SPHERICAL AND PRACTICAL ASTRONOMY

as Applied to Geodesy

By the same author

Introduction to Satellite Geodesy	1964
Gravimetric and Celestial Geodesy *A Glossary of Terms* with John D.Rockie	1966

SPHERICAL AND PRACTICAL ASTRONOMY

as Applied to Geodesy

IVAN I. MUELLER

Professor, Department of Geodetic Science
The Ohio State University

With a contribution by

HEINRICH EICHHORN

Chairman, Department of Astronomy
University of South Florida

FREDERICK UNGAR PUBLISHING CO.
NEW YORK

Printed in the United States of America

Library of Congress Catalog Card No. 68-31453

In memory of

Dr. ISTVÁN RÉDEY

Professor of Geodesy
Technical University of Budapest

PREFACE

The art and science of determining the precise location of a terrestrial point from measurements on natural celestial bodies is the principal subject of this book. Comprehensive treatments of this topic published in English in the past include certain government agency manuals, rather limited in scope, books written mainly for the use of the land surveyor, limited to simple methods of low accuracy, and the 'practical astronomy' books from Chauvenet (1891) to Nassau (1931) which are probably the most comprehensive but for the most part obsolete. Recent advances in pertinent instrumentation, in computational methods due to the use of electronic computors, and new astronomic discoveries and the resulting new concepts and conventions made it necessary to prepare this text. The book is an outgrowth of lecture material presented in courses at The Ohio State University over a period of years. The aim of the book is to bring together an up-to-date balanced selection and treatment of the subject that can serve as an introduction yet is sufficiently detailed to be also used as a teaching text and reference work.

A review on the geometric and physical properties of the earth is given in the second chapter of the book. This is followed by three chapters which deal with topics commonly known as spherical astronomy–subjects such as the various celestial coordinate systems and their relation to each other; the variations of star coordinates due to precession, nutation, polar motion, aberration, parallax, refraction, and proper motion; the different time systems including ephemeris and atomic times. The next three chapters describe the various star catalogues and their accuracies, the instrumentation as used in the observatories and in the field, international time coordination and dissemination. The final chapters deal with topics known as 'positional astronomy,' treating first- and second-order position determination methods. The extent of these last sections varies rather arbitrarily; methods popular in the U.S. are discussed in great detail, while others are treated in a shorter form. References and numerical examples are included throughout.

Background required from the reader includes the basic rules of calculus, matrix algebra and general physics, the elements of geodetic surveying, and descriptive astronomy. Lack of knowledge of matrix operations, however, will not hinder understanding since each problem is treated both with and without the use of matrices.

It is a pleasant duty of the author to acknowledge his debt for comments on the manuscript first of all to Professor H.K. Eichhorn of the University of South Florida who also wrote Chapter 6; also to Mr. Georges Blaha of The Ohio State University, Professor Karl Ramsayer of the Technical University of Stuttgart, and Dr. A.R. Robbins, Reader at Oxford University. The manuscript was typed, in a form suitable for direct reproduction, by Mrs. Jeanne C. Preston and Mrs. Irene B. Tesfai whose excellent work is reflected in the appearance of the book. The illustrations were drawn by Mr. John R. Miller. The index was compiled by Mr. C. R. Schwarz. Graduate students performed the burdensome tasks of proofreading (Messrs. G. Blaha, E. J. Krakiwsky, C.L. Noren, J. P. Reilly, C. R. Schwarz, and T. J. Thomason) and of computing the numerical examples (names are listed within the examples). The author wishes to express his special appreciation to all of them for their careful voluntary work. It is also a pleasure to express appreciation to the publisher for his efforts in making the book available to the public.

Credit for certain illustrations, tables, and other material is due the following organizations: Aeroflex Laboratories, Inc., Woodside, N. Y., for Fig. 8.7; Askania-Werke, Berlin, GFR, for Figs. 7.17, 7.18; Bureau International de l'Heure, Paris, for Tables 4.7, 5.6, 8.3–8.14; Hewlett-Packard, Palo Alto, California, for Fig. 8.4; H. M. Nautical Almanac Office, Royal Greenwich Observatory, Sussex, for Fig. 5.10, and Tables 4.1, 4.2, 5.1, 5.3; Kern Instruments, Inc., Port Chester, N. Y., for Figs. 7.11–7.14, 7.24; Sky and Telescope, Cambridge, Mass., for Fig. 12.11 and Table 12.1; Tracor Inc., Austin, Texas, for Fig. 8.6; U.S. Army Map Service, Washington, D. C., for Figs. 12.13–12.16; U.S. Coast and Geodetic Survey, Rockville, Md., for Figs. 8.18–8.21, 11.2, and Tables 4.8–4.10; U.S. National Bureau of Standards, Boulder, Colo., for Figs. 8.3, 8.12, 8.13, 8.15, 8.16; U.S. Naval Observatory, Washington, D. C., for Figs. 7.32, 7.36, 7.37, 8.8, 11.7, 12.17, and Tables 4.3, 4.4, 4.11, 4.12, 5.4, 5.5, 5.7, 8.16–8.18; VEB Carl Zeiss-Jena, GDR, for Figs. 7.15, 7.16, 7.30; VEB Carl Zeiss-Oberkochen, GFR, for Fig. 7.29; Wild Heerbrugg Instruments, Inc., Farmingdale, N. Y., for Figs. 7.2–7.10, 7.23, 8.5, 8.25.

Department of Geodetic Science
The Ohio State University
Columbus, Ohio
April, 1968

I.I.M.

CONTENTS

1. INTRODUCTION 1

2. THE EARTH FROM THE GEODETIC POINT OF VIEW 3

2.1 The Gravity Field of the Earth, the Geoid 3
2.2 The Normal Gravity Field of the Earth, the Spheroid, Earth-Spherop, Earth and Reference Ellipsoids, and Sphere 8
2.3 Fundamental Definitions Related to the Geodetic Reference Surfaces 11

2.31 The Sphere 11
2.32 The Ellipsoid 13
2.33 The Geopotential Surface 17

2.4 Absolute and Relative Positions, the Geodetic Datum, the Deflection of the Vertical 21

References 28

3. THE CELESTIAL SPHERE AND ITS COORDINATE SYSTEMS 29

3.1 Fundamental Definitions 29
3.2 Celestial Coordinate Systems 32

3.21 Horizon System 33
3.22 Hour Angle System 33
3.23 Right Ascension System 34
3.24 Ecliptic System 34
3.25 Summary 35

3.3 Transformation of Coordinates 37

3.31 Conversions Between the Horizon and Hour Angle Systems 37
3.32 Conversions Between the Hour Angle and Right Ascension Systems 38
3.33 Conversions Between the Right Ascension and Ecliptic Systems 39
3.34 Transformation by Matrices 41
3.35 Numerical Examples 44
3.36 Differential Relations 50

3.4 Special Star Positions 51

3.41 Circumpolar and Equatorial Stars 51
3.42 Visibility, Rising, Setting 52
3.43 Culmination or Transit 54
3.44 Prime Vertical Crossing 56

3.45 Elongation 56
3.46 Numerical Example 57

References 58

4. VARIATIONS IN THE CELESTIAL COORDINATES 59

4.1 Variations Due to the Motion of the Coordinate Systems 59

4.11 General Precession 62

4.111 Rigorous Transformation 62
4.112 Approximate Transformation 65

4.12 Astronomic Nutation 68

4.121 Nutational Parameters 68
4.122 Transformation from the Mean to the True Celestial System 71
4.123 Besselian Transformation for General Precession and Nutation 75
4.124 Short-Period Nutation 79

4.13 Polar Motion 80

4.131 Motion of the Pole, the International Polar Motion Service 80
4.132 Transformation from the Celestial to the Terrestrial System 84
4.133 Polar Motion in Latitude, Longitude, and Azimuth 86

4.2 Variations Due to Physical Effects 88

4.21 Aberration 91

4.211 General Law of Aberration 91
4.212 Circular Annual Aberration 92
4.213 The Effect of the Ellipticity of the Earth's Orbit on Annual Aberration 94
4.214 Diurnal Aberration 96

4.22 Parallax 98

4.221 Annual Parallax 98
4.222 Geocentric Parallax 102

4.23 Astronomic Refraction 103

4.231 Rigorous Theory 103
4.232 Approximate Astronomic Refraction and Refraction Tables 106

4.3 Variations Due to the Proper Motion of Stars 109

4.31 The Effect of Time on the Components of Proper Motion 109
4.32 The Effect of Precession on the Components of Proper Motion 111
4.33 The Effect of Proper Motion on the Mean Place of a Star 113

4.331 Transformation by Components 113
4.332 Transformation by Time Series 114
4.333 Transformation by Matrices 114

4.4 Summary of the Reduction of Star Positions, and Secondary Effects 115

4.41 Reduction of the Catalogued Mean Place of a Star from One Epoch to Another 118

4.411 Rigorous Reduction 118
4.412 Conventional Reduction 118

4.42 Reduction of the Apparent Place of a Star from Its Mean Position 119

4.421 Conventional Reduction 119
4.422 Rigorous Matrix Reduction 120
4.423 Trigonometric Reduction 123
4.424 Numerical Examples 124

4.43 Reduction of the Observed Place of a Star to Its Apparent Position 133

References 134

5. TIME SYSTEMS 137

5.1 Sidereal Time System 139

5.11 Sidereal Epoch: Apparent and Mean Sidereal Times, the Equation of the Equinox 139
5.12 Sidereal Interval: Mean Sidereal Day and the Period of Rotation 140
5.13 Sidereal Calendar: Greenwich Sidereal Date and Day Numbers 141

5.2 Universal (Solar) Time System 141

5.21 The Motion of the Sun 141

5.211 Annual Motion, Kepler's Laws 141

5.212 Diurnal Motion 145
5.213 Actual Motion 145

5.22 Universal (Solar) Time 145

5.221 Solar Epoch: Apparent (True) and Mean Solar, Universal, and Zonal Times; the Equation of Time 145
5.222 Solar Interval: Mean Solar Day, Tropical, Sidereal and Julian Years 150

5.23 Solar Calendars 152

5.231 Civil Calendars: Julian and Gregorian 152
5.232 Astronomic Calendars: Astronomic and Julian Dates; Julian Day Numbers 153

5.3 Conversion of Time: Sidereal-Universal 153

5.31 Conversion of Epoch 153
5.32 Conversion of Interval 156
5.33 Numerical Examples 158

5.4 Irregularities of the Rotational (Universal and Sidereal) Time Systems 163

5.41 Classification 163
5.42 Rotational Time Systems UT0, UT1, and UT2 163
5.43 Seasonal Variations in the Rotational Velocity of the Earth 164
5.44 Secular and Irregular Variations in the Rotational Velocity of the Earth 165

5.5 Ephemeris Time System 166

5.51 Ephemeris Epoch and Interval 168
5.52 The Fictitious Mean Sun, the Ephemeris Meridian 169
5.53 Ephemeris Calendars 172

5.6 Atomic Time Systems 174

5.61 Atomic Time Epoch 175
5.62 Atomic Time Interval 176
5.63 Relation Between Atomic and Ephemeris Times 176

References 177

6. STAR CATALOGUES 179

6.1 Fundamental Definitions and Classification 179

6.11 Requirements 179
6.12 Types of Catalogues 180

6.2 Original Observation Catalogues 182

6.21 Absolute Star Catalogues 182

6.211 Determination of Absolute Declination 182
6.212 Determination of Absolute Right Ascension 183

6.22 Relative Star Catalogues 186

6.221 Zone Observations 186
6.222 Photographically Determined Relative Star Positions 187

6.3 Catalogues Defining a Fundamental System 189

6.31 Fundamental Star Catalogues: The Principle 189

6.311 Systematic Errors Affecting Star Positions 189
6.312 The Establishment of a Catalogue System 190
6.313 Definition of a Fundamental System 190
6.314 Reduction Between Systems 192

6.32 The German Series of Fundamental Catalogues 193

6.321 The FC 193
6.322 The NFK 194
6.323 The FK3 195
6.324 The FK4, FK4 Sup, and the 'Apparent Places of Fundamental Stars' 195

6.33 The General Catalogue (GC) 202

6.331 The Scope of the GC 202
6.332 The GC System 204
6.333 Description of the GC 208

6.34 The N30 Catalogue 209

6.341 History and Establishment of the System 209
6.342 The Accuracy of the N30 210

6.4 Relationship Between Catalogue Systems 211

6.41 Representation of the Systematic Differences 211
6.42 Sources for Reduction Tables to the GC 213
6.43 Systematic Differences Between the Fundamental Systems 213
6.44 Brosche's Method 214

6.5 Lists of Star Catalogues 215

6.51 Old Lists 215

6.52 The 'Geschichte des Fixsternhimmels' (GFH) 216
6.53 The 'Index der Sternörter' 217
6.54 The 'Astronomischer Jahresbericht' 217

6.6 The AG-Type Catalogues 218

6.61 The AGK1 and Its Extensions 218

6.611 The Original AGK1 218
6.612 The South American Extensions 218

6.62 The Photographic Zone Catalogues 219

6.621 The AGK2 219
6.622 The AGK3 224
6.623 The Yale Catalogues 225
6.624 The 'Cape Photographic Catalogue for 1950.0' (CPC) 229

6.63 The Smithsonian Astrophysical Observatory 'Star Catalog' 231

6.631 History and Original Purpose 231
6.632 Description of the Catalogue 232
6.633 The Sources 235
6.634 Using the SAOC 236

6.7 The Astrographic Catalogue (AC) 237

6.71 History 237
6.72 Photography 237
6.73 Measurements 238
6.74 Reduction 239
6.75 The Zones of the AC 240

6.8 Future Work 241

References 242

7. OPTICAL INSTRUMENTS FOR ASTRONOMIC POSITION, AZIMUTH, AND TIME DETERMINATIONS 249

7.1 First-Order (Precision) Instruments 250

7.11 The Wild T4 Universal Theodolite 254

7.111 Horizontal System 254
7.112 Vertical System 255
7.113 Telescope 257
7.114 Telescope Eyepiece and Impersonal Micrometer 258
7.115 Levels 259

7.116 Electric System 259

7.12 The Kern DKM3-A Universal Theodolite 260

7.121 Vertical Axis 260
7.122 Precisely Graduated Reading Circles and Eyepiece 262
7.123 Telescope 263
7.124 Telescope Eyepiece and Impersonal Micrometer 263
7.125 Levels 265

7.13 The Zeiss (Jena) Theo-003 Universal Theodolite 266

7.131 Precisely Graduated Reading Circles 267
7.132 Telescope and Automatic Tilt Compensator 268

7.14 Universal Telescopes 269
7.15 Miscellaneous Instruments 271
7.16 Adjustment of the Instruments 273

7.161 General Adjustment 274
7.162 Vertical Axis Adjustment 275
7.163 Horizontal Axis Adjustment 275
7.164 Striding Level Adjustment 275
7.165 Horizontal Collimation Adjustment 275
7.166 Vertical Collimation Adjustment 276
7.167 Thread Adjustment 276
7.168 Focusing 277
7.169 Miscellaneous Comments 277

7.17 Instrument Calibration 278

7.171 Conventional Level Calibration 278
7.172 Level Calibration with the Universal Theodolite 282
7.173 Impersonal Micrometer Calibration for Lost Motion 284
7.174 Impersonal Micrometer Calibration for Mean Width of Contact Strips 285
7.175 Impersonal Micrometer Calibration for Equatorial Value of One Drum Revolution 286

7.2 Second-Order (Geodetic) Instruments 288

7.21 Geodetic Theodolites 289
7.22 Prismatic Astrolabes 291
7.23 Pendulum Astrolabes 294

7.3 Observatory (High-Precision) Instruments 296

7.31 Photographic Zenith Tube (PZT) 298
7.32 Danjon-OPL Impersonal Astrolabe 301
7.33 Markowitz Moon Camera 303

7.4 Equations of the Optical Instruments 305

7.41 Equation for Horizontal Angle and Azimuth Measurement 306
7.42 Equation for Vertical Angle Measurement 308
7.43 Equation for Transit Time Measurement 310
7.44 Determination of the Constants of the Instrumental Equations 312

7.441 Inclination of the Horizontal Axis (b) 312
7.442 Horizontal Collimation (c) 313
7.443 Horizontal Index Correction (ΔA) 314
7.444 Instrumental Zenith Point (Z) 314
7.445 Meridian (Azimuth) Setting Error (a) 315
7.446 Inclinations of the Horizontal and Vertical Axes (i' and i) 316

References 316

8. TIMEKEEPING AND TIME DISSEMINATION 319

8.1 Timekeeping Instruments 319

8.11 Primary Frequency Standards 321
8.12 Clocks and Chronometers 326

8.121 Mechanical 327
8.122 Quartz Crystal 328
8.123 Atomic 334

8.2 Time and Frequency Dissemination and Coordination 336

8.21 Bureau International de l'Heure 336

8.211 Circulars of the BIH 337
8.212 Frequency Offsets and Step Adjustments 338
8.213 The Mean Observatory (the Greenwich Mean Astronomic Meridian) 343
8.214 The Bulletin Horaire 345
8.215 Recent Changes in the Procedures of the BIH 351

8.22 Time Services 354

8.221 Radio Time Signals 354
8.222 Transmissions Controlled by the U.S. National Bureau of Standards 355

8.223 Standard Time and Frequency Broadcasts 361
8.224 Loran-C and Omega Transmissions 361
8.225 International Coordination of Time and Frequency Services 369

8.23 Corrections to Time Signals 371

8.231 Corrections for the Propagation Delay of HF Signals 371
8.232 Correction for the Propagation Delay of LF and VLF Signals 372
8.233 Corrections to the Emitted Time Signals for Epoch Reference 373

8.3 Time Receiving and Comparison 381

8.31 High Precision Time Comparisons Using HF Transmissions 383

8.311 Comparison by Tick Phasing Adjustment 383
8.312 Comparison with Stroboscopic Flashes 384
8.313 Comparison with Delay Counter 385

8.32 High Precision Frequency Comparisons Using LF/VLF Transmissions 386
8.33 First-Order Time Comparisons 391
8.34 Second-Order Time Comparisons 396

References 397

9. DETERMINATION OF ASTRONOMIC AZIMUTH 401

9.1 Second-Order Azimuth Determination 402

9.11 Azimuth by Star Hour Angles 402

9.111 Azimuth by the Hour Angle of Polaris 404
9.112 Azimuth by the Hour Angles of Stars Near Elongation 409
9.113 Azimuth by the Hour Angles of Stars Near Culmination 414
9.114 Effect of Random Observational Errors 414

9.12 Azimuth by Star Altitudes 416
9.13 Miscellaneous Methods of Azimuth Determination 422

9.131 Azimuth by Equal Altitudes 422
9.132 Azimuth by the Hour Angles of Almucantar Crossings 422
9.133 Azimuth by the Rate of Change of Zenith Distance 423

9.2 First-Order Azimuth Determination 423

9.21 Azimuth by the Hour Angle of Polaris: Direction Method 425
9.22 Azimuth by the Hour Angle of Polaris: Micrometer Method 426

9.3 Corrections to the Observed Azimuth 428

9.31 Correction for the Motion of the Pole 429
9.32 Correction for the Curvature of the Plumb Line 430
9.33 Correction for the Skew of the Normal 432
9.34 Correction for the Eccentricity of the Station 432
9.35 Correction for the Eccentricity of the Reference Mark 434

References 435

10. DETERMINATION OF ASTRONOMIC LATITUDE 437

10.1 Second-Order Latitude Determination 438

10.11 Latitude by Meridian Zenith Distances 438
10.12 Latitude by Circum-Meridian Zenith Distances 440
10.13 Latitude by the Zenith Distance of Polaris and of a South Star 444

10.131 Latitude by the Zenith Distance of Polaris 444
10.132 Latitude by the Combined Zenith Distances of Polaris and the South Star 445

10.14 Latitude by Equal Zenith Distances of Two Stars 456
10.15 Latitude by the Azimuths of a Star Pair Near Elongation 456
10.16 Miscellaneous Methods of Second-Order Latitude Determination 457

10.161 Latitude by Ex-Meridian Zenith Distances 457
10.162 Latitude by Equal Zenith Distances of One Star 458
10.163 Latitude by the Hour Angles of Stars in the Prime Vertical Plane 458
10.164 Approximate Latitude by the Rate of Change of the Zenith Distance Near the Prime Vertical 458

10.2 First-Order Latitude Determination 459

10.21 The Horrebow-Talcott Method 459

10.211 Latitude Equation: Corrections to the Observed Zenith Distance Differences 460
10.212 Observing List 462

10.213 Observations 463
10.214 Computations 464

10.22 Alternative Methods for First-Order Latitude Determination 469

10.221 The Pevtsov Method 469
10.222 The Sterneck Method 470

10.3 Corrections to the Observed Latitude 471

10.31 Correction for the Motion of the Pole 471
10.32 Correction for the Curvature of the Plumb Line 471
10.33 Correction for the Eccentricity of the Station 472

10.4 Simultaneous Determination of Latitude and Azimuth 472

References 474

11. DETERMINATION OF ASTRONOMIC LONGITUDE AND TIME 476

11.1 Time and Longitude 476

11.11 Longitude Equations 476
11.12 Principles of Determining Local Sidereal Time 478

11.2 Second-Order Longitude Determination 481

11.21 Longitude by Zenith Distances Measured Near the Prime Vertical 481

11.211 Observing List 481
11.212 Observations 482
11.213 Computations 483

11.22 Longitude by Transit Times 485

11.221 Longitude by Meridian Transit Times 492
11.222 Longitude by Transit Times Through the Vertical Plane of Polaris 496

11.23 Longitude by Equal Zenith Distances 497
11.24 Longitude by Horizontal Angles 502

11.3 First-Order Longitude Determination 504

11.31 The Mayer Method 504

11.311 Longitude Equation: Corrections to the Observed Meridian Transit Time 505
11.312 Observing List 507
11.313 Observations 508
11.314 Longitude Computations 509

11.32 Alternative Methods for First-Order Longitude Determination 516

11.321 The Doellen Method 516
11.322 The Tsinger Method 517

11.4 Corrections to the Observed Longitude 518

11.41 Corrections for the Motion of the Pole 519
11.42 Correction for the Curvature of the Plumb Line 520
11.43 Correction for the Eccentricity of the Station 520

11.5 Simultaneous Determination of Longitude and Latitude 520

11.51 Latitude and Longitude by Ex-Meridian Zenith Distances 520

11.511 Latitude-Longitude Equation 521
11.512 Star Selection 522
11.513 Observations 523
11.514 Least Squares Adjustment 524
11.515 Graphical Adjustment: Position Lines 525

11.52 Latitude and Longitude by Equal Zenith Distances 531

11.521 Star Selection 531
11.522 Observations 534
11.523 Computations 535

11.53 Observatory Longitude (Time) and Latitude Determinations 535

11.531 Observations with the Danjon-OPL Astrolabe 536
11.532 Observation with the PZT 538

11.6 Simultaneous Determination of Longitude, Latitude, and Azimuth 541

11.61 Longitude, Latitude, and Azimuth from Horizontal Directions 541

11.611 The Fundamental Equation 541
11.612 Star Selection 542
11.613 Observation 543
11.614 Computations 543

11.62 Alternative Methods of Determining Longitude, Latitude, and Azimuth Simultaneously 544

11.621 Determination from Two Stars, at Altitude Equal to Declination 544

11.622 Determination from Two Ex-Meridian Stars 545
11.623 Determination from Meridian Transits 545

11.7 Determination of Ephemeris Time 545

11.71 Determination of the Direction of the Moon 546
11.72 Interpolation of Ephemeris Time 547

References 548

12. SOLAR ECLIPSES AND OCCULTATIONS 551

12.1 Introduction 551

12.11 Definitions 551
12.12 Two Different Points of View 551

12.2 Fundamentals of the Theory of Solar Eclipses 552

12.21 General Prediction 552
12.22 Condition of the Beginning or Ending of a Solar Eclipse at a Given Place on Earth 555

12.221 Position of the Shadow Axis at Any Given Time 558
12.222 Distance of a Given Observation Place from the Shadow Axis at a Given Time 559
12.223 Radius of the Shadow 562
12.224 Fundamental Equation of the Theory of Eclipses 564
12.225 Besselian Elements of the Eclipse 564

12.3 Prediction of a Solar Eclipse for a Given Place 565

12.31 Time of the Contact 565
12.32 The Position Angle 568
12.33 Time and Degree of Maximum Obscuration 570
12.34 Correction for Refraction 570

12.4 Prediction of an Occultation 572

12.41 General Method 572
12.42 Method of the 'American Ephemeris and Nautical Almanac' and of the 'Astronomical Ephemeris' 578

12.421 Tables of 'Sky and Telescope' 579
12.422 Computer-Printed Tables of H.M. Nautical Almanac Office 580

12.43 Limits of an Occultation 581
12.44 Prediction of the Isolimb of an Occultation 583

12.5 Geodetic Applications of Occultations 585

12.51 Position Determination 586

12.511 Effect of Errors in the Assumed Data on the Predicted Time of Contact 586
12.512 On the Possibility of Determining the Errors in the Assumed Data 589

12.52 Determination of the Equatorial Semidiameter of the Earth and the Parallax of the Moon 594

12.6 Observations of Occultations 598

12.61 Optical Methods 598
12.62 Photoelectric Method 598

12.7 Major Factors Affecting the Accuracy of the Occultation Results 601

12.71 Topography of the Lunar Limb 601
12.72 The Libration 603
12.73 Coordinates of the Moon 604
12.74 Parallax of the Moon 605

References 605

INDEX 608

LIST OF ILLUSTRATIONS

2.1 Attraction of a Unit Mass 4
2.2 The Sphere 12
2.3 Azimuth on the Sphere 13
2.4 The Ellipsoid 14
2.5 The Meridian Ellipse 15
2.6 Azimuths on the Ellipsoid 17
2.7 The Geopotential Surface 18
2.8 Astronomic Meridians and Parallels 19
2.9 Astronomic Latitudes 20
2.10 Absolute and Relative Geodetic Coordinates 22
2.11 Major Geodetic Datums 24
2.12 Undulations of the Geoid 27

3.1 The Celestial Sphere 30
3.2 The Celestial Sphere 31
3.3 The Horizon System 32

3.4 The Hour Angle System 33
3.5 The Right Ascension System 34
3.6 The Ecliptic System 35
3.7 Relations Between the Horizon and Hour Angle Systems 37
3.8 The Astronomic Triangle 38
3.9 Relations Between Hour Angle and Right Ascension Systems 39
3.10 Relations Between the Right Ascension and Ecliptic Systems 40
3.11 The Ecliptic Triangle 41
3.12 Circumpolar and Equatorial Stars 52
3.13 Stars at the Equator and the Pole 53
3.14 Rising and Setting Limits of Stars 53
3.15 The Zenith Distances of Stars 55
3.16 Stars at the Prime Vertical 56

4.1 Lunisolar Precession (Cause) 60
4.2 Lunisolar Precession (Effect) 61
4.3 General Precession 62
4.4 Annual General Precession 65
4.5 Schematic General Precession and Nutation 70
4.6 Lunar and Solar Orbits 71
4.7 Nutation 75
4.8 Polar Motion 82
4.9 The Motion of the True Celestial and the Average Terrestrial Poles in the Terrestrial System of 1900–05 83
4.10 True Celestial and Average (Mean) Terrestrial Systems 85
4.11 Polar Motion in Azimuth 88
4.12 Aberration (Schematic) 88
4.13 Geocentric Parallax (Schematic) 89
4.14 Annual Parallax (Schematic) 90
4.15 Astronomic Refraction (Schematic) 90
4.16 Aberration of Light 91
4.17 Annual Aberration in Case of Circular Orbit 92
4.18 Effect of Orbital Ellipticity on Annual Aberration 95
4.19 Diurnal Velocity 97
4.20 Diurnal Aberration 98
4.21 Annual Parallax 100
4.22 Geocentric Parallax 103
4.23 Astronomic Refraction 104
4.24 Proper Motion 112
4.25 Mean and Apparent Places of the Star GC 23487 117

5.1 Sidereal Times 139
5.2 Orbital Velocities of the Earth 143
5.3 Apparent Solar Motion 146
5.4 Solar Times 148

5.5 Equation of Time and of the Equinox for 0^h UT, 1966 149
5.6 Standard Time Zones 151
5.7 Solar and Sidereal Days 158
5.8 Seasonal Variation in the Rotational Velocity of the Earth, 1967 165
5.9 Secular and Irregular Variations in the Rotational Velocity of the Earth 167
5.10 General Trend of ΔT, 1635-1967 172
5.11 Astronomic Time Systems 173

7.1 Distinctive Features of First-Order Optical Instruments 251
7.2 Wild T4 Universal Theodolite 253
7.3 Reading the Horizontal Circle on the Wild T4 254
7.4 Horizontal Setting Circle on the Wild T4 255
7.5 Reading the Vertical Circle on the Wild T4 256
7.6 Reading the Vertical Setting Circle on the Wild T4 256
7.7 Optical System of the Wild T4 Telescope 257
7.8 Telescope Eyepiece and Impersonal Micrometer of the Wild T4 258
7.9 Suspension Level of the Wild T4 260
7.10 Twin Horrebow Levels of the Wild T4 260
7.11 Kern DKM3-A Universal Theodolite 261
7.12 Reading the Vertical and Horizontal Circles on the Kern DKM3-A 262
7.13 Optical System of the Kern DKM3-A Telescope 263
7.14 Telescope Eyepiece and Impersonal Micrometer of the Kern DKM3-A 264
7.15 Zeiss (Jena) Theo-003 Universal Theodolite 268
7.16 Optical System of the Zeiss (Jena) Theo-003 Telescope 269
7.17 Askania AP 70 Universal Telescope 270
7.18 Askania Theodolite TPR 272
7.19 Level Trier 279
7.20 Dependence of Level Sensitivity on Bubble Position and Length 281
7.21 Dependence of Level Tilt on the Inclination of the Alidade 283
7.22 Lost Motion in the Impersonal Micrometer 284
7.23 Wild T3 Geodetic Theodolite with Astrolabe Attachment 290
7.24 Kern DKM3 Geodetic Theodolite 291
7.25 Principle of Prismatic Astrolabes 293
7.26 View Through the Astrolabe Eyepiece 293
7.27 Cooke 45° Prismatic Astrolabe 294
7.28 Nušl-Frič Circumzenithal 295
7.29 Zeiss (Oberkochen) Ni2 Pendulum Astrolabe 296
7.30 Zeiss (Jena) Pendulum Astrolabe 297
7.31 Principle of Photographic Zenith Tube 298

7.32 Photographic Zenith Tube of the U.S. Naval Observatory 299
7.33 The Images of a Star on the PZT Plate 300
7.34 Schematic Cross Section of the Danjon Astrolabe 301
7.35 Image Formation in a Double Symmetrical Wollaston Prism 302
7.36 Danjon Impersonal Prismatic Astrolabe at the U.S. Naval Observatory 303
7.37 Markowitz Dual Rate Moon Camera at the U.S. Naval Observatory 304
7.38 The Effect of Residual Instrumental Adjustment Errors on Horizontal and Vertical Angle Measurements 306
7.39 The Effect of Residual Instrumental Adjustment Errors on Transit Time Measurements 311

8.1 Stability of Atomic and Quartz Crystal Oscillators 322
8.2 Schematic Diagram of a Cesium Beam Frequency Standard 323
8.3 The NBS-III Primary Cesium Beam Frequency Standard at the National Bureau of Standards, Boulder, Colorado 324
8.4 Hewlett-Packard Cesium Beam Frequency Standard, Model 5061A 325
8.5 Nardin Mechanical Sidereal Chronometer with Electric Contacts 328
8.6 Tracor Quartz Crystal Chronometer, Model Sulzer-A5 330
8.7 Newtek Quartz Crystal Chronometer, Model Chronofax-102, with Sample Printout 333
8.8 The Atomic Clock System at the U.S. Naval Observatory, Washington, D.C. 335
8.9 Deviation of UTC from UT2, 1965-1966 343
8.10 Variation of the BIH Average Terrestrial Pole in the CIO System, 1958-1967 351
8.11 The International ONOGO System 355
8.12 Time and Frequency Measurement Console, and the NBS-A and UTC (NBS) Clocks at the National Bureau of Standards in Boulder, Colorado 356
8.13 NBS Time and Frequency Facilities (1967) 357
8.14 Station WWV and the 2.5 MHz Transmitter 358
8.15 Chart of Time Code Transmissions from WWV 360
8.16 Hourly Broadcast Schedule of the NBS Radio Stations (1967) 362
8.17 Standard Frequency/Time Services, Loran-C and Omega Stations 367
8.18 Drift Rate of the Sulzer-A5 Quartz Crystal Oscillator M5A-8 389
8.19 Record of VLF Phase Comparisons, VLF (18.6 kHz): Station Oscillator (6059-2) (Sample) 391
8.20 Drift Rate and Clock Error Curves for the Station Oscillator (6059-2) Using 18.6 kHz Signals 392

8.21 Comparison of Drift Rate Curves for the Station Oscillator (6059-2) Using 18.6 kHz and 23.4 kHz Signals 392
8.22 First-Order Time Comparison Equipment with One-Pen Drum Chronograph 393
8.23 Favag Two-Pen Tape Chronograph 393
8.24 Schematic Diagram of First-Order Time Comparison Equipment 394
8.25 Chronograph Record Scaling Fan 394

9.1 Corrections to the Observed Azimuth for the Curvature of the Plumb Line and the Skew of the Normal 430
9.2 Correction to the Observed Azimuth for the Eccentricity of the Station 433
9.3 Correction to the Observed Azimuth for the Eccentricity of the Reference Mark 434

10.1 Reduction of Zenith Distances to the Meridian 461
10.2 Correction to the Observed Latitude for the Eccentricity of the Station 472

11.1 Hookup Diagram of Auxiliary Equipment for Longitude Work 479
11.2 Star Chart for Selection of Longitude Pairs ($\Phi = 40^\circ$) 488
11.3 Longitude and Latitude by Position Lines 530
11.4 Local Hour Angles and Azimuths for 30° Zenith Distance ($\Phi = 40^\circ$) 532
11.5 Azimuths for 30° Zenith Distance ($\Phi = 40^\circ$) 533
11.6 Declination Limits for Stars at 30° Zenith Distance During an Observation Period ($\Phi = 40^\circ$) 533
11.7 Photograph Taken by the Markowitz Dual Rate Moon Camera 546

12.1 The Moon and the Sun in Conjunction in Longitude 553
12.2 Simplified Diagram of the Moon and the Sun at Conjunction 553
12.3 The Shadow of the Moon 556
12.4 Relation Between the Different Cartesian Coordinate Systems 558
12.5 The Position of the Observer 561
12.6 The Projection of the Observer and of the Shadow on the Apparent Plane 562
12.7 The Geometry of the Shadow 563
12.8 The Relative Polar Coordinates of the Center of the Shadow with Respect to the Observer 567
12.9 The Position Angle 569
12.10 Effect of Refraction 571
12.11 Central Stations for Occultation Predictions in North America 579

12.12 Diagram of the Path of Occulted Stars Observed at Williston Observatory 593
12.13 Diagram of the Photoelectric Occultation Equipment 599
12.14 Record from the Photoelectric Occultation Equipment 599
12.15 Successful Equal Limb Lines, Pacific Program, U.S. Army Map Service, 1954-1964 600
12.16 Nippon-Kogaku Occultation Telescope 601
12.17 Watts Lunar Limb-Correction and Libration Chart (Sample) 602

LIST OF TABLES

2.1 Mass Coefficients of the Earth 8
2.2 Parameters of Reference Ellipsoids 10
2.3 Major Geodetic Datums 25

3.1 Celestial Coordinate Systems 36
3.2 Cartesian Celestial Coordinate Systems 42

4.1 Equatorial Precessional Elements for Reduction to 1950.0 or Other Epochs 64
4.2 Series for the Nutation 72
4.3 Sample Page of the AENA: Besselian Day Numbers, 1967 77
4.4 Sample Page of the AENA: Independent Day Numbers, 1967 78
4.5 Latitude Observatories of the International Polar Motion Service 81
4.6 Coordinates of the Average Terrestrial Pole in the System of 1900–05 82
4.7 Correction for Diurnal Aberration 99
4.8 Mean Refraction Angle, Δz_R^m 107
4.9 Pressure Correction Factor, C_B 108
4.10 Temperature Correction Factor, C_T 108
4.11 Sample Page of the AENA: Second-Order Day Number, J, 1967 121
4.12 Sample Page of the AENA: Second-Order Day Number, J′, 1967 122

5.1 Greenwich Sidereal Day Number, 1950-1999 142
5.2 The Seasons 146
5.3 Julian Day Numbers, 1950-1999 154
5.4 Sample Page of the AENA: Calendar, 1967 155
5.5 Sample Page of the AENA: Universal and Sidereal Times, 1967 157

5.6	BIH Coefficients for Seasonal Variation	164
5.7	ΔT, Reduction from Universal Time to Ephemeris Time	171
6.1	Standard Errors of the FK4 System	197
6.2	Sample Pages of the FK4	200
6.3	Sample Page of the APFS	203
6.4	Sample Pages of the GC	206
6.5	Standard Deviations of Positions in the N30	211
6.6	The Zones of the AGK1 and Their South American Counterparts	220
6.7	The Yale Zone Catalogues	226
6.8	The Zones of the 'Cape Photographic Catalogue'	231
6.9	Sample Page of the SAO Catalog	234
6.10	Sources for SAOC Data	236
6.11	The Zones of the Astrographic Catalogue	241
7.1	Universal Theodolite Specifications	267
7.2	Universal Telescope Specifications	271
7.3	Geodetic and One-Second Theodolite Specifications	292
8.1	Characteristics of Primary Atomic Frequency Standards (1967)	326
8.2	Characteristics of Secondary Quartz Crystal Frequency Standards or Chronometers (1967)	331
8.3	Circular 'A' of the BIH (Sample)	339
8.4	Circular 'B/C' of the BIH (Sample)	340
8.5	Circular 'D' of the BIH (Sample)	341
8.6	Corrections to Universal Time Due to the Changes in the Meridian of the Mean Observatory, 1931-1958	344
8.7	Corrections to Universal Time Due to the Changes in the Average Terrestrial Pole, 1931-1963	345
8.8	Bulletin Horaire: Table 1, Coordinates of the Pole and Time Corrections UT1-UT0 (Sample)	346
8.9	Bulletin Horaire: Table 3, List of Participating Stations (Sample)	347
8.10	Bulletin Horaire: Table 6, Atomic Time (Sample)	348
8.11	Bulletin Horaire: Table 7, Coordinated Time UTC (Sample)	349
8.12	Coordinates of the Average Terrestrial Pole Used by the BIH in the System of the Conventional International Origin, and the Resulting Corrections to Universal Time, 1955-1967	352
8.13	Coordinated Standard Time and Frequency Broadcasts (1967)	363
8.14	Noncoordinated Standard Time and Frequency Broadcasts (1967)	366
8.15	Loran-C Networks (1967)	368

8.16 U.S. Naval Observatory Extrapolated UT2-UTC (USNO) Corrections (Sample) 375
8.17 U.S. Naval Observatory Preliminary Emission Times and Provisional Coordinates of the Pole (Sample) 376
8.18 U.S. Naval Observatory Time Signals Bulletin (Sample) 377
12.1 'Sky and Telescope' Occultation Predictions for 1968 (Sample) 580

LIST OF NUMERICAL EXAMPLES

3.1 Conversion from the Right Ascension to the Hour Angle System 45
3.2 Conversion from the Hour Angle to the Right Ascension System 45
3.3 Conversion from the Hour Angle to the Horizon System 46
3.4 Conversion from the Horizon to the Hour Angle System 47
3.5 Conversion from the Right Ascension to the Ecliptic System 49
3.6 Conversion from the Ecliptic to the Right Ascension System 49
3.7 Special Star Positions 57
4.1 Conventional Reduction of the Mean Place of a Star and Its Proper Motion from One Epoch to Another 124
4.2 Conventional Reduction of the Apparent Place of a Star from Its Mean Position Using Besselian Day Numbers and Star Constants 126
4.3 Conventional Reduction of the Apparent Place of a Star from Its Mean Position Using Independent Day Numbers 128
4.4 Reduction of the Apparent Place of a Star from Its Mean Position Using Matrices 129
5.1 Conversion of True Solar to Apparent Sidereal Time Without Conversion Tables 159
5.2 Conversion of True Solar to Apparent Sidereal Time with Tables 160
5.3 Conversion of Apparent Sidereal to True Solar Time Without Conversion Tables 161
5.4 Conversion of Apparent Sidereal to True Solar Time with Tables 162
7.1 Level Sensitivity Determination 280
7.2 Lost Motion Determination 286
7.3 Determination of the Mean Width of Contact Strips 287

7.4 Equatorial Drum Value Determination 288

8.1 Referencing an Emitted Time Signal to the BIH System 381
8.2 Referencing an Emitted Time Signal to the USNO System 382
8.3 High Precision Time Comparison with Portable Quartz Crystal Oscillators and VLF Signal Monitoring 387
8.4 First-Order Time Comparison with HF Signal Monitoring 395

9.1 Azimuth by the Hour Angle of Polaris 407
9.2 Azimuth by Hour Angle of Stars Near Elongation 411
9.3 Azimuth by Star Altitudes 419
9.4 Horizontal Angle by the Micrometer Method (Sample) 429

10.1 Latitude by Meridian Zenith Distances 440
10.2 Latitude by the Combined Zenith Distances of Polaris and a South Star 447
10.3 Latitude by the Horrebow-Talcott Method 464

11.1 Longitude by Zenith Distances Measured Near the Prime Vertical 485
11.2 Longitude by Meridian Transit Times 494
11.3 Longitude by the Prime Vertical Transit Times of Equal Altitude Stars 500
11.4 Longitude by Mayer's Method 509
11.5 Longitude and Latitude by Ex-Meridian Zenith Distances 525

12.1 General Prediction of a Solar Eclipse 555
12.2 Variations in the Besselian Coordinates 566
12.3 Prediction of the Immersion of 24 Psc, January 20, 1961 575
12.4 Position of Williston Observatory, Massachusetts, from Occultations 591

1 INTRODUCTION

The geographic position of a terrestrial point, as interpreted by map makers is its position relative to a geodetic reference surface, which is used in mapping by tradition as a substitute for the real surface of the earth. The most often used reference surfaces are the sphere, the ellipsoid (bi- or triaxial), and the equipotential surface of the gravity field of the earth (geopotential surface or geop). The position is usually defined by curvilinear coordinates, such as latitude, longitude, and height above the reference surface. It is customary to distinguish between spherical, geodetic (ellipsoidal), and astronomic (natural or gravitational) coordinates, depending on whether the sphere, the ellipsoid, or the geopotential surface is used as the reference surface, respectively. According to this interpretation the term 'geographic coordinate' is a general one including the three types of coordinates mentioned above.

Geodesy in the history of the sciences is one of the oldest. It has both scientific and practical missions. The major task of scientific geodesy is the determination of the size, shape, and gravity field of the earth. Using the results obtained, practical geodesy carries out the measurements and computations necessary for the accurate mapping of the earth's surface. The main mission of practical geodesy is the determination of geodetic and astronomic coordinates of fixed terrestrial points. The geodetic coordinates are traditionally determined by length and/or direction measurements (triangulation, trilateration, traversing), leveling, and gravity observations, practiced within the framework of geodetic surveying. The astronomic coordinates (excluding height) of fixed terrestrial points are determined from direction and/or time observations on stars.

The art and science of determining geographic positions and directions between neighboring stations (azimuths) from measurements on natural celestial bodies is generally termed geodetic astronomy. Direct measurements on artificial satellites for geographic position determination are handled in satellite (cosmic) geodesy. The position determination of moving objects (ships, airplanes, etc.) is treated in navigation.

The chapters which follow treat the details needed to practice geodetic astronomy; however, the requirements for satellite geodesy and navigation are also kept in mind in those sections which are relevant to all three of these disciplines. The discussion is restricted to applications on the northern hemisphere throughout except when noted.

2 THE EARTH FROM THE GEODETIC POINT OF VIEW

2.1 The Gravity Field of the Earth, the Geoid

The gravitational potential of the earth of M mass is given by

$$V = G \int_M \frac{dm}{s}, \tag{2.1}$$

where G is the Newtonian gravitational constant ($6.673 \pm 0.003 \times 10^{-8}\, cm^3\, g^{-1} s^{-2}$) and s is the distance between an attracting mass element, dm, and the attracted unit mass. The symbol M indicates that the integration is extended over the whole mass M. The solution is investigated in a geocentric rectangular coordinate system u, v, w, where the axis w coincides with the rotation axis of the earth and is positive to the north, the axis u is in the equatorial plane oriented in an arbitrary direction, and the axis v is perpendicular to u and w as shown in Fig. 2.1. The position of the unit mass m in this coordinate system is given either by the rectangular coordinates $\bar{u}, \bar{v}, \bar{w}$, or by its geocentric spherical coordinates φ', λ', and r. The position of the mass element, dm, is given by the rectangular coordinates u, v, w. Its position with respect to the attracted point is sufficiently defined by the polar coordinates ρ and ψ.

From the figure it is evident that

$$s^2 = \rho^2 + r^2 - 2 r\rho \cos\psi = r^2 \left[1 + \left(\frac{\rho}{r}\right)^2 - 2\frac{\rho}{r}\cos\psi\right],$$

or

$$\frac{1}{s} = \frac{1}{r}(1+t)^{-\frac{1}{2}},$$

where

$$t = \left(\frac{\rho}{r}\right)^2 - 2\frac{\rho}{r}\cos\psi\ . \tag{2.2}$$

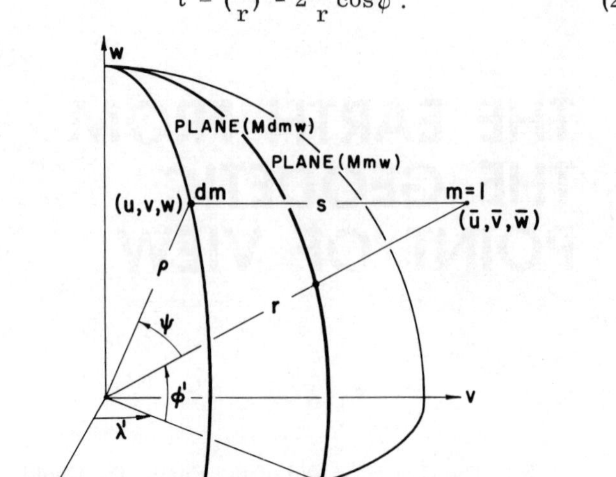

Fig. 2.1 Attraction of a unit mass

With binomial expansion,

$$\frac{1}{s} = \frac{1}{r}\left(1 - \frac{1}{2}t + \frac{3}{8}t^2 - \frac{5}{16}t^3 + \frac{35}{128}t^4 - \ldots\right).$$

Substituting this in equation (2.1) and indicating the integration term by term

$$V = \frac{G}{r}\int_M dm - \frac{G}{r}\int_M \frac{t}{2}\,dm + \frac{G}{r}\int_M \frac{3t^2}{8}\,dm - \frac{G}{r}\int_M \frac{5t^3}{16}\,dm + \ldots,$$

or

$$V = V_0 + V_1 + V_2 + V_3 + \ldots = \sum_{n=0}^{\infty} V_n \tag{2.3}$$

where, substituting the value of t from (2.2),

$$\left.\begin{aligned} V_0 &= \frac{G}{r}\int_M dm\,, \\ V_1 &= \frac{G}{r^2}\int_M \rho\cos\psi\,dm\,, \\ V_2 &= \frac{G}{2r^3}\int_M \rho^2\,(3\cos^2\psi - 1)\,dm\,, \\ &\ \ \vdots \end{aligned}\right\} . \tag{2.4}$$

From the infinite series (2.3), the first term is

$$V_0 = \frac{G}{r}\int_M dm = \frac{GM}{r} . \qquad (2.5)$$

This is the attraction potential of a sphere of mass M. If the earth were of spherical mass distribution, all other terms in (2.4) would be zero.

From the figure it is seen that

$$\cos\psi = \frac{u\bar{u} + v\bar{v} + w\bar{w}}{\rho r} ,$$

and therefore the second term, V_1, becomes

$$V_1 = \frac{G}{r^3}(\bar{u}\int_M udm + \bar{v}\int_M vdm + \bar{w}\int_M wdm) . \qquad (2.6)$$

As is known from classical mechanics, the integrals in this expression are zero because the origin of the coordinate system was chosen to be at the center of gravity. Therefore, in our system

$$V_1 = 0 .$$

The computation of the term V_2 is a little more space consuming. From Fig.2.1 it is evident that

$$\rho^2 = u^2 + v^2 + w^2 ,$$

and

$$\cos^2\psi = \frac{(u\bar{u} + v\bar{v} + w\bar{w})^2}{\rho^2 r^2} .$$

Substituting these into the expression for V_2 in (2.4)

$$V_2 = \frac{G}{2r^5}\Big[\bar{u}^2\int_M (2u^2 - v^2 - w^2)\, dm + \bar{v}^2\int_M (2v^2 - w^2 - u^2)\, dm + \\ + \bar{w}^2\int_M (2w^2 - u^2 - v^2)\, dm + 6\bar{u}\bar{v}\int_M uv\, dm + \\ + 6\bar{v}\bar{w}\int_M vw\, dm + 6\bar{w}\bar{u}\int_M wu\, dm\Big] ,$$

or by denoting the moments of inertia with

$$A = \int_M (v^2 + w^2)\, dm, \qquad D = \int_M vw\, dm,$$

$$B = \int_M (u^2 + w^2)\, dm, \qquad E = \int_M wu\, dm,$$

$$C = \int_M (u^2 + v^2)\, dm, \qquad F = \int_M uv\, dm,$$

the following is obtained:

$$V_2 = \frac{1}{2}\frac{G}{r^5}\Big[\bar{u}^2(B + C - 2A) + \bar{v}^2(C + A - 2B) + \bar{w}^2(A + B - 2C) + + 6\bar{v}\bar{w}D + 6\bar{w}\bar{u}E + 6\bar{u}\bar{v}F\Big] . \tag{2.7}$$

The quantities A, B, and C are the moments of inertia of the earth about the axes u, v, w respectively. If the axis w is the principal (maximum) inertia axis of the earth, the moments D and E are zero.

Transforming the rectangular coordinates $\bar{u}, \bar{v}, \bar{w}$ into spherical coordinates by means of

$$\begin{aligned} \bar{u} &= r\cos\varphi'\cos\lambda' , \\ \bar{v} &= r\cos\varphi'\sin\lambda' , \\ \bar{w} &= r\sin\varphi' , \end{aligned}$$

the following expression is obtained from (2.7):

$$V_2 = \frac{G}{r^3}\left\{\frac{2C - (A+B)}{4}(1 - 3\sin^2\varphi') + (3E\cos\lambda' + 3D\sin\lambda')\sin\varphi'\cos\varphi' + \left[\frac{3}{4}(B - A)\cos 2\lambda' + \frac{3}{2}F\sin 2\lambda'\right]\cos^2\varphi'\right\} \cdot \tag{2.8}$$

Similar expressions may be derived for V_3, V_4, etc.

The potential of gravity W is the sum of the gravitational (attraction) potential and the potential of the centrifugal force, thus

$$W = V + V' , \tag{2.9}$$

where the gravitational potential V may be taken from (2.3), while the potential of the centrifugal force is given by [Bomford, 1962, p. 389]

$$V' = \frac{\omega_e^2 r^2}{2}\cos^2\varphi' , \tag{2.10}$$

where the notation corresponds to Fig. 2.1, and ω_e is the rotational velocity of the earth.

By substituting expressions (2.5), (2.6), and (2.8) into (2.3), and the result with (2.10) into (2.9), the potential function of the earth's gravity (geopotential function) is obtained:

$$\begin{aligned} W = \frac{GM}{r} + \frac{G}{r^3}\Big\{&\frac{2C - (A+B)}{4}(1 - 3\sin^2\varphi') + \\ &+ (3E\cos\lambda' + 3D\sin\lambda')\sin\varphi'\cos\varphi' + \\ &+ \left[\frac{3}{4}(B - A)\cos 2\lambda' + \frac{3}{2}F\sin 2\lambda'\right]\cos^2\varphi'\Big\} + \\ &+ \ldots + \frac{\omega_e^2 r^2}{2}\cos^2\varphi' \cdot \end{aligned} \tag{2.11}$$

This equation may also be written in the following more general form of a spherical harmonic expansion [Mueller, 1964, pp. 181 - 186, and 354]:

$$W = \frac{GM}{r}\left\{\sum_{n=0}^{\infty}\sum_{m=0}^{n}\left(\frac{a_e}{r}\right)^n (C_{nm}\cos m\lambda' + S_{nm}\sin m\lambda')\,P_{nm}(\sin\varphi') + \right.$$

$$\left. + \frac{\omega_e^2 r^3}{2GM}\cos^2\varphi'\right\} \tag{2.12}$$

where a_e is the equatorial radius of the earth; C_{nm} and S_{nm} are mass-coefficients depending mostly on the distribution of masses inside the earth (e.g., on the moments of inertia); and $P_{nm}(\sin\varphi')$ is the associated function of Legendre (or Legendrian) of nth degree and mth order to be computed from the following equation:

$$P_{nm}(\sin\varphi') = \sqrt{(2n+1)\frac{k_m(n-m)!}{(n+m)!}}\;\frac{d^m P_n(\sin\varphi')}{d(\sin\varphi')^m}\cos^m\varphi'$$

where $k_m = 1$ when $m = 0$, while $k_m = 2$ for $m \neq 0$, and

$$P_n(\sin\varphi') = \frac{1}{2^n n!}\,\frac{d^n}{d(\sin\varphi')^n}(\sin^2\varphi' - 1)^n .$$

A recent set of mass-coefficients as determined from observations of artificial satellites is given in Table 2.1 [Gaposchkin, 1966].

The main properties of the function W are that together with its first derivatives, it is single valued and continuous, its second derivatives exist, and within and on the boundaries of a certain region it satisfies the equation of Poisson [Bomford, 1962, p. 394],

$$\frac{\partial^2 W}{\partial u^2} + \frac{\partial^2 W}{\partial v^2} + \frac{\partial^2 W}{\partial w^2} = 2\omega_e^2 - 4\pi G\sigma$$

where σ is the density of the matter at the point in question.

The equipotential surface, W = constant, is called a geopotential surface or a geop. The geop which most nearly coincides with the undisturbed mean surface of the oceans is the geoid ($W = W_g$), customarily used in classical geodesy as the shape of the earth in first approximation (theoretical shape of the earth).

TABLE 2.1 Mass Coefficients of the Earth

n	m	C_{nm}	S_{nm}	n	m	C_{nm}	S_{nm}
2	0	-484.174	.000	7	7	.055	.096
2	2	2.379	-1.351	8	0	.065	.000
3	0	.963	.000	8	1	-.075	.065
3	1	1.936	.266	8	2	.026	.039
3	2	.734	-.538	8	3	-.037	.004
3	3	.561	1.620	8	4	-.212	-.012
4	0	.550	.000	8	5	-.053	.118
4	1	-.572	-.469	8	6	-.017	.318
4	2	.330	.661	8	7	-.0087	.031
4	3	.851	-.190	8	8	-.248	.102
4	4	-.053	.230	9	0	.012	.000
5	0	.063	.000	9	1	.117	.012
5	1	-.079	-.103	9	2	-.004	.035
5	2	.631	-.232	9	9	-.065	.0909
5	3	-.520	.007	10	0	.012	.000
5	4	-.265	.064	10	1	.105	-.126
5	5	.156	-.592	10	2	-.105	-.042
6	0	-.179	.000	10	3	-.065	.030
6	1	-.047	-.027	10	4	-.074	-.111
6	2	.069	-.366	11	0	-.063	.000
6	3	-.054	.031	11	1	-.053	.015
6	4	-.044	-.518	12	0	.071	.000
6	5	-.313	-.458	12	1	-.163	-.071
6	6	-.040	-.155	12	2	-.103	-.0051
7	0	.086	.000	13	0	.022	.000
7	1	.197	.156	13	12	-.058	.048
7	2	.364	.163	13	13	-.075	.010
7	3	.250	.018	14	0	-.033	.000
7	4	-.152	-.102	14	1	-.015	.0053
7	5	.076	.054	15	12	-.062	.058
7	6	-.209	.063	15	13	-.063	-.066
				15	14	.0083	-.0201

Multiply all values by 10^{-6}

2.2 The Normal Gravity Field of the Earth, the Spheroid, Earth-Spherop, Earth and Reference Ellipsoids, and Sphere

Equation (2.11) or (2.12) expresses the geopotential function of the earth in the coordinate system of Fig. 2.1. In the case of rotational symmetry in mass distribution about the axis w, in (2.11) the moment A = B, and F = 0. If the axis w is the principal inertia axis of the earth, the moments D and E are zero. These conditions are the equivalent of making in equation (2.12) the mass-coefficients $S_{nm} = 0$, and $C_{nm} = 0$ (for $m > 0$). In addition, if it is assumed that the distribution of masses is symmetrical about the plane of the equator and that in expression (2.12) the terms with degrees higher than four are zero, the concept of the normal earth is reached. The geopotential function of the normal earth is called spheropotential function and is denoted by U. Thus

$$U = \left\{ \frac{GM}{r} \sum_{n=0,2,4} \left(\frac{a_e}{r}\right)^n C_{no} P_{no} (\sin\varphi') + \frac{\omega^2 r^3}{2\,GM} \cos^2\varphi' \right\} . \quad (2.13)$$

The equipotential surface, U = constant, is called spheropotential surface or spherop. The spherop which possesses the same constant as the geoid, i.e., upon which $U = W_g$, is the spheroid. The spherop whose volume is the same as that of the geoid may be called the earth-spherop. The surface of the spheroid or that of the earth-spherop is used in geodesy as the theoretical shape of the earth in second approximation. It may be shown that, neglecting terms smaller than the second power of the flattening, the surface of the earth-spherop is identical to that of a rotational ellipsoid (earth-ellipsoid), which is used as a third approximation to the theoretical shape of the earth.

The flattening f of the earth-ellipsoid may be computed from [Mueller, 1964, p.357]

$$f = \frac{a_e - b_e}{a_e} = 0.001730342 - \frac{3\sqrt{5}}{2} C_{20} - \frac{15}{8} C_{40} \quad (2.14)$$

where b_e is the polar semi-diameter. The size of the earth-ellipsoid is defined by the equal volume condition mentioned above.

It may be shown that the gravity of the normal earth, or, as it is called, the normal gravity, γ, on the surface of the earth-spherop (or approximately on the surface of the earth-ellipsoid of the same flattening) is [Mueller, 1964, p.360]

$$\gamma = -\frac{\partial U}{\partial r} = \gamma_e (1 + \beta \sin^2\varphi + \beta_1 \sin^2 2\varphi) \quad (2.15)$$

where the normal gravity at the equator, γ_e, is

$$\gamma_e = \frac{GM}{a_e^2} (1 + f - \frac{3}{2} q - \frac{3}{7} qf + f^2) , \quad (2.16)$$

and

$$\begin{aligned} \beta &= \frac{5}{2} \bar{q} - f - \frac{17}{14} \bar{q} f + \frac{15}{4} \bar{q}^2 , \\ \beta_1 &= \frac{1}{8} f^2 - \frac{5}{8} \bar{q} f , \end{aligned} \quad (2.17)$$

$$\tan \varphi = \frac{\tan \varphi'}{(1-f)^2} \;, \tag{2.18}$$

$$\bar{q} = \frac{\omega_e^2 a_e^3}{GM} (1-f) = q(1-f) \;. \tag{2.19}$$

Equations (2.15) and (2.17), relating normal gravity and flattening, are known as the theorem of Clairaut.

Using the recent determinations of C_{20}, C_{40}, GM, and γ_e, equations (2.14) and (2.16) yield the following parameters and standard errors for

TABLE 2.2 Parameters of Reference Ellipsoids

Name	Year	a_e	1/f
Bouguer, Maupertuis	1738	6,397,300 m	216.8
Delambre	1800	6,375,653	334.0
Walbeck	1819	6,376,896	302.78
Everest	1830	7,276.345	300.8017
Airy	1830	7,563.396	299.324964
Bessel	1841	7,397.155	299.152813
Clarke	1858	8,293.645	294.26
Pratt	1863	8,245	295.3
Clarke[1]	1866	8,206.4	294.978698
Clarke (modified)	1880	8,249.145	293.4663
Hayford	1906	8,283	297.8
Helmert	1907	8,200	298.3
Hayford (International Ellipsoid)	1909-10	8,388	297.0
Heiskanen	1926	8,397	297.0
Krasovsky	1936	8,210	298.6
Krasovsky[2]	1940	8,245	298.3
Jeffreys	1948	8,099	297.10
Ledersteger	1951	8,298	297.0
Hough	1956	8,260	297.0
Fischer	1960	8,155	298.3
Australian "165"	1962	8,165	298.3
Kaula[3]	1964	8,160	298.25
Veis	1964	8,169	298.25

[1] Defined by the values of a_e and b_e = 6,356,583.8 m

[2] Originally a triaxial ellipsoid (equatorial ellipticity = 1/30,000; direction of the longest axis, $\lambda = 15°$ E)

[3] Recommended by the International Astronomical Union, 1964.

the earth-ellipsoid [Rapp, 1967]:

$$1/f = 298.25 \pm 0.02,$$

$$a_e = 6{,}378{,}157.5 \pm 10.8\text{m} .$$

On the corresponding earth-spherop

$$\gamma_e = 978.0284 \pm 0.0022\,\text{cm s}^{-2} ,$$

$$W_s = 6{,}263{,}678.8 \pm 10.6 \times 10^5\text{cm}^2\text{s}^{-2} ,$$

$$GM = 3.986013 \pm 0.000010 \times 10^2\text{cm}^3\text{s}^{-2} .$$

Since the size and shape of the earth-ellipsoid have not been known until quite recently, the practice has been to substitute various similar but arbitrarily defined reference ellipsoids (see Table 2.2). The dimensions, shapes, centers, and orientations of these reference ellipsoids usually differ only slightly from those of the more recently determined earth-ellipsoid. These are used as reference surfaces in geodetic calculations by the different countries.

For those geodetic computations where low accuracy is sufficient, the ellipsoid may be substituted by a sphere serving as the fourth approximation to the theoretical shape of the earth. The radius of the sphere having the volume (mass) of the ellipsoid is

$$R = a_e \sqrt[3]{1-f} . \tag{2.20}$$

2.3 Fundamental Definitions Related to the Geodetic Reference Surfaces

2.31 The Sphere. The sphere is a closed surface upon which all points are of the same distance from the center, O (Fig.2.2). The distance of any surface point from the center is the spherical radius, R.

The intersection of any plane with the sphere always results in a circle. Every plane containing the center O intersects the sphere in a great circle. Let the sphere rotate about an axis containing the center. The great circle perpendicular to the rotation axis is the spherical equator. All other circles parallel to the equator are the spherical parallels. The great circles whose generating planes contain the rotation axis are the spherical meridians. The intersection points of these meridians are the spherical poles. It is customary to distinguish between the north and the south spherical pole.

Let point P be an arbitrary surface point on the sphere. The line OP is perpendicular to the surface and is called the spherical normal. The angle b between this normal and a plane perpendicular to the rotation axis, e.g., the equator, is called the spherical latitude of P. It is taken from 0° to 90° with a positive sign on the northern half of the sphere and with a negative sign on the southern half. Points of the same spherical latitude are situated on a spherical parallel.

One of the spherical meridians is designated as the starting (zero) meridian. The angle ℓ between the plane of this zero meridian and the spherical meridian plane of P, measured in a plane perpendicular to the rotation axis, e.g., in the equatorial plane, is the spherical longitude of P. It is reckoned positive to the east from the zero meridian to 360°. Points of the same spherical longitude are situated on a spherical meridian.

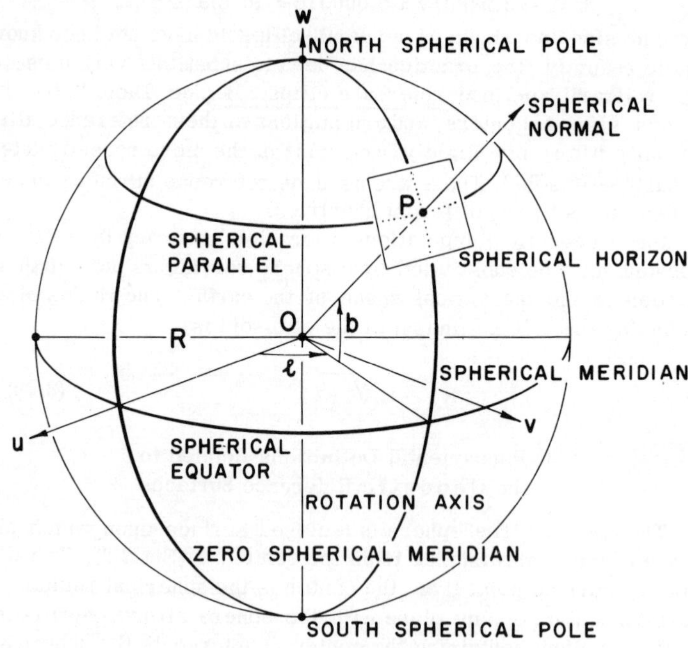

Fig. 2.2 The sphere

Planes containing the spherical normal of a point are called spherical normal planes. Their intersections with the sphere are the spherical normal sections, which are always great circles. The plane perpendicular to the spherical normal at P, thus tangent to the sphere, is the spherical horizon of P.

Let Q be an arbitrary surface point on the sphere different from P (Fig. 2.3). The angle a between the spherical meridian plane of P and its normal plane through Q, measured in the spherical horizon at P, is the spherical azimuth of Q. It is reckoned from the north side of the meridian and is measured clockwise from 0° to 360°.

Let point S be a surface point on the sphere different from P and Q. The great circles PQ, PS, and QS form a spherical triangle.

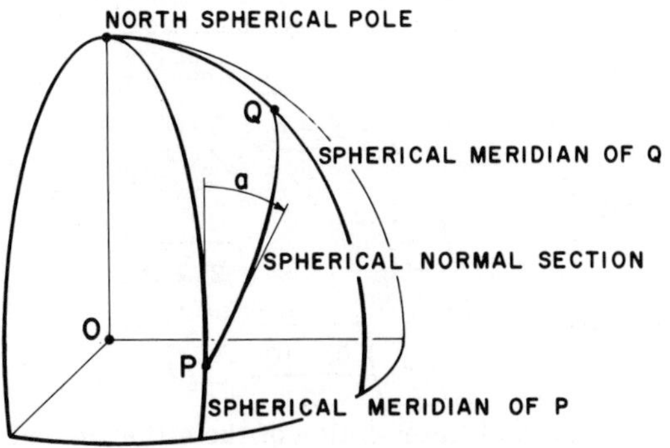

Fig. 2.3 Azimuth on the sphere

A Cartesian coordinate system, u, v, w, is now introduced with the origin at the center of the sphere. The axis w coincides with the rotation axis and points to the north. The axis u is the intersection of the zero spherical meridian and the spherical equator, while the axis v forms a right-handed system with u and w (Fig. 2.2). The rectangular coordinates of P in this sytem are

$$\left.\begin{aligned} u &= R\cos b \cos \ell \\ v &= R\cos b \sin \ell \\ w &= R\sin b \end{aligned}\right\} . \tag{2.21}$$

2.32 The Ellipsoid. The quadric surface on which all curves of intersection with planes are ellipses is called an ellipsoid. From the geodetic point of view, it is customary to distinguish between rotational, or biaxial, and triaxial ellipsoids. The rotational ellipsoid is generated by rotating an ellipse about its shortest (minor) axis. It is defined by two parameters, e.g., by the lengths of the shortest and longest axes. The triaxial ellipsoid is defined by three parameters, e.g., by the lengths of three mutually perpendicular axes. From here on only the rotational ellipsoid is used since it is more frequently used as a geodetic reference surface than the triaxial one. It will be called simply the 'ellipsoid.'

When the ellipsoid is defined by the lengths of its longest and shortest axes, a_e and b_e respectively, its equation in the coordinate system shown in Fig. 2.4 and defined later is

$$\frac{u^2 + v^2}{a_e^2} + \frac{w^2}{b_e^2} = 1 . \tag{2.22}$$

It is also customary to define the ellipsoid by the length of the long axis a_e and by one of the following auxiliary quantities:

Flattening,

$$f = \frac{a_e - b_e}{a_e} . \tag{2.23}$$

First eccentricity,

$$e = \sqrt{\frac{a_e^2 - b_e^2}{a_e^2}} = \sqrt{2f - f^2} . \tag{2.24}$$

Second eccentricity,

$$e' = \sqrt{\frac{a_e^2 - b_e^2}{b_e^2}} . \tag{2.25}$$

In geodesy the curve of intersection of the ellipsoid and a plane perpendicular to the rotation axis and containing the ellipsoid's center is called the geodetic equator. All other circles on the ellipsoid parallel to the equator are the geodetic parallels. The ellipses whose generating planes contain the rotation axis are the geodetic meridians. The two points in which the meridians intersect are the geodetic poles. It is customary to distinguish between the north and the south geodetic pole.

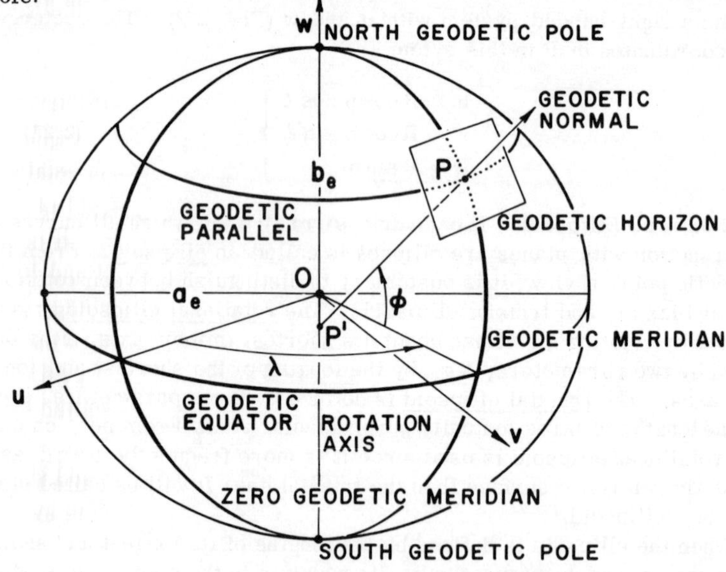

Fig. 2.4 The ellipsoid

Let point P be an arbitrary surface point on the ellipsoid. The line PP′ is perpendicular to the surface and is called geodetic normal. The angle φ between this normal and a plane perpendicular to the rotation axis, e.g., the geodetic equator, is called the geodetic latitude of P. It is taken from 0° to 90° with a positive sign on the northern half of the ellipsoid and with a negative sign on the southern half. Points of the same geodetic latitude are situated on a geodetic parallel.

One of the geodetic meridians is designated as the starting (zero) meridian. The angle λ, between the plane of this zero meridian and the geodetic meridian plane of P, measured in a plane perpendicular to the rotation axis, e.g., in the geodetic equatorial plane, is the geodetic longitude of P. It is reckoned positive to the east from the zero meridian to 360°. Points of the same geodetic longitude are situated on a geodetic meridian.

The geodetic latitude and longitude are commonly known as geodetic coordinates, and they define the position of a point on the surface of the ellipsoid.

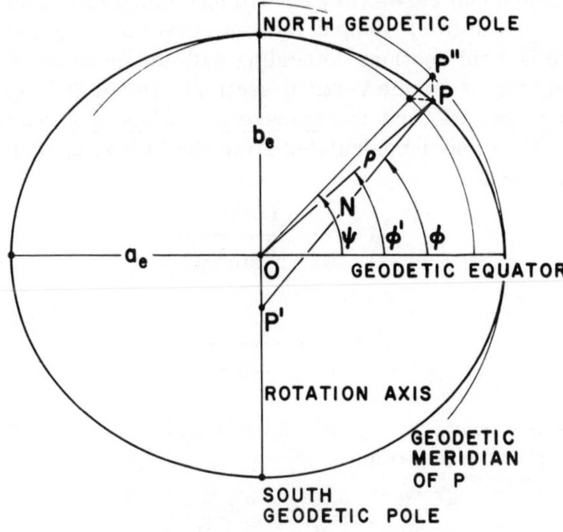

Fig. 2.5 The meridian ellipse

For computational purposes two special latitudes are introduced in Fig. 2.5 showing the meridian plane of point P. The line OP is the geocentric radius of P. The angle φ', between the radius and the geodetic equator is the geocentric latitude. The angle ψ between the line OP'' and the equator is called the reduced latitude. The point P'' is obtained by projecting P parallel to the rotation axis to the auxiliary circle of radius a_e. The relations between the three latitudes are [Bomford, 1962, p. 496]

$$\left.\begin{aligned} \tan\varphi' &= \frac{b_e}{a_e}\tan\psi &&= \frac{b_e^2}{a_e^2}\tan\varphi \\ \tan\varphi' &= (1-f)\tan\psi &&= (1-f)^2\tan\varphi \\ \tan\varphi' &= \sqrt{1-e^2}\tan\psi &&= (1-e^2)\tan\varphi \end{aligned}\right\} . \tag{2.26}$$

The length ρ of the geocentric radius is given by

$$\rho = \frac{a_e\sqrt{1-e^2}}{\sqrt{1-e^2\cos^2\varphi'}} . \tag{2.27}$$

Planes containing the geodetic normal of a point are called geodetic normal planes. Their intersections with the ellipsoid are the geodetic normal sections. The radii of curvature of these normal sections, unlike in the spherical case, vary according to the directions of the normal sections. The two principal radii of curvature at a given point are those of the normal sections coinciding with the geodetic meridian, and perpendicular to it (prime vertical section), called the meridional radius of curvature, M, and the transverse radius of curvature, N, respectively. They may be calculated from the following relations [Bomford, 1962, p.496]:

$$M = \frac{a_e(1-e^2)}{(1-e^2\sin^2\varphi)^{\frac{3}{2}}} , \tag{2.28}$$

$$N = \frac{a_e}{(1-e^2\sin^2\varphi)^{\frac{1}{2}}} . \tag{2.29}$$

The radius of curvature, R_α, of an arbitrary geodetic normal section is given by Euler's theorem:

$$\frac{1}{R_\alpha} = \frac{\cos^2\alpha}{M} + \frac{\sin^2\alpha}{N} \tag{2.30}$$

where α is the azimuth of the normal section at the point in question. The normal section azimuth from the point P to an arbitrary point Q on the ellipsoid is defined as the angle between the plane of the geodetic meridian of P and the geodetic normal plane at P, containing Q, measured in the geodetic horizon of P (plane perpendicular to the geodetic normal at P). The azimuth is reckoned from the north side of the meridian and is measured clockwise from 0° to 360°. This definition of the normal section azimuth requires some explanation. On the sphere the spherical normal plane of point P through point Q and that of point Q through P are identical; thus their curves of intersection on the sphere coincide. On the ellipsoid the situation is different. The geodetic normal sections originating at points P and Q, and terminating at Q and P respectively, do not coincide (Fig. 2.6). Thus at any of the two points there are two intersecting geodetic normal sections which in fact may lead to some confusion as to which normal section the azimuth should be measured.

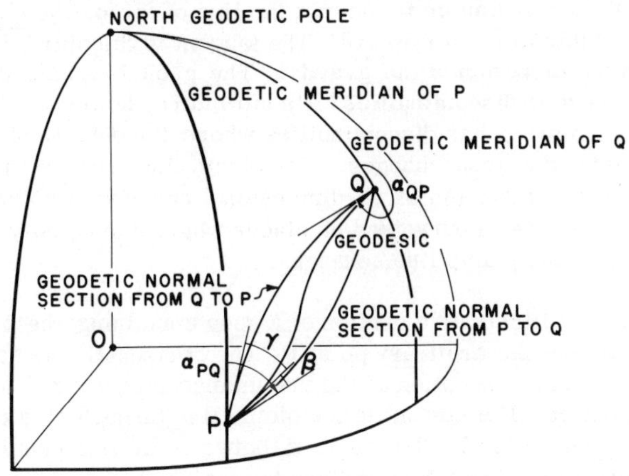

Fig. 2.6 Azimuths on the ellipsoid

To avoid the possibility of confusion, the term geodetic azimuth is introduced. It is defined as the angle between the geodetic meridian and the tangent to the surface curve of shortest distance between the two points, the geodesic. The geodesic divides the angle between the normal sections in the proportion $\beta:\gamma = 1:2$. The angle β is very small ($\beta_{maximum} = 0\text{''}014$ for a normal section 100 km long), but it increases with the length of the line.

Finally a Cartesian coordinate system u, v, w is introduced with the origin at the center of the ellipsoid (Fig. 2.4). The axis w coincides with the rotation axis of the ellipsoid and points to the north. The axis u is the intersection of the zero geodetic meridian and the geodetic equator, while the axis v forms a right-handed system with u and w. The rectangular coordinates of point P in this system may be calculated from one of the following sets of equations:

$$\left.\begin{aligned} u &= N\cos\varphi\cos\lambda = \rho\cos\varphi'\cos\lambda = a_e\cos\psi\cos\lambda \\ v &= N\cos\varphi\sin\lambda = \rho\cos\varphi'\sin\lambda = a_e\cos\psi\sin\lambda \\ w &= N(1-e^2)\sin\varphi = \rho\sin\varphi' = b_e\sin\psi \end{aligned}\right\} \quad (2.31)$$

2.33 The Geopotential Surface. The geopotential surface or the geop is defined by the equation

$$W(\varphi', \lambda', r) = W(u, v, w) = \text{constant} \qquad (2.32)$$

where the function $W(\varphi', \lambda', r)$ is given by equation (2.11) or (2.12). Since the shape of the geop depends mostly on the irregular distribution of masses in the earth's interior, it cannot be described by a closed mathematical formula such as the ellipsoid's; thus, in the conventional sense, it is considered a physical rather than a mathematical surface. Its main properties which follow from each other are that the potential W is constant on it and that the line of force, the plumb line, is perpendicular to it everywhere. The tangent to the plumb line coincides with the direction of the gravity. The geop is a smooth surface with no breaks or discontinuities. Its curvature, however, due to the equation of Poisson, has discontinuities where the density of the mass intersected by the geop changes. The plumb line has the same properties; thus it is a smooth three-dimensional curve with no breaks and with discontinuities in curvature at places where the density of mass, intersected by the plumb line, changes.

Let Fig. 2.7 illustrate a section of a geop containing the tangent to the plumb line at the arbitrary point P. The horizontal line at the center is the line of intersection of the instantaneous equatorial plane with the geop-section. The tangent to the plumb line through P is called the astronomic normal of P. The angle Φ between the astronomic normal and the plane of the instantaneous equator (which is perpendicular to the instantaneous axis of rotation) is the astronomic latitude of P. It is taken from 0° to 90° with a positive sign on the northern half of the geop and with a negative sign on the southern half.

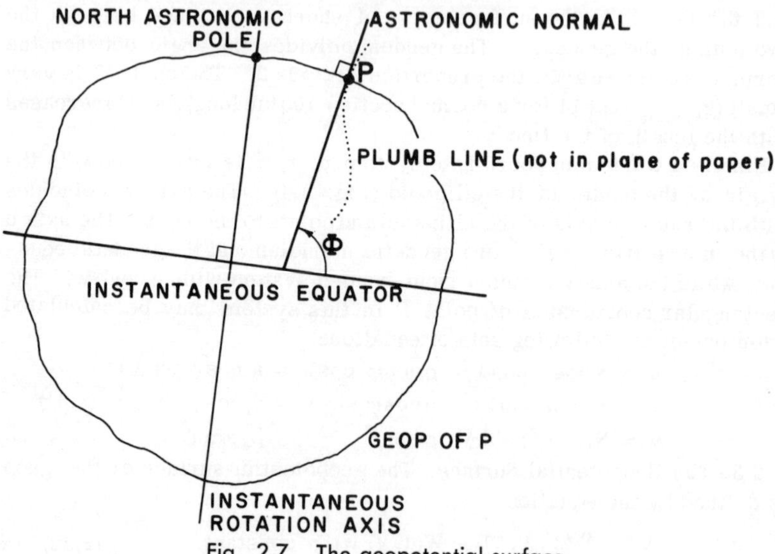

Fig. 2.7 The geopotential surface

The plane containing the astronomic normal at P and parallel to the instantaneous rotation axis of the earth is the astronomic meridian plane of P. The meridian plane cannot be defined as previously (the plane containing both the normal and the rotation axis) because the astronomic normal does not necessarily intersect the rotation axis. The angle Λ, between some designated zero (Greenwich mean) astronomic meridian plane and that of the point P, measured in the plane of the instantaneous equator, is the astronomic longitude of P. It is reckoned positive to the east from the zero meridian to 360°. Points of the same astronomic latitude or longitude are situated on the same astronomic parallel and meridian respectively. These curves are neither circles nor ellipses, as on the sphere and on the ellipsoid, but are irregular (Fig. 2.8) space curves.

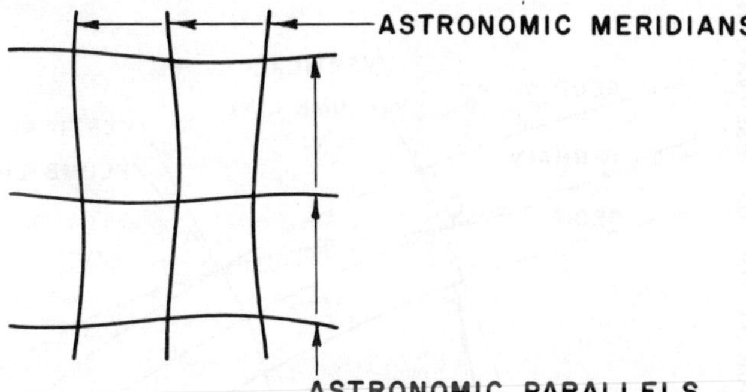

Fig. 2.8 Astronomic meridians and parallels

The astronomic poles are defined as the points of intersection of the instantaneous rotation axis with the geop. It is customary to distinguish between the north and the south astronomic pole. It follows from the definition of the astronomic latitude that it is not a necessity that at the astronomic poles the astronomic latitude be ±90°. Similarly, it is possible that the astronomic latitude be ±90° at a point different from the astronomic poles. This depends on the shape of the geop, i.e., on the distribution of masses.

The plane perpendicular to the astronomic normal at P is the astronomic horizon. The planes containing the astronomic normal at P are called the astronomic normal planes of P.

Let Q be an arbitrary point different from P. The astronomic azimuth of Q is the angle between the astronomic meridian plane of P and the astronomic normal plane of P through Q, measured in the astronomic horizon of P. It is reckoned from the north side of the meridian clockwise from 0° to 360°.

In practice the geop to which the astronomic system described above is most often referred is the one containing the terrestrial point where the observation takes place. In this case the astronomic quantities (latitude, longitude, azimuth, poles, etc.) should bear the prefix 'observed' or 'instantaneous.' Thus one may speak about observed or instantaneous astronomic latitude, longitude, azimuth, etc., all referring to the geop of the observation point. In this work, in order to simplify the terminology, these will be called simply 'astronomic' quantities. Whenever these are referred to a different geop a note is made to this effect. Exceptions to this rule are the astronomic normal and the astronomic normal plane which when referred to the observer's geop are called vertical and vertical plane respectively. According to this definition,

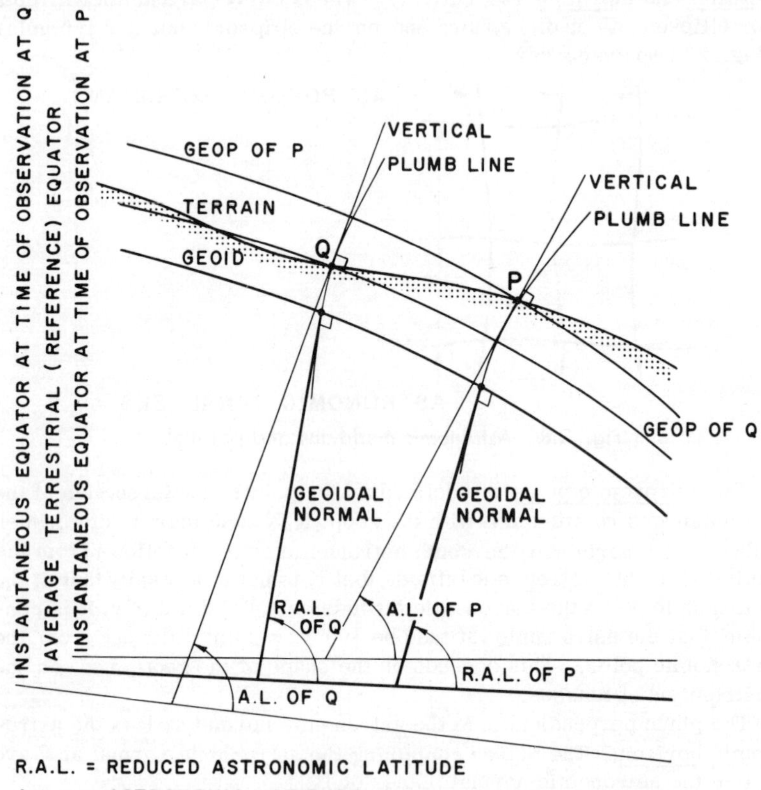

Fig. 2.9 Astronomic latitudes

for example, the astronomic latitude is the angle between the observer's vertical and the instantaneous equatorial plane.

Evidently this definition of the astronomic quantities has a few disadvantages. First, since the various observation points are not likely to be located on the same geop, the quantities will refer to different reference surfaces; thus the astronomic meridians and parallels lose their meanings. Secondly, since the rotation axis of the earth changes its position with respect to the solid body of the earth (polar motion, see section 4.13) and since the observations are not likely to be carried out at the same time, they will refer to different instantaneous rotation axes and equators (Fig. 2.9). To avoid this difficulty the astronomic quantities may also be referred to the geoid, which is a common reference surface for all observers, and to a designated average (reference) rotation axis and corresponding equator of the earth (see section 4.131). In this case they are usually called '<u>ordinary</u>,' '<u>average</u>,' or '<u>reduced</u>' quantities. The last designation will be used here; thus, for example, the reduced astronomic latitude is the angle between the geoidal normal and the average terrestrial (reference) equator. The reduced astronomic quantities differ from the astronomic ones by corrections for the polar motion and for the curvature of the plumb line (see sections 4.133, 9.32, 10.32, and 11.42).

2.4 Absolute and Relative Positions, the Geodetic Datum, the Deflection of the Vertical

As shown previously, the position of a point on a reference surface may be defined by its astronomic, geodetic, or other coordinates. The geodetic coordinates may refer to a reference ellipsoid (Table 2.2) of arbitrary orientation and position, or to one whose orientation is still arbitrary but whose center is at the center of gravity of the earth and whose minor axis coincides with its average rotation axis. In the first case the position is called <u>relative</u>, in the second case <u>absolute</u>. The geodetic coordinates in the absolute (geocentric) system will be called absolute- or <u>geocentric-geodetic coordinates</u> (not to be confused with the commonly used geocentric coordinates defined by equation 2.26).

The customary astronomic coordinates are the reduced astronomic latitude and longitude, Φ and Λ respectively, and the height of the observer above the geoid measured along the plumb line. This height is called the <u>orthometric height</u>, here denoted by H, and is obtained from spirit leveling and en route gravity observations [Baeschlin, 1960]. The quantites Φ, Λ, and H define the position of the observer with respect to the geoid and average rotation axis of the earth.

The customary geodetic coordinates are the geodetic latitude and longitude, φ and λ respectively, and the height of the observer above the reference ellipsoid, measured along the geodetic normal. This height is called the <u>geodetic height</u> and is denoted by h (Fig. 2.10). The quantities φ, λ, and h define the position of the observer with respect to a reference ellipsoid.

The geodetic coordinates are determined from direction and/or

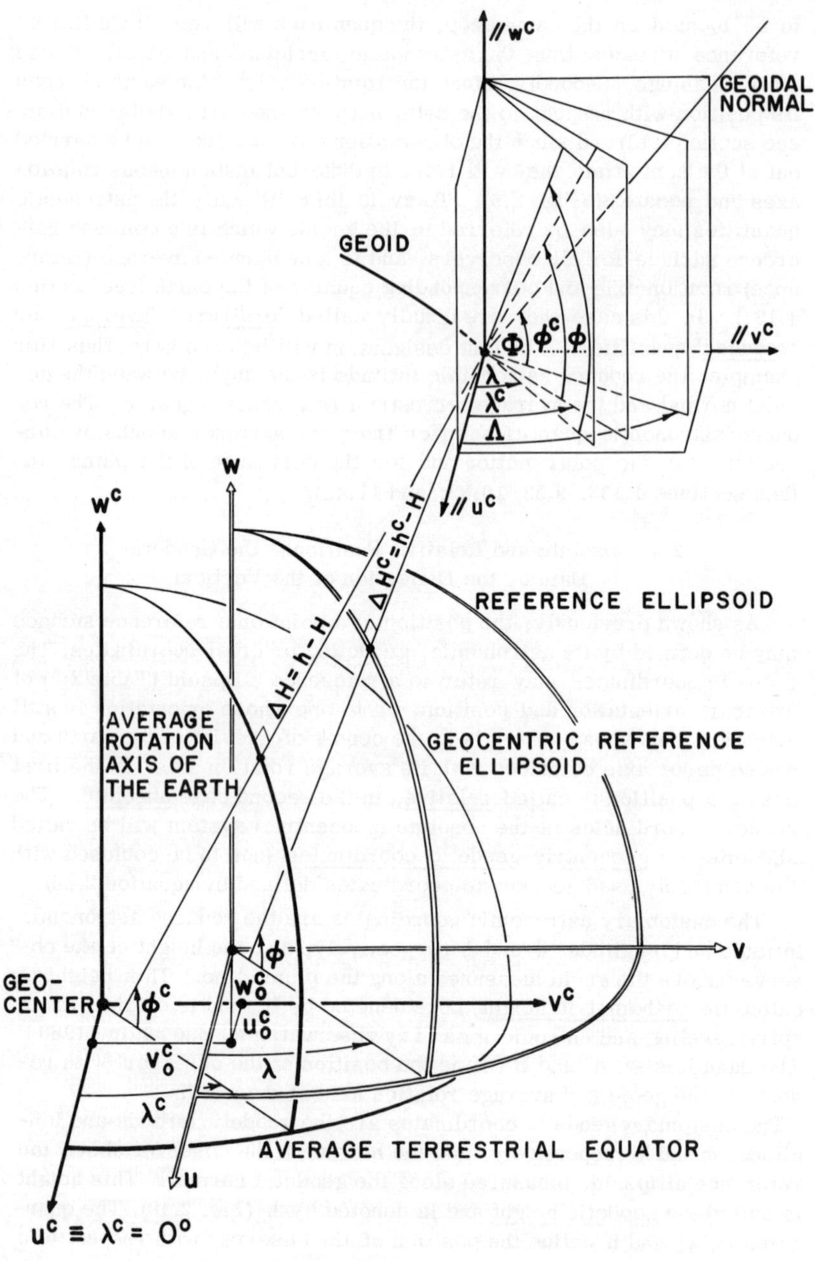

Fig. 2.10 Absolute and relative geodetic coordinates

length measurements observed on the surface, reduced to the ellipsoid, and are assumed to be adjusted using the Laplace condition $(A-\alpha)=(\Lambda-\lambda)\sin\varphi$ (A and α are the astronomic and geodetic azimuths respectively) to keep control on angular distortions within the net [Bomford, 1962, pp. 90 and 134-136]. The position of the reference ellipsoid with respect to the geoid and the average rotation axis of the earth is defined by the geodetic coordinates of the initial point of the triangulation or trilateration, φ_o, λ_o, and h_o, or by the differences between the astronomic and the geodetic coordinates there:

$$\left.\begin{aligned}\Delta\Phi_o &= \Phi_o - \varphi_o\\ \Delta\Lambda_o &= \Lambda_o - \lambda_o\\ \Delta H_o &= h_o - H_o\end{aligned}\right\}. \tag{2.33}$$

It should be noted that the elevations h and H are measured along different directions (geodetic normal and plumb line respectively); therefore, theoretically, they cannot be directly combined. For practical purposes, however, the effect of the different directions may be neglected. The quantities $\Delta\Phi_o$, $\Delta\Lambda_o$, and ΔH_o, together with the parameters of the reference ellipsoid a_e and f, define the size, the shape, and the position of the center of the reference ellipsoid. The orientation of the reference ellipsoid with respect to the geoid is given by the condition that its minor axis should be parallel to the average rotation axis of the earth, i. e., by also measuring an astro-azimuth at the initial point and applying the Laplace equation to obtain the corresponding geodetic azimuth. This condition together with the quantities $\Delta\Phi_o$, $\Delta\Lambda_o$, ΔH_o, a_e, and f define a geodetic datum. The major geodetic datums are compiled in Fig. 2.11, and Table 2.3 lists the locations of the corresponding initial points and the reference ellipsoids used.

The geocentric geodetic coordinates are denoted by φ^c, λ^c, and h^c (Fig. 2.10). Thus a geocentric geodetic datum is defined by $\Delta\Phi_o^c$, $\Delta\Lambda_o^c$, ΔH_o^c, a_e, and f, where

$$\left.\begin{aligned}\Delta\Phi_o^c &= \Phi_o - \varphi_o^c\\ \Delta\Lambda_o^c &= \Lambda_o - \lambda_o^c\\ \Delta H_o^c &= h_o^c - H_o\end{aligned}\right\} \tag{2.34}$$

and the index o refers again to the initial point.

The relative Cartesian coordinates u, v, w of a point may be computed from

$$\left.\begin{aligned}u &= (N+h)\cos\varphi\cos\lambda\\ v &= (N+h)\cos\varphi\sin\lambda\\ w &= [N(1-e^2)+h]\sin\varphi\end{aligned}\right\}. \tag{2.35}$$

The axis u is in the zero geodetic meridian plane, v points to $\lambda=90°$ in the equatorial plane, and w is parallel to the average rotation axis of the earth (Fig. 2.10). In the equations, N is the transverse radius of curvature to be computed from equation (2.29).

The absolute (geocentric) Cartesian coordinates u^c, v^c, w^c, of the point

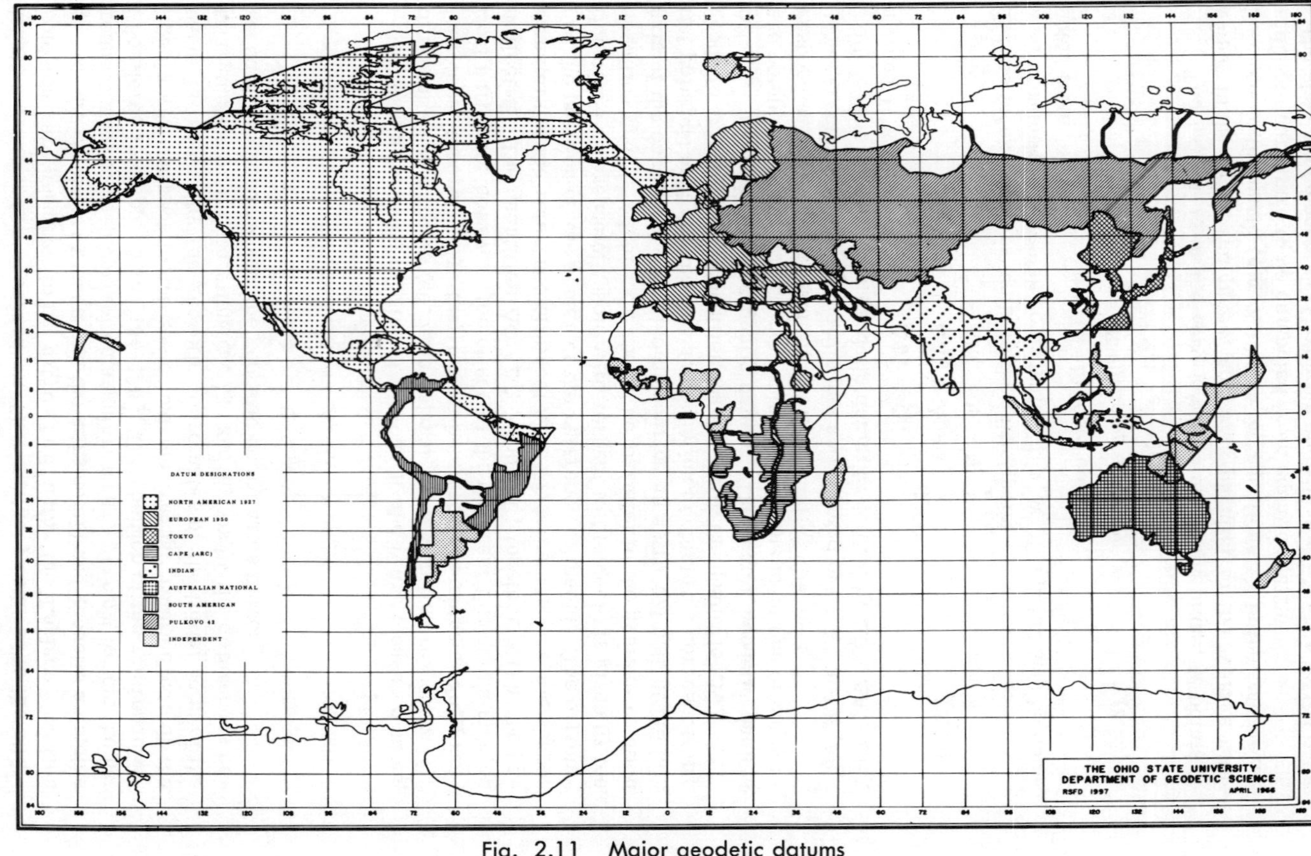

Fig. 2.11 Major geodetic datums

TABLE 2.3 Major Geodetic Datums

Name[1]	Reference Ellipsoid	Initial Point: Location	Latitude	Longitude
Argentinian	International	Campo Inchauste, Argentina	-35° 58'	-62° 10'
Australian National	Kaula	Grundy, Australia	-25° 54'	134° 33'
Cape (Arc)	Clarke 1880	Buffelsfontein South Africa	-34° 00'	25° 31'
European 1950	International	Potsdam, Germany	52° 23'	13° 04'
Indian	Everest	Kalianpur, India	24° 07'	77° 39'
North American 1927	Clarke 1866	Meades Ranch, Kansas	39° 14'	-98° 32'
Pulkovo 42	Krasovsky	Pulkovo, U. S. S. R.	59° 46'	30° 19'
South American	International	La Canoa, Venezuela	08° 34'	-63° 52'
Tokyo	Bessel	Tokyo, Japan	35° 39'	139° 45'

[1] For the location of these datums see Fig. 2.11.

may be calculated from equations (2.35) substituting φ^c, λ^c and h^c for φ, λ and h respectively.

The conversion of the relative coordinates to absolute may be done by computing first the geocentric coordinates of the center of the reference ellipsoid and adding them to the coordinates of the point. The geocentric coordinates of the ellipsoidal center are given with sufficient accuracy by [Veis, 1960, p. 101]:

$$\left.\begin{aligned} u_e^c &= \delta\Phi_o M_o \sin\varphi_o \cos\lambda_o + \delta\Lambda_o N_o \cos\varphi_o \sin\lambda_o \\ &\quad + \delta H_o \cos\varphi_o \cos\lambda_o \\ v_e^c &= \delta\Phi_o M_o \sin\varphi_o \sin\lambda_o - \delta\Lambda_o N_o \cos\varphi_o \cos\lambda_o \\ &\quad + \delta H_o \cos\varphi_o \sin\lambda_o \\ w_e^c &= -\delta\Phi_o M_o \cos\varphi_o + \delta H_o \sin\varphi_o \end{aligned}\right\} \tag{2.36}$$

where

$$\delta\Phi_o = \Delta\Phi_o^c - \Delta\Phi_o = \varphi_o - \varphi_o^c \,,$$
$$\delta\Lambda_o = \Delta\Lambda_o^c - \Delta\Lambda_o = \lambda_o - \lambda_o^c \,,$$
$$\delta H_o = \Delta H_o^c - \Delta H_o = h_o^c - h_o \,,$$

and M is the meridional radius of curvature to be computed from equation (2.28). The index o again refers to the initial point. With these the geocentric Cartesian coordinates of the point are

$$\left.\begin{aligned} u^c &= u_e^c + u \\ v^c &= v_e^c + v \\ w^c &= w_e^c + w \end{aligned}\right\} . \tag{2.37}$$

It is seen from equations (2.36) that the shift of the center of the reference ellipsoid to the geocenter, i. e., placing it in the absolute system, is the equivalent of changing the initial quantities $\Delta\Phi_o$, $\Delta\Lambda_o$, and ΔH_o to $\Delta\Phi_o^c$, $\Delta\Lambda_o^c$, and ΔH_o^c respectively.

The quantities $\Delta\Phi_o$ and $\Delta\Lambda_o \cos\varphi_o$ are called the deflection of the vertical components in the meridian and in the prime vertical plane of the initial point respectively. In the relative system these quantities are the relative or astro-geodetic deflections. In the absolute (geocentric) system they ($\Delta\Phi_o^c$, $\Delta\Lambda_o^c \cos\varphi_o^c$) are the absolute deflections. The quantities appearing in (2.36), i. e., $\delta\Phi_o$ and $\delta\Lambda_o \cos\varphi_o$, are the differences between the components of the absolute and the relative deflections at the initial point. Depending on whether the quantities Φ and Λ are astronomic or reduced-astronomic, the deflection components may be termed observed or reduced respectively. Both types could be either absolute or relative.

As far as the elevations are concerned, the difference between the ellipsoidal and geoidal heights ($\Delta H_o = h_o - H_o$) is the geoid undulation which is referred either to the relative (ΔH_o) or to the absolute (ΔH_o^c) reference ellipsoid.

For the deflection components and the geoid undulation, the following general notation will be used:

$$\left.\begin{aligned} \xi &= \Delta\Phi &&= \Phi - \varphi \\ \eta &= \Delta\Lambda \cos\varphi &&= (\Lambda - \lambda)\cos\varphi \\ \zeta &= \Delta H &&= h - H \end{aligned}\right\} . \tag{2.38}$$

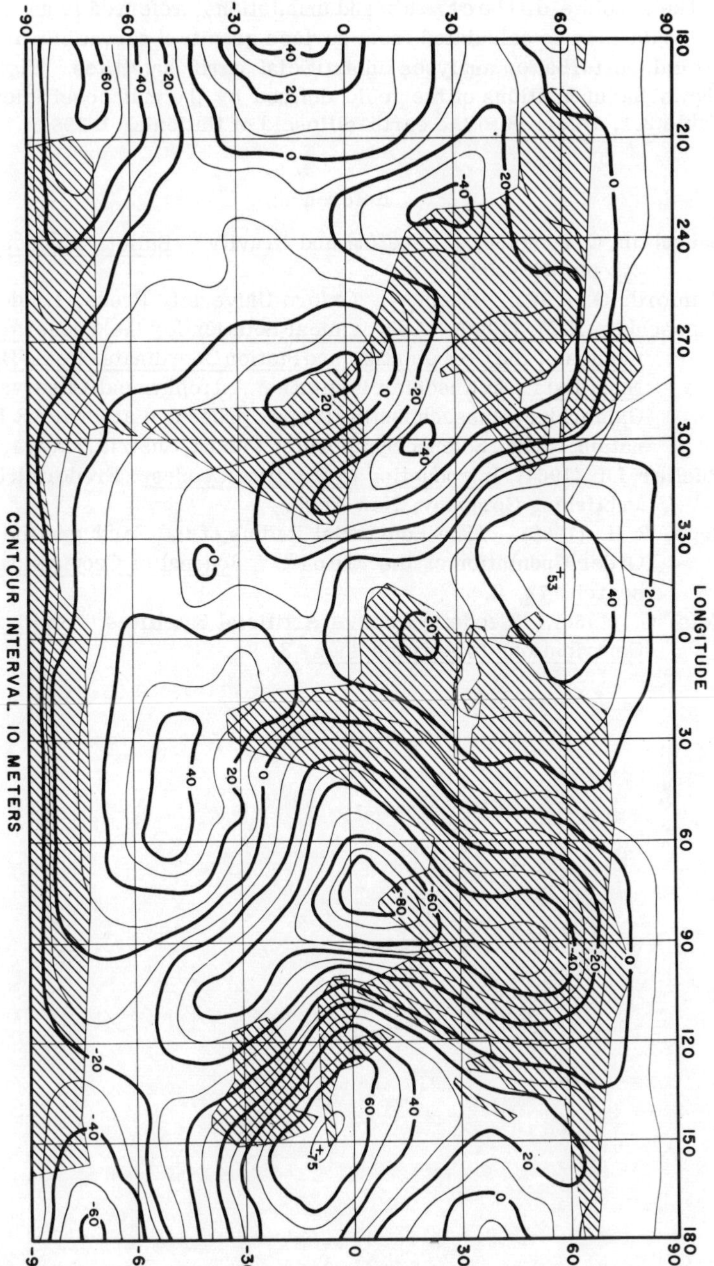

Fig. 2.12 Undulations of the geoid

The absolute deflections and undulations, referred to an earth-ellipsoid, may be calculated from surface gravity observations or from orbital perturbation analyses on artificial earth satellites. Fig. 2.12 shows the undulations of the geoid defined by the mass coefficients of Table 2.1, referred to the earth-ellipsoid of flattening 1/298.3.

References

Baeschlin, C. F. (1960). "Leveling and Gravity." Bulletin Géodésique, 57, pp. 247-298.

Bomford, G. (1962). Geodesy. Oxford University Press, London.

Gaposchkin, E. M. (1966). A Dynamical Solution for the Tesseral Harmonics of the Geopotential, and Station Coordinates Using Baker-Nunn Data. Smithsonian Institution Astrophysical Observatory, Cambridge, Massachusetts. (Paper presented at COSPAR International Space Science Symposium, Vienna, Austria, May 14, 1966.)

Mueller, I. I. (1964). Introduction to Satellite Geodesy. Frederick Ungar Publishing Company, New York.

Rapp, R. H. (1967). "The Equatorial Radius of the Earth and the Zero Order Undulation of the Geoid." Journal of Geophysical Research, 72, 2.

Veis, G. (1960). "Geodetic Uses of Artificial Satellites." Smithsonian Contributions to Astrophysics, 3, 9.

3 THE CELESTIAL SPHERE AND ITS COORDINATE SYSTEMS

3.1 Fundamental Definitions

Almost all celestial objects are at a distance many times the radius of the earth from which we make our observations. The objects of astronomical observation being at practically infinite distances thus give the impression of being located on the celestial sphere, on which, as a two-dimensional manifold, a suitably chosen pair of coordinates is sufficient to uniquely define a place, or position.

The rotation axis of the earth intersects this sphere at the north and south celestial poles (Fig. 3.1). The plane perpendicular to the rotation axis and containing the center is the celestial equator. A great circle containing the poles, thus perpendicular to the celestial equator, is called an hour circle. A small circle parallel to the celestial equator is termed a celestial parallel (of declination). The celestial poles, equator, parallels, and the hour circles are, for the time being, imagined to be fixed on the celestial sphere.

The vertical of the observer intersects the celestial sphere at two points. The point above the observer is the zenith, and opposite it is the nadir. The plane perpendicular to the vertical and containing the observer (i. e., the center of the earth) is the celestial horizon. The plane containing the zenith, and perpendicular to the horizon, is a vertical plane. A small circle parallel to the celestial horizon is termed an almucantar.

The observer's vertical plane containing the poles is the celestial meridian. It contains the hour circle through the zenith. The points on the celestial sphere where the celestial meridian intersects the celestial horizon are called north and south points. The vertical plane perpendicular to the celestial meridian is designated as the prime vertical.

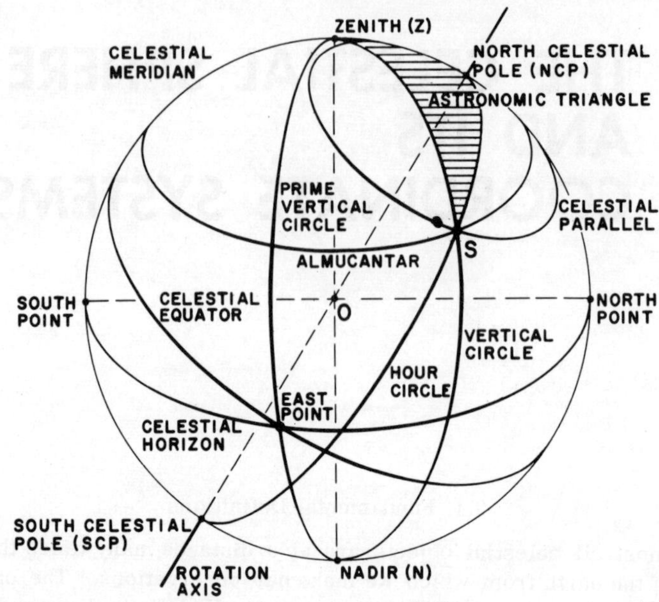

Fig. 3.1 The celestial sphere

Its intersection points on the celestial sphere with the celestial horizon are the east and west points. They are situated on the line of intersection of the celestial horizon and the celestial equator. The zenith, nadir, vertical planes, almucantars, the celestial horizon and meridian, the prime vertical, and their associated quantities may be imagined fixed relative to the earth (observer); thus, due to the rotation of the earth, they continuously change their position on the celestial sphere, i.e., with respect to the stars.

Let S be an arbitrary point on the celestial sphere. The celestial meridian and the hour and vertical circles through S form a spherical triangle, the astronomic triangle of S. Its vertices are the zenith, the north celestial pole, and the point S.

Owing to the rotation of the earth about its axis from west to east, an observer views the celestial sphere with its stars (and the celestial equator, parallels and hour circles) as if it were rotating from east to west. He also sees how the stars change their positions continuously with respect to the quantities fixed relative to the earth (zenith, celestial meridian, horizon, prime vertical, etc.).

The quantities previously described are all related either to the observer (vertical, horizon, vertical planes, etc.) or to the rotation of the earth (rotation axis, equator, hour circles, etc.). There are also some other important features on the celestial sphere which are related to the revolution of the earth about the sun or, in the reversed concept, to the apparent motion of the sun about the earth. The most important of these is the ecliptic, which is defined in terms of the plane containing the center of the sun, the barycenter of the earth-moon system, and the inertial heliocentric velocity vector of the earth-moon system's barycenter. The plane of the ecliptic is this plane, except that it is freed from the periodic perturbations caused by Venus and Jupiter. The ecliptic is always within about 2" of the sun's apparent path about the earth. The line perpendicular to the ecliptic and containing the center of the earth intersects the celestial sphere in the ecliptic poles (Fig. 3.2). It is customary to distinguish between north and south ecliptic poles. A plane parallel to the ecliptic intersects the celestial sphere in the ecliptic parallel (of latitude), while one perpendicular to it and containing the ecliptic poles intersects the celestial sphere in the ecliptic meridian (circle of ecliptic longitude).

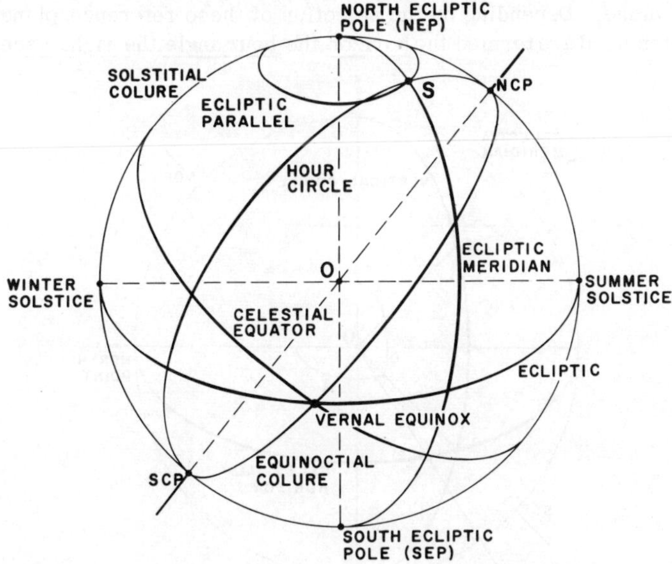

Fig. 3.2 The celestial sphere

The ecliptic intersects the celestial equator in a line connecting the equinoxes. The vernal equinox is that point of intersection near which the apparent sun crosses the celestial equator from south to north in the spring. The sun crosses the celestial equator from north to south in the fall near the autumnal equinox. The acute angle between the celestial equator and the ecliptic is termed the obliquity of the ecliptic.

The two opposite points on the ecliptic 90° from either equinox are termed solstices and are near the points where the sun reaches its greatest angular distance from the celestial equator, north (summer solstice) or south (winter solstice).

Two special hour circles at right angles to each other, passing, one through the celestial poles and the equinoxes, the other through the poles and the solstices, are called the equinoctial and the solstitial colures respectively. The ecliptic poles lie on the solstitial colure.

3.2 Celestial Coordinate Systems

In geodetic astronomy celestial coordinate systems are used to define the positions (directions) of objects (stars) on the celestial sphere. Customarily, the direction is expressed in terms of two perpendicular components or curvilinear coordinates. One component is reckoned from a primary reference plane and is measured perpendicular to it, the other from a secondary reference plane and is measured in the primary plane. Depending on the selection of these reference planes the systems used are termed the horizon, the hour angle, the right-ascension

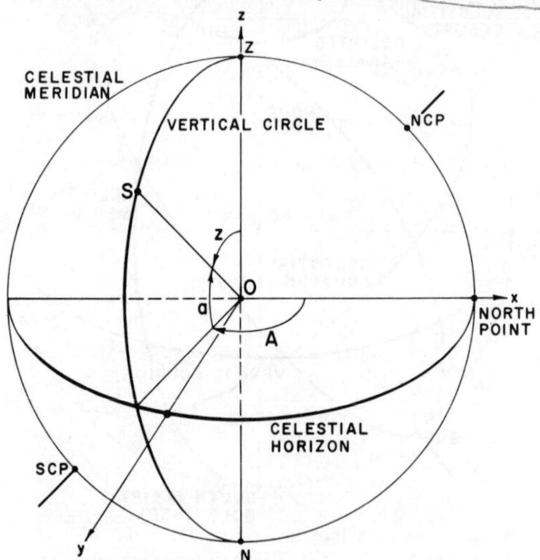

Fig. 3.3 The horizon system

and the ecliptic systems. All of these may have their origin of coordinates at the observer (topocentric) or at the center of the earth (geocentric). Since the dimensions of the earth are negligibly small compared to the distances to the stars, no distinction is necessary in geodetic astronomy between the topocentric and the geocentric systems.

In addition to the four systems mentioned above and described in the following sections, other curvilinear systems are also used in astronomy with origins at the center of the sun (heliocentric) or at the center of mass of the solar system (barycentric); they have various primary and secondary reference planes. In geodetic astronomy, however, these systems are not used and, therefore, not treated here.

3.21 Horizon System. In the horizon system the primary reference plane is the celestial horizon; the secondary is the observer's celestial meridian (Fig. 3.3). Let S be an arbitrary point on the celestial sphere. Its direction is defined by the parameters altitude and azimuth.

The altitude a is the angle between the direction OS and the celestial horizon measured from 0° to 90° in the plane of the vertical circle through S. It is taken with a positive sign above the horizon and with a negative sign below. Its complementary angle, $z = 90° - a$, is called the zenith distance.

The azimuth A is the angle between the vertical plane of S and the celestial meridian of the observer measured from the direction of the north point to the east in the celestial horizon from 0° to 360°.

3.22 Hour Angle System. In the hour angle system the primary reference plane is the celestial equator; the secondary is the hour circle containing the zenith (the observer's celestial meridian) (Fig. 3.4). Let

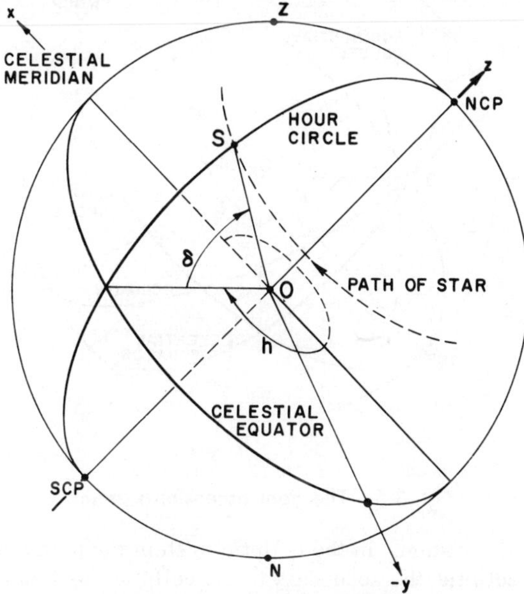

Fig. 3.4 The hour angle system

S be an arbitrary point on the celestial sphere. Its direction is given by the parameters declination and hour angle.

The declination δ is the angle between the direction OS and the celestial equator, measured from 0° to 90° in the plane of the hour circle through S. It is taken with a positive sign on the northern half of the celestial sphere and with a negative sign on the southern half. The seldom used complement of the declination is termed the polar distance.

The hour angle h is the angle between the hour circle of S and that of the observer measured from that half of the latter which contains the zenith, to the west (in the direction of the star's apparent daily motion) in the plane of the celestial equator from 0^h to 24^h or, on rare occasions, from 0° to 360°.

3.23 Right Ascension System. In the right ascension system the primary reference plane is also the celestial equator, but the secondary is the equinoctial colure (Fig. 3.5). Let S be an arbitrary point on the celestial sphere. Its direction is given by the parameters declination and right ascension.

The right ascension is the angle between the hour circle of S and the equinoctial colure, measured from the vernal equinox ♈ to the east in the plane of the celestial equator from 0^h to 24^h or, on rare occasions, from 0° to 360°.

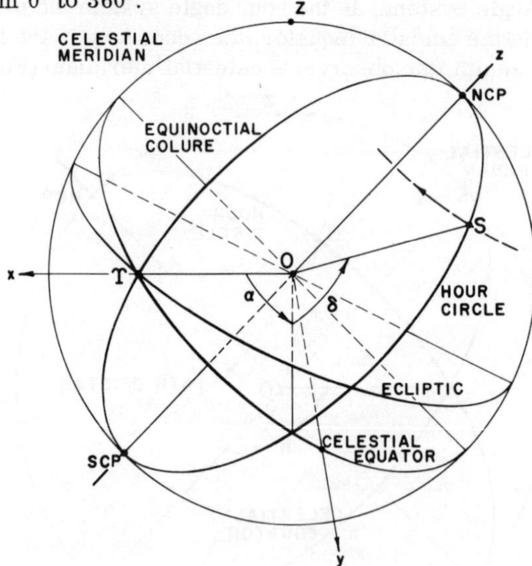

Fig. 3.5 The right ascension system

3.24 Ecliptic System. In the ecliptic system the primary reference plane is the ecliptic; the secondary is the ecliptic meridian of the equinox (Fig. 3.6). Let S be an arbitrary point on the celestial sphere. Its direction is given by the parameters (ecliptic) latitude and (ecliptic) longitude.

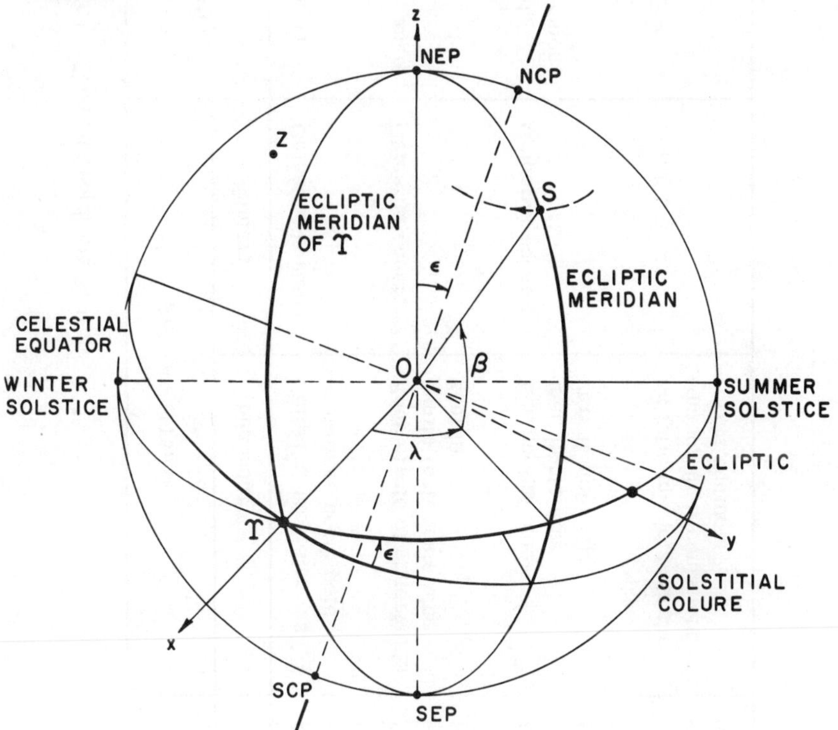

Fig. 3.6 The ecliptic system

The (ecliptic) latitude β is the angle between the direction OS and the ecliptic measured in the ecliptic meridian of S from 0° to 90°. It is taken with a positive sign north from the ecliptic and with a negative sign south. The complement of the latitude is the seldom used ecliptic polar distance.

The (ecliptic) longitude λ is the angle between the ecliptic meridian of S and that of the equinox measured from the vernal equinox ♈, to the east, in the ecliptic from 0° to 360°.

The acute angle between the ecliptic and the celestial equator, the obliquity of the ecliptic, is usually denoted by ϵ.

3.25 Summary. The most important characteristics of the coordinate systems described above are summarized in Table 3.1.

TABLE 3.1 Celestial Coordinate Systems

System	Reference Plane		Parameters Measured from the Reference Plane	
	Primary	Secondary	Primary	Secondary
Horizon	Celestial horizon	Celestial meridian (half containing north pole)	Altitude $-90° \leq a \leq +90°$ (+ toward zenith)	Azimuth $0° \leq A \leq 360°$ (+ east)
Hour angle	Celestial equator	Hour circle of observer's zenith (half containing zenith)	Declination $-90° \leq \delta \leq +90°$ (+ north)	Hour angle $0^h \leq h \leq 24^h$ $0° \leq h \leq 360°$ (+ west)
Right ascension	Celestial equator	Equinoctial colure (half containing vernal equinox)	Declination $-90° \leq \delta \leq +90°$ (+ north)	Right ascension $0^h \leq \alpha \leq 24^h$ $0° \leq \alpha \leq 360°$ (+east)
Ecliptic	Ecliptic	Ecliptic meridian of the equinox (half containing vernal equinox)	(Ecliptic) Latitude $-90° \leq \beta \leq +90°$ (+ north)	(Ecliptic) Longitude $0° \leq \lambda \leq 360°$ (+ east)

3.3 Transformation of Coordinates

3.31 Conversions Between the Horizon and Hour Angle Systems. Let Fig. 3.7 represent a point S on the northern half of the celestial sphere where the horizon and hour angle systems are shown together.

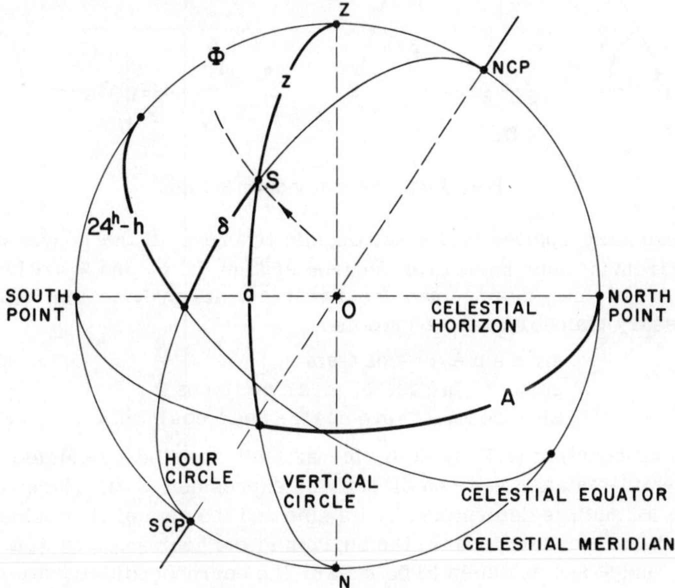

Fig. 3.7 Relations between the horizon and hour angle systems

As has been mentioned earlier, the spherical triangle with vertices at the pole (NCP), zenith (Z), and S is the astronomic triangle which forms an important link between the horizon and the hour angle systems. It is evident that if the point S is west of the celestial meridian (Fig. 3.8 a), the angles in the astronomic triangle at NCP, Z, and S are the hour angle h; the explement of the azimuth, $360° - A$; and the parallactic angle p (the angle between the hour circle and the vertical circle through S) respectively. If the point S is east of the celestial meridian (Fig. 3.8b), then the corresponding angles are the explement of the hour angle, $24^h - h$; the azimuth A; and the parallactic angle of S, p. The sides of the astronomic triangle in both cases are: celestial pole to zenith—the complement of the astronomic latitude or the co-latitude, $90° - \Phi$; celestial pole to S—the complement of the declination or the polar distance, $90° - \delta$; zenith to S—the zenith distance z, or the complement of altitude, $90° - a$.

The relations between the parameters of the horizon and the hour angle systems may be derived by means of the basic laws of spherical

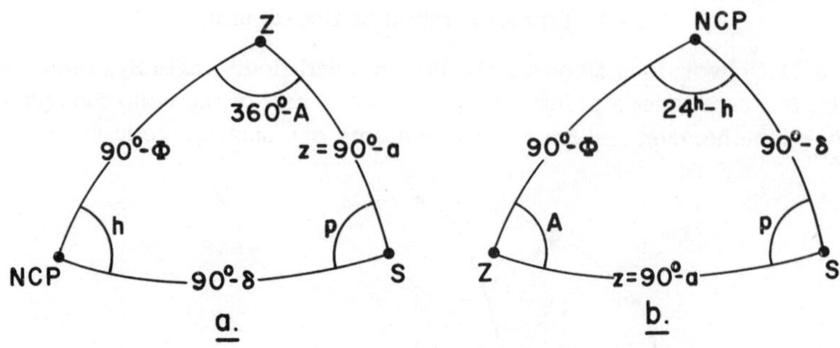

Fig. 3.8 The astronomic triangle

trigonometry applied to the astronomic triangle. If the conversion is made from the hour angle to the horizon system (h, δ, and Φ are known), the laws of sine, cosine, and the five elements provide the following necessary transformation equations:

$$\sin z \sin A = -\cos\delta \sin h, \qquad (3.1)$$
$$\cos z = \sin\delta \sin\Phi + \cos\delta \cos h \cos\Phi, \qquad (3.2)$$
$$\sin z \cos A = \sin\delta \cos\Phi - \cos\delta \cos h \sin\Phi . \qquad (3.3)$$

Dividing equation (3.1) by (3.3), the azimuth may be calculated while the zenith-distance is given directly by expression (3.2). The quadrant of the azimuth is determined by its sine and the sign of its cosine.

If the conversion is from the horizon to the hour angle system (i.e., z, A, and Φ are assumed to be known), the corresponding equations are

$$\cos\delta \sin h = -\sin z \sin A, \qquad (3.4)$$
$$\sin\delta = \cos z \sin\Phi + \sin z \cos A \cos\Phi, \qquad (3.5)$$
$$\cos\delta \cos h = \cos z \cos\Phi - \sin z \cos A \sin\Phi . \qquad (3.6)$$

Dividing equation (3.4) by (3.6), the hour angle may be calculated; the declination is given directly by expression (3.5). The quadrant of the hour angle is determined by its sine and the sign of its cosine.

3.32 Conversions Between the Hour Angle and Right Ascension Systems. From the descriptions of the hour angle and right ascension systems in sections 3.22 and 3.23, it is apparent that both use the equator as the primary reference plane; thus the declination δ is a common parameter. The conversion, therefore, is restricted to the hour angle and the right ascension.

Let Fig. 3.9 represent the celestial sphere viewed from the north celestial pole. From this view the celestial equator is a circle, and the hour circles are radial straight lines. From the figure it is evident that

$$ST = h + \alpha, \qquad (3.7)$$

where ST is the measure of the local sidereal time (see section 5.11), defined as the hour angle of the vernal equinox.

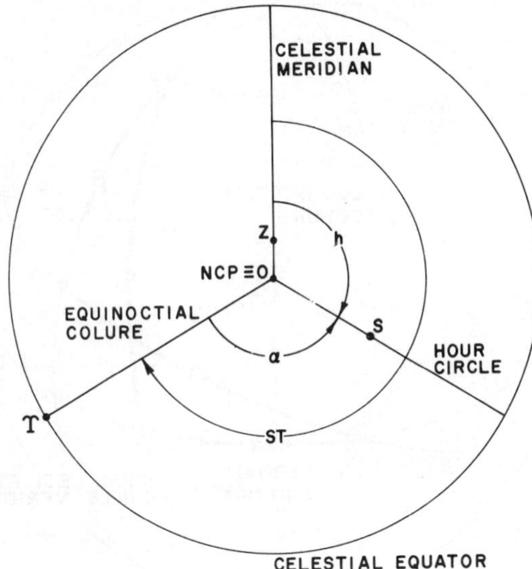

Fig. 3.9 Relations between hour angle and right ascension systems

Assuming that the local sidereal time ST is known, the conversion formulas from the hour angle to the right ascension systems are

$$\begin{aligned} \delta &= \delta, \\ \alpha &= ST - h, \end{aligned} \tag{3.8}$$

while the parameters of the hour angle system may be calculated from those of the right ascension system by means of the following expressions

$$\begin{aligned} \delta &= \delta, \\ h &= ST - \alpha. \end{aligned} \tag{3.9}$$

3.33 Conversions Between the Right Ascension and Ecliptic Systems. Let Fig. 3.10 represent a point S on the northern half of the celestial sphere where the right ascension and ecliptic systems are shown together. It is useful to note immediately that the right ascension of the north ecliptic pole (NEP) is 270° (18^h). Conversely, the longitude of the north celestial pole (NCP) is 90°. Considering these, it is evident that in the spherical triangle NEP-NCP-S the angles at NEP and at NCP are $90° - \lambda$ and $90° + \alpha$ respectively (Figs. 3.10 and 3.11a). If the point S is on the other side of the solstitial colure, the corresponding

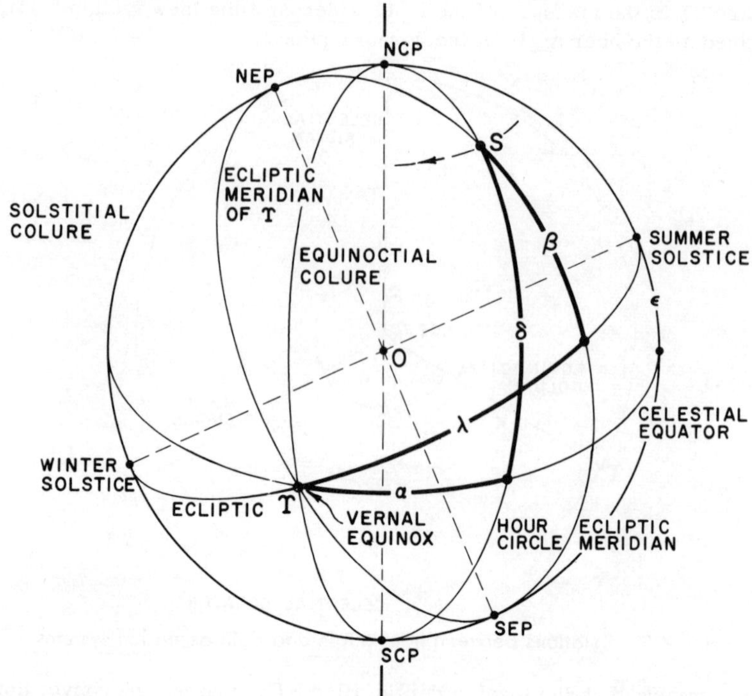

Fig. 3.10 Relations between the right ascension and ecliptic systems

angles are $\lambda - 90^\circ$ and $270^\circ - \alpha$ respectively (Fig. 3.11b). The sides of the triangles in both cases are celestial pole to ecliptic pole—the obliquity of the ecliptic ϵ; celestial pole to S—the polar distance, $90^\circ - \delta$; ecliptic pole to S—ecliptic polar distance, $90^\circ - \beta$.

The relations between the parameters of the ecliptic and the right ascension systems may be derived by applying the laws of sine, cosine, and the five elements of spherical trigonometry on any of the triangles shown in Fig. 3.11. The resulting sets of equations are

$$\cos\delta \cos\alpha = \cos\beta \cos\lambda , \tag{3.10}$$
$$\sin\delta = \cos\beta \sin\lambda \sin\epsilon + \sin\beta \cos\epsilon , \tag{3.11}$$
$$\cos\delta \sin\alpha = \cos\beta \sin\lambda \cos\epsilon - \sin\beta \sin\epsilon , \tag{3.12}$$

$$\cos\beta \cos\lambda = \cos\delta \cos\alpha , \tag{3.13}$$
$$\sin\beta = -\cos\delta \sin\alpha \sin\epsilon + \sin\delta \cos\epsilon , \tag{3.14}$$
$$\cos\beta \sin\lambda = \cos\delta \sin\alpha \cos\epsilon + \sin\delta \sin\epsilon . \tag{3.15}$$

If the ecliptic parameters ϵ, λ, and β are given, α may be calculated by dividing equation (3.12) by (3.10). Expression (3.11) provides δ.

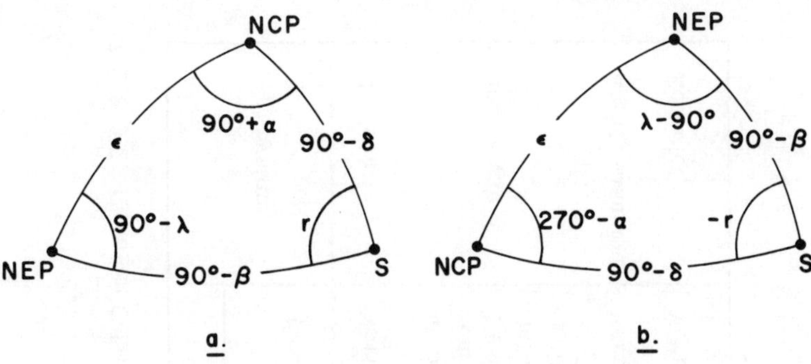

Fig. 3.11 The ecliptic triangle

If the parameters of the right ascension system α, δ, and ϵ are given, λ may be computed by dividing expression (3.15) by (3.13). Equation (3.14) provides β.

The quadrant of α or λ may be determined by checking the corresponding sines and the signs of the cosines.

3.34 Transformation by Matrices. A more general system of transformation, especially useful for electronic machine computations, may be substituted for that described in sections 3.31 - 3.33.

Assume a general Cartesian coordinate system x, y, z with its origin at the center of the earth. The axis z is perpendicular to the primary reference plane (see Table 3.1) of any celestial system and is positive towards the primary pole (see Table 3.2). The axis x is the intersection of the primary and secondary planes and is positive towards the secondary pole. The axis y is perpendicular to both and is generated by rotating the axis x through a positive angle of ninety degrees corresponding to the sign conventions followed in the various celestial systems. Following this system it is evident that the horizon and hour angle systems are left-handed, and the right ascension and ecliptic systems are right-handed (Figs. 3.3 - 3.6). Denoting the curvilinear coordinates measured in the primary reference plane and perpendicular to it with the general symbols μ and ν respectively, the Cartesian coordinates of a point on the celestial sphere (radius = 1) may be calculated from

$$\begin{pmatrix} x \\ y \\ z \end{pmatrix}_{\mu,\nu} = \begin{pmatrix} \cos\nu \cos\mu \\ \cos\nu \sin\mu \\ \sin\nu \end{pmatrix} . \qquad (3.16)$$

Conventional orthogonal rotation matrices of 3×3 dimensions, $\mathbf{R}_1(\theta)$, $\mathbf{R}_2(\theta)$, and $\mathbf{R}_3(\theta)$, may be used to rotate the general system by the angle

TABLE 3.2 Cartesian Celestial Coordinate Systems

System	Orientation of the Positive Axis			μ	ν	Left or Right-handed
	x (Secondary pole)	y	z (Primary pole)			
Horizon	North point	$A = 90^\circ$	Zenith	A	a	left
Hour angle	Intersection of the zenith's hour circle with the celestial equator on the zenith's side.	$h = 90^\circ = 6^h$	North celestial pole	h	δ	left
Right ascension	Vernal equinox	$\alpha = 90^\circ = 6^h$	North celestial pole	α	δ	right
Ecliptic	Vernal equinox	$\lambda = 90^\circ$	North ecliptic pole	λ	β	right

θ about the axes x, y, and z respectively [Goldstein, 1950, p. 109]:

$$\mathbf{R}_1(\theta) = \begin{pmatrix} 1 & 0 & 0 \\ 0 & \cos\theta & \sin\theta \\ 0 & -\sin\theta & \cos\theta \end{pmatrix}, \tag{3.17}$$

$$\mathbf{R}_2(\theta) = \begin{pmatrix} \cos\theta & 0 & -\sin\theta \\ 0 & 1 & 0 \\ \sin\theta & 0 & \cos\theta \end{pmatrix}, \tag{3.18}$$

$$\mathbf{R}_3(\theta) = \begin{pmatrix} \cos\theta & \sin\theta & 0 \\ -\sin\theta & \cos\theta & 0 \\ 0 & 0 & 1 \end{pmatrix}. \tag{3.19}$$

These are consistent with a right-handed coordinate system and positive signs for counterclockwise rotation as viewed looking toward the origin from the positive end of the rotation axis.

Coordinates in the hour angle system may be calculated from those in the horizon system by means of rotating the latter with $\theta = 180°$ about the axis z, and then with $\theta = -(90° - \Phi)$ about its axis y:

$$\begin{pmatrix} x \\ y \\ z \end{pmatrix}_{h,\delta} = \mathbf{R}_2(\Phi - 90°)\, \mathbf{R}_3(180°) \begin{pmatrix} x \\ y \\ z \end{pmatrix}_{A,a}. \tag{3.20}$$

The hour angle system may be converted into the right ascension system by changing the left-handed system into the right-handed by means of reflection of the coordinate axes [Goldstein, 1950, p. 122] and applying a rotation of $\theta = -ST$ about the axis z:

$$\begin{pmatrix} x \\ y \\ z \end{pmatrix}_{\alpha,\delta} = \mathbf{R}_3(-ST) \cdot \begin{pmatrix} 1 & 0 & 0 \\ 0 & -1 & 0 \\ 0 & 0 & 1 \end{pmatrix} \begin{pmatrix} x \\ y \\ z \end{pmatrix}_{h,\delta}. \tag{3.21}$$

The ecliptic system may be obtained from the right ascension system by a single rotation of $\theta = \epsilon$ about the axis x:

$$\begin{pmatrix} x \\ y \\ z \end{pmatrix}_{\lambda,\beta} = \mathbf{R}_1(\epsilon) \begin{pmatrix} x \\ y \\ z \end{pmatrix}_{\alpha,\delta}. \tag{3.22}$$

The notation in these equations is the same as in sections 3.31-3.33.

The inverse solution of the equations above may be easily obtained by applying the following set of rules of orthogonal-matrix algebra: if $x = \mathbf{R}x'$, then $x' = \mathbf{R}^{-1}x$; also $(\mathbf{R}_i\mathbf{R}_j)^{-1} = \mathbf{R}_j^{-1}\mathbf{R}_i^{-1}$; and $\mathbf{R}^{-1}(\theta) = \mathbf{R}(-\theta)$. With these, the inverse relationships between the different systems are

$$\begin{pmatrix} x \\ y \\ z \end{pmatrix}_{A,a} = \mathbf{R}_3(-180^\circ)\ \mathbf{R}_2(90^\circ - \Phi) \begin{pmatrix} x \\ y \\ z \end{pmatrix}_{h,\delta} , \tag{3.23}$$

$$\begin{pmatrix} x \\ y \\ z \end{pmatrix}_{h,\delta} = \begin{pmatrix} 1 & 0 & 0 \\ 0 & -1 & 0 \\ 0 & 0 & 1 \end{pmatrix} \mathbf{R}_3(\mathrm{ST}) \begin{pmatrix} x \\ y \\ z \end{pmatrix}_{\alpha,\delta} , \tag{3.24}$$

$$\begin{pmatrix} x \\ y \\ z \end{pmatrix}_{\alpha,\delta} = \mathbf{R}_1(-\epsilon) \begin{pmatrix} x \\ y \\ z \end{pmatrix}_{\lambda,\beta} . \tag{3.25}$$

After the conversion, the curvilinear coordinates μ and ν may be calculated from

$$\begin{aligned} \mu &= \tan^{-1}\frac{y}{x} , \\ \nu &= \tan^{-1}\frac{z}{(x^2+y^2)^{\frac{1}{2}}} = \sin^{-1} z . \end{aligned} \tag{3.26}$$

In geodetic astronomy generally $0^\circ \leq \nu \leq 90^\circ$ and $0^\circ \leq \mu \leq 360^\circ$; thus the quadrant of the latter must be determined from the signs of x and y, realizing that actually $x = p\cos\mu$ and $y = p\sin\mu$ (p is a constant).

3.35 Numerical Examples. The conversions described in sections 3.31 - 3.34 are illustrated in the following examples in which the basic data for a star (General Catalogue #23487) is as follows:

$$\begin{aligned} \alpha &= 17^h20^m36^s\!.622 = 260^\circ09'09''\!.33, \\ \delta &= 40^\circ00'08''\!.04, \\ \mathrm{ST} &= 6^h10^m16^s\!.550 = 92^\circ34'08''\!.25, \\ \Phi &= 60^\circ00'00''\!.00, \\ \epsilon &= 23^\circ26'43''\!.06. \end{aligned}$$

	α	δ	ST	Φ	ϵ
sin	-0.98526672	0.64281747	0.99899499	0.86602540	0.39787329
cos	-0.17102479	0.76601939	-0.04482176	0.50000000	0.91744037

(The data above reflects the apparent position of the star referred to the true equator and equinox on January 9, 1967, $4^h30^m10^s\!.000$ UT (Universal Time). The ST is local apparent sidereal time at $\Lambda = 277^\circ00'00''\!.00$. For the explanation of these terms see Chapters 4 and 5.)

EXAMPLE 3.1

Conversion from the Right Ascension to the Hour Angle System

1. Trigonometric conversion (equation (3.9)):

$$\begin{array}{lll} ST = & & 06^h 10^m 16^s.550 \\ \alpha = & & 17^h 20^m 36^s.622 \\ \hline h & = ST - \alpha = & 12^h 49^m 39^s.928 \; \| \\ \delta & = \delta & = 40^\circ 00' 08''.04 \; \| \end{array}$$

2. Matrix conversion (equations (3.16), (3.24) and (3.26)):

$$\begin{pmatrix} x \\ y \\ z \end{pmatrix}_{\alpha,\delta} = \begin{pmatrix} \cos\delta \cos\alpha \\ \cos\delta \sin\alpha \\ \sin\delta \end{pmatrix} = \begin{pmatrix} -0.13100831 \\ -0.75473341 \\ 0.64281747 \end{pmatrix}$$

$$\mathbf{R}_3(ST) = \begin{pmatrix} \cos(ST) & \sin(ST) & 0 \\ -\sin(ST) & \cos(ST) & 0 \\ 0 & 0 & 1 \end{pmatrix} = \begin{pmatrix} -0.04482176 & 0.99899499 & 0 \\ -0.99899499 & -0.04482176 & 0 \\ 0 & 0 & 1 \end{pmatrix}$$

$$\begin{pmatrix} x \\ y \\ z \end{pmatrix}_{h,\delta} = \begin{pmatrix} 1 & 0 & 0 \\ 0 & -1 & 0 \\ 0 & 0 & 1 \end{pmatrix} \mathbf{R}_3(ST) \begin{pmatrix} x \\ y \\ z \end{pmatrix}_{\alpha,\delta} = \begin{pmatrix} -0.74810287 \\ -0.16470512 \\ 0.64281747 \end{pmatrix}$$

$$h = \tan^{-1}\frac{y}{x} = \tan^{-1} 0.22016373 = 12^\circ 24' 58''.92 + 180^\circ 00' 00''.00 =$$

$$= 192^\circ 24' 58''.92 = 12^h 49^m 39^s.928 \; \|$$

(180° is added because both x and y are negative.)

$$\delta = \sin^{-1} z = \sin^{-1} 0.64281747 = 40^\circ 00' 08''.04 \; \|$$

EXAMPLE 3.2

Conversion from the Hour Angle to the Right Ascension System

Data from Example 3.1:

$$h = 12^h 49^m 39^s.928 = 192^\circ 24' 58''.92$$

$$\delta = 40^\circ 00' 08''.04$$

$$\sin h = -0.21501430; \; \cos h = -0.97661090$$

1. Trigonometric conversion (equation (3.8)):

$$\begin{array}{lll} ST & = & 6^h 10^m 16\overset{s}{.}550 \\ h & = & 12^h 49^m 39\overset{s}{.}928 \\ \hline \alpha & = ST - h = & 17^h 20^m 36\overset{s}{.}622 \; \| \\ \delta & = \delta = & 40^\circ 00' 08''\!.04 \; \| \end{array}$$

2. Matrix conversion (equations (3.21) and (3.26)):

$$\begin{pmatrix} x \\ y \\ z \end{pmatrix}_{h,\delta} = \begin{pmatrix} \cos\delta \cos h \\ \cos\delta \sin h \\ \sin\delta \end{pmatrix} = \begin{pmatrix} -0.74810289 \\ -0.16470512 \\ 0.64281747 \end{pmatrix}$$

$$\mathbf{R}_3(-ST) = \begin{pmatrix} \cos(ST) & -\sin(ST) & 0 \\ \sin(ST) & \cos(ST) & 0 \\ 0 & 0 & 1 \end{pmatrix} = \begin{pmatrix} -0.04482176 & -0.99899499 & 0 \\ 0.99899499 & -0.04482176 & 0 \\ 0 & 0 & 1 \end{pmatrix}$$

$$\begin{pmatrix} x \\ y \\ z \end{pmatrix}_{\alpha,\delta} = \mathbf{R}_3(-ST) \begin{pmatrix} 1 & 0 & 0 \\ 0 & -1 & 0 \\ 0 & 0 & 1 \end{pmatrix} \begin{pmatrix} x \\ y \\ z \end{pmatrix}_{h,\delta} = \begin{pmatrix} -0.13100830 \\ -0.75473341 \\ 0.64281747 \end{pmatrix}$$

$$\alpha = \tan^{-1} \frac{y}{x} = \tan^{-1} 5.76095873 = 80^\circ 09' 09''\!.33 + 180^\circ 00' 00''\!.00 =$$
$$= 260^\circ 09' 09''\!.33 = 17^h 20^m 36\overset{s}{.}622 \; \|$$

(180° is added because both x and y are negative.)

$$\delta = \sin^{-1} z = \sin^{-1} 0.64281747 = 40^\circ 00' 08''\!.04 \; \|$$

EXAMPLE 3.3

Conversion from the Hour Angle to the Horizon System

Data from Example 3.1:

$$h = 12^h 49^m 39\overset{s}{.}928 = 192^\circ 24' 58''\!.92$$
$$\delta = 40^\circ 00' 08''\!.04$$
$$\sin h = -0.21501430; \; \cos h = -0.97661090$$

1. Trigonometric conversion (equations (3.1) - (3.3)):

$$\cos z = \sin\delta \sin\Phi + \cos\delta \cos h \cos\Phi = 0.18264482$$

$$z = 79°28'34''.14 \;\|$$

$$a = 90° - z = 10°31'25''.86 \;\|$$

$$\tan A = \frac{-\cos\delta \sin h}{\sin\delta \cos\Phi - \cos\delta \cos h \sin\Phi} = \frac{0.16470512}{0.96928484} = 0.16992437$$

$$A = 9°38'37''.80 \;\|$$

2. Matrix conversion (equations (3.23) and (3.26)):

$$\begin{pmatrix} x \\ y \\ z \end{pmatrix}_{h,\delta} = \begin{pmatrix} \cos\delta \cos h \\ \cos\delta \sin h \\ \sin\delta \end{pmatrix} = \begin{pmatrix} -0.74810289 \\ -0.16470512 \\ 0.64281747 \end{pmatrix}$$

$$\mathbf{R}_2(90° - \Phi) = \begin{pmatrix} \sin\Phi & 0 & -\cos\Phi \\ 0 & 1 & 0 \\ \cos\Phi & 0 & \sin\Phi \end{pmatrix} = \begin{pmatrix} 0.86602540 & 0 & -0.50000000 \\ 0 & 1 & 0 \\ 0.50000000 & 0 & 0.86602540 \end{pmatrix}$$

$$\mathbf{R}_3(-180°) = \begin{pmatrix} \cos 180° & -\sin 180° & 0 \\ \sin 180° & \cos 180° & 0 \\ 0 & 0 & 1 \end{pmatrix} = \begin{pmatrix} -1 & 0 & 0 \\ 0 & -1 & 0 \\ 0 & 0 & 1 \end{pmatrix}$$

$$\begin{pmatrix} x \\ y \\ z \end{pmatrix}_{A,a} = \mathbf{R}_3(-180°)\, \mathbf{R}_2(90° - \Phi) \begin{pmatrix} x \\ y \\ z \end{pmatrix}_{h,\delta} = \begin{pmatrix} 0.96928484 \\ 0.16470512 \\ 0.18264481 \end{pmatrix}$$

$$A = \tan^{-1}\frac{y}{x} = \tan^{-1} 0.16992437 = 9°38'37''.80 \;\|$$

$$a = \sin^{-1} z = \sin^{-1} 0.18264481 = 10°31'25''.86 \;\|$$

EXAMPLE 3.4

Conversion from the Horizon to the Hour Angle System

Data from Example 3.3:

$$A = 9°38'37''.80$$

$$a = 10°31'25''.86$$

$$z = 90° - a = 79°28'34''.14$$

	A	a	z
sin	0.16752302	0.18264480	0.98317896
cos	0.98586816	0.98317896	0.18264480

1. Trigonometric conversion (equations (3.4) - (3.6)):

$$\tan h = \frac{-\sin z \sin A}{\cos z \cos \Phi - \sin z \cos A \sin \Phi} = \frac{-0.16470511}{-0.74810288} = 0.22016372$$

$$h = 12^\circ 24' 58''.92 + 180^\circ 00' 00''.00 = 192^\circ 24' 58''.92 \|$$

(180° is added because both the numerator and the denominator are negative.)

$$\sin \delta = \cos z \sin \Phi + \sin z \cos A \cos \Phi = 0.64281745$$

$$\delta = 40^\circ 00' 08''.04 \|$$

2. Matrix conversion (equations (3.20) and (3.26)):

$$\begin{pmatrix} x \\ y \\ z \end{pmatrix}_{A,a} = \begin{pmatrix} \cos a \cos A \\ \cos a \sin A \\ \sin a \end{pmatrix} = \begin{pmatrix} 0.96928483 \\ 0.16470511 \\ 0.18264480 \end{pmatrix}$$

$$\mathbf{R}_3(180^\circ) = \begin{pmatrix} \cos 180^\circ & \sin 180^\circ & 0 \\ -\sin 180^\circ & \cos 180^\circ & 0 \\ 0 & 0 & 1 \end{pmatrix} = \begin{pmatrix} -1 & 0 & 0 \\ 0 & -1 & 0 \\ 0 & 0 & 1 \end{pmatrix}$$

$$\mathbf{R}_2(\Phi - 90^\circ) = \begin{pmatrix} \sin \Phi & 0 & \cos \Phi \\ 0 & 1 & 0 \\ -\cos \Phi & 0 & \sin \Phi \end{pmatrix} = \begin{pmatrix} 0.86602540 & 0 & 0.50000000 \\ 0 & 1 & 0 \\ -0.50000000 & 0 & 0.86602540 \end{pmatrix}$$

$$\begin{pmatrix} x \\ y \\ z \end{pmatrix}_{h,\delta} = \mathbf{R}_2(\Phi - 90^\circ)\, \mathbf{R}_3(180^\circ) \begin{pmatrix} x \\ y \\ z \end{pmatrix}_{A,a} = \begin{pmatrix} -0.74810288 \\ -0.16470511 \\ 0.64281745 \end{pmatrix}$$

$$h = \tan^{-1} \frac{y}{x} = \tan^{-1} 0.22016372 = 12^\circ 24' 58''.92 + 180^\circ 00' 00''.00 =$$

$$= 192^\circ 24' 58''.92 \|$$

(180° is added because both x and y are negative.)

$$\delta = \sin^{-1} z = \sin^{-1} 0.64281745 = 40^\circ 00' 08''.04 \|$$

EXAMPLE 3.5

Conversion from the Right Ascension to the Ecliptic System

1. Trigonometric conversion (equations (3.13) - (3.15)):

$$\sin\beta = -\cos\delta \sin\alpha \sin\epsilon + \sin\delta \cos\epsilon = 0.89003496$$
$$\beta = 62°52'39''.51 \,\|$$

$$\tan\lambda = \frac{\cos\delta \sin\alpha \cos\epsilon + \sin\delta \sin\epsilon}{\cos\delta \cos\alpha} = \frac{-0.43666300}{-0.13100831} = 3.33309391$$

$$\lambda = 73°17'58''.64 + 180°00'00''.00 = 253°17'58''.64 \,\|$$

(180° is added because both the numerator and the denominator are negative.)

2. Matrix conversion (equations (3.22) and (3.26)):

$$\begin{pmatrix} x \\ y \\ z \end{pmatrix}_{\alpha,\delta} = \begin{pmatrix} \cos\delta \cos\alpha \\ \cos\delta \sin\alpha \\ \sin\delta \end{pmatrix} = \begin{pmatrix} -0.13100831 \\ -0.75473341 \\ 0.64281747 \end{pmatrix}$$

$$\mathbf{R}_1(\epsilon) = \begin{pmatrix} 1 & 0 & 0 \\ 0 & \cos\epsilon & \sin\epsilon \\ 0 & -\sin\epsilon & \cos\epsilon \end{pmatrix} = \begin{pmatrix} 1 & 0 & 0 \\ 0 & 0.91744037 & 0.39787329 \\ 0 & -0.39787329 & 0.91744037 \end{pmatrix}$$

$$\begin{pmatrix} x \\ y \\ z \end{pmatrix}_{\lambda,\beta} = \mathbf{R}_1(\epsilon) \begin{pmatrix} x \\ y \\ z \end{pmatrix}_{\alpha,\delta} = \begin{pmatrix} -0.13100831 \\ -0.43666300 \\ 0.89003496 \end{pmatrix}$$

$$\lambda = \tan^{-1}\frac{y}{x} = \tan^{-1} 3.33309391 = 73°17'58''.64 + 180°00'00''.00 =$$
$$= 253°17'58''.64 \,\|$$

(180° is added because both x and y are negative.)

$$\beta = \sin^{-1} z = \sin^{-1} 0.89003496 = 62°52'39''.51 \,\|$$

EXAMPLE 3.6

Conversion from the Ecliptic to the Right Ascension System

Data from Example 3.5:

$$\beta = 62°52'39''.51$$
$$\lambda = 253°17'58''.64$$

	β	λ
sin	0.89003497	-0.95782060
cos	0.45589226	-0.28736683

1. Trigonometric conversion (equations (3.10) - (3.12)):

$$\sin\delta = \cos\beta\,\sin\lambda\,\sin\epsilon + \sin\beta\,\cos\epsilon = 0.64281747$$

$$\delta = 40^\circ\,00'\,08''\!.04 \;\|$$

$$\tan\alpha = \frac{\cos\beta\,\sin\lambda\,\cos\epsilon - \sin\beta\,\sin\epsilon}{\cos\beta\,\cos\lambda} = \frac{-0.75473340}{-0.13100831} = 5.7609582$$

$$\alpha = 80^\circ\,09'\,09''\!.33 + 180^\circ 00'\,00''\!.00 = 260^\circ\,09'\,09''\!.33 \;\|$$

(180° is added because both the numerator and the denominator are negative.)

2. Matrix conversion (equations (3.25) and (3.26)):

$$\begin{pmatrix} x \\ y \\ z \end{pmatrix}_{\lambda,\beta} = \begin{pmatrix} \cos\beta\,\cos\lambda \\ \cos\beta\,\sin\lambda \\ \sin\beta \end{pmatrix} = \begin{pmatrix} -0.13100831 \\ -0.43666300 \\ 0.89003497 \end{pmatrix}$$

$$\mathbf{R}_1(-\epsilon) = \begin{pmatrix} 1 & 0 & 0 \\ 0 & \cos\epsilon & -\sin\epsilon \\ 0 & \sin\epsilon & \cos\epsilon \end{pmatrix} = \begin{pmatrix} 1 & 0 & 0 \\ 0 & 0.91744037 & -0.39787329 \\ 0 & 0.39787329 & 0.91744037 \end{pmatrix}$$

$$\begin{pmatrix} x \\ y \\ z \end{pmatrix}_{\alpha,\delta} = \mathbf{R}_1(-\epsilon)\begin{pmatrix} x \\ y \\ z \end{pmatrix}_{\lambda,\beta} = \begin{pmatrix} -0.13100831 \\ -0.75473340 \\ 0.64281743 \end{pmatrix}$$

$$\alpha = \tan^{-1}\frac{y}{x} = \tan^{-1}5.7609582 = 80^\circ 09'\,09''\!.33 + 180^\circ 00'\,00''\!.00 =$$

$$= 260^\circ 09'\,09''\!.33 \;\|$$

(180° is added because both x and y are negative.)

$$\delta = \sin^{-1}z = \sin^{-1}0.64281743 = 40^\circ\,00'\,08''\!.04 \;\|$$

3.36 Differential Relations. Supplementary to the problem of coordinate transformation is that of finding the effect of small changes in the values of one pair of coordinates upon the values of the other. For this purpose the angles p ($\le 90^\circ$) and r ($-90^\circ \le r \le 90^\circ$) in Figs. 3.8 and 3.11 are useful.

Applying the laws of sine and cosine of spherical trigonometry to the triangles shown in Figs. 3.8 and 3.11 yields the following results: From Fig. 3.8,

$$\sin p = \mp \sin A \cos \Phi \sec \delta = \pm \sin h \operatorname{cosec} z \cos \Phi, \quad (3.27)$$
$$\cos p = \sin \Phi \operatorname{cosec} z \sec \delta - \tan \delta \cot z, \quad (3.28)$$

where the upper sign applies to $0^h < h < 12^h$, and the lower sign to $12^h < h < 24^h$; from Fig. 3.11,

$$\sin r = \cos \lambda \sec \delta \sin \epsilon = \cos \alpha \sec \beta \sin \epsilon, \quad (3.29)$$
$$\cos r = \sec \delta \sec \beta \cos \epsilon - \tan \delta \tan \beta. \quad (3.30)$$

Differentiating equations (3.1) - (3.3) and using the quantities above, yields the following differential relations between the horizon and right ascension systems:

$$dA = (\cos p \cos \delta dh + \cos \Phi \sin h \operatorname{cosec} z d\delta + \cos z \sin A d\Phi) \operatorname{cosec} z, \quad (3.31)$$
$$dz = -\sin A \cos \Phi dh - \cos p d\delta - \cos A d\Phi, \quad (3.32)$$

and, conversely,

$$dh = (\cos p \sin z dA + \cos \Phi \sin h \operatorname{cosec} z dz + \sin \delta \sin h d\Phi) \sec \delta, \quad (3.33)$$
$$d\delta = \cos \Phi \sin h dA - \cos p dz + \cos h d\Phi. \quad (3.34)$$

Differentiating equations (3.10) - (3.12) yields the differential relations between the right ascension and the ecliptic systems:

$$d\alpha = (\cos r \cos \beta d\lambda - \sin r d\beta - \sin \delta \cos \alpha d\epsilon) \sec \delta, \quad (3.35)$$
$$d\delta = \sin r \cos \beta d\lambda + \cos r d\beta + \sin \alpha d\epsilon, \quad (3.36)$$

and, conversely,

$$d\lambda = (\cos r \cos \delta d\alpha + \sin r d\delta + \sin \beta \cos \lambda d\epsilon) \sec \beta, \quad (3.37)$$
$$d\beta = -\sin r \cos \delta d\alpha + \cos r d\delta - \sin \lambda d\epsilon. \quad (3.38)$$

3.4 Special Star Positions

3.41 Circumpolar and Equatorial Stars. If the observer is situated somewhere between the equator and the north pole, the behavior of a star as seen by him may be classified as being one of six types (Fig. 3.12). Stars in the first group are always above the horizon and they never cross the prime vertical. The second group contains stars which, for observers situated at latitudes higher than 45°, are also above the horizon but cross the prime vertical. Stars in these two groups would be visible all the time if darkness would prevail in the sky. They are called northern circumpolar stars. Stars in the third group rise above and set below the celestial horizon spending more time above than below. Stars on the celestial equator constitute the fourth group, being as long above the horizon as below. In the fifth group, stars are above the horizon for a shorter period of time than below. The third, fourth, and fifth groups constitute the equatorial stars. The never rising stars, i.e., those not visible to the observer, form the sixth group, the southern circumpolar stars.

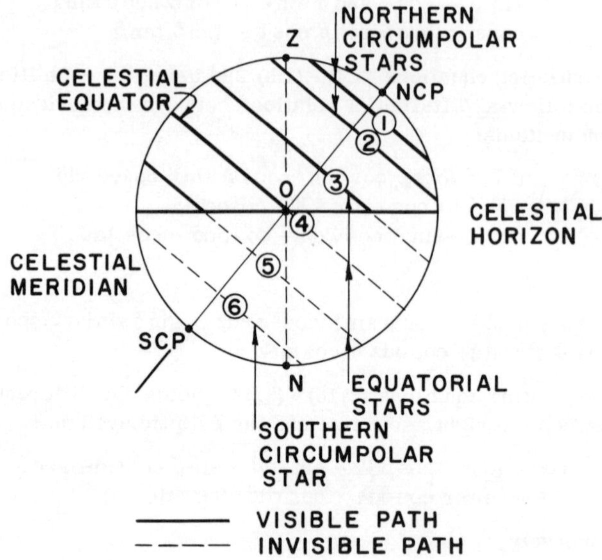

Fig. 3.12 Circumpolar and equatorial stars

If the observer is situated on the equator, he sees half of the path of all the stars (Fig. 3.13a). Each star spends equal time above and below the horizon. If, however, he is at the north (south) pole, stars on the northern (southern) half of the celestial sphere are always above his horizon. They move around the zenith in circles (Fig. 3.13b). He cannot observe stars situated on the southern (northern) half of the celestial sphere.

3.42 Visibility, Rising, Setting. Let Fig. 3.14 show the celestial meridian of an observer situated at astronomic latitude Φ with the paths of northern (n) and southern (s) circumpolar stars reaching but not crossing the celestial horizon. It is evident that for northern circumpolar stars,

$$\delta \geq 90° - \Phi,$$

for southern circumpolar stars,

$$\delta \leq \Phi - 90°,$$

and for equatorial stars,

$$90^\circ - \Phi > \delta > \Phi - 90^\circ . \tag{3.39}$$

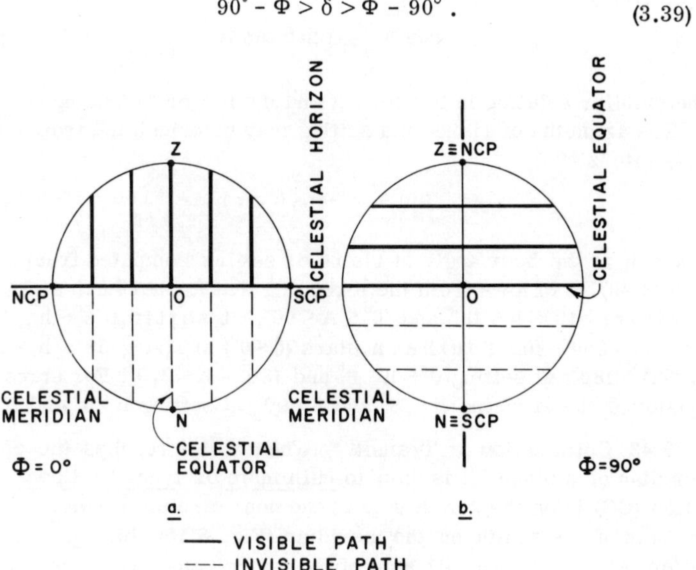

Fig. 3.13 Stars at the equator and the pole

Since the northern circumpolar stars do not set and the southern ones do not rise, expression (3.39) is the condition for rising and setting. For example, if the observer is situated at $\Phi = +40^\circ$, the condition for a star to rise and set is $50^\circ > \delta > -50^\circ$. If $\delta \geq 50^\circ$ the star does not set, while if $\delta \leq -50^\circ$ it does not rise.

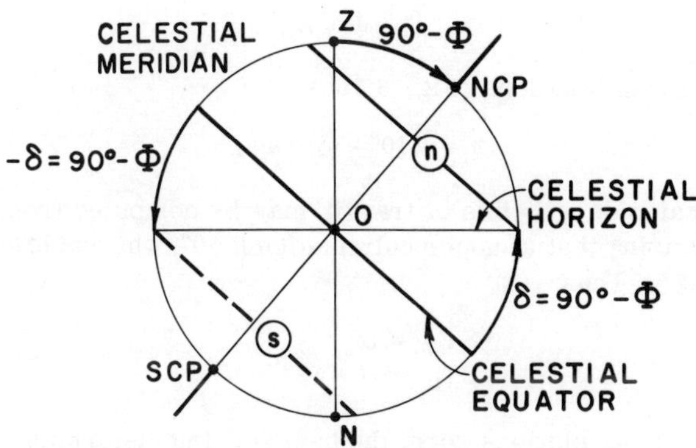

Fig. 3.14 Rising and setting limits of stars

The hour angle of rising or setting may be calculated from equation (3.2) with $z = 90°$,

$$\cos h = -\tan\delta \tan\Phi. \tag{3.40}$$

The smaller solution is for setting and the larger for rising.

The azimuths of rising and setting may be calculated from equation (3.1) with $z = 90°$,

$$\sin A = -\cos\delta \sin h, \tag{3.41}$$

where h is the hour angle of rising or setting computed from expression (3.40). It follows from the foregoing that for northern stars ($\delta > 0°$) at rising $12^h < h < 18^h$ and $0° < A < 90°$, at setting $6^h < h < 12^h$ and $270° < A < 360°$; for southern stars ($\delta < 0°$) at rising $18^h < h < 24^h$ and $90° < A < 180°$, at setting $0^h < h < 6^h$ and $180° < A < 270°$. For stars on the equator ($\delta = 0°$) at rising $h = 18^h$ and $A = 90°$, at setting $h = 6^h$ and $A = 270°$.

3.43 Culmination or Transit. When a star reaches the celestial meridian of a place, it is said to culminate or transit. Upper culmination (UC) is on the zenith side of the hour circle. It can occur north or south of the zenith on the meridian (Fig. 3.15a, b). Lower culmination (LC) is on the nadir side of the hour circle, and it occurs always north of the zenith (Fig. 3.15c).

The zenith distance of the star at culmination may be calculated for the various cases as follows: For upper culmination north of the zenith (Fig. 3.15a),

$$z = \delta - \Phi; \tag{3.42}$$

for upper culmination south of the zenith (Fig. 3.15b),

$$z = \Phi - \delta; \tag{3.43}$$

for all lower culminations (Fig. 3.15c),

$$z = 180° - (\delta + \Phi). \tag{3.44}$$

The local sidereal time of transit may be computed from equation (3.7) recognizing that at upper culmination $h = 0^h$, while at lower culmination $h = 12^h$. Thus

$$\begin{aligned} ST_{UC} &= \alpha, \\ ST_{LC} &= \alpha + 12^h. \end{aligned} \tag{3.45}$$

These important relations form the basis of time determinations to be described later.

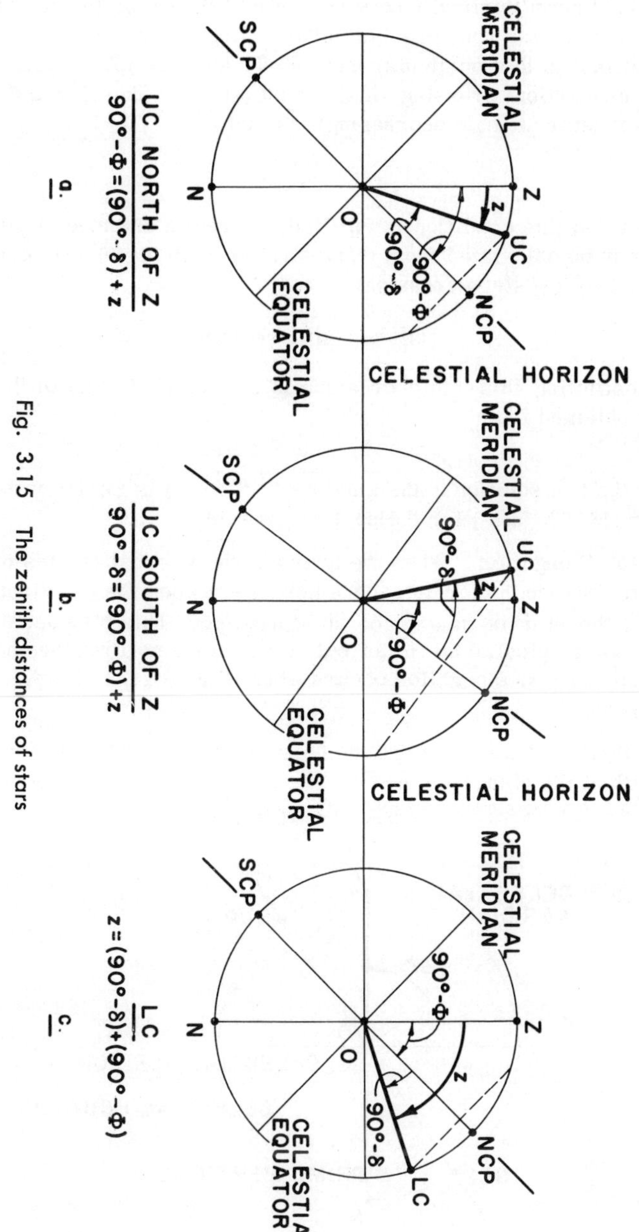

Fig. 3.15 The zenith distances of stars

3.44 Prime Vertical Crossing. Fig. 3.16 shows the celestial meridian with a star reaching but not crossing the prime vertical. It is evident that in this particular case $\delta = \Phi$, and, therefore, the condition of prime vertical crossing, in the east (star altitude increasing) or in the west (star altitude decreasing), is that

$$0^\circ < \delta < \Phi . \tag{3.46}$$

Provided this condition is fulfilled, the zenith distance of the crossings may be calculated from equation (3.5), with $A = 90^\circ$ (eastern crossing) or 270° (western crossing),

$$\cos z = \sin\delta \operatorname{cosec}\Phi . \tag{3.47}$$

Substituting this into expression (3.2), the hour angle of the crossing is obtained from

$$\cos h = \cot\Phi \tan\delta . \tag{3.48}$$

Of the two solutions, the smaller $(0^h < h < 6^h)$ is for the western, the larger $(18^h < h < 24^h)$ for the eastern crossing.

3.45 Elongation. When the hour and the vertical circles through a star are perpendicular to each other, i.e., when the parallactic angle $p = 90^\circ$, the star is said to be at elongation. This may occur on both sides of the celestial meridian, but only to stars not crossing the prime vertical. Thus, elongation occurs when (Fig. 3.16)

$$\delta > \Phi . \tag{3.49}$$

If this condition is fulfilled, from equation (3.27) with $p = 90^\circ$, the azimuth of the elongation is

$$\sin A = \cos\delta \sec\Phi . \tag{3.50}$$

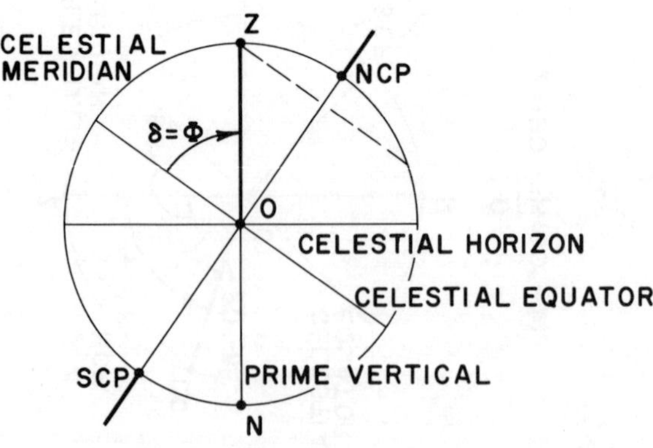

Fig. 3.16 Stars at the prime vertical

Of the two solutions, the smaller ($0° < A < 90°$) is for the eastern elongation. The azimuth of the western elongation is 360° minus the smaller solution.

From equation (3.28) with $p = 90°$ the zenith distance of both elongations is

$$\cos z = \sin\Phi \operatorname{cosec}\delta . \qquad (3.51)$$

The hour angle of the elongation is obtained by substituting (3.51) into equation (3.2),

$$\cos h = \cot\delta \tan\Phi . \qquad (3.52)$$

Of the two solutions, the smaller ($0^h < h < 6^h$) is for the western and the larger ($18^h < h < 24^h$) for the eastern elongation.

3.46 Numerical Example. Let an observer be situated at $\Phi = +40°$. The declinations of the stars A, B, and C are as follows:

$$\delta_A = +55°00'00''.00 ,$$
$$\delta_B = +35°00'00''.00 ,$$
$$\delta_C = -30°00'00''.00 .$$

According to section 3.42, A is a northern circumpolar star ($\delta > 90° - \Phi$); B and C are equatorial ($90° - \Phi > \delta > \Phi - 90°$). Conditions (3.46) and (3.49) show that star B crosses the prime vertical ($0° < \delta < \Phi$), and that A can reach elongation ($\delta > \Phi$). In Example 3.7, equations (3.40)-(3.52) yield the coordinates of the stars in the hour angle and horizon systems.

EXAMPLE 3.7

Special Star Positions

Special Position	Star	h	A	z
Rising	A	not rising	not rising	not rising
	B	$15^h36^m04^s.102$	38°31'28''.80	90°00'00''.00
	C	$22^h04^m05^s.584$	115°11'37''.79	90°00'00''.00
Prime vertical crossing (east)	A	no crossing	no crossing	no crossing
	B	$21^h46^m14^s.687$	90°00'00''.00	26°48'58''.55
	C	not visible	not visible	not visible
Elongation (east)	A	$20^h23^m55^s.898$	48°28'56''.24	38°18'25''.82
	B	no elongation	no elongation	no elongation
	C	no elongation	no elongation	no elongation
Upper culmination	A	$0^h00^m00^s.000$	0°00'00''.00	15°00'00''.00
	B	$0^h00^m00^s.000$	180°00'00''.00	5°00'00''.00
	C	$0^h00^m00^s.000$	180°00 00''.00	70°00'00''.00

Special Star Positions
(continued)

Special Position	Star	h	A	z
Prime vertical crossing (west)	A	no crossing	no crossing	no crossing
	B	$2^h 13^m 45^s.314$	270°00'00".00	26°48 58".55
	C	not visible	not visible	not visible
Elongation (west)	A	$3^h 36^m 04^s.102$	311°31'03".76	38°18'25".82
	B	no elongation	no elongation	no elongation
	C	no elongation	no elongation	no elongation
Setting	A	no setting	no setting	no setting
	B	$8^h 23^m 55^s.898$	321°28 31".20	90°00'00".00
	C	$1^h 55^m 54^s.416$	204°48'22".21	90°00'00".00
Lower culmination	A	$12^h 00^m 00^s.000$	0°00'00".00	85°00'00".00
	B	not visible	not visible	not visible
	C	not visible	not visible	not visible

References

Goldstein, H. (1950). Classical Mechanics. Addison-Wesley Publising Company, Reading, Massachusetts.

4 VARIATIONS IN THE CELESTIAL COORDINATES

In the previous discussion it has been assumed that the coordinates of stars in the right ascension and ecliptic systems are constant with respect to time, and that in the hour angle and horizon systems they change periodically only due to the rotation of the earth. Observations, however, show that all celestial coordinates are also subject to other slower and small variations due to the following factors:

(1) The motion of the coordinate systems with respect to the stars (precession and nutation) and with respect to the solid earth (polar motion).

(2) The apparent displacements in the directions of stars due to physical phenomena (refraction, aberration, and parallax).

(3) The relative motion of stars with respect to each other in space (proper motion).

The main effects causing these variations and the methods used in their numerical evaluation are described in this section.

4.1 Variations Due to the Motion of the Coordinate Systems

Variations in this group are due to the gravitational action of the extra-terrestrial bodies of the solar system, chiefly that of the sun and the moon, on the equatorial bulge and the orbital elements of the earth.

Let, for example, Fig. 4.1a show a schematic section of the earth with its equatorial bulge. The sun, situated in the direction S, attracts the opposite bulges with forces F_1 and F_2, where $F_1 > F_2$. The centrifugal forces at the gravity centers of the bulges, due to the revolution of the earth around the sun, are C_1 and C_2, where $C_1 < C_2$. The resultant forces of the attractions and centrifugal forces are R_1 and R_2, pointing in the opposite directions as shown in Fig. 4.1b.

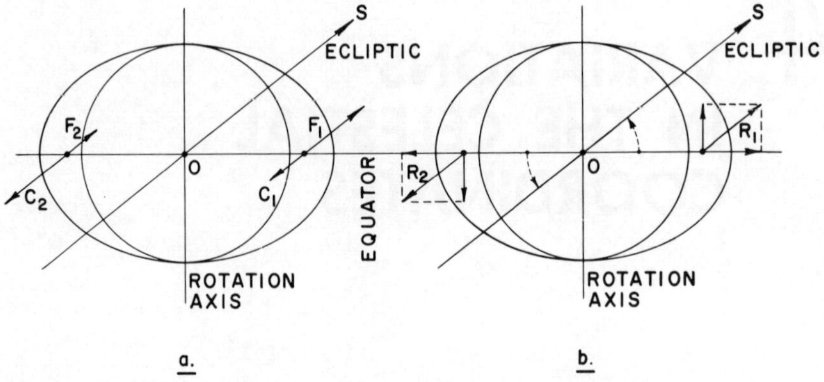

Fig. 4.1 Lunisolar precession (cause)

Let the forces R_1 and R_2 be resolved into components which are perpendicular and parallel respectively to the equator. The perpendicular components cause a moment which tries to rotate the equator into the ecliptic. This moment, combined with the rotational momentum of the earth, results mainly in a motion similar to that of an ordinary top termed the 'precession.'

The actual motion combined with the effects of the moon and the planets is more complicated. The effects of the sun and the moon are resolved into two components. The first component, the lunisolar precession, explained in principle above, moves the celestial pole around the ecliptic pole with a period of about 25,800 years and an amplitude equal to the obliquity of the ecliptic, about 23°.5 (Fig. 4.2), resulting in the westerly motion of the equinox on the equator of about 50".3 a year. The second component, termed the astronomic nutation, is due partly to the periodic motion of the earth around the sun and of the moon around the earth, partly to the fact that the moon's orbit does not coincide with the ecliptic, and partly to other causes. The nutation is a relatively short periodic motion of the celestial pole superimposed on the luni-solar precession with a maximum amplitude of about 9" and a main period of about 18.6 years (Fig. 4.5).

The planets affect the position of the mean orbital plane of the earth, the ecliptic. This effect, the planetary precession, consists of a slow rotation of the ecliptic (0".5 a year) about a slowly moving axis of rotation ($\lambda \cong 174°$), and it results in the westerly motion of the equinox of about 12".5 per century and a decrease in the obliquity of the ecliptic of about 47" per century.

The lunisolar and planetary precessions normally are considered together and are termed general precession.

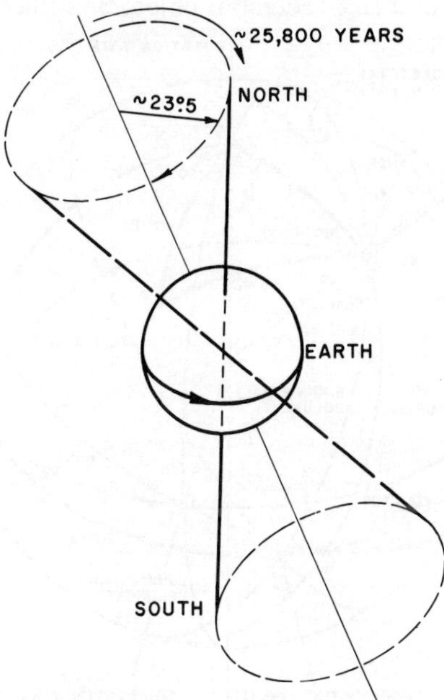

Fig. 4.2 Lunisolar precession (effect)

It follows that since both the general precession and the nutation affect the positions of the celestial equator and the ecliptic, thus the equinoxes, the orientation of most primary and secondary reference planes defining the various celestial coordinate systems described in Chapter 3, changes with time.

In addition to these factors, the relative motion of the solid mass of the earth with respect to the rotation axis (polar motion) makes the geographic coordinates of the observer change in time. Polar motion is due partly to free-force precession (Eulerian motion) disturbed by the elastic yielding of the earth and partly to the continuous redistribution of mass in the meteorological and geophysical processes.

These various factors are discussed below in some detail as they affect the coordinates in the right ascension system. Coordinate changes in the other systems may be obtained by the use of the transformation equations or the differential relationships derived in the previous Chapter. The exception to this is the polar motion whose effect is evaluated for the coordinates of the observer.

4.11 General Precession

4.111 Rigorous Transformation. The effect of general precession on the coordinates of a fixed celestial object S is illustrated in Fig. 4.3.

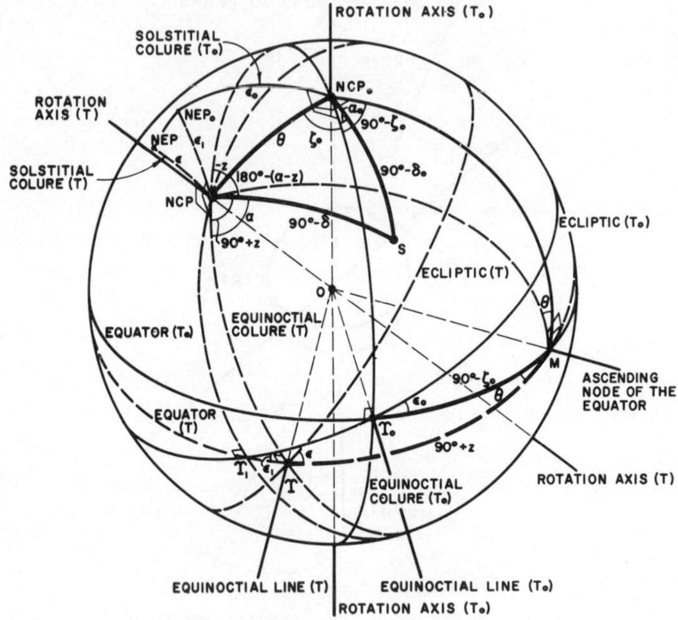

Fig. 4.3 General precession

The coordinates of S, at the initial epoch T_0, are α_0 and δ_0 in a system defined by the north celestial pole NCP_0 (or, equivalently, by the celestial equator at T_0), and by the vernal equinox Υ_0. At this initial epoch the north pole of the ecliptic is at NEP_0. The symbols α, δ, NCP, Υ, and NEP denote the respective coordinates and positions of these points at a later time T. Although at any instant the NCP moves, due to the luni-solar precession, in a direction perpendicular to the solstitial colure, the arc NCP_0-NCP is not perpendicular either to the colure at T_0 or at T. Due to planetary precession, NEP is itself in motion along a curve which is always convex to the solstitial colure. The combined motion is defined by means of three angles ζ_0, z, and θ (precessional elements) and the corresponding values of the obliquity of the ecliptic ϵ. The angle $90°-\zeta_0$ is the right ascension of the ascending node of the equator at T on the initial equator at T_0 measured in the initial system (the arc Υ_0M); $90°+z$ is the right ascension of the node in the new system (the arc ΥM); and θ is the inclination of the equator at T with respect to the initial equator.

Simon Newcomb derived constants, based partly on theoretical consideration but mainly on observation, from which, after neglecting small secular changes in the coefficients, the following expressions for ζ_0, z, and θ may be deduced [Nautical Almanac Offices, 1961, pp. 29-31]:

$$\left.\begin{aligned} \zeta_0 &= (2304''.250 + 1''.396\,t_0)\,t + 0''.302\,t^2 + 0''.018\,t^3 \\ z &= \zeta_0 + 0''.791\,t^2 + 0''.001\,t^3 \\ \theta &= (2004''.682 - 0''.853\,t_0)\,t - 0''.426\,t^2 - 0''.042\,t^3 \end{aligned}\right\} , \quad (4.1)$$

where the initial epoch is $T_0 = 1900.0 + t_0$, the final epoch $T = 1900.0 + t_0 + t$ (see section 5.53). The quantities t_0 and t are measured in tropical centuries (see section 5.222). It can be verified that ζ_0, z, θ for initial epoch T_0 and interval t are identically equal to $-z$, $-\zeta_0$, $-\theta$ respectively for epoch T_0+t and interval $-t$. The precessional elements are tabulated in table 2.1 of the Explanatory Supplement for years 1900 to 1980 at intervals of one year, for reduction to 1950.0, but expressions are given so that the table may be used for reductions between any two epochs (part of this table is reproduced in Table 4.1) [Nautical Almanac Offices, 1961, pp. 32-33]. The annual volumes of the American Ephemeris and Nautical Almanac also contain similar tables (Table III) for reduction to the date of the particular issue [U. S. Naval Observatory, Annual].

Positions referred to the reference system defined by NCP and NEP, constantly in motion due to general precession, are designated as 'mean positions' referred to the 'mean celestial equinox and equator of epoch T.' Similarly, the pole moving about NEP due to general precession is designated as the 'mean celestial pole.'

By use of the spherical triangle NCP_0-S-NCP in Fig. 4.3, the mean celestial coordinates of S at epoch T may be calculated from those at the initial epoch T_0 by means of the following set of equations based on the well-known rules of spherical trigonometry:

$$\left.\begin{aligned} \cos\delta \sin(\alpha - z) &= \cos\delta_0 \sin(\alpha_0 + \zeta_0) \\ \cos\delta \cos(\alpha - z) &= \cos\theta \cos\delta_0 \cos(\alpha_0 + \zeta_0) - \sin\theta \sin\delta_0 \\ \sin\delta &= \sin\theta \cos\delta_0 \cos(\alpha_0 + \zeta_0) + \cos\theta \sin\delta_0 \end{aligned}\right\} . \quad (4.2)$$

The same results may be obtained by means of the rotation matrices of section 3.34. Using the system introduced there, from Fig. 4.3 it is evident that a positive rotation of the line $O\Upsilon_0$ through the angle $90° - \zeta_0$ about the 3 axis (O-NCP_0), and one of O-NCP_0 through θ about the 1 axis (OM) produces results identical to those in (4.2) noting that $\sin(\alpha - z) = \cos[\alpha - (90° + z)]$ and that $\cos(\alpha - z) = -\sin[\alpha - (90° + z)]$. Thus

$$\begin{pmatrix} x \\ y \\ z \end{pmatrix}_{\alpha-(90°+z),\,\delta} = \mathbf{R}_1(\theta)\,\mathbf{R}_3(90° - \zeta_0) \begin{pmatrix} x \\ y \\ z \end{pmatrix}_{\alpha_0,\,\delta_0} .$$

TABLE 4.1 Equatorial Precessional Elements for Reduction to 1950.0 or Other Epochs

Date t_0	ζ_0	z	$\sin\theta$	$\cos\theta - 1$	M	$N = \theta$	$N = \theta$
	s	s	Unit = 10^{-8}		s	s	"
1950·0	0·000	0·000	0	0	0·000	0·000	0·00
1951	− 1·537	− 1·537	− 9717	0	− 3·073	− 1·336	− 20·04
1952	3·073	3·073	1 9434	− 2	6·147	2·672	40·08
1953	4·610	4·610	2 9151	4	9·220	4·008	60·13
1954	6·147	6·147	3 8867	8	12·293	5·345	80·17
1955·0	− 7·683	− 7·683	− 4 8584	− 12	− 15·367	− 6·681	− 100·21
1956	9·220	9·220	5 8301	17	18·440	8·017	120·25
1957	10·757	10·757	6 8017	23	21·513	9·353	140·30
1958	12·294	12·293	7 7734	30	24·587	10·689	160·34
1959	13·830	13·830	8 7450	38	27·660	12·025	180·38
1960·0	− 15·367	− 15·367	− 9 7167	− 47	− 30·734	− 13·361	− 200·42
1961	16·904	16·903	10 6883	57	33·807	14·698	220·46
1962	18·441	18·440	11 6600	68	36·881	16·034	240·50
1963	19·977	19·977	12 6316	80	39·954	17·370	260·55
1964	21·514	21·513	13 6032	93	43·028	18·706	280·59
1965·0	− 23·051	− 23·050	− 14 5749	− 106	− 46·101	− 20·042	− 300·63
1966	24·588	24·587	15 5465	121	49·175	21·378	320·67
1967	26·125	26·123	16 5181	137	52·248	22·714	340·71
1968	27·662	27·660	17 4897	153	55·322	24·050	360·75
1969	29·199	29·197	18 4613	171	58·395	25·386	380·79
1970·0	− 30·736	− 30·733	− 19 4330	− 189	− 61·469	− 26·722	− 400·83
1971	32·273	32·270	20 4046	208	64·543	28·058	420·87
1972	33·809	33·807	21 3762	229	67·616	29·394	440·92
1973	35·346	35·344	22 3477	250	70·690	30·730	460·96
1974	36·883	36·880	23 3193	272	73·764	32·066	481·00
1975·0	− 38·420	− 38·417	− 24 2909	− 295	− 76·837	− 33·402	− 501·04
1976	39·957	39·954	25 2625	319	79·911	34·738	521·08
1977	41·494	41·491	26 2341	344	82·985	36·074	541·12
1978	43·031	43·027	27 2056	370	86·059	37·410	561·16
1979	44·568	44·564	28 1772	397	89·133	38·746	581·20
1980·0	− 46·106	− 46·101	− 29 1488	− 424	− 92·206	− 40·082	− 601·24

These values are for the reduction from the epoch t_0, in the left-hand argument column, to the epoch 1950·0. For reduction from 1950·0 to t_0 enter the table with t_0 as argument, reverse the signs of all respondents except $\cos\theta - 1$, and interchange ζ_0 and z.

For reduction from the epoch $t_0 + \Delta t$ to 1950·0 + Δt, and vice versa, take out values from the table using argument t_0, and multiply:

$$\zeta_0,\ z,\ M \text{ by } (1 + 0{\cdot}0000\ 06\ \Delta t)$$

and

$$N,\ \theta,\ \sin\theta \text{ by } (1 - 0{\cdot}0000\ 04\ \Delta t).$$

Over the range of the table $\tan \tfrac{1}{2}\theta$ can be taken as $\tfrac{1}{2}\sin\theta$.

Notation: $t_0 = T_0$

$\Delta t = t$

A third negative rotation of OM through the angle $90°+z$ about the 3 axis (O-NCP) completes the transformation from the system defined by $NCP_\circ$ and $\Upsilon_\circ$ into the one defined by NCP and Υ:

$$\begin{pmatrix} x \\ y \\ z \end{pmatrix}_{\alpha,\delta} = \mathbf{P} \begin{pmatrix} x \\ y \\ z \end{pmatrix}_{\alpha_\circ,\delta_\circ} , \tag{4.3}$$

where

$$\begin{aligned} \mathbf{P} &= \mathbf{R}_3(-z-90°)\,\mathbf{R}_1(\theta)\,\mathbf{R}_3(90°-\zeta_\circ) \\ &= \mathbf{R}_3(-z)\,\mathbf{R}_2(\theta)\,\mathbf{R}_3(-\zeta_\circ) \ . \end{aligned} \tag{4.4}$$

Equations (4.2) and (4.3) are rigorous and either of them may be used to compute the effect of general precession on the right ascension and declination of a fixed object. Equation (4.3) is particularly useful for machine computations.

4.112 Approximate Transformation. During a short interval of time (e.g., a year) the motions due to general precession are relatively small; thus both the diagram and the equations of the previous section may be simplified considerably. Let Fig. 4.4 illustrate the situation in the area of the vernal equinox on the celestial sphere over the interval of exactly one year. For such a short interval the mean equators (and ecliptics) at the initial and final epochs can be considered parallel to each other. The most convenient precessional elements for this case are the annual luni-solar precession in longitude ψ' and the annual planetary precession in right ascension λ'. These, together with the corresponding obliquity of the ecliptic, define the precessional motion.

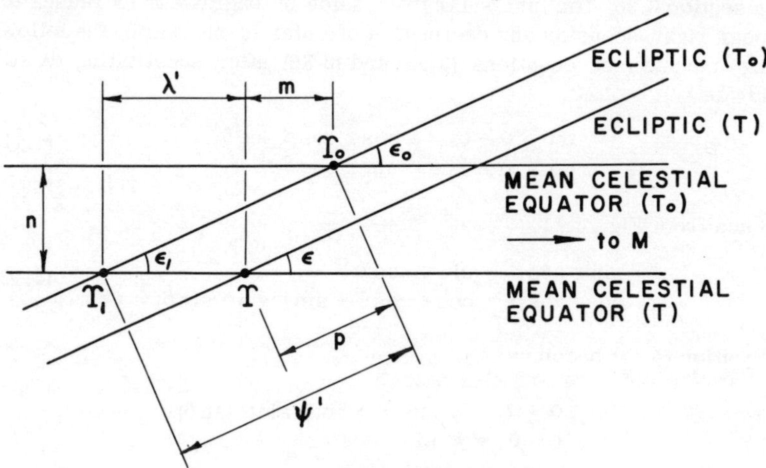

Fig. 4.4 Annual general precession

The expressions for these elements are deduced from Newcomb's discussion [Newcomb, 1897]:

$$\left.\begin{aligned} \psi' &= 50''.3708 + 0''.0050\,t \\ \lambda' &= 0''.1247 - 0''.0188\,t \\ \epsilon &= 23°27'08''.26 - 46''.845\,t - 0''.0059\,t^2 + 0''.00181\,t^3 \end{aligned}\right\}, \qquad (4.5)$$

where t is in tropical centuries from 1900.0 (see sections 5.222 and 5.53). Derived quantities from Fig. 4.4 are the following:

— Annual general precession in longitude,

$$p = \psi' - \lambda' \cos \epsilon_1 = 50''.2564 + 0''.0222\,t\,. \qquad (4.6)$$

— Annual general precession in right ascension,

$$m = \psi' \cos \epsilon_1 - \lambda' = 3^s.07234 + 0^s.00186\,t = 46''.0851 + 0''.0279\,t. \qquad (4.7)$$

— Annual general precession in declination,

$$n = \psi' \sin \epsilon_1 = 1^s.33646 - 0^s.00057\,t = 20''.0468 - 0''.0085\,t\,. \qquad (4.8)$$

— Longitude of the axis of rotation of the ecliptic (the ascending node of the ecliptic at the time t on the ecliptic at 1900.0),

$$\Pi = 173°57'.06 + 54'.77\,t\,. \qquad (4.9)$$

— Annual rate of rotation of the ecliptic,

$$\pi = 0''.4711 - 0''.0007\,t\,. \qquad (4.10)$$

The effect of the annual general precession on the mean coordinates of a star may be computed from the differential relationships derived in section 3.36. The luni-solar precession in longitude ψ' increases the mean right ascension and declination of a star in one year by the following amounts (see equations (3.35) and (3.36) after substituting $d\lambda = \psi'$, $d\beta = d\epsilon = 0$):

$$\begin{aligned} d\alpha &= \alpha - \alpha_0 = \psi' \cos r \cos\beta \sec\delta\,, \\ d\delta &= \delta - \delta_0 = \psi' \sin r \cos\beta\,. \end{aligned} \qquad (4.11)$$

Since from Fig. 3.11

$$\begin{aligned} \sin r \cos\beta &= \sin\epsilon \cos\alpha\,, \\ \cos r \cos\beta &= \cos\epsilon \cos\delta + \sin\epsilon \sin\alpha \sin\delta\,, \end{aligned} \qquad (4.12)$$

equation (4.11) becomes

$$\begin{aligned} \alpha - \alpha_0 &= \psi'(\cos\epsilon + \sin\epsilon \sin\alpha \tan\delta)\,, \\ \delta - \delta_0 &= \psi' \sin\epsilon \cos\alpha\,. \end{aligned}$$

The planetary precession decreases only the right ascension by λ' per year; thus the effect of the annual general precession is

$$\Delta\alpha_P = \alpha - \alpha_0 = (\psi'\cos\epsilon - \lambda') + \psi'\sin\epsilon\,\sin\alpha\,\tan\delta\,,$$
$$\Delta\delta_P = \delta - \delta_0 = \psi'\sin\epsilon\,\cos\alpha\,.$$

The same effect over a period shorter than a year, using the notation of expressions (4.7) and (4.8), is

$$\begin{aligned}\Delta\alpha_P &= \alpha - \alpha_0 = (m + n\,\sin\alpha\,\tan\delta)(T - T_0)\,,\\ \Delta\delta_P &= \delta - \delta_0 = n\,\cos\alpha\,(T - T_0)\,.\end{aligned} \tag{4.13}$$

where $\Delta\alpha_P$ and $\Delta\delta_P$ are the corrections for general precession, T_0 is the initial epoch, T is the final epoch expressed in tropical years. The quantities m and n vary slowly with the time and are given in the American Ephemeris and Nautical Almanac at the end of the sun's tables for the year of the volume.

A more accurate version of equation (4.13), with averaged terms, may also be used in place of equation (4.2) or (4.3) over a period of time during which the precessional elements of equation (4.1) vary linearly. Thus

$$\begin{aligned}\Delta\alpha_P &= M + N\sin\tfrac{1}{2}(\alpha+\alpha_0)\,\tan\tfrac{1}{2}(\delta+\delta_0)\,,\\ \Delta\delta_P &= N\cos\tfrac{1}{2}(\alpha+\alpha_0)\,,\end{aligned} \tag{4.14}$$

where

$$\begin{aligned}M &= \overline{m}\,(T - T_0) \cong \zeta_0 + z\,,\\ N &= \overline{n}\,(T - T_0) \cong \theta\,,\end{aligned} \tag{4.15}$$

$\overline{m}$ and $\overline{n}$ are the average values of m and n over the interval $(T-T_0)$ and are computed from (4.7) and (4.8) for the epoch $(T+T_0)/2$.

Finally, it is also possible and often convenient to calculate the general precession from the Taylor series,

$$\begin{aligned}\Delta\alpha_P &= \alpha - \alpha_0 = \frac{d\alpha_0}{dt}t + \frac{1}{2}\frac{d^2\alpha_0}{dt^2}t^2 + \frac{1}{6}\frac{d^3\alpha_0}{dt^3}t^3 + \ldots\,,\\ \Delta\delta_P &= \delta - \delta_0 = \frac{d\delta_0}{dt}t + \frac{1}{2}\frac{d^2\delta_0}{dt^2}t^2 + \frac{1}{6}\frac{d^3\delta_0}{dt^3}t^3 + \ldots\,,\end{aligned} \tag{4.16}$$

which is almost never carried beyond the third term, and where $t = T - T_0$. The first derivatives are identical to equation (4.13) for $T - T_0 = 1$ year; thus

$$\begin{aligned}\frac{d\alpha_0}{dt} &= m + n\,\sin\alpha_0\tan\delta_0\,,\\ \frac{d\delta_0}{dt} &= n\,\cos\alpha_0\,.\end{aligned} \tag{4.17}$$

The second derivatives are

$$\begin{aligned}\frac{d^2\alpha_\circ}{dt^2} &= \frac{dm}{dt} + (1 + 2\tan^2\delta_\circ)\frac{n^2}{2}\sin 2\alpha_\circ + (\frac{dn}{dt}\sin\alpha_\circ + m\,n\cos\alpha_\circ)\tan\delta_\circ\,, \\ \frac{d^2\delta_\circ}{dt^2} &= -m\,n\sin\alpha_\circ + \frac{dn}{dt}\cos\alpha_\circ - n^2\sin^2\alpha_\circ\tan\delta_\circ\,.\end{aligned} \tag{4.18}$$

The third derivatives are

$$\left.\begin{aligned}\frac{d^3\alpha_\circ}{dt^3} &= \frac{mn}{2} + \frac{3}{2}mn^2\cos 2\alpha_\circ + \frac{3n}{2}\frac{dn}{dt}\sin 2\alpha_\circ + [(2n^2 - m^2 + 3n^2\cos 2\alpha_\circ)n\sin\alpha_\circ + \\ &\quad + (2m\frac{dn}{dt} + n\frac{dm}{dt})\cos\alpha_\circ]\tan\delta_\circ + (3n\frac{dn}{dt}\sin 2\alpha_\circ + 3mn^2\cos 2\alpha_\circ)\tan^2\delta_\circ + \\ &\quad + [2n^3\sin\alpha_\circ(1 + 2\cos 2\alpha_\circ)]\tan^3\delta_\circ \\ \frac{d^3\delta_\circ}{dt^3} &= -m^2n\cos\alpha_\circ - n\frac{dm}{dt}\sin\alpha_\circ - n^2\sin^2\alpha_\circ\cos\alpha_\circ(1 + 3n\tan^2\delta_\circ) - \\ &\quad - (\frac{3}{2}m\,n^2\sin 2\alpha_\circ + 3n\frac{dn}{dt}\sin^2\alpha_\circ)\tan\delta_\circ\end{aligned}\right\} \tag{4.19}$$

Since the second and third derivatives are very small quantities (except near the poles), it is customary to compute their contribution over time intervals of 10^2 and 10^6 years respectively and to tabulate the following terms:

Annual variation in α,	AV_α	$= \frac{d\alpha_\circ}{dt}$,
Annual variation in δ,	AV_δ	$= \frac{d\delta_\circ}{dt}$,
Secular variation in α,	SV_α	$= 100\frac{d^2\alpha_\circ}{dt^2}$,
Secular variation in δ,	SV_δ	$= 100\frac{d^2\delta_\circ}{dt^2}$,
Third term in α,	3^{rd}_α	$= \frac{10^6}{6}\frac{d^3\alpha_\circ}{dt^3}$,
Third term in δ,	3^{rd}_δ	$= \frac{10^6}{6}\frac{d^3\delta_\circ}{dt^3}$.

With this notation the effect of general precession may be computed from

$$\begin{aligned}\Delta\alpha_p &= AV_\alpha(T - T_\circ) + \frac{SV_\alpha}{200}(T - T_\circ)^2 + \frac{3^{rd}_\alpha}{10^6}(T - T_\circ)^3\,, \\ \Delta\delta_p &= AV_\delta(T - T_\circ) + \frac{SV_\delta}{200}(T - T_\circ)^2 + \frac{3^{rd}_\delta}{10^6}(T - T_\circ)^3\,.\end{aligned} \tag{4.20}$$

4.12 Astronomic Nutation

4.121 Nutational Parameters. Astronomic nutation is due partly to the fact that the orbit of the earth is not circular; thus its distance to the sun undergoes periodic changes and, consequently, the sun's influence on the motions of the mean pole and the mean equinox is not quite uniform. The elliptic character of the moon's orbit produces similar effects. The combined motion which is superimposed on general precession is called nutation in longitude and is denoted by the symbol $\Delta\psi$.

The main term of astronomic nutation is produced by the non-coincidence of the moon's orbit with the ecliptic in conjunction with the retrograde motion of the lunar nodes. This results in a periodic change in the obliquity of the ecliptic termed nutation in obliquity, denoted by $\Delta\epsilon$.

The astronomic nutation (from now on called simply 'nutation') is not to be confused with the true nutation appearing as a force-free precession (Eulerian motion) of the earth's rotation axis about its principal moment of inertia axis, which is part of the polar motion described in section 4.13 [Goldstein, 1950, pp. 159-163, 174-175].

Positions referred to the reference system defined by the equator and ecliptic in motion due to general precession and nutation (in longitude and in obliquity) are designated as 'true positions' referred to the 'true equinox and celestial equator of epoch T.' Similarly, the pole having a somewhat irregular circular motion about the mean pole due to nutation is called the 'true celestial pole' (Fig. 4.5).

The theory and the numerical series upon which the nutation is now based may be found in full detail in [Woolard, 1953]. The first six terms of the expressions for $\Delta\psi$ and $\Delta\epsilon$ are

$$\left.\begin{aligned}\Delta\psi &= -(17''\!.2327 + 0''\!.01737\,t)\sin\Omega + (0''\!.2088 + 0''\!.00002\,t)\sin 2\Omega + \\ &\quad + 0''\!.0045\sin(2\omega_M + \Omega) - 0''\!.0010\sin 2\omega_M - 0''\!.0004\sin(2\omega_s - \\ &\quad - \Omega) - 0''\!.0003\sin 2(\omega_M + \Omega) + \ldots \\ \Delta\epsilon &= (9''\!.2100 + 0''\!.00091\,t)\cos\Omega - (0''\!.0904 - 0''\!.0004\,t)\cos 2\Omega - \\ &\quad - 0''\!.0024\cos(2\omega_M + \Omega) + 0''\!.0002\cos(2\omega_s - \Omega) + 0''\!.0002\cos 2(\omega_M + \\ &\quad + \Omega) + (0''\!.5522 - 0''\!.00029\,t)\cos 2\lambda_s + \ldots\end{aligned}\right\} \qquad (4.21)$$

where t is the time interval measured from 1900 January $0^{d}\!.5$ ET in Julian centuries of 36525 days (see sections 5.222, 5.232, and 5.53), Ω is the longitude of the mean ascending node of the lunar orbit on the ecliptic measured from the mean equinox of date, ω_M is the 'argument' of the point where the moon is nearest to the earth (i.e., the lunar perigee), ω_s is the mean longitude of the solar perigee measured from the mean equinox of date, and λ_s is the geometric mean longitude of the sun measured from the mean equinox of date. The new terms are illustrated in Fig. 4.6.

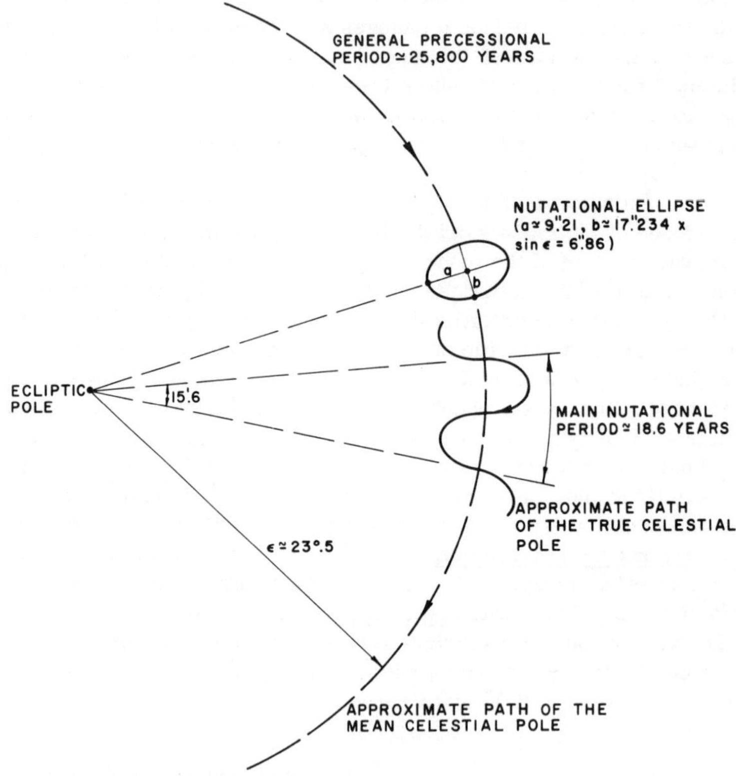

Fig. 4.5 Schematic general precession and nutation

In Woolard's development there are 69 terms in the expression for $\Delta\psi$ and 40 in that for $\Delta\epsilon$. Nutation is customarily divided into long-period and short-period terms; the latter, consisting of terms with periods of less than 35 days are denoted by $d\psi$ and $d\epsilon$ respectively. Both long- and short-period terms are listed in Table 4.2 taken from [Nautical Almanac Offices, 1961, pp. 44-45]. The principal term in longitude has an amplitude of 17''.2327 and in obliquity 9''.21, the period being 6798 days (18.62 years). The amplitude 9''.21 is known as the constant of nutation. In the table the nutational series are given with the following fundamental arguments:

$$
\begin{aligned}
l &= \lambda_M - \Omega - \omega_M, \\
l' &= \lambda_S - \omega_S, \\
F &= \lambda_M - \Omega, \\
D &= \lambda_M - \lambda_S, \\
\Omega &= \Omega.
\end{aligned}
$$

The procedure to be followed when using the table is first to compute the fundamental arguments from formulas given at the top of the table. These terms, multiplied by the factors listed in each period-row and summed, constitute the argument of the sine or cosine term whose coefficients are given in the last two columns. The sine terms summed for each period give the nutation in longitude, while the sum of the cosine terms is the nutation in obliquity.

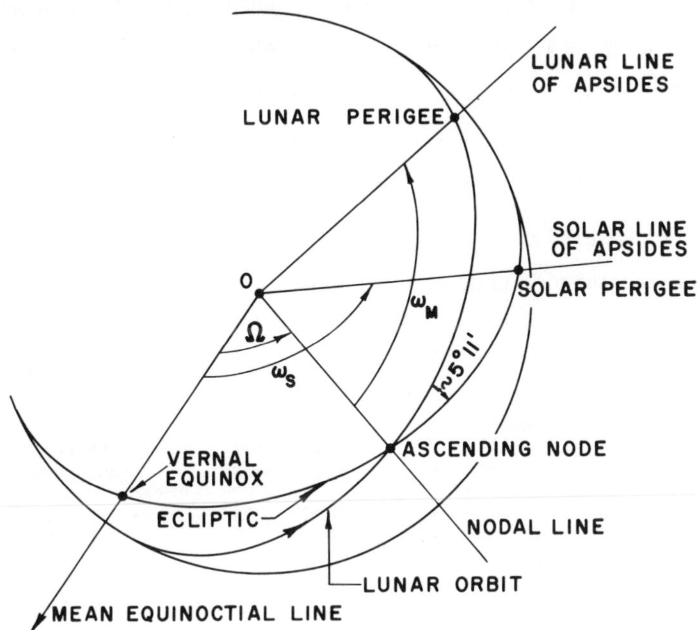

Fig. 4.6 Lunar and solar orbits

Values of the nutational parameters $\Delta\psi$ and $\Delta\epsilon$, have been calculated for each day from 1900 to 2000. Those for 1952-1959 have been included in the 'Improved Lunar Ephemeris 1952 - 1959' [Eckert et al., 1954]. Since 1960 the parameters are given to 0".001 in the 'American Ephemeris and Nautical Almanac,' $\Delta\psi$ in the ninth column in the sun's tables, and $\Delta\epsilon = -B$ in the table for Besselian Day Numbers. The short-period terms $d\psi$ and $d\epsilon$ may also be found in the same place.

4.122 Transformation from the Mean to the True Celestial System. Since the nutation in longitude $\Delta\psi$ and the nutation in obliquity $\Delta\epsilon$ are small quantities, their effect on the mean coordinates α and δ may be calculated by means of equations (3.35) and (3.36). Substituting $d\lambda = \Delta\psi$

TABLE 4.2 **Series for the Nutation**

The fundamental arguments are:

$$l = 296^\circ.10460\,8 + 13^\circ.06499\,24465d + 0^\circ.00068\,90\bar{d}^2 + 0^\circ.00000\,0295\bar{d}^3$$

$$l' = 358^\circ.47583\,3 + 0^\circ.98560\,02669d - 0^\circ.00001\,12\bar{d}^2 - 0^\circ.00000\,0068\bar{d}^3$$

$$F = 11^\circ.25088\,9 + 13^\circ.22935\,04490d - 0^\circ.00024\,07\bar{d}^2 - 0^\circ.00000\,0007\bar{d}^3$$

$$D = 350^\circ.73748\,6 + 12^\circ.19074\,91914d - 0^\circ.00010\,76\bar{d}^2 + 0^\circ.00000\,0039\bar{d}^3$$

$$\Omega = 259^\circ.18327\,5 - 0^\circ.05295\,39222d + 0^\circ.00015\,57\bar{d}^2 + 0^\circ.00000\,0046\bar{d}^3$$

where the fundamental epoch is 1900 January $0^d.5$ ET = JED 241 5020.0 and

d is measured in days,

$\bar{d}$ is measured in units of 10 000 days.

Period (days)	ARGUMENT Multiple of l	l'	F	D	Ω	LONGITUDE Coefficient of sine argument		OBLIQUITY Coefficient of cosine argument	
						Unit = 0″·0001			
6798					+1	−172327	−173·7 T	+92100	+9·1 T
3399					+2	+ 2088	+ 0·2 T	− 904	+0·4 T
1305	−2		+2		+1	+45		−24	
1095	+2		−2			+10			
6786		−2	+2	−2	+1	− 4		+ 2	
1616	−2		+2		+2	− 3		+ 2	
3233	+1	−1		−1		− 2			
183			+2	−2	+2	−12729	−1·3 T	+5522	−2·9 T
365		+1				+ 1261	−3·1 T		
122		+1	+2	−2	+2	− 497	+1·2 T	+ 216	−0·6 T
365		−1	+2	−2	+2	+ 214	−0·5 T	− 93	+0·3 T
178			+2	−2	+1	+ 124	+0·1 T	− 66	
206	+2			−2		+45			
173			+2	−2		−21			
183		+2				+16	−0·1 T		
386		+1			+1	−15		+ 8	
91		+2	+2	−2	+2	−15	+0·1 T	+ 7	
347		−1			+1	−10		+ 5	
200	−2			+2	+1	− 5		+ 3	
347		−1	+2	−2	+1	− 5		+ 3	
212	+2			−2	+1	+ 4		− 2	
120		+1	+2	−2	+1	+ 3		− 2	
412	+1			−1		− 3			

TABLE 4.2 (continued)

Period (days)	ARGUMENT Multiple of l	l'	F	D	☊	LONGITUDE Coefficient of sine argument		OBLIQUITY Coefficient of cosine argument	
						Unit = 0″·0001			
13·7			+2		+2	−2037	−0·2T	+884	−0·5T
27·6	+1					+675	+0·1T		
13·6			+2		+1	−342	−0·4T	+183	
9·1	+1		+2		+2	−261		+113	−0·1T
31·8	+1			−2		−149			
27·1	−1		+2		+2	+114		−50	
14·8				+2		+60			
27·7	+1				+1	+58		−31	
27·4	−1				+1	−57		+30	
9·6	−1		+2	+2	+2	−52		+22	
9·1	+1		+2		+1	−44		+23	
7·1			+2	+2	+2	−32		+14	
13·8	+2					+28			
23·9	+1		+2	−2	+2	+26		−11	
6·9	+2		+2		+2	−26		+11	
13·6			+2			+25			
27·0	−1		+2		+1	+19		−10	
32·0	−1			+2	+1	+14		−7	
31·7	+1			−2	+1	−13		+7	
9·5	−1		+2	+2	+1	−9		+5	
34·8	+1	+1		−2		−7			
13·2		+1	+2		+2	+7		−3	
9·6	+1			+2		+6			
14·8				+2	+1	−6		+3	
14·2		−1	+2		+2	−6		+3	
5·6	+1		+2	+2	+2	−6		+3	
12·8	+2		+2	−2	+2	+6		−2	
14·7				−2	+1	−5		+3	
7·1			+2	+2	+1	−5		+3	
23·9	+1		+2	−2	+1	+5		−3	
29·5				+1		−4			
15·4		+1		−2		−4			
29·8	+1	−1				+4			
26·9	+1		−2			+4			
6·9	+2		+2		+1	−4		+2	
9·1	+1		+2			+3			
25·6	+1	+1				−3			
9·4	+1	−1	+2		+2	−3			
13·7	−2				+1	−2			
32·6	−1		+2	−2	+1	−2			
13·8	+2				+1	+2			
9·8	−1	−1	+2	+2	+2	−2			
7·2		−1	+2	+2	+2	−2			
27·8	+1				+2	−2			
8·9	+1	+1	+2		+2	+2			
5·5	+3		+2		+2	−2			

and $d\epsilon = \Delta\epsilon$ yields

$$\Delta\alpha_N = \alpha_T - \alpha = \Delta\psi \cos r \cos\beta \sec\delta - \Delta\epsilon \tan\delta \cos\alpha,$$
$$\Delta\delta_N = \delta_T - \delta = \Delta\psi \sin r \cos\beta + \Delta\epsilon \sin\alpha,$$

or using expression (4.12)

$$\begin{aligned} \Delta\alpha_N &= \alpha_T - \alpha = \Delta\psi(\cos\epsilon + \sin\epsilon \sin\alpha \tan\delta) - \Delta\epsilon \tan\delta \cos\alpha, \\ \Delta\delta_N &= \delta_T - \delta = \Delta\psi \sin\epsilon \cos\alpha + \Delta\epsilon \sin\alpha. \end{aligned} \tag{4.22}$$

where α, δ are the mean and α_T, δ_T the true right ascensions and declinations respectively at epoch T, and $\Delta\alpha_N$, $\Delta\delta_N$ are the corrections for nutation.

By means of relations (4.7) and (4.8), the expressions above may also be written in the following form:

$$\begin{aligned} \Delta\alpha_N &= \frac{\Delta\psi}{\psi'}(m + n \sin\alpha \tan\delta) + \frac{\Delta\psi}{\psi'}\lambda' - \Delta\epsilon \tan\delta \cos\alpha, \\ \Delta\delta_N &= \frac{\Delta\psi}{\psi'} n \cos\alpha + \Delta\epsilon \sin\alpha. \end{aligned} \tag{4.23}$$

An approximate transformation of the direction numbers x, y, z for machine computations may be obtained by means of the rotation matrices of section 3.34. Using the system introduced there, from Fig. 4.7 it is evident that a negative rotation of the mean equinoctial line about the mean celestial pole at epoch T (axis 3) by the angle $\Delta\mu = \Delta\psi \cos\epsilon$, a second positive rotation about the axis $\alpha = 90° - \Delta\mu$ (axis 2) by the angle $\Delta\nu = \Delta\psi \sin\epsilon$, and a third negative rotation about the true equinoctial line (axis 1) by the angle $\Delta\epsilon$, produces the following results:

$$\begin{pmatrix} x \\ y \\ z \end{pmatrix}_{\alpha_T, \delta_T} = \mathbf{N} \begin{pmatrix} x \\ y \\ z \end{pmatrix}_{\alpha, \delta}, \tag{4.24}$$

where

$$\mathbf{N} = \mathbf{R}_1(-\Delta\epsilon)\,\mathbf{R}_2(\Delta\nu)\,\mathbf{R}_3(-\Delta\mu) \cong \begin{pmatrix} 1 & -\Delta\mu & -\Delta\nu \\ \Delta\mu & 1 & -\Delta\epsilon \\ \Delta\nu & \Delta\epsilon & 1 \end{pmatrix}. \tag{4.25}$$

Second-order terms arising in the above simplification of the nutation matrix $\mathbf{N}$ are given below [Scott and Hughes, 1964]. If needed, they should be added term by term to the matrix on the right side of equations (4.25)

$$\begin{pmatrix} -\Delta\psi^2/2 & 0 & 0 \\ -\Delta\epsilon\,\Delta\nu & -\frac{1}{2}(\Delta\epsilon^2 + \Delta\mu^2) & -\frac{1}{2}\,\Delta\nu\,\Delta\mu \\ +\Delta\epsilon\,\Delta\mu & -\frac{1}{2}\,\Delta\nu\,\Delta\mu & -\frac{1}{2}(\Delta\epsilon^2 + \Delta\nu^2) \end{pmatrix}.$$

The maximum value of any of these terms is about 4×10^{-9} radians.

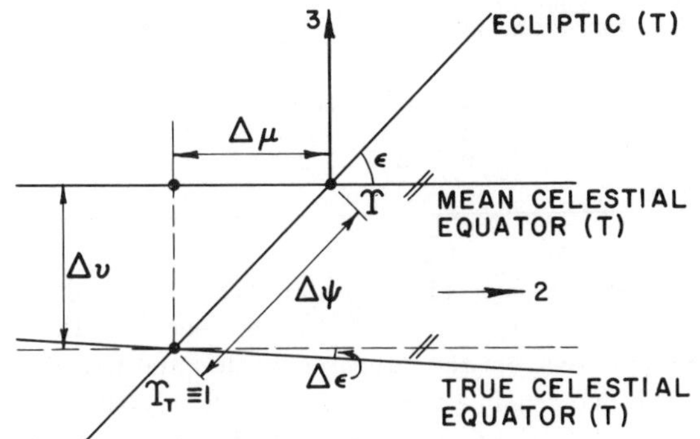

Fig. 4.7 Nutation

Successive rotations about the mean equinoctial line (axis 1) by ϵ, about the mean ecliptic pole (axis 3) by $-\Delta\psi$, and about the true equinoctial line (axis 1) by $-(\epsilon+\Delta\epsilon)$ provide the following rigorous nutation matrix which does not require second-order terms:

$$\mathbf{N} = \mathbf{R}_1(-\epsilon-\Delta\epsilon)\,\mathbf{R}_3(-\Delta\psi)\,\mathbf{R}_1(\epsilon) \; . \qquad (4.26)$$

4.123 Besselian Transformation for General Precession and Nutation. In practice, a position (α_T, δ_T) of a star referred to the true equinox and equator of any particular date is usually found by first determining the mean coordinates (α_0, δ_0) at the beginning (or at the end) of the current Besselian year (see section 5.53) by means of such equations as (4.20), then adding (subtracting) the further precession to date and the nutation at the date by combining equations (4.13) and (4.23) as follows:

$$\alpha_T - \alpha_0 = (m + n \sin\alpha \tan\delta)\left(\frac{\Delta\psi}{\psi'} + t\right) + \lambda'\frac{\Delta\psi}{\psi'} - \Delta\epsilon \cos\alpha \tan\delta \, ,$$

$$\delta_T - \delta_0 = n \cos\alpha \left(\frac{\Delta\psi}{\psi'} + t\right) + \Delta\epsilon \sin\alpha ,$$

or introducing Bessel's notation

$$\begin{aligned} \alpha_T - \alpha_0 &= Aa + Bb + E, \\ \delta_T - \delta_0 &= Aa' + Bb'. \end{aligned} \qquad (4.27)$$

The Besselian day numbers

$$\begin{aligned} A &= nt + n\frac{\Delta\psi}{\psi'} = nt + \Delta\psi \sin\epsilon, \\ B &= -\Delta\epsilon \, , \\ E &= \lambda'\frac{\Delta\psi}{\psi'} \, , \end{aligned}$$

depend only on the obliquity of the ecliptic and on the parameters of general precession and nutation, while the Besselian star constants

$$
\begin{aligned}
a &= \frac{m}{n} + \sin\alpha \tan\delta , \\
b &= \cos\alpha \tan\delta , \\
a' &= \cos\alpha , \\
b' &= -\sin\alpha ,
\end{aligned}
$$

depend mainly on the position of the star. In the above equations, $t = T - T_0$ in tropical years, T_0 and T are the epoch of the beginning of the Besselian year nearest the date in question and that of the date respectively. The Besselian day numbers are tabulated for each day in the 'American Ephemeris and Nautical Almanac,' a sample page of which is shown in Table 4.3, and in other ephemerides. Care should be exercised since in some cases the tabulated values include the long-period nutation only (see section 4.124). The Besselian star constants are so nearly constant for a particular star that the same values can be used over an extended interval of time.

The Besselian method as outlined above is particularly useful when an extended ephemeris is required, i.e., when positions of a star are necessary for many dates. If the coordinates are needed at a few isolated dates only, the system of independent day numbers is more advantageous. In this system

$$
\begin{aligned}
\alpha_T - \alpha_0 &= f + g \sin (G+\alpha) \tan\delta , \\
\delta_T - \delta_0 &= g \cos (G+\alpha) ,
\end{aligned}
\tag{4.28}
$$

where the independent day numbers are

$$
\begin{aligned}
f &= \frac{m}{n} A + E = mt + \Delta\psi \cos\epsilon , \\
g \sin G &= B , \\
g \cos G &= A .
\end{aligned}
$$

They are also tabulated for each day in the 'American Ephemeris and Nautical Almanac' on the page opposite the Besselian day numbers; a sample page is shown in Table 4.4.

In electronic machine computations the Besselian transformation of the coordinates of a star from the mean equator and equinox of the beginning of the Besselian year, nearest to the date in question, to the true equator and equinox of the date may also be performed rapidly by the following successive rotations [Scott and Hughes, 1964]:

$$
\begin{pmatrix} x \\ y \\ z \end{pmatrix}_{\alpha_T, \delta_T} = \mathbf{B} \begin{pmatrix} x \\ y \\ z \end{pmatrix}_{\alpha_0, \delta_0} , \tag{4.29}
$$

where

$$
\mathbf{B} = \mathbf{R}_3(-f)\, \mathbf{R}_2(A)\, \mathbf{R}_1(B) \tag{4.30}
$$

TABLE 4.3 Sample Page of the AENA

258

BESSELIAN DAY NUMBERS, 1967

FOR 0^h EPHEMERIS TIME

Date	*A*	*B*	*C*	*D*	*E*	dψ	dε	τ	S.T.
	″	″	″	″	(0^s·0001)	(0″·001)			h
Jan. 0	− 4·447	−6·166	− 2·885	+20·212	−16	+214	+ 9	−0·0028	6·6
1	4·381	6·229	3·214	20·155	16	+184	+ 60	− ·0001	6·7
2	4·335	6·278	3·542	20·091	16	+106	+ 96	+ ·0026	6·7
3	4·299	6·303	3·870	20·021	16	+ 3	+108	·0054	6·8
4	4·262	6·305	4·196	19·945	16	− 98	+ 96	·0081	6·9
5	− 4·214	−6·285	− 4·522	+19·862	−16	−169	+ 61	+0·0108	6·9
6	4·148	6·252	4·847	19·772	16	−195	+ 13	·0136	7·0
7	4·061	6·219	5·170	19·676	16	−167	− 36	·0163	7·1
8	3·956	6·196	5·492	19·574	16	− 94	− 76	·0191	7·1
9	3·841	6·192	5·812	19·465	16	+ 6	− 97	·0218	7·2
10	− 3·725	−6·210	− 6·130	+19·349	−16	+111	− 97	+0·0245	7·3
11	3·616	6·248	6·446	19·226	15	+198	− 77	·0273	7·3
12	3·521	6·302	6·760	19·097	15	+251	− 42	·0300	7·4
13	3·443	6·362	7·072	18·962	15	+261	− 1	·0327	7·5
14	3·384	6·423	7·381	18·821	15	+226	+ 40	·0355	7·5
15	− 3·339	−6·475	− 7·688	+18·673	−15	+157	+ 72	+0·0382	7·6
16	3·304	6·516	7·991	18·519	15	+ 62	+ 92	·0410	7·6
17	3·275	6·540	8·292	18·360	15	− 43	+ 95	·0437	7·7
18	3·244	6·548	8·590	18·195	16	−144	+ 82	·0464	7·8
19	3·205	6·541	8·884	18·024	16	−224	+ 53	·0492	7·8
20	− 3·154	−6·526	− 9·176	+17·848	−16	−273	+ 15	+0·0519	7·9
21	3·088	6·504	9·464	17·667	16	−280	− 29	·0546	8·0
22	3·003	6·487	9·749	17·480	15	−242	− 69	·0574	8·0
23	2·903	6·482	10·031	17·288	15	−162	− 98	·0601	8·1
24	2·792	6·494	10·310	17·091	15	− 53	−109	·0629	8·2
25	− 2·679	−6·531	−10·585	+16·889	−15	+ 63	− 96	+0·0656	8·2
26	2·574	6·591	10·857	16·683	15	+159	− 60	·0683	8·3
27	2·488	6·667	11·126	16·472	15	+210	− 9	·0711	8·4
28	2·425	6·746	11·391	16·256	15	+203	+ 46	·0738	8·4
29	2·386	6·815	11·653	16·035	15	+140	+ 90	·0766	8·5
30	− 2·361	−6·860	−11·912	+15·809	−15	+ 41	+110	+0·0793	8·6
31	2·339	6·878	12·168	15·578	15	− 63	+103	·0820	8·6
Feb. 1	2·308	6·872	12·420	15·342	15	−142	+ 72	·0848	8·7
2	2·259	6·851	12·668	15·101	15	−177	+ 26	·0875	8·8
3	2·191	6·826	12·913	14·856	15	−161	− 24	·0902	8·8
4	− 2·106	−6·808	−13·154	+14·605	−15	− 99	− 67	+0·0930	8·9
5	2·008	6·807	13·391	14·349	15	− 6	− 93	·0957	9·0
6	1·909	6·827	13·623	14·089	14	+ 95	− 98	·0985	9·0
7	1·814	6·866	13·851	13·824	14	+185	− 84	·1012	9·1
8	1·732	6·922	14·075	13·554	14	+245	− 53	·1039	9·2
9	− 1·665	−6·987	−14·294	+13·280	−14	+267	− 13	+0·1067	9·2
10	1·617	7·053	14·508	13·001	14	+245	+ 29	·1094	9·3
11	1·585	7·113	14·717	12·719	14	+183	+ 64	·1121	9·4
12	1·564	7·161	14·921	12·432	14	+ 94	+ 88	·1149	9·4
13	1·551	7·193	15·120	12·141	15	− 11	+ 96	·1176	9·5
14	− 1·537	−7·208	−15·314	+11·847	−15	−114	+ 87	+0·1204	9·6
15	− 1·519	−7·209	−15·503	+11·549	−15	−203	+ 64	+0·1231	9·6

Notation: $\tau = t = T - T_0$

TABLE 4.4 Sample Page of the AENA

INDEPENDENT DAY NUMBERS, 1967 259

FOR 0h EPHEMERIS TIME

Date	f	g	G	h	H	i	f′	g′	G′
	s	″	h m s	″	h m s	″	(0s·0001)	(0″·001)	h m
Jan. 0	−0·6836	7·602	15 36 48	20·417	23 27 30	−1·251	+131	86	23 36
1	·6735	7·615	15 39 31	20·410	23 23 46	1·394	+113	95	21 23
2	·6664	7·629	15 41 30	20·401	23 20 00	1·536	+ 65	105	19 35
3	·6609	7·629	15 42 49	20·392	23 16 14	1·678	+ 2	108	18 02
4	·6553	7·610	15 43 46	20·382	23 12 28	1·820	− 60	104	16 32
5	−0·6479	7·567	15 44 38	20·370	23 08 42	−1·961	−103	91	14 49
6	·6378	7·503	15 45 45	20·357	23 04 54	2·102	−119	79	12 38
7	·6244	7·427	15 47 25	20·344	23 01 07	2·242	−102	76	10 06
8	·6084	7·351	15 49 46	20·330	22 57 19	2·382	− 57	85	7 45
9	·5907	7·287	15 52 45	20·314	22 53 30	2·520	+ 4	97	5 55
10	−0·5728	7·242	15 56 10	20·297	22 49 41	−2·658	+ 68	107	4 22
11	·5561	7·219	15 59 46	20·278	22 45 52	2·795	+121	110	2 57
12	·5415	7·219	16 03 14	20·258	22 42 02	2·931	+154	108	1 31
13	·5296	7·234	16 06 19	20·238	22 38 11	3·067	+160	104	0 02
14	·5205	7·260	16 08 52	20·217	22 34 21	3·201	+138	98	22 24
15	−0·5136	7·285	16 10 53	20·194	22 30 29	−3·334	+ 96	95	20 44
16	·5083	7·306	16 12 27	20·170	22 26 38	3·465	+ 38	95	19 00
17	·5038	7·314	16 13 36	20·146	22 22 47	3·596	− 26	97	17 19
18	·4990	7·308	16 14 35	20·121	22 18 55	3·725	− 88	100	15 40
19	·4931	7·284	16 15 35	20·095	22 15 03	3·852	−137	104	14 03
20	−0·4853	7·248	16 16 49	20·069	22 11 10	−3·979	−167	110	12 31
21	·4751	7·200	16 18 25	20·042	22 07 17	4·104	−171	115	11 02
22	·4622	7·148	16 20 38	20·015	22 03 24	4·228	−148	119	9 37
23	·4468	7·102	16 23 30	19·987	21 59 30	4·350	− 99	117	8 13
24	·4297	7·069	16 26 56	19·960	21 55 36	4·471	− 32	111	6 44
25	−0·4123	7·059	16 30 47	19·932	21 51 42	−4·590	+ 39	99	5 02
26	·3963	7·076	16 34 40	19·905	21 47 47	4·708	+ 97	87	2 54
27	·3830	7·116	16 38 09	19·878	21 43 51	4·825	+128	84	0 25
28	·3734	7·169	16 40 55	19·850	21 39 55	4·940	+124	93	22 01
29	·3674	7·221	16 42 49	19·822	21 35 59	5·053	+ 86	106	20 07
30	−0·3635	7·255	16 44 02	19·794	21 32 01	−5·166	+ 25	111	18 34
31	·3602	7·265	16 44 52	19·767	21 28 02	5·277	− 39	106	17 05
Feb. 1	·3554	7·249	16 45 44	19·739	21 24 02	5·386	− 87	91	15 28
2	·3480	7·214	16 47 00	19·711	21 20 02	5·493	−108	75	13 21
3	·3375	7·169	16 48 49	19·684	21 16 01	5·600	− 98	68	10 38
4	−0·3244	7·126	16 51 15	19·655	21 11 58	−5·704	− 61	78	8 02
5	·3095	7·097	16 54 16	19·627	21 07 55	5·807	− 4	93	6 06
6	·2942	7·089	16 57 31	19·598	21 03 51	5·908	+ 58	105	4 36
7	·2796	7·102	17 00 48	19·569	20 59 47	6·006	+113	112	3 15
8	·2670	7·135	17 03 49	19·540	20 55 41	6·104	+150	111	1 54
9	−0·2568	7·183	17 06 23	19·511	20 51 34	−6·198	+163	107	0 28
10	·2494	7·236	17 08 21	19·481	20 47 27	6·291	+150	102	22 54
11	·2445	7·287	17 09 45	19·452	20 43 20	6·382	+112	97	21 15
12	·2414	7·330	17 10 43	19·421	20 39 12	6·470	+ 57	96	19 32
13	·2394	7·358	17 11 20	19·391	20 35 03	6·557	− 7	96	17 50
14	−0·2372	7·370	17 11 51	19·362	20 30 54	−6·641	− 70	98	16 10
15	−0·2344	7·367	17 12 24	19·332	20 26 44	−6·723	−124	103	14 34

Transformation from the mean place of any epoch T_O to the true place of epoch T may be also performed rapidly by means of electronic computers from the combination of equations (4.3) and (4.24) as follows:

$$\begin{pmatrix} x \\ y \\ z \end{pmatrix}_{\alpha_T,\delta_T} = \mathbf{N}\,\mathbf{P} \begin{pmatrix} x \\ y \\ z \end{pmatrix}_{\alpha_O,\delta_O} \tag{4.31}$$

where $\mathbf{P}$ is from (4.4) and $\mathbf{N}$ from (4.25) or (4.26).

4.124 Short-Period Nutation. As has been mentioned earlier, the Besselian day numbers in some ephemerides do not contain the short-period nutation terms. In other cases, such as when star positions are tabulated at intervals longer than one day, the short-period terms are omitted because of the difficulty of interpolation at such intervals (e.g., ten-day tables of the 'Apparent Places of Fundamental Stars,' [Astronomisches Rechen Institut, Heidelberg, Annual]). In these cases the effect of the short-period terms must be calculated separately and applied to the values obtained using the tables.

For single reductions the system of independent day numbers should be used, the necessary corrections to be added being

$$\begin{aligned} d\alpha_N &= f' + g' \sin (G' + \alpha) \tan \delta, \\ d\delta_N &= g' \cos (G' + \alpha), \end{aligned} \tag{4.32}$$

where

$$\begin{aligned} f' &= d\psi \cos \epsilon, \\ g' \sin G' &= -d\epsilon, \\ g' \cos G' &= d\psi \sin \epsilon. \end{aligned}$$

For ephemeris type work where a number of reductions are to be made, the above formulas may be written in the following form:

$$\begin{aligned} d\alpha_N &= d\alpha(\psi)\, d\psi + d\alpha(\epsilon)\, d\epsilon, \\ d\delta_N &= d\delta(\psi)\, d\psi + d\delta(\epsilon)\, d\epsilon, \end{aligned} \tag{4.33}$$

where

$$\begin{aligned} d\alpha(\psi) &= \cos \epsilon + \sin \alpha \tan \delta \sin \epsilon, \\ d\alpha(\epsilon) &= -\cos \alpha \tan \delta, \\ d\delta(\psi) &= \cos \alpha \sin \epsilon, \\ d\delta(\epsilon) &= \sin \alpha. \end{aligned}$$

These coefficients for each star usually are given in those ephemerides where the short-period nutational corrections need to be applied. The quantities $d\psi$, $d\epsilon$, f', g', and G' are tabulated in the 'American Ephemeris and Nautical Almanac' for each day in the tables of the Besselian and independent day numbers (see Tables 4.3 and 4.4). The quantities $d\psi$ and $d\epsilon$ may also be found in table 1 of the 'Apparent Places of Fundamental Stars.'

4.13 Polar Motion

4.131 Motion of the Pole, the International Polar Motion Service. Polar motion is the motion of the true celestial pole ('instantaneous' rotation axis of the earth) with respect to a reference point fixed to the earth's crust. The reference point is usually selected to be near the average position of the true pole over a certain time interval and is called the 'average' or 'mean' terrestrial pole of that interval. The absolute geodetic and the reduced astronomic coordinates are all referred to this average (mean) terrestrial pole and to its corresponding equator (see sections 2.33 and 2.4). The terrestrial Cartesian coordinate system u, v, w thus has its origin at the center of the earth; its axis u is oriented towards the Greenwich mean astronomic meridian of Λ (reduced) $= 0°$ (see section 8.213), the axis w towards the average north terrestrial pole; the axis v is perpendicular to both and forms a right-handed coordinate system (Fig. 4.10).

In the literature the term 'variation of latitude' is often used in lieu of polar motion. Although the magnitude of the variation in the pole's position is generally determined from latitude observations, the expression is a poor choice since both longitude and azimuth also vary due to the motion of the pole.

The earliest evidence that a motion of the true pole actually occurs was obtained by F. Küstner at Berlin, from observations beginning in 1884. His results led the International Association of Geodesy to organize observations at several stations within the framework of the International Latitude Service (ILS), established in 1899. Within a few years these observations confirmed the existence of polar motion, but they also showed that it could not be in accordance with Euler's theory of motion for rigid bodies (see section 4.121). The correct interpretation was obtained first from S.C. Chandler's analysis of the accumulated data and from Newcomb's dynamical explanation of the results. From the discussion of the observations, Chandler established that the motion is the resultant of two components; one is the revolution of the true pole around the principal moment of inertia axis, counterclockwise as viewed from the north, with a period of about 1.2 years (Chandler period), and the other is a revolution in the same direction with an annual period [Chandler, 1891, p. 65; 1892, p. 97]. Newcomb recognized that the first type of motion could be explained by the effects of elastic yielding of the earth and a consequent displacement of the maximum moment of inertia axis [Woolard, 1953, p. 157]. The further disturbance of the theoretical rigid Eulerian motion by the continuous redistribution of mass in meteorological and geophysical processes produces the annual component. This complex motion and its variations are only incompletely understood at present.

The ILS was reorganized in 1962 into the International Polar Motion

Service (IPMS) following resolutions of the XIIth General Assembly of the International Astronomic Union (IAU) at Berkeley in 1961. The Central Bureau of the IPMS is located at the Mizusawa Observatory, Japan, and its activities are directed by an international scientific council [Yumi, 1964]. In principle, the IPMS continues the work of the ILS.

Another service, the Rapid Latitude Service (RLS) was established by action of the General Assembly of the IAU at Dublin in 1955 and put under the direction of the Bureau International de l'Heure (BIH) in Paris. The function of the RLS used to be to predict the coordinates of the instantaneous pole and to provide time corrections on a nearly current basis. This work has become part of the routine of the BIH, and the designation 'RLS' is no longer used (see also sections 5.4 and 8.211).

The motion of the pole is usually determined from frequent and uninterrupted latitude observations at so-called latitude observatories. At present there are about thirty such stations located about the world, some of them determining time and latitude simultaneously. Five of these stations, located at nearly the same latitude, are designated IPMS stations (see Table 4.5). The other stations primarily participate in the BIH program described in section 8.21. The advantage of having stations located at nearly the same latitude is that the derived coordinates of the pole can be freed of errors in star positions. The exact method of reduction is not given here. An outline can be found in [Markowitz, 1961, pp. 29-39], and details in [Yumi, 1964 and 1965].

TABLE 4.5 Latitude Observatories of the International Polar Motion Service

Station	Longitude	Latitude
Carloforte, Italy	8°18'44"	39°08'08".941
Gaithersburg, Maryland	-77 11 57	39 08 13 .202
Kitab, U.S.S.R.	66 52 51	39 08 01 .850
Mizusawa, Japan	141 07 51	39 08 03 .602
Ukiah, California	-123 12 35	39 08 12 .096

Both the IPMS and the BIH publish the positions of the true celestial pole with respect to a designated average terrestrial pole in terms of the Cartesian coordinates x_P and y_P, expressed in seconds of arc. The origin of the system is at the designated terrestrial pole; x_P is positive south towards Greenwich and y_P towards the west (Fig. 4.8).

Over a period of about six years the periodic motion of the true pole with respect to the earth's crust averages out. The IPMS determines the position of the average pole from the observational data accumulated over six years. This position is usually referred to as the average (mean) terrestrial pole of this period. The fact that determina-

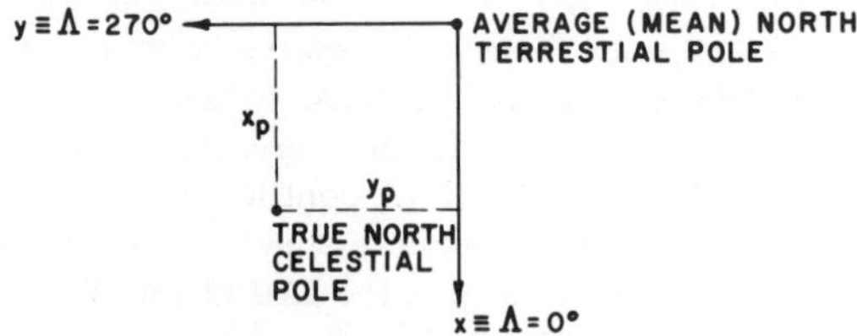

Fig. 4.8 Polar motion

tions of the average poles of consecutive six year intervals show a displacement of 0".003 to 0".006 per year in the direction of about 285° longitude (Fig. 4.9) has led to much debate on the question of the secular motion of the pole. The apparent secular motion of the average pole is attributed to various causes, including the actual secular motion of the pole, possible horizontal crustal displacements, and possible changes in the direction of the verticals at the observing stations. In recent years the secular motion of the average pole seems to be confirmed, and, in addition, a random (libration-type) motion appears to be likely [Markowitz, 1960 and 1967].

The question of the epoch to which the terrestrial coordinate system should refer was settled by the general assembly of the International Union of Geodesy and Geophysics, at Helsinki, in 1960. It was decided that the average position of the true celestial pole during the period between 1900 and 1905 shall be adopted as the terrestrial pole. This position actually is defined by the adoption of the latitudes of the IPMS stations as listed in Table 4.5 as absolute constants. The coordinate system associated with this pole is usually called the new system of the period 1900-05 or (misleadingly) of the mean epoch 1903.

The coordinates of consecutive terrestrial poles for the epochs from 1903 to 1962 are listed in Table 4.6 relative to the system of 1900-05 (see also Fig. 4.9).

TABLE 4.6 Coordinates of the Average Terrestrial Pole in the System of 1900–1905

Mean epoch	x_o	y_o
1903	0".000	0".000
1909	-0.007	0.043
1915	0.001	0.076
1927	0.039	0.080
1932	0.027	0.130
1938	0.031	0.139
1952	0.074	0.142
1957	0.071	0.178
1962	0.060	0.218

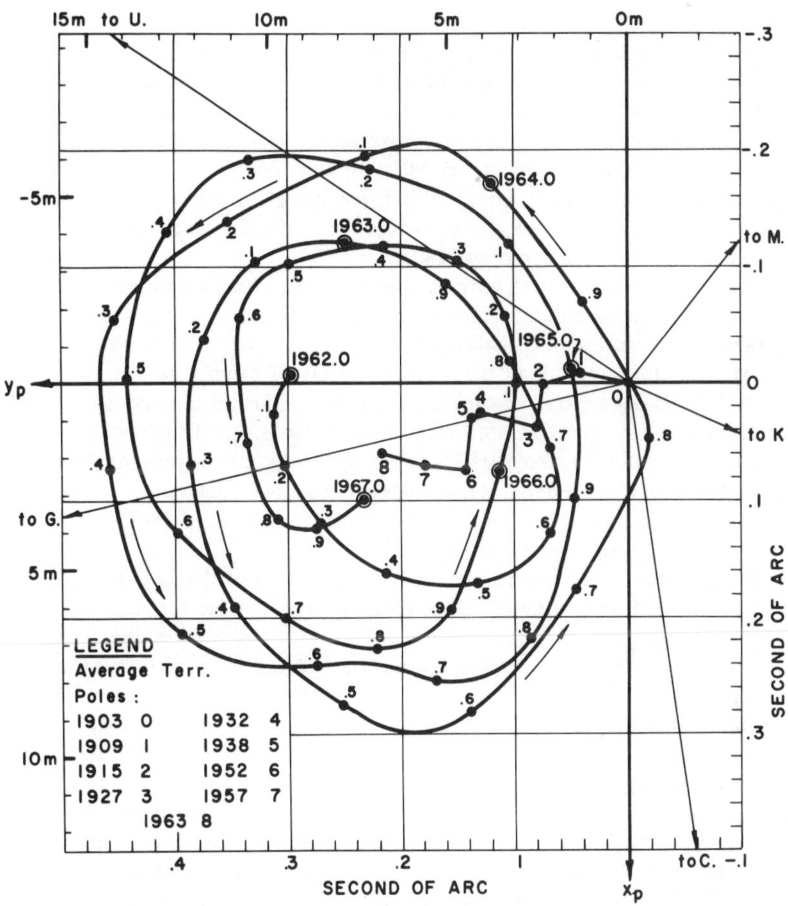

I.P.M.S. Stations:

C. CARLOFORTE, ITALY
G. GAITHERSBURG, MARYLAND
K. KITAB, U.S.S.R.
M. MIZUSAWA, JAPAN
U. UKIAH, CALIFORNIA

Fig. 4.9 The motion of the true celestial and the average terrestrial poles in the terrestrial system of 1900–1905.

A plot of the motion of the true celestial pole between 1962.0 and 1967.0 in the new system 1900-05, based on IPMS data, is shown in Fig. 4.9.

Since 1963 preliminary coordinates of the true celestial pole, based on the observations at the five IPMS stations, are published by the Central Bureau of the IPMS in its 'Monthly Notes of the International Polar Motion Service.' The data is given for a period of about 0.15 years, three months in arrears. Final coordinates, referring to the new system 1900-05, are published annually in the 'Annual Report of the International Polar Motion Service' (first volume for the year 1962). The coordinates x_P and y_P are given at intervals of 0.05 year and are published about two years in arrears.

The BIH also publishes both preliminary and final coordinates of the true celestial pole based on the observations of some thirty cooperating stations. Preliminary coordinates at ten-day intervals are distributed in their monthly circular 'B/C' for a period of twenty to thirty days, about two weeks in arrears for interpolated coordinates and about ten weeks ahead for extrapolated ones (see Table 8.4). Final coordinates at five-day intervals are published in the monthly circular 'D' (see Table 8.5) for a period of about thirty days, one month in arrears, and also at ten-day intervals in the 'Bulletin Horaire' (Serie J) about one year in arrears (see section 8.214).

Since 1959, in accordance with the resolution of the IAU at Moscow in 1958, the BIH coordinates have been referred to the moving 'average pole of the epoch.' This reference point is calculated as the average position of the true pole during the current Chandler period (1.2 years). From the geodetic point of view, satisfactory annual values for the coordinates of this BIH pole, referred to the IPMS 1900-05 fixed pole are available in Table 8.12 or in [Yumi and Wako, 1966, p. 66].

It is expected that from 1968 the BIH coordinates will also be referred to the IPMS pole of 1900-05 which should resolve much of the confusion arising from the present difference in the definition of the IPMS and the BIH poles. For more information on polar motion and on the work of the BIH, see sections 5.4 and 8.2.

4.132 Transformation from the Celestial to the Terrestrial System. The coordinate systems involved are illustrated in Fig. 4.10 where the true celestial (x, y, z) and the average (mean) terrestrial (u, v, w) Cartesian systems are shown together. The position of the axis z with respect to w is given by the parameters of polar motion x_P and y_P, while the angle between x and u is defined by the Greenwich hour angle of x (true vernal equinox), the Greenwich apparent sidereal time (GAST, see section 5.11).

The terrestrial direction cosines or direction numbers of a point may be obtained from those in the true system by rotating the axis x about z (axis 3) with a positive GAST, then the axis z about the new x (axis 1) with a negative y_P, and finally the new axis z about the new y (axis 2) with a negative x_P. Thus, with the notation of section 3.34,

$$\begin{pmatrix} u \\ v \\ w \end{pmatrix}_{\ell,b} = \mathbf{S} \begin{pmatrix} x \\ y \\ z \end{pmatrix}_{\alpha_T,\delta_T}, \tag{4.34}$$

where

$$\mathbf{S} = \mathbf{R}_2(-x_P)\,\mathbf{R}_1(-y_P)\,\mathbf{R}_3(\text{GAST}), \tag{4.35}$$

and the quantities ℓ and b are the spherical longitude and latitude of the point respectively.

Using equation (4.31) gives the transformation from the mean celestial system at epoch T_0 to the terrestrial system at epoch T as

$$\begin{pmatrix} u \\ v \\ w \end{pmatrix}_{\ell,b} = \mathbf{S\,N\,P} \begin{pmatrix} x \\ y \\ z \end{pmatrix}_{\alpha_0,\delta_0}, \tag{4.36}$$

where the matrices **P**, **N**, and **S** are to be computed from equations (4.4), (4.25) or (4.26), and (4.35) respectively.

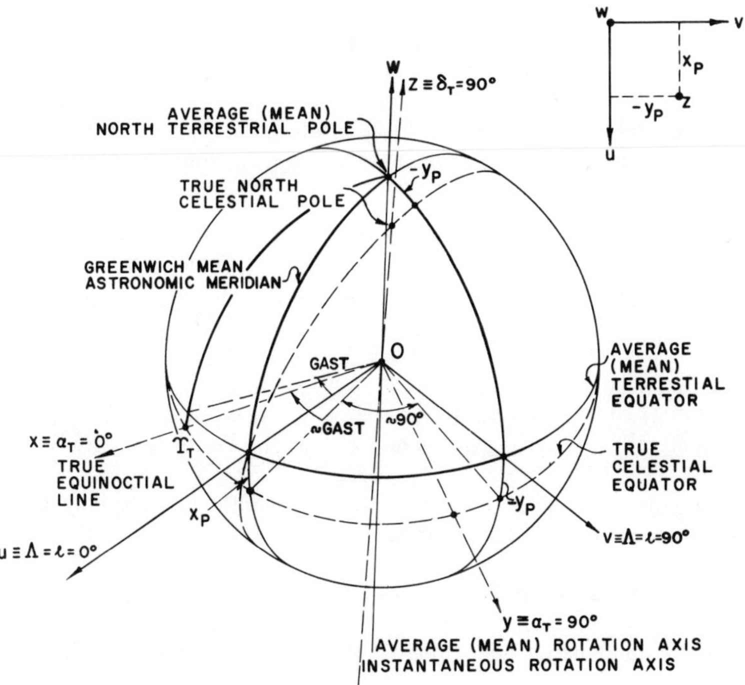

Fig. 4.10 True celestial and average (mean) terrestrial systems

4.133 Polar Motion in Latitude, Longitude, and Azimuth. It has been mentioned in section 2.4 that the observed astronomic latitude, longitude, and azimuth at a given place on the earth change in time due to the fact that the true pole moves with respect to the solid earth. In order to make the results of the observations time independent, it is customary, as explained earlier, to refer the observed quantities to an average terrestrial pole of some defined period (e.g., that of 1900-05) by adding a correction for polar motion to the observed quantities.

Let Φ_T, Λ_T, and A_T denote the observed astronomic latitude, longitude, and azimuth respectively. These quantities refer to a coordinate system which can be generated from the true celestial system by a single positive rotation of GAST about the axis z (Fig. 4.10). Thus, assuming a spherical earth for this derivation, with the notation of section 3.34,

$$\begin{pmatrix} u \\ v \\ w \end{pmatrix}_{\Lambda_T, \Phi_T} = \mathbf{R}_3(\text{GAST}) \begin{pmatrix} x \\ y \\ z \end{pmatrix}_{\alpha_T, \delta_T} .$$

Comparing this expression with equations (4.34) and (4.35) it is evident that

$$\begin{pmatrix} u \\ v \\ w \end{pmatrix}_{\Lambda, \Phi} = \mathbf{R}_2(-x_P)\ \mathbf{R}_1(-y_P) \begin{pmatrix} u \\ v \\ w \end{pmatrix}_{\Lambda_T, \Phi_T} . \tag{4.37}$$

Recognizing that $\cos x_P \cong \cos y_P \cong 1$, $\sin x_P \cong x_P$, $\sin y_P \cong y_P$, and neglecting products $x_P y_P$, appropriate substitutions yield the following set of equations:

$$\begin{pmatrix} \cos\Phi \cos\Lambda \\ \cos\Phi \sin\Lambda \\ \sin\Phi \end{pmatrix} = \begin{pmatrix} 1 & 0 & x_P \\ 0 & 1 & -y_P \\ -x_P & y_P & 1 \end{pmatrix} \begin{pmatrix} \cos\Phi_T \cos\Lambda_T \\ \cos\Phi_T \sin\Lambda_T \\ \sin\Phi_T \end{pmatrix} ,$$

or

$$\left.\begin{aligned} \cos\Phi \cos\Lambda &= \cos\Phi_T \cos\Lambda_T + x_P \sin\Phi_T \\ \cos\Phi \sin\Lambda &= \cos\Phi_T \sin\Lambda_T - y_P \sin\Phi_T \\ \sin\Phi &= -x_P \cos\Phi_T \cos\Lambda_T + y_P \cos\Phi_T \sin\Lambda_T + \sin\Phi_T \end{aligned}\right\} . \tag{4.38}$$

The effect of polar motion on the astronomic latitude may be derived by means of the last equation, from which

$$\sin\Phi = \sin\Phi_T + \cos\Phi_T\,(y_P \sin\Lambda_T - x_P \cos\Lambda_T).$$

On the other hand, according to Taylor's expansion,

$$\sin\Phi = \sin\Phi_T + (\Phi - \Phi_T)\cos\Phi_T + \ldots ;$$

consequently,

$$\sin\Phi_T + (\Phi - \Phi_T)\cos\Phi_T = \sin\Phi_T + \\ + \cos\Phi_T\,(y_P \sin\Lambda_T - x_P\cos\Lambda_T).$$

Thus the correction to be applied to the observed latitude is

$$\Delta\Phi_P = \Phi - \Phi_T = y_P\sin\Lambda_T - x_P\cos\Lambda_T\,. \qquad (4.39)$$

The effect of polar motion on the longitude of the station may be derived from the first equation in (4.38) by substituting the following Taylor expansions:

$$\sin\Phi_T = \sin\Phi + \cos\Phi\,(x_P\cos\Lambda_T - y_P\sin\Lambda_T),$$
$$\cos\Phi_T = \cos\Phi - \sin\Phi\,(x_P\cos\Lambda_T - y_P\sin\Lambda_T),$$
$$\cos\Lambda_T = \cos\Lambda + \Delta\Lambda_P\sin\Lambda_T,$$

where $\Delta\Lambda_P = \Lambda - \Lambda_T$. Thus, dropping the subscripts for Λ and neglecting the small products x_P^2, $x_P y_P$,

$$\cos\Phi\cos\Lambda = x_P\sin\Phi + x_P(x_P\cos\Lambda - y_P\sin\Lambda) + [\cos\Phi - \\ -\sin\Phi(x_P\cos\Lambda - y_P\sin\Lambda)](\cos\Lambda + \Delta\Lambda_P\sin\Lambda)$$
$$= x_P\sin\Phi + \cos\Phi\cos\Lambda - \sin\Phi\,(x_P\cos^2\Lambda - \\ -y_P\sin\Lambda\cos\Lambda) + \Delta\Lambda_P\cos\Phi\sin\Lambda - \\ -\sin\Phi\sin\Lambda\,(x_P\cos\Lambda - y_P\sin\Lambda)\,\Delta\Lambda_P.$$

Substituting $\cos^2\Lambda = 1 - \sin^2\Lambda$, and neglecting the small products $x_P\Delta\Lambda_P$ and $y_P\Delta\Lambda_P$,

$$\cos\Phi\cos\Lambda = x_P\sin\Phi\sin^2\Lambda + y_P\sin\Phi\sin\Lambda\cos\Lambda + \\ + \cos\Phi\cos\Lambda + \Delta\Lambda_P\cos\Phi\sin\Lambda,$$

from which the correction to the observed longitude is

$$\Delta\Lambda_P = \Lambda - \Lambda_T = -(x_P\sin\Lambda + y_P\cos\Lambda)\tan\Phi. \qquad (4.40)$$

The effect of polar motion on the astronomic azimuth may be derived from Fig. 4.11, where z is the true celestial pole and w is the average terrestrial pole. The symbol P denotes the station where the azimuth A_T to a station Q is observed. The angle A is the azimuth referred to the terrestrial system, and $\Delta A_P = A - A_T$. From the figure it is evident that

$$\sin\Delta A_P = -\frac{\sin d\,\sin(\theta + \Lambda)}{\cos\Phi_T}\,.$$

Since d and ΔA_P are small angles, this equation may be written as follows:

$$\Delta A_P = -d\sin(\theta + \Lambda)\sec\Phi_T = -(d\sin\theta\cos\Lambda + d\cos\theta\sin\Lambda)\sec\Phi_T,$$

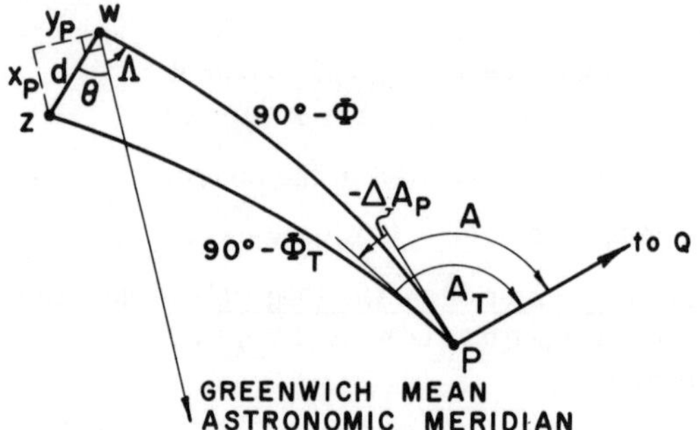

Fig. 4.11 Polar motion in azimuth

but since

$$d \cos\theta = x_P,$$
$$d \sin\theta = y_P,$$

the correction to the observed azimuth is

$$\Delta A_P = A - A_T = -(x_P \sin\Lambda + y_P \cos\Lambda) \sec \Phi_T. \tag{4.41}$$

4.2 Variations Due to Physical Effects

Variations of celestial coordinates in this group include effects from aberration, parallax, and astronomic refraction.

The aberration is the apparent displacement of a celestial object caused by the finite velocity of light propagation in combination with the relative motion of the observer and the object (Fig. 4.12). Plane-

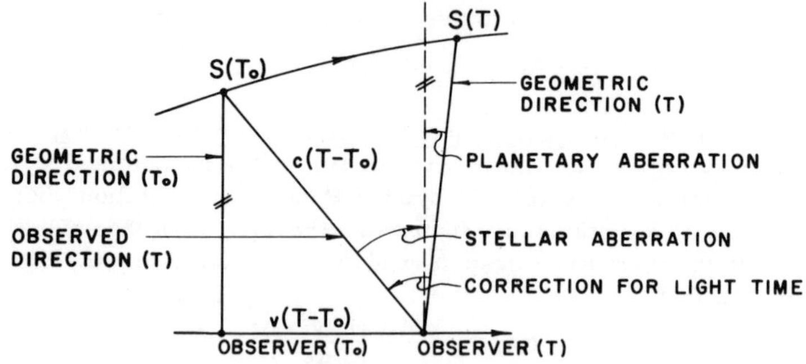

Fig. 4.12 Aberration (schematic)

tary aberration is the angular displacement of the geometric direction between the object S and the observer at the instant of light emission T_o from the geometric direction at the instant of observation T. It has two components: stellar aberration due to the inertial velocity of the observer at the instant of observation v, and the correction for light time due to the motion of the object in the inertial frame of reference during the interval while the light was propagating. Stellar aberration has three components: diurnal aberration due to the rotation of the earth; annual aberration due to the revolution of the earth around the center of mass of the solar system; and secular aberration due to the motion of the solar system in space around the center of our galaxy. In geodetic astronomy, when the objects of observations are stars, the corrections for light-time are of no importance and may be assumed to be zero, and, since the secular aberration is practically constant for each star and therefore ignored, only the diurnal and annual aberrations are considered.

The parallactic displacement is the angle between the directions of a celestial object as seen from the observer and from some standard point of reference. The geocentric parallactic displacement is the angle at the object A between the direction to the observer and the direction to the center of the earth (Fig. 4.13). To reduce an observed or topo-

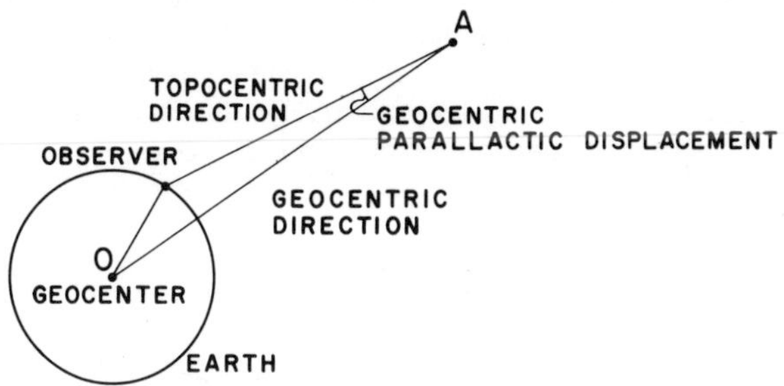

Fig. 4.13 Geocentric parallax (schematic)

centric direction to the geocenter, a correction to this effect must be applied. The annual or stellar parallactic displacement is the angle subtended at an object by the radius of the earth's orbit. A correction for this effect must be applied to a geocentric direction to reduce it to the center of mass of the solar system (Fig. 4.14). In geodetic astronomy, when the objects of observations are stars whose distances from the earth are very large, the effect of the geocentric parallax is always neglected, while the correction for annual parallax must sometimes be applied.

The astronomic refraction is the apparent displacement of the celestial object outside the atmosphere that results from light rays being bent in passing through the atmosphere. This results in all objects appearing higher above the horizon than they would appear if there were no refraction (Fig. 4.15). The magnitude of this displacement depends upon the zenith distance of the object and upon atmospheric conditions (temperature, barometric pressure, etc.).

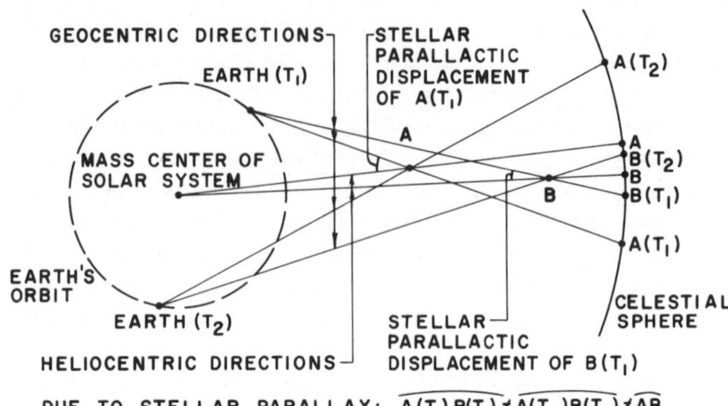

Fig. 4.14 Annual parallax (schematic)

None of the above-mentioned phenomena affects the frame of reference, but cause changes in the direction in which the celestial object is observed. The corresponding corrections, described below, must therefore be calculated with respect to a particular reference system and applied to the object's position in the same system.

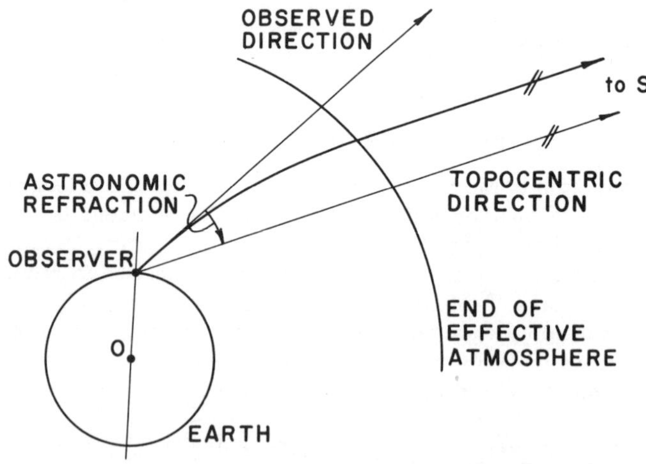

Fig. 4.15 Astronomic refraction (schematic)

4.21 Aberration

4.211 General Law of Aberration. Let S in Fig. 4.16 be the position of a star at the moment when a ray of light leaves it and O that of the observer at the instant of light detection. Let OO′ be the direction in which the observer is moving at this moment and v his velocity. Let the line SS′ be parallel to OO′ and of such length that

$$SS' : SO = v : c,$$

where c is the velocity of light. The law of aberration is that the star S will be seen by the observer in the displaced direction OS′. Stated in a general form the law is the following:

> The apparent direction of an object seen by an observer in motion is displaced from the true direction in which it would be seen if the observer were motionless by an amount equal in linear measure to the observer's motion at a constant speed during the time interval the light propagated from the object to the observer. The direction of the displacement is that of the observer's motion at the moment of the observation [Newcomb, 1906].

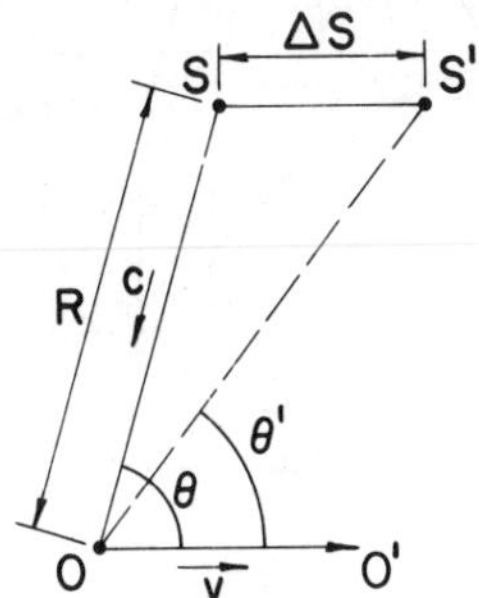

Fig. 4.16 Aberration of light

Thus in linear measure

$$\Delta S = v\frac{R}{c} = \varkappa R,$$

where

$$\varkappa = \frac{\Delta S}{R} = \frac{v}{c}.$$

From Fig. 4.16 it is also evident that

$$R \sin(\theta - \theta') = \Delta S \sin\theta'.$$

Since the angle $\theta - \theta'$ is small, the angular displacement is

$$\Delta\theta = \theta - \theta' = \frac{\Delta S}{R}\sin\theta' = \varkappa \sin\theta', \qquad (4.42)$$

where $\varkappa$, expressed in seconds of arc, is called the constant of aberration.

4.212 Circular Annual Aberration. As mentioned previously, the annual aberration is due to the revolution of the earth around the center of mass of the solar system. Let Fig. 4.17a illustrate the orbit of the earth, first assumed to be circular and around the sun. The instantaneous orbital velocity-vector in this case points to the direction F of $\lambda_S - 90^\circ$ celestial longitude (λ_S is the longitude of the sun). Let Fig. 4.17b show the situation on the celestial sphere, where O is the observ-

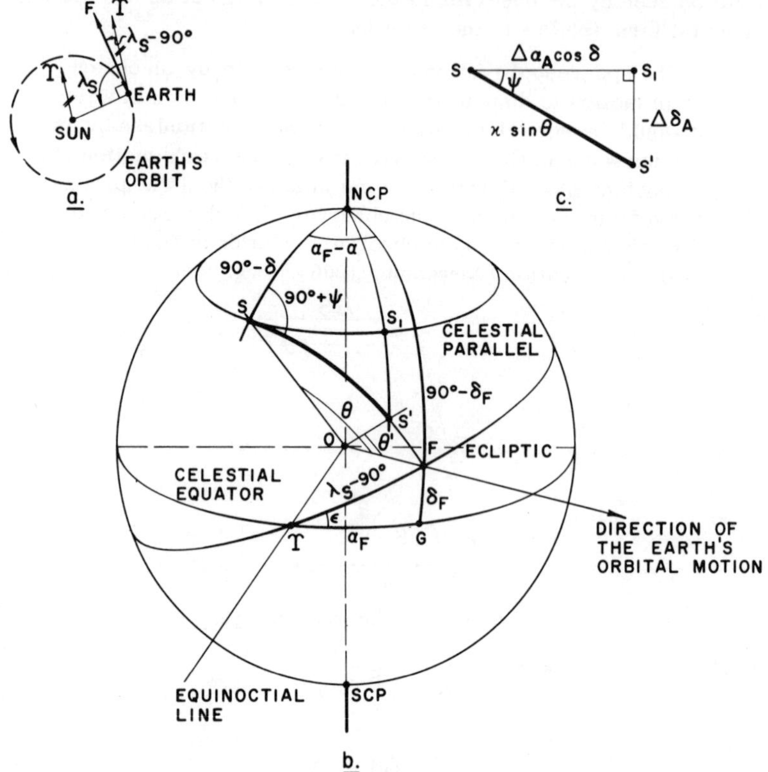

Fig. 4.17 Annual aberration in case of circular orbit

er moving in the F direction, and OS and OS′ are the actual and displaced directions of the celestial object S respectively. According to the general law of aberration (the displacement is in the direction of the velocity of the observer), the direction OS′ lies in the plane SOF. The actual angular displacement according to equation (4.42) is

$$SS' = \theta - \theta' \cong \varkappa \sin\theta,$$

where the adopted value for the constant of aberration is $\varkappa = 20''.4958$. This number is the equivalent to a light-time of $499^s.012$ for the astronomic unit [Fricke, W. et al., 1965].

The displacement in right ascension and in declination, $\Delta\alpha_A$ and $\Delta\delta_A$ respectively, may be determined from the triangle $SS'S_1$ (Fig. 4.17c):

$$\begin{aligned} \Delta\alpha_A &= \alpha' - \alpha = \varkappa \sin\theta \cos\psi \sec\delta, \\ \Delta\delta_A &= \delta' - \delta = -\varkappa \sin\theta \sin\psi, \end{aligned} \tag{4.43}$$

where the primed quantities denote the coordinates of the displaced position S'.

From the triangle S-NCP-F, it is evident that

$$\begin{aligned} \sin\theta \cos\psi &= \cos\delta_F \sin(\alpha_F - \alpha), \\ -\sin\theta \sin\psi &= \sin\delta_F \cos\delta - \cos\delta_F \sin\delta \cos(\alpha_F - \alpha). \end{aligned} \tag{4.44}$$

On the other hand, from the triangle F♈G,

$$\left.\begin{aligned} \sin\lambda_S &= \cos\alpha_F \cos\delta_F \\ -\cos\lambda_S \sin\epsilon &= \sin\delta_F \\ -\cos\lambda_S \cos\epsilon &= \sin\alpha_F \cos\delta_F \end{aligned}\right\}. \tag{4.45}$$

The aberrational corrections (to be applied to the true positions to get the displaced ones) are obtained by substituting equations (4.45) into (4.44), and then the result into (4.43):

$$\begin{aligned} \Delta\alpha_A &= -\varkappa \sec\delta (\cos\alpha \cos\lambda_S \cos\epsilon + \sin\alpha \sin\lambda_S), \\ \Delta\delta_A &= -\varkappa [\cos\lambda_S \cos\epsilon (\tan\epsilon \cos\delta - \sin\alpha \sin\delta) + \\ &\quad + \cos\alpha \sin\delta \sin\lambda_S]. \end{aligned}$$

These formulas are usually simplified by means of the Besselian day numbers C and D, referring to aberration, and star constants c, d, c′ and d′, to

$$\begin{aligned} \Delta\alpha_A &= \alpha' - \alpha = Cc + Dd, \\ \Delta\delta_A &= \delta' - \delta = Cc' + Dd', \end{aligned} \tag{4.46}$$

where

$$\begin{aligned} C &= -\varkappa \cos\epsilon \cos\lambda_S, \\ D &= -\varkappa \sin\lambda_S, \\ c &= \cos\alpha \sec\delta, \\ d &= \sin\alpha \sec\delta, \\ c' &= \tan\epsilon \cos\delta - \sin\alpha \sin\delta, \\ d' &= \cos\alpha \sin\delta. \end{aligned}$$

These relations are particularly useful when positions of the same star are required for many dates. If the coordinates are needed for a few isolated times only, the use of independent aberrational day numbers, H, h, and i, is more advantageous.

In this system

$$\Delta\alpha_A = h \sin (H + \alpha) \sec \delta,$$
$$\Delta\delta_A = i \cos \delta + h \cos (H + \alpha) \sin \delta, \tag{4.47}$$

where

$$h \cos H = -\varkappa \sin \lambda_S = D,$$
$$h \sin H = -\varkappa \cos \epsilon \cos \lambda_S = C,$$
$$i = -\varkappa \sin \epsilon \cos \lambda_S = C \tan \epsilon.$$

Both types of aberrational day numbers are tabulated for each day in the 'American Ephemeris and Nautical Almanac.' Sample pages are shown in Tables 4.3 and 4.4.

It may be shown that the aberrational terms -D, C, and C tan ϵ, expressed in radians, may be regarded as displacements of the direction numbers x, y, z. The displaced direction of an object due to circular annual aberration may be calculated from [Scott and Hughes, 1964]

$$\begin{pmatrix} x \\ y \\ z \end{pmatrix}_{\alpha',\delta'} = \begin{pmatrix} x \\ y \\ z \end{pmatrix}_{\alpha,\delta} + \begin{pmatrix} -D \\ C \\ C \tan \epsilon \end{pmatrix} . \tag{4.48}$$

The day numbers in the second term on the right side of the equation should be referred to the true coordinate system of the epoch to which the quantities α, δ are referred also. The second matrix may be calculated from the tabulated day numbers, normally referred to the mean coordinate system of the beginning of the Besselian year nearest to the epoch (e.g., in Table 4.3), by rotating the quantities -D, C, and C tan ϵ, expressed in radians, with the matrix **B** of equation (4.30). Thus

$$\begin{pmatrix} x \\ y \\ z \end{pmatrix}_{\alpha',\delta'} = \begin{pmatrix} x \\ y \\ z \end{pmatrix}_{\alpha,\delta} + \mathbf{B} \begin{pmatrix} -D \\ C \\ C \tan \epsilon \end{pmatrix}^*$$

where the asterisk denotes that the day numbers in the vector are as they are tabulated for the beginning of the nearest Besselian year, in radians.

4.213 The Effect of the Ellipticity of the Earth's Orbit on Annual Aberration. So far, it has been assumed that the earth's orbit is circular; thus its orbital velocity vector is perpendicular to its radius to the center of mass of the solar system (Fig. 4.17a). In the near actual situation shown in Fig. 4.18a, where the center of mass of the solar system is replaced by the sun, the velocity T may be divided into the components F perpendicular to the radius, and f parallel to the minor semi-axis of the orbit. It may be shown that both F and f can be considered constant along the orbit and that $f = eF$, the quantity e being the eccentricity of the orbit (see section 5.211). The direction of f is defined by the celestial longitude $90° + \bar{\omega}_s$, where $\bar{\omega}_s$ is the longitude of the perihelion (point of shortest distance to the sun) of the earth's orbit. (It differs from the longitude of the solar perigee ω_s by 180°.)

The effect of the velocity component F on the annual aberration has been described in the previous section and is given as a correction to the coordinates by any of the equations (4.46)-(4.48). Let Fig. 4.18b illustrate the aberrational displacement of the point S to S′ on account of the component f. Comparing Fig. 4.17b to 4.18b, and recognizing that $f = eF$, it is evident that equations (4.46)-(4.48) may be used to calculate the effect of the orbital ellipticity with the substitution $\lambda_s = \bar{\omega}_s + 180°$ (since $\lambda_s - 90° = 90° + \bar{\omega}_s$) and by multiplying with the factor e. The resulting

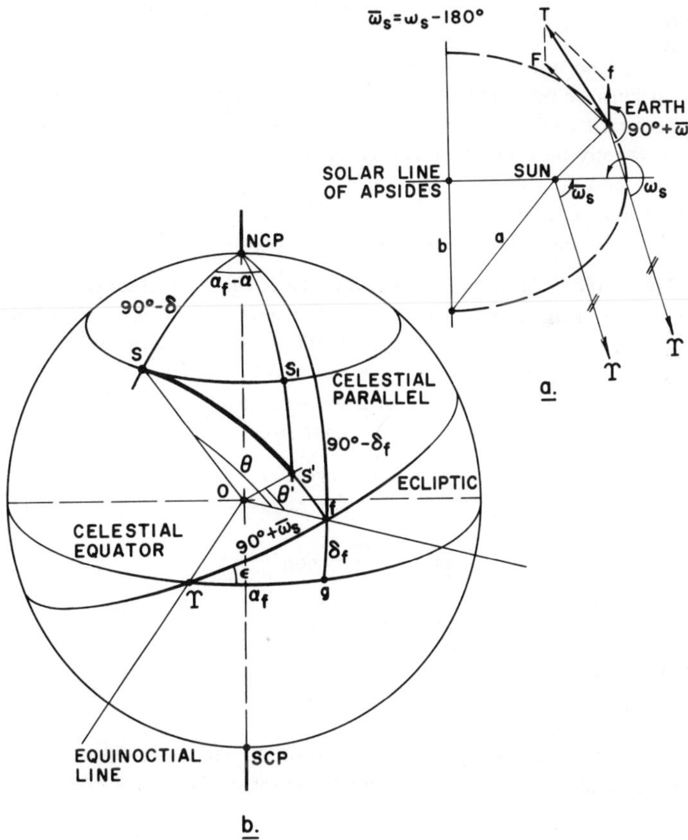

Fig. 4.18 Effect of orbital ellipticity on annual aberration

equations are

$$\begin{aligned}\Delta\alpha_{\bar{A}} &= c\,\Delta C + d\,\Delta D,\\ \Delta\delta_{\bar{A}} &= c'\Delta C + d'\Delta D,\end{aligned} \tag{4.49}$$

where

$$\begin{aligned}\Delta C &= e\varkappa\cos\bar{\omega}_s\cos\epsilon,\\ \Delta D &= e\varkappa\sin\bar{\omega}_s,\end{aligned}$$

and c, d, c′, d′ are the same as in equation (4.46). In the independent system

$$\begin{aligned}\Delta\alpha_{\bar{A}} &= \Delta h\sin(\Delta H+\alpha)\sec\delta,\\ \Delta\delta_{\bar{A}} &= \Delta i\cos\delta + \Delta h\cos(\Delta H+\alpha)\sin\delta,\end{aligned} \tag{4.50}$$

where

$$\begin{aligned}\Delta h\cos\Delta H &= e\varkappa\sin\bar{\omega}_s = \Delta D,\\ \Delta h\sin\Delta H &= e\varkappa\cos\bar{\omega}_s\cos\epsilon = \Delta C,\\ \Delta i &= e\varkappa\sin\epsilon\cos\bar{\omega}_s = \Delta C\tan\epsilon.\end{aligned}$$

It should be noted that since these corrections, commonly known as the e-terms or perigee terms of aberration, depend only on very slowly varying quantities ($\bar{\omega}_s$, e, ϵ, α, δ, all independent of the longitude of the sun), they may be considered constant throughout a year for any particular star and they change very slowly during the centuries. Since the eccentricity of the orbit is $e = 0.016726$, the maximum effect of the ellipticity of the earth's orbit on annual aberration, $e\varkappa$, is only about 0ʺ.343.

4.214 Diurnal Aberration. In addition to the orbital motion of the observer around the center of mass of the solar system, there is also the diurnal motion due to the rotation of the earth about its axis. Let ρ denote the geocentric radius of the observer, φ' his geocentric latitude, and ω_e the rotational velocity of the earth (Fig. 4.19). The linear velocity of the observer towards the east is

$$v = \omega_e\rho\cos\varphi'.$$

With

$$\omega_e = \frac{2\pi}{86164}\,\sec^{-1},$$

$$\rho = 6378\ \text{km},$$

the velocity is

$$v = 0.464\,\rho\cos\varphi'\ \text{km sec}^{-1} \tag{4.51}$$

where ρ is in units of the earth's equatorial radius (6378 km).

Using equation (4.51) yields the constant of diurnal aberration

$$k = \frac{v}{c}\operatorname{cosec}1'' = 0''.320\,\rho\cos\varphi' = 0^{s}.0213\,\rho\cos\varphi'. \tag{4.52}$$

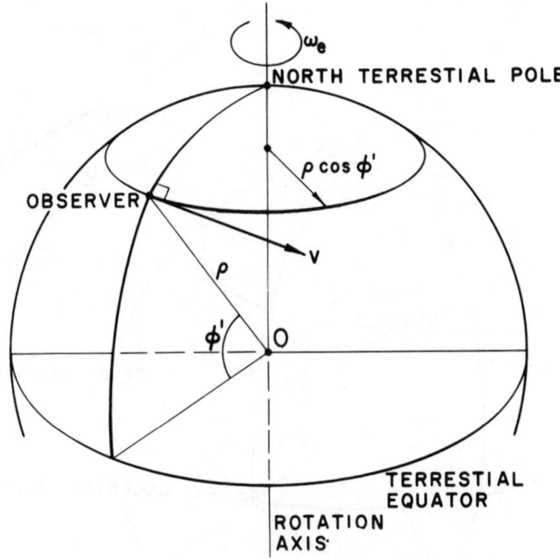

Fig. 4.19 Diurnal velocity

Let Fig. 4.20a illustrate the displacement of the object S to S$'$ owing to the velocity of the observer moving at O in the direction E (east). The total displacement according to the law of aberration is

$$SS' = k \sin \theta.$$

The displacement in right ascension and declination $\Delta\alpha_D$ and $\Delta\delta_D$ may be determined from the triangle SS$'$S$_1$ (Fig. 4.20b):

$$\begin{aligned} \Delta\alpha_D \cos \delta &= k \sin \theta \cos \psi , \\ -\Delta\delta_D &= k \sin \theta \sin \psi , \end{aligned} \tag{4.53}$$

From the triangle S-NCP-E it is evident that

$$\begin{aligned} \sin \theta \cos \psi &= \cos h, \\ \sin \theta \sin \psi &= -\sin \delta \sin h . \end{aligned} \tag{4.54}$$

The corrections for diurnal aberration are obtained by substituting equations (4.52) and (4.54) into (4.53). Thus

$$\begin{aligned} \Delta\alpha_D &= \alpha' - \alpha = 0^s\!.02132\, \rho \cos \varphi' \cos h \sec \delta, \\ \Delta\delta_D &= \delta' - \delta = 0''\!.3198\, \rho \cos \varphi' \sin h \sin \delta , \end{aligned} \tag{4.55}$$

where the primed quantities again refer to the displaced position and ρ is in units of the equatorial radius of the earth.

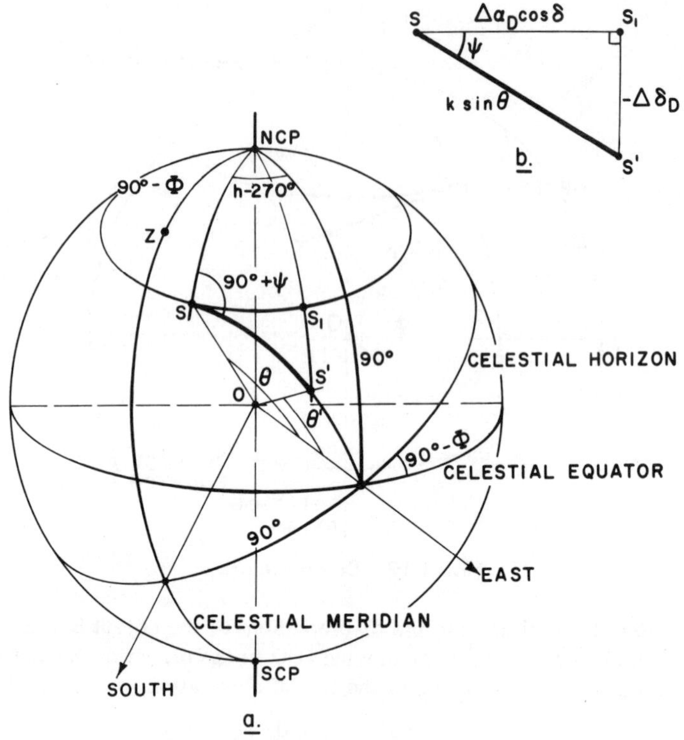

Fig. 4.20 Diurnal aberration

From equation (4.55) it is obvious that at transit

$$\Delta\alpha_D = \pm 0^s\!.02132\, \rho \cos\varphi' \sec\delta,$$
$$\Delta\delta_D = 0''\!.0000,$$

where the positive sign corresponds to upper culmination ($h = 0^h$), the negative to lower culmination ($h = 12^h$). Thus, owing to diurnal aberration, the stars seem to transit later than they actually do. In astronomical observations the diurnal aberration is usually subtracted from the observed (displaced) parameters (e.g., from time in case of meridian observations) instead of being added to the true right ascension and declination. Values of the diurnal aberration at meridian crossings are tabulated in Table 4.7.

4.22 Parallax

4.221 Annual Parallax. Owing to the revolution of the earth around the center of mass of the solar system, the geocentric direction to a

TABLE 4.7 Correction for Diurnal Aberration

Dec. \ Lat.	0°	10°	20°	30°	35°	40°	45°	50°	52°	54°	56°	58°	60°
						Unit $0^s.001$							
°													
0	21	21	20	18	17	16	15	14	13	13	12	11	11
5	21	21	20	19	18	16	15	14	13	13	12	11	11
10	22	21	20	19	18	17	15	14	13	13	12	11	11
15	22	22	21	19	18	17	16	14	14	13	12	12	11
20	23	22	21	20	19	17	16	15	14	13	13	12	11
25	24	23	22	20	19	18	17	15	14	14	13	12	12
30	25	24	23	21	20	19	17	16	15	14	14	13	12
35	26	26	24	23	21	20	18	17	16	15	15	14	13
40	28	27	26	24	23	21	20	18	17	16	16	15	14
45	30	30	28	26	25	23	21	19	19	18	17	16	15
50	33	33	31	29	27	25	23	21	20	20	19	18	17
52	35	34	33	30	28	27	25	22	21	20	19	18	17
54	36	36	34	31	30	28	26	23	22	21	20	19	18
56	38	38	36	33	31	29	27	24	23	22	21	20	19
58	40	40	38	35	33	31	28	26	25	24	23	21	20
60	43	42	40	37	35	33	30	27	26	25	24	23	21
62	45	45	43	39	37	35	32	29	28	27	25	24	23
64	49	48	46	42	40	37	34	31	30	29	27	26	24
66	52	52	49	45	43	40	37	34	32	31	29	28	26
68	57	56	54	49	47	44	40	37	35	33	32	30	28
70	62	61	59	54	51	48	44	40	38	37	35	33	31
71	66	65	62	57	54	50	46	42	40	39	37	35	33
72	69	68	65	60	57	53	49	44	43	41	39	37	35
73	73	72	69	63	60	56	52	47	45	43	41	39	36
74	77	76	73	67	63	59	55	50	48	45	43	41	39
75	82	81	77	71	68	63	58	53	51	48	46	44	41
76	88	87	83	76	72	68	62	57	54	52	49	47	44
77	95	93	89	82	78	73	67	61	58	56	53	50	47
78	103	101	96	89	84	79	73	66	63	60	57	54	51
79	112	110	105	97	92	86	79	72	69	66	63	59	56
					Unit $0^s.01$								
° ′													
80 00	12	12	12	11	10	9	9	8	8	7	7	7	6
81 00	14	13	13	12	11	10	10	9	8	8	8	7	7
82 00	15	15	14	13	13	12	11	10	9	9	9	8	8
83 00	18	17	16	15	14	13	12	11	11	10	10	9	9
84 00	20	20	19	18	17	16	14	13	13	12	11	11	10
85 00	24	24	23	21	20	19	17	16	15	14	14	13	12
85 30	27	27	26	24	22	21	19	17	17	16	15	14	14
86 00	31	30	29	26	25	23	22	20	19	18	17	16	15
86 30	35	34	33	30	29	27	25	22	22	21	20	19	17
87 00	41	40	38	35	33	31	29	26	25	24	23	22	20
87 30	49	48	46	42	40	37	35	31	30	29	27	26	24
88 00	61	60	57	53	50	47	43	39	38	36	34	32	31
88 10	67	66	63	58	55	51	47	43	41	39	37	35	33
88 20	73	72	69	64	60	56	52	47	45	43	41	39	37
88 30	82	80	77	71	67	62	58	52	50	48	46	43	41
88 40	92	90	86	79	75	70	65	59	56	54	51	49	46
88 50	105	103	98	91	86	80	74	67	65	62	59	56	52
89 00	122	120	115	106	100	94	86	79	75	72	68	65	61

This correction is to be *subtracted* from the observed time of transit for transits above pole, and *added* to the time of transit for transits below pole.

certain star changes with respect to an inertial system. In order to take this variation in the star's coordinates properly into account, the geocentric direction must be reduced to the center of mass of the solar system. For most practical work, however, it is adequate to reduce to the center of the sun.

Fig. 4.21a shows the earth's orbit, regarded here as circular (the error produced by this assumption is about 1%). The geocentric direction to a star A at a certain moment is defined by the angle θ', the heliocentric direction by θ. The angle Π subtended at A by the radius of the orbit, a, is called the annual or stellar parallax. If d denotes the distance between the sun and the point A, then from the figure it is evident that

$$\sin \Pi = \frac{a}{d}, \tag{4.56}$$

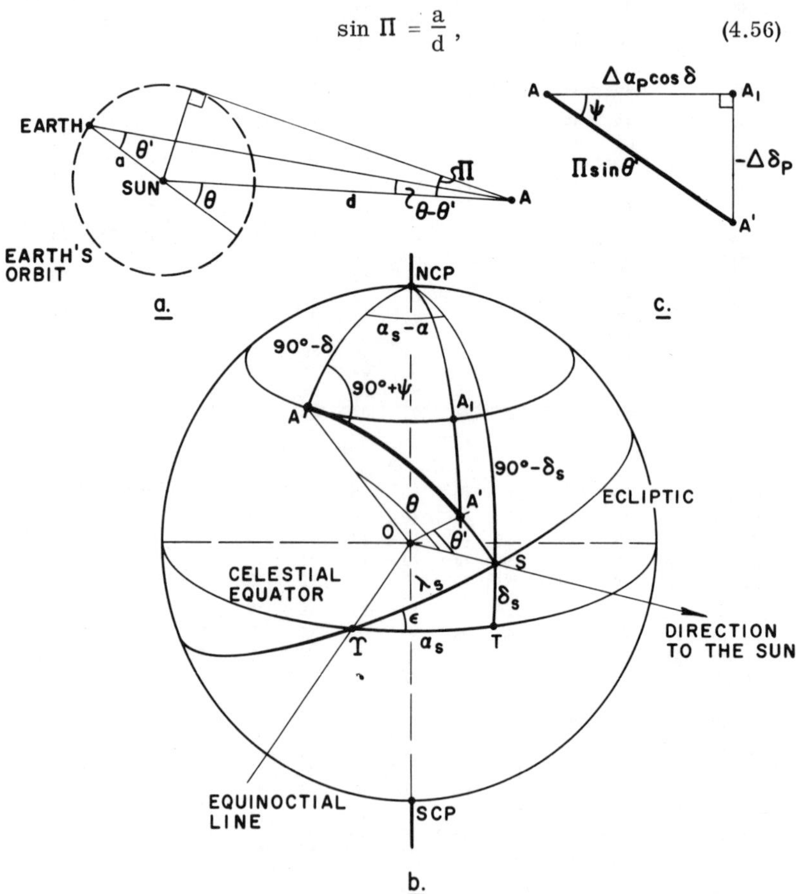

Fig. 4.21 Annual parallax

and

$$\sin(\theta - \theta') = \frac{a}{d} \sin \theta'. \tag{4.57}$$

Since both Π and $\theta - \theta'$ are small angles, the reduction to heliocentric direction is

$$\theta - \theta' \cong \Pi \sin \theta. \tag{4.58}$$

Let Fig. 4.21b illustrate the effect of the annual parallax on the celestial sphere. The geocentric direction to the star is OA′, while the heliocentric is OA. The arc AA′ is equal to $\theta - \theta'$ as given by equation (4.58). The effect of annual parallax on the right ascension and declination, $\Delta\alpha_P$ and $\Delta\delta_P$ respectively, may be derived from the triangle AA′A$_1$ (Fig. 4.21c):

$$\begin{aligned} \Delta\alpha_P &= \alpha' - \alpha = \Pi \sin\theta \cos\psi \sec\delta, \\ \Delta\delta_P &= \delta' - \delta = -\Pi \sin\theta \sin\psi. \end{aligned} \tag{4.59}$$

It is evident from the triangle A-NCP-S that

$$\begin{aligned} \sin\theta \cos\psi &= \cos\delta_s \sin(\alpha_s - \alpha), \\ -\sin\theta \sin\psi &= \sin\delta_s \cos\delta - \cos\delta_s \sin\delta \cos(\alpha_s - \alpha), \end{aligned} \tag{4.60}$$

and, from the triangle S♈T, that

$$\left.\begin{aligned} \cos\lambda_s &= \cos\delta_s \cos\alpha_s \\ \sin\lambda_s \sin\epsilon &= \sin\delta_s \\ \sin\lambda_s \cos\epsilon &= \cos\delta_s \sin\alpha_s \end{aligned}\right\}. \tag{4.61}$$

The annual parallax correction, to be applied to the heliocentric directions, is obtained by substituting (4.61) into (4.60) and then the result into (4.59):

$$\begin{aligned} \Delta\alpha_P &= \alpha' - \alpha = \Pi(\cos\alpha \cos\epsilon \sin\lambda_s - \sin\alpha \cos\lambda_s) \sec\delta, \\ \Delta\delta_P &= \delta' - \delta = \Pi(\cos\delta \sin\epsilon \sin\lambda_s - \cos\alpha \sin\delta \cos\lambda_s - \\ &\qquad - \sin\alpha \sin\delta \cos\epsilon \sin\lambda_s), \end{aligned}$$

where the primed quantities denote the geocentric directions.

The equations above may be written in the following simpler form

$$\begin{aligned} \Delta\alpha_P &= \Pi(Yc - Xd), \\ \Delta\delta_P &= \Pi(Yc' - Xd'), \end{aligned} \tag{4.62}$$

where

$$\begin{aligned} X &= \cos\delta_S \cos\alpha_S = \cos\lambda_S, \\ Y &= \cos\delta_S \sin\alpha_S = \sin\lambda_S \cos\epsilon \end{aligned}$$

are the equatorial Cartesian coordinates of the sun, tabulated in the

'American Ephemeris and Nautical Almanac,' and c, c′, d, d′, are the star constants of equation (4.46).

The conversion from heliocentric to geocentric coordinates may also be effected through the direction numbers x, y, z by the operation

$$\begin{pmatrix} x \\ y \\ z \end{pmatrix}_{\alpha',\delta'} = \begin{pmatrix} x \\ y \\ z \end{pmatrix}_{\alpha,\delta} + \begin{pmatrix} -C \sec \epsilon \\ -D \cos \epsilon \\ -D \sin \epsilon \end{pmatrix} (\Pi/\varkappa), \tag{4.63}$$

where C, D are the aberration day numbers of equation (4.46) referred, as in equation (4.48), to the true coordinate system of the epoch in question; and $\varkappa$ is the constant of aberration 20″.4958.

Corrections for the annual parallax may be included with the aberration terms to be applied to the true directions as follows from equations (4.46) and (4.62):

$$\begin{aligned} \Delta\alpha_{A+P} &= (C + \Pi Y)c + (D - \Pi X)\, d\ , \\ \Delta\delta_{A+P} &= (C + \Pi Y)c' + (D - \Pi X)\, d'\ . \end{aligned} \tag{4.64}$$

Heliocentric positions referred to the true equator and equinox of the date to which the combined annual aberration and parallax corrections have been applied are referred to as apparent (geocentric) positions. The correction due to the annual parallax is very small (for the nearest star, $\Pi = 0''.76$) and is currently neglected when $\Pi < 0''.01$.

4.222 Geocentric Parallax. Owing to the rotation of the earth about its axis the topocentric direction from the observer to a celestial object changes. In order to eliminate this periodic variation in the object's geocentric coordinates, the topocentric direction must be reduced to the center of the earth.

Fig. 4.22 shows the earth with the observer situated at ρ distance from the center of the earth. The topocentric direction to the object A is the topocentric zenith distance z'; the geocentric direction is the geocentric zenith distance z. The angle π at the celestial body between the direction to the observer and the direction to the center of the earth is the geocentric parallax. From the figure it is evident that

$$z' = z + \pi, \tag{4.65}$$

and

$$\sin \pi = \frac{\rho}{r} \sin z'. \tag{4.66}$$

If the body is on the horizon, $z' = 90°$, thus

$$\sin \pi = \frac{\rho}{r},$$

the parallax is called horizontal parallax. If, in addition, the observer is at the equator, thus $\rho = a_e$,

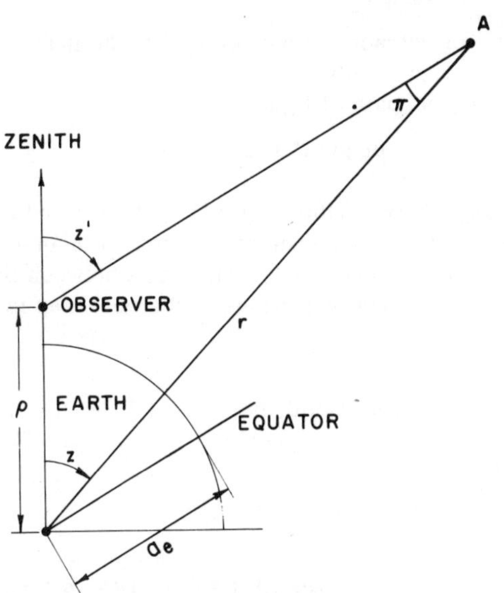

Fig. 4.22 Geocentric parallax

$$\sin \pi_0 = \frac{a_e}{r},$$

which is called equatorial horizontal parallax.

By use of the formulas (3.27), (3.28), (3.33), and (3.34) with $dA = d\Phi = 0$, and $dz = \pi$, the effect of geocentric parallax on the right ascension and declination of an object may be derived:

$$\begin{aligned} \Delta\alpha_G &= \alpha' - \alpha = -\pi \sin h \operatorname{cosec} z \cos \Phi \sec \delta, \\ \Delta\delta_G &= \delta' - \delta = -\pi (\sin \Phi \operatorname{cosec} z \sec \delta - \tan \delta \cot z), \end{aligned} \tag{4.67}$$

where the primed quantities refer to the topocentric directions. The correction is usually subtracted from the observed quantities rather than added to the geocentric right ascension and declination.

In the case of stars the effect of geocentric parallax is always negligibly small.

4.23 Astronomic Refraction

4.231 Rigorous Theory. As a light ray passes through the atmosphere of the earth, the variation of air density along the path causes a continuous change in the direction of propagation. In addition, scattering and absorption by the gases of the atmosphere and by dust particles cause attenuation and change of spectral composition. The complexity of atmospheric refraction makes its determination exceedingly difficult. The general effect is that the light rays bend downward; thus a celestial object appears to be at a higher altitude than it is in reality (Fig. 4.23).

From the law of refraction, assuming that the index of refraction μ is radially symmetric ($\mu = \mu(r)$ where r is the distance from the center of the earth) along the path of light

$$\mu r \sin \zeta = \text{constant} \tag{4.68}$$

where ζ is the angle between the direction of the ray and the radius vector r, i.e., the angle of incidence at P situated in an arbitrary atmospheric layer [Smart, 1960, p.63]. The angle between the direction of the ray at P and the geocentric radius to the observer is

$$\bar{z}_P = z_P + \zeta$$

where z_P is the geocentric zenith distance of P. At the observer, where $r = \rho$, $\mu = \mu_0$, $z_P = 0$, the observed zenith distance is $z' = \zeta_0$. The constant in equation (4.68) is equal to

$$\mu_0 \rho \sin \zeta_0 = \bar{k} ;$$

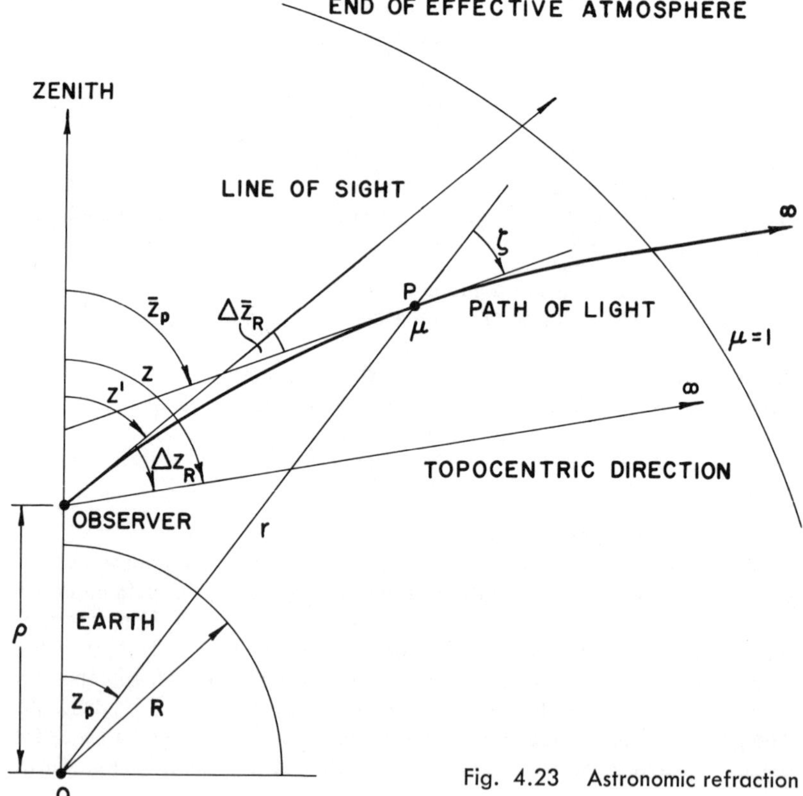

Fig. 4.23 Astronomic refraction

thus equation (4.68) may be written in the following form:

$$\mu r \tan \zeta = \bar{k} \sec \zeta = \bar{k}(1+\tan^2\zeta)^{\frac{1}{2}},$$

from which

$$\tan \zeta = \frac{\bar{k}}{(\mu^2 r^2 - \bar{k}^2)^{\frac{1}{2}}} . \tag{4.69}$$

The total amount of refraction from the point P to the observer is

$$\Delta \bar{z}_R = \bar{z}_P - z';$$

thus

$$d\Delta \bar{z}_R = d\bar{z}_P = dz_P + d\zeta . \tag{4.70}$$

Differentiating equation (4.68) yields

$$d(\mu r) \sin \zeta + \mu r \cos \zeta \, d\zeta = 0,$$

from which

$$d\zeta = -\tan \zeta \frac{d(\mu r)}{\mu r} .$$

From differential geometry

$$dz_P = \tan \zeta \frac{dr}{r} ; \tag{4.71}$$

therefore by (4.70)

$$d\Delta \bar{z}_R = -\tan \zeta \left[\frac{d(\mu r)}{\mu r} - \frac{dr}{r}\right] = -\tan \zeta \frac{d\mu}{\mu} .$$

Substituting equation (4.69) gives the result

$$d\Delta \bar{z}_R = -\frac{\bar{k}}{(\mu^2 r^2 - \bar{k}^2)^{\frac{1}{2}}} \frac{d\mu}{\mu} . \tag{4.72}$$

The integration of this equation gives the astronomic refraction

$$\Delta z_R = z - z' = \rho \mu_0 \sin z' \int_1^{\mu_0} (r^2 \mu^2 - \rho^2 \mu_0^2 \sin^2 z)^{-\frac{1}{2}} \frac{d\mu}{\mu} . \tag{4.73}$$

The correction Δz_R is called astronomic refraction angle, since it is to be applied to astronomic observations involving objects outside the atmosphere. If the object were inside the effective atmosphere, the correction would be called atmospheric refraction angle.

The astronomic refraction correction to be applied to the observed right ascension and declination may be calculated from equation (3.27), (3.28), (3.33), and (3.34) by substituting $dA = d\Phi = 0$ and $dz = \Delta z_R$; thus,

$$\Delta\alpha_R = \alpha - \alpha' = -\Delta z_R \sin h \operatorname{cosec} z' \cos\Phi \sec\delta,$$
$$\Delta\delta_R = \delta - \delta' = -\Delta z_R (\sin\Phi \operatorname{cosec} z' \sec\delta - \tan\delta \cot z'), \tag{4.74}$$

where the primed quantities refer to the observed (displaced) and the nonprimed refer to the topocentric directions.

4.232 Approximate Astronomic Refraction and Refraction Tables. Since our knowledge of the physical state of the atmosphere is insufficient to know the relation between the two variables μ and r, the integral (4.73) is treated by approximate methods, usually through developments into series [Baldini, 1963;Chauvenet, 1863,pp. 127-172; Garfinkel, 1944; Newcomb, 1906, pp. 173-224; Oterma, 1960; Willis, 1941]. The convergence of the series determines the upper limit of z', to which the formula is still applicable. General forms of these series are

$$\Delta z_R = A_1 \tan z' (1 - A_2 \sec^2 z' + A_3 \sec^4 z' - A_4 \sec^6 z' + \ldots), \tag{4.75}$$

or

$$\Delta z_R = B_1 \tan z' - B_2 \tan^3 z' + B_3 \tan^5 z' - B_4 \tan^7 z' + \ldots, \tag{4.76}$$

where the values A_i and B_i are coefficients depending on the physical state of the atmosphere. In geodetic astronomy, restricting equation (4.76) to the first two terms provides satisfactory results to moderate zenith distances,up to $z'=75°$. If the zenith distance is smaller than 60°, the first term alone is often sufficient,depending on the accuracy desired.

Table 4.8 contains the 'mean' astronomic refraction angles Δz_R^m for the standard condition; barometric pressure 760 mm, temperature 10°C, and relative humidity 60% [Hoskinson and Duerksen, 1952, p. 173]. For zenith distances from 0° to 85° the table was computed by means of the Willis method. The remainder of the table was taken from the Pulkovo tables (see below).

If the atmospheric conditions are not normal, then the mean angle of refraction must be corrected for deviations in temperature and barometric pressure. A suitable form for such a reduction, for example, is

$$\Delta z_R = \Delta z_R^m C_B C_T, \tag{4.77}$$

where the coefficients C_B and C_T depend on barometric pressure and on temperature respectively. In connection with the standard atmospheric conditions mentioned above

$$C_B = B/760, \qquad C_T = 283/(273+T)$$

where B is the actual barometric pressure in millimeters, and T is the actual temperature in centigrades. The values C_B and C_T may be found in Tables 4.9 and 4.10 [Hoskinson and Duerksen, 1952, pp. 174-175].

Most available refraction tables should be used generally for moderate zenith distances (up to 75°). For observations made near the horizon, special tables of refraction have been prepared. Among the earliest tables were the ones constructed by Bessel based, at low altitudes, entirely on observations. These,and later tables based upon them,were in general use until in 1870 the Pulkovo Observatory Tables were published [Orlov, 1956]. The Pulkovo tables in successive editions, and

TABLE 4.8 Mean Refraction Angle, Δz_R^m

[Pressure=760 *mm*; Temperature=10° *C*; Relative humidity=60%]

z	00′	10′	20′	30′	40′	50′	60′
°	″	″	″	″	″	″	″
0	0.0	0.2	0.3	0.5	0.7	0.8	1.0
1	1.0	1.2	1.3	1.5	1.7	1.9	2.0
2	2.0	2.2	2.4	2.5	2.7	2.9	3.0
3	3.0	3.2	3.4	3.5	3.7	3.9	4.0
4	4.0	4.2	4.4	4.6	4.7	4.9	5.1
5	5.1	5.2	5.4	5.6	5.7	5.9	6.1
6	6.1	6.3	6.4	6.6	6.8	6.9	7.1
7	7.1	7.3	7.5	7.6	7.8	8.0	8.1
8	8.1	8.3	8.5	8.7	8.8	9.0	9.2
9	9.2	9.3	9.5	9.7	9.9	10.0	10.2
10	10.2	10.4	10.6	10.7	10.9	11.1	11.3
11	11.3	11.4	11.6	11.8	12.0	12.1	12.3
12	12.3	12.5	12.7	12.8	13.0	13.2	13.4
13	13.4	13.5	13.7	13.9	14.1	14.3	14.4
14	14.4	14.6	14.8	15.0	15.2	15.3	15.5
15	15.5	15.7	15.9	16.1	16.2	16.4	16.6
16	16.6	16.8	17.0	17.2	17.3	17.5	17.7
17	17.7	17.9	18.1	18.3	18.4	18.6	18.8
18	18.8	19.0	19.2	19.4	19.6	19.8	19.9
19	19.9	20.1	20.3	20.5	20.7	20.9	21.1
20	21.1	21.3	21.5	21.7	21.8	22.0	22.2
21	22.2	22.4	22.6	22.8	23.0	23.2	23.4
22	23.4	23.6	23.8	24.0	24.2	24.4	24.6
23	24.6	24.8	25.0	25.2	25.4	25.6	25.8
24	25.8	26.0	26.2	26.4	26.6	26.8	27.0
25	27.0	27.2	27.4	27.6	27.8	28.0	28.2
26	28.2	28.4	28.7	28.9	29.1	29.3	29.5
27	29.5	29.7	29.9	30.1	30.4	30.6	30.8
28	30.8	31.0	31.2	31.4	31.7	31.9	32.1
29	32.1	32.3	32.5	32.8	33.0	33.2	33.4
30	33.4	33.6	33.9	34.1	34.3	34.6	34.8
31	34.8	35.0	35.2	35.5	35.7	35.9	36.2
32	36.2	36.4	36.6	36.9	37.1	37.4	37.6
33	37.6	37.8	38.1	38.3	38.6	38.8	39.0
34	39.0	39.3	39.5	39.8	40.0	40.3	40.5
35	40.5	40.8	41.0	41.3	41.5	41.8	42.1
36	42.1	42.3	42.6	42.8	43.1	43.3	43.6
37	43.6	43.9	44.1	44.4	44.7	44.9	45.2
38	45.2	45.5	45.8	46.0	46.3	46.6	46.9
39	46.9	47.1	47.4	47.7	48.0	48.3	48.6
40	48.6	48.8	49.1	49.4	49.7	50.0	50.3
41	50.3	50.6	50.9	51.2	51.5	51.8	52.1
42	52.1	52.4	52.7	53.0	53.3	53.6	54.0
43	54.0	54.3	54.6	54.9	55.2	55.5	55.9
44	55.9	56.2	56.5	56.9	57.2	57.5	57.9
45	57.9	58.2	58.5	58.9	59.2	59.6	59.9
46	59.9	60.2	60.6	61.0	61.3	61.7	62.0
47	62.0	62.4	62.8	63.1	63.5	63.9	64.2
48	64.2	64.6	65.0	65.4	65.7	66.1	66.5
49	66.5	66.9	67.3	67.7	68.1	68.5	68.9
50	68.9	69.3	69.7	70.1	70.6	71.0	71.4
51	71.4	71.8	72.2	72.7	73.1	73.6	74.0
52	74.0	74.4	74.9	75.3	75.8	76.2	76.7
53	76.7	77.2	77.6	78.1	78.6	79.1	79.5
54	79.5	80.0	80.5	81.0	81.5	82.0	82.5
55	82.5	83.0	83.5	84.1	84.6	85.1	85.6
56	85.6	86.2	86.7	87.3	87.8	88.4	88.9
57	88.9	89.5	90.1	90.7	91.2	91.8	92.4
58	92.4	93.0	93.6	94.2	94.8	95.5	96.1
59	96.1	96.7	97.4	98.0	98.6	99.3	100.0
60	100.0	100.6	101.3	102.0	102.7	103.4	104.1
61	104.1	104.8	105.5	106.3	107.0	107.7	108.5
62	108.5	109.2	110.0	110.8	111.6	112.4	113.2
63	113.2	114.0	114.8	115.6	116.5	117.3	118.2
64	118.2	119.0	119.9	120.8	121.7	122.6	123.5
65	123.5	124.5	125.4	126.4	127.4	128.3	129.3
66	129.3	130.3	131.4	132.4	133.4	134.5	135.6
67	135.6	136.7	137.8	138.9	140.0	141.2	142.3
68	142.3	143.5	144.7	145.9	147.2	148.4	149.7
69	149.7	151.0	152.3	153.6	155.0	156.4	157.8
70	157.8	159.2	160.6	162.1	163.6	165.1	166.6
71	166.6	168.2	169.7	171.4	173.0	174.7	176.3
72	176.3	178.1	179.8	181.6	183.4	185.3	187.2
73	187.2	189.1	191.0	193.0	195.1	197.1	199.2
74	199.2	201.4	203.6	205.8	208.1	210.4	212.8
75	212.8	215.2	217.7	220.2	222.8	225.5	228.2
76	228.2	230.9	233.7	236.6	239.6	242.6	245.7
77	245.7	248.9	252.1	255.4	258.9	262.3	265.9
78	265.9	269.6	273.4	277.2	281.2	285.3	289.5
79	289.5	293.8	298.2	302.8	307.5	312.3	317.3
80	317.3	322.4	327.7	333.2	338.8	344.6	350.6
81	350.6	356.8	363.2	369.8	376.6	383.7	391.1
82	391.1	398.7	406.6	414.8	423.3	432.1	441.3
83	441.3	450.9	460.9	471.2	482.0	493.3	505.1
84	505.1	517.4	530.3	543.8	558.0	572.8	588.4
85	588.4	604.4	621.6	639.7	658.8	678.9	700.2
86	700.2	722.7	746.6	771.8	798.7	827.2	857.6
87	857.6	890.0	924.7	961.6	1,001.3	1,043.9	1,089.7
88	1,089.7	1,138.9	1,192.0	1,249.2	1,311.4	1,378.6	1,452.0
89	1,452.0	1,531.7	1,618.8	1,714.0	1,818.4	1,933.1	2,059.5

TABLE 4.9 Pressure Correction Factor, C_B

(Apply to mean refraction in Table 4.8)

$$\Delta z_R = \Delta z_R^m C_B C_T$$

Barometer		C_B	Barometer		C_B	Barometer		C_B	Barometer		C_B	Barometer		C_B
Inches	*mm*		*Inches*	*mm*		*Inches*	*mm*		*Inches*	*mm*		*Inches*	*mm*	
20.0	508	0.670	22.4	569	0.749	24.8	630	0.829	27.2	691	0.909	29.6	752	0.989
20.1	511	0.673	22.5	572	0.752	24.9	632	0.832	27.3	693	0.912	29.7	754	0.992
20.2	513	0.676	22.6	574	0.755	25.0	635	0.835	27.4	696	0.916	29.8	757	0.996
20.3	516	0.679	22.7	576	0.759	25.1	637	0.838	27.5	699	0.920	29.9	759	0.999
20.4	518	0.682	22.8	579	0.762	25.2	640	0.842	27.6	701	0.923	30.0	762	1.003
20.5	521	0.685	22.9	582	0.766	25.3	643	0.846	27.7	704	0.926	30.1	765	1.007
20.6	523	0.688	23.0	584	0.770	25.4	645	0.849	27.8	706	0.929	30.2	767	1.010
20.7	526	0.692	23.1	587	0.773	25.5	648	0.853	27.9	709	0.933	30.3	770	1.013
20.8	528	0.696	23.2	589	0.776	25.6	650	0.856	28.0	711	0.936	30.4	772	1.016
20.9	531	0.699	23.3	592	0.779	25.7	653	0.859	28.1	714	0.939	30.5	775	1.020
21.0	533	0.703	23.4	594	0.783	25.8	655	0.862	28.2	716	0.942	30.6	777	1.023
21.1	536	0.706	23.5	597	0.786	25.9	658	0.866	28.3	719	0.946	30.7	780	1.026
21.2	538	0.709	23.6	599	0.789	26.0	660	0.869	28.4	721	0.949	30.8	782	1.029
21.3	541	0.712	23.7	602	0.792	26.1	663	0.872	28.5	724	0.953	30.9	785	1.033
21.4	544	0.716	23.8	605	0.796	26.2	665	0.875	28.6	726	0.956	31.0	787	1.036
21.5	546	0.719	23.9	607	0.799	26.3	668	0.879	28.7	729	0.959			
21.6	549	0.722	24.0	610	0.803	26.4	671	0.882	28.8	732	0.963			
21.7	551	0.725	24.1	612	0.806	26.5	673	0.885	28.9	734	0.966			
21.8	554	0.729	24.2	615	0.809	26.6	676	0.889	29.0	737	0.970			
21.9	556	0.732	24.3	617	0.813	26.7	678	0.892	29.1	739	0.973			
22.0	559	0.735	24.4	620	0.816	26.8	681	0.896	29.2	742	0.976			
22.1	561	0.739	24.5	622	0.820	26.9	683	0.899	29.3	744	0.979			
22.2	564	0.742	24.6	625	0.823	27.0	686	0.902	29.4	747	0.983			
22.3	566	0.746	24.7	627	0.826	27.1	688	0.905	29.5	749	0.986			

TABLE 4.10 Temperature Correction Factor, C_T

(Apply to mean refraction in Table 4.8)

$$\Delta z_R = \Delta z_R^m C_B C_T$$

Temperature		C_T	Temperature		C_T	Temperature		C_T	Temperature		C_T	Temperature		C_T
Fahrenheit	Centigrade		Fahrenheit	Centigrade		Fahrenheit	Centigrade		Fahrenheit	Centigrade		Fahrenheit	Centigrade	
°	°		°	°		°	°		°	°		°	°	
−25	−31.7	1.172	8	−13.3	1.089	41	5.0	1.018	74	23.3	0.955	107	41.7	0.900
−24	−31.1	1.169	9	−12.8	1.087	42	5.6	1.016	75	23.9	0.953	108	42.2	0.899
−23	−30.6	1.166	10	−12.2	1.085	43	6.1	1.014	76	24.4	0.952	109	42.8	0.897
−22	−30.0	1.164	11	−11.7	1.082	44	6.7	1.012	77	25.0	0.950	110	43.3	0.895
−21	−29.4	1.161	12	−11.1	1.080	45	7.2	1.010	78	25.6	0.948	111	43.9	0.894
−20	−28.9	1.158	13	−10.6	1.078	46	7.8	1.008	79	26.1	0.946	112	44.4	0.892
−19	−28.3	1.156	14	−10.0	1.076	47	8.3	1.006	80	26.7	0.945	113	45.0	0.891
−18	−27.8	1.153	15	− 9.4	1.073	48	8.9	1.004	81	27.2	0.943	114	45.6	0.890
−17	−27.2	1.151	16	− 8.9	1.071	49	9.4	1.002	82	27.8	0.941	115	46.1	0.888
−16	−26.7	1.148	17	− 8.3	1.069	50	10.0	1.000	83	28.3	0.939	116	46.7	0.886
−15	−26.1	1.145	18	− 7.8	1.067	51	10.6	0.998	84	28.9	0.938	117	47.2	0.885
−14	−25.6	1.143	19	− 7.2	1.064	52	11.1	0.996	85	29.4	0.936	118	47.8	0.884
−13	−25.0	1.140	20	− 6.7	1.062	53	11.7	0.994	86	30.0	0.934	119	48.3	0.882
−12	−24.4	1.138	21	− 6.1	1.060	54	12.2	0.992	87	30.6	0.933	120	48.9	0.881
−11	−23.9	1.135	22	− 5.6	1.058	55	12.8	0.990	88	31.1	0.931	121	49.4	0.880
−10	−23.3	1.133	23	− 5.0	1.056	56	13.3	0.988	89	31.7	0.929	122	50.0	0.878
− 9	−22.8	1.130	24	− 4.4	1.054	57	13.9	0.986	90	32.2	0.928	123	50.6	0.877
− 8	−22.2	1.128	25	− 3.9	1.051	58	14.4	0.985	91	32.8	0.926	124	51.1	0.876
− 7	−21.7	1.125	26	− 3.3	1.049	59	15.0	0.983	92	33.3	0.924	125	51.7	0.874
− 6	−21.1	1.123	27	− 2.8	1.047	60	15.6	0.981	93	33.9	0.923	126	52.2	0.873
− 5	−20.6	1.120	28	− 2.2	1.045	61	16.1	0.979	94	34.4	0.921	127	52.8	0.871
− 4	−20.0	1.118	29	− 1.7	1.043	62	16.7	0.977	95	35.0	0.919	128	53.3	0.870
− 3	−19.4	1.115	30	− 1.1	1.041	63	17.2	0.975	96	35.6	0.917	129	53.9	0.868
− 2	−18.9	1.113	31	− 0.6	1.039	64	17.8	0.973	97	36.1	0.916	130	54.4	0.867
− 1	−18.3	1.111	32	0.0	1.036	65	18.3	0.972	98	36.7	0.914			
0	−17.8	1.108	33	+ 0.6	1.034	66	18.9	0.970	99	37.2	0.912			
+ 1	−17.2	1.106	34	1.1	1.032	67	19.4	0.968	100	37.8	0.911			
2	−16.7	1.103	35	1.7	1.030	68	20.0	0.966	101	38.3	0.909			
3	−16.1	1.101	36	2.2	1.028	69	20.6	0.964	102	38.9	0.908			
4	−15.6	1.099	37	2.8	1.026	70	21.1	0.962	103	39.4	0.906			
5	−15.0	1.096	38	3.3	1.024	71	21.7	0.961	104	40.0	0.905			
6	−14.4	1.094	39	3.9	1.022	72	22.2	0.959	105	40.6	0.903			
7	−13.9	1.092	40	4.4	1.020	73	22.8	0.957	106	41.1	0.902			

other tables such as the Greenwich Observatory Tables, have remained in widespread use [Cowell, 1900]. Meanwhile, many tables have been constructed from more recent theoretical developments taking into account new discoveries in the atmosphere [Willis, 1941; Garfinkel, 1944]. The latest development in this area replaces the tables by an electronic computer routine and extends the validity of the formulas for both astronomic and atmospheric refractions to the entire domain $0 \leq z \leq 180°$ [Garfinkel, 1967].

The refraction correction is usually directly added to the observed zenith distance rather than subtracted from the topocentric right ascension and declination.

4.3 Variations Due to the Proper Motion of Stars

When from direct observations of the apparent positions of the stars their mean places are reduced, it is found that the differences in these mean places between different dates do not agree with those which arise solely from the precession expressed by either equations (4.2) or (4.3). Each star appears to have a small motion of its own designated as its proper motion. This motion (i. e., change of direction) is the resultant of the actual motion of the star in space and of its apparent motion due to the changing direction arising from the motion of the sun with the planets.

4.31 The Effect of Time on the Components of Proper Motion

The proper motion which takes place in an arbitrary direction can be divided into a component in the direction of the star (radial component) and one perpendicular to it (tangential component). Let μ denote the annual tangential component of the proper motion (from here on 'proper motion'), in units of seconds of arc per year, generally assumed to be constant for many years. Let S_0 denote the mean position of a star at an epoch T_0 referred to the mean equator and equinox of the same epoch (Fig. 4.24). Let S_1 be the star's position after an interval of $t = T - T_0$ years. The arc S_0S_1 is the proper motion of the star during the interval t. Let μ_0^a and μ_0^d denote the components of the proper motion at S_0 in the direction of the celestial parallel and perpendicular to it at the epoch T_0, referred to the mean equator and equinox of the same epoch respectively. The components may be computed from the following expressions [Newcomb, 1906, p. 263]:

$$\begin{aligned} \mu_0^a &= \mu \sin \psi_0 \sec \delta_0 , \\ \mu_0^d &= \mu \cos \psi_0 , \end{aligned} \qquad (4.78)$$

where ψ_0 is the azimuth of the direction in which the proper motion takes place. The same components t time later at S_1 are

$$\begin{aligned} \mu_1^a &= \mu \sin \psi_1 \sec \delta_1 , \\ \mu_1^d &= \mu \cos \psi_1 . \end{aligned} \qquad (4.79)$$

It is thus clear from equations (4.78) and (4.79) that the components of proper motion at S_0 and S_1 are not the same, i.e., they alter with respect to the epoch for which they are defined, even though the mean equator and equinox of T_0 are the same and the total proper motion μ is constant.

By the sine formula of spherical trigonometry applied to the triangle S_0-NCP$_0$-S_1,

$$\sin\psi_1 \cos\delta_1 = \sin\psi_0 \cos\delta_0 .$$

Writing $\psi_0 + \Delta\psi_1$ for ψ_1 and $\delta_0 + \Delta\delta_1$ for δ_1, and regarding the quantities $\Delta\psi_1$ and $\Delta\delta_1$ as small angles, the equation above yields

$$(\sin\psi_0 + \Delta\psi_1 \cos\psi_0)(\cos\delta_0 - \Delta\delta_1 \sin\delta_0) = \sin\psi_0 \cos\delta_0 ,$$

from which, neglecting second-order infinitesimal quantities,

$$\Delta\psi_1 \cos\psi_0 \cos\delta_0 = \Delta\delta_1 \sin\psi_0 \sin\delta_0 . \tag{4.80}$$

Since μt is a very small angle, from the triangle S_0S_1U it is evident that

$$\Delta\delta_1 = S_0U = \delta_1 - \delta_0 = \mu t \cos\psi_0 ; \tag{4.81}$$

hence from (4.80) and (4.81)

$$\Delta\psi_1 = \mu t \sin\psi_0 \tan\delta_0 . \tag{4.82}$$

With similar approximations, equation (4.79) gives

$$\mu_1^a = \mu \frac{\sin\psi_0 + \Delta\psi_1 \cos\psi_0}{\cos\delta_0 - \Delta\delta_1 \sin\delta_0} = \mu \sec\delta_0 (\sin\psi_0 +$$

$$+ \Delta\psi_1 \cos\psi_0)(1 + \Delta\delta_1 \tan\delta_0) ,$$

or, again neglecting small quantities of the second order,

$$\mu_1^a = \mu \sin\psi_0 \sec\delta_0 + \mu\Delta\psi_1 \cos\psi_0 \sec\delta_0 + \mu\Delta\delta_1 \sin\psi_0 \sec\delta_0 \tan\delta_0 .$$

From this by (4.78), (4.81), and (4.82) the change in the right ascension component of proper motion during the interval $t = T - T_0$ is

$$\Delta\mu_1^a = \mu_1^a - \mu_0^a = 2(T - T_0)\, \mu_0^a \mu_0^d \tan\delta_0 .$$

Similarly for the declination component

$$\mu_1^d = \mu(\cos\psi_0 - \Delta\psi_1 \sin\psi_0);$$

thus

$$\Delta\mu_1^d = \mu_1^d - \mu_0^d = -\mu\Delta\psi_1 \sin\psi_0 = -\mu^2 t \sin^2\psi_0 \tan\delta_0 ,$$

which yields through (4.78),

$$\Delta\mu_1^d = \mu_1^d - \mu_0^d = -(T - T_0)(\mu_0^a)^2 \sin\delta_0 \cos\delta_0 .$$

Another small effect of time on the components of proper motion is a variation due to a progressive change brought about by the radial motion of the star, in the projection of its actual motion on the surface of the celestial sphere. The combined secular effects of both tangential and radial variations on spherical coordinates are [Scott and Hughes, 1964],

$$\Delta\mu_1^a = \mu_1^a - \mu_0^a = (2\mu_0^a\mu_0^d \tan\delta_0 - 2\Pi nV\, \mu_0^a \sin 1''/a)(T - T_0), \qquad (4.83)$$
$$\Delta\mu_1^d = \mu_1^d - \mu_0^d = [-(\mu_0^a)^2 \sin\delta_0 \cos\delta_0 - 2\Pi nV \mu_0^d \sin 1''/a](T - T_0), \quad (4.84)$$

where Π is the annual parallax in seconds of arc, n the number of seconds per year, V the radial velocity in kilometers per second, and a the radius of the earth's orbit. The numerical value of $2n \sin 1''/a$ is 2.047×10^{-6}. Unfortunately, the radial velocity and the annual parallax occur only in the combination ΠV. Since both are usually not known, the terms containing them are usually not considered at all, though their contribution to $\Delta\mu_1$ would be just as significant as that of the other terms in the expressions above.

4.32 The Effect of Precession on the Components of Proper Motion

The components of proper motion also change due to the motion of the mean celestial pole NCP_0-NCP on account of the luni-solar precession (and unnoticeably also due to nutation) occurring during the interval $t = T - T_0$ (Fig. 4.24).

Again let μ_0^a and μ_0^d denote the components of proper motion of the star S_0 at the epoch T_0 referring to the mean equator and equinox of the same epoch. The symbols μ^a and μ^d denote the same quantities referring to the mean equator and equinox at epoch T. Thus

$$\begin{aligned} \mu^a &= \mu \sin\psi \sec\delta, \\ \mu^d &= \mu \cos\psi. \end{aligned} \qquad (4.85)$$

The quantities μ_0^a and μ_0^d are given by equation (4.78). From the spherical triangle NCP_0-S_0-NCP,

$$\sin\Delta\psi = \sin(NCP_0 - NCP) \sin\alpha_0 \sec\delta .$$

Since both $\Delta\psi$ and the arc NCP_0-NCP $= \theta \cong nt$ (see Fig. 4.3 and equation (4.15)) are small,

$$\Delta\psi = nt \sin\alpha_0 \sec\delta_0 , \qquad (4.86)$$

in which δ has been substituted by δ_0 without any loss of accuracy.

Let $\Delta\delta = \delta - \delta_0$ be represented by the arc NCP_0-V; then

$$\Delta\delta \cong \theta \cos \alpha_0 = nt \cos \alpha_0 \, . \tag{4.87}$$

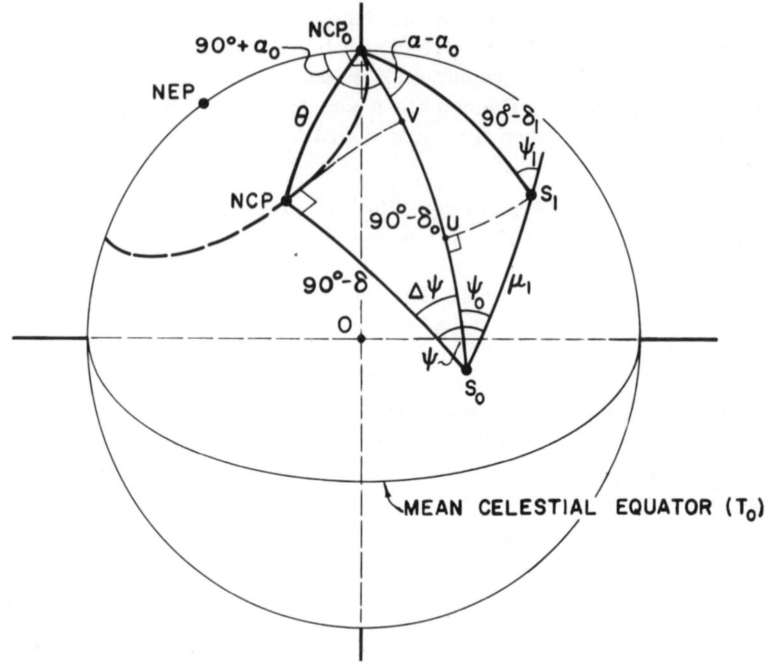

Fig. 4.24 Proper motion

From (4.85) as before

$$\mu^a = \mu \frac{\sin(\psi_0 + \Delta\psi)}{\cos(\delta_0 + \Delta\delta)} = \mu \frac{\sin\psi_0 + \Delta\psi \cos\psi_0}{\cos\delta_0 (1 - \Delta\delta \tan\delta_0)} =$$

$$= \mu \sec\delta_0 (\sin\psi_0 + \Delta\psi \cos\psi_0)(1 + \Delta\delta \tan\delta_0),$$

or, neglecting quantities of the second order,

$$\mu^a = \mu \sec\delta_0 \sin\psi_0 + \mu\Delta\delta \sec\delta_0 \sin\psi_0 \tan\delta_0 + \\ + \mu\,\Delta\psi \sec\delta_0 \cos\psi_0 \, ,$$

which by (4.78), (4.86), and (4.87) yields the correction to the right ascension component of proper motion due to the effect of precession during the period $t = T - T_0$:

$$\Delta\mu^a = \mu^a - \mu_0^a = n(T - T_0)(\mu_0^a \cos\alpha_0 \tan\delta_0 + \\ + \mu_0^d \sin\alpha_0 \sec^2\delta_0). \tag{4.88}$$

Similarly

$$\mu^{d} = \mu(\cos\psi_0 - \Delta\psi \sin\psi_0)$$
$$= \mu\cos\psi_0 - \mu nt \sin\alpha_0 \sec\delta_0 \sin\psi_0;$$

thus

$$\Delta\mu^{d} = \mu^{d} - \mu_0^{d} = -n(T - T_0)\mu_0^{a}\sin\alpha_0 . \quad (4.89)$$

Equations (4.88) and (4.89) give the components of the proper motion at the epoch T when the components for epoch T_0 are known or vice versa. Both corrections are very small but need to be taken into account when the proper motion is large.

4.33 The Effect of Proper Motion on the Mean Place of a Star

Let α_0, δ_0, and μ_0^{a}, μ_0^{d} denote the respective right ascension, declination, and proper motion components of a star at epoch T_0, referred to the mean equator and equinox of the same epoch; α, δ, and μ^{a}, μ^{d}, the same quantities the interval $t = T - T_0$ later, i.e., referred to the mean equator and equinox of epoch T. Possibilities of transforming one set of parameters to the other are the following: (1) transformation by components, (2) transformation by time series, (3) rigorous matrix transformation. These methods are briefly described below.

4.331 Transformation by Components. The quantities α, δ, μ^{a}, and μ^{d} may be calculated from α_0, δ_0, μ_0^{a}, and μ_0^{d} by the following steps:

(1) Compute the change of the proper motion components $\Delta\mu_1^{a}$ and $\Delta\mu_1^{d}$ for the interval $T - T_0$, by means of equations (4.83) and (4.84).

(2) Compute the components of the proper motion at the epoch T in the mean reference frame of epoch T_0 by

$$\mu_1^{a} = \mu_0^{a} + \Delta\mu_1^{a} ,$$
$$\mu_1^{d} = \mu_0^{d} + \Delta\mu_1^{d} .$$

(3) Compute the star's position at the epoch T in the mean reference frame of epoch T_0 by averaging as follows:

$$\alpha_1 = \alpha_0 + \tfrac{1}{2}(\mu_0^{a} + \mu_1^{a})(T - T_0) = \alpha_0 + (\mu_0^{a} + \Delta\mu_1^{a}/2)(T - T_0),$$
$$\delta_1 = \delta_0 + \tfrac{1}{2}(\mu_0^{d} + \mu_1^{d})(T - T_0) = \delta_0 + (\mu_0^{d} + \Delta\mu_1^{d}/2)(T - T_0).$$

(4) Compute the star's position α and δ at the epoch T in the mean reference frame of epoch T by equations (4.2), (4.3), or (4.13), (4.14), or (4.20), depending on the length of the interval $T - T_0$. The inputs in these equations are the quantities α_1 and δ_1 obtained in the previous step.

(5) If required also compute the components of the proper motion in the mean reference frame of epoch T using equations (4.88) and (4.89):

$$\mu^{a} = \mu_1^{a} + \Delta\mu^{a} ,$$
$$\mu^{d} = \mu_1^{d} + \Delta\mu^{d} .$$

4.332 Transformation by Time Series. It is also possible, and often convenient to calculate the effect of general precession and proper motion on the mean place of a star from the Taylor series (4.16), or from equation (4.20), but with the following substitutions [Doolittle, 1895]:

$$\left.\begin{aligned}\frac{d\alpha_0}{dt} &= \left(\frac{d\alpha_0}{dt}\right)_P + \mu_0^a \\ \frac{d\delta_0}{dt} &= \left(\frac{d\delta_0}{dt}\right)_P + \mu_0^d\end{aligned}\right\} \quad (4.90)$$

$$\left.\begin{aligned}\frac{d^2\alpha_0}{dt^2} &= \left(\frac{d^2\alpha_0}{dt^2}\right)_P + \frac{2\Delta\mu^a + \Delta\mu_1^a}{T-T_0} = \left(\frac{d^2\alpha_0}{dt^2}\right)_P + 2n(\mu_0^a \cos\alpha_0 \tan\delta_0 + \\ &+ \mu_0^d \sin\alpha_0 \sec^2\delta_0) + 2\mu_0^a\,\mu_0^d \tan\delta_0 \\ \frac{d^2\delta_0}{dt^2} &= \left(\frac{d^2\delta_0}{dt^2}\right)_P + \frac{2\Delta\mu^d + \Delta\mu_1^d}{T-T_0} = \left(\frac{d^2\delta_0}{dt^2}\right)_P - 2n\mu_0^a \sin\alpha_0 + \\ &+ \tfrac{1}{2}(\mu_0^a)^2 \sin 2\delta_0\end{aligned}\right\} \quad (4.91)$$

$$\left.\begin{aligned}\frac{d^3\alpha_0}{dt^3} &= \left(\frac{d^3\alpha_0}{dt^3}\right)_P + 2\mu_0^a(\mu_0^d)^2 + 3\frac{dn}{dt}\mu_0^d \sin\alpha_0 + 3n\mu_0^d(m + 2\mu_0^a)\cos\alpha_0 + \\ &+ 3\mu_0^a n^2 \cos 2\alpha_0 - 2(\mu_0^a)^3 \sin^2\delta_0 + 3\{[-2(\mu_0^a)^2 + 2(\mu_0^d)^2 - \\ &- m\mu_0^a]n \sin\alpha_0 + \frac{dn}{dt}\mu_0^a \cos\alpha_0 + 2n^2\,\mu_0^d \sin 2\alpha_0\} \tan\delta_0 + \\ &+ 3(2\mu_0^a\,\mu_0^d + \frac{dn}{dt}\mu_0^d \sin\alpha_0 + 4\mu_0^a\,\mu_0^d\, n \cos\alpha_0 + \\ &+ 2\mu_0^a\, n^2 \cos 2\alpha_0) \tan^2\delta_0 + 6\mu_0^d n\,(\mu_0^d \sin\alpha_0 + n \sin 2\alpha_0)\tan^3\delta_0 \\ \frac{d^3\delta_0}{dt^3} &= \left(\frac{d^3\delta_0}{dt^3}\right)_P - (\mu_0^a)^2\mu_0^d(2m + 3\mu_0^a)\,\frac{dn}{dt}\sin\alpha_0 - [3(\mu_0^a)^2 + \\ &+ 3m\mu_0^a]\, n\cos\alpha_0 - 3n^2\mu_0^d \sin^2\alpha_0 - 2(\mu_0^a)^2\mu_0^d \sin^2\delta_0 - \\ &- (6n\mu_0^a\,\mu_0^d \sin\alpha_0 + 3\mu_0^a n^2 \sin 2\alpha_0)\tan\delta_0 - \\ &- 3n^2\mu_0^d \sin^2\alpha_0 \tan^2\delta_0\end{aligned}\right\} \quad (4.92)$$

where the derivatives in parentheses with the subscript P are the precessional variations expressed by equations (4.17) - (4.19). The other notations correspond to those of section 4.112.

4.333 Transformation by Matrices. Using the system introduced in section 3.34, the notation of Fig. 4.24, and the experience gained with matrix rotations in sections 4.111, 4.122, and 4.123, it is evident that the transformations described in section 4.331 or 4.332 may also be performed by the following rotations:

$$\begin{pmatrix} x \\ y \\ z \end{pmatrix}_{\alpha,\delta} = \mathbf{P}\,\mathbf{M} \begin{pmatrix} x \\ y \\ z \end{pmatrix}_{\alpha_0,\delta_0}, \qquad (4.93)$$

where $\mathbf{P}$ is from equation (4.4), and

$$\mathbf{M} = \mathbf{R}_3(-\alpha_0)\,\mathbf{R}_2(\delta_0 - 90^\circ)\,\mathbf{R}_3(\psi_0)\,\mathbf{R}_2(\mu t)\,\mathbf{R}_3(-\psi_0)\,\mathbf{R}_2(90^\circ - \delta_0)\,\mathbf{R}_3(\alpha_0).$$

It may be easily shown that

$$\mathbf{M} \begin{pmatrix} x \\ y \\ z \end{pmatrix}_{\alpha_0,\delta_0} = \begin{pmatrix} \sin\delta_0\cos\alpha_0 & -\sin\alpha_0 & \cos\delta_0\cos\alpha_0 \\ \sin\delta_0\sin\alpha_0 & \cos\alpha_0 & \cos\delta_0\sin\alpha_0 \\ -\cos\delta_0 & 0 & \sin\delta_0 \end{pmatrix} \begin{pmatrix} -\cos\psi_0\sin\mu t \\ \sin\psi_0\sin\mu t \\ \cos\mu t \end{pmatrix} \qquad (4.94)$$

where $t = T - T_0$, and

$$\mu = \sqrt{(\mu_0^a\cos\delta_0)^2 + (\mu_0^d)^2}\,,$$

$$\sin\psi_0 = \frac{\mu_0^a\cos\delta_0}{\mu} = \frac{\mu_0^a\cos\delta_0}{[(\mu_0^a\cos\delta_0)^2 + (\mu_0^d)^2]^{\frac{1}{2}}}\,,$$

$$\cos\psi_0 = \frac{\mu_0^d}{\mu} = \frac{\mu_0^d}{[(\mu_0^a\cos\delta_0)^2 + (\mu_0^d)^2]^{\frac{1}{2}}}\,.$$

Equation (4.94) gives the mean place of the star in terms of direction numbers at the epoch T referred to the mean equator and equinox at T_0.

The above transformations are completely rigorous provided μ is constant, i.e., the effect of the radial velocity V described in section 4.31 is negligible. If this is not the case, the quantity μ in expression (4.94) should be substituted by

$$\mu + \frac{1}{2}\frac{d\mu}{dt}\,t$$

where

$$\frac{d\mu}{dt} = -\frac{2\pi V \mu n \sin 1''}{a}\,,$$

and the notation corresponds to that of equation (4.83) or (4.84). For an illustration see Example 4.4.

4.4. Summary of the Reduction of Star Positions, and Secondary Effects

In order to be able to distinguish between the variable coordinates of stars or the elements of the celestial sphere (pole, equinox, equator, etc.) referred to a particular coordinate system in motion, the prefixes observed, apparent, true, and mean are used conventionally.

The observed place of a celestial object is its actual position as determined by means of direct readings on some instrument corrected for systematic instrumental errors (e. g., dislevelment, collimation).

The apparent place of an object is its geocentric position, referred to the true equator and equinox of date when the object actually is observed. It differs from the observed place by the effects of astronomic refraction, diurnal aberration, and geocentric parallax. It is the place of the object as viewed by an imaginary observer situated at the center of the earth without an atmosphere, in a coordinate system affected by precession and nutation.

The true place of an object is its barycentric (practically heliocentric) position, referred to the true equator and equinox of date when the object actually is observed. It differs from the apparent place by the effects of annual aberration and annual parallax. It is the place of the object as viewed by an imaginary observer situated at the center of mass of the solar system (practically at the center of the sun) in a coordinate system affected by precession and nutation.

The mean place of an object is its barycentric (practically heliocentric) position, referred to some specified mean equator and equinox, generally that of the beginning of a Besselian year (see section 5.53). It differs from the true place by the effect of nutation. It is the place of an object as viewed by an imaginary observer situated at the center of mass of the solar system (practically at the center of the sun) in a coordinate system affected by precession.

It follows that if the prefixes are applied to nonvisible elements (e.g., equinox, equator, pole, etc.), then there is no difference between the apparent and the true place. For example, the apparent and the true equinoxes are identical moving points on the celestial sphere, representing the intersections of the equator and ecliptic continuously changing their place due to the precession and nutation.

As an illustration of the variation of star coordinates, Fig. 4.25 shows the variations of the mean and apparent right ascensions and declinations of the star No. 23487 from the General Catalogue between January 1, 1966, and January 1, 1969.

The star places are usually listed in various star catalogues and ephemerides (see Chapter 6). The former usually contain mean places referred to some designated epoch, the latter apparent places at certain time intervals (e.g., every day, every tenth day). In all star catalogues the effect of the ellipticity of the earth's orbit on annual aberration (section 4.213), called the e-terms of aberration, are included in the mean places. Thus

$$\text{Catalogued Mean Place} = \text{Mean Place (as defined previously)} + \text{e-terms of Aberration},$$

where the e-terms are those of equations (4.49) or (4.50). The rigor-

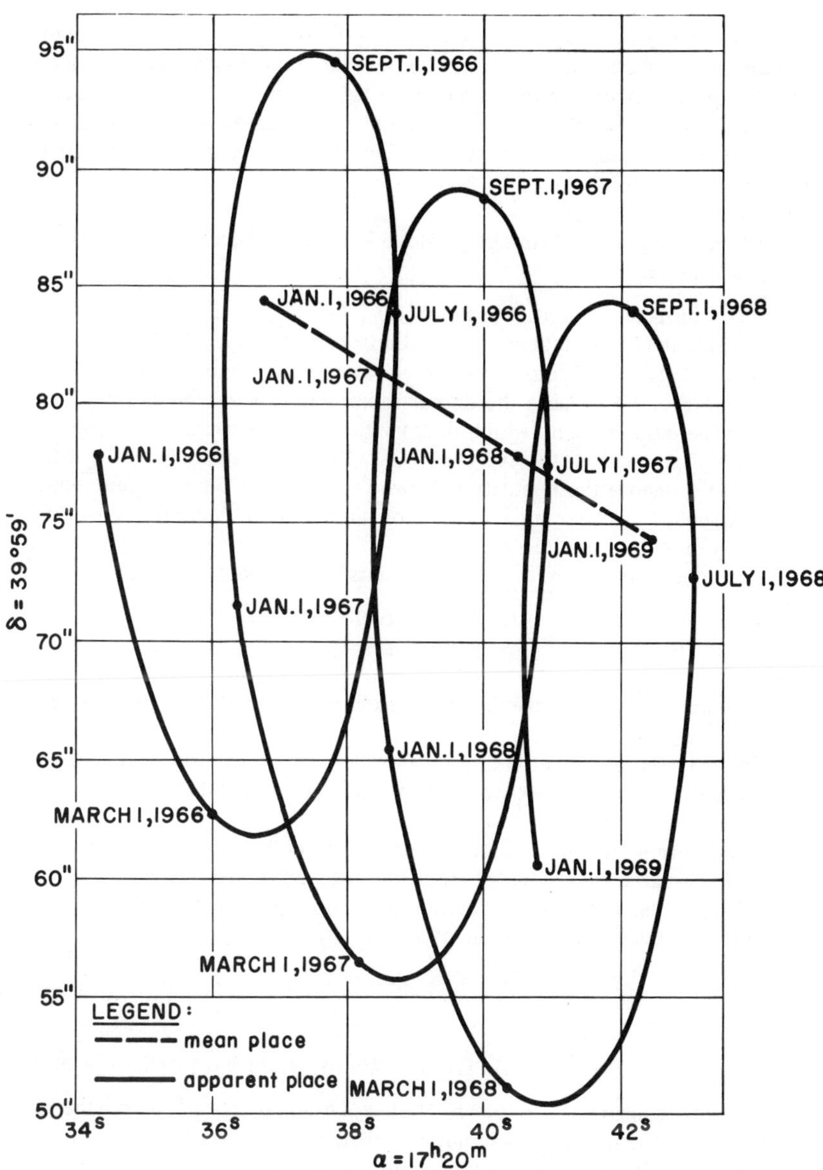

Fig. 4.25 Mean and apparent places of the star GC 23487

ous calculation of a meanplace for an epoch T to be summarized below necessitates the removal of the e-terms at the epoch of the catalogue T_o before applying precession and proper motion. The updated e-terms for the epoch T should then be put back if the catalogued mean place for this epoch is desired. If this procedure is not applied, i.e., the e-terms are handled as part of precession, a small error in the maximum order of 0".002 per century is introduced in both right ascension and declination.

4.41 Reduction of the Catalogued Mean Place of a Star from One Epoch to Another

4.411 Rigorous Reduction. Let α_o^c, δ_o^c, and μ_o^a, μ_o^d denote the respective catalogued right ascension, declination, and proper motion components of a star referred to the mean equator and equinox of the catalogue epoch T_o; α_o and δ_o denote the actual mean right ascension and declination for the same epoch, which differs from the previous ones by the e-terms of aberration for the epoch T_o. The same symbols without the subscript o denote the quantity referred to the mean equator and equinox of epoch T. The correct steps of the rigorous reduction from the epoch T_o to T then are the following:

(1) Remove the e-terms from the catalogued positions

$$\alpha_o = \alpha_o^c - (\Delta\alpha_{\bar{A}})_o \ ,$$
$$\delta_o = \delta_o^c - (\Delta\delta_{\bar{A}})_o \ ,$$

where $(\Delta\alpha_{\bar{A}})_o$ and $(\Delta\delta_{\bar{A}})_o$ are from equation (4.49) or (4.50) computed for the epoch T_o.

(2) Apply the precession and proper motion for the interval $T-T_o$ as described in section 4.331 (in step (4) use equation (4.2) or (4.3)) or in 4.333.

If the star's proper motion is zero, or not known, the precession may be computed by means of equation (4.2) or (4.3).

(3) Apply the e-terms to the positions obtained in step (2); α, δ, to obtain the catalogued positions α^c and δ^c,

$$\alpha^c = \alpha + \Delta\alpha_{\bar{A}} \ ,$$
$$\delta^c = \delta + \Delta\delta_{\bar{A}} \ ,$$

where $\Delta\alpha_{\bar{A}}$ and $\Delta\delta_{\bar{A}}$ are again from (4.49) or (4.50) but for the epoch T.

4.412 Conventional Reduction. Let α_o, δ_o, and μ_o^a, μ_o^d denote the respective catalogued right ascension, declination, and proper motion components of a star referred to the mean equator and equinox of the catalogue epoch T_o; the same symbols without the subscript o are the same quantities referred to the mean equator and equinox of epoch T. The reduction from the epoch T_o to T, correct for an interval of about 100 years, may be calculated as explained in section 4.332. If the proper motion components are zero, or not known, equation (4.20) may be used to compute precession. For an illustration see Example 4.1.

4.42 Reduction of the Apparent Place of a Star from Its Mean Position

4.421 Conventional Reduction. In the method of conventional reduction, the apparent place of a star at epoch T, α' and δ', is computed from its mean place α_0 and δ_0 at the catalogue epoch T_0 in four steps as follows:

(1) Compute the mean place α and δ of the star for the beginning of the Besselian year nearest to the epoch T using the method outlined in section 4.412.

(2) Apply the precession for the remaining fraction of the year, nutation, and annual aberration by using the Besselian or independent day numbers and star constants, according to equations (4.27) and (4.46), or (4.28) and (4.47). Thus, to the first order

$$\left.\begin{aligned} \alpha' &= \alpha + Aa + Bb + Cc + Dd + E \\ &= \alpha + f + g\sin(G+\alpha)\tan\delta + h\sin(H+\alpha)\sec\delta \\ \delta' &= \delta + Aa' + Bb' + Cc' + Dd' \\ &= \delta + g\cos(G+\alpha) + h\cos(H+\alpha)\sin\delta + i\cos\delta \end{aligned}\right\} \qquad (4.95)$$

where α' and δ' are the apparent coordinates at the epoch for which the Besselian or independent day numbers are calculated; α and δ are the mean coordinates of the star referred to the beginning of the Besselian year nearest to this epoch. The same reduction may be also performed by means of matrices according to equations (4.29) and (4.48):

$$\begin{pmatrix} x \\ y \\ z \end{pmatrix}_{\alpha',\delta'} = \mathbf{B}\left[\begin{pmatrix} -D \\ C \\ C\tan\epsilon \end{pmatrix} + \begin{pmatrix} x \\ y \\ z \end{pmatrix}_{\alpha,\delta}\right]. \qquad (4.96)$$

where C, D, and ϵ are as tabulated and

$$\mathbf{B} = \mathbf{R}_3(-f)\,\mathbf{R}_2(A)\,\mathbf{R}_1(B)\ .$$

The apparent right ascension and declination is then computed from equation (3.26) after changing the notation appropriately. Thus

$$\begin{aligned} \alpha' &= \arctan\frac{y}{x}\ , \\ \delta' &= \arctan\frac{z}{(x^2+y^2)^{\frac{1}{2}}}\ , \end{aligned} \qquad (4.97)$$

where the direction numbers x, y, z are from the left side of the matrix relation (4.96).

For precise calculations, second-order terms should be included [Nautical Almanac Offices, 1961, pp. 152-153]. The most advantageous method of doing this is to introduce additional day numbers J and J′ which can be tabulated in the ephemerides (e.g., in the 'American Ephemeris and Nautical Almanac') to give additional reductions in the form:

$$\begin{aligned} &\text{in right ascension} && + J\tan^2\delta, \\ &\text{in declination} && + J'\tan\delta. \end{aligned} \tag{4.98}$$

The full expressions for J and J′ are

$$\left.\begin{aligned} J &= [g\sin(G+\alpha) \pm h\sin(H+\alpha)][g\cos(G+\alpha) \pm h\cos(H+\alpha)] \\ &= [(A+D)\sin\alpha + (B\pm C)\cos\alpha][(A\pm D)\cos\alpha - (B\pm C)\sin\alpha] \\ J' &= -\tfrac{1}{2}[g\sin(G+\alpha) \pm h\sin(H+\alpha)]^2 \\ &= -\tfrac{1}{2}[(A\pm D)\sin\alpha + (B\pm C)\cos\alpha]^2 \end{aligned}\right\}, \tag{4.99}$$

the upper signs being used for positive declinations and the lower signs for negative declinations. The notation corresponds to that of sections 4.123 and 4.212. Tables 4.11 and 4.12 are sample pages showing the tabulated second-order day numbers.

The effect of the second-order terms is reduced by the fact that the expansions of day numbers are not used for an interval longer than half a year. The maximum error introduced by this method of reduction (with second-order terms) is 0.″003. When no second-order correction is applied, the error is about 0.″01 at $\delta = 70°$ increasing about 0.″002 with every 2° in declination.

(3) Apply corrections for annual parallax, if applicable, by using equation (4.62) or (4.63).

(4) Apply proper motion for the remaining fraction of the year, if applicable, by proportioning the annual proper motion components, referred to the beginning of the appropriate Besselian year (not to the original epoch T_0). For illustration see Examples 4.2 and 4.3.

4.422 Rigorous Matrix Reduction. The present availability of electronic computers has now made possible the transformation from mean to apparent place by rigorous formulas that, opposed to the method described in the previous section, involve very little expansion in series.

Let α_0^c, δ_0^c, and μ_0^a, μ_0^d denote the respective catalogued right ascension, declination, and proper motion components referred to the mean equator and equinox of the catalogue epoch T_0; let α_0 and δ_0 denote the actual mean right ascension and declination for the same epoch. The same quantities primed (α', δ') refer to the true equator and equinox of epoch T and denote the apparent places. The correct steps of the rigorous reduction, from the mean place of epoch T_0 to the apparent place of epoch T, are the following:

(1) Remove the e-terms of annual aberration from the catalogued mean positions by means of equation (4.49) or (4.50) as follows:

TABLE 4.11 Sample Page of the AENA

SECOND-ORDER DAY NUMBER *J*, 1967

FOR NORTHERN DECLINATIONS
FOR 0^h EPHEMERIS TIME

Date \ R.A.		0^h 12^h	1^h 13^h	2^h 14^h	3^h 15^h	4^h 16^h	5^h 17^h	6^h 18^h	7^h 19^h	8^h 20^h	9^h 21^h	10^h 22^h	11^h 23^h	12^h 24^h
								J ($0^s.00001$)						
Jan.	0	−5	−3	0	+3	+5	+5	+5	+3	0	−3	−5	−5	−5
	10	−6	−5	−2	+1	+4	+6	+6	+5	+2	−1	−4	−6	−6
	20	−8	−7	−4	0	+3	+6	+8	+7	+4	0	−3	−6	−8
	30	−8	−9	−6	−3	+2	+6	+8	+9	+6	+3	−2	−6	−8
Feb.	9	−8	−10	−8	−5	0	+4	+8	+10	+8	+5	0	−4	−8
	19	−7	−10	−10	−8	−3	+2	+7	+10	+10	+8	+3	−2	−7
Mar.	1	−5	−9	−11	−10	−6	0	+5	+9	+11	+10	+6	0	−5
	11	−3	−8	−11	−11	−8	−3	+3	+8	+11	+11	+8	+3	−3
	21	0	−6	−10	−11	−10	−5	0	+6	+10	+11	+10	+5	0
	31	+2	−3	−8	−11	−11	−8	−2	+3	+8	+11	+11	+8	+2
Apr.	10	+5	−1	−6	−10	−11	−9	−5	+1	+6	+10	+11	+9	+5
	20	+7	+2	−4	−8	−10	−10	−7	−2	+4	+8	+10	+10	+7
	30	+8	+4	−1	−6	−9	−10	−8	−4	+1	+6	+9	+10	+8
May	10	+8	+5	+1	−4	−7	−9	−8	−5	−1	+4	+7	+9	+8
	20	+8	+6	+3	−2	−5	−8	−8	−6	−3	+2	+5	+8	+8
	30	+7	+6	+4	0	−3	−6	−7	−6	−4	0	+3	+6	+7
June	9	+5	+6	+4	+2	−1	−4	−5	−6	−4	−2	+1	+4	+5
	19	+4	+5	+4	+3	0	−2	−4	−5	−4	−3	0	+2	+4
	29	+2	+3	+4	+3	+1	−1	−2	−3	−4	−3	−1	+1	+2
July	9	+1	+2	+3	+3	+2	+1	−1	−2	−3	−3	−2	−1	+1
June	29	+6	+14	+19	+18	+13	+4	−6	−14	−19	−18	−13	−4	+6
July	9	+2	+10	+16	+17	+14	+7	−2	−10	−16	−17	−14	−7	+2
	19	−1	+7	+13	+15	+14	+8	+1	−7	−13	−15	−14	−8	−1
	29	−3	+4	+10	+13	+13	+9	+3	−4	−10	−13	−13	−9	−3
Aug.	8	−4	+1	+7	+10	+11	+9	+4	−1	−7	−10	−11	−9	−4
	18	−5	−1	+4	+7	+9	+8	+5	+1	−4	−7	−9	−8	−5
	28	−5	−2	+1	+5	+7	+7	+5	+2	−1	−5	−7	−7	−5
Sept.	7	−5	−3	0	+2	+4	+5	+5	+3	0	−2	−4	−5	−5
	17	−4	−3	−1	0	+2	+3	+4	+3	+1	0	−2	−3	−4
	27	−2	−3	−2	−1	+1	+2	+2	+3	+2	+1	−1	−2	−2
Oct.	7	−1	−2	−2	−1	−1	0	+1	+2	+2	+1	+1	0	−1
	17	0	−1	−1	−1	−1	−1	0	+1	+1	+1	+1	+1	0
	27	+1	0	0	−1	−1	−1	−1	0	0	+1	+1	+1	+1
Nov.	6	+1	+1	+1	0	0	−1	−1	−1	−1	0	0	+1	+1
	16	+1	+2	+2	+2	+1	0	−1	−2	−2	−2	−1	0	+1
	26	0	+2	+2	+3	+2	+1	0	−2	−2	−3	−2	−1	0
Dec.	6	−1	+1	+3	+4	+4	+3	+1	−1	−3	−4	−4	−3	−1
	16	−3	−1	+2	+4	+5	+4	+3	+1	−2	−4	−5	−4	−3
	26	−5	−3	+1	+4	+6	+6	+5	+3	−1	−4	−6	−6	−5
	36	−7	−5	−1	+3	+6	+7	+7	+5	+1	−3	−6	−7	−7

The quantity J is given in this table in units of $0^s.00001$, and is to be multiplied by $\tan^2\delta_0$ to give the second-order correction in the calculation of the apparent right ascension of a star.

TABLE 4.12 Sample Page of the AENA

SECOND-ORDER DAY NUMBER *J'*, 1967 279

FOR NORTHERN DECLINATIONS

FOR 0^h EPHEMERIS TIME

R.A. / Date		0^h / 12^h	1^h / 13^h	2^h / 14^h	3^h / 15^h	4^h / 16^h	5^h / 17^h	6^h / 18^h	7^h / 19^h	8^h / 20^h	9^h / 21^h	10^h / 22^h	11^h / 23^h	12^h / 24^h
								J' (0″·0001)						
Jan.	0	−2	−1	0	−1	−2	−4	−6	−7	−8	−7	−6	−4	−2
	10	−4	−1	0	0	−1	−3	−6	−8	−9	−9	−8	−6	−4
	20	−6	−3	−1	0	−1	−2	−5	−8	−10	−11	−11	−9	−6
	30	−9	−5	−2	0	0	−2	−4	−8	−11	−13	−13	−11	−9
Feb.	9	−11	−8	−4	−1	0	−1	−3	−7	−10	−13	−14	−14	−11
	19	−13	−10	−6	−3	0	0	−2	−5	−9	−13	−15	−15	−13
Mar.	1	−13	−12	−8	−4	−1	0	−1	−4	−8	−12	−15	−16	−13
	11	−15	−14	−11	−6	−3	0	0	−2	−6	−11	−14	−17	−15
	21	−17	−16	−12	−8	−4	−1	0	−1	−4	−9	−13	−16	−17
	31	−17	−16	−14	−10	−6	−2	0	0	−3	−7	−11	−15	−17
Apr.	10	−16	−17	−15	−12	−8	−4	−1	0	−1	−5	−9	−13	−16
	20	−14	−16	−15	−13	−9	−5	−2	0	−1	−3	−7	−11	−14
	30	−12	−14	−14	−13	−10	−6	−3	−1	0	−2	−5	−8	−12
May	10	−9	−12	−13	−13	−11	−8	−4	−1	0	−1	−3	−6	−9
	20	−7	−10	−12	−12	−11	−8	−5	−2	0	0	−1	−4	−7
	30	−5	−8	−9	−10	−10	−8	−5	−3	−1	0	−1	−2	−5
June	9	−3	−5	−7	−8	−9	−8	−6	−3	−1	0	0	−1	−3
	19	−2	−3	−5	−6	−7	−7	−6	−4	−2	−1	0	0	−2
	29	−1	−2	−3	−4	−5	−5	−5	−4	−2	−1	0	0	−1
July	9	0	−1	−2	−3	−3	−4	−4	−2	−2	−1	−1	0	0
June	29	−1	−5	−11	−19	−25	−29	−28	−24	−18	−10	−4	0	−1
July	9	0	−3	−8	−15	−21	−25	−26	−23	−18	−12	−5	−1	0
	19	0	−1	−5	−11	−17	−21	−23	−22	−18	−12	−6	−2	0
	29	0	0	−3	−8	−13	−17	−20	−20	−17	−12	−7	−3	0
Aug.	8	−1	0	−2	−5	−9	−13	−16	−17	−15	−12	−7	−3	−1
	18	−1	0	−1	−3	−6	−9	−12	−13	−13	−11	−7	−4	−1
	28	−2	0	0	−1	−4	−6	−9	−10	−11	−9	−7	−4	−2
Sept.	7	−2	−1	0	0	−2	−4	−6	−7	−8	−8	−6	−4	−2
	17	−2	−1	0	0	−1	−2	−3	−4	−5	−6	−5	−4	−2
	27	−3	−2	−1	0	0	−1	−1	−2	−3	−4	−4	−3	−3
Oct.	7	−2	−2	−1	0	0	0	0	−1	−2	−2	−3	−3	−2
	17	−2	−2	−1	−1	0	0	0	0	0	−1	−1	−2	−2
	27	−1	−2	−2	−2	−1	−1	0	0	0	0	0	−1	−1
Nov.	6	−1	−1	−2	−2	−2	−2	−1	−1	0	0	0	0	−1
	16	0	−1	−2	−2	−3	−3	−3	−2	−1	−1	0	0	0
	26	0	0	−1	−2	−3	−4	−4	−4	−3	−2	−1	0	0
Dec.	6	0	0	−1	−2	−3	−5	−6	−6	−5	−4	−2	−1	0
	16	−1	0	0	−1	−3	−5	−7	−7	−7	−6	−4	−2	−1
	26	−2	0	0	−1	−3	−5	−7	−9	−9	−8	−7	−4	−2
	36	−4	−1	0	0	−2	−5	−8	−10	−11	−11	−9	−7	−4

The quantity *J'* is given in this table in units of 0″·0001, and is to be multiplied by tan δ_0 to give the second-order correction in the calculation of the apparent declination of a star.

$$\alpha_0 = \alpha_0^c - (\Delta\alpha_{\bar{A}})_0 ,$$
$$\delta_0 = \delta_0^c - (\Delta\delta_{\bar{A}})_0 ,$$

where the quantities in parentheses are the e-terms at the epoch T_0.

(2) Compute the apparent place from the mean place using the direction numbers

$$\begin{pmatrix} x \\ y \\ z \end{pmatrix}_{\alpha_0,\delta_0} = \begin{pmatrix} \cos\delta_0 \cos\alpha_0 \\ \cos\delta_0 \sin\alpha_0 \\ \sin\delta_0 \end{pmatrix} ,$$

in the following combination of equations (4.3), (4.24), (4.48), (4.63), and (4.93):

$$\begin{pmatrix} x \\ y \\ z \end{pmatrix}_{\alpha',\delta'} = \begin{pmatrix} -D \\ C \\ C\tan\epsilon \end{pmatrix} - \begin{pmatrix} C\sec\epsilon \\ D\cos\epsilon \\ D\sin\epsilon \end{pmatrix} (\Pi/\varkappa) + \mathbf{NPM} \begin{pmatrix} x \\ y \\ z \end{pmatrix}_{\alpha_0,\delta_0} , \quad (4.100)$$

where

$$\mathbf{M} = \mathbf{R}_3(-\alpha_0)\, \mathbf{R}_2(\delta_0 - 90°)\, \mathbf{R}_3(\psi_0)\, \mathbf{R}_2 (\mu t + $$
$$+ \tfrac{1}{2} \frac{d\mu}{dt} t^2)\, \mathbf{R}_3(-\psi_0)\, \mathbf{R}_2(90° - \delta_0)\, \mathbf{R}_3(\alpha_0) , \quad (4.94)$$
$$\mathbf{N} = \mathbf{R}_1(-\epsilon-\Delta\epsilon)\, \mathbf{R}_3(-\Delta\psi)\, \mathbf{R}_1(\epsilon), \quad (4.26)$$
$$\mathbf{P} = \mathbf{R}_3(-z)\, \mathbf{R}_2(\theta)\, \mathbf{R}_3(-\zeta_0), \quad (4.4)$$

C and D are the Besselian day numbers at epoch T, ϵ is the obliquity of the ecliptic at the same epoch, Π is the parallax of the star, and $\varkappa$ is the constant of annual aberration. The first two vectors on the right side of the equation may be calculated by first setting them up from the day numbers, tabulated for the beginning of the Besselian year nearest to the epoch T, in radians, and then multiplying them with the matrix $\mathbf{B}$ of equation (4.30) (see comments after equations (4.48) and (4.63) and Example 4.4).

(3) Compute the apparent right ascension and declination from equation (4.97) using the direction numbers obtained from equation (4.100).

(4) Apply the e-terms of annual aberration computed from equations (4.49) or (4.50) for the epoch T, to the positions obtained in step (3).

If the interval $T-T_0$ is shorter than 100 years, steps (1) and (4) may be excluded from the reduction (maximum error about 0.″002 in both coordinates). For illustration see Example 4.4.

4.423 Trigonometric Reduction. An accurate reduction method using only trigonometric relations, applicable for longer intervals than the conventional method, consists of the following steps:

(1) Compute the mean place of the star for the epoch T, α and δ, from the mean place at the catalogue epoch T_0 as outlined in section 4.411.

(2) Compute the true place of the star for the epoch T, α_T and δ_T, from the mean place at epoch T obtained in step (1), by correcting it for nutation by means of equation (4.22) or (4.23).

(3) Compute the apparent place of the star referred to the true equator and equinox of epoch T, α' and δ', from its true place obtained in step (2), by adding corrections for annual (circular) aberration and annual parallax (when necessary). Thus

$$\alpha' = \alpha_T + \Delta\alpha_A + \Delta\alpha_P = \alpha_T + \Delta\alpha_{A+P},$$
$$\delta' = \delta_T + \Delta\delta_A + \Delta\delta_P = \delta_T + \Delta\delta_{A+P},$$

where $\Delta\alpha_A$, $\Delta\delta_A$ are the corrections for annual (circular) aberration given by equation (4.46) or (4.47); $\Delta\alpha_P$, $\Delta\delta_P$ are those for annual parallax given by (4.62); $\Delta\alpha_{A+P}$, $\Delta\delta_{A+P}$ are for the combined annual aberration and parallax as given by (4.64).

If the catalogued mean place does not contain the e-terms of aberration, i.e., when steps (1) and (3) in section 4.411 are omitted, these terms, given by equation (4.49) or (4.50), should be added to the corrections above. Thus, in this case

$$\alpha' = \alpha_T + \Delta\alpha_{A+P} + \Delta\alpha_{\bar{A}},$$
$$\delta' = \delta_T + \Delta\delta_{A+P} + \Delta\delta_{\bar{A}}.$$

(4) Second-order terms may be added according to equation (4.98) in section 4.421.

4.424 Numerical Examples. The following examples illustrate the various methods of reducing the mean place of a star to its apparent position. The star selected is No. 23487 from the 'General Catalogue of Stars for 1950.0' (GC). The final epoch of reduction is

$$\begin{aligned} T &= 23^h30^m10^s.000 \text{ EST (Eastern Standard Time), January 8, 1967} \\ &= 4^h30^m10^s.000 \text{ UT (Universal Time), January 9, 1967} \\ &= 1967 \text{ January } 9^d.188033 \text{ ET (Ephemeris Time)} = 1967.02231 \end{aligned}$$
(see section 5.53).

In the examples sufficient references are made to the tables used in the calculations. Interpolations between the tabulated values are to be made by use of first and second differences, unless noted.

EXAMPLE 4.1

Conventional Reduction of the Mean Place of a Star
and Its Proper Motion from One Epoch to Another
(GC No. 23487 from 1950.0 to 1967.0)

From equations (4.16) and (4.20) (with (4.90)–(4.92)),

$$\alpha = \alpha_0 + AV_\alpha (T - T_0) + \frac{SV_\alpha}{200} (T-T_0)^2 + \frac{3^{rd}_\alpha}{10^6} (T - T_0)^3 ,$$

$$\delta = \delta_0 + AV_\delta (T - T_0) + \frac{SV_\delta}{200} (T-T_0)^2 + \frac{3^{rd}_\delta}{10^6} (T - T_0)^3 .$$

From the GC the following data is obtained:

$\alpha_0 = 17^h 20^m 05^s.048$, $\delta_0 = 40°01'21''.12$,
$AV_\alpha = 1^s.9686$, $AV_\delta = -3''.543$,
$SV_\alpha = 0^s.0035$, $SV_\delta = 0''.284$,
$3^{rd}_\alpha = -0^s.001$, $3^{rd}_\delta = 0''.02$,
$\mu_0^a = 0^s.0004$, $\mu_0^d = -0''.070$.

In accordance with the above equation the computation of the mean place of the star for the epoch 1967.0, referred to the mean equator and equinox at the same epoch, is as follows ($T-T_0$=1967.0 - 1950.0=17 years):

α_0	=	$17^h 20^m 05^s.048$	δ_0	=	$40°01'21''.12$
$AV_\alpha (T-T_0)$	=	33.4662	$AV_\delta (T-T_0)$	=	- 60.231
$SV_\alpha \frac{(T-T_0)^2}{200}$	=	0.0051	$SV_\delta \frac{(T-T_0)^2}{200}$	=	0.410
$3^{rd}_\alpha \frac{(T-T_0)^3}{10^6}$	=	0.0000	$3^{rd}_\delta \frac{(T-T_0)^3}{10^6}$	=	0.000
$\alpha_{1967.0}$	=	$17^h 20^m 38^s.519(+3)$ ‖	$\delta_{1967.0}$	=	$40°00'21''.30(-1)$ ‖

The last figures in parentheses, if added to the next place (which is zero), give the number before rounding (e.g., 0.519(+3) = 0.5193 or 0.30(−1) = 0.299). The results are correct within one unit in the last place as given. To get the figure correct to the last place (e.g., to $0^s.001$ or $0''.01$), one more figure is needed for the values of AV than given in the GC.

The components of the annual proper motion for the epoch 1967.0, referred to the mean equator and equinox at the same epoch, may be computed by means of equations (4.83) - (4.84) and (4.88) - (4.89) as follows:

$$\mu^a = \mu_0^a + [2\mu_0^a \mu_0^d \tan\delta_0 + \bar{n}(\mu_0^a \cos\alpha_0 \tan\delta_0 + \mu_0^d \sin\alpha_0 \sec^2\delta_0)](T-T_0),$$
$$\mu^d = \mu_0^d - [(\mu_0^a)^2 \sin\delta_0 \cos\delta_0 + \bar{n}\mu_0^a \sin\alpha_0](T-T_0)$$

where $T-T_0$ = 1967.0 - 1950.0 = 17 years, $\bar{n}$ from equation (4.8) for the mean epoch $(T_0+T)/2$ is

$$\bar{n} = 20''.0468 - 0''.0085\,[(T_0+T)/2 - 1900.0]^{CENT} = 20''.0418,$$
$$\sin\alpha_0 = 0.98487,\ \cos\alpha_0 = 0.17329,\ \sin\delta_0 = 0.64388,\ \cos\delta_0 = 0.76579.$$

In the equations, μ^a and μ_0^a are to be in seconds of time, μ^d, μ_0^d and $\bar{n}$ in seconds or arc, and $T - T_0$ in years, thus

$$
\begin{aligned}
&\mu_0^a && = 0^s\!.0004 \\
&2\mu_0^a \mu_0^d (T - T_0) \tan\delta_0 \sin 1'' && = 0.0000 \\
&\bar{n}(\mu_0^a \cos\alpha_0 \tan\delta_0 + \frac{1}{15}\mu_0^d \sin\alpha_0 \sec^2\delta_0)(T - T_0) \sin 1'' && = \underline{0.0000} \\
&\mu^a_{1967.0} && = 0^s\!.0004 \,\|
\end{aligned}
$$

$$
\begin{aligned}
&\mu_0^d && = -0''\!.070 \\
&-(15\,\mu_0^a)^2 (T - T_0) \sin\delta_0 \cos\delta_0 \sin 1'' && = 0.000 \\
&-\bar{n}(15\mu_0^a)(T - T_0) \sin\alpha_0 \sin 1'' && = \underline{0.000} \\
&\mu^d_{1967.0} && = -0''\!.070 \,\|
\end{aligned}
$$

EXAMPLE 4.2

Conventional Reduction of the Apparent Place of a Star from Its Mean Position Using Besselian Day Numbers and Star Constants (GC No. 23487 from Its Mean Place at 1967.0 to Its Apparent Place at 1967.02231)

According to equations (4.95) and (4.98)

$$
\begin{aligned}
\alpha' &= \alpha + \mu^a\tau + Aa + Bb + Cc + Dd + E + J\tan^2\delta, \\
\delta' &= \delta + \mu^d\tau + Aa' + Bb' + Cc' + Dd' + J'\tan\delta,
\end{aligned}
$$

where α, δ, are the mean coordinates of the star at the beginning of the Besselian year T_0, nearest to the epoch T at which the apparent coordinates α' and δ' are sought; μ^a and μ^d are the proper motion components at T_0; A, B, C, D, E are the Besselian day numbers as tabulated in the 'American Ephemeris and Nautical Almanac,' pages 258–272 (see Table 4.3); a, b, c, d, a′, b′, c′, d′ are the Besselian star constants as defined by equations (4.27) and (4.46); J and J′ are the second-order day numbers as tabulated in the same publication on pages 278–281 (see Tables 4.11 and 4.12); and $\tau = T - T_0$ in years. The annual parallax Π is assumed to be zero.

From Example 4.1 the mean place of the star for 1967.0 is

$$
\begin{aligned}
\alpha &= 17^h 20^m 38^s\!.519(+3), & \delta &= 40^\circ 00' 21''\!.30(-1), \\
\mu^a &= 0^s\!.0004, & \mu^d &= -0''\!.070, \\
\sin\alpha &= -0.98529031, & \sin\delta &= 0.64286671, \\
\cos\alpha &= -0.17088888, & \cos\delta &= 0.76597806.
\end{aligned}
$$

From equations (4.7) and (4.8), or from the 'American Ephemeris and Nautical Almanac,' 1967, p. 50, the annual precessional constants are

$$m = 3^s.07359,$$
$$n = 1^s.33608.$$

From the 'American Ephemeris and Nautical Almanac,' 1967, pp.258, 278 and 279 (Tables 4.3, 4.11 and 4.12) the Besselian and second-order day numbers for 1967 January $9^d.188033$ ET (or τ = 1967.02231 - 1967.0 = 0.02231), from interpolation with first and second differences, are

$$A = -3''.8190,$$
$$B = -6''.1938,$$
$$C = -5''.8720,$$
$$D = 19''.4437,$$
$$E = -0^s.0016,$$
$$J = 0^s.00006,$$
$$J' = -0''.0003.$$

From equation (4.5), or from the 'American Ephemeris and Nautical Almanac,' 1967, p. 50, the obliquity of the ecliptic is

$$\epsilon = 23°26'36''.87,$$
$$\tan\epsilon = 0.43364190.$$

From equations (4.27) and (4.46) the Besselian star constants (for right ascension divided by 15 for conversion from arc to time) are

$$a = \frac{1}{15}\left(\frac{m}{n} + \sin\alpha\tan\delta\right) = 0.098235, \quad a' = \cos\alpha = -0.170889,$$

$$b = \frac{1}{15}\cos\alpha\tan\delta = -0.009562, \quad b' = \sin\alpha = 0.985290,$$

$$c = \frac{1}{15}\cos\alpha\sec\delta = -0.014873, \quad c' = \tan\epsilon\cos\delta - \sin\alpha\sin\delta = 0.965570,$$

$$d = \frac{1}{15}\sin\alpha\sec\delta = -0.085754, \quad d' = \cos\alpha\sin\delta = -0.109859.$$

With this data the apparent place of the star for epoch 1967.02231, referred to the true equator and equinox at that epoch, is computed as follows:

α	=	$17^h20^m38^s.5193$	δ	=	$40°00'21''.299$
$\mu^a\tau$	=	0.0000	$\mu^d\tau$	= -	0.002
Aa	= -	0.3752	Aa$'$	= +	0.653
Bb	= +	0.0592	Bb$'$	= -	6.103
Cc	= +	0.0873	Cc$'$	= -	5.670
Dd	= -	1.6674	Dd$'$	= -	2.136
E	= -	0.0016			
$J\tan^2\delta$	=	0.0000	$J'\tan\delta$	=	0.000
α'	=	$17^h20^m36^s.622$	δ'		$40°00'08''.04$

EXAMPLE 4.3

Conventional Reduction of the Apparent Place of a Star from Its Mean Position Using Independent Day Numbers (GC No. 23487 from Its Mean Place at 1967.0 to Its Apparent Place at 1967.02231)

According to equations (4.95) and (4.98),

$$\alpha' = \alpha + \mu^a\tau + f + g\sin(G+\alpha)\tan\delta + h\sin(H+\alpha)\sec\delta + J\tan^2\delta,$$
$$\delta' = \delta + \mu^d\tau + g\cos(G+\alpha) + h\cos(H+\alpha)\sin\delta + i\cos\delta + J'\tan\delta,$$

where α, δ are the mean coordinates of the star at the beginning of the Besselian year T_0, nearest to the epoch T at which the apparent coordinates α' and δ' are sought; μ^a and μ^d are the proper motion components at T_0; f, g, h, i, G, H are the independent day numbers as tabulated in the 'American Ephemeris and Nautical Almanac,' pp. 259-273 (see Table 4.4); J and J' are the second-order day numbers as tabulated in the same publication on pp. 278-281 (see Tables 4.11 and 4.12); and $\tau = T - T_0$ in years. The annual parallax Π is assumed to be zero.

From Example 4.1 the mean place of the star for 1967.0 is

$$\alpha = 17^h20^m38\overset{s}{.}519(+3), \quad \delta = 40°00'21\overset{''}{.}30(-1),$$
$$\mu^a = 0\overset{s}{.}0004, \quad \mu^d = -0\overset{''}{.}070,$$
$$\sin\delta = 0.64286671, \quad \cos\delta = 0.76597806.$$

From the 'American Ephemeris and Nautical Almanac,' 1967, the independent and second-order day numbers for 1967 January $9\overset{d}{.}188033$ ET (or $\tau = 1967.02231 - 1967.0 = 0.02231$), from interpolation with first and second differences (third differences in G), are

$$f = -0\overset{s}{.}5873, \quad G = 15^h53^m21\overset{s}{.}980,$$
$$g = 7\overset{''}{.}2768, \quad H = 22^h52^m46\overset{s}{.}940,$$
$$h = 20\overset{''}{.}3108, \quad J = 0\overset{s}{.}00006,$$
$$i = -2\overset{''}{.}5459, \quad J' = -0\overset{''}{.}0003.$$

It follows that

$$G+\alpha = 9^h14^m00\overset{s}{.}499 = 138°30'7\overset{''}{.}485, \quad H+\alpha = 16^h13^m25\overset{s}{.}459 = 243°21'21\overset{''}{.}885,$$
$$\sin(G+\alpha) = 0.6625929, \quad \sin(H+\alpha) = -0.8938107,$$
$$\cos(G+\alpha) = -0.7489798, \quad \cos(H+\alpha) = -0.4484444.$$

With this data the apparent place of the star for epoch 1967.02231, referred to the true equator and equinox at that epoch, is computed as follows:

α	=	$17^h 20^m 38^s.5193$	δ	=	$40°00'21''.299$
$\mu^a T$	=	0.0000	$\mu^d T$	= -	0.002
f	= -	0.5873	$g \cos(G+\alpha)$	= -	5.450
$\frac{1}{15} g \sin(G+\alpha) \tan\delta$	=	0.2698	$h \cos(H+\alpha) \sin\delta$	= -	5.855
$\frac{1}{15} h \sin(H+\alpha) \sec\delta$	= -	1.5800	$i \cos\delta$	= -	1.950
$J \tan^2\delta$	=	0.0000	$J' \tan\delta$	=	0.000
α'	=	$17^h 20^m 36^s.622$ ‖	δ'	=	$40°00'08''.04$ ‖

EXAMPLE 4.4

Reduction of the Apparent Place of a Star from Its Mean Position Using Matrices (GC No. 23487 from Its Mean Place at 1950.0 to Its Apparent Place at 1967.02231)

From equations (4.30), (4.48) and (4.100)

$$\begin{pmatrix} x \\ y \\ z \end{pmatrix}_{\alpha',\delta'} = \mathbf{B} \begin{pmatrix} -D \\ C \\ C \tan\epsilon \end{pmatrix} + \mathbf{NPM} \begin{pmatrix} x \\ y \\ z \end{pmatrix}_{\alpha_0,\delta_0},$$

where the matrices in the notation explained in section 3.34 are as follows:

$$\mathbf{B} = \mathbf{R}_3(-f)\, \mathbf{R}_2(A)\, \mathbf{R}_1(B), \tag{4.30}$$
$$\mathbf{N} = \mathbf{R}_1(-\epsilon - \Delta\epsilon)\, \mathbf{R}_3(-\Delta\psi)\, \mathbf{R}_1(\epsilon), \tag{4.26}$$
$$\mathbf{P} = \mathbf{R}_3(-z)\, \mathbf{R}_2(\theta)\, \mathbf{R}_3(-\zeta_0), \tag{4.4}$$

$$\mathbf{M}\begin{pmatrix} x \\ y \\ z \end{pmatrix}_{\alpha_0,\delta_0} = \begin{pmatrix} \sin\delta_0\cos\alpha_0 & -\sin\alpha_0 & \cos\delta_0\cos\alpha_0 \\ \sin\delta_0\sin\alpha_0 & \cos\alpha_0 & \cos\delta_0\sin\alpha_0 \\ -\cos\delta_0 & 0 & \sin\delta_0 \end{pmatrix}\begin{pmatrix} -\cos\psi_0\sin\mu t \\ \sin\psi_0\sin\mu t \\ \cos\mu t \end{pmatrix}. \tag{4.94}$$

In the above expressions α_0, δ_0 are the mean coordinates of the star at the catalogue epoch T_0; α', δ' are the apparent coordinates at the desired epoch T, $t=T-T_0$ years later; A, B, C, D are the Besselian day numbers interpolated with first and second differences from the tabulated values in the 'American Ephemeris and Nautical Almanac,' pp. 258-272, for the epoch T; f is the independent day number computed similarly from the same publication; ϵ is the mean obliquity of the ecliptic at the epoch T, as computed from equation (4.5); $\Delta\epsilon$ and $\Delta\psi$ are the nutations in obliquity and in longitude at the epoch T; z, θ, and ζ_0 are the precessional elements as given in equation (4.1) for the interval t; μ and ψ_0 are the proper motion and its azimuth at the epoch T_0.

The basic steps of the calculations are the following:

1. The effect of proper motion on the mean place. From the GC for star No. 23487:

$$\alpha_0 = 17^h 20^m 05^s.048, \qquad \delta_0 = 40°01'21''.12,$$
$$\mu_0^a = 0^s.0004, \qquad \mu_0^d = -0''.070,$$
$$\sin\alpha_0 = -0.984871435, \qquad \sin\delta_0 = 0.643088825,$$
$$\cos\alpha_0 = -0.173286646, \qquad \cos\delta_0 = 0.765791586.$$

From equation (3.16) the mean direction cosines of the star at 1950.0, referred to the mean equator and equinox of 1950.0, are the following:

$$\begin{pmatrix} x \\ y \\ z \end{pmatrix}_{\alpha_0,\delta_0} = \begin{pmatrix} \cos\delta_0 \cos\alpha_0 \\ \cos\delta_0 \sin\alpha_0 \\ \sin\delta_0 \end{pmatrix} = \begin{pmatrix} -0.132701455 \\ -0.754206258 \\ 0.643088825 \end{pmatrix}.$$

Other elements necessary to use equation (4.94) are

$$\mu = [(15\mu_0^a \cos\delta_0)^2 + (\mu_0^d)^2]^{\frac{1}{2}} = 0''.07015064,$$
$$\sin\psi_0 = \frac{15\mu_0^a}{\mu}\cos\delta_0 = 0.06549833,$$
$$\cos\psi_0 = \frac{\mu_0^d}{\mu} = -0.99785262,$$

$$t = T - T_0 = 1967.02231 - 1950.0 = 17.02231 \text{ years},$$
$$\mu t = 1''.1941,$$
$$\sin\mu t = 0.00000579,$$
$$\cos\mu t = 1.00000000.$$

With these, equation (4.94) becomes

$$\mathbf{M}\begin{pmatrix} x \\ y \\ z \end{pmatrix}_{\alpha_0,\delta_0} = \begin{pmatrix} -0.111438706 & 0.984871435 & -0.132701455 \\ -0.633359814 & -0.173286646 & -0.754206258 \\ -0.765791586 & 0 & 0.643088825 \end{pmatrix} \times$$
$$\times \begin{pmatrix} 0.000005777 \\ 0.000000379 \\ 1.000000000 \end{pmatrix} = \begin{pmatrix} -0.132701726 \\ -0.754209983 \\ 0.643084401 \end{pmatrix}.$$

These are the mean direction numbers of the star at 1967.02231, referred to the mean equator and equinox of 1950.0.

2. The effect of precession. The elements of precession from equation (4.1), with $t_0 = 1950.0 - 1900.0 = 0.5$ tropical century, and $t = 1967.02231 - 1950.0 = 0.1702231$ tropical century, are

$$\zeta_0 = (2304''.250 + 1''.396\, t_0)t + 0''.302\, t^2 + 0''.018\, t^3 = 392''.3642,$$
$$z = \zeta_0 + 0''.791\, t^2 + 0''.001\, t^3 = 392''.3871,$$
$$\theta = (2004''.682 - 0''.853\, t_0)\, t - 0''.426\, t^2 - 0''.042\, t^3 = 341''.1580.$$

With these elements the rotational matrices necessary to use equation (4.4), from equations (3.18) and (3.19), are

$$\mathbf{R}_3(-z) = \begin{pmatrix} \cos z & -\sin z & 0 \\ \sin z & \cos z & 0 \\ 0 & 0 & 1 \end{pmatrix} = \begin{pmatrix} 0.999998186 & -0.001902347 & 0 \\ 0.001902347 & 0.999998186 & 0 \\ 0 & 0 & 1 \end{pmatrix},$$

$$\mathbf{R}_2(\theta) = \begin{pmatrix} \cos\theta & 0 & -\sin\theta \\ 0 & 1 & 0 \\ \sin\theta & 0 & \cos\theta \end{pmatrix} = \begin{pmatrix} 0.999998628 & 0 & -0.001653976 \\ 0 & 1 & 0 \\ 0.001653976 & 0 & 0.999998628 \end{pmatrix},$$

$$\mathbf{R}_3(-\zeta_0) = \begin{pmatrix} \cos\zeta_0 & -\sin\zeta_0 & 0 \\ \sin\zeta_0 & \cos\zeta_0 & 0 \\ 0 & 0 & 1 \end{pmatrix} = \begin{pmatrix} 0.999998186 & -0.001902236 & 0 \\ 0.001902236 & 0.999998186 & 0 \\ 0 & 0 & 1 \end{pmatrix}.$$

Thus the precessional matrix (equation (4.4)) is

$$\mathbf{P} = \begin{pmatrix} 0.999991381 & -0.003804573 & -0.001653973 \\ 0.003804573 & 0.999992753 & -0.000003146 \\ 0.001653973 & -0.000003146 & 0.999998628 \end{pmatrix}$$

3. <u>The effect of nutation and the true place of the star</u>. The elements of nutation as interpolated from the tables for the sun, p. 18 of the 'American Ephemeris and Nautical Almanac,' 1967, for January $9^{d}.188033$ ET, are

$$\Delta\psi = -10''.7232,$$
$$\epsilon_T = \epsilon + \Delta\epsilon = 23°26'43''.055.$$

From equation (4.5) with $t = 1967.02231 - 1900.0 = 0.6702231$ tropical century,

$$\epsilon = 23°27'08''.26 - 46''.845\,t - 0''.0059t^2 + 0''.00181t^3 = 23°26'36''.861\ .$$

The necessary rotational matrices, from equations (3.17) and (3.19), are

$$\mathbf{R}_1(-\epsilon_T) = \begin{pmatrix} 1 & 0 & 0 \\ 0 & \cos\epsilon_T & -\sin\epsilon_T \\ 0 & \sin\epsilon_T & \cos\epsilon_T \end{pmatrix} = \begin{pmatrix} 1 & 0 & 0 \\ 0 & 0.917440384 & -0.397873265 \\ 0 & 0.397873265 & 0.917440384 \end{pmatrix},$$

$$\mathbf{R}_3(-\Delta\psi) = \begin{pmatrix} \cos\Delta\psi & -\sin\Delta\psi & 0 \\ \sin\Delta\psi & \cos\Delta\psi & 0 \\ 0 & 0 & 1 \end{pmatrix} = \begin{pmatrix} 1.000000000 & 0.000051988 & 0 \\ -0.000051988 & 1.000000000 & 0 \\ 0 & 0 & 1 \end{pmatrix},$$

$$\mathbf{R}_1(\epsilon) = \begin{pmatrix} 1 & 0 & 0 \\ 0 & \cos\epsilon & \sin\epsilon \\ 0 & -\sin\epsilon & \cos\epsilon \end{pmatrix} = \begin{pmatrix} 1 & 0 & 0 \\ 0 & 0.917452338 & 0.397845713 \\ 0 & -0.397845713 & 0.917452338 \end{pmatrix}.$$

Thus the nutational matrix (equation (4.26)) is

$$\mathbf{N} = \begin{pmatrix} 1.000000000 & 0.000047697 & 0.000020683 \\ -0.000047696 & 0.999999998 & -0.000030033 \\ -0.000020685 & 0.000030033 & 0.999999998 \end{pmatrix}.$$

The true direction numbers of the star at 1967.02231,referred to the true equator and equinox at that epoch, from the results of steps 1, 2, and 3 are

$$\begin{pmatrix} x \\ y \\ z \end{pmatrix}_{\alpha,\delta} = \mathbf{NPM} \begin{pmatrix} x \\ y \\ z \end{pmatrix}_{\alpha_0,\delta_0} = \begin{pmatrix} -0.130917481 \\ -0.754724476 \\ 0.642846446 \end{pmatrix}.$$

4. The effect of aberration. The Besselian and independent day numbers, from the 'American Ephemeris and Nautical Almanac,' 1967, pp. 258-259 (see Tables 4.3 and 4.4), for January $9^{\mathrm{d}}.188033$ ET, are

$$\begin{aligned} A &= -3''.8190 &&= -0.000018515 \text{ radian}, \\ B &= -6''.1938 &&= -0.000030028 \text{ radian}, \\ C &= -5''.8720 &&= -0.000028468 \text{ radian}, \\ D &= 19''.4437 &&= 0.000094265 \text{ radian}, \\ f &= -0^{\mathrm{s}}.5873 = -8''.8095 &&= -0.000042709 \text{ radian}. \end{aligned}$$

The necessary rotation matrices for equation (4.30) from equations (3.17) - (3.19), recognizing that due to the small magnitudes of the day numbers $\sin\theta \simeq \theta$ and $\cos\theta \simeq 1$, are

$$\mathbf{R}_3(-f) = \begin{pmatrix} 1 & -f & 0 \\ f & 1 & 0 \\ 0 & 0 & 1 \end{pmatrix} = \begin{pmatrix} 1 & 0.000042709 & 0 \\ -0.000042709 & 1 & 0 \\ 0 & 0 & 1 \end{pmatrix},$$

$$\mathbf{R}_2(A) = \begin{pmatrix} 1 & 0 & -A \\ 0 & 1 & 0 \\ A & 0 & 1 \end{pmatrix} = \begin{pmatrix} 1 & 0 & 0.000018515 \\ 0 & 1 & 0 \\ -0.000018515 & 0 & 1 \end{pmatrix},$$

$$\mathbf{R}_1(B) = \begin{pmatrix} 1 & 0 & 0 \\ 0 & 1 & B \\ 0 & -B & 1 \end{pmatrix} = \begin{pmatrix} 1 & 0 & 0 \\ 0 & 1 & -0.000030028 \\ 0 & 0.000030028 & 1 \end{pmatrix}.$$

Thus the matrix $\mathbf{B}$ of equation (4.30) is

$$\mathbf{B} = \begin{pmatrix} 1.000000000 & 0.000042709 & 0.000018515 \\ -0.000042709 & 1.000000000 & -0.000030028 \\ -0.000018515 & 0.000030028 & 1.000000000 \end{pmatrix}.$$

The aberrational vector of equation (4.48) is

$$\begin{pmatrix} -D \\ C \\ C\tan\epsilon \end{pmatrix} = \begin{pmatrix} -0.000094265 \\ -0.000028468 \\ -0.000012345 \end{pmatrix} .$$

Thus the effect of aberration on the true direction numbers at 1967.02231 is

$$\mathbf{B}\begin{pmatrix} -D \\ C \\ C\tan\epsilon \end{pmatrix} = \begin{pmatrix} -0.000094266 \\ -0.000028464 \\ -0.000012344 \end{pmatrix} .$$

5. <u>The apparent place of the star</u>. The apparent direction numbers of the star at 1967.02231, referred to the true equator and equinox at that epoch, from the results of steps 3 and 4 are

$$\begin{pmatrix} x \\ y \\ z \end{pmatrix}_{\alpha',\delta'} = \mathbf{B}\begin{pmatrix} -D \\ C \\ C\tan\epsilon \end{pmatrix} + \begin{pmatrix} x \\ y \\ z \end{pmatrix}_{\alpha,\delta} = \begin{pmatrix} -0.131011747 \\ -0.754752940 \\ 0.642834102 \end{pmatrix} .$$

Using these numbers, the apparent right ascension and declination from equation (4.97) are

$$\alpha' = \arctan\frac{y}{x} = 17^h 20^m 36\overset{s}{.}621 \; \| ,$$

$$\delta' = \arctan\frac{z}{(x^2+y^2)^{\frac{1}{2}}} = 40°\,00'\,08\overset{''}{.}04 \; \| .$$

4.43 Reduction of the Observed Place of a Star to Its Apparent Position

Let α^O and δ^O denote the observed coordinates of a star at epoch T, and α' and δ' its apparent position at the same epoch. The reduction from the observed place to the apparent place consists of corrections for diurnal aberration, geocentric parallax (if applicable), and astronomic refraction applied as follows:

$$\alpha' = \alpha^O - \Delta\alpha_D - \Delta\alpha_G + \Delta\alpha_R ,$$
$$\delta' = \delta^O - \Delta\delta_D - \Delta\delta_G + \Delta\delta_R ,$$

where $\Delta\alpha_D$ and $\Delta\delta_D$, $\Delta\alpha_G$ and $\Delta\delta_G$, $\Delta\alpha_R$ and $\Delta\delta_R$ are corrections for diurnal aberration, geocentric parallax, and astronomic refraction, to be computed from equations (4.55), (4.67), and (4.74) respectively.

As has been mentioned earlier, in geodetic astronomy these corrections usually are not applied to the star's equatorial coordinates as shown above, but rather to the actually observed parameters (e.g., zenith distance, time of transit). Geocentric parallax is always neglected for stars.

References

Astronomisches Rechen-Institut, Heidelberg. (Annual). Apparent Places of Fundamental Stars Containing the 1535 Stars in the Fourth Fundamental Catalogue (FK4). Verlag G. Braun, Karlsruhe.

Baldini, A. A. (1963). "Formulas for Computing Atmospheric Refraction for Objects Inside or Outside the Atmosphere." GIMRADA Research Note, 8.

Chandler, S. (1891). "On the Variation of Latitude." The Astronomical Journal, 11, p. 83.

Chandler, S. (1892). "On the Variation of Latitude." The Astronomical Journal, 12, p. 17.

Chauvenet, W. (1863). A Manual of Spherical and Practical Astronomy. Philadelphia. Reprinted by Dover Publications, Inc., New York, 1960.

Cowell, P.H. (1900). "Refraction Tables Arranged for Use at the Royal Observatory, Greenwich." Greenwich Observations for 1898, Appendix I, London.

Doolittle, C. L. (1895). A Treatise on Practical Astronomy. John Wiley and Sons, Inc., New York.

Eckert, W. J., R. Jones, and H. K. Clark. (1954). Improved Lunar Ephemeris 1952-1959, U.S.Naval Observatory, Washington, D.C.

Fricke, W., D.Brouwer, J.Kovalevsky, A.A.Mikhailov, and G.A. Wilkins. (1965). "Report to the Executive Committee of the Working Group on the System of Astronomical Constants." Bulletin Géodésique, 75, pp. 59-67.

Fricke, W. and A. Kopff. (1963). Fourth Fundamental Catalogue, Veröff. d. Astronomischen Rechen-Instituts, 10, Heidelberg.

Garfinkel, B. (1944). "An Investigation in the Theory of Astronomical Refraction." The Astronomical Journal, 50, 8.

Garfinkel, B. (1967). "Astronomical Refraction in a Polytropic Atmosphere." The Astronomical Journal, 72, 2.

Goldstein, H. (1950). Classical Mechanics. Addison-Wesley Publishing Company, Reading, Massachusetts.

Hoskinson, A. J. and J. A. Duerksen. (1952). "Manual of Geodetic Astronomy-Determination of Longitude, Latitude, and Azimuth." U. S. Coast and Geodetic Survey, Special Publication, 237. Government Printing Office, Washington, D. C.

Markowitz, W. (1960). "Latitude and Longitude and the Secular Motion of the Pole." Methods and Techniques in Geophysics (S.K. Runcorn, ed.), 1, pp. 325-361, Interscience Publishers, New York.

Markowitz, W. (1961). "International Determination of the Total Motion of the Pole." Bulletin Géodésique, 59, pp. 29-41.

Markowitz, W. (1967). "Current Astronomical Observations for Studying Continental Drift, Polar Motion, and the Rotation of the Earth." Continental Drift, Secular Motion of the Pole, and Rotation of the Earth (Wm. Markowitz and B. Guinot, eds.). D. Reidel Publishing Co., Dordrecht, Holland.

Nautical Almanac Offices of the United Kingdom and the United States of America. (1961). Explanatory Supplement to the Astronomical Ephemeris and the American Ephemeris and Nautical Almanac. H.M. Stationery Office, London. (The introduction of the IAU system of astronomical constants (1964) requires changes in this reference. They are listed in the supplement of the "American Ephemeris and Nautical Almanac," 1968.)

Newcomb, S. (1897). "A New Determination of the Precessional Constant with the Resulting Precessional Motions." Astronomical Papers of the American Ephemeris, VIII, Part 1.

Newcomb, S. (1906). A Compendium on Spherical Astronomy. The Macmillan Co., New York. Reprinted by Dover Publications, Inc., New York, 1960.

Oterma, L. (1960). "Computing the Refraction for the Väisälä Astronomical Method of Triangulation." Astronomia-Optika Institucio, Universitato de Turku, Informo, 20, Turku.

Orlov, B.A. (1956). Refraction Tables of Pulkova Observatory, Fourth edition, Pulkovo.

Porter, J.C. and D.H. Sadler. (1953). "The Accurate Calculation of Apparent Places of Stars." Monthly Notices of the Royal Astronomical Society, 113, pp. 455-467.

Scott, F.P. and J.A. Hughes. (1964). "Computation of Apparent Places for the Southern Reference Star Program." The Astronomical Journal, 69, pp. 368-371.

Smart, W.M. (1960). Spherical Astronomy. Fourth edition. Cambridge University Press, London.

Stoyko, A. (1964). "La réduction de heures définitives dans un système uniforme et au pôle moyen de l'époque." Bulletin Horaire, Serie H, 2, pp. 41-44.

U.S. Naval Observatory, Nautical Almanac Office. (Annual). The American Ephemeris and Nautical Almanac. U.S. Government Printing Office, Washington, D.C. (The same publication is also issued by H.M. Nautical Almanac Office under the title "The Astronomical Ephemeris" in London.)

Willis, J. E. (1941). "A Determination of Astronomical Refraction from Physical Data." Transactions of the American Geophysical Union, pp. 324-336.

Woolard, E. W. (1953). "Theory of the Rotation of the Earth Around Its Center of Mass." The Astronomical Journal, 58, 2; also Astronomical Papers Prepared for the Use of the American Ephemeris, XV, Part I.

Yumi, S. (1964). Annual Report of the International Polar Motion Service for 1962. Central Bureau of the International Polar Motion Service, Mizusawa.

Yumi, S. (1965). Annual Report of the International Polar Motion Service for 1963. Central Bureau of the International Polar Motion Service, Mizusawa.

Yumi, S. and Y. Wako. (1966). "On the Secular Motion of the Mean Pole." Publications of the International Latitude Observatory of Mizusawa, V, 1.

5 TIME SYSTEMS

As has been mentioned in the preceding chapter, the celestial coordinates are subject to change with respect to time. In order to further discuss these variations and their effect on the observations and calculations in geodetic astronomy, it is necessary to become familiar with the various types of time systems presently in use.

First, it is necessary to distinguish between two aspects of time, namely, the epoch and the interval. The main purpose of the epoch is to define with precision the moment or the instant of the occurrence of some phenomenon or observation. The time interval is the time elapsed between two epochs measured in units of some time scale.

A fundamental necessity of any system of time measurement is the establishment of a relationship between the adopted measure of time (year, month, day, hour, minute, second, decimals of second) and some observable physical phenomenon which is either repetitive and countable or continuous and measurable, or both. Those systems which are based on observable astronomic phenomena (e.g., star transits, occultations) are the astronomic time systems. In all systems the most important practical requirement is that the phenomenon on which the time is based should be free, or could be freed, from short periodic irregularities to permit interpolation or extrapolation by man-made timekeeping devices.

In geodetic astronomy there are three systems of time measurements, each related to some natural observable phenomenon:

(1) Sidereal and universal (solar) times based on the diurnal rotation of the earth.

(2) Ephemeris time which is the independent variable in the equations of motion of dynamics. Its measure is defined by the orbital motion of the earth about the sun.

(3) Atomic time based on the electromagnetic oscillations produced by the quantum transition of an atom.

The universal and sidereal times are equivalent forms of time which are related to each other by rigorous formulas; thus one defines the other. The use of one instead of the other is purely a matter of convenience. Research in several areas utilizes this measure, such as

— star position determinations for catalogue constructions,
— astronomic position determinations in geodesy and in navigation,
— determination of the variations in the rotation of the earth, including variations in the position of the rotation axis and in the rotational velocity.

Universal or sidereal time is determined from observations on stars (usually transits) and can be traced as far back in the past as such observations are available.

Ephemeris and atomic times are independent systems. They are used mostly in observations and calculations which are periodic or repetitive in nature where the uniformity of time intervals plays an important role, such as in connection with the observations of artificial satellites, in celestial mechanics, and for ephemeris purposes. Ephemeris time or an equivalent thereof must be used where the availability of the epoch and interval is not imminent and when Newton's equations of motion enter the considerations, such as in (celestial) mechanics. Atomic time may be used in other cases, mostly for physical applications, where the need is an increasing relative precision in the measurement of short intervals of time.

Ephemeris time is determined from observations of a body in the solar system (usually of the moon; solar eclipses and occultations in the past, moon camera observations at present) and can be traced as far back in the past as such observations are available (seventeenth century). The rotational (universal and sidereal) and the ephemeris times are the only astronomic time systems.

The rotational time scales vary with respect to ephemeris and atomic times because of variations in the speed of rotation of the earth. The relation between the universal time and ephemeris or atomic time cannot be predicted; hence it must be determined empirically through observations. The present time scales of ephemeris and atomic times may be regarded as practically equivalent.

In the past some of the time systems described in the following sections had a somewhat different interpretation; some of the present designations were not used and others disappeared from general usage. There were, for example, important changes introduced and internationally accepted in 1925 and again in 1960. Therefore, if references are used or work is performed with material originating before these dates, caution should be exercised and proper references consulted, e.g., [Nautical Almanac Offices, 1961, pp. 88-95].

5.1 Sidereal Time System

5.11 Sidereal Epoch: Apparent and Mean Sidereal Times, the Equation of the Equinox

Sidereal time is directly related to the rotation of the earth; equal intervals of angular motion correspond to equal intervals of sidereal time.

The sidereal epoch is numerically measured by the hour angle of the vernal equinox. The local hour angle of the true vernal equinox is termed apparent sidereal time (AST). When the hour angle is referred to the Greenwich mean astronomic meridian (see section 8.213), it is called Greenwich apparent sidereal time (GAST). The local hour angle of the mean vernal equinox is termed mean sidereal time (MST). When the hour angle is referred to the Greenwich mean astronomic meridian, it is called Greenwich mean sidereal time (GMST).

The difference between AST and MST is termed the equation of the equinox (Eq.E). Thus

$$\text{Eq. E} = \text{AST} - \text{MST} = \text{GAST} - \text{GMST}. \qquad (5.1)$$

The equation of the equinox is due to the nutation. It varies with rather short periods, its maximum value reaching about 1^s (in 1966 $\text{Eq. E}_{max} = -0^s.949$ on April 22, see Fig. 5.5). It is tabulated in the 'American Ephemeris and Nautical Almanac' in the tables for universal and sidereal times (see Table 5.5).

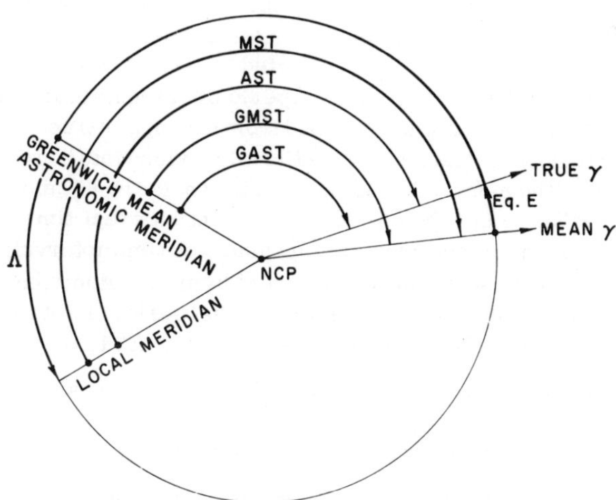

Fig. 5.1 Sidereal times

The above relations are illustrated in Fig. 5.1, from which it is evident that

$$MST = GMST + \Lambda, \tag{5.2}$$
$$AST = GAST + \Lambda, \tag{5.3}$$

or that

$$\Lambda = MST - GMST = AST - GAST \tag{5.4}$$

where Λ is the reduced astronomic longitude with the positive sign to the east. Sidereal time is conventionally measured in hours, minutes, and seconds, so that longitude in the above equation is measured in time at the rate of one hour to 15°. Thus, $24^h = 360°$, $1^h = 15°$, $1^m = 15'$, and $1^s = 15''$. There are tables included in the 'Ephemeris' which facilitate conversions from arc to time and from time to arc. The latter conversion is not necessary if trigonometric tables with time arguments are available [H.M. Nautical Almanac Office, 1958].

Since the hour angle of the vernal equinox is equal to the right ascension of an object at the instant of its upper transit over the meridian from which the hour angle is reckoned, the sidereal time is usually determined by observing transits of stars. In this case, allowance must be made for the variation in the position of the meridian due to the motion of the true celestial pole with respect to the solid body of the earth and should also (when possible) be made for short periodic irregularities in the rate of rotation of the earth. With this understanding, the definition of the sidereal epoch as given above is precise (see Fig. 4.10).

5.12 Sidereal Interval: Mean Sidereal Day and the Period of Rotation

The fundamental unit of sidereal interval is the mean sidereal day, defined as the interval between two successive upper transits of the mean vernal equinox over some meridian (in the rest of Chapter 5 it is understood that the meridian is not affected by polar motion and that the rotation of the earth is free from short periodic irregularities, unless noted). The mean sidereal day is reckoned from 0^h at upper transit which is known as sidereal noon, and it is conventionally divided into hours, minutes, and seconds. The mean sidereal interval is denoted by (S); thus $1^d(S) = 24^h(S)$, $1^h(S) = 60^m(S)$, and $1^m(S) = 60^s(S)$. Apparent sidereal time, because of its variable rate, is not used as a measure of time interval.

Since the mean equinox is affected by precession, the mean sidereal day is about $0\overset{s}{.}0084$ shorter than the actual period of rotation of the earth. More precisely [Nautical Almanac Offices, 1961, p. 73],

$$1^d(S) = (0.999999902907 - 59 \times 10^{-12}t) \text{ period of rotation} \tag{5.5}$$

or

$$1 \text{ period of rotation} = [1.0 + (97093 + 59t) \times 10^{-12}]^d(S) \tag{5.6}$$

where t is the number of ephemeris centuries of 36525 ephemeris days since 1900 January $0\overset{d}{.}5$ ET (see sections 5.222, 5.232, and 5.53).

5.13 Sidereal Calendar: Greenwich Sidereal Date and Day Numbers

In order to facilitate the tabulation of certain quantities with arguments in sidereal time, the concepts of the Greenwich sidereal date (GSD) and Greenwich sidereal day number have been introduced.

In certain applications it is convenient to express the epoch as a certain number of days and a fraction of a day after a fundamental epoch sufficiently in the past to precede the historical period. According to this concept the Greenwich sidereal date is defined as the number of mean sidereal days that have elapsed on the Greenwich mean astronomic meridian since the beginning of the sidereal day that was in progress at Greenwich noon on Jan. 1, 4713 B.C. The integral part of the GSD is the Greenwich sidereal day number, a means of numbering consecutively successive sidereal days beginning at the instants of upper transit of the mean vernal equinox over the Greenwich meridian. The nonintegral part of GSD is simply the GMST expressed usually in fractions of a mean sidereal day.

The Greenwich sidereal day numbers are tabulated for the beginning of the first sidereal day each month, from 1900 to 1999, in [Nautical Almanac Offices, 1961, pp. 440-441]. A sample page is shown in Table 5.1.

Analogous to this system are the Julian dates and day numbers, and the Julian ephemeris dates and day numbers used in the solar and ephemeris time systems respectively, which are described later.

5.2 Universal (Solar) Time System

5.21 The Motion of the Sun

The apparent motion of the sun as seen by an observer on the earth is due partly to the annual revolution of the earth around the sun and partly to the diurnal rotation of the earth about its axis.

5.211 Annual Motion, Kepler's Laws. The annual motion is closely governed by the first two of the three laws of Kepler which, applied to the earth, are summarized below.

Kepler's first law states that the orbit of a planet around the sun is an ellipse, the position of the sun being at a focus of the ellipse. Let Fig. 5.2 illustrate the approximate orbit of the earth: the ellipse lying in the ecliptic with the sun at the focus S. Let point E represent the earth at a certain time. Its position is defined by the radius vector r and the angle $\theta = \lambda_s - 180^\circ$ (λ_s is the longitude of the sun). The equation of the ellipse is known to be [Smart, 1960, p. 108]

$$r = \frac{p}{1 + e\cos(\theta - \bar{\omega}_s)} \tag{5.7}$$

where the mean radius

$$p = \frac{b^2}{a} = a(1 - e^2) \cong 149.56 \times 10^6 \text{km}, \tag{5.8}$$

TABLE 5.1 Greenwich Sidereal Day Number, 1950–1999

	OF DAY COMMENCING AT UNIVERSAL TIME:											
Year	Jan. 0·73	Feb. 0·64	Mar. 0·57	Apr. 0·48	May 0·40	June 0·32	July 0·23	Aug. 0·15	Sept.† 0·07	Oct. 0·98	Nov. 0·90	Dec. 0·81
1950	243 9945	9976	*0004	*0035	*0065	*0096	*0126	*0157	*0188	*0219	*0250	*0280
1951	244 0311	0342	0370	0401	0431	0462	0492	0523	0554	0585	0616	0646
1952	0677	0708	0737	0768	0798	0829	0859	0890	0921	0952	0983	1013
1953	1044	1075	1103	1134	1164	1195	1225	1256	1287	1318	1349	1379
1954	1410	1441	1469	1500	1530	1561	1591	1622	1653	1684	1715	1745
1955	244 1776	1807	1835	1866	1896	1927	1957	1988	2019	2050	2081	2111
1956	2142	2173	2202	2233	2263	2294	2324	2355	2386	2417	2448	2478
1957	2509	2540	2568	2599	2629	2660	2690	2721	2752	2783	2814	2844
1958	2875	2906	2934	2965	2995	3026	3056	3087	3118	3149	3180	3210
1959	3241	3272	3300	3331	3361	3392	3422	3453	3484	3515	3546	3576
1960	244 3607	3638	3667	3698	3728	3759	3789	3820	3851	3882	3913	3943
1961	3974	4005	4033	4064	4094	4125	4155	4186	4217	4248	4279	4309
1962	4340	4371	4399	4430	4460	4491	4521	4552	4583	4614	4645	4675
1963	4706	4737	4765	4796	4826	4857	4887	4918	4949	4980	5011	5041
1964	5072	5103	5132	5163	5193	5224	5254	5285	5316	5347	5378	5408
1965	244 5439	5470	5498	5529	5559	5590	5620	5651	5682	5713	5744	5774
1966	5805	5836	5864	5895	5925	5956	5986	6017	6048	6079	6110	6140
1967	6171	6202	6230	6261	6291	6322	6352	6383	6414	6445	6476	6506
1968	6537	6568	6597	6628	6658	6689	6719	6750	6781	6812	6843	6873
1969	6904	6935	6963	6994	7024	7055	7085	7116	7147	7178	7209	7239
1970	244 7270	7301	7329	7360	7390	7421	7451	7482	7513	7544	7575	7605
1971	7636	7667	7695	7726	7756	7787	7817	7848	7879	7910	7941	7971
1972	8002	8033	8062	8093	8123	8154	8184	8215	8246	8277	8308	8338
1973	8369	8400	8428	8459	8489	8520	8550	8581	8612	8643	8674	8704
1974	8735	8766	8794	8825	8855	8886	8916	8947	8978	9009	9040	9070
1975	244 9101	9132	9160	9191	9221	9252	9282	9313	9344	9375	9406	9436
1976	9467	9498	9527	9558	9588	9619	9649	9680	9711	9742	9773	9803
1977	244 9834	9865	9893	9924	9954	9985	*0015	*0046	*0077	*0108	*0139	*0169
1978	245 0200	0231	0259	0290	0320	0351	0381	0412	0443	0474	0505	0535
1979	0566	0597	0625	0656	0686	0717	0747	0778	0809	0840	0871	0901
1980	245 0932	0963	0992	1023	1053	1084	1114	1145	1176	1207	1238	1268
1981	1299	1330	1358	1389	1419	1450	1480	1511	1542	1573	1604	1634
1982	1665	1696	1724	1755	1785	1816	1846	1877	1908	1939	1970	2000
1983	2031	2062	2090	2121	2151	2182	2212	2243	2274	2305	2336	2366
1984	2397	2428	2457	2488	2518	2549	2579	2610	2641	2672	2703	2733
1985	245 2764	2795	2823	2854	2884	2915	2945	2976	3007	3038	3069	3099
1986	3130	3161	3189	3220	3250	3281	3311	3342	3373	3404	3435	3465
1987	3496	3527	3555	3586	3616	3647	3677	3708	3739	3770	3801	3831
1988	3862	3893	3922	3953	3983	4014	4044	4075	4106	4137	4168	4198
1989	4229	4260	4288	4319	4349	4380	4410	4441	4472	4503	4534	4564
1990	245 4595	4626	4654	4685	4715	4746	4776	4807	4838	4869	4900	4930
1991	4961	4992	5020	5051	5081	5112	5142	5173	5204	5235	5266	5296
1992	5327	5358	5387	5418	5448	5479	5509	5540	5571	5602	5633	5663
1993	5694	5725	5753	5784	5814	5845	5875	5906	5937	5968	5999	6029
1994	6060	6091	6119	6150	6180	6211	6241	6272	6303	6334	6365	6395
1995	245 6426	6457	6485	6516	6546	6577	6607	6638	6669	6700	6731	6761
1996	6792	6823	6852	6883	6913	6944	6974	7005	7036	7067	7098	7128
1997	7159	7190	7218	7249	7279	7310	7340	7371	7402	7433	7464	7494
1998	7525	7556	7584	7615	7645	7676	7706	7737	7768	7799	7830	7860
1999	245 7891	7922	7950	7981	8011	8042	8072	8103	8134	8165	8196	8226

† There are two transits of the first point of Aries on one mean solar day near September 22.

the eccentricity of the orbit

$$e = \sqrt{\frac{a^2 - b^2}{a^2}} \cong 0.016726, \tag{5.9}$$

the major semiaxis

$$a \cong 149.60 \times 10^6 \,\text{km},$$

the minor semiaxis

$$b \cong 149.58 \times 10^6 \,\text{km},$$

the longitude of the perihelion (point A where the earth is nearest to the sun, as opposed to point B, the aphelion) $\overline{\omega}_s \cong 102^\circ$.

When $\theta = \overline{\omega}_s$ (earth at perihelion), equation (5.7) yields

$$r_p = \frac{p}{1+e} \cong 147.1 \times 10^6 \,\text{km}, \tag{5.10}$$

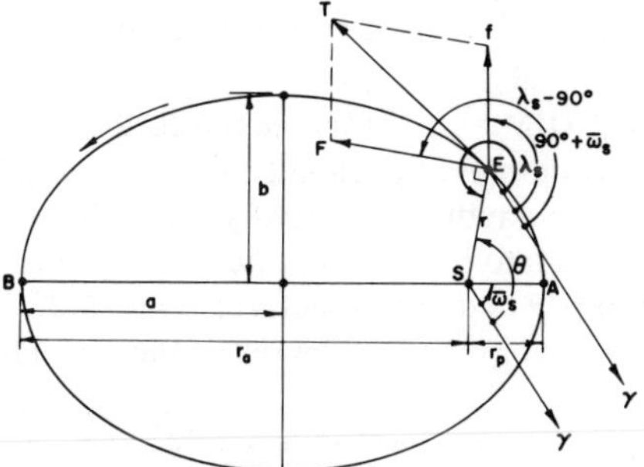

Fig. 5.2 Orbital velocities of the earth

while when $\theta = 180^\circ + \overline{\omega}_s$ (earth at aphelion),

$$r_a = \frac{p}{1-e} \cong 152.1 \times 10^6 \,\text{km}. \tag{5.11}$$

It is obvious that

$$r_p + r_a = \frac{p}{1+e} + \frac{p}{1-e} = \frac{2p}{1-e^2} = 2a.$$

Kepler's second law states that the radius vector r sweeps over equal areas during equal times. Thus, from the element of the area, $dA = \frac{1}{2} r^2 \, d\theta$, it follows that

$$\frac{dA}{dt} = \frac{1}{2} r^2 \frac{d\theta}{dt} = \frac{1}{2} h \tag{5.12}$$

where h, a constant, is twice the areal velocity, defined as dA/dt. It may be determined by applying equation (5.12) for a full revolution. Since the whole area of the ellipse $ab\pi$ is described by the radius vector during the interval defined by the period P, the constant h is given by

$$h = \frac{2ab\pi}{P} = \frac{2\pi a^2 (1 - e^2)^{\frac{1}{2}}}{P} , \qquad (5.13)$$

or, introducing the mean angular velocity n, termed the mean motion,

$$n = \frac{2\pi}{P} \cong 1.99106 \times 10^{-7} \text{ radians s}^{-1}, \qquad (5.14)$$

equation (5.13) becomes

$$h = na^2(1-e^2)^{\frac{1}{2}} \cong 4.4553 \times 10^9 \text{km}^2 \text{s}^{-1}. \qquad (5.15)$$

Kepler's third law states that the square of the period of revolution around the sun is inversely proportional to the cube of the major semi-axis of the orbit. The mathematical form of this law, using equations (5.8) and (5.15), is

$$n^2 a^3 = \frac{h^2}{p} = G(M+m) \cong 13.2718 \times 10^{10} \text{km}^3 \text{s}^{-2} \qquad (5.16)$$

where the quantity G(M + m) is termed the heliocentric gravitational constant (G is the Newtonian gravitational constant, M and m are the masses of the sun and of the earth respectively). Let the earth's tangential velocity T be resolved into components perpendicular to the radius vector and to the major axis F and f respectively (Fig. 5.2). It may be shown that these components are given by [Smart, 1960, p. 110]:

$$F = \frac{h}{p} = na(1 - e^2)^{-\frac{1}{2}} \cong 29.79 \text{ km s}^{-1}, \qquad (5.17)$$

$$f = e\frac{h}{p} = nae(1 - e^2)^{-\frac{1}{2}} \cong .50 \text{ km s}^{-1}; \qquad (5.18)$$

thus, they are constant anywhere along the orbit. The directions of the velocity components F and f are evidently defined by longitudes $\lambda_s - 90°$ and $90° + \bar{\omega}_s$ respectively. It follows that the maximum velocity of the earth occurs at perihelion and is

$$T_{max} = F + f = \frac{h}{p}(1+e) \cong 30.29 \text{ km s}^{-1}; \qquad (5.19)$$

its minimum velocity occurs at aphelion and is

$$T_{min} = F - f = \frac{h}{p}(1-e) \cong 29.29 \text{ km s}^{-1}. \qquad (5.20)$$

Anywhere else the tangential velocity is between these values. It may be calculated from [Smart, 1960, p. 109]

$$T = [n^2 a^3 (\frac{2}{r} - \frac{1}{a})]^{\frac{1}{2}}. \qquad (5.21)$$

The annual apparent motion of the sun as seen by an observer on the earth is the reversed annual motion of the earth around the sun as explained in the foregoing.

5.212 Diurnal Motion. The apparent diurnal motion of the sun is governed by the rotation of the earth about its axis; thus the sun may be viewed as a regular star moving uniformly along the celestial parallels as described in section 3.4.

5.213 Actual Motion. The actual motion of the sun is the combination of its annual nonuniform motion on the ecliptic and its nearly uniform diurnal motion on a celestial parallel, as described above, the result being a nonuniform spiral-like motion as shown in Fig. 5.3. Around March 21 the sun is close to the vernal equinox; thus it moves near the equator with zero declination. After March 21 its declination increases until it reaches the summer solstice around June 22 where the declination is equal to the obliquity of the ecliptic or about $23\overset{\circ}{.}5$. From approximately June 22 to September 23 the declination decreases to zero, when the sun is close to the autumnal equinox, and thus it moves again near the equator. After September 23 the declination becomes negative and decreases until the sun reaches the winter solstice around December 22. From here on the negative declinations increase until approximately March 21 when the sun is again near the equator at the vernal equinox and the whole cycle is repeated. From the figure it is evident that at $0^\circ < \Phi < 90^\circ - \epsilon$ between December 22 and June 22 the length of the day increases and between June 22 and December 22 it decreases. It is also obvious that the length of the day between March 21 and September 23 is longer than 12^h, while between September 23 and March 21 it is shorter than 12^h. These relations are summarized in Table 5.2.

If the observer is at the arctic (antarctic) circle, $\pm\Phi = 90^\circ - \epsilon \cong 66\overset{\circ}{.}5$, he finds that on June 21 (December 21) the sun does not set and on December 21 (June 21) it does not rise. If he is between the arctic (antarctic) circle and the north (south) pole, he finds that as the absolute value of the latitude increases, the number of days when the sun does not set or rise increases symmetrically around June 21 (December 21) and December 21 (June 21) respectively. At the north (south) pole ($\Phi = \pm 90^\circ$) he finds the sun above the horizon all the time between March 21 and September 21 (September 21 and March 21) and below for the rest of the year.

On the equator ($\Phi = 0^\circ$) the length of the day (and night) is exactly 12^h throughout the entire year.

5.22 Universal (Solar) Time

5.221 Solar Epoch: Apparent (True) and Mean Solar, Universal, and Zonal Times; the Equation of Time. The epoch of the apparent or true solar time (TT) at any place is defined as

$$TT = h_s + 12^h \qquad (5.22)$$

TABLE 5.2 The Seasons

Approximate Epoch	The place of the sun	Coordinates of the sun				Name of the season (Northern Hemisphere)	Approximate length of the season
		λ_s	β_s	α_s	δ_s		
March 21	vernal equinox	0°	0°	0^h	0°	spring	93^d
June 22	summer solstice	90°	0°	6^h	+23°.5	summer	93^d
Sept 23	autumnal equinox	180°	0°	12^h	0°	autumn	90^d
Dec 22	winter solstice	270°	0°	18^h	-23°.5	winter	89^d
March 21	vernal equinox	0°	0°	0^h	0°		

where h_S denotes the local hour angle of the true sun. The 12^h is added for convenience so that 0^hTT should occur at night (at lower transit), conforming with civil timekeeping. The true solar time may be determined, for example, from altitude and azimuth observations on the sun and converting the properly reduced results to its hour angle.

Kepler's second law implies that the motion of the true sun is nonuniform on the ecliptic (strictly speaking it does not even occur in the ecliptic). Due partly to this and partly to the fact that it does not move along the celestial equator, its hour angle does not increase uniformly. Since the uniformity of the physical phenomenon associated with a time system is a prerequi-

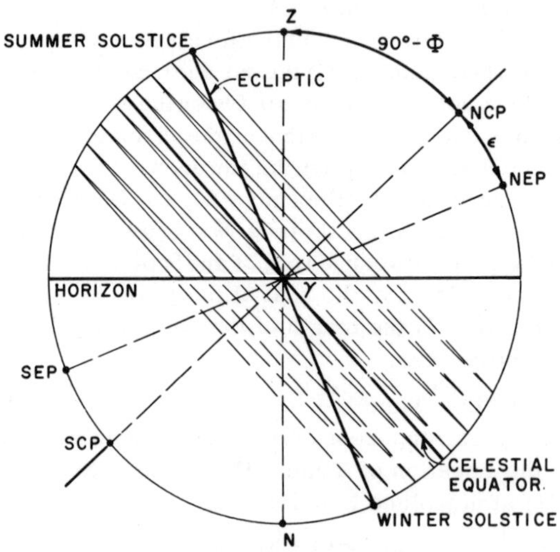

Fig. 5.3 Apparent solar motion

site, it is not advisable to use the true sun as a means for precise time-keeping. It is replaced by a fiducial point called the 'fictitious sun,' which is characterized by a uniform sidereal motion along the equator at a rate which differs from the mean rate of the annual motion of the sun along the ecliptic only by the amount of a slight secular acceleration of the sun. Relative to any meridian, this point has a diurnal motion in hour angle virtually the same as the average diurnal motion of the sun and with only very slight inequalities due to the variations of the local meridian (polar motion) and variations of the rate of rotation of the earth. The time system known as <u>mean solar time</u> (MT) is defined from this fiducial point by the following relation:

$$MT = h_M + 12^h \qquad (5.23)$$

where h_M is the local hour angle of the fictitious sun.

If the hour angles of the true and fictitious suns are referred to the Greenwich mean astronomic meridian, the resulting times are termed <u>Greenwich apparent (true) solar time</u> (GTT) and <u>universal time</u> (UT) respectively. Thus

$$GTT = h_S^G + 12^h, \qquad (5.24)$$
$$UT = h_M^G + 12^h. \qquad (5.25)$$

These and the previous relations are shown in Fig. 5.4. From that figure it is evident that,

$$TT = GTT + \Lambda, \qquad (5.26)$$
$$MT = UT + \Lambda, \qquad (5.27)$$
$$h_S = h_S^G + \Lambda, \qquad (5.28)$$
$$h_M = h_M^G + \Lambda, \qquad (5.29)$$

or that

$$\Lambda = TT - GTT = MT - UT = h_S - h_S^G = h_M - h_M^G. \qquad (5.30)$$

The right ascension of the fictitious sun, whose motion thus defines the mean solar time system, is given by the expression [Newcomb, 1895]:

$$\alpha_M = 18^h 38^m 45^s\!.836 + 8{,}640{,}184^s\!.542\, t_M + 0^s\!.0929\, t_M^2 \qquad (5.31)$$

where t_M is the number of Julian centuries of 36525 mean solar days (see section 5.222) which, at midnight beginning the day, have elapsed since the <u>standard epoch of UT</u>:

1900 January 0.5 UT (see section 5.232).

The difference between the true and mean solar times (see Fig. 5.4) at a given instant is termed the <u>equation of time</u> (Eq. T). It is tabulated in the almanacs for navigators and surveyors who use the sun for time determination of limited accuracy (e.g., Eq. T for 1965-1966 is tabulated in Circular No. 97 of the U. S. Naval Observatory; tables for future years

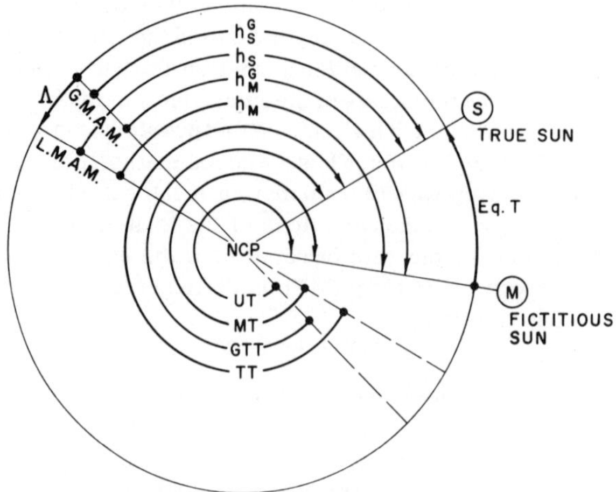

Fig. 5.4 Solar times

appear in its annual publication 'Astronomical Phenomena'). Thus

$$\text{Eq. T} = \text{TT} - \text{MT} = \text{GTT} - \text{UT} = h_S - h_M = h_S^G - h_M^G\,. \tag{5.32}$$

The variation of the Eq. T for one year is illustrated in Fig. 5.5.

Since the fictitious sun is not actually observable, mean solar time in the navigational and surveyor practice is ascertained by determining the true solar time and correcting it with the tabulated equation of time:

$$\text{MT} = \text{TT} - \text{Eq. T}, \tag{5.33}$$

or

$$\text{UT} = \text{GTT} - \text{Eq. T} = \text{TT} - \Lambda - \text{Eq. T}. \tag{5.34}$$

For geodetic purposes sun observations are not sufficiently accurate; therefore a definition is formulated as a relation to sidereal time, and in practice mean solar time is obtained through this intermediary by observing the diurnal motion of the stars (see section 5.31).

In civil timekeeping confusion would arise if every place would use the mean solar time appropriate to its meridian. To avoid this, a uniform system of timekeeping, the standard or zone time (ZT), is chosen, corresponding to the mean solar time of a particular meridian (the standard meridian) which is uniformly used within certain limits of longitude or geographical area. Within these areas the same ZT is kept and not the local MT of each place. The world is divided into twenty-four zones each having a width of fifteen degrees (one hour) of longitude, in each of which the same standard

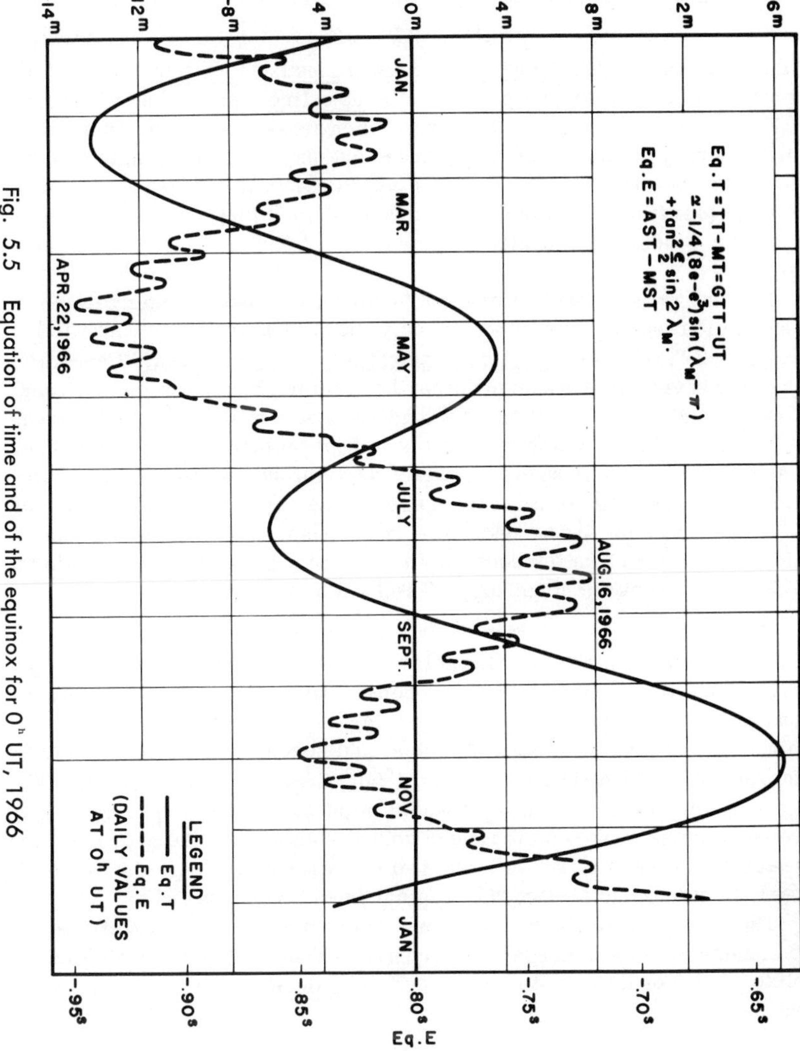

Fig. 5.5 Equation of time and of the equinox for 0^h UT, 1966

time is kept (sometimes the border of a zone is slightly modified to make it correspond to certain geographical boundaries). The meridian of Greenwich is taken as the center of the system and of zone 0. Zones to the east are numbered $\Delta Z = -1$, -2, etc. and those to the west $\Delta Z = +1$, $+2$, etc., according to the number of hours to be added to ZT to obtain UT. Thus,

$$UT = ZT + \Delta Z \qquad (5.35)$$

where ΔZ is called the zonal correction or zonal description. The twelfth zone is divided into two parts by the date line, that to the west being $\Delta Z = -12$ and that to the east $\Delta Z = +12$. When crossing this line on a westerly course, the date must be advanced one day. Zone time may also be indicated by letters. UT is Z (zero) and the zones to the east are lettered A to M (omitting J) and those to the west N to Y (see Fig. 5.6). Zones may also be identified locally by names, e.g., Atlantic, Eastern, Mountain, etc.

In many geographical areas the ZTs for winter and summer differ for economic reasons. The latter, which is usually one hour in advance of the regular ZT, is termed daylight saving time or summer time. Generally, daylight saving time is kept in the northern hemisphere between April and October and in the southern hemisphere between October and March.

Using the ZT system the same time is kept on land and sea throughout each zone (except during the periods of daylight saving time), and in different zones the times differ from one another by an integral number of hours, the minutes and seconds in all zones remaining the same. Exceptions to this are in a few countries (e.g., Iran, India, Venezuela) where ZT differs from the neighboring ZTs by 0.5 or 1.5 hours.

5.222 Solar Interval: Mean Solar Day, Tropical, Sidereal and Julian Years. The basic interval in the solar time system is the mean solar day, defined as the interval between two consecutive transits of the fictitious sun over a meridian (MT or UT system). The mean solar day conventionally is divided into hours, minutes, and seconds. The mean solar interval is denoted by (M); thus $1^d(M) = 24^h(M)$, $1^h(M) = 60^m(M)$, $1^m(M) = 60^s(M)$.

The time required by the fiducial point (fictitious sun) to make two consecutive passages on the mean vernal equinox, or the average time required by the true sun to make two consecutive passages on the true vernal equinox, is termed the tropical (mean solar) year.

The time required by the fictitious sun to make a complete circuit on the equator, or the average time required by the true sun to make a complete circuit on the ecliptic, is termed the sidereal year. With sufficient accuracy,

$$\begin{aligned} 1 \text{ tropical year} &= 365^d\!.24219879 \text{ (M)} \\ &= 365^d 05^h 48^m 45^s\!.9754 \text{ (M)}, \end{aligned} \qquad (5.36)$$

$$\begin{aligned} 1 \text{ sidereal year} &= 365^d\!.25636042 \text{ (M)} \\ &= 365^d 06^h 09^m 09^s\!.5403 \text{ (M)}. \end{aligned} \qquad (5.37)$$

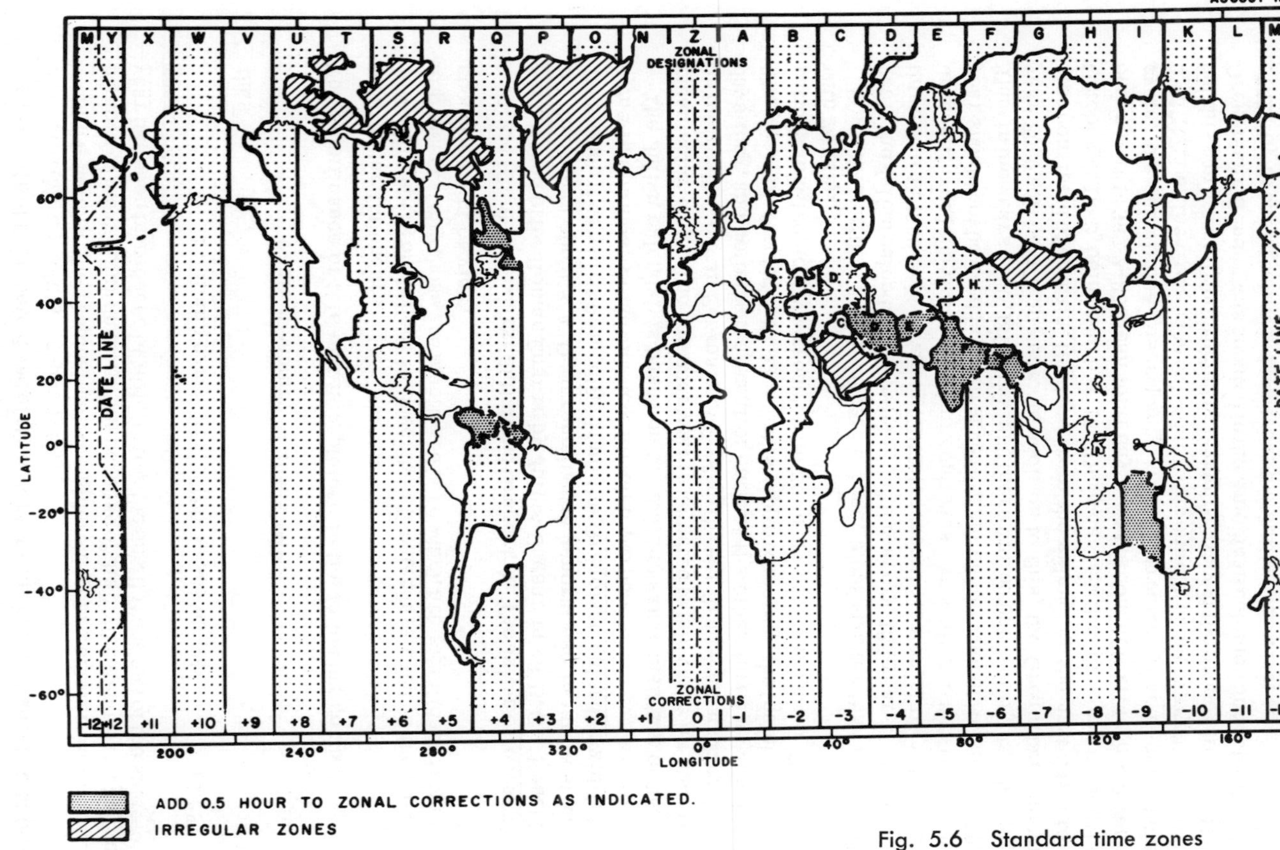

Fig. 5.6 Standard time zones

The length of the tropical and of the sidereal year changes about $-0^s.53$(M) and $+0^s.01$(M) a century respectively.

Since neither the tropical nor the sidereal year contains an integral number of mean solar days, for the purpose of the civil calendar the Julian year has been introduced by eliminating the small fractions of the sidereal year,

$$1 \text{ Julian year} = 365^d.25(\mathrm{M}) = 365^d 06^h(\mathrm{M}); \tag{5.38}$$

thus

$$1 \text{ Julian century} = 36525^d(\mathrm{M}). \tag{5.39}$$

The significance of this unit is explained in the following sections.

5.23 Solar Calendars

No attempt is made here to describe all calendars used in the past or at present. Only those are mentioned here which are significant from the geodetic point of view.

5.231 Civil Calendars: Julian and Gregorian. The Julian calendar was established in the Roman Empire by Julius Caesar in 46 B. C. Reaching its final form about 8 A. D. it remained in general use in the west until 1582 when it was further modified into the Gregorian calendar which has come into almost worldwide use for civil purposes.

The Julian calendar is based on the Julian year consisting of 365.25 mean solar days. The calendar year is adjusted to this value by inserting an intercalary day in February every fourth year (leap year) making it 366 days long, and defining the length of each of the other three as 365 days. The year is divided into 12 months (approximating the number of lunations in a tropical year) each with 30 or 31 days except for February with 28 (29).

Since the length of a tropical year is about $0^d.008$ shorter than the Julian year, the Julian calendar was gaining one day every 125 years. Realizing this error in 1582, Pope Gregory XIII introduced his intercalary rule which dropped three leap years each four centuries. According to this rule, leap year should not be used in the first year of any century not divisible by 400. Thus in the years 1700, 1800, and 1900 there was no intercalary day, but the year 2000 will be a leap year. In addition to this, the Gregorian reform also consisted of omitting ten days from the calendar, the next day after October 4, 1582, being October 15. The mean length of the Gregorian calendar year is 365.2425 mean solar days. At the completion of the 400 year cycle, the cumulative discrepancy with the tropical year is only about $2^h 53^m$(M), i.e., the Gregorian calendar gains one day about every 3300 years, but this can be easily corrected by dropping a day at the proper time. The Gregorian calendar was adopted in the American colonies on September 3, 1752, by dropping eleven days from the Julian calendar.

For the description of other calendars see the appropriate section in [Nautical Almanac Offices, 1961, pp. 407-422].

5.232 Astronomic Calendars: Astronomic and Julian Dates; Julian Day Numbers. The Gregorian calendar is also used in astronomy or in other sciences with deviation only in notation from the civilian practice. The astronomic year commences at 0^hMT (UT) on December 31 of the previous year, which epoch is denoted by the astronomic date January $0^d.0$ MT (UT). According to this notation, for example,

$$1966 \text{ January } 1\ 12^h \text{ UT} \equiv 1966 \text{ January } 1^d.5 \text{ UT},$$
$$1966 \text{ December } 31\ 18^h \text{ UT} \equiv 1967 \text{ January } 0^d.75 \text{ UT}.$$

As has been mentioned earlier, in certain applications it is convenient to express the epoch as a certain number of days and a fraction of a day after a fundamental epoch sufficiently in the past to precede the historical period. The number which denotes a day in this continuous count in the solar time system is the Julian day number. The Julian date (JD) commences with 0 Julian day number at 12^hUT on January 1, 4713 B.C. The JD thus denotes the length of time that has elapsed at 12^hUT on the day designated since this epoch. The Julian date of the standard epoch of UT, for example, is

$$1900 \text{ January } 0^d.5 \text{ UT} = \text{JD } 2{,}415{,}020.0.$$

Julian day numbers are listed for the last day of every month from 1900 to 1999, commencing at 12^hUT, in [Nautical Almanac Offices, 1961, pp. 438-439]. The calendar sections of the 'American Ephemeris and Nautical Almanac' list the Julian dates for every day of the current year. Sample pages are shown in Tables 5.3 and 5.4 respectively. Note that the Julian day begins at noon.

5.3 Conversion of Time: Sidereal - Universal

5.31 Conversion of Epoch

The fundamental relations upon which the conversion of epoch is based are equations (3.9) and (5.23). Using these equations it is evident that

$$\text{MT} = 12^h + h_M = 12^h + \text{MST} - \alpha_M,$$

or

$$\text{MST} = \text{MT} + \alpha_M - 12^h,$$

from which, with equation (5.31),

$$\text{MT} = \text{MST} - 6^h38^m45^s.836 - 8{,}640{,}184^s.542\, t_M - 0^s.0929\, t_M^2, \qquad (5.40)$$
$$\text{MST} = \text{MT} + 6^h38^m45^s.836 + 8{,}640{,}184^s.542\, t_M + 0^s.0929\, t_M^2. \qquad (5.41)$$

These two equations define the epoch of MST in terms of MT and the reverse at any meridian.

TABLE 5.3 Julian Day Numbers, 1950–1999

	OF DAY COMMENCING AT GREENWICH NOON ON:												
Year		Jan. 0	Feb. 0	Mar. 0	Apr. 0	May 0	June 0	July 0	Aug. 0	Sept. 0	Oct. 0	Nov. 0	Dec. 0
1950	243	3282	3313	3341	3372	3402	3433	3463	3494	3525	3555	3586	3616
1951		3647	3678	3706	3737	3767	3798	3828	3859	3890	3920	3951	3981
1952		4012	4043	4072	4103	4133	4164	4194	4225	4256	4286	4317	4347
1953		4378	4409	4437	4468	4498	4529	4559	4590	4621	4651	4682	4712
1954		4743	4774	4802	4833	4863	4894	4924	4955	4986	5016	5047	5077
1955	243	5108	5139	5167	5198	5228	5259	5289	5320	5351	5381	5412	5442
1956		5473	5504	5533	5564	5594	5625	5655	5686	5717	5747	5778	5808
1957		5839	5870	5898	5929	5959	5990	6020	6051	6082	6112	6143	6173
1958		6204	6235	6263	6294	6324	6355	6385	6416	6447	6477	6508	6538
1959		6569	6600	6628	6659	6689	6720	6750	6781	6812	6842	6873	6903
1960	243	6934	6965	6994	7025	7055	7086	7116	7147	7178	7208	7239	7269
1961		7300	7331	7359	7390	7420	7451	7481	7512	7543	7573	7604	7634
1962		7665	7696	7724	7755	7785	7816	7846	7877	7908	7938	7969	7999
1963		8030	8061	8089	8120	8150	8181	8211	8242	8273	8303	8334	8364
1964		8395	8426	8455	8486	8516	8547	8577	8608	8639	8669	8700	8730
1965	243	8761	8792	8820	8851	8881	8912	8942	8973	9004	9034	9065	9095
1966		9126	9157	9185	9216	9246	9277	9307	9338	9369	9399	9430	9460
1967		9491	9522	9550	9581	9611	9642	9672	9703	9734	9764	9795	9825
1968	243	9856	9887	9916	9947	9977	*0008	*0038	*0069	*0100	*0130	*0161	*0191
1969	244	0222	0253	0281	0312	0342	0373	0403	0434	0465	0495	0526	0556
1970	244	0587	0618	0646	0677	0707	0738	0768	0799	0830	0860	0891	0921
1971		0952	0983	1011	1042	1072	1103	1133	1164	1195	1225	1256	1286
1972		1317	1348	1377	1408	1438	1469	1499	1530	1561	1591	1622	1652
1973		1683	1714	1742	1773	1803	1834	1864	1895	1926	1956	1987	2017
1974		2048	2079	2107	2138	2168	2199	2229	2260	2291	2321	2352	2382
1975	244	2413	2444	2472	2503	2533	2564	2594	2625	2656	2686	2717	2747
1976		2778	2809	2838	2869	2899	2930	2960	2991	3022	3052	3083	3113
1977		3144	3175	3203	3234	3264	3295	3325	3356	3387	3417	3448	3478
1978		3509	3540	3568	3599	3629	3660	3690	3721	3752	3782	3813	3843
1979		3874	3905	3933	3964	3994	4025	4055	4086	4117	4147	4178	4208
1980	244	4239	4270	4299	4330	4360	4391	4421	4452	4483	4513	4544	4574
1981		4605	4636	4664	4695	4725	4756	4786	4817	4848	4878	4909	4939
1982		4970	5001	5029	5060	5090	5121	5151	5182	5213	5243	5274	5304
1983		5335	5366	5394	5425	5455	5486	5516	5547	5578	5608	5639	5669
1984		5700	5731	5760	5791	5821	5852	5882	5913	5944	5974	6005	6035
1985	244	6066	6097	6125	6156	6186	6217	6247	6278	6309	6339	6370	6400
1986		6431	6462	6490	6521	6551	6582	6612	6643	6674	6704	6735	6765
1987		6796	6827	6855	6886	6916	6947	6977	7008	7039	7069	7100	7130
1988		7161	7192	7221	7252	7282	7313	7343	7374	7405	7435	7466	7496
1989		7527	7558	7586	7617	7647	7678	7708	7739	7770	7800	7831	7861
1990	244	7892	7923	7951	7982	8012	8043	8073	8104	8135	8165	8196	8226
1991		8257	8288	8316	8347	8377	8408	8438	8469	8500	8530	8561	8591
1992		8622	8653	8682	8713	8743	8774	8804	8835	8866	8896	8927	8957
1993		8988	9019	9047	9078	9108	9139	9169	9200	9231	9261	9292	9322
1994		9353	9384	9412	9443	9473	9504	9534	9565	9596	9626	9657	9687
1995	244	9718	9749	9777	9808	9838	9869	9899	9930	9961	9991	*0022	*0052
1996	245	0083	0114	0143	0174	0204	0235	0265	0296	0327	0357	0388	0418
1997		0449	0480	0508	0539	0569	0600	0630	0661	0692	0722	0753	0783
1998		0814	0845	0873	0904	0934	0965	0995	1026	1057	1087	1118	1148
1999	245	1179	1210	1238	1269	1299	1330	1360	1391	1422	1452	1483	1513

TABLE 5.4 Sample Page of the AENA

CALENDAR, 1967 3

Day of Month	JULY		AUGUST		SEPTEMBER		OCTOBER		NOVEMBER		DECEMBER	
	Day of Week	Julian Date	Day of Week	Julian Date	Day of Week	Julian Date	Day of Week	Julian Date	Day of Week	Julian Date	Day of Week	Julian Date
		2439		2439		2439		2439		2439		2439
1·0	S.	672·5	Tu.	703·5	F.	734·5	**S.**	764·5	W.	795·5	F.	825·5
2·0	**S.**	673·5	W.	704·5	S.	735·5	M.	765·5	Th.	796·5	S.	826·5
3·0	M.	674·5	Th.	705·5	**S.**	736·5	Tu.	766·5	F.	797·5	**S.**	827·5
4·0	Tu.	675·5	F.	706·5	M.	737·5	W.	767·5	S.	798·5	M.	828·5
5·0	W.	676·5	S.	707·5	Tu.	738·5	Th.	768·5	**S.**	799·5	Tu.	829·5
6·0	Th.	677·5	**S.**	708·5	W.	739·5	F.	769·5	M.	800·5	W.	830·5
7·0	F.	678·5	M.	709·5	Th.	740·5	S.	770·5	Tu.	801·5	Th.	831·5
8·0	S.	679·5	Tu.	710·5	F.	741·5	**S.**	771·5	W.	802·5	F.	832·5
9·0	**S.**	580·5	W.	711·5	S.	742·5	M.	772·5	Th.	803·5	S.	833·5
10·0	M.	681·5	Th.	712·5	**S.**	743·5	Tu.	773·5	F.	804·5	**S.**	834·5
11·0	Tu.	682·5	F.	713·5	M.	744·5	W.	774·5	S.	805·5	M.	835·5
12·0	W.	683·5	S.	714·5	Tu.	745·5	Th.	775·5	**S.**	806·5	Tu.	836·5
13·0	Th.	684·5	**S.**	715·5	W.	746·5	F.	776·5	M.	807·5	W.	837·5
14·0	F.	685·5	M.	716·5	Th.	747·5	S.	777·5	Tu.	808·5	Th.	838·5
15·0	S.	686·5	Tu.	717·5	F.	748·5	**S.**	778·5	W.	809·5	F.	839·5
16·0	**S.**	687·5	W.	718·5	S.	749·5	M.	779·5	Th.	810·5	S.	840·5
17·0	M.	688·5	Th.	719·5	**S.**	750·5	Tu.	780·5	F.	811·5	**S.**	841·5
18·0	Tu.	689·5	F.	720·5	M.	751·5	W.	781·5	S.	812·5	M.	842·5
19·0	W.	690·5	S.	721·5	Tu.	752·5	Th.	782·5	**S.**	813·5	Tu.	843·5
20·0	Th.	691·5	**S.**	722·5	W.	753·5	F.	783·5	M.	814·5	W.	844·5
21·0	F.	692·5	M.	723·5	Th.	754·5	S.	784·5	Tu.	815·5	Th.	845·5
22·0	S.	693·5	Tu.	724·5	F.	755·5	**S.**	785·5	W.	816·5	F.	846·5
23·0	**S.**	694·5	W.	725·5	S.	756·5	M.	786·5	Th.	817·5	S.	847·5
24·0	M.	695·5	Th.	726·5	**S.**	757·5	Tu.	787·5	F.	818·5	**S.**	848·5
25·0	Tu.	696·5	F.	727·5	M.	758·5	W.	788·5	S.	819·5	M.	849·5
26·0	W.	697·5	S.	728·5	Tu.	759·5	Th.	789·5	**S.**	820·5	Tu.	850·5
27·0	Th.	698·5	**S.**	729·5	W.	760·5	F.	790·5	M.	821·5	W.	851·5
28·0	F.	699·5	M.	730·5	Th.	761·5	S.	791·5	Tu.	822·5	Th.	852·5
29·0	S.	700·5	Tu.	731·5	F.	762·5	**S.**	792·5	W.	823·5	F.	853·5
30·0	**S.**	701·5	W.	732·5	S.	763·5	M.	793·5	Th.	824·5	S.	854·5
31·0	M.	702·5	Th.	733·5			Tu.	794·5			**S.**	855·5

The Julian Day begins at noon.

The fraction of the year, τ, measured from the beginning of the Besselian solar year, is given on pages 258–272. For the first half of the year, on pages 258–264, it is measured from 1967·0 or 1967 January $1^{d}{\cdot}041$; for the second half of the year, on pages 266–272, it is measured from 1968·0 or 1968 January $1^{d}{\cdot}283$.

For practical computations, equations (5.40) and (5.41) are not convenient unless the quantity $6^h 38^m 45^s.836 + 8,640,184^s.542\, t_M^2 + 0^s.0929\, t_M$ is tabulated. It is obvious that this quantity corresponds to MST at 0^hMT or to -MT at 0^h MST (to the hour angle of the mean vernal equinox or to the MT of its upper transit, referred to a certain meridian, respectively). In the 'American Ephemeris and Nautical Almanac' these quantities are tabulated for the Greenwich meridian (Greenwich hour angle of the mean vernal equinox at 0^h UT, and the UT of its upper transit with arguments JD and GSD respectively) for every day in the tables of universal and sidereal times. In these tables the corresponding apparent times are also given. A sample page is shown in Table 5.5. In order that these tables may be utilized, the conversion is generally executed at Greenwich by computing the Greenwich mean times from the local mean times and then retransferring to the local meridian.

The conversion from mean to apparent times is done by applying the equation of the equinox (in the sidereal system) or the equation of time (in the solar system) according to equations (5.1) or (5.32) respectively.

The conversion between the Julian date and the Greenwich sidereal date with adequate approximation to give proper day numbers may be performed by the following expressions:

$$\mathrm{GSD} = 0.671 + 1.0027379093\ \mathrm{JD}, \tag{5.42}$$

$$\mathrm{JD} = -0.669 + 0.9972695664\ \mathrm{GSD}. \tag{5.43}$$

The first coefficients on the right side of these equations stem from the initial difference between JD and GSD; the second coefficients from the ratio, sidereal to solar interval, explained in the next section.

5.32 Conversion of Interval

In section 5.222 the length of a mean solar day of 86400^s (M) has been defined as the interval between two consecutive transits of the fictitious sun over a meridian. Based on equation (5.40), this definition is the equivalent of saying that the length of a mean solar day is the interval between two instants at which the mean vernal equinox reaches the tabulated hour angles (at 0^h MT) for two consecutive dates. It follows that the hour angle, which the equinox describes during one mean solar day, consists of a complete circuit of 24^h(S) plus an excess angle β which at 12^hUT is equal to the rate of change of MST during a mean solar day (Fig. 5.7). From equation (5.41) this rate of change per day is

$$\frac{d(\mathrm{MST})}{dt_M} = \frac{8,640,184^s.542 + 0^s.1858\, t_M}{36525}$$

$$= 236^s.5553605 + 0^s.000005087\, t_M.$$

Thus the interval of mean sidereal time in a mean solar day is

TABLE 5.5 Sample Page of the AENA

UNIVERSAL AND SIDEREAL TIMES, 1967

Date 0ʰ U.T.		Julian Date	Sidereal Time H.A. of First Point of Aries		Equation of Equi-noxes	G.S.D. 0ʰ S.T.	Universal Time Transit of First Point of Aries		
			Apparent	Mean				Apparent	Mean
		2439	h m s	s	s	2446		d h m s	s
Jan.	0	490·5	6 35 53·152	53·827	−0·675	171·0	Jan.	0 17 21 15·794	15·122
	1	491·5	6 39 49·709	50·382	·673	172·0		1 17 17 19·885	19·213
	2	492·5	6 43 46·263	46·937	·675	173·0		2 17 13 23·978	23·303
	3	493·5	6 47 42·815	43·493	·677	174·0		3 17 09 28·071	27·394
	4	494·5	6 51 39·368	40·048	·680	175·0		4 17 05 32·164	31·484
	5	495·5	6 55 35·922	36·603	−0·681	176·0		5 17 01 36·253	35·575
	6	496·5	6 59 32·479	33·159	·680	177·0		6 16 57 40·340	39·665
	7	497·5	7 03 29·040	29·714	·675	178·0		7 16 53 44·423	43·756
	8	498·5	7 07 25·603	26·270	·667	179·0		8 16 49 48·505	47·846
	9	499·5	7 11 22·167	22·825	·658	180·0		9 16 45 52·586	51·937
	10	500·5	7 15 18·732	19·380	−0·648	181·0		10 16 41 56·668	56·028
	11	501·5	7 19 15·296	15·936	·640	182·0		11 16 38 00·752	00·118
	12	502·5	7 23 11·857	12·491	·634	183·0		12 16 34 04·838	04·209
	13	503·5	7 27 08·416	09·046	·630	184·0		13 16 30 08·927	08·299
	14	504·5	7 31 04·972	05·602	·630	185·0		14 16 26 13·018	12·390
	15	505·5	7 35 01·526	02·157	−0·631	186·0		15 16 22 17·112	16·480
	16	506·5	7 38 58·078	58·712	·634	187·0		16 16 18 21·206	20·571
	17	507·5	7 42 54·630	55·268	·638	188·0		17 16 14 25·300	24·661
	18	508·5	7 46 51·182	51·823	·642	189·0		18 16 10 29·394	28·752
	19	509·5	7 50 47·734	48·379	·644	190·0		19 16 06 33·485	32·842
	20	510·5	7 54 44·289	44·934	−0·645	191·0		20 16 02 37·575	36·933
	21	511·5	7 58 40·846	41·489	·643	192·0		21 15 58 41·662	41·023
	22	512·5	8 02 37·406	38·045	·639	193·0		22 15 54 45·746	45·114
	23	513·5	8 06 33·968	34·600	·632	194·0		23 15 50 49·829	49·204
	24	514·5	8 10 30·532	31·155	·623	195·0		24 15 46 53·910	53·295
	25	515·5	8 14 27·097	27·711	−0·614	196·0		25 15 42 57·993	57·386
	26	516·5	8 18 23·660	24·266	·606	197·0		26 15 39 02·077	01·476
	27	517·5	8 22 20·220	20·821	·601	198·0		27 15 35 06·165	05·567
	28	518·5	8 26 16·777	17·377	·600	199·0		28 15 31 10·257	09·657
	29	519·5	8 30 13·330	13·932	·603	200·0		29 15 27 14·352	13·748
	30	520·5	8 34 09·880	10·488	−0·607	201·0		30 15 23 18·447	17·838
	31	521·5	8 38 06·431	07·043	·612	202·0		31 15 19 22·542	21·929
Feb.	1	522·5	8 42 02·982	03·598	·616	203·0	Feb.	1 15 15 26·634	26·019
	2	523·5	8 45 59·537	60·154	·617	204·0		2 15 11 30·724	30·110
	3	524·5	8 49 56·094	56·709	·615	205·0		3 15 07 34·811	34·200
	4	525·5	8 53 52·654	53·264	−0·610	206·0		4 15 03 38·895	38·291
	5	526·5	8 57 49·216	49·820	·604	207·0		5 14 59 42·979	42·381
	6	527·5	9 01 45·778	46·375	·597	208·0		6 14 55 47·063	46·472
	7	528·5	9 05 42·340	42·930	·591	209·0		7 14 51 51·149	50·563
	8	529·5	9 09 38·899	39·486	·586	210·0		8 14 47 55·236	54·653
	9	530·5	9 13 35·457	36·041	−0·585	211·0		9 14 43 59·327	58·744
	10	531·5	9 17 32·011	32·597	·586	212·0		10 14 40 03·420	02·834
	11	532·5	9 21 28·563	29·152	·589	213·0		11 14 36 07·515	06·925
	12	533·5	9 25 25·113	25·707	·594	214·0		12 14 32 11·612	11·015
	13	534·5	9 29 21·662	22·263	·601	215·0		13 14 28 15·709	15·106
	14	535·5	9 33 18·211	18·818	−0·607	216·0		14 14 24 19·805	19·196
	15	536·5	9 37 14·761	15·373	−0·613	217·0		15 14 20 23·900	23·287

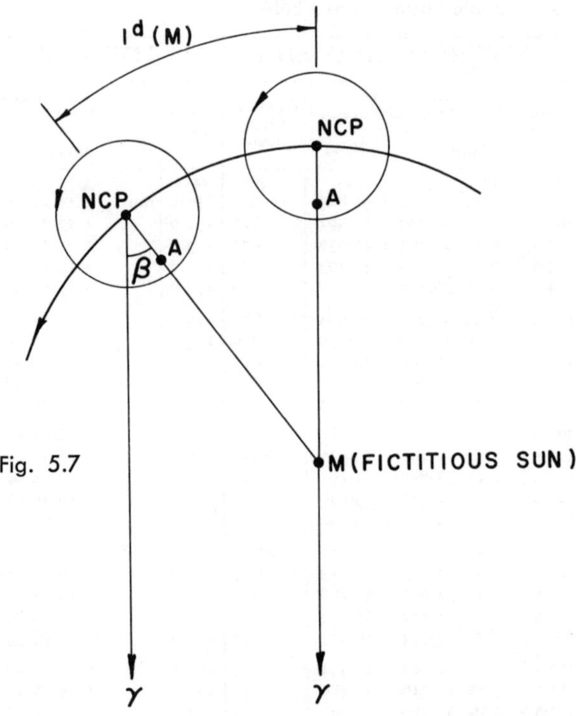

Fig. 5.7

$$1^d(M) = 86400^s(M) = [86636\overset{s}{.}5553605 + 0\overset{s}{.}000005087t_M](S).$$

The ratio of a sidereal day of 86400^s (S) to this interval thus is

$$\frac{1^d(S)}{1^d(M)} = 0.997269566414 - 0.586 \times 10^{-10}t_M = \frac{(S)}{(M)}. \tag{5.44}$$

Inversely

$$\frac{1^d(M)}{1^d(S)} = 1.002737909265 + 0.589 \times 10^{-10}t_M = \frac{(M)}{(S)}. \tag{5.45}$$

By use of these ratios, neglecting the inappreciable secular variations, the following equivalent measures of intervals are obtained:

$$\begin{aligned}
1^d\,(S) &= 23^h56^m04\overset{s}{.}09054\ (M), & 1^d(M) &= 24^h03^m56\overset{s}{.}55536\ (S),\\
1^h\,(S) &= 59^m50\overset{s}{.}17044\ (M), & 1^h(M) &= 1^h00^m09\overset{s}{.}85647\ (S),\\
1^m\,(S) &= 59\overset{s}{.}83617\ (M), & 1^m(M) &= 1^m00\overset{s}{.}16427\ (S),\\
1^s\,(S) &= 0\overset{s}{.}99727\ (M), & 1^s(M) &= 1\overset{s}{.}00273\ (S).
\end{aligned}$$

5.33 Numerical Examples

Examples 5.1 - 5.4 were computed to illustrate the computational steps necessary in the various time conversions described in sections 5.31 and 5.32. Examples 5.1 and 5.3 show conversions when conversion tables are not available, while the other examples use such tables. In all examples

EXAMPLE 5.1

Conversion of True Solar to Apparent Sidereal Time Without Conversion Tables

1	TT	$15^h33^m19^s.000$	December 14, 1966
2	Λ	-5 32 09.303	
3	GTT = TT - Λ	21 05 28.303	
4	Eq. T at 0^hUT	05 41.970	From U.S. Naval Obs. Circular No. 97
5	Approximate UT (first)	20 59 46.333	(3) - (4)
6	Δ(Eq. T) for approximate UT (first)	-25.090	From U.S. Naval Obs. Circular No. 97
7	Approximate UT (second)	21 00 11.423	(5) - (6)
8	Δ(Eq. T) for approximate UT (second)	-25.099	From U.S. Naval Obs. Circular No. 97
9	UT = GTT - Eq. T	21 00 11.432	(3) - (4) - (8)
10	JD at 1966 Dec. 14, $21^h00^m11^s.432$ UT	2439474.375132	From AENA, 1966, p. 17
11	JD at 1900 January $0^d.5$ UT	2415020.0	Fundamental epoch of UT
12	t_M (mean solar days)	24454.375132	(10) - (11)
13	t_M (Julian centuries)	0.6695243020	(12)/36525
14	$(GMST)_o = 6^h38^m45^s.836$	$6^h38^m45^s.836$	$(GMST)_o$ = GMST at 0^hUT, from
	$+ 8640184^s.542t_M$	22 53 33.525	equation (5.41)
	$+ 0^s.0929t_M^2$	0.042	
15	GMST = UT + $(GMST)_o$	26 32 30.835	
16	Eq. E	-0.732	From AENA, 1966, p. 17
17	GAST = GMST + Eq. E	26 32 30.103	
18	Λ	-5 32 09.303	
19	AST = GAST + Λ	$21^h00^m20^s.800$	December 14, 1966

EXAMPLE 5.2

Conversion of True Solar to Apparent Sidereal Time with Tables

1	TT	$15^h 33^m 19^s.000$	December 14, 1966
2	Λ	-5 32 09.303	
3	GTT = TT - Λ	21 05 28.303	
4	Eq. T at 0^h UT	05 41.970	From U. S. Naval Obs. Circular No. 97
5	Approximate UT (first)	20 59 46.333	(3) - (4)
6	Δ(Eq.T) for approx. UT (first)	-25.090	From U. S Naval Obs. Circular No. 97
7	Approximate UT (second)	21 00 11.423	(5) - (6)
8	Δ(Eq.T) for approx. UT (second)	-25.099	From U. S. Naval Obs. Circular No. 97
9	UT = GTT - Eq.T	21 00 11.432	(3) - (4) - (8)
10	GAST at 1966 Dec. 14, 0^h UT	5 28 51.646	From AENA, 1966, p. 17
11	21^h(S) - 21^h(M)	3 26.986	Conversion of mean solar to mean
	$11^s.432$(S) - $11^s.432$(M)	0.031	sidereal interval
12	Δ(Eq. E) for UT	0.008	From AENA, 1966, p. 17
13	GAST	26 32 30.103	(9) + (10) + (11) + (12)
14	Λ	-5 32 09.303	
15	AST = GAST + Λ	$21^h 00^m 20^s.800$	December 14, 1966

EXAMPLE 5.3

Conversion of Apparent Sidereal to True Solar Time Without Conversion Tables

1	AST	$21^h 00^m 20^s.800$	December 14, 1966
2	Λ	-5 32 09.303	
3	GAST = AST - Λ	26 32 30.103	
4	Eq. E at 1966 December 14, $26^h 32^m 30^s.103$ GAST	-0.732	From AENA, 1966, p. 17
5	GMST = GAST - Eq. E	26 32 30.835	
6	GSD at 1966 Dec. 14, $26^h 32^m 30^s.835$ GMST	2446154.105913	
7	JD = 0.9972695664 × GSD	2439475.044133	Equation (5.43)
	-0.669	-0.669	
	=	2439474.375133	
8	JD at 1900 January $0^d.5$ UT	2415020.0	
9	t_M (mean solar days)	24454.375133	(7) - (8)
10	t_M (Julian centuries)	0.6695243020	(9)/36525
11	$(UT)_\circ = -6^h 38^m 45^s.836$	$-6^h 38^m 45^s.836$	$(UT)_\circ$ = UT at 0^h GMST, from
	$-8640184^s.542 t_M$	-22 53 33.525	equation (5.40)
	$-0^s.0929 t_M^2$	0.042	
12	UT = GMST + $(UT)_\circ$	21 00 11.432	(5) + (11)
13	Eq. T	05 16.871	From U. S. Naval Obs. Circular No. 97
14	GTT = UT + Eq. T	21 05 28.303	
15	Λ	-5 32 09.303	
16	TT = GTT + Λ	$15^h 33^m 19^s.000$	December 14, 1966

EXAMPLE 5.4

Conversion of Apparent Sidereal to True Solar Time with Tables

1	AST	$21^h 00^m 20^s.800$	December 14, 1966
2	Λ	-5 32 09.303	
3	GAST = AST - Λ	26 32 30.103	
4	GAST at 0^h UT	5 28 51.646	From AENA, 1966, p. 17
5	Apparent Sidereal Interval	21 03 38.457	(3) - (4)
6	Δ(Eq. E) for $21^h 03^m 38^s.457$	-0.008	From AENA, 1966, p. 17
7	Mean Sidereal Interval	21 03 38.449	(5) + (6)
8	$21^h 03^m$(M) - $21^h 03^m$(S) $38^s.419$(M) - $38^s.419$(S)	-03 26.912 -0.105	Conversion of mean sidereal to mean solar interval
9	UT	21 00 11.432	(7) + (8)
10	Eq.T	05 16.871	From U.S. Naval Obs. Circular No. 97
11	GTT = UT + Eq.T	21 05 28.303	
12	Λ	-5 32 09.303	
13	TT = GTT + Λ	$15^h 33^m 19^s.000$	December 14, 1966

the initial data is

$$\Lambda = -5^h32^m09^s.303,$$
$$TT = 15^h33^m19^s.000.$$

In practice similar conversions occur frequently with the possible exception that instead of TT, the zonal time ZT is given (or sought after). In this case the first (or last) steps in Examples 5.1 and 5.2 (or 5.3 and 5.4), leading to (or from) the computation of UT, should simply be replaced by UT = ZT + ΔZ (or ZT = UT - ΔZ).

The abbreviation AENA in all examples stands for the 'American Ephemeris and Nautical Almanac.'

5.4 Irregularities of the Rotational (Universal and Sidereal) Time Systems

5.41 Classification

As has been stated earlier, the sidereal and universal time systems are based on the rotation of the earth, and their relation to each other is defined by equations (5.40), (5.41), (5.44), and (5.45). Irregularities in the rotation of the earth thus will affect both of these equivalent forms of time. These irregularities are of two different characters:

(1) Variations in the position of the axis of rotation (true celestial pole) with respect to the solid earth, known as the motion of the pole.

(2) Variations in the earth's rotation speed.

The latter irregularities are of three types: (a) seasonal or periodic variation, more or less reproducible from year to year, probably due to meteorological causes and to earth tides; (b) secular decrease, essentially due to dissipative tidal forces; (c) irregular fluctuations, possibly connected with solar activities. Details on the reasons for these irregularities may be found in [Munk and MacDonald, 1960] and in [Marsden and Cameron, 1966].

The effect of polar motion on universal time is determined from observations coordinated and analyzed by the Bureau International de l'Heure (BIH), as discussed in sections 4.13 and 8.21. The effect of polar motion on time (longitude) $\Delta\Lambda_P$, is given by equation (4.40).

An effort is also made by the BIH to predict and determine the seasonal variations in the rotational speed and its effect on universal time, $\Delta\Lambda_S$ (see sections 5.43 and 8.21).

5.42 Rotational Time Systems UT0, UT1, and UT2

For the foregoing reasons it is obvious that distinction must be made between universal time as deduced directly from observations and its corrected values for polar motion and seasonal variation. The numbers 0, 1, and 2 are used conventionally for this purpose, in the following sense:

$$\left.\begin{aligned} UT0 &= \text{UT as deduced directly from observations} \\ UT1 &= UT0 + \Delta\Lambda_P \\ UT2 &= UT1 + \Delta\Lambda_S = UT0 + \Delta\Lambda_P + \Delta\Lambda_S \end{aligned}\right\} . \quad (5.46)$$

Thus UT1 is universal time corrected for polar motion and it represents the true angular rotation of the earth. The time UT2 is universal time corrected for both polar motion and periodic variations in the rotational velocity of the earth; however, it is still affected by secular and irregular variations. It is the best approximation to a uniform universal time at present. The system UT1 representing the actual rotation of the earth has the greatest importance in geodesy when observations are referred to a certain epoch. Since UT2 is a nearly uniform time, it should be used for purposes that require a relatively uniform measure.

The distinctions which were made above for the UT may be made naturally for MT, MST, or for GMST as well. In this sense one may speak about the systems MT0, MT1, MT2, or MST0, MST1, MST2, or GMST0, GMST1, GMST2, etc., where the numbers 1 and 2 represent the corrected versions of the observed (0) times, similar to equations (5.46).

5.43 Seasonal Variations in the Rotational Velocity of the Earth

The seasonal variation was first reliably determined by N. Stoyko in 1937 at the BIH. He found from an analysis of time observations that the amplitude of the seasonal variation is about $0^s.060$ during the course of a year. Later investigators found similar results which led to the assumption that the seasonal variations are repetitive.

The correction for the seasonal variation may be written in the form [Markowitz, 1958, p. 28]

$$\Delta\Lambda_S = a \sin 2\pi t + b \cos 2\pi t + c \sin 4\pi t + d \cos 4\pi t \qquad (5.47)$$

where a, b, c, and d are empirical constants and t is the fraction of the tropical year from the beginning of the Besselian year (see section 5.53). The coefficients are determined from observed differences between astronomical observations for UT0 and precision atomic clocks marking a uniform time. The BIH has been instructed by the General Assembly of the IAU at Dublin in 1955 to determine and publish in advance the coefficients so that they may be used by all coordinated time services in the determination of UT2. The coefficients adopted since 1956 are listed in Table 5.6.

TABLE 5.6 BIH Coefficients for Seasonal Variation

Years	a	b	c	d
1956-1961	$+0^s.022$	$-0^s.017$	$-0^s.007$	$+0^s.006$
1962-1968	+0.022	-0.012	-0.006	+0.007

The BIH also distributes the values $\Delta\Lambda_S = UT2 - UT1$ at five-day intervals for the whole year in advance, in the first issue of the 'Bulletin Horaire' each year and in its annual circular 'A.' A sample is shown in Table 8.3 (note that $\Delta T_S = \Delta\Lambda_S$). The seasonal variation is illustrated graphically in Fig. 5.8. The published values are calculated from equation (5.47) with the adopted coefficients in Table 5.6.

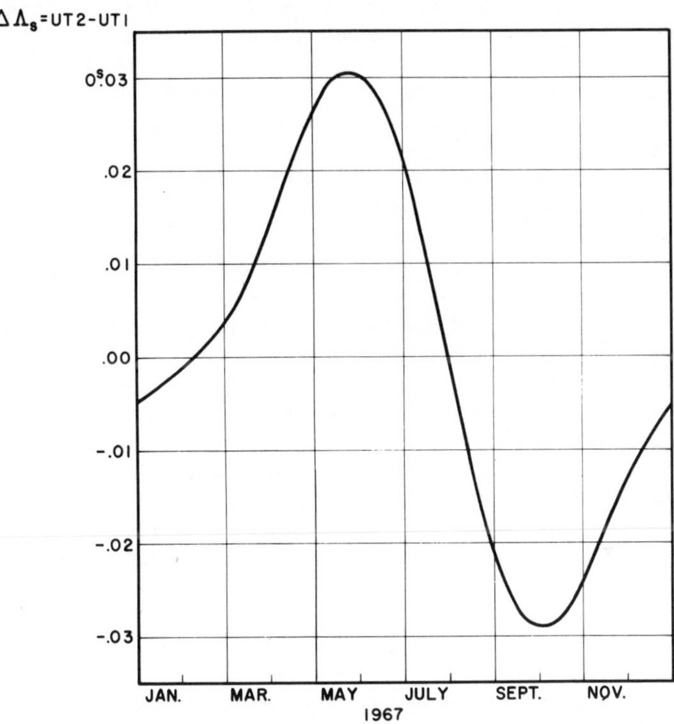

Fig. 5.8 Seasonal variation in the rotational velocity of the earth, 1967

Contrary to the effect of polar motion, which is different for each station, the correction $\Delta\Lambda_S$ is the same for all stations.

Equation (5.47) removes variations of annual and semiannual period. There are other periodic variations of $27^d.55$ and of $13^d.66$ periods in the rotation speed probably due to earth tides induced by the moon. In practice, these terms are eliminated either by smoothing the observations over a period of about two months, or by adding a correction term depending on the longitude of the moon's node [Markowitz, 1962, p. 242].

5.44 Secular and Irregular Variations in the Rotational Velocity of the Earth

Since the secular and irregular variations in the rotational velocity are present in UT2, they cannot be determined by star observations. The best way to recognize them is to compare UT2 with time provided by clocks keeping uniform time. This is currently done using atomic clocks. An example of such a comparison is shown in Fig. 5.9 with a solid line (x), where UT2 is compared to the atomic time system of the U.S. Naval Observatory, A.1 (see section 5.6). The figure indicates, for instance, that a deceleration took place approximately between 1955.5 and 1957.75, followed by an acceleration until about 1962.1, and a deceleration since that time. The figure is based on monthly means of UT2 - A.1, corrected for a linear term, as determined from observations at Washington, D.C., and Richmond, Florida. For other comparisons see [Stoyko, 1966; Jung, 1966].

Though these types of comparisons greatly improve our knowledge of the irregular fluctuation, they really should be made with a time scale based on a system that has been observed in the past, such as a gravitational (ephemeris) time scale defined from the motion of the sun, the moon, or some planet. Such a comparison would also yield the absolute values of the differences, not just the relative variations. This type of comparison is discussed in section 5.52.

A note should also be made of the fact that at present there are about forty observatories in the world which determine UT2. Consistency in the determination is assured by adherence to international agreements. Each observatory calculates UT2 according to equation (5.46) and thus determines its own UT2 system. The BIH smoothes out the discrepancies by forming a so-called 'mean' observatory and therewith a unique international UT2 system. This work is discussed in more detail in section 8.2.

5.5 Ephemeris Time System

The failure of the rotational (universal and sidereal) time as a uniform time scale led astronomers to a new system in which the purely gravitational motion of a celestial body in the solar system can be used to define the new time scale. Since the apparent motion of all bodies in the solar system is a combination of the motion of the earth and that of the body, it was simplest to consider as the basis of the new time system the apparent motion of the sun itself, in particular, the variation of its geometric longitude. Of several theories of the variation of the coordinates of the sun, the one given by Newcomb was selected since it was the most widely used [Newcomb, 1895].

The resulting time is called ephemeris time (ET) and, assuming the solar theory to be flawless, it is theoretically a strictly uniform measure of time in the Newtonian sense. It is used for purposes that require such a measure, e.g., in celestial mechanics. It is the indepen-

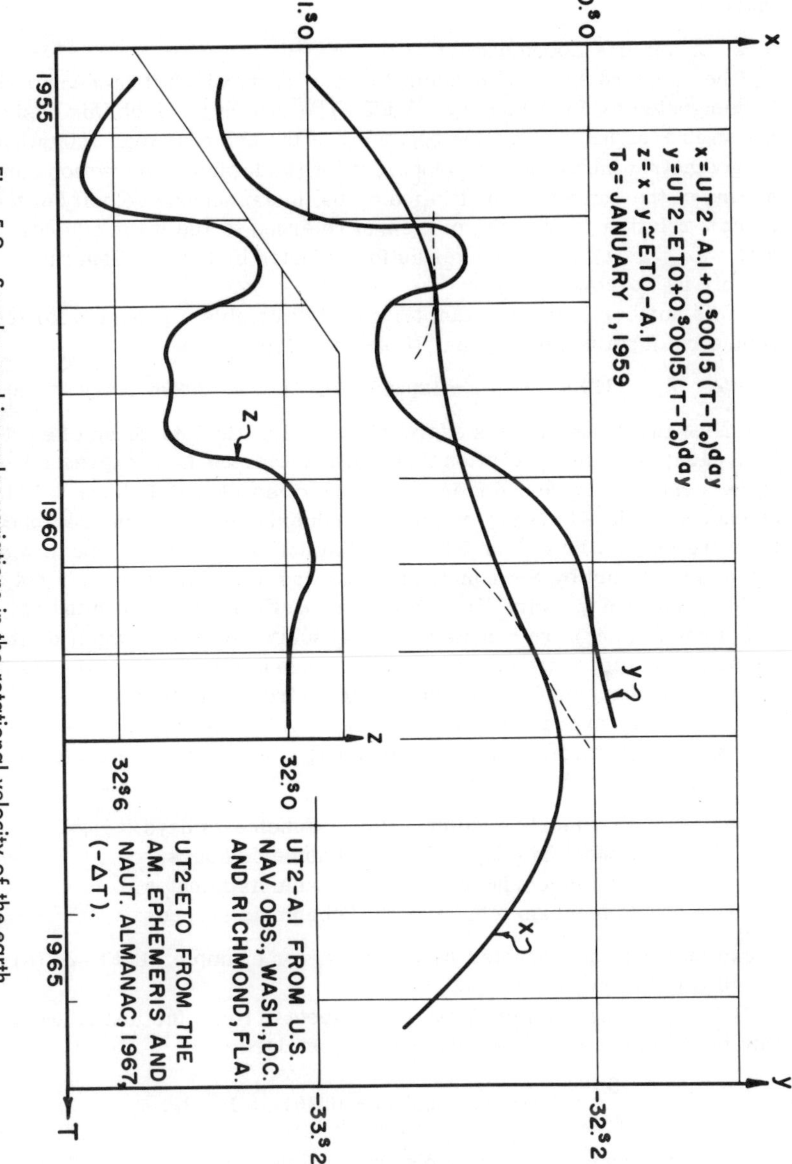

Fig. 5.9 Secular and irregular variations in the rotational velocity of the earth

dent variable in the orbital theories of the earth, moon, and planets, and the argument for the fundamental tables in the astronomical ephemerides.

5.51 Ephemeris Epoch and Interval

The standard epoch of ephemeris time, from which it is measured, is designated as 1900 January $0^{d}.5$ ET. The instant to which this designation is assigned is near the beginning of the calendar year 1900, when the geometric mean longitude (apparent longitude plus a correction consisting of the light-time multiplied by the instantaneous velocity of the earth relative to the sun) of the sun referred to the mean equinox of date was $279^{\circ}41'48''.04$ (IAU resolution, adopted by the Xth General Assembly in Moscow, 1958).

This number is the constant term of Newcomb's expression for the geometric longitude of the sun [Newcomb, 1895]:

$$\lambda_E = 279^{\circ}41'48''.04 + 129,602,768''.13t_E + 1''.089t_E^2 . \qquad (5.48)$$

The first two terms in this expression are identical to those of equation (5.31) with the exception that while there the t_M is expressed in Julian centuries reckoned from the standard epoch of UT, here t_E is to be expressed in strictly uniform ephemeris centuries and reckoned from the standard epoch of ET. The distinction between t_M and t_E was not pointed out by Newcomb, since he was not aware that t_M is not a strictly uniform quantity. The first term and the coefficient in the second term in (5.48) are considered as absolute constants which define the origin and measure of ET. A future possible revision of the solar theory may change the coefficients of the third term slightly.

The primary unit of ephemeris time interval is the length of the tropical century at the standard epoch of ET ($t_E = 0$), termed the ephemeris century. It is divided as follows:

1 Ephemeris century = 36525 Ephemeris days,
1 Ephemeris day = 24 Ephemeris hours,
1 Ephemeris hour = 60 Ephemeris minutes,
1 Ephemeris minute = 60 Ephemeris seconds.

These intervals are denoted by (E); thus, for example, $1^{d}(E) = 24^{h}(E) = 1440^{m}(E) = 86400^{s}(E)$.

Differentiating equation (5.48) with respect to time, the instantaneous rate of change of λ_E per second of ET is obtained,

$$\frac{d\lambda_E}{dt_E} = 0''.0410686389744 + 6''.9017 \times 10^{-10} t_E .$$

During one tropical year the sun's longitude increases by $2\pi = 1,296,000''$. Hence the length of the tropical year expressed as a function of t_E in ephemeris seconds is

$$1{,}296{,}000'' \Big/ \frac{d\lambda_E}{dt_E} = 31{,}556{,}925.^{s}9747\,(E) - 0.^{s}53032 t_E\,(E).$$

The first term again is an absolute constant; the coefficient of the second term, however, might be changed. To provide a constant unit of time, the relation between the second and the tropical year has been defined for the fundamental epoch, i. e., for $t_E = 0$. Thus the duration of the ephemeris second, as adopted in 1956 by the International Committee of Weights and Measures, is the fraction 1/31,556,925.9747 of the tropical year at 1900 January $0.^{d}5$ ET.

The adopted duration of the ephemeris second multiplied by 60, 3600, 86400, etc., defines the length of the respective ephemeris minute, hour, day, etc.

The ET at any instant may be determined by comparing the observed positions of the sun, moon, or planets with the ephemeris of the body where the argument is the measure of time defined by Newcomb's formula. The ET is the value of the argument for which the ephemeris position is the same as the observed (properly reduced) position. In practice, observations of the moon are the most effective since its geocentric motion is much greater than those of other bodies, though its motion due to tidal effects is not purely gravitational. It follows that the precision with which ET may be determined depends on the errors of observation and on the accuracy of the ephemeris, i.e., on the solar and lunar theories. The revision of these theories may change the practical measure of ET (designated ET0 when determined from lunar observations), which thus contains errors and is not strictly uniform. The difference ET - ET0 has a systematic character, but its exact nature and magnitude are still unknown at present.

Since the necessary observational reductions and data analysis require a great deal of data and time, ET is made available only after a delay of several years. For this reason ET is the time appropriate for purposes of which neither the epoch nor the unit of time interval need be made on a current basis.

5.52 The Fictitious Mean Sun, the Ephemeris Meridian

As is evidenced by the previous section, ephemeris time is independent of the rotation of the earth; thus it is unsuitable for the calculation of quantities which depend on the rotation. In spite of this, in order to facilitate better understanding of the nature of ephemeris time and of the practical calculations, two fictitious terms are introduced here which will enable the reader to view the ephemeris time system from the same point of view that has been used when discussing the rotational times. The two terms are the fictitious mean sun and the ephemeris meridian.

The fictitious mean sun is analogous to the fictitious sun which defines the solar time system. Its right ascension is given by Newcomb's expression,

$$\alpha_E = 18^h 38^m 45^s.836 + 8{,}640{,}184^s.542 t_E + 0^s.0929 t_E^2, \tag{5.49}$$

where t_E is the number of ephemeris centuries of strictly uniform length elapsed since 1900 January $0^d.5$ ET. This equation differs from (5.31) only in that t was expressed in Julian centuries of variable length reckoned from the standard epoch of UT. Thus, while the motion of the fictitious sun is not strictly uniform, that of the fictitious mean sun is. It should be noted that since Newcomb assumed that the length of the Julian century is strictly uniform, he actually derived expression (5.49) and not (5.31). It may be shown that

$$\alpha_E - \alpha_M = 0.002738\ \Delta T \tag{5.50}$$

where

$$\Delta T = ET - UT \tag{5.51}$$

at any instant. The correction ΔT may be calculated by determining ET through inverse interpolation in the ephemeris of the sun, moon, or planets, as described in the previous section, and deducting the recorded UT of the observation. Only in recent years have reasonably accurate values of ΔT become available, but fair values may be determined back to the beginning of the nineteenth century, and approximate estimates may be made back into the seventeenth century. The annual values of ΔT are tabulated in the 'American Ephemeris and Nautical Almanac' to $0^s.01$, starting at 1901 and ending with the current year (Table 5.7). For years up to 1948 inclusive, they are taken from Brouwer's smoothed values [Brouwer, 1952]; for the later years definitive values are given up to about five years before the current year, provisional and extrapolated values for the years following to $0^s.1$ and $1^s.0$ respectively. Thus basically ET is defined by a table of corrections to be added to UT (which is easily accessible and known with a much greater precision: $0^s.002$ opposed to $0^s.05$ - $0^s.2$).

Fig. 5.10 shows the general trend of ΔT = ET - UT (actually ET0 - UT2) based on the supplemented values of Brouwer. In Fig. 5.9 the solid line (y) illustrates the variations since 1955, in the sense UT2 - ET0, and with a linear term removed, for possible comparison with the solid line (x). The line z in the figure is the difference $z = x - y = ET0 - A.1 + \Delta$, where Δ is the probably very small difference in UT2 between the values of the U. S. Naval Observatory and those of the BIH (see Examples 8.1 and 8.2).

The ephemeris meridian is defined as the meridian where the Greenwich mean astronomic meridian would have been if the earth had rotated uniformly at the rate implicit in the definition of ephemeris time. It is located at

TABLE 5.7 ΔT, Reduction from Ephemeris Time to Universal Time

ΔT

REDUCTION FROM UNIVERSAL TIME TO EPHEMERIS TIME

Add to Universal Time

	s	d		s	d		s	d
1901.5	− 2.54	−.000029	1926.5	+22.72	+.000263	1951.5	+29.66	+.000343
1902.5	− 1.13	−.000013	1927.5	22.82	.000264	1952.5	30.29	.000351
1903.5	+ 0.35	+.000004	1928.5	22.92	.000265	1953.5	30.96	.000358
1904.5	1.80	.000021	1929.5	23.05	.000267	1954.5	31.09	.000360
1905.5	3.26	.000038	1930.5	23.18	.000268	1955.5	31.59	.000366
1906.5	+ 4.69	+.000054	1931.5	+23.34	+.000270	1956.5	+32.06	+.000371
1907.5	6.11	.000071	1932.5	23.50	.000272	1957.5	31.82	.000368
1908.5	7.51	.000087	1933.5	23.60	.000273	1958.5	32.69	.000378
1909.5	8.90	.000103	1934.5	23.64	.000274	1959.5	33.05	.000383
1910.5	10.28	.000119	1935.5	23.63	.000273	1960.5	33.16	.000384
1911.5	+11.64	+.000135	1936.5	+23.58	+.000273	1961.5	+33.59	+.000389
1912.5	12.95	.000150	1937.5	23.63	.000273	1962.5	34.08	.000394
1913.5	14.18	.000164	1938.5	23.76	.000275	1963.5	34.2	.00040
1914.5	15.31	.000177	1939.5	23.99	.000278	1964.5	35	. . .
1915.5	16.39	.000190	1940.5	24.30	.000281	1965.5	35	. . .
1916.5	+17.37	+.000201	1941.5	+24.71	+.000286	1966.5	+36	. . .
1917.5	18.27	.000211	1942.5	25.15	.000291	1967.5	36	. . .
1918.5	19.08	.000221	1943.5	25.61	.000296	1968.5		
1919.5	19.83	.000230	1944.5	26.08	.000302	1969.5		
1920.5	20.48	.000237	1945.5	26.57	.000308	1970.5		
1921.5	+21.06	+.000244	1946.5	+27.08	+.000313	1971.5		
1922.5	21.56	.000250	1947.5	27.61	.000320	1972.5		
1923.5	21.97	.000254	1948.5	28.15	.000326	1973.5		
1924.5	22.29	.000258	1949.5	28.94	.000335	1974.5		
1925.5	+22.55	+.000261	1950.5	+29.42	+.000341	1975.5		

$$\Lambda_{EM} = 1.002738\ \Delta T. \tag{5.52}$$

Thus, its position varies with ΔT with respect to the Greenwich mean astronomic meridian. The two coincided at some date between 1900 and 1905 (when $\Delta T = 0$).

By substituting the fictitious mean sun and the ephemeris meridian inplace of the respective fictitious sun and Greenwich mean astronomic meridian, phenomena depending on the rotation of the earth maybe handled in the ephemeris time system the same way as in the universal time system. Thus, analogous to section 5.221,

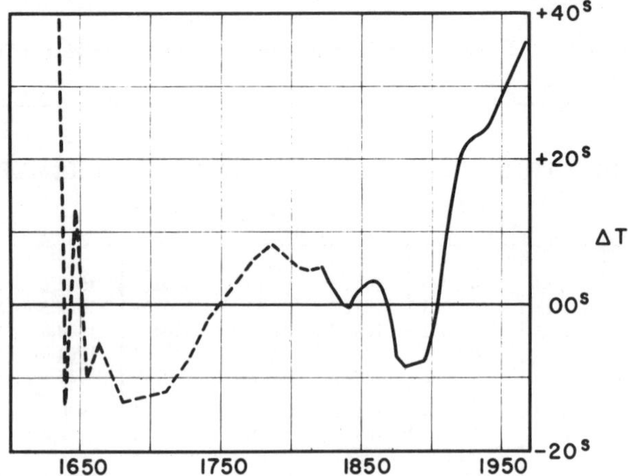

Fig. 5.10 General trend of ΔT, 1635–1967

$$h_E = h_E^E + \Lambda^E , \tag{5.53}$$
$$ET = h_E^E + 12^h , \tag{5.54}$$
$$Eq.\,ET = h_S^E - h_E^E \tag{5.55}$$

where h_E is the local hour angle of the fictitious mean sun, h_E^E is the e-ephemeris hour angle of the fictitious mean sun, Eq. ET is the equation of ephemeris time.

Using the same logic it is also appropriate to introduce the ephemeris mean (apparent) sidereal time EMST (EAST), analogous to the Greenwich mean (apparent) sidereal time GMST (GAST),

$$EMST = \alpha_E + h_E^E , \tag{5.56}$$
$$EAST = EMST + Eq.\,E. \tag{5.57}$$

These relations and others are illustrated in Fig. 5.11 from which it is evident that

$$ET - UT = EMST - \alpha_E - (GMST - \alpha_M) = \Lambda_{EM} - (\alpha_E - \alpha_M);$$

thus, with (5.50) and (5.52),

$$ET - UT = 1.002738\,\Delta T - 0.002738\,\Delta T = \Delta T.$$

5.53 Ephemeris Calendars

The Julian day numbers and the Julian date may be conveniently applied to ephemeris time. In this case, the Julian ephemeris day number will represent the number of ephemeris days that have elapsed at 12^hET preceding the epoch in question since 12^hET January 1, 4713 B. C. The proper term for this type of count is Julian ephemeris date (JED).

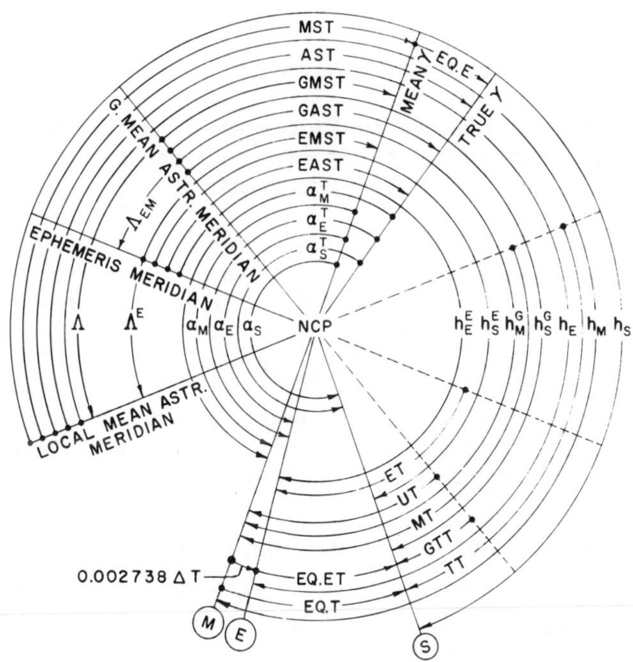

AST	APPARENT SIDEREAL TIME	α	MEAN RIGHT ASCENSION
MST	MEAN SIDEREAL TIME	α^T	TRUE RIGHT ASCENSION
ET	EPHEMERIS TIME	γ	VERNAL EQUINOX
MT	MEAN SOLAR TIME	h	HOUR ANGLE
UT	UNIVERSAL TIME	E	EPHEMERIS
TT	TRUE SOLAR TIME	G	GREENWICH
EQ.T	EQUATION OF TIME	NCP	NORTH CEL. POLE (PROJECTED)
EQ.E	EQUATION OF THE EQUINOX	G	REFERS TO G. MEAN ASTR. MER.
EQ.ET	EQUATION OF EPHEMERIS TIME	M	REFERS TO FICTITIOUS SUN
Λ	LONGITUDE	S	REFERS TO TRUE SUN
Λ_{EM}	LONGITUDE OF EPHEMERIS MERIDIAN = 1.002738 ΔT	E	REFERS TO FICTITIOUS MEAN SUN (SUPERSCRIPT E TO EPHEMERIS MER.)

Fig. 5.11 Astronomic time systems

Thus, the standard epoch of ephemeris time is

$$1900 \text{ January } 0^{d}.5 \text{ ET} = \text{JED } 2{,}415{,}020.0.$$

Generally,

$$\text{JED} = \text{JD} + \Delta\text{T}. \tag{5.58}$$

The Julian ephemeris day begins at 12^{h}ET, analogous to the Julian day which begins at 12^{h}UT.

The concept of the astronomic date described in section 5.232 for the solar time may also be conveniently applied for the ephemeris time with no change. Thus, for example, 1966 January 1 12^{h}ET = 1966 January $1^{d}.5$ ET, etc.

Another calendar frequently used in astronomy is the Besselian. The beginning of the Besselian year is the instant when the right ascension of the fictitious mean sun α_{E}, affected by aberration and measured from the mean equinox, is precisely

$$\alpha_{E} = 18^{h}40^{m}.$$

This instant always occurs near the beginning of the calendar year and is denoted by the notation .0 after the year. For example, the Besselian year in 1966 and 1967 commences at the following Besselian dates (BD):

$$1966.0 = 1966 \text{ January } 0^{d}.799 \text{ ET} = \text{JED } 2439126.299,$$
$$1967.0 = 1967 \text{ January } 1^{d}.041 \text{ ET} = \text{JED } 2439491.541.$$

The ephemeris astronomic date of the beginning of the Besselian years 1900 - 1999 is tabulated in [Nautical Almanac Offices, 1961, pp. 434-435]. The figure for the current year is also given in the 'American Ephemeris and Nautical Almanac,' Calendar (see Table 5.4).

The length of the Besselian year is equivalent with the period of one complete circuit of the fictitious mean sun in right ascension and is shorter than the (tropical) ephemeris year of $365^{d}.24219879$(E) by the amount $0^{s}.148\,t$, where t denotes the time in centuries after 1900. This insignificant difference is usually ignored [Nautical Almanac Offices, 1961, p. 30].

5.6 Atomic Time Systems

Atomic time (AT) is based on the electromagnetic oscillations produced by the quantum transition of an atom. It is obtained by associating a high precision quartz crystal clock with an atomic standard of frequency. The basis of any atomic time system is the adopted atomic standard which controls the frequency of the quartz crystal. Thus each atomic clock provides its own time system. Continuous coordination and comparison between the different systems provide an internationally accepted atomic standard which at present (1967) is the transition frequency of the cesium-133 atom.

There are several atomic time systems in operation. Some of the most important ones at present are the following:

A.1 is the system adopted by the U. S. Naval Observatory in Washington, D. C. It is based on the operation of cesium beam oscillators at laboratories located about the world as follows: U. S. Naval Observatory, Washington, D. C. and Richmond, Florida; U. S. Naval Research Laboratory, Washington, D. C.; U. S. National Bureau of Standards, Boulder, Colorado; National Physical Laboratory, Teddington, G.B.; and Laboratoire du CNET, Bagneux, France. The system is maintained through the use of several (around sixteen) Hewlett-Packard Model 5060A portable cesium beam oscillators (see section 8.13). The frequencies of these oscillators are regularly compared to the laboratory systems mentioned above.

NBS-A is the atomic time system maintained by the U. S. National Bureau of Standards, Boulder, Colorado. It is based on the operation of a laboratory type cesium beam standard, the NBS-III. The system is maintained through the use of five portable atomic oscillators, which are regularly compared to NBS-III.

A3 is the atomic time scale adopted by the Bureau International de l'Heure, Paris, France, based on the weighted means of cesium beams operated at the National Physical Laboratory, Teddington, G. B.; U. S. National Bureau of Standards, Boulder, Colorado; National Research Council of Canada, Ottawa; U. S. Naval Observatory, Washington, D. C.; Radio Research Laboratory, Tokyo, Japan; Research Institute for National Defense, Stockholm, Sweden; Paris Observatory, France; and the Republic Observatory, Johannesburg, South Africa (NBS carries a weight of five, NRC two, and the others one).

Standards of frequency and atomic clocks are discussed in more detail in section 8.1.

5.61 Atomic Time Epoch

The fundamental epoch of atomic time depends on the initial reading of the particular atomic clock and is, therefore, different for each of the systems mentioned above. The adopted initial epoch, for instance, for the system A.1 is $0^h0^m0^s$ UT2 on January 1, 1958, at which instant A.1 was $0^h0^m0^s$.

Other atomic time systems may have different initial epochs. The present (1967) epoch-differences of the systems mentioned, from direct comparisons made by transportable clocks, are the following [Bodily et al., 1967]:

$$\text{A.1} - (\text{NBS-A}) \approx -10.9 \text{ ms},$$
$$\text{A.1} - \text{A3} \approx 34.4 \text{ ms}$$

(positive difference means that A.1 is early).

The difference in the epochs does not cause practical difficulties since there is, at present, no real requirement for a definite atomic epoch. The atomic time is chiefly used as a measure of interval.

5.62 Atomic Time Interval

The fundamental unit of atomic time interval is the second, which was adopted by the thirteenth conference of the International Committee of Weights and Measures in Paris in October, 1967.

The wording of the new definition is:

> The standard to be employed is the transition between two hyperfine levels $F = 4$, $m_F = 0$ and $F = 3$, $m_F = 0$ of the fundamental state $^2S_{\frac{1}{2}}$ of the atom of cesium-133 undisturbed by external fields and the value 9,192,631,770 Hertz is assigned.

(The symbol F denotes the energy level, m_F the magnetic field.)

This definition was based on a previous experiment, conducted jointly by the U. S. Naval Observatory and the National Physical Laboratory, Teddington, where it was found that the frequency of cesium-133 at zero magnetic field at 1957.0 was 9,192,631,770 ±20 cycles per ephemeris second [Markowitz et al., 1958; Essen et al., 1958]. The epoch is stated because the atomic and gravitational time scales may diverge due to cosmic causes.

Other atomic time interval units, denoted by (A), are obtained from the fundamental second the conventional way. Thus, $1^d(A) = 24^h(A) = 1440^m(A) = 86400^s(A)$, etc.

5.63 Relation Between Atomic and Ephemeris Times

The difference between the atomic and ephemeris times is calculated by determining ET through inverse interpolation in an ephemeris, as mentioned in section 5.51, and subtracting it from the recorded AT of the observation.

The relation between atomic time and ephemeris time was found to be [Markowitz, 1959],

$$AT - ET = a + bt + ct^2 \qquad (5.59)$$

where t is the time from the epoch when

$$AT = ET + a.$$

The coefficient 'a' thus determines the epoch of atomic time in relation to ephemeris time. The coefficient 'b' is the division ratio adopted for the atomic resonator. The coefficient 'c' is a cosmic constant whose value may be zero or of the order of $0^s.001$ per year.

The degree to which $1^s(A)$ represents $1^s(E)$ depends mainly upon the accuracy of 'b', i. e., upon the adopted frequency of cesium. There is a question also whether 'c' is negligible or can have a significant effect.

At present 1^s(A) corresponds to 1^s(E) to 2×10^{-9}. This means that the ephemeris time intervals for most practical purposes can be replaced very efficiently with a system of atomic time.

The epoch of an atomic system may differ considerably from the observed ephemeris epoch. Curve 'z' of Fig. 5.9 illustrates the difference ET0 - A.1. If ET0 could be determined as accurately as A.1, the difference z ('a' in Equation (5.59)) would be a constant $32^s.15$.

The atomic time systems have the great advantage of being immediately available through time signals and special correction bulletins (see section 8.23).

References

Bodily, L. N. and R. C. Hyatt. (1967). "Flying Clock Comparisons Extended to East Europe, Africa and Australia." Hewlett-Packard Journal, 19, 4.

Brouwer, D. (1952). "A Study of the Changes in the Rate of Rotation of the Earth." Astronomical Journal, 57, pp. 125-146.

Essen, L., J. V. L. Parry, W. Markowitz, and R. G. Hall. (1958). "Variation in the Speed of Rotation of the Earth Since June 1955," Nature, 181, p. 1054.

H. M. Nautical Almanac Office. (1958). Seven-figure Trigonometric Tables for Every Second of Time. H. M. Stationery Office, London.

Jung, Karl. (1966). "Zur Änderung der Rotationsgeschwindigkeit der Erde," Zeitschrift für Vermessungwesen, 91, 10.

Markowitz, W. (1958). "Variations in Rotation of the Earth, Results Obtained with the Dual-Rate Moon Camera and Photographic Zenith Tubes." Proceedings of the Symposium on the Rotation of the Earth and Atomic Time Standards (Brouwer, ed.), pp. 26-33, Moscow.

Markowitz, W. (1959). Astronomical and Atomic Times. U. S. Naval Observatory, March 9, Washington, D. C.

Markowitz, W. (1962). "The Atomic Time Scale." Institute of Radio Engineers, Transactions, I-11, 3-4, pp. 239-242.

Markowitz, W., R. G. Hall, L. Essen, and J. V. L. Parry. (1958). "Frequency of Cesium in Terms of Ephemeris Time," Physical Review Letters, 1, 3.

Marsden, B. G. and A. G. W. Cameron (eds.). (1966). The Earth-Moon System. Plenum Press, New York.

Munk, W. H. and G. J. F. MacDonald. (1960). The Rotation of the Earth, A Geophysical Discussion. Cambridge University Press, London.

Nautical Almanac Offices of the United Kingdom and the United States of America. (1961). Explanatory Supplement to the Astronomical Ephemeris and the American Ephemeris and Nautical Almanac

H. M. Stationery Office, London. (The introduction of the IAU system of astronomical constants (1964) requires changes in this reference. They are listed in the supplements to the 1968 volumes of the Astronomical Ephemeris and of the American Ephemeris and Nautical Almanac.)

Newcomb, S. (1895). "Tables of the Motion of the Earth on Its Axis and Around the Sun." Astronomical Papers Prepared for the Use of the American Ephemeris and Nautical Almanac, VI, Part I.

Smart, W. M. (1960). Spherical Astronomy, Fourth Edition. Cambridge University Press, London.

Stoyko, A. (1966). "Die Erdrotation verlangsamt sich jetzt," Die Umschau, 66, 1, pp. 26-29.

6 STAR CATALOGUES

6.1 Fundamental Definitions and Classification

6.11 Requirements

The term 'star catalogue' somewhat loosely describes what one may term 'catalogue of accurate star positions.' As mentioned earlier, in contrast to a complete set of coordinates which fully define the point at which an object is located, the term 'position' here refers to a set of data which defines only the direction to an object from a certain point.

For recording star positions in catalogues, the coordinate system is the mean heliocentric right ascension system, described in sections 3.23 and 4.111, for a certain epoch T_0, which is usually chosen to be the beginning of a Besselian year. (In recent times, only 1900.0, 1925.0, 1950.0, and 2000.0 are used.)

For every star listed, a complete catalogue must contain its right ascension and declination; their time derivatives with respect to an (ideally) nonmoving system, i.e., the proper motions; and finally the epoch to which this data refers. Also it is necessary to know the epoch of the coordinate system and the constant of precession employed in the computation of the motions. For completely rigorous reductions, the distances and radial velocities of the stars listed should also be available. It could also be useful to know for every star the covariance matrix (or an equivalent) of all this data. All of this is almost never completely available but is given only in the form of standard deviations of position and of proper motion. For the sake of completeness, catalogues usually contain additional data which has strictly nothing to do with position information, such as current number, magnitude, spectrum, cross references to other catalogues, etc. The data as listed above constitutes what is termed a 'complete' catalogue. From this, the stars' coordinates and their time derivatives can be computed for any epoch, referred to an inertial system or the right ascension system at some (not necessarily the same) epoch.

Only a complete catalogue is really useful for the purposes of geodetic astronomy, which is mostly the empirical establishment of directions with respect to a well-defined coordinate system by directly observing these directions with respect to stars whose positions at the epochs of the observations are known. These positions can usually be computed from those at some other epoch if only the proper motions are known, i.e., even if the distances and radial velocities of the stars are not known. The proper motions, however, differ from star to star and are, if not individually known, only stochastically predictable as functions of the star's position, magnitude, and spectral type. In contrast to this, as shown in Chapter 4, the transformation from one coordinate system to another involves only precession and nutation. The effects of these on any star's known coordinates are completely predictable but, of course, only with an accuracy corresponding to that of the pertinent parameters (constant of precession, etc.).

In order to be useful for geodesy, a star catalogue must also contain a sufficiently large number of stars to serve whatever purpose is intended. The construction of such a catalogue is no small task and it has taxed the industry and ingenuity of astrometrists (positional astronomers) for over two centuries.

6.12 Types of Catalogues

To establish a star position by observation, one must be able to realize the directions to which the axes of the coordinate system point, and one must be able to measure accurately the direction defined by the position of the star with respect to this system. The first task can be solved either by determining the orientation of the coordinate system directly or by measuring the star position with respect to the positions of a set of other relatively close stars whose positions with respect to the system in question are known. Positions obtained in the first way, i.e., by referring them directly to the coordinate axes whose determination is an integral part of the observing process, are termed 'absolute' positions. Extensive lists of such positions are called <u>absolute catalogues</u>. Positions obtained in the second way are termed 'relative.' Extensive lists of such positions are known as <u>relative catalogues</u>. The various ways in which absolute and relative positions are determined will be discussed later.

The proper motions cannot be determined instantly as can the positions. As a matter of fact, they cannot be determined at all by direct observations in the strict sense of the term, but must be calculated from observations that were made at different epochs. Since the directly observed positions are usually referred to the true right ascension system of the observation epoch and thus change from one epoch to the next not only because of proper motion but also due to precession and nutation, all observations that are used for the *derivation* of a proper motion must be reduced to the same coordinate system. (This illustrates the mutual dependence of proper motions and the constant of precession.) After elimination of the influence of the motion of the coordinate system on the positions at the various epochs, the proper motions can be computed. For this purpose, one needs positions observed at least at two *epochs* sufficiently long apart from each other (which makes accurate old star positions especially valuable since the weight of a proper motion is proportional to the square of the time interval between the initial and the final epoch), because otherwise the star might not have moved far enough to produce a measurable proper motion, and differences in the measured positions will be due to unavoidable errors of measurement rather than proper motion.

The determination of a proper motion from positions at only two epochs is, of course, a minimum situation. In the interest of higher accuracy and to provide crosschecks against errors, proper motions are often determined from a least squares adjustment of several (and sometimes all available) independently observed positions. The positions and proper motions resulting from such an adjustment may be referred to any epoch, but are often referred to the same epoch as is the underlying coordinate system, or to that epoch at which there is no correlation between position and proper motion.

Catalogues which contain positions that are not the direct results of independent (absolute or relative) observations, but were obtained (usually together with proper motions) by combining information from several original observation catalogues, are called *compilation catalogues*. These vary widely in scope. For the compilation of some, all available absolute and the better of the relative independent observations are utilized for the determination of the best possible positions and proper motions of a selected number of stars. These are the *fundamental catalogues*. Other compilation catalogues list positions and proper motions for a large number of stars, calculated from only two independently (and usually relatively) determined positions. Some catalogues contain, in addition to the newly and independently determined positions, proper motions that *were* determined by comparison of these positions with those determined at another epoch and listed in another catalogue. Such catalogues *are* not regarded as compilation catalogues, since the primary information contained in them, namely the positions, were independently observed; the determination of proper motions must always involve data from at least two different sources, and thus a compilation process.

6.2 Original Observation Catalogues

6.21 Absolute Star Catalogues

The determination of absolute positions involves first finding the directions of the axes of the right ascension system and then determining the positions of the stars with respect to these axes.

In practice, absolute right ascensions are much harder to determine than absolute declinations.

6.211 Determination of Absolute Declination. Since declination is the complement of polar distance, an instrument capable of measuring the polar distance of an object is suitable for the determination of absolute declinations. One such instrument is the so-called 'meridian circle' (see Fig. 7.1). Ideally, it consists of a telescope whose optical axis is exactly perpendicular to the only axis on which the instrument can be turned, which lies in the plane of the horizon in the east-west line. Consequently, the optical axis of the instrument is always in the plane of the local celestial meridian. The axis of rotation carries a graduated circle which allows the reading of the angle between the telescope's optical axis and an index-mark fixed to the immovable parts of the instrument. Since the celestial pole is by definition always in the local meridian, there will be one circle reading corresponding to the position of this pole. The angle between the pole and a star can be measured only when the star crosses the meridian, since only then can the meridian circle be pointed at the star. The difference between the circle reading corresponding to the star at its meridian passage and that corresponding to the pole is the star's polar distance. The circle reading corresponding to the pole is in principle easily established by taking the mean of the circle readings on any circumpolar star at its upper and its lower culmination.

At locations on the earth's southern hemisphere, there are no stars circumpolar with respect to the north celestial pole. In this case, the circle reading corresponding to the south celestial pole would have to be established. The reading corresponding to the north celestial pole is then diametrically opposite on the circle.

It is obvious from this discussion that the same principles can be applied if the optical axis of the instrument is restrained to move in any plane which contains the direction to the celestial poles even if this plane is not that of the meridian. The reasons why the plane of the meridian is chosen are purely practical ones. In practice, the determination of absolute declinations is by no means as simple as just described, because it is impossible to construct and set up an instrument that conforms exactly to the theoretical model. Furthermore, effects of refraction and of residual stellar aberration render the direction in which an object is seen different from that to the apparent position. Add to this the complications introduced by systematic errors caused by the subjectivity of the observing process and the effect of polar motion and it becomes apparent that the accurate determination of absolute declinations (i. e., with a standard deviation in the order of 0''.2 for a single observation) that are reasonably free from systematic errors is extremely difficult.

The direct measurement of the angle between the directions to the pole and to the star is not the only principle available for the determination of absolute declination. Several other methods which utilize the general relationships between altitude and azimuth on the one hand and hour angle and declination on the other, in which the latitude of the observer appears as a parameter,have been developed and successfully applied. Appropriately arranged observations such as those of transit times of stars through the prime vertical, or through a certain almucantar, both on the eastern and western hemisphere of the sky, allow the direct computation of the declination from equations from which the unknown latitude can be eliminated. Some of these methods, in particular Danjon's very successful prism astrolabe, are discussed in sections 7.32 and 11.531.

6.212 Determination of Absolute Right Ascension. The accurate determination of absolute right ascensions is probably the most difficult problem in positional astronomy. It requires finding the direction to the vernal equinox, defined in turn as the ascending node of the ecliptic on the equator. Since the ecliptic is (roughly speaking) defined as the apparent path of the sun in the sky,it follows that a direct determination of the location of the ecliptic must at least indirectly be based on observations of the sun.

The orientation of the ecliptic with respect to an inertial system is determined by two Eulerian angles. Because of planetary precession, these angles are functions of time, for which at present only approximation polynomials are available. The approximations of these functions contain no periodic terms, since the ecliptic is defined in such a way that its orientation with respect to an inertial system is not subject to oscillations.

Suppose that an equatorial system at a certain epoch T_0 is chosen as the inertial system. Since the direction to the vernal equinox is not known, the direction of its axis x is set arbitrarily. In this system the position of the ecliptic is defined by the parameters

$$\begin{aligned} \Omega &= \Omega_0 + \Omega_1 t + \Omega_2 t^2 + \Omega_3 t^3, \\ \epsilon &= \epsilon_0 + \epsilon_1 t + \epsilon_2 t^2 + \epsilon_3 t^3, \end{aligned} \tag{6.1}$$

where t is the time interval elapsed since the epoch T_0, Ω is the angle between the axis x and the vernal equinox, and ϵ is the obliquity of the ecliptic. The plane of the ecliptic is obtained by removing the periodic oscillations from the plane which contains the center of the sun, the barycenter of the earth-moon system and the velocity vector of the earth-moon system. These periodic oscillations are due to the influence of Venus and Jupiter. Provided the relative orientations and velocities of the bodies in the planetary system are known (and they are), the values of Ω_1, Ω_2, Ω_3, ϵ_1, ϵ_2, and ϵ_3 may be calculated from celestial mechanics, leaving only Ω_0 and ϵ_0 to be determined from direct observations. The terms Ω_0 and ϵ_0 may be found very simply by observing the position of the sun on at least two occasions if the directions can be measured in (or at least reduced to) the above-defined inertial system. This is possible, since the sun is observed with respect to the local hour angle system, which can be transformed to the inertial system by a rotation on the common axis z through time. In principle, the process is as follows: It is assumed that the sun's apparent ecliptic latitude is always zero (which is strictly not true); then from equations (3.10)-(3.12)

$$\left.\begin{aligned} \cos\delta_s \cos\alpha_s &= \cos\lambda_s \\ \cos\delta_s \sin\alpha_s &= \sin\lambda_s \cos\epsilon \\ \sin\delta_s &= \sin\lambda_s \sin\epsilon \end{aligned}\right\}, \tag{6.2}$$

so that one may calculate α_s from δ_s by

$$\sin\alpha_s = \tan\delta_s \cot\epsilon \tag{6.3}$$

when ϵ is known. The absolute determination of the sun's right ascension is thus reduced to that of the sun's declination which can be carried out, for instance, by means of a meridian circle as described above. If at the same time the hour angle h_s of the sun is accurately measured (this is always the case when the declination is measured at the instant of meridian transit since then the hour angle is 0^h or 12^h), the apparent sidereal time AST can be computed from AST $= \alpha_s + h_s$. This time is used to set a sidereal clock from which the sidereal time can thereafter be read. Since an object's right ascension equals the sidereal time at which its upper culmination occurs, the absolute right ascensions of objects can then simply be determined by recording the sidereal times of their meridian passages in upper culmination.

In order that this procedure can be followed, ϵ must be known. This also may be obtained by observing the sun's declination during the solstices, when $\alpha_s = \lambda_s = 6^h$ or 18^h, $\cos\alpha_s = \cos\lambda_s = 0$, and $\sin\alpha_s = \sin\lambda_s = \pm 1$; thus

$$\sin\delta_s = \pm\sin\epsilon\ . \tag{6.4}$$

The practical difficulties are tremendous. Consider alone that the sun is a daytime object, has a finite diameter, is much brighter than any star and radiates a considerable amount of heat. All this, and many other circumstances not mentioned, will tend to produce systematic errors which can be eliminated only by extremely laborious procedures, if they can be eliminated at all.

Fortunately, it is not necessary to use the sun directly as a reference object. Any other object whose motion with respect to the ecliptic system is known will also define the ecliptic. Observations of a planet or an asteroid for which a theory of motion is available that will accurately predict its position with respect to the ecliptic will also determine the orientation of the ecliptic.

Systematic observations for this purpose of objects in the planetary system are on the program of only a very few observatories; in the United States, for instance, only on that of the U.S. Naval Observatory at Washington, DC.

The right ascension system (or any inertial system) thus cannot be defined by statics and kinematics alone, but can only be realized experimentally by comparing the results of the integration of certain equations of motion with observations in a well-defined but not necessarily inertial system.

The tedium and the difficulties associated with the determination of completely absolute right ascensions have led to the consequence that they are rather seldom determined as such. What are often termed 'absolute right ascensions' or sometimes more correctly 'semi-absolute right ascensions' are in fact only absolute right ascension differences. They can in principle be obtained very simply by recording the sidereal time differences between the meridian passages of the several stars by means of a meridian circle or some other device in connection with an accurate clock. Ideally, the time indicated by this clock must be strictly proportional to the 'rotation angle' of the earth at the observing station. It is also important, of course, that the rate of the clock be accurately known, i.e., that the indicated sidereal time of two successive upper culminations of the star differ exactly by the amount by which the right ascension has changed during one day. Since systematic instrumental errors, originating in the meridian circle as well as in the clock, can be eliminated (or rather minimized) only very laboriously and with difficulties, the determination of even absolute right ascension differences will be carried out only when relative positions would not serve the purpose intended. In practice, right ascensions (or rather right ascension differences) are often determined which are absolute only with respect to one coordinate, which means that they will not contain systematic errors that depend on δ, while they may contain systematic errors that depend on α, or vice versa.

Experience has shown that the determination of absolute positions requires about ten times as much time as the determination of strictly relative positions.

6.22 Relative Star Catalogues

If all right ascensions and declinations were determined absolutely, i.e., without reference to any previously known star positions, only a very few star positions would ever become available. It is much more efficient to determine star positions relatively, i.e., with respect to already known positions of other stars (termed 'reference stars') when these are available. These 'relative' star positions will, of course, be affected by systematic errors which will differ from those of the reference stars only because of the uncertainty in the determination of the reduction constants.

6.221 Zone Observations. The classical technique of observing relative star positions is 'in zones' with a meridian instrument. Since about 1925, many of the positions which used to be observed in zone observations have been determined by photographic techniques. Relative meridian positions are at this time observed practically only for stars that are to serve later as reference stars in conjunction with the photographic determination of relative star positions (see section 6.222).

Relative (zone) observations start with the selection of all the 'field' or 'zone' stars whose positions are to be determined. The stars are to be situated within a strip ('zone') of a few degrees width in declination, and a length of several hours in right ascension, corresponding to the duration of a normal observing session. As the stars cross the meridian, readings of the clock and of the altitude (or declination) circle are recorded. Interspersed among the observations of the field stars are observations of 'reference' or 'fundamental' stars whose positions are already known. There are usually not enough fundamental stars available within the zone, so that they will have to be chosen from outside. From the circle readings associated with the reference stars, and from their declinations, one obtains basically a relationship, which, when applied to the readings corresponding to the field stars, gives the declinations of the latter.

Since the observations are made in narrow zones only a small section of the circle is involved. Thus the influence of division errors is far less pernicious than on absolutely determined declinations which involve the entire circle. Likewise, refraction enters only in the form of the difference of the refraction for reference and field stars, i.e., as 'differential refraction,' which can be computed with considerably higher accuracy than the entire refraction needed for the absolute determination of declinations. The uncertainities in the accurate computation of refraction are generally considered to be the main obstacle to achieving a higher accuracy in absolutely determined declinations.

The clock readings at the meridian passages of the reference stars are used in connection with their known right ascensions to calculate the collimation error and the azimuth of the instrument as well as the clock correction and the clock rate during the observation of the zone. (The clock correction is what must be added to the clock reading to obtain sidereal time; the rate is the time derivative of this quantity.) Within the time span during which a certain zone was observed, the clock rate is assumed to be constant. After the quantities necessary for the computation of positions from the crude readings have in this way become known, the positions of the field stars can be calculated by means of the well-known formula of T. Mayer or another equivalent formula (see section 7.43).

6.222 Photographically Determined Relative Star Positions. The photographic method of determining relative star positions is in principle similar. Assume that a photograph is made of a region in the sky with a camera for which the geometry of image formation is known. This means that the rectangular Cartesian coordinates of known star images can be predicted on the photographic plate provided that the camera's focal length and the position of the 'tangential point' at whose image the camera's optical axis penetrates the photographic plate are also known. This point is chosen as the origin of a rectangular coordinate system ξ, η, whose axis η points toward the (fictitious) image of the north pole. The rectangular coordinates of star images in this idealized coordinate system are usually termed 'standard coordinates.' The gnomonic projection is most frequently used as a model for setting up relations between the standard coordinates of the stars and their positions. For this projection, the standard coordinates ξ and η, expressed in radians, may be calculated from the star positions α, δ and the position α_0, δ_0 of the tangential point as follows:

$$\xi = \frac{\cos\delta \sin(\alpha-\alpha_0)}{\sin\delta \sin\delta_0 + \cos\delta \cos\delta_0 \cos(\alpha-\alpha_0)},$$
$$\eta = \frac{\sin\delta \cos\delta_0 - \cos\delta \sin\delta_0 \cos(\alpha-\alpha_0)}{\sin\delta \sin\delta_0 + \cos\delta \cos\delta_0 \cos(\alpha-\alpha_0)}. \tag{6.5}$$

The coordinates of the stellar images measured on the photographic plate, x and y, will differ from the standard coordinates, since the plane of the plate will be tilted against the assumed plane of projection; the origin with respect to which the images are measured will not coincide with the tangential point; the axis y will not exactly point toward the image of the north pole; and the scale of the projection, being a function of the instrument's focal length, will not be known exactly.

If it is assumed that the camera furnishes a mapping of the celestial sphere onto a plane exactly in the same way as a pinhole camera would, i.e., if the camera produces an ideal gnomonic projection, the relationship between the images' measured rectangular coordinates x, y and their standard coordinates is given by a projective transformation with an orthogonal matrix, provided that the apparent refracted α and δ of the stars were used in (6.5). Thus we have

$$\left.\begin{aligned} \Xi &= f[\alpha_{11}(x-x_0) + \alpha_{12}(y-y_0) + \alpha_{13}] \\ H &= f[\alpha_{21}(x-x_0) + \alpha_{22}(y-y_0) + \alpha_{23}] \\ Z &= \alpha_{31}(x-x_0) + \alpha_{32}(y-y_0) + \alpha_{33} \end{aligned}\right\} ; \quad \begin{aligned} \xi &= \frac{\Xi}{Z} \\ \eta &= \frac{H}{Z} \end{aligned} \tag{6.6}$$

where six independent relationships exist between the α_{ik}. Since, however, it is the mean α and δ of the comparison stars which are normally available for the computation of the ξ and η, the actual relationship between the standard and the measured coordinates will deviate from (6.6) by the effects of refraction and aberration. A comprehensive discussion of this is given in [König, 1933].

There it is also shown that in most cases which involve small areas of the sky (up to $2° \times 2°$, say) the exact orthogonal projective relationship (6.6) may be replaced by a nonorthogonal affine relationship of the form

$$\begin{aligned} \xi &= Ax + By + C, \\ \eta &= Dx + Ey + F, \end{aligned} \tag{6.7}$$

where ξ, η are the standard coordinates calculated from the stars' mean position; x, y are the measured coordinates of their images; and A, B, C, D, E, F are Turner's plate constants. The equations (6.7) may evidently be used only when the spherical position of the tangential point and its location on the plate are sufficiently well-known. For a full discussion again see [König, 1933].

Note that the number of independent constants in (6.7) is the same as in (6.6), namely, six. The transition from (6.6) to (6.7) specializes the projective transformation to an affine one (because it is assumed that the data concerning the tangential point is sufficiently well-known) but generalizes from an orthogonal to a nonorthogonal relationship so that the effects of refraction and aberration (and other sources) may be properly accommodated without going through a complicated special analysis.

The plate constants can be found from equations (6.6) or (6.7) in which the values of the standard coordinates, computed from (6.5) with respect to the tangential point at α_0, δ_0, are used on the left-hand side, and the corresponding measured rectangular coordinates on the right. If more than three stars are used, the system is over-determined and the plate constants are computed by a statistical adjustment. Once the plate constants have in this way become known, equations (6.6) or (6.7) can be used to calculate the standard coordinates from the measured coordinates, and equations (6.5) can be inverted to give α and δ in terms of ξ, η, α_0, and δ_0.

6.3 Catalogues Defining a Fundamental System

6.31 Fundamental Star Catalogues: The Principle

6.311 Systematic Errors Affecting Star Positions. As mentioned above, an ideal star catalogue should contain positions, proper motions, distances and radial velocities so that it can define a coordinate system not only for the epoch at which the stars were observed (which in general is not the same for all stars) but for a reasonably long time interval, the length of which depends on the accuracy with which the positions, proper motions and the constant of precession are known. Since proper motions cannot be observed instantly, such a catalogue must be the result of a compilation. Fundamental catalogues aim at providing the best obtainable positional information for a selected number of stars distributed over the entire sky.

The question arises: Which are the systematically and individually most accurate positions and proper motions obtainable for a certain star at a certain time?

One might be tempted to subject all published independently determined positions of a certain star to a statistical adjustment, and compute therefrom its position and proper motion for a certain epoch. Two reasons, one practical and the other theoretical, can be invoked against this procedure. First, the number of published star positions is enormous, and the inclusion of all of them for the purpose of compiling a general catalogue would require an amount of labor and of computer time which is at present not available for this purpose. Second, one must consider that star positions contained in certain lists (catalogues) will be affected not only by accidental but also by systematic errors. As explained above, the determination of absolute star positions also involves the determination of the orientation of the reference frame and is based on a number of other idealizations such as a clock that is perfectly synchronous with the earth's rotation, perfect knowledge of the effects of refraction, a perfectly adjusted ideal instrument, etc. Deviations from the ideal conditions are unavoidable; and the parameters, which are used in the correction equations that reduce the raw observations to what they would have been if they had been made under ideal circumstances, must themselves be calculated from observations and are thus affected by errors. With respect to the parameters these errors are mostly accidental, but they will propagate themselves as systematic errors into the positions. Another important source for hidden systematic errors in the positions are inadequate or unrealistic models for the correction equations. No published star position may, therefore, ever be regarded as free from systematic errors. This is especially true for relative star positions, since they will not only be affected by the systematic errors of the reference stars, but also by the inadequacies and inaccuracies of the reduction to that coordinate system which is defined by the reference stars.

Before the available star position material is therefore subjected to an adjustment from which zero epoch positions, and proper motions are obtained, an effort must be made to remove as many as possible of the systematic errors of the published directly observed positions.

6.312 The Establishment of a Catalogue System. If one assumes that all observers of absolute star positions succeeded in minimizing all systematic errors, the best possible values for star positions and proper motions are obtained by adjusting (i.e., taking the means of) all the available absolute observations. Clearly, it would be wrong to include at this stage any relative observations, since they share (by definition) the systematic errors of their reference stars, and one would thus in fact assign an unduly high weight to the position of these.

After this first stage, the residuals of the positions given in the individual catalogues (including those of relative positions) with respect to the adjusted means of the absolutely observed star positions can be established. It will normally be found that these residuals show a systematic trend that is usually correlated with the position itself, and mostly also with the stars' magnitudes and sometimes their spectra. From these residuals one may construct tables or formulas which represent the systematic deviations of the individual catalogue positions from the averaged ones. These tables can then be used for the reduction of an individual catalogue to the system of the averaged catalogue.

The construction of a fundamental catalogue starts with the judicious selection of those absolute catalogues which are to be incorporated. As a rule, catalogues will be left out only if one has reasons to suspect that the observations were affected by systematic errors which could have been avoided had reasonable care been exercised. (In practice, catalogues based on observations made before a certain date were also excluded from some fundamental catalogues because it was felt that an attempt to remove the systematic errors from the positions recorded in them had little chance of success.) Average positions and proper motions are then computed, as indicated above, usually with different weight given to the information from different catalogues mostly according to the judgment of the investigator. Next incorporated are the positions given in selected relative catalogues, after the differences between these positions and their averages in the absolute catalogues are removed as far as possible. A list of positions and proper motions of stars obtained in the manner just described which cover with a more or less homogeneous density most of the sky constitutes a fundamental catalogue. The coordinate system to which the positions in the catalogue are referred is called a fundamental system.

6.313 Definition of a Fundamental System. A fundamental system must not be confused with an inertial system, although it is the best available approximation to one (note, however, the complications introduced by an erroneous constant of precession [Clemence, 1966]).

An explicit definition of a fundamental system beyond the somewhat trivial statement that it is 'all the information contained in the positions and proper motions in a given fundamental catalogue' has never been given, and the exact nature of a fundamental system must therefore be defined operationally.

For this purpose one might proceed as follows: Three coordinates of two stars (e.g., two declinations and one right ascension) are in principle sufficient to define the equator with the point of zero right ascension on it, thus a coordinate system. These coordinates, even in a fundamental catalogue, are affected by systematic errors so that the axis of the coordinate system defined by the stars' positions in one region of the sky will not quite coincide with the axes of the system defined by the stars in another region. Since more than three star coordinates overdetermine a system, the axes of the coordinate system must be fitted to the star positions by some kind of an adjustment. The same sample of stars will yield the directions of the axes somewhat differently if a different mathematical model is chosen in the adjustment. The directions of the axes will also change somewhat if the sample of stars is changed even only slightly. Therefore, it can be concluded that the system of a fundamental catalogue is defined only with respect to a sample of positions from the catalogue and with respect to a specific adjustment model.

It would, of course, be an ideal situation if the system defined by a fundamental catalogue were well and uniquely defined for every point in the sky, and independent of a model. For this to be possible, the regions in which the axes of the system are fitted to the star samples would have to be infinitely small and would have to contain an infinite number of stars. Of course, there can be no such regions with an infinite number of fundamental stars anywhere in the sky. The best one can hope for, in principle, is to obtain uniquely defined estimates of the positions of the coordinate axes of the system for every point in the sky. This could also be achieved by a procedure that involves all stars in the fundamental catalogue but cuts their influence down by a weight factor that diminishes according to a function of increasing distance from the point at which the system is to be defined. One could then agree upon a model for the fit and thus render the definition such that it yields a unique result. If such a definition were available, it would be clear that it could hardly ever be applied in practice, since the entire sky is almost never available for the comparison of two systems. This illustrates the situation that the reduction of the system of one catalogue to that of another must always be an estimation, as long as perfectly accurate star positions themselves are unavailable.

A fundamental system is thus defined as a set of positions and proper motions, both for a certain epoch, with respect to the coordinate system at a certain epoch and to a constant of precession; or the equivalent thereof (e.g., a set of two positions each at two different epochs with respect to the coordinate system at different epochs). As a rule, positions and coordinate systems in fundamental catalogues are referred to the same epoch, or epochs, which are then, of course, the same for every star in the catalogue.

6.314 Reduction Between Systems. To reduce a position from the fundamental system A to the system B, there must be available a table (or formula) giving the systematic differences $\Delta\alpha$ and $\Delta\delta$ of the positions as referred to A and B respectively at a certain epoch, dependent on the relevant arguments (usually α and δ), and a table giving the systematic differences $\Delta\mu^a$ and $\Delta\mu^d$ of the proper motions as listed in systems A and B respectively. If, then, at the epoch T_o the differences between the coordinates of the same star referred to A and to B are the tabulated $\Delta\alpha(T_o)$ and $\Delta\delta(T_o)$ in right ascension and declination respectively, the differences at the epoch T will be given by

$$\begin{aligned} \Delta\alpha(T) &= \Delta\alpha(T_o) + (T-T_o)\Delta\mu^a, \\ \Delta\delta(T) &= \Delta\delta(T_o) + (T-T_o)\Delta\mu^d. \end{aligned} \tag{6.8}$$

The systematic difference between the positions of a star with respect to one and the other coordinate systems depends thus on the epoch to which this position refers.

The systematic differences between two systems at a certain epoch and the systematic differences between the proper motions thus define the systematic differences between the positions in two systems at any epoch. The systematic differences between the positions at any two epochs would serve the same purpose. For reasons of accuracy, however, these epochs should be chosen as far apart as practicable.

The need for a unified fundamental system became more and more obvious as the amount of absolutely determined positions grew, and since the second half of the nineteenth century, astrometrists have devoted a considerable portion of their time to establishing one. That which is by international agreement generally adopted at this time, and which represented at the time of its publication about the best available approximation to an inertial system (except for the imperfect value of the constant of precession) is that of the FK4, the fourth of a series of fundamental catalogues established by various generations of astronomers of the German 'Astronomisches Rechen-Institut' [Fricke and Kopff, 1963].

Also in use (and in a sense not competing with the FK4) are the systems of the 'General Catalogue' by Benjamin Boss and that of the N30 catalogue by R. A. Morgan; but although these catalogues each define a system, they are not fundamental catalogues in the strict sense of the word [Boss, 1937; Morgan, 1952].

6.32 The German Series of Fundamental Catalogues

6.321 The FC. Around 1860, when astronomers realized the desirability of systematically and efficiently measuring the positions of all stars on the entire accessible sky, complete to a certain magnitude, the need for a list of positions of fundamental stars which were to serve as reference stars for the individual determinations also became clear.

When the German (but international in membership) 'Astronomische Gesellschaft' therefore resolved the inception of the observations which have long since led to the establishment of the 'Catalog der Astronomischen Gesellschaft,' commonly called the AGK1 (see section 6.61), they entrusted Artur Auwers who was then the most respected German astrometrist with the establishment of a list of reference stars.

The result of his efforts, called Fundamental-Catalog, abbreviated as FC, was issued in two parts covering in the first list the mean positions and proper motions of 539 stars between declination $-10°$ and the north celestial pole, and in the second, 83 stars between the declinations $-10°$ and $-32°$ [Auwers, 1879 and 1883].

In detail, the list contains a current number, the (visual) magnitude, the weighted average of the epochs of the positions that were used for the formation of the catalogued right ascension, the same for the declination, position for epoch and coordinate system at 1875.0, annual proper motions and the first three terms of the precession series, based on the precession constant of Struve-Peters (to which the proper motions also refer) and some other data which are no longer relevant.

The weighted average of the epochs of the individual observations from which the positions were computed are important for the following reason. Assume the standard deviation of a coordinate (say α) at epoch T_0 to be $m_\alpha(T_0)$ and that of the corresponding proper motion component to be m_μ, then the formula

$$m_\alpha^2(T) = m_\alpha^2(T_0) + (T-T_0)^2 m_\mu^2, \tag{6.9}$$

will give the mean error of the right ascension at T only if α and μ are completely uncorrelated. Since they are as a rule the results of a statistical adjustment and are obtained as solutions from the same system of equations, α and μ will generally be correlated and formula (6.9) will yield wrong results, unless T_0 is the weighted average of the epochs of the various observations that were used to calculate positions and proper motions at the standard epoch, the epoch at which the estimates obtained for position and proper motion are not correlated. The epoch T_0 then is necessary to calculate the mean errors of the position components at any epoch from those at T_0 and the errors of the proper motion. For the FC, however, these latter errors are given only for the southern part so that listing of the T_0 in the northern part of the catalogue is somewhat pointless.

It is remarkable that only one series of absolute observations in right ascension and declination each (both made at Pulkovo on different instruments) was used to establish the system of the positions in the FC, while the proper motion system was established using two series of absolute observations made at Greenwich about 100 years apart. The final values of the positions and proper motions were derived by combining the data of altogether six (for the northern part) and fourteen additional catalogues, after these had been reduced to the system previously described.

In assigning weights to the catalogues, Auwers was guided not only by the preliminary adjustment residuals, but relied a great deal on his personal judgement which of necessity was not always infallible.

6.322 The NFK. Auwers kept working on the improvement of the accidental and systematic accuracy of the positions published in the FC and on the extension of the lists to the south so that they would eventually cover the entire sky. The results of these investigations were published in a series of papers [Auwers, 1889, 1897, 1898, 1904, 1905] and finally combined in the 'Neuer Fundamentalkatalog' [Peters, 1907], known as the NFK.

The number of available absolute and relative star positions had increased considerably since the publication of the FC, and the system of the NFK represents a much better approximation to an inertial system than that of the FC, mainly because it is in essence a genuine representation of all the absolute observations which were then available. Besides the catalogues of absolute positions, about fifty relative catalogues were incorporated into the results documented in the NFK. Positions and proper motions were determined by a procedure that mixed averaging and least squares, which at the time represented an optimum balance between computing economy and accuracy.

The weights of the relative catalogues were mainly obtained from investigating the residuals (after reduction to the NFK system), but Auwers' judgement was still the main determinant for assigning weights to the fundamental catalogues and a major factor in fixing the system. Tables of weights of the catalogues are given in various papers by Auwers, particularly in [Auwers, 1900 and 1903]. Peters was mainly responsible for compiling the material provided by Auwers and for referring it from Struve's to Newcomb's constant of precession which had been internationally adopted in the meantime.

The NFK contains current numbers, names (when assigned) and magnitudes of 925 stars. Their positions and proper motions are listed for the epoch and mean coordinate system of both 1925.0 and 1950.0. Some peripheral information is also given, but data for the computation of the errors of the positions at a certain epoch are missing. The average epochs of the catalogues used for the determination of a star's position, which are not listed but are generally in the neighborhood of 1877, can be found in [Auwers, 1904 and 1905].

6.323 The FK3. In the late nineteen hundred and twenties, the Astronomische Gesellschaft resolved to redetermine the positions contained in the AGK1 (see section 6.611) for the northern hemisphere by photographic means. This enterprise resulted in the AGK2 (see section 6.621) and necessitated the establishment of a revised system of fundamental star positions. The catalogue by which it is defined was to become known as the 'Dritter Fundamental-Katalog des Berliner Astronomischen Jahrbuchs,' abbreviated FK3. The overall direction of the work was in the hands of August Kopff; A. Kahrstedt supervised the establishment of the right ascension system and K. Heinemann that of the declinations [Kopff, 1937 and 1938, Kahrstedt, 1937; Heinemann, 1937].

Seventy-seven catalogues (21 of them containing absolute positions) had become available for the revision of the NFK. This, as was the revision of the FC, was carried out in two steps. First, the positions of the stars were individually improved within the NFK system; and, second, the system of positions and proper motions of the NFK was corrected. The details are rather complicated and cumbersome involving a great deal of personal judgment. Characteristic of the FK3 is that data contained in catalogues with an observing epoch before 1845 was almost not used; the investigators felt that the attempt to remove systematic errors from them would be futile and only spoil the results. Furthermore, the FK3 reflects the assumption that the system of positions of the NFK is essentially free from errors at the average of the epochs of the catalogues which contributed to the NFK, which is about 1877.

The star density in the NFK increases noticeably toward the poles. In order to provide a more uniform star density all over the sky, Kopff selected some 600 'Zusatzsterne' (additional stars) [Kopff and Nowacki, 1934]. Their positions and proper motions were computed only after the catalogues in which they occur were reduced to the FK3; the 'Zusatzsterne,' therefore, do not contribute to the revision of the NFK system.

Since 1935, up to the adoption of the FK4 in 1964, the FK3 has been the basis for the ephemerides of the stars in the various national and international almanacs, especially the 'Apparent Places of Fundamental Stars' (see below).

6.324 The FK4, FK4 Sup, and the 'Apparent Places of Fundamental Stars.' The 'Fourth Fundamental Catalogue,' or the FK4, was published in 1963 by the Astronomisches Rechen-Institut at Heidelberg [Fricke and Kopff, 1963]. As in analogous situations before, the immediate stimulus for the revision of the FK3, or rather its replacement by the FK4, was the plan for the establishment of the AGK3 (see section 6.622).

The observations of relative and absolute star positions that had accumulated in the meantime made it mandatory that the AGK3 be reduced to the best system then available which was no longer that of the FK3.

The establishment of the FK4 followed in its main aspects that of the FK3 and is described in [Kopff, Nowacki and Strobel, 1964] for the declinations, and in [Gliese, 1963] for the right ascensions.

First, the individual positions of the stars within the system of the FK3 were improved in the following way. A selection from the catalogues which had become available after the work on the FK3 was finished was compared with the FK3 and reduced to the FK3 system, thus providing material for the improvement of the individual stars' positions and proper motions within the system of the FK3. These positions were then compared with those available for the same stars in absolute catalogues. The system of the positions in declination was established using catalogues which are absolute in declination, some having had their equator point (the zero point of their declination system) determined by including observations of the sun or of major and minor planets. Although for the correction of the equator point (which may be regarded as a constant bias afflicting all declinations) and the establishment of the proper motion system, absolute catalogues with epochs ranging from 1846 to 1956 were used; the overall system of the declinations was derived only from absolute catalogues with an epoch after 1900. The epoch of the declination system of the FK4 (i.e., the average weighted epoch of those catalogues that determined the system of the declinations in the FK4, in contrast to the individual positions) is about 1925, with slight variations depending on the declination zone under consideration.

The system of the positions in right ascension was derived from absolute observations covering the period from 1918 to 1956, with an average of 1935, whereas observations obtained on only four instruments (the two meridian circles at the U. S. Naval Observatory in Washington, the one at Pulkovo, and the one at the Cape of Good Hope), covering the period from 1897 to 1956, were used for the derivation of the proper motion system. In particular, the system of the proper motions in right ascension south of $\delta = -20°$ is based on observations at the Cape meridian circle alone.

Absolute proper motions were, of course, also obtained from the adjustment of the positions in all catalogues of absolute right ascension used to establish the FK4 system. Yet these were not used, but were replaced, as explained above, by proper motions obtained on the four instruments under the assumption that the homogeneity of material obtained quasi-differentially on one instrument only would be systematically more accurate than proper motions obtained from observations on a large number of instruments. Only the future will decide whether a less exclusive procedure for the determination of the right ascension proper motions would have led to more accurate results.

The standard errors of the system of the FK4 are listed in Table 6.1; the zero epoch of the right ascension system is 1935, and that of the declination system, 1925.

TABLE 6.1 Standard Errors of the FK4 System

δ	$m_\alpha \cos\delta$	$100 m_{\mu^a} \cos\delta$	m_δ	$100 m_{\mu^d}$
$>+80°$	$0^s\!.001$	$0^s\!.010$	$0''\!.017$	$0''\!.07$
+70	.002	.008	.016	.07
+50	.003	.010	.016	.05
+20	.002	.008	.015	.06
0	.001	.006	.014	.05
-20	.002	.012	.020	.08
-50	.004	.017	.030	.10
-75	.009	.024	.040	.13

These quantities do not vary susceptibly with right ascension. In order to get the accuracy with which the FK4 system will represent the ideal system at an epoch T, the standard errors have to be calculated from the formulas

$$(m_\alpha \cos\delta)_T^2 = (m_\alpha \cos\delta)_{T_0}^2 + \left(\frac{T-1935}{100}\right)^2 (m_{\mu^a} \cos\delta)^2,$$
$$(m_\delta^2)_T = (m_\delta^2)_{T_0} + \left(\frac{T-1925}{100}\right)^2 m_{\mu^d}^2 . \qquad (6.10)$$

If, for example, data from Table 6.1 is used to compute the standard errors with which the FK4 system represents the ideal system in 1970 at a declination of -75°, equation (6.10) yields, in right ascension

$$(m_\alpha \cos\delta)^2 = (0^s\!.009)^2 + \left(\frac{35}{100}\right)^2 (0.024)^2 = 0.000151,$$

so that at 1970

$$m_\alpha \cos\delta = 0^s\!.012;$$

and in declination

$$m_\delta^2 = (0''\!.040)^2 + \left(\frac{45}{100}\right)^2 (0''\!.13)^2 = 0.00502,$$

so that at 1970

$$m_\delta = 0''\!.072.$$

Two things must be well understood in this connection: (1) The standard errors thus calculated are those of the system only, not those of the star positions. (2) These standard errors may not be interpreted as the standard errors of the difference between the FK4 system and an inertial system, but must rather be taken as a direct measure for the probable difference between the actual and the ideal FK4 system. Even the latter would not be an inertial system because it is well known that Newcomb's constant of precession, with respect to which all proper motions in the FK4 have been computed, requires a correction of about one second of arc per century. This means that even the ideal FK4 system (or any fundamental system) will change its orientation with respect to an inertial system by exactly the amount of the correction to be applied to the used constant of precession to make it the physically correct one. This circumstance may in the future have to be taken into account when geodetic information is extracted from the dynamic properties of satellite orbits which were established by observing the satellite against a background of stars whose positions are in the FK4 system.

More detailed discussions concerning the errors of the FK4 system are given in the above-mentioned papers by Gliese and by Kopff, Nowacki and Strobel.

The body of the FK4 gives the following information: On the left-hand page (see sample in Table 6.2) are the FK3 number and name of the star, its magnitude and spectrum from the Henry Draper catalogue, mean α, $d\alpha/dT$ and $\frac{1}{2}(d^2\alpha/dT^2)$ (the latter referred to the century as a unit and taking proper motion into account) for equinox and epoch 1950.0 as well as 1975.0, followed by the centennial μ^a (there denoted μ), referred to Newcomb's constant of precession and its time derivative $d\mu^a/dT$ for the same epochs and equinoxes. Next listed is Ep. (α), the weighted average of the epochs (minus 1900) of those catalogues whose right ascensions were used to derived the FK4. These epochs vary from star to star and are, of course, not to be confused with the epoch of the FK4 system described earlier. The column m(α) contains the standard deviation in units of $0^s.001$ of the star's right ascension at the epoch Ep. (α) within the system; under the heading m(μ) is listed the standard deviation of the centennial proper motion in right ascension in units of $0^s.001$.

The right-hand page repeats the FK3 number and lists the quantities pertaining to the declinations analogous to the right ascension information; only the errors are given in units of $0''.01$. Two additional columns give the star's number in the GC and the N30 catalogues (see sections 6.33-6.34).

The main body of the catalogue is followed by special tables for the components of double stars, a short treatment of the foreshortening effects on FK4 stars, tables for the systematic differences FK4-FK3, and a list of catalogues from which data was incorporated into the FK4.

The standard deviation (M_δ) at epoch t (in tropical years after 1900) of a star's declination within the FK4 system in units of 0".01 is obtained by

$$M_\delta^2 = m^2(\delta) + \left(\frac{t - Ep(\delta)}{100}\right)^2 m^2(\mu), \tag{6.11}$$

with a completely analogous formula in right ascension. The right ascension of Star No. 166 (δ Men) will, for example, have a standard deviation M_α within the FK4 system at 1980, as follows:

$$M_\alpha^2 = 0.0183^2 + \left(\frac{80 - 12.25}{100}\right)^2 \times 0.098^2 = 0.00464,$$

so that at 1980

$$M_\alpha = 0^s.068.$$

When the standard error E of the position with respect to the ideal fundamental system rather than with respect to the actual fundamental system is required, it must be computed from

$$E^2 = m^2 + M^2, \tag{6.12}$$

where m is the standard error of the system at this epoch computed from (6.10); while M is the standard deviation of the position within the system from equation (6.11).

After the construction of the FK4, it became apparent that there was a need for additional stars in a fundamental catalogue. To achieve this aim, 1,987 stars not in the FK4 were selected all over the sky within the magnitude range of the fundamental stars. Most of these were already in the N30 (see section 6.34). They had originally not been included in the FK4 because there were no, or not enough, observations of their absolute positions available. Information concerning these stars was published as the 'Preliminary Supplement to the Fourth General Catalogue,' abbreviated FK4 Sup, and contains number, spectrum, position in the FK4 system at epoch and for the equinox 1950.0, the proper motions for the same epoch and equinox, the star's numbers in the GC and N30 (when included there), as well as remarks, usually pertaining to duplicity or variability [Astronomisches Rechen-Institut, Heidelberg, 1963].

The FK4 Sup positions of the N30 stars are N30 positions to which systematic reductions to the FK4 system have been applied; for these, the data concerning average epoch of the incorporated catalogues and standard deviations is identical with that given in the N30. The positions of the non-N30 stars in the FK4 Sup are those of the GC systematically corrected for reduction FK4 minus GC, and individually improved by the incorporation of more recent observations made by F. P. Scott at the U. S. Naval Observatory. For these stars, data concerning the

TABLE 6.2 A Left-Hand Page of the FK4

EQUINOX AND EPOCH 1950.0 AND 1975.0 65

No.	δ	$\frac{d\delta}{dT}$	$\frac{1}{2}\frac{d^2\delta}{dT^2}$	μ'	$\frac{d\mu'}{dT}$	Ep. (δ)	m (δ)	m (μ')	GC	N30
1328	+ 7 56′ 47″.16 + 7 48 35.22	−1968″.91 −1966.61	+ 4″.54 + 4.66	+ 0″.03 + 0.02	−0″.02 −0.02	17.60	3.2	15	17346	2941
481	−59 24 56.53 −59 33 8.14	−1967.80 −1965.07	+ 5.37 + 5.56	− 1.69 − 1.69	−0.01 −0.01	11.87	4.4	18	17374	2947
1329	−24 34 46.08 −24 42 56.13	−1961.46 −1958.94	+ 4.97 + 5.11	+ 3.89 + 3.89	−0.01 −0 01	21.98	5.1	32	17380	2948
1330	+ 3 50 43.38 + 3 42 32.19	−1965.96 −1963.53	+ 4.79 + 4.91	− 0.74 − 0.74	0.00 0.00	14.18	2.6	13	17381	2949
1331	−33 43 37.32 −33 51 47.70	−1962.86 −1960.15	+ 5.34 + 5.49	− 2.34 − 2.34	−0.01 −0.01	14.47	4.1	25	17433	2959
1332	+27 48 44.74 +27 40 35.19	−1959.45 −1956.95	+ 4.96 + 5.06	− 1.32 − 1.32	0.00 0.00	15.43	2.4	11	17455	2963
1333	+17 20 43.66 +17 12 34.24	−1958.97 −1956.40	+ 5.10 + 5.21	− 1.69 − 1.69	0.00 0.00	22.40	3.8	16	17464	2965
482	−39 54 26.31 −40 2 35.48	−1958.14 −1955.24	+ 5.74 + 5.89	− 2.65 − 2.65	+0.02 +0.02	11.78	5.0	23	17489	2969
1334	−17 45 59.76 −17 54 8.02	−1954.42 −1951.61	+ 5.54 + 5.67	− 0.28 − 0.28	−0.01 −0.01	23.96	4.6	32	17506	2971
1335	− 9 16 3 92 − 9 24 12.42	−1955.38 −1952.60	+ 5.50 + 5.63	− 2.00 − 2 00	−0.01 −0.01	07.51	2.8	11	17516	2974
483	+56 13 51.14 +56 5 42.88	−1954.23 −1951.84	+ 4.75 + 4.83	− 1.02 − 1.01	+0.04 +0.04	08.42	2.1	8	17518	2976
484	+ 3 40 7.60 + 3 31 58.81	−1956.53 −1953.79	+ 5.42 + 5.54	− 5.79 − 5.82	−0.11 −0.11	10.27	2.0	8	17543	2979
486	+65 42 33.68 +65 34 25.64	−1953.26 −1951.03	+ 4.44 + 4.50	− 3.35 − 3.35	0.00 0.00	16.63	2.9	12	17554	2982
485	+38 35 16.82 +38 27 11.07	−1944.31 −1941.72	+ 5.13 + 5.22	+ 5.20 + 5.18	−0.07 −0.07	01.02	2.2	7	17557	2983
1336	− 3 32 33.43 − 3 40 38.57	−1942.08 −1939.06	+ 5 96 + 6.09	+ 0.34 + 0.34	−0.01 −0.01	15.49	3.3	14	17631	2990
487	−71 16 47.18 −71 24 52.12	−1941.82 −1937.66	+ 8.16 + 8.45	− 3.16 − 3.10	+0.22 +0.23	12.33	4.1	20	17672	2996
488	+11 13 38.90 +11 5 35.54	−1934.94 −1931.92	+ 5.98 + 6.09	+ 1.73 + 1.71	−0.07 −0.07	13.03	1.9	8	17687	2999
1337	+36 3 57.55 +35 55 56.34	−1926.35 −1923.33	+ 5.99 + 6.08	+ 1.70 + 1.70	−0.01 −0.01	19.48	3.4	14	17751	3007
1338	+45 32 7.73 +45 24 6.81	−1925.15 −1922.23	+ 5.80 + 5.88	+ 2.37 + 2.37	−0.01 −0.01	23.83	3.6	16	17758	3008
1339	+21 25 16.61 +21 17 14.11	−1931.58 −1928.42	+ 6.27 + 6.37	− 4.76 − 4.77	−0.02 −0.02	22.45	4.6	14	17767	3010
489	−49 38 19.89 −49 46 21.40	−1927.90 −1924.13	+ 7.44 + 7.62	− 1.23 − 1.23	−0.01 −0.01	12.79	5.6	25	17773	3011
1340	−53 11 33.36 −53 19 34.94	−1928.24 −1924.38	+ 7.64 + 7.82	− 3.21 − 3.21	−0.02 −0.02	13.86	6.1	35	17783	3014
490	− 5 16 21.11 − 5 24 21.17	−1921.98 −1918.46	+ 6.96 + 7.09	− 3 66 − 3.66	−0.01 −0.01	09.18	1.8	7	17828	3021
491	+38 45 51.03 +38 37 53.05	−1913.49 −1910.34	+ 6.24 + 6.33	+ 3.81 + 3.80	−0.03 −0.03	15.56	3.0	13	17835	3022
1341	−26 17 10.71 −26 25 8.79	−1914.17 −1910.41	+ 7.44 + 7.59	+ 0.12 + 0.11	−0.02 −0.02	22.19	4.5	31	17861	3026
492	+28 7 52.03 +28 0 16.22	−1824.82 −1821.61	+ 6.36 + 6.45	+ 87.88 + 87.81	−0.28 −0.28	13.08	2.2	9	17874	3027
493	−67 37 48.79 −67 45 45.26	−1908.28 −1903.45	+ 9.53 + 9.81	− 1.67 − 1.68	−0.03 −0.03	15.54	6.0	29	17927	3035
1342	−31 14 31.61 −31 22 27.59	−1905.94 −1901.84	+ 8.13 + 8.28	− 5.54 − 5.54	+0.01 +0.01	13.58	3.8	23	17968	3041
1343	−43 42 57.34 −43 50 51.98	−1900.68 −1896.40	+ 8.48 + 8.66	− 0.89 − 0.89	0.00 0.00	16.92	6.2	35	17978	3044
1344	+ 5 43 57.99 + 5 36 4.33	−1896.54 −1892.76	+ 7.51 + 7.62	+ 1.13 + 1.13	0.00 0.00	20.27	3.0	15	17995	3047

5

TABLE 6.2 (continued) A Right-Hand Page of the FK4

No.	Name	Mag.	Sp.	α	$\frac{d\alpha}{dT}$	$\frac{1}{2}\frac{d^2\alpha}{dT^2}$	μ	$\frac{d\mu}{dT}$	Ep. (α)	m (α)	m (μ)
1328	32 d² Vir	5.24	A 5	$12^h 43^m 5^s.423$ 12 44 21.200	+ 303^s.103 + 303.111	+ 0^s.014 + 0.021	− 0^s.738 − 0.738	+0^s.001 +0.001	18.29	1.9	11
481	β Cru	1.50	B 1	12 44 47.037 12 46 14.929	+ 350.719 + 352.423	+ 3.386 + 3.430	− 0.504 − 0.506	−0.007 −0.007	13.01	5.6	23
1329	332 G. Hya	6.29	B 9	12 45 13.841 12 46 33.650	+ 319.001 + 319.470	+ 0.935 + 0.944	− 0.310 − 0.310	−0.002 −0.002	23.11	3.5	21
1330	35 Vir	6.66	M o	12 45 18.552 12 46 34.936	+ 305.507 + 305.569	+ 0.121 + 0.128	− 0.056 − 0.056	0.000 0.000	15.58	1.6	7
1331	143 G. Cen	5.01	A o	12 47 57.957 12 49 19.449	+ 325.637 + 326.297	+ 1.313 + 1.324	− 0.223 − 0.223	−0.001 −0.001	15.79	3.4	20
1332	31 Com	5.07	G o	12 49 15.862 12 50 28.881	+ 292.191 + 291.962	− 0.462 − 0.453	− 0.101 − 0.101	+0.001 +0.001	20.00	1.7	8
1333	32 Com	6.53	K 5	12 49 43.043 12 50 57.612	+ 298.324 + 298.228	− 0.197 − 0.189	− 0.020 − 0.020	0.000 0.000	25.29	2.6	13
482	150 G. Cen	4.34	A 5	12 50 39.510 12 52 2.708	+ 332.380 + 333.206	+ 1.644 + 1.659	+ 0.553 + 0.554	+0.005 +0.005	14.79	4.5	22
1334	52 G. Crv	6.84	A o	12 51 21.719 12 52 40.914	+ 316.597 + 316.962	+ 0.726 + 0.734	− 0.245 − 0.245	−0.001 −0.001	23.34	3.2	22
1335	ψ Vir	4.91	M 3	12 51 44.944 12 53 2.979	+ 312.020 + 312.261	+ 0.479 + 0.486	− 0.188 − 0.188	0.000 0.000	09.45	1.9	8
483	ε UMa	1.68	A o p	12 51 50.084 12 52 55.960	+ 263.830 + 263.178	− 1.315 − 1.293	+ 1.314 + 1.309	−0.018 −0.018	16.20	2.4	9
484	δ Vir	3.66	M o	12 53 4.986 12 54 20.549	+ 302.215 + 302.292	+ 0.149 + 0.156	− 3.145 − 3.144	+0.003 +0.003	11.37	1.2	5
486	8 Dra	5.27	F o	12 53 29.454 12 54 29.037	+ 238.712 + 237.958	− 1.525 − 1.493	− 0.141 − 0.140	+0.006 +0.006	21.64	3.8	19
485	α CVn *sq*	2.90	A o p	12 53 41.478 12 54 51.575	+ 280.566 + 280.212	− 0.713 − 0.702	− 2.011 − 2.008	+0.014 +0.014	07.87	1.7	6
1336	44 Vir	5.88	A o	12 57 4.803 12 58 22.100	+ 309.101 + 309.272	+ 0.338 + 0.345	− 0.265 − 0.265	0.000 0.000	18.90	2.1	10
487	δ Mus	3.63	K 2	12 58 47.973 13 0 31.686	+ 412.983 + 416.734	+ 7.429 + 7.577	+ 5.601 + 5.641	+0.161 +0.163	11.90	9.8	49
488	ε Vir	2.95	K o	12 59 41.210 13 0 55.867	+ 298.629 + 298.624	− 0.014 − 0.008	− 1.868 − 1.867	+0.003 +0.003	14.39	1.1	5
1337	14 CVn	5.11	B 9	13 3 24.273 13 4 34.356	+ 280.479 + 280.184	− 0.594 − 0.584	− 0.268 − 0.268	+0.001 +0.001	25.07	2.7	11
1338	Grb 1956 CVn	5.72	K o	13 3 37.463 13 4 44.879	+ 269.872 + 269.457	− 0.835 − 0.821	− 0.145 − 0.145	+0.001 +0.001	30.01	3.1	15
1339	39 Com	6.04	F 5	13 3 55.003 13 5 8.081	+ 292.372 + 292.254	− 0.239 − 0.232	− 0.524 − 0.523	+0.003 +0.003	30.19	3.3	13
489	ξ^2 Cen	4.40	B 3	13 3 58.753 13 5 26.493	+ 350.352 + 351.566	+ 2.418 + 2.440	− 0.296 − 0.297	−0.003 −0.003	14.31	5.4	29
1340	177 G. Cen	5.96	B 9	13 4 39.192 13 6 8.516	+ 356.598 + 357.999	+ 2.789 + 2.817	− 0.436 − 0.437	−0.004 −0.004	12.93	7.5	43
490	ϑ Vir	4.46	A o	13 7 21.476 13 8 39.168	+ 310.667 + 310.872	+ 0.408 + 0.414	− 0.232 − 0.232	0.000 +0.001	10.21	1.1	5
491	17 CVn	6.05	F o	13 7 45.560 13 8 54.380	+ 275.438 + 275.126	− 0.629 − 0.618	− 0.627 − 0.626	+0.003 +0.003	21.93	2.2	11
1341	342 G. Hya	6.48	A 3	13 8 55.695 13 10 17.362	+ 326.400 + 326.935	+ 1.064 + 1.073	− 0.478 − 0.479	−0.002 −0.002	22.16	3.2	20
492	β Com	4.32	G o	13 9 32.444 13 10 42.404	+ 279.933 + 279.751	− 0.366 − 0.359	− 6.049 − 6.048	+0.005 +0.005	16.11	1.5	7
493	η Mus	4.95	B 8	13 11 49.986 13 13 32.072	+ 406.848 + 409.847	+ 5.951 + 6.044	− 0.581 − 0.584	−0.011 −0.011	14.78	11.1	60
1342	195 G. Cen	5.36	K o	13 14 6.074 13 15 29.495	+ 333.359 + 334.006	+ 1.289 + 1.299	+ 0.279 + 0.280	+0.003 +0.003	12.89	3.2	21
1343	196 G. Cen	5.87	A 3 p	13 14 19.307 13 15 46.429	+ 347.989 + 348.989	+ 1.992 + 2.008	− 0.046 − 0.046	0.000 0.000	15.52	6.2	36
1344	σ Vir	5.01	M o	13 15 4.685 13 16 20.435	+ 302.963 + 303.041	+ 0.154 + 0.160	− 0.048 − 0.048	0.000 0.000	23.56	2.0	10

epochs and the standard deviations is at this time not available in the printed literature.

The FK4 provides, of course, only the mean positions for the two equinoxes and epochs 1950.0 and 1975.0. For many purposes of the observer's practice, apparent positions are required at odd epochs. So that these may be readily available, the International Astronomical Union since 1941 has sponsored an annual volume, entitled 'Apparent Places of Fundamental Stars,' which was published from 1941 through 1959 by the Royal Greenwich Observatory, and from 1960 on by the Astronomisches Rechen-Institut at Heidelberg. These volumes give annually the apparent places of the 1,535 FK3 (FK4) stars at ten-day intervals. Up to 1963, the positions were referred to the FK3 system; since 1964 the FK4 system was adopted.

The apparent places were computed using those values for the fundamental astronomical constants which were adopted at the 'Conférence Internationale des Étoiles Fondamentales' that was held in Paris in 1896. Starting with the 1968 volume, however, the new constant of annual aberration of 20″.496, internationally agreed upon in 1964, will be adopted [Fricke, Brouwer et al., 1965].

The short-period nutation terms are not included in the published positions; neither are the e-terms of the annual aberration (see section 4.213). Starting with the 1960 volume, the effects of annual aberration are no longer calculated by a closed formula, but by numerical differentiation from the actual velocity of the earth's center with respect to the barycenter of the solar system. Nutation is calculated from Woolard's new formulas which consider all terms with an amplitude of at least 0″.0002 [Woolard, 1953]. The influence of annual parallax is included for those 721 stars whose annual parallax, as listed in [Jenkins, 1952], is at least equal to 0″.010.

Besides the apparent positions (see Table 6.3), the volumes contain auxiliary tables for the Besselian day numbers, short-period terms of nutation, conversion of time, interpolation, and diurnal aberration.

6.33 The General Catalogue (GC)

6.331 The Scope of the GC. More or less concurrently with the work on the FK3 in Germany, a group of American astronomers at the Dudley Observatory in Albany, New York, under the leadership of Benjamin Boss, was working on the establishment of the 'General Catalogue of 33,342 Stars for the Epoch 1950,' abbreviated the General Catalogue or the GC [Boss, 1937]. (This is the name that is generally accepted in the astronomical community. Some users have recently called it the 'Boss Catalogue.' Since this term could apply to several other catalogues it should be avoided.) According to the introduction

> ...the General Catalogue contains the standard positions and proper motions of all stars brighter than seventh

TABLE 6.3 Sample Page of the APFS

196 **APPARENT PLACES OF STARS, 1967**

AT UPPER TRANSIT AT GREENWICH

No.	1327		1328		481		1329	
Name	Y Canum Venat.		32 Virginis		β Crucis		332 G. Hydrae	
Mag. Spect.	4.8 to 6.0	N3	5.24	A5	1.50	B1	6.29	B9
U.T.	R.A.	Dec.	R.A.	Dec.	R.A.	Dec.	R.A.	Dec.
	h m 12 43	° ′ +45 36	h m 12 43	° ′ + 7 50	h m 12 45	° ′ −59 30	h m 12 46	° ′ −24 40
d	s +429	″ −205	s +341	″ −225	s +560	″ −116	s +363	″ −192
I 0.3	34.664 +435	61.19 −158	56.259 +341	71.77 −213	45.135 +556	10.79 −167	07.109 +362	08.86 −213
I 10.2	35.099 +426	59.61 −101	56.600 +330	69.64 −193	45.691 +533	12.46 −217	07.471 +349	10.99 −231
I 20.2	35.525 +401	58.60 − 46	56.930 +309	67.71 −166	46.224 +495	14.63 −256	07.820 +325	13.30 −238
I 30.2	35.926 +369	58.14 + 10	57.239 +283	66.05 −138	46.719 +449	17.19 −290	08.145 +296	15.68 −241
II 9.1	36.295 +323	58.24 + 66	57.522 +247	64.67 −105	47.168 +390	20.09 −316	08.441 +259	18.09 −237
II 19.1	36.618 +270	58.90 +113	57.769 +209	63.62 − 71	47.558 +327	23.25 −331	08.700 +219	20.46 −227
III 1.1	36.888 +217	60.03 +157	57.978 +170	62.91 − 41	47.885 +264	26.56 −342	08.919 +180	22.73 −215
III 11.1	37.105 +156	61.60 +192	58.148 +130	62.50 − 9	48.149 +194	29.98 −344	09.099 +137	24.88 −197
III 21.0	37.261 +100	63.52 +215	58.278 + 92	62.41 + 17	48.343 +131	33.42 −336	09.236 +101	26.85 −177
III 31.0	37.361 + 47	65.67 +231	58.370 + 58	62.58 + 39	48.474 + 69	36.78 −327	09.337 + 65	28.62 −157
IV 9.9	37.408 − 6	67.98 +235	58.428 + 25	62.97 + 57	48.543 + 7	40.05 −307	09.402 + 30	30.19 −134
IV 19.9	37.402 − 47	70.33 +229	58.453 − 1	63.54 + 69	48.550 − 47	43.12 −283	09.432 + 3	31.53 −112
IV 29.9	37.355 − 86	72.62 +216	58.452 − 24	64.23 + 79	48.503 − 98	45.95 −256	09.435 − 22	32.65 − 89
V 9.9	37.269 −119	74.78 +194	58.428 − 46	65.02 + 82	48.405 −148	48.51 −220	09.413 − 47	33.54 − 65
V 19.9	37.150 −142	76.72 +166	58.382 − 61	65.84 + 82	48.257 −186	50.71 −183	09.366 − 64	34.19 − 42
V 29.8	37.008 −163	78.37 +133	58.321 − 74	66.66 + 80	48.071 −224	52.54 −143	09.302 − 82	34.61 − 20
VI 8.8	36.845 −176	79.70 + 95	58.247 − 86	67.46 + 73	47.847 −255	53.97 − 97	09.220 − 97	34.81 + 4
VI 18.8	36.669 −182	80.65 + 54	58.161 − 91	68.19 + 65	47.592 −276	54.94 − 52	09.123 −106	34.77 + 25
VI 28.8	36.487 −187	81.19 + 14	58.070 − 98	68.84 + 56	47.316 −294	55.46 − 5	09.017 −115	34.52 + 47
VII 8.7	36.300 −183	81.33 − 30	57.972 − 98	69.40 + 43	47.022 −299	55.51 + 43	08.902 −118	34.05 + 67
VII 18.7	36.117 −175	81.03 − 71	57.874 − 96	69.83 + 30	46.723 −295	55.08 + 87	08.784 −117	33.38 + 83
VII 28.7	35.942 −163	80.32 −112	57.778 − 91	70.13 + 16	46.428 −283	54.21 +130	08.667 −113	32.55 + 99
VIII 7.7	35.779 −143	79.20 −153	57.687 − 80	70.29 − 2	46.145 −255	52.91 +169	08.554 −101	31.56 +110
VIII 17.6	35.636 −119	77.67 −188	57.607 − 65	70.27 − 20	45.890 −217	51.22 +200	08.453 − 83	30.46 +115
VIII 27.6	35.517 − 90	75.79 −224	57.542 − 46	70.07 − 39	45.673 −170	49.22 +227	08.370 − 61	29.31 +118
IX 6.6	35.427 − 51	73.55 −255	57.496 − 17	69.68 − 62	45.503 −105	46.95 +244	08.309 − 29	28.13 +113
IX 16.5	35.376 − 10	71.00 −281	57.479 + 13	69.06 − 84	45.398 − 37	44.51 +250	08.280 + 7	27.00 +102
IX 26.5	35.366 + 37	68.19 −306	57.492 + 49	68.22 −109	45.361 + 41	42.01 +249	08.287 + 49	25.98 + 87
X 6.5	35.403 + 92	65.13 −322	57.541 + 91	67.13 −134	45.402 +128	39.52 +235	08.336 + 97	25.11 + 63
X 16.5	35.495 +145	61.91 −332	57.632 +133	65.79 −157	45.530 +210	37.17 +210	08.433 +145	24.48 + 35
X 26.4	35.640 +203	58.59 −339	57.765 +178	64.22 −181	45.740 +296	35.07 +178	08.578 +194	24.13 + 3
XI 5.4	35.843 +260	55.20 −334	57.943 +222	62.41 −200	46.036 +375	33.29 +134	08.772 +243	24.10 − 34
XI 15.4	36.103 +311	51.86 −321	58.165 +260	60.41 −217	46.411 +441	31.95 + 86	09.015 +283	24.44 − 71
XI 25.4	36.414 +358	48.65 −303	58.425 +295	58.24 −229	46.852 +499	31.09 + 33	09.298 +321	25.15 −108
XII 5.3	36.772 +395	45.62 −271	58.720 +321	55.95 −232	47.351 +539	30.76 − 27	09.619 +346	26.23 −144
XII 15.3	37.167 +420	42.91 −234	59.041 +337	53.63 −232	47.890 +559	31.03 − 81	09.965 +362	27.67 −174
XII 25.3	37.587 +435	40.57 −189	59.378 +345	51.31 −222	48.449 +568	31.84 −137	10.327 +367	29.41 −202
XII 35.2	38.022 +432	38.68 −136	59.723 +339	49.09 −206	49.017 +552	33.21 −188	10.694 +360	31.43 −222
Mean Place	34.974	73.40	56.951	72.58	46.757	30.90	08.098	19.38
sec δ, tan δ	+1.430	+1.022	+1.009	+0.138	+1.971	−1.698	+1.100	−0.459
dα(ψ), dδ(ψ)	+0.056	−0.39	+0.060	−0.39	+0.070	−0.39	+0.064	−0.39
dα(ε), dδ(ε)	+0.067	−0.19	+0.009	−0.19	−0.111	−0.20	−0.030	−0.20
Dble. Trans.	April 3		April 3		April 3		April 3	

> magnitude, extending from the north to the south pole, and some thousands of additional fainter stars promising to yield reasonably accurate proper motions.

The astronomers who constructed the General Catalogue have achieved this aim by incorporating positions from almost 250 individual catalogues, two of which were observed (at Albany, N.Y., 20,811 stars; at San Luis, Argentina, 15,333 stars) especially for the purpose of contributing position material to be incorporated in the GC [Boss et al., 1931; Boss and Boss, 1928].

In contrast to the German fundamental catalogues, the GC aimed not only at establishing a fundamental system by giving the positions and proper motions of a carefully selected number of frequently and fundamentally observed stars, but also to calculate for all stars brighter than seventh visual magnitude the best positions and proper motions that could be derived from the existing transit observations. Furthermore, whenever there was enough raw material for establishing accurate positional data for a fainter star this star was included. The results fill five volumes.

6.332 The GC System. The system of the GC resulted from a correction to the system of the Preliminary General Catalogue (PGC) [Boss, 1910] which was originally established by carefully combining data from 27 high-quality catalogues, nine of which were considered absolute [Boss, 1903]. Thirty-six additional catalogues, of which eight were absolute, served for the correction of the PGC system, so that one may say that the GC system is based on 53 catalogues, of which 17 were considered absolute. This is fewer than the number of catalogues incorporated to derive the system of the FK3. For details regarding the establishment of the PGC system, the above-mentioned papers must be consulted.

In establishing the systematic differences between two catalogues, the German School assumes that the various terms in the corrections are of no particular analytical mathematical form; the functions representing them are, therefore, not defined by formulas but given in the form of tables. The investigators of the American School assume that, for example, right ascension dependent corrections can be expressed by a Fourier series broken off with the 2α terms, so that

$$\Delta\alpha_\alpha = a + b \sin\alpha + c \cos\alpha + d \sin 2\alpha + e \cos 2\alpha, \qquad (6.13)$$

which is the general expression that is supposed to completely represent that part of the systematic right ascension difference between the two systems which depends on the right ascension itself.

The fact that competent investigators have tenaciously clung to their mutually contradictory practices, insisting that it was superior to the alternative, indicates that the question for the 'truest' function describing a systematic difference is not an easy one to answer. The German School insists that representation of a systematic difference by a model of the form (6.13) may not leave enough flexibility to represent all the significant features of the $\Delta\alpha$ function. The American School argues that the empirically found values for $\Delta\alpha$ at certain arguments cannot be sufficiently accurately determined to define a function more complicated than the simple terminated Fourier series, a form which is moreover suggested by certain physical and geometric considerations.

There is merit in both viewpoints. It has recently been shown that providing for too flexible a model may introduce systematic errors, just as much as not providing for a sufficiently flexible model may leave significant systematic differences unrepresented [Eichhorn, Googe, and Gatewood, 1967]. This paper also established a criterion for deciding which of two models will better represent an empirical function.

A breakthrough in the treatment of catalogue comparisons may recently have been achieved by Brosche who replaced the Fourier series model with the more flexible development of the difference between two systems as a series in spherical harmonics. The series is terminated when objective statistical tests show that the inclusion of more terms will not produce a significantly more accurate representation [Brosche, 1966]. This is a compromise between the established practices, and likely an improvement over both. Brosche's method, however, could not have been applied when the FK3 or the GC were compiled, even if it had then been known, because it requires a tremendous amount of numerical calculations which can be carried out efficiently only when an electronic computer is available.

Benjamin Boss, in deriving the corrections of the various incorporated catalogues to the GC system, assumed (as did Auwers and Kopff) that the corrections could always be written in the following form:

$$\begin{aligned} \Delta\alpha &= \Delta\alpha_\alpha + \Delta\alpha_\delta + \Delta\alpha_m , \\ \Delta\delta &= \Delta\delta_\alpha + \Delta\delta_\delta , \end{aligned} \tag{6.14}$$

where $\Delta\alpha_\alpha$ is a correction to the right ascension which depends on right ascension alone; and the other terms are explained analogously (m denotes the magnitude). The observed (and tabulated) systematic differences between the GC system and that of the individual catalogues were assumed to be of the form (6.14), where $\Delta\alpha_\alpha$ was restricted to the model shown in equation (6.13). The five coefficients a, b, c, d, e were found by a least squares solution. No analytical model was enforced on $\Delta\delta_\delta$, while $\Delta\alpha_m$ was restricted to be linear in the magnitudes. It is interesting that the models for the total correction $\Delta\alpha$ and $\Delta\delta$, as given by (6.14) are completely justifiable geometrically and physically if one allows for the influence of error sources of <u>known type</u> on star positions obtained by

TABLE 6.4 A Left-Hand Page of the GC

38 A GENERAL CATALOGUE OF STARS FOR 1950

No.	Draper No.	Mag.	Type	R. A. 1950			Epoch	An. Var.	Sec. Var.	$3^{d}t$	μ and 100Δμ		Prob. Errors αEp	100μ	α50
				h	m	s		s	s	s	s		″	″	″
17351	110979	8.1	G5	12	43	24.270	04.1	+3.1339	+.0091	+.009	+.0206	0	.11	.80	.38
17352	110956	4.86	B3			29.590	10.0	+3.4450	+.0586	+.046	-.0050	0	.09	.48	.21
17353	55°5216	9.17				30.682	07.4	+3.4427	+.0585	+.046	-.0076	- 1	.15	1.3:	.59
17354	111041	9.0	F8			42.547	02.6	+2.7723	-.0238	+.025	+.0019	0	.13	.54	.29
17355	111028	5.86	Ko			50.057	89.8	+3.0479	-.0006	+.010	+.0186	0	.06	.19	.13
17356	111020	7.91	Ko			54.629	99.8	+3.3016	+.0411	+.027	-.0046	0	.11	.95	.49
17357	111043	7.67	Fo			55.248	10.3	+3.0283	-.0005	+.010	-.0022	0	.10	.52	.23
17358	111031	6.89	G5			55.298	00.4	+3.1059	+.0104	+.010	-.0193	0	.09	.64	.33
17359	111044	8.13	Ko			58.566	03.3	+3.0744	+.0046	+.009	-.0013	0	.11	.56	.28
17360	111032	5.98	Ko		44	3.719	96.4	+3.2383	+.0254	+.015	-.0010	0	.11	.71	.40
17361	111066	6.72	F5			5.277	97.6	+2.9484	-.0078	+.012	-.0089	+ 1	.07	.41	.22
17362	111059	8.48	K2			6.263	04.8	+3.0730	+.0045	+.009	-.0020	0	.12	.63	.30
17363	111067	5.33	K2			8.902	07.2	+2.9964	-.0041	+.010	+.0006	0	.048	.22	.11
17364	111096	7.9	G5			23.666	01.8	+3.1018	+.0083	+.010	-.0054	0	.06	.52	.26
17365	111153	7.90	F8			27.032	98.6	+2.8290	-.0189	+.019	-.0005	0	.11	.47	.26
17366	111133	6.39	B9			29.867	96.8	+3.0473	+.0012	+.009	+.0022	0	.10	.45	.26
17367	111132	7.17	Ko			30.640	00.5	+3.0310	-.0003	+.010	+.0001	0	.11	.55	.29
17368	111105	7.34	A2			38.904	05.3	+3.5201	+.0692	+.060	+.0022	0	.14	.90	.44
17369	111117	9.68	Ko			39.790	02.6	+3.2194	+.0228	+.014	-.0029	0	.15	.87	.44
17370	111163	6.68	Ko			40.534	04.9	+2.9999	-.0036	+.010	+.0002	0	.11	.65	.32
17371	111164	6.05	A3			42.601	00.8	+3.0200	-.0018	+.010	+.0028	0	.06	.24	.13
17372	111156	7.22	G5			42.949	07.6	+3.1543	+.0148	+.011	-.0068	0	.13	1.06	.47
17373	111102	7.19	A5			44.251	02.9	+3.4853	+.0637	+.052	-.0031	0	.13	.80	.40
17374	111123	1.50	B1			47.055	03.8	+3.5069	+.0677	+.058	-.0054	- 1	.036	.16	.08
17375	111199	6.26	F8			58.191	98.4	+3.1007	+.0075	+.009	-.0001	0	.036	.24	.13
17376	111160	7.50	B8		45	6.065	06.5	+3.5150	+.0678	+.057	+.0013	0	.13	.95	.43
17377	111270	5.83	A5			11.251	87.7	+2.5608	-.0338	+.043	+.0025	0	.08	.36	.24
17378	111161	7.71	A2			11.301	99.5	+3.6783	+.0997	+.111	-.0074	- 1	.15	.83	.45
17379	111214	6.86	Ko			13.341	97.7	+3.1904	+.0188	+.012	-.0036	0	.10	.73	.39
17380	111226	6.29	B9			13.817	07.5	+3.1894	+.0187	+.012	-.0038	0	.06	.50	.22
17381	111239	6.66	Ma			18.568	02.4	+3.0552	+.0024	+.009	-.0004	0	.023	.13	.06
17382	111271	7.11	Fo			22.984	05.7	+2.9154	-.0106	+.013	-.0074	0	.12	.80	.37
17383	111235	6.98	Ao			29.476	08.4	+3.3292	+.0378	+.024	-.0025	0	.10	.69	.30
17384	111272	6.93	Ko			31.174	99.7	+2.9841	-.0051	+.011	+.0022	0	.10	.80	.42
17385	111306	6.77	Fo			31.342	95.3	+2.7434	-.0233	+.025	-.0108	+ 1	.09	.32	.20
17386	111285	7.31	G5			31.810	98.3	+2.9510	-.0078	+.012	-.0029	0	.08	.50	.27
17387	111335	5.67	K5			32.218	93.9	+2.4506	-.0365	+.053	+.0006	0	.05	.25	.15
17388	111111	7.20	A2			35.537	98.4	+4.6441	+.3549	+.969	+.0046	+ 3	.13	.80	.43
17389	111262	7.49	K2			35.683	96.1	+3.2348	+.0246	+.015	-.0038	0	.11	.75	.42
17390	111308	6.43	Ao			44.068	00.6	+3.0048	-.0024	+.010	-.0033	0	.08	.28	.16
17391	111295	5.80	G5			46.035	07.8	+3.1996	+.0207	+.013	-.0107	0	.048	.36	.16
17392	111307	7.85	Ma			48.898	99.5	+2.9781	-.0053	+.011	-.0008	0	.09	.54	.29
17393	111318	7.74	Ko			49.754	07.1	+2.9113	-.0111	+.013	-.0046	0	.11	.83	.38
17394	111347	8.02	F8			56.799	08.4	+2.9080	-.0113	+.013	-.0051*	0	.13	.90	.40
17395	111381	7.03	G5			57.664	05.8	+2.7224	-.0251	+.027	-.0012	0	.14	1.00	.46
17396	111283	7.60	Ao			58.766	02.6	+3.6513	+.0918	+.096	-.0014	0	.12	.60	.31
17397	111420	7.32	K5		46	5.235	03.3	+2.2980	-.0377	+.064	+.0096	- 2	.10	.64	.32
17398	111443	7.60	K2			19.155	10.9	+2.5933	-.0318	+.039	+.0058	- 1	.14	.83	.36
17399	111398	7.07	G5			20.880	98.4	+3.0304	-.0018	+.010	+.0160	0	.09	.48	.26
17400	111395	6.39	G5			20.938	11.8	+2.9226	-.0079	+.012	-.0249	+ 1	.07	.41	.17

TABLE 6.4 (continued) A Right-Hand Page of the GC

POSITIONS AND PROPER MOTIONS OF STARS FOR 1950														39
No.	Decl. 1950			Epoch	An. Var.	Sec. Var.	3dt	μ' and 100 Δμ'		Prob. Errors δEp.	100μ'	δ50	Remarks	
	°	′	″		″	″	″	″		″	″	″		
17351	- 9	2	14.26	02.2	-19.905	+.095	+.17	-.221	+ 1	.10	.73	.36		
17352	-56	12	56.58	03.9	-19.721	+.103	+.23	-.039	0	.08	.36	.18	B5s I	
17353	-56	13	47.64	03.2	-19.714	+.103	+.23	-.031	0	.16	1.13	.55	I	
17354	+50	5	45.69	89.6	-19.672	+.085	+.12	+.008	0	.11	.32	.23	A 8661 9.5 5″.8 207°	
17355	+ 9	49	8.48	89.3	-20.129	+.093	+.16	-.452	0	.05	.16	.11	K1 33Vir	
17356	-47	0	26.26	95.9	-19.697	+.101	+.21	-.021	0	.11	.73	.41		
17357	+ 9	33	31.33	06.4	-19.702	+.092	+.16	-.027	0	.10	.45	.22		
17358	-11	32	24.34	99.7	-19.623	+.094	+.17	+.052	- 1	.08	.61	.32		
17359	- 0	32	54.89	99.9	-19.674	+.094	+.16	.000	0	.11	.48	.26		
17360	-33	2	32.06	96.2	-19.711	+.098	+.19	-.037	0	.11	.73	.41	I	
17361	+24	25	15.63	96.4	-19.890	+.090	+.15	-.217	0	.06	.34	.19		
17362	- 0	23	22.86	01.7	-19.686	+.094	+.16	-.013	0	.12	.50	.27		
17363	+16	51	0.65	04.6	-19.672	+.092	+.15	.000	0	.048	.19	.10	K4 27Com	
17364	- 7	31	34.68	99.9	-19.716	+.095	+.17	-.048	0	.06	.42	.22		
17365	+43	25	29.87	95.0	-19.714	+.088	+.13	-.047	0	.10	.40	.24		
17366	+ 6	13	26.82	93.4	-19.715	+.094	+.16	-.049	0	.09	.39	.24		
17367	+ 9	20	12.92	97.8	-19.685	+.094	+.16	-.019	0	.10	.47	.26		
17368	-59	48	29.87	00.6	-19.686	+.108	+.24	-.023	0	.13	.69	.36		
17369	-29	56	22.25	97.6	-19.697	+.099	+.19	-.034	0	.14	.61	.34		
17370	+15	51	52.72	06.1	-19.695	+.093	+.15	-.032	0	.11	.73	.34		
17371	+12	13	51.76	94.2	-19.687	+.094	+.16	-.025	0	.05	.21	.13	Aon 34Vir	
17372	-18	43	52.75	03.8	-19.659	+.097	+.18	+.003	0	.13	.90	.44		
17373	-58	1	29.26	99.8	-19.698	+.107	+.24	-.036	0	.12	.71	.38		
17374	-59	24	56.95	99.4	-19.687	+.108	+.24	-.026	0	.039	.16	.09	B1s βCru I	
17375	- 6	1	42.65	97.2	-19.708	+.096	+.17	-.050	0	.039	.23	.12	F5	
17376	-59	19	14.91	01.8	-19.679	+.109	+.24	-.023	0	.12	.67	.35		
17377	+63	3	12.73	84.5	-19.660	+.082	+.10	-.006	0	.08	.30	.21	A4n	
17378	-66	51	30.83	95.3	-19.697	+.113	+.28	-.043	0	.13	.75	.43	I 8.8 13″ 97°	
17379	-24	44	31.97	92.9	-19.683	+.099	+.18	-.029	0	.11	.67	.40	6.96	
17380	-24	34	46.44	08.0	-19.624	+.099	+.18	+.030	0	.06	.46	.20		
17381	+ 3	50	43.21	03.8	-19.661	+.096	+.16	-.008	0	.024	.13	.06	M4 35Vir	
17382	+29	48	15.52	06.7	-19.664	+.092	+.14	-.013	0	.11	.83	.38		
17383	-44	26	46.13	04.5	-19.667	+.104	+.21	-.018	0	.09	.51	.25		
17384	+19	6	33.69	99.8	-19.686	+.094	+.15	-.038	0	.09	.78	.40		
17385	+50	25	45.60	94.1	-19.670	+.087	+.12	-.022	0	.08	.29	.18		
17386	+24	22	4.63	97.3	-19.645	+.093	+.15	+.003	0	.08	.41	.23		
17387	+67	3	46.69	94.3	-19.656	+.079	+.09	-.008	0	.047	.18	.11	K5 7Dra	
17388	-80	25	47.46	95.6	-19.652	+.142	+.54	-.005	0	.11	.75	.42		
17389	-32	3	7.61	96.2	-19.663	+.102	+.19	-.016	0	.10	.73	.40		
17390	+13	49	33.39	94.4	-19.680	+.095	+.15	-.036	0	.07	.20	.13	Aon 28Com	
17391	-27	19	26.60	07.7	-19.716	+.101	+.19	-.071	0	.05	.26	.12		
17392	+19	35	41.38	98.1	-19.656	+.095	+.15	-.012	0	.08	.44	.24		
17393	+30	40	8.00	07.5	-19.646	+.092	+.14	-.003	0	.11	.83	.37		
17394	+31	2	21.88	09.5	-19.607	+.093	+.14	+.034*	0	.12	.95	.40		
17395	+52	43	33.15	06.5	-19.645	+.087	+.12	-.004	0	.14	1.06	.48		
17396	-65	19	14.33	95.4	-19.651	+.114	+.27	-.010	0	.11	.54	.31	I 9.2 8″.6 9°	
17397	+71	13	10.83	06.5	-19.704	+.076	+.07	-.066	0	.10	.87	.39		
17398	+61	5	33.64	04.4	-19.648	+.084	+.10	-.014	0	.13	.56	.29		
17399	+12	22	14.59	95.6	-19.774	+.098	+.16	-.140	0	.08	.40	.23		
17400	+25	6	50.22	09.6	-19.750	+.093	+.14	-.116	- 1	.07	.40	.18	G6	

transit observations. Many decades of experience have since shown that in transit observation there will occur systematic errors that cannot be split into components in the way indicated in (6.14), and no definite causes have yet been pinpointed which produce the actually observed systematic errors in transit observations.

6.333 Description of the GC. The first volume of the GC (in the following abbreviated as GCI) contains a description of the various steps that led to its compilation, as well as a history of the project. The detailed description of the various columns in the catalogue proper (pages 44-53) must be regarded as m a n d a t o r y reading for anyone using the GC. This is followed by the appendices. Volumes II through V contain the mean positions; annual, secular variation and the third term (with p r o p e r motions; see equations (4.20 and (4.91)); proper motions for equinox and epoch 1950.0 of the 33,342 stars; and peripheral data (numbers, magnitudes, spectra)(see Table 6.4). Also listed for every star are the epoch (i.e., the weighted average of the epochs from which the published epoch was derived) and the probable errors of the star position at the epoch, and at 1950.0, and that of the proper motion components. Note that the published positions refer to the epoch 1950; the epochs are to be used only for calculating the errors of the positions at epochs other than the one published, as shown by formula (6.9).

Appendix I of the first volume gives the ephemerides of the polar stars. For these, the series development of precession (equation (4.20)), broken off after the third term, will no longer give accurate results. Appendix II treats peculiarities of proper motion, mostly concerning binary stars. The systematic corrections to be applied to reduce the positions contained in about 250 catalogues to the system of the GC are given in Appendix III.

The standard deviations of star positions in the GC vary, of course, from star to star even for the same epoch and depend mainly on the number of incorporated catalogues in which the star appeared. On the average, however, they are about 0".15 at the epoch in both coordinates; but due to the errors of the proper motions, they rise to at least an average of 0".70 at the epoch 1970 [Schlesinger and Barney, 1939]. Thus at the time of this writing (early 1967), the GC is rapidly losing its usefulness and becoming a work of mainly historical interest. The deviations of the system defined by the GC from the ideal fundamental system (based on Newcomb's constant of precession) are increasing, and a much better approximation of the ideal fundamental system is now available through the FK4.

Although tables and even formulas for the reduction of the GC system to that of the FK4 have been derived [Brosche, Nowacki and Strobel, 1964; Brosche, 1966], the accidental errors of the positions have by now become so large (because of the uncertainties of the proper motions and the long time interval between now and the average epoch of the incorporated catalogues) that positions of better accidental accuracy are available from other sources.

At the present time, the GC will be used to advantage only in two types of applications, namely, either when star positions are not required to have an accidental dispersion of less than a second of arc, or when only the proper motions (systematically corrected, of course, for reduction to the FK4 or another system) are used for transferring accurate star positions of more recent origin than those in the GC to a different, usually even more recent epoch. There are still stars for which the proper motion as listed in the GC is the most accurate available in print.

6.34 The N30 Catalogue

6.341 History and Establishment of the System. It may at this point be well to recall that one of the aims of *celestial* mechanics is the practical verification of the laws on which mechanics is based, namely, Newton's postulates of motion, which have so far not been modified at all, and Newton's law of gravitation, which has suffered a minor (but nevertheless significant) modification through the general theory of relativity.

The experimental verification of the laws of motion can be carried out only when an inertial system is available, and it is in no mean part toward providing this that the efforts of the catalogue astronomers are directed.

Systematic errors in the star catalogues which define the system in which the planetary positions are observed, will produce apparent discrepancies between the computed and the observed planetary positions and complicate the verification of the laws of mechanics.

It had become apparent about one decade after the GC was published that it was affected by known systematic errors. Also, a number of excellent catalogues had then become available which had not appeared in time to be incorporated into the GC. Therefore, D. Brouwer of the Yale University Observatory and G.M. Clemence, then of the U.S. Naval Observatory at Washington, D.C. who, of the scientists in the U.S. were most concerned with this particular aspect of celestial mechanics, suggested the correction of the GC system, by whatever means were then available.

This work was carried out as a joint enterprise of the Yale University Observatory and the U. S. Naval Observatory under the direction of H. R. Morgan and published under the title 'Catalog of 5268 Standard Stars, 1950, Based on the Normal System N30' [Morgan, 1952].

Over seventy catalogues, twenty of them absolute and thirteen partially absolute were incorporated into the whole project.

Observations of the sun and the planets, as well as the absolute catalogues, were utilized for the corrections of the equinox and the equator point, while the absolute catalogues only (according to their suitability) were used for the correction of the GC system.

The $\Delta\alpha_\alpha$ and $\Delta\delta_\alpha$ were represented by means of formulas of type (6.13). The corrections in right ascension were presumed to be independent of the declination, while the corrections to the declination system were allowed to depend on declination through the equator point correction. $\Delta\alpha_\delta$ and $\Delta\delta_\delta$ as well as magnitude corrections in α and δ were applied in tabular form.

From there on the procedure was as follows: The absolute and the relative positions from those star catalogues which had not been incorporated into the GC were combined into normal positions by taking their weighted averages (not allowing for proper motions) after the catalogues had been reduced to the corrected GC, i.e., N30 system. These positions constitute the system of N30 positions at the mean epoch of the non-GC catalogues, which is on the average about 1930.

The proper motions were determined by comparing these new 'normal' positions with the GC positions at 1900 (about the average mean epoch of the GC stars) which had been systematically corrected only by the application of $\Delta\alpha_\alpha$ and $\Delta\delta_\alpha$, i.e., no corrections varying with declination were applied.

6.342 The Accuracy of the N30. The N30 catalogue was rather severely criticized by A. Kopff who supervised the work that led to the FK3 and also the initial steps in the compilation of the FK4 [Kopff, 1954]. Now that the FK4 is available, it is fair to say that Kopff's criticism was partly justified. While the right ascension (position and proper motion) systems of the FK4 and the N30 agree better than those of the FK4 and the FK3, the N30 declinations (positions and proper motions) do not agree systematically on the average better with the FK4 than those of the FK3, although the N30 is based on more recent material than the FK3. Whether the N30 or the FK3 represents a superior system has become an idle question since the FK4 has become available.

The N30 is at the present time valuable mainly as a source of highly accurate positions of bright stars, many of which will not be found in the FK4 and FK4 Sup. (As mentioned above, positions of those stars in the FK4 Sup which occur in the N30 are only N30 positions to which the systematic corrections FK4–N30 have been applied.) The systematic corrections of the system of the N30 to that of the FK4 are readily available [Brosche, Nowacki, and Strobel, 1964].

The standard deviations of the positions listed in the N30 vary considerably depending on the number of star catalogues from which they were compiled. Table 6.5 gives the accidental standard deviations of the N30 data.

TABLE 6.5 Standard Deviations of Positions in the N30

No. of Stars	Epoch 1932	Epoch 1970	Standard Deviation of Annual Proper Motion
100	0".039	0".060	0".0012
200	48	77	16
200	48	87	19
300	61	113	25
500	71	154	36
1000	77	160	37
2200	104	247	59
700	134	291	68

From this, one can see that the majority of star positions listed in the N30 will have a standard deviation of about 0".2 in either coordinate for the epoch 1970. The uncertainity of the system must, of course, be taken into account separately. Since it is to be assumed that the N30 positions will be used in the system of the FK4, the uncertainty of the system will be composed of that of the FK4 system and the uncertainty of the reduction of the N30 to the FK4 system.

6.4 Relationship Between Catalogue Systems

6.41 Representation of the Systematic Differences

The systematic errors in the positions contained in a catalogue are usually of such a nature that the coordinate system which is represented by the catalogued positions of stars within a limited region in the sky varies very slightly (i. e., represents a somewhat different coordinate system) depending on which region in the sky one considers (see section 6.313). The positions in a catalogue not displaying this feature would be self-consistent. The frame of reference defined by these positions could be brought to coincidence with an ideal system by an orthogonal transformation, i. e., a rotation.

It is safe to say that in practice no system is ever self-consistent, but some are closer to this ideal than others.

The frame of reference defined by the FK4 defines the best easily accessible approximation to a self-consistent system as well as the most error free system in general, and one will, therefore, often need means for the reduction of positions from some system to that of the FK4, or, when required, generally the relationship between any two systems.

In principle, the systematic differences $\Delta\alpha$ and $\Delta\delta$ between the positions as given in two catalogues will be of the form

$$\Delta\alpha = \Delta\alpha_{\alpha,\delta,m} ,$$
$$\Delta\delta = \Delta\delta_{\alpha,\delta,m} ,$$

where the notation corresponds to that of equation (6.14). In order to be fully general, one would have to admit the spectrum and also perhaps other parameters as quantities on which the systematic difference between two systems depends, but this has so far not yet been done anywhere and there appears at this time no evidence that would make it necessary.

For the description of the systematic differences between two systems of position which are represented all over the sky, one would, therefore, expect a set of double entry tables for different magnitudes.

Fortunately, in all practical cases (except one; see section 6.44) so far treated, it was found that the systematic differences can be written in the form

$$\begin{aligned} \Delta\alpha &= \Delta\alpha_{\alpha,\delta} + \Delta\alpha_{m} , \\ \Delta\delta &= \Delta\delta_{\alpha,\delta} + \Delta\delta_{m} , \end{aligned} \tag{6.15}$$

i.e., the magnitude dependent parts of the systematic differences are themselves position independent. Thus, the systematic differences may be displayed in the form of a double entry table plus a single entry table which allows one to take the magnitude dependence into account.

The form in which the systematic differences are given can often be specialized further. It is clear that for narrow zones of declination, one will be able to express the corrections in a form given by equations (6.14); thus

$$\begin{aligned} \Delta\alpha &= \Delta\alpha_{\alpha} + \Delta\alpha_{\delta} + \Delta\alpha_{m} , \\ \Delta\delta &= \Delta\delta_{\alpha} + \Delta\delta_{\delta} + \Delta\delta_{m} . \end{aligned}$$

In defining the relationship between two fundamental systems, one must take account of the fact that such a system, by definition, consists of a set of positions and a set of proper motions (or the equivalent thereof). Since the proper motions vary only very slowly with time, at least for declinations which are not too high, one may in practice regard the systematic differences between the proper motions systems of two fundamental catalogues as time invariant.

If $\Delta\alpha(T_0)$ is a systematic difference between the right ascensions of two fundamental catalogues at the standard epoch T_0, and $\Delta\mu$ is the corresponding systematic difference between the proper motions, we have seen that the systematic differences $\Delta\alpha(T)$ at some other epoch T is given by equations (6.8):

$$\Delta\alpha(T) = \Delta\alpha(T_0) + (T - T_0)\Delta\mu .$$

The same is true for declinations. This is a very important point when star positions from a catalogue in a system A are to be reduced to a system B. If the catalogue in which the star occurs is in the system A (i.e., if the positions in it show no significant systematic differences against the A positions) as recorded in catalogue A, and if the star's position is listed for the epoch T, the systematic difference between the A and B at the same epoch T must be used as calculated from equation (6.8).

6.42 Sources for Reduction Tables to the GC

Unfortunately, tables of systematic corrections of the individual catalogues to the system of the FK4 are not yet available in the printed literature. The most comprehensive collections of tables for the reduction of catalogues positions to a fundamental system are those tables for the reduction of position from various catalogues to the system of the GC which are described below.

Appendix III of the first volume of the GC gives tables containing $\Delta\alpha_\alpha$, $\Delta\alpha_\delta$, and $\Delta\alpha_m$; $\Delta\delta_\alpha$ and $\Delta\delta_\delta$ for 238 catalogues; as well as tables from which one may calculate the weight of a position in a particular catalogue, depending on its magnitude and on how many times the star was observed. The unit weights in these tables (as well as all weights in the GC) correspond to a probable error of 0''.30 in $\alpha \cos \delta$ and in δ, thus to a standard error of 0''.445. The tables for the corrections follow the form of equations (6.8) and give corrections in the sense GC minus catalogue.

Another source is the collection of tables in [Gyllenberg, 1948], in which reduction tables are provided for catalogues not considered in Appendix III of the GC, either because the compilers of the GC regarded them to be of such inferior accuracy that their reduction to the system of the GC did not seem to be worthwhile or because the catalogues appeared after the work on the GC had been completed. The latest publication date of any catalogue for which reduction tables are given is 1948. Gyllenberg also provides tables for computing weights of positions, exactly as in the GC. His correction tables were derived essentially by a graphical method (at that time and place the only way in which a single individual could have accomplished a task of such magnitude). He represented the corrections $\Delta\alpha_\alpha$ and $\Delta\delta_\alpha$ as sine curves and the terms $\Delta\delta_\delta$ and $\Delta\alpha_\delta$ as straight lines or smoothed lines of simple curvature; $\Delta\alpha_m$ was usually assumed to be a linear function of magnitude.

There are no other recent sources where collections of corrections to the system of some fundamental catalogue have been published. For catalogues that were published after the GC, systematic differences with respect to the GC can often be found in the introductions or appendices.

At the present time, one will usually want star positions in the system of the FK4, yet no systematic corrections to the FK4 system have been published for any catalogue although they were derived and applied at the Heidelberg Astronomisches Rechen-Institut as part of the work that led to the formation of the FK4. Their eventual publication by the Rechen-Institut is contemplated.

6.43 Systematic Differences Between the Fundamental Systems

Since almost all published corrections are to the GC system, systematic differences between the GC and the FK4 are necessary to reduce positions further from the GC system to that of the FK4. Other systematic differences one is likely to need are those between FK4 and FK3, and FK4 and N30.

Tables for the systematic differences FK4-N30 and FK4-GC are given in [Brosche, Nowacki and Strobel, 1964] and tables for FK4-FK3 are contained on pages 131-134 of [Fricke and Kopff, 1963].

In the literature there are also other widely scattered comparisons between the GC, FK3, and N30, but the tables mentioned above seem to be more accurate than any of the previously published ones. Relationships between any two of the catalogue systems of the GC, N30, FK3 may be derived from these tables.

6.44 Brosche's Method

Corrections FK4-GC which are slightly different from those given in [Brosche, Nowacki and Strobel, 1964] may be found in [Brosche, 1966]. This paper presents in various aspects a radical deviation from the classical method of establishing the systematic difference between two catalogue systems. As indicated before (see section 6.322), Brosche develops the systematic differences between two systems as a series in spherical harmonics which is cut off when a significance test shows that higher order terms would not bring about a significant improvement in the accuracy of the representation of the individually observed differences at selected points.

Another field in which Brosche has completely abandoned the established practice is the treatment of magnitude equations. As indicated by formula (6.15), it was up to the publication of Brosche's paper always tacitly assumed that the so-called magnitude equation, i. e., the magnitude dependent part of the systematic difference between two catalogue systems is independent of the position.

Brosche applied his method to make a comparison FK4-GC separately for stars over and under fifth magnitude. A comparison of the coefficients of the same spherical harmonic for the group of bright stars and the group of faint stars definitively ruled out the classical model (6.15) and demonstrated that the general corrections would have to be of the form $\Delta\alpha_{\alpha,\delta,m}$ and $\Delta\delta_{\alpha,\delta,m}$. Brosche discusses this and shows how a general expression for these functions of three arguments could be expanded in a series of spherical harmonics. For practical applications he recommends, however, that the magnitude equation be assumed linear in the magnitude, i. e., the corrections to be of the form

$$\begin{aligned} \Delta\alpha &= \Delta_0\alpha_{\alpha,\delta} + (m-m_0)\Delta_1\alpha_{\alpha,\delta}, \\ \Delta\delta &= \Delta_0\delta_{\alpha,\delta} + (m-m_0)\Delta_1\delta_{\alpha,\delta}, \end{aligned} \tag{6.16}$$

(m_0 being a suitably chosen zero point for the magnitudes) where in practice the Δ_0 and Δ_1 will be developed in series of spherical harmonics (and analogously for the proper motions). At this time (early 1967) no results of catalogue comparisons based on this have yet been published.

It is interesting that the model (6.16) for catalogue system comparisons was implicitly postulated by [Kopff, 1954] who notices that the systematic differences between the N30 and the GC show different behavior, depending on whether the stars which also occur in the FK3 are compared or those stars, common to N30 and GC, which are not in the FK3. It is easy to see that these are essentially two magnitude groups since the FK3 contains mostly bright stars. Kopff, of course, was looking for the source of this error definitely outside the FK3.

The model (6.16) for the systematic differences between catalogues is at this time probably the most sophisticated one which can be supported by the available material. When tables of systematic catalogue corrections based on this model are published, one may expect the residual magnitude equation to be removed from many catalogues and, in general, the 'true physical system' of right ascensions and declinations to become better defined. Until then one will have to use the material which has so far been published.

Experience has shown that 'error models' based on purely physical and geometrical considerations, as for instance (6.13) and (6.14), had to be abandoned in the end in favor of more general formulas which are actually pure interpolation formulas. It turns out that no matter how accurately an investigator designs an error model that takes all sources of errors into account which he could conceive of, there are always those errors which nobody suspected and which are not covered by the model used.

6.5 Lists of Star Catalogues

6.51 Old Lists

Star positions have been observed and published in more or less extensive lists (i.e., catalogues) since at least several hundred years B.C. Since about 1700 the telescope has been employed in the service of obtaining accurate star positions, and the meridian circle emerged as the instrument par excellence for the determination of star positions. During the second half of the nineteenth century astronomers in all countries felt the need for organization in the activity of determining star positions, and (as remarked above) the Astronomische Gesellschaft (AG) organized the observation of the now famous zone catalogues, of which the third generation is about to be published.

However, there are now many hundreds of catalogues, of which the AG catalogues are only a small part, and the position of any star would be listed in many of them. Around 1900 an enterprise was started which, now essentially in its third generation, lists for almost every star that was at any time included in a catalogue (except the Astrographic Catalogue, see section 6.7) all catalogues in which it occurred and its number in that catalogue.

The first step toward this goal was the compilation of lists of star catalogues. The first of these was published by Knobel and contains some useful information on pretelescopic catalogues [Knobel, 1877]. Based in part on this list, Ristenpart appended a list of all star catalogues published up to 1900 to his article 'Sterncataloge und-Karten' in W. Valentiner's 'Handwörterbuch der Astronomie' [Ristenpart, 1901a]. This article was subsequently published in somewhat extended format as a monograph [Ristenpart, 1901b]. Essentially, this list contains the author, number of star positions given in the catalogue, the equinox to which the positions refer, and the accurate title.

6.52 The 'Geschichte des Fixsternhimmels' (GFH)

It also seems that Ristenpart was the first to have had the idea that all the information contained in all star catalogues should be used to derive one general catalogue of positions and proper motions, and that this catalogue should be kept up to date by immediately including the information contained in any new catalogue as it became available. Ristenpart, then an assistant astronomer at the Heidelberg Observatory, communicated these ideas to Auwers, who in turn obtained in 1898 from the Royal Prussian Academy of Sciences an appropriation of the funds necessary to make Ristenpart's proposals a reality.

The realization of Ristenpart's ideas in their entirety was a far too ambitious enterprise for the pre-desk calculator (and even for the pre-computer) era. All that was evenutally done and published under the title 'Geschichte des Fixsternhimmels' (GFH) was a list of all star positions observed on transit type instruments up to 1900 arranged star by star [Preussische Akademie der Wissenschaften, 1922-]. Most of the positions were precessed to 1875.0, using the Struve-Peters value for the precession. (The exceptions are a few hours of right ascension in the southern hemisphere.) The completed work takes up 48 volumes, each containing the information for stars in one hour of right ascension, with one volume each for every right ascension hour in the northern and in the southern hemispheres. All of the GFH has appeared except for one volume referring to the southern hemisphere, the publication of which is imminent.

The volumes 1 through 24 containing the information for the northern hemisphere were published from 1922 to 1936, and by 1943 the first seven volumes containing southern hemisphere stars were published. After World War II part of the material had been removed from Berlin to Bonn, and the edition of the rest of the work was accomplished by two different agencies, one each in West and in East Germany.

The GFH is mainly valuable because it contains (almost) all star positions observed up to 1900 (excepting especially positions obtained photographically). The positions were taken unaltered from the original catalogue, except that obvious errors were corrected (when they were detected) and except that they were precessed (on the basis of the Struve-Peters constant) to 1875.0. It has, therefore, become unnecessary to consult the original catalogues, which is a tremendous advantage since these are often extremely hard to find.

The systematic corrections to the GC system (see section 6.42) are available for many catalogues incorporated in the GFH so that the positions can be processed and utilized in modern investigations.

One of the by-products of the GFH is Ristenpart's list of errata in the star catalogues of the eighteenth and nineteenth centuries, the most extensive such list which has ever appeared [Ristenpart, 1909].

6.53 The 'Index der Sternörter'

Although the presently available automatic data processing equipment would facilitate enormously the continuation of the GFH to include catalogues published recently, this apparently is not being done.

The task accomplished by the GFH is, in a very restricted way, being continued by the 'Index der Sternörter.' The first Index, now called 'Index I,' was originally compiled under the direction of R. Schorr [Schorr and Kruse, 1928]. It contains, tabulated against the star number, only the catalogue numbers (no positions) of stars referred to those catalogues which contain the stars' originally observed right ascensions and/or declinations. The 281 catalogues covered in the Index I cover the period 1900 to 1925 and are coded by number and listed in an appendix. One volume each contains material pertaining to the northern and the southern hemispheres. There is a certain amount of overlap with the GFH.

The Index must thus be used in conjunction with the original catalogues if actual position information is required.

Schorr's work was continued by A. Kahrstedt who issued the 'Index II' which provides an index to those catalogues that were published between 1925 and 1960 [Kahrstedt, 1961-1966]. As in the Index I, photographic positions were included only if they are referred to a system of reference stars which were observed on meridian circles (or equivalent instruments). It is a measure for the increased production in the field when one considers that the Index I could be published in two volumes, while nine volumes were necessary for the Index II, without the inclusion of the extensive Astrographic Catalogue (see section 6.7).

6.54 The 'Astronomischer Jahresbericht'

If a complete record of originally and independently observed positions of a certain star is required, one will have to search the 'Astronomischer Jahresbericht' (AJB) for catalogues that may possibly contain a newly observed position of the star. The AJB, published annually (at present) by the Astronomisches Rechen-Institut, Heidelberg, contains a listing, as complete as humanly possible, of the astronomical literature published in a particular year. References to catalogues will be found there under the heading, 'Sternkataloge, Sternkarten.'

6.6 The AG-Type Catalogues

6.61 The AGK1 and Its Extensions

6.611 The Original AGK1. Although the mode of observing star positions by means of meridian circles became well established and standardized during the eighteenth and the first half of the nineteenth century, catalogues of star positions were observed rather haphazardly favoring mostly bright stars. Observers paid little regard to which stars others had in their observing lists; neither did they pay any attention to complete coverage with respect to magnitudes.

In August of 1867, F. W. Argelander proposed to the Astronomische Gesellschaft (then only two years old) that it should organize a concerted effort directed at the observation of all stars down to the ninth magnitude. The positions thus obtained should be observed at meridian circles, relative to a homogeneous system of about one-half thousand reference stars. The observations were originally planned only for stars with declinations between +80° and -2°, but in 1887 it was agreed to observe in the same manner the stars between -2° and -23°, the latter part being designated as the 'Zweite Abteilung' (second division) of the whole enterprise, which had been named 'Catalog der Astronomischen Gesellschaft' and is now commonly known as the AGK1 [Astronomische Gesellschaft, 1890-1912; Gonnesiat, 1924].

6.612 The South American Extensions. Late in the nineteenth century, the South American astronomers at Cordoba and at LaPlata decided to continue observations following the plan of the AGK1 from -23° to -82° and thus to complete the whole enterprise from pole to pole (with the exception of the polar caps).

The last South American observations were made as late as 1935, when elsewhere the observation of zone catalogues on meridian circles had been abandoned in favor of the more accurate and, above all, more efficient photographic methods.

Especially on the northern hemisphere, the time will soon have come when all AG stars will have been observed photographically at least twice (in the AGK2 and the AGK3, to be discussed below) so that some astrometrists feel that the less accurate AGK1 has completely lost its value, especially as the various zones are afflicted by different systematic errors which may be difficult to remove. Expert opinions are divided, however, and many regard the AGK1 as a document of considerable permanent value.

In any case, it is a monument to the industry and dedication of the scores of astronomers who contributed to the first organized international effort of the astronomical community.

Table 6.6 shows the zones of the AGK1 and their South American counterparts. The columns are mostly self-explanatory. $E(\alpha)$ and $E(\delta)$ are the average standard errors of catalogue positions. These vary greatly, of course, from star to star depending on how often the star in question was observed. The following comments are indicated in the 'Remarks' column:

(1) The Dorpat zone was never quite finished and is given only in the form of differences against Berlin C.

(2) The instrument was transported from Helsingfors to Gotha while the observations were going on. Observations marked by an R in the catalogue were obtained on a refractor.

(3) The data refers not to the originally published catalogue, but to the new reduction in [Mönnichmeyer, 1909].

(4) The catalogue is preceded by a very detailed introduction by A. Auwers which is of great interest for the history of the AGK1.

(5) The introductions to the Cordoba catalogues are either very scanty or completely lacking.

(6) The catalogue appeared too late to be included in the GC, and no systematic corrections to the GC system for Cordoba D are available in Gyllenberg's collection either. The catalogue is supposed to be in the system of the GC, and an Appendix gives a comparison of Cordoba D and the GC. From this, reduction tables could be constructed.

(7) Only the first part of this catalogue is so far available. A comparison with the GC, from which reduction tables could be constructed, is given in the Appendix.

6.62 The Photographic Zone Catalogues

6.621 The AGK2. The pioneer work by Schlesinger had shown that positions of an accuracy comparable to and better than that of excellent meridian observations could be obtained from photographic plates taken with wide angle ($5^\circ \times 5^\circ$ to $15^\circ \times 15^\circ$) medium focal length (about 200 cm) cameras [Schlesinger and Hudson, 1916]. One of Schlesinger's great merits is to have realized much of the great potential accuracy of the photographic-astrometric technique which had not been fully exploited during the work on the Astrographic Catalogue (see section 6.7).

In 1921, R. Schorr and F. Cohn proposed to the Astronomische Gesellschaft what essentially amounted to a repetition of the AGK1 by photographic means. A study commission, appointed in the same year, reported in 1924 the following guidelines which then were followed in the course of the work:

(1) All published positions should be in the same fundamental system. (This requirement led to the revision of the NFK and the establishment of the FK3.)

TABLE 6.6 The Zones of the AGK1 and Their South American Counterparts

Declination Zone From	To	Observed At	Coord. System	E(α)	E(δ)	Average No. of Obs.	Average Epoch	No. of Stars	Published By (if other than AG)	Remarks (See Text)
+80°	+75°	Kasan	1875.0	0″.45	0″.66	4.4	1875	4281		
+75°	+70°	Berlin (C)	1905.0	.29	.35	2.3	1905	3461		
+75°	+70°	Dorpat	1905.0	.45	.71		1875			1
+70°	+65°	Christiania	1875.0	.45	.89	2.9	1875	3949		
+65°	+55°	Helsingfors and Gotha	1875.0	.86	.71	2.2	1876	14680		2
+55°	+50°	Harvard	1875.0	.57	.73	3.1	1877	8627		
+50°	+40°	Bonn	1875.0	1.23	.74	2.6	1875	18457		3
+40°	+35°	Lund	1875.0	.60	.47	2.7	1881	11415		
+35°	+30°	Leiden	1875.0	.51	.54	2.2	1873	10209		
+30°	+25°	Cambridge (E)	1875.0	.70	.45	4.0	1879	14441		
+25°	+20°	Berlin (B)	1875.0	.29	.29	2.7	1881	9208		
+20°	+15°	Berlin (A)	1875.0	.48	.57	2.5	1870	9789		4
+15°	+10°	Leipzig I	1875.0	.77	.56	2.3	1880	9547		
+10°	+ 5°	Leipzig II	1875.0	.56	.50	2.3	1885	11875		
+ 5°	+ 1°	Albany	1875.0	.56	.52	2.3	1880	8241		
+ 1°	− 2°	Nicolayev	1875.0	.66	.93	2.3	1885	5954		
− 2°	− 6°	Strassburg	1900.0	.45	.45	2.9	1891	8204		
− 6°	−10°	Wien-Ottakring	1900.0	.42	.42	2.2	1895	8468		
−10°	−14°	Harvard	1900.0	.62	.68	3.2	1892	8337		
−14°	−18°	Washington	1900.0	.72	.51	2.2	1895	8824		
−18°	−23°	Algiers	1900.0	.50	.31	4.3	1892	9997	[Gonnesiat, 1924]	

TABLE 6.6 (continued)

Declination Zone From	To	Observed At	Coord. System	E(α)	E(δ)	Average No. of Obs.	Average Epoch	No. of Stars	Published By (if other than AG)	Remarks (See Text)
-22°	-27°	Cordoba (A)	1900.0	0″.63	0″.63	2.5	1895	15975	Cordoba Res., Vol. 22	5
-27°	-32°	Cordoba (B)	1900.0	.54	.54	2.7	1896	15200	Cordoba Res., Vol. 23	5
-32°	-37°	Cordoba (C)	1900.0	.57	.57	3.0	1897	12757	Cordoba Res., Vol. 24	
-37°	-47°	Cordoba (D)	1950.0	.37	.38	2.7	1935	16610	Cordoba Res., Vol. 38	5,6
-47°	-52°	LaPlata (F)	1935.0	.50	.50	3.7	1935	4828	LaPlata Pub., Vol. 13	
-52°	-57°	LaPlata (A)	1925.0	.58	.58	2.4	1915	7412	LaPlata Pub., Vol. 5	
-57°	-63°	LaPlata (B)	1925.0	.36	.34	3.5	1916	7792	LaPlata Pub., Vol. 7	
-62°	-66°	LaPlata (C)	1925.0	.59	.47	2.7	1920	4412	LaPlata Pub., Vol. 8	
-66°	-72°	LaPlata (D)	1925.0	.53	.47	3.0	1920	4513	LaPlata Pub., Vol. 9	
-72°	-82°	LaPlata (E)	1925.0	.55	.48	2.7	1929	22486	LaPlata Pub., Vol. 10, No. 1	7

(2) A sufficient number of reference stars were to be observed at several meridian circles, contemporaneously with the photographic plates and reduced to the common fundamental system.

(3) The plates were designed to be obtained at a small number of observatories on astrographs of identical design.

The positions of the reference stars (about one per square degree) were observed at a series of meridian circles, mostly in Germany. Every star was observed on at least two different instruments. The observations were sent to the Astronomisches Rechen-Institut, then located at Berlin-Dahlem and under the direction of A. Kopff. There they were reduced to the system of the FK3 and compiled into one homogeneous catalogue, the AGK2R (the R denoting 'reference')[Astronomisches Rechen-Institut, Berlin-Dahlem, 1943].

The photographic plates were exposed at the observatories at Bonn (for declination between +20° and -2°), Hamburg-Bergedorf (+70° to +20°, later also +90° to +70°) and Pulkovo (+90° to +70°) on Zeiss Astrographs specially designed and built for this purpose. These instruments are equipped with four-component objectives of 110 mm effective opening and 2060 mm focal length, so that one millimeter in the focal plane at the tangential point corresponds to 100 seconds of arc. The exposed plates have a useful field of 5°×5°. Most of the plates were obtained during the years 1929 and 1930.

In 1937, the Pulkovo astronomers informed their German colleagues that the work on the region assigned to them was to be published in Pulkovo. The Bergedorf and Bonn observatories thus published the AGK2 in fifteen volumes for declinations between +90° and -2° since, with remarkable foresight, the plate coverage had been extended to the pole [Schorr and Kohlschütter, 1951-1958].

The Pulkovo observatory did indeed publish the results of their efforts [Belyavsky, 1947]. The standard deviations of these positions turned out to be 0".28 in α and 0".24 in δ, considerably higher than those of their German counterparts, although the procedures employed for photography, measurement, and reduction were practically identical with those at the German observatories. The explanation for this discrepancy may perhaps be found in the optical system, since all three observatories complain about having had to send the optics back to the manufacturer for further adjustment, and Pulkovo complains that Zeiss never succeeded in adjusting the instrument to their complete satisfaction.

The plates were taken generally in a corner-in-center overlap pattern, so that the sky is covered twice and every plate has one quadrant each of its area in common with four other plates. Some generous margins occasionally provided for multiple overlap. In order to be able to choose a reduction model with as few plate constants as possible, the location of the tangential point α_0, δ_0 (see section 6.222) was determined experimentally for every plate, the radial distortion of the system was determined once and for all (but separately for the Bonn and the Bergedorf camera, of course) by means of suitably arranged exposures of star pairs. The nonlinearly distorting influence of refraction was eliminated from the measurements before adjustment, and differential aberration will not produce effects which are large enough to warrant special efforts to remove them from plates of this size. These precautions permitted the investigators to regard the standard coordinates as obtainable from the measured coordinates by a general affine relationship (see equation (6.7)) involving six Turner-type plate constants. No efforts were thus made to arrange the measurements and preliminary reductions in such a way that the standard coordinates could be obtained from the measured coordinates by a similarity transformation (which would have involved only four linear plate constants). Detailed introductions to the AGK2 zones can be found in Vol. I, referring to the Bergedorf part, and in Vol. II, describing the Bonn part.

The volumes themselves, listing the information for the stars in the order of increasing mean right ascension for equator and equinox 1950.0, are divided into zones of 5° width in declination. Within each zone they give current number (newly assigned for the purposes of this catalogue), photographic magnitude (newly determined from the plates themselves), mean right ascension and declination (1950.0) for the average epoch of the contributing plates (hopefully) in the system of the FK3, and the average epoch Ep of the plates. Also given are the first two (and in high declinations the first three) coefficients of the development of the precession as a power series (analogous to formula (4.16)) without proper motion. The precession terms, formulas for the use of which are given at the beginning of every 5°-wide zone, were computed with Newcomb's constant of precession. In this age of electronic computers they have lost much of their usefulness. Also given is the number of plates from which each position was derived. In the Bonn zones this information is given under 'Bemerkungen' (remarks), and only if it differs from 2. The 'Bemerkungen' inform about other items of interest, such as whether the position is taken from the FK3, in which case it would refer to the epoch 1950. FK3 positions override positions measured on the plates which, incidentally, makes a comparison of the AGK2 with the FK3 very difficult.

The standard deviation of a listed catalogue position derived from two plates is quoted for the Hamburg zones as

$$[0''.145 + 0''.061\,(m - 9.12)^2]$$

in both coordinates (m is the photographic magnitude), and, for the Bonn zones, 0''.155 in right ascension and 0''.185 in declination.

At the end of every 5° zone, there is a list of (mostly double) stars which for some reason were not measured on the photographic plates and for which positions were obtained at the meridian circle of the Berlin-Babelsberg Observatory. These lists of 'Babelsberger Meridian-beobachtungen' are arranged quite analogously to the main body of the catalogue.

The positions in the polar cap, i.e., those of stars with declinations larger than 85°, are given in the form of polar standard coordinates ($X = \cos\delta \cos\alpha \csc 1''$; $Y = \cos\delta \sin\alpha \csc 1''$) with the precession terms dX/dt, $\frac{1}{2}(d^2X/dt^2)$ and $\frac{1}{6}(d^3X/dt^3)$, and analogous ones for Y.

After publication, careful investigations of the residuals (against the reference star positions) and a comparison of the Bonn and Hamburg positions in the small overlapping zone showed that the published positions are affected by various kinds of systematic errors. The Bonn positions especially are affected by tangential distortion, caused by imperfect alignment of the components of the objective. Both the Bonn and the Hamburg positions are afflicted with errors that depend on magnitude and spectrum as well as on position on the plate and spectrum. These systematic errors can be partly removed by referring to the diagrams on page E4 in Volume 10 and the tables on page 38 of Volume 11. Those systematic errors which also depend on the image position on the plate cannot be applied easily since the user has no means of ascertaining where the images of a certain star were situated on the plates.

6.622 The AGK3. Unlike the Yale Catalogues (section 6.623) and the Cape Catalogues (section 6.624), the AGK2 contains no information on proper motions although they could have been derived from comparisons with the various zones of the AGK1. In the judgment of the astronomers responsible for the setup of the AGK2, the AGK1 is not sufficiently free from systematic errors nor are its various zones sufficiently homogeneous to serve as a basis for the derivations of reliable proper motions.

In order to get proper motions for the AGK2 stars, the (then) director of the Hamburg-Bergedorf Observatory, O. Heckmann initiated a plan for the repetition of the AGK2 with the aim of providing material for the derivation of highly reliable proper motions for which a standard deviation of 0".008 could be expected. All the photographic work was to be done at Hamburg, and the reference stars were observed on meridian circles in a program with wide international participation.

At the time of this writing, the observations at the Hamburg-Bergedorf Observatory have been carried out according to plans. A complete set of overlapping 5°×5° plates has been taken of the northern sky for the mean epoch of 1958. The measurements have been completed; final solution is expected when the reference star program, AGK3R, is finished. The latter is under the supervision of F. P. Scott at the U. S. Naval Observatory who has collected all the meridian observations and is responsible for the compilation of the catalogue

[Scott, 1967]. The work on the photographic plates is under the supervision of W. Dieckvoss and H. Kox. The proper motions in the AGK3 will be based on AGK2 positions from which all known systematic errors have been removed. The end of the work on the AGK3 is in sight, and the catalogue should be in print in 1968.

6.623 The Yale Catalogues. Of all the photographic zone catalogues, the Yale catalogues are the first and the most extensive and ambitious enterprise, and the most heterogeneous one. Five different cameras have been used to secure the needed plates, the measuring and reduction procedures were changed several times, and reference stars that represented a considerable variety of coordinate systems were used to derive the plate constants. The plate epochs range from 1914 to 1956, and more than one generation of Yale astrometrists were busy with the work on the Yale zones. The study of the introductions to the zones, in chronological order, is highly recommended to anyone who is interested in the development of the art of efficiently computing star positions from measurements on photographic plates. There he will find many 'firsts' in astrometric technique.

The feasibility of using wide angle and medium focal length photographic cameras for constructing catalogues of star positions with about the same scope as that of the AGK1, namely, completeness to about ninth visual magnitude (corresponding to an average star density of about ten stars per square degree) was demonstrated by Frank Schlesinger who started his experiments in 1913 when he was the director of the Allegheny Observatory of the University of Pittsburgh [Schlesinger and Hudson, 1916]. Soon afterwards he was appointed to the directorship of the observatory of Yale University.

All but the more recent Yale catalogues have the name of Dr. Ida Barney on the cover, at least as one of the coauthors. It was she who was responsible for carrying out Schlesinger's plans. She not only supervised the measurements and computations (in itself an almost unimaginably huge task, considering that most of it was done before the advent of electronic computers), but actively participated in them herself. The volumes of catalogues that appeared while she was actively working on the project are a monument to her industry, professional competence and dedication.

After Miss Barney's retirement, Dr. Dorrit Hoffleit assumed the responsibility for the production of the Yale catalogues which, in Schlesinger's own words, were eventually to cover the whole sky. As one can see from Table 6.7, gaps still exist only between declinations $+85^\circ$ to $+60^\circ$, $+50^\circ$ to $+30^\circ$, and -50° to -60°.

Unlike the AGK2 (and the AGK3), the Yale catalogues were not constructed in one massive, homogeneous effort. The individual catalogues were published as work on the various zones was finished, and the mea-

TABLE 6.7 The Yale Zone Catalogues

Declination Zone		Plate Epoch	Coordinate System	Standard Deviation	Published in Yale Transactions	Remarks
From	To					
+90°	+85°	1951	1950.0	0″.13	26/I	
+60°	+55° I	1916	1875.0	0.22	7	
+55°	+50° I	1916	1875.0	0.22	4	
+60°	+55° II	1947	1950.0	0.14	27	
+55°	+50° II	1947	1950.0	0.14	26/II	
+30°	+25°	1928	1875.0	0.19	9	
+25°	+20°	1928	1875.0	0.19	10	
+30°	+25° r		1950.0	0.14	24	2
+25°	+20° r		1950.0	0.14	25	2
+20°	+15°	1940	1950.0	0.17	18	
+15°	+10°	1940	1950.0	0.17	19	
+10°	+ 9°	1940	1950.0	0.17	22/II	
+ 9°	+ 5°	1936	1950.0	0.17	22/I	
+ 5°	+ 1°	1936	1950.0	0.17	20	
+ 2°	+ 1°	1914	1875.0	0.24	5	
+ 1°	− 2° I	1914	1875.0	0.19	5	
+ 1°	− 2° II	1937	1950.0	0.17	21	
− 2°	− 6°	1933	1950.0	0.16	17	
− 6°	−10°	1933	1950.0	0.16	16	
−10°	−14°	1933	1950.0	0.17	11	
−14°	−18°	1933	1950.0	0.17	12/I	

TABLE 6.7 (continued)

Declination Zone		Plate Epoch	Coordinate System	Standard Deviation	Published in Yale Transactions	Remarks
From	To					
-18°	-20°	1933	1950	0".17	12/II	
-20°	-22°	1933	1950	0.16	13/I	
-22°	-27°	1933	1950	0.16	14	
-27°	-30°	1933	1950	0.16	13/II	
-30°	-35°	1955	1950	0.35	28	
-35°	-40°	1955	1950	0.35		1, 3
-40°	-50°	1942	1950	0.37		1, 3
-60°	-70°	1942	1950			1, 3
-70°	-90°	1956	1950			1

Remarks:

(1) Not yet published.

(2) These volumes contain positions which are essentially the means of the Yale positions published in YT 9 and 10 the corresponding volumes of the AGK2.

(3) Measurements concluded.

suring and reduction methods were constantly changed and improved as time went on.

This procedure was facilitated by the belt pattern in which the plates were taken. Double coverage of the sky was achieved by an edge-in-center-line overlap pattern within the belts. Thus, every plate covers a region that is also partially covered by two other plates, and there is no overlap in declination (except for some slight edge-on-edge overlap on the boundary region between two zones (i.e., belts). It stands to reason that as a consequence of this, the right ascensions in the Yale zone catalogues are systematically more reliable than the declinations. If, however, a corner-in-center overlap pattern had been followed (as, for instance, in the Astrographic Catalogue and the AGK2) the Yale zones could not have been treated each independently.

A detailed and accurate description of the procedures employed for the work on and characteristics of each individual Yale zone catalogue would fall beyond the scope of this work and would be of little interest for the average user. Therefore, only a general description will be given from which details concerning individual zones may deviate somewhat. Exhaustive discussions will be found in the introductions to the various zones in the respective volumes of the 'Transactions of the Astronomical Observatory of Yale University,' or, for short, the 'Yale Transactions' (YT).

The plate format was usually in the neighborhood of $10^\circ \times 10^\circ$, and the focal length such that one millimeter on the plates corresponded closely to 100 seconds of arc. The exposure time was such that a ninth magnitude star appeared as a well-blackened disk. Since the planeness of the emulsion is quite critical for astrometric work at such wide angle fields, the glass plates were frequently checked for planeness before they were coated with the emulsion.

Reference stars, desired with a density of about one per square degree, were usually not available in the existing catalogues and as a rule were especially observed at meridian circles more or less at the same time the plates were exposed. The fundamental systems to which the reference stars were referred thus vary considerably from zone to zone. However, tables for the reduction to the GC system of the positions (and proper motions) as published in the various zones were given in [Barney, 1951] for those zones that appeared in YT volumes up to No. 22. These tables supersede those given in the introductions to the volumes.

The plates were measured on long screw measuring machines usually in one position only, but with various measurers working on the same plate. The plates for one zone were measured on a fully automatic measuring machine.

The model for the computation of standard coordinates ξ, η (see section 6.222) from the measured coordinates x, y was usually of the form

$$\xi = \Sigma a_{ik} x^i y^k, \qquad \eta = \Sigma b_{ik} x^i y^k,$$

where in no case terms in x and y of higher than third order (i. e., with $i+k>3$) were included; but even within this restriction, some of the coefficients a_{ik} and b_{ik} were arbitrarily assumed equal to zero. The third-order terms (if included at all) were usually not allowed to vary from plate to plate within a given zone and usually not carried in the original least squares adjustment which was, of course, carried out for every plate individually. They were determined instead from an investigation of the residuals after the original adjustment of the entire zone.

Particularly remarkable is the fact that it was found necessary to include in the formulas correction terms that depend on magnitude and color, often in connection with the position of the image on the plate. This is expressed by the fact that some of the coefficients a_{ik} and b_{ik} depend on magnitude and on color. In almost all zones the effective focal length of the camera (described essentially by the terms a_{10} and b_{01}) depended on magnitude (coma effect) and color (color magnification error), and in one zone there is a systematic effect which may be described as a color dependent coma effect. In this case, for example, one will have to allow a_{10} and b_{01} to be of the form

$$a_{10} = A_0 + A_1 m + A_2 c + A_3 cm,$$
$$b_{01} = B_0 + B_1 m + B_2 c + B_3 cm,$$

where the A_i and B_i are constants; m and c are magnitude and color equivalent respectively.

One particularly ingenious scheme, namely the use of a coarse objective grating for the determination of the coma effect (which Schlesinger calls magnitude distortion) was used for the first time in the Yale zone catalogues. Details can be found in the introduction to YT, Vol. 9.

As a rule the published catalogues include the AGK1 number (when assigned); magnitude; spectrum (when available); mean right ascension and declination for the average of the epochs of the plates, referred to the coordinate systems at 1875 or 1950; proper motions that have been derived by comparison with old catalogues (notably the AGK1); annual and secular variations, and the third term of the precession series, frequently computed with the Struve-Peters precession constant but always without the effects of proper motion. Table 6.7 gives the pertinent information (when available) for those Yale zone catalogues which have either already appeared or are being worked on.

6.624 The 'Cape Photographic Catalogue for 1950.0' (CPC). A few years after the work on the AGK2 was begun, the Royal Observatory at the Cape of Good Hope embarked on a very similar and hardly less ambitious program, namely, the accurate determination of the positions of all stars brighter than ninth magnitude from declination -30° to the celestial south pole by means of photography.

The camera was of the same general type as that used for the observation of the AGK2 and of the Yale zones. It was equipped with a five-element objective lens of about four inches effective aperture. The focal length of near 80 inches produces a scale of one millimeter on the plate (near the focal point) corresponding to 103".2. Two zones were observed on plates with a field of 4°.8 × 5°.0, the others on plates which rendered a field of 4° × 4°. During the period from 1934 to 1961, the plates were measured on a one-screw measuring machine that was equipped with a projection device. The sky is essentially covered doubly by an 'edge-in-center-line' overlap pattern (as in the Yale zones).

The positions of the reference stars were determined on the meridian circles of the Cape Observatory fairly contemporaneously with the exposures of the plates. The distribution of the reference stars over the sky is quite uniform: There are on the average four reference stars for every five square degrees. (This is very close to the density of the GC.)

The transformations of the measured coordinates of the star images to standard coordinates were performed assuming an affine (Turner type, i.e., six linear plate constant) relationship (see equation (6.7)) which was eventually further sophisticated as will be discussed below.

It is interesting to note that the linear plate constants were not determined by a rigorous least squares adjustment, but by the so-called Dyson-Christie method. In this method an average of the reference star data in every quadrant of the plate is taken, and the six plate constants are then obtained from the eight equations following from the data provided by the four 'normal points,' again by a process of averaging. This procedure is designed to save computing labor by avoiding the heavy arithmetic that is characteristic for any least squares adjustment. Naturally, the plate constants obtained in this way will typically be less accurate than those that would have resulted from a least squares adjustment. The high accuracy of the published positions, however, has vindicated the method employed and demonstrates that apparently any reasonable adjustment method is capable of providing high quality results as long as the original data is of high quality. In the case of the CPC, these are the positions of the reference stars and the measured rectangular coordinates of the star images.

Fortunately, the advent and ready availability of electronic computers have eliminated the need for and even the desirability of adjustment methods designed to avoid the computing tedium of least squares at the price of even only a slight loss of accuracy.

The residuals from the preliminary Turner type six plate constant adjustments were then in every zone averaged for representative points on the plate; and from this data, corrections to the standard coordinates ξ and η were determined (zone by zone) in the form

$$\Delta\xi = \Sigma a_{ik} x^i y^k, \quad \Delta\eta = \Sigma b_{ik} x^i y^k,$$

where $0 \leq i+k \leq 3$. The coordinates were corrected, moreoever, for the coma effect (essentially a magnitude dependence of the effective focal length) and reduced to the FK3 (using the GC as an intermediary because of the low density of the positions in the FK3).

Table 6.8 gives a general survey over the CPC [Jackson and Stoy, 1954-1966]. The catalogues contain the following data: serial number, DM number, i.e., Cordoba Durchmusterung (CD) or Cape Photographic Durchmusterung (CPD) number, mean right ascension and declination for the mean of the plate epochs referred to the coordinate system of the FK3 at 1950.0, precession terms, secular proper motions in the FK3 system (when available) and their probable errors (except in Vol. 17), and photographic and photovisual magnitudes (as well as color index). The photometric data was specially determined for this catalogue. Epochs and remarks are on the bottom of the pages, a 'GC' indicates a position supplied by the GC.

Recently plates were also taken in the zone -40° to -52° and their reduction has been started. At present, for these zone catalogues of positions and proper motions with essentially the same coverage as the CPC are used; they were derived from the Cape section of the Astrographic Catalogue [Gill and Hough, 1923; Spencer Jones and Jackson, 1936].

TABLE 6.8 The Zones of the 'Cape Photographic Catalogue'

Declination Zone		Plates Taken	Standard Deviation		Published in Cape Observatory Annals Volume
From	To		α	δ	
-30°	-35°	1930-1933	0.″22	0.″22	17
-35°	-40°	1935-1937	0.″21	0.″15	18
-40°	-52°	A.C.			
-52°	-56°	1938	0.″18	0.″15	19
-56°	-64°	1945-1947	0.″18	0.″15	20
-64°	-80°				
-80°	-90°				

6.63 The Smithsonian Astrophysical Observatory 'Star Catalog'

6.631 History and Original Purpose. Even before the first artificial satellite was launched, the Smithsonian Astrophysical Observatory (SAO) at Cambridge, Massachusetts was given the responsibility for the photographic tracking of these objects. This was to be done with cameras especially designed for this purpose (the Baker-Nunn Super Schmidt cameras). It had been originally estimated that the highest precision that would ever be required of observed satellite positions would correspond to a standard deviation of 2".0 in either coordinate, and the entire system of observing and measuring the satellite positions photographically was geared to this figure. (Needless to say, the accuracy requirements for satellite positions have since risen tremendously. This brings to mind a dictum by E. Hertzsprung: 'We don't know what data the next generation of astronomers will want. But we do know that they will want them more accurately.')

Even so, it soon became apparent that the reductions of the measurements were seriously hampered by the lack of a sufficiently numerous and homogeneous source of suitable reference star positions. The available accurate star positions (in the northern hemisphere mostly from the AGK2 and the Yale catalogues, and in the southern hemisphere from the Yale catalogues and the CPC) were several decades old, and proper motions were not available in all zones. Besides, the positions and proper motions from the available catalogues were referred to a variety of systems although this would hardly have been harmful.

Thus, the SAO in 1959 started the compilation of what was originally intended to be a catalogue of reference star positions for the photographic determination of positions of artificial satellites, containing in no region of the sky less than four stars per square degree and giving positions and proper motions with such accuracy that no coordinate would have an average standard deviation of over 1" by 1970. As far as the existing material has permitted it, this goal was attained and even overshot to a mean standard deviation of 0".5 [Haramundanis, 1967].

The SAO catalogue of 258,997 star positions is strictly a compilation catalogue in which no information is found that was not already available [Smithsonian Astrophysical Observatory, 1966]. Its value consists in presenting this information in the form in which it is most likely to be used.

The actual work of compiling the catalogue essentially followed guidelines set up by H. Eichhorn and was under the general supervision of G. Veis.

6.632 Description of the Catalogue. The SAO catalogue consists of four volumes. It contains for each star the following information (see sample page in Table 6.9): current number, photographic and visual magnitude, mean α and δ for coordinate system and epoch 1950.0 to $0^s.001$ and 0".01 respectively, the annual proper motions μ in α and μ' in δ for the epoch and coordinate system 1950.0. μ is given (as is customary) in units of $0^s.0001$ and μ' in units of 0".001 both on the great circle. The standard deviation (σ) in units of 0".01 refers to the position as a whole at the epoch 1950. This is the dispersion of the distances between the unknown actual and the catalogued positions at epoch 1950. No other catalogue contains this information and, unfortunately, the quantity was calculated with a wrong formula (quoted on page xiv of the introduction, No. (10)) which tends to give values which are too large. Critical users will, therefore, regard the column giving σ only for the purpose of getting an approximate idea concerning the accuracy of the position at the epoch 1950. In investigations that are sensitive to this quantity, it may (if all the pertinent information is available) be calculated from rigorous formulas as given in [Eichhorn, 1962] and described below. Suppose the proper motion components were (as is the case for almost all proper motions in the SAO catalogue) obtained by dividing the position components of the star, referred to the same coordinate system but at two different epochs t_1 and t_2, by the elapsed time interval $t_2 - t_1$. If $m_{\alpha 1}$ is the standard deviation of the right ascension at epoch t_1, one obtains for the standard deviation $m_\alpha(t)$ of a right ascension for the epoch t which was obtained by linear inter- or extrapolation between α_1 and α_2

$$m_\alpha^2(t) = \frac{1}{t_2 - t_1}\left[m_{\alpha 1}^2 (t - t_2)^2 + m_{\alpha 2}^2 (t - t_1)^2\right], \qquad (6.17)$$

and an analogous formula for $m_\delta(t)$, so that one finally obtains

$$m(t) = \left[m_\alpha^2(t) \sec^2 \delta + m_\delta^2(t)\right]^{\frac{1}{2}}, \qquad (6.18)$$

if m_α was given on the parallel, or

$$m(t) = \left[m_\alpha^2(t) + m_\delta^2(t)\right]^{\frac{1}{2}}, \qquad (6.19)$$

if m_α was given on the great circle. It is the latter, somewhat unconventional alternative which was chosen for indicating the standard deviations of the right ascension proper motions (σ_μ) in the SAO Catalog.

The SAO Catalog also contains the following additional information: the seconds of time (accurate to 0.001) and minutes of the mean right ascension, and seconds of arc (to 0".01) of the mean declination at the 'original' epoch (see below); the standard deviations of these quantities given for both coordinates in units of 0".01 on the great circle; the 'original epoch' in tropical years after 1900. The latter indicates completely different things depending on the source. For those positions that were taken from the FK4 and the GC, it is the mean epoch of the catalogues which were incorporated to give the fundamental positions. Only for these will formula (10) for 'the standard deviation of the position at 1950.0,' quoted on page xiv of the introduction give correct answers. For those positions that were adapted from some other catalogue, 'original epoch' is the epoch of the catalogue from which the position was taken. The proper motions, however, were obtained by comparison of a modern with an old catalogue, whether this proper motion was already available in the source catalogue (as, for instance, in the case of the Yale and the Cape Photographic Catalogues) or whether it had to be newly derived at the SAO. The positions for 1950 are, therefore, just as dependent on the earlier as on the

TABLE 6.9 Sample Page of the SAO Catalog

240200			EPOCH 1950							ORIGINAL EPOCH						SP.	SOURCE		-50° 12H
NUMBER	MAGNITUDES m_{pg}	m_v	α_{1950} h m s	μ s	σ_μ ".001	δ_{1950} ° ' "	μ' "	$\sigma_{\mu'}$ ".001	σ_{1950} "	α_2 s	σ ".01	ep.	δ_2 "	σ' ".01	ep.		CAT.	STAR NUMBER	DM NUMBER
1	8.9	8.8W	12 41 30.753	0.0004	06	-58 46 2.51	-0.010	06	0.14	30.751	14	46.4	2.48	14	46.4	A2	C0	3907	P -58 4466
2	9.2	9.0W	41 35.687	-0.0055	15	-54 4 22.56	-0.013	15	0.24	35.750	15	38.4	22.41	18	38.4	F8	C9	4781	P*-53 5309
3		6.8T	41 41.802	-0.0086	14	-57 0 47.53	-0.024	10	0.63	42.245	21	03.9	46.30	19	98.4	A0	GC	17320K	P*-56 5415
4		7.4T	41 49.713	-0.0031	14	-53 48 42.39	-0.005	11	0.64	49.859	21	02.9	42.16	19	98.6	K0	GC	17323	P -53 5313
5	10.21		41 59.989	0.0016	05	-52 29 7.20	-0.008	05	0.18	59.971	15	38.4	7.10	18	38.4	F5 O	C9	4783	P -52 5846
6		8.1T	42 0.960	-0.0006	17	-52 28 56.23	-0.008	10	0.67	0.986	21	04.9	55.82	17	00.5	F2	GC	17327K	P -52 5847
7	9.1	9.0W	42 5.092	-0.0051	08	-52 41 59.18	-0.006	08	0.19	5.150	15	38.4	59.10	18	38.4	A2	C9	4785	P -52 5849
8		7.0T	42 10.491	-0.0088	13	-59 48 9.07	-0.008	09	0.58	10.947	19	03.8	8.67	18	98.4	K2	GC	17331	P -59 4418
9	10.2E	9.9*	42 13.955	-0.0024	12	-51 43 31.34	0.005	12	0.65	14.070	31	01.8	31.55	31	01.8	A2	CZ	10652	P -51 5463
10	9.3	9.5W	42 15.480	-0.0014	07	-56 32 57.54	0.010	07	0.14	15.485	14	46.4	67.58	14	46.4	B8	C0	3910	P -56 5413
11	8.5	8.8W	42 17.531	-0.0053	07	-56 3 19.98	0.014	07	0.18	17.592	15	38.4	20.15	18	38.4	B9	C9	4786	P -55 5208
12	10.2	9.9W	42 20.191	-0.0062	15	-55 27 8.17	0.058	15	0.24	20.264	15	38.4	8.85	18	38.4	A5	C9	4787	P -55 5209
13	8.6E	8.2Z	42 20.671	-0.0021	12	-51 33 53.86	0.009	12	0.81	20.766	63	04.3	54.25	57	04.3	K2	CZ	10653	P -51 5467
14	8.7	7.9W	42 22.204	-0.0227	06	-57 4 58.33	-0.153	06	0.14	22.285	14	46.4	57.78	14	46.4	K0	C0	3911	P -56 5415
15	10.6		42 22.215	-0.0074	15	-52 42 50.66	0.021	15	0.24	22.301	15	38.4	50.90	18	38.4	F2	C9	4788	P -52 5853
16	9.0	9.1W	42 26.179	-0.0027	07	-52 24 8.84	-0.011	07	0.18	26.211	15	38.4	8.70	18	38.4	A2	C9	4789	P -51 5469
17	9.5	9.3W	42 30.072	-0.0008	11	-52 42 48.83	-0.036	11	0.21	30.080	15	38.4	48.40	18	38.4	F2	C9	4790	P -52 5858
18	9.6E	8.8*	42 35.516	-0.0012	12	-50 49 19.68	-0.002	12	0.64	35.574	31	02.7	19.57	28	02.7	K0	CZ	10655	P -50 5449
19		8.3T	42 41.557	0.0015	12	-53 35 17.37	0.008	12	0.62	41.486	21	02.6	17.79	19	99.8	F2	GC	17340	P -53 5317
20	9.5	9.5W	42 51.579	-0.0040	06	-57 56 14.56	0.005	06	0.14	51.594	14	46.4	14.58	14	46.4	A5	C0	3912	P -57 5686
21	9.4E	8.5Z	42 57.882	-0.0019	12	-51 26 14.62	-0.014	12	0.81	57.971	63	04.3	13.97	57	04.3	A0	CZ	10662	P -51 5477
22	8.2	8.0W	42 57.972	-0.0005	06	-58 13 32.49	-0.004	06	0.14	57.973	14	46.4	32.48	14	46.4	F0	C0	3913	P -57 5690
23	9.8	9.4W	43 1.736	-0.0056	12	-52 58 27.44	-0.037	12	0.22	1.800	15	38.4	27.00	18	38.4	F5	C9	4792	P -52 5868
24	9.6	8.7W	43 2.807	-0.0022	07	-57 0 50.96	0.007	07	0.14	2.815	14	46.4	50.98	14	46.4	K0	C0	3914	P -56 5421
25	9.5	9.0W	43 4.517	-0.0038	12	-53 20 36.96	0.013	12	0.22	4.560	15	38.4	37.10	18	38.4	G0	C9	4793	P -52 5869
26	9.7E	9.0Z	43 4.668	-0.0159	12	-51 10 37.37	-0.026	12	0.70	5.395	45	04.3	36.17	45	04.3	G5	CZ	10665	P -50 5453
27	9.6	8.5W	43 11.191	-0.0067	07	-56 56 37.54	-0.017	07	0.14	11.215	14	46.4	37.48	14	46.4	K2	C0	3915	P -56 5423
28	9.4	9.5W	43 12.014	-0.0063	07	-56 12 21.66	-0.025	07	0.14	12.036	14	46.4	21.57	14	46.4	A5	C0	3916	P -55 5214
29		6.7T	43 13.658	-0.0042	09	-54 20 20.61	-0.005	07	0.37	13.828	14	09.5	20.40	12	05.2	K0	GC	17349	P -53 5319
30	10.1		43 26.762	-0.0067	13	-56 27 52.54	-0.018	13	0.15	26.786	14	46.4	52.47	14	46.4	B8	C0	3917	P -56 5428
31			43 26.930	-0.0005	06	-58 51 5.82	-0.012	06	0.14	26.931	14	46.4	5.78	14	46.4	G5	C0	3918V	P*-58 4490
32	9.7	9.7W	43 27.771	0.0009	12	-53 8 33.42	-0.018	12	0.22	27.761	15	38.4	33.20	18	38.4	A0	C9	4795	P -52 5876
33	9.2	8.3W	43 28.628	-0.0016	07	-58 18 6.63	-0.013	07	0.14	28.633	14	46.4	6.58	14	46.4	G5	C0	3919	P -57 5694
34	10.1E	9.5Z	43 28.766	-0.0037	12	-50 35 40.81	-0.054	12	0.70	28.936	45	04.3	38.36	45	04.3	G5	CZ	10672	P -50 5459
35		4.9T	43 29.575	-0.0049	07	-56 12 56.12	-0.031	05	0.30	29.800	14	10.0	54.70	12	03.9	B3	GC	17352K	P*-55 5215
36		9.2T	43 30.665	-0.0075	20	-56 13 47.18	-0.023	17	0.85	31.015	22	07.4	46.12	24	03.2		GC	17353K	P -55 5216
37	9.6	9.6W	43 38.575	-0.0041	07	-59 55 51.13	-0.015	07	0.14	38.590	14	46.4	51.08	14	46.4	A0	C0	3690	P -59 4434
38	9.8	10.1W	43 39.305	-0.0072	07	-59 4 41.05	-0.020	07	0.14	39.331	14	46.4	40.98	14	46.4	B3	C0	3923	P -58 4494
39	9.9	9.3W	43 41.066	-0.0047	14	-52 0 30.86	-0.030	14	0.23	41.121	15	38.4	30.50	18	38.4	G0	C9	4796	P -51 5485
40	9.8	9.5W	43 49.192	-0.0016	12	-53 21 13.77	0.012	12	0.22	49.210	15	38.4	13.90	18	38.4	F8	C9	4797	P -52 5880
41	10.2	9.7W	44 1.841	-0.0052	12	-53 21 49.10	0.001	12	0.22	1.900	15	38.4	49.10	18	38.4	G0	C9	4798	P -52 5884
42	9.3E	9.0Z	44 7.761	-0.0026	12	-50 55 57.67	-0.009	12	0.70	7.881	45	04.3	57.23	45	04.3	K0	CZ	10680	P -50 5465
43	9.3	9.2W	44 10.601	-0.0027	07	-59 15 2.35	-0.020	07	0.14	10.611	14	46.4	2.28	14	46.4	F8	C0	3924	P -58 4504
44	9.9	10.2W	44 12.543	-0.0010	13	-56 20 11.98	-0.002	13	0.15	12.546	14	46.4	11.97	14	46.4	A0	C0	3925	P -55 5225
45	10.1	9.1W	44 15.547	-0.0003	09	-52 9 52.13	-0.019	09	0.20	15.551	15	38.4	51.90	18	38.4	G5	C9	4799	P -51 5490
46	9.5	9.2W	44 16.041	-0.0121	07	-52 38 42.00	0.027	07	0.18	16.181	15	38.4	42.30	18	38.4	F2	C9	4800	P -52 5887
47	8.8	8.6W	44 18.449	-0.0010	07	-54 18 34.10	-0.017	07	0.18	18.460	15	38.4	33.91	18	38.4	F0	C9	4801	P -53 5328
48	9.2	9.3W	44 33.947	-0.0012	07	-54 11 12.29	-0.007	07	0.18	33.960	15	38.4	12.21	18	38.4	A2	C9	4803	P -53 5329
49	8.7	8.9W	44 35.096	-0.0013	06	-52 47 57.48	0.011	06	0.18	35.110	15	38.4	57.60	18	38.4	A2	C9	4804	P -52 5889
50	9.9	9.0W	44 35.658	-0.0046	07	-57 10 5.90	-0.033	07	0.14	35.675	14	46.4	5.78	14	46.4	G5	C0	3926	P*-56 5434
51	9.8	9.8W	12 44 35.940	-0.0026	04	-54 40 9.56	-0.012	04	0.17	35.970	15	38.4	9.42	18	38.4	A0	C9	4805	P -54 5331
52	9.4	9.4W	44 37.229	-0.0020	07	-55 41 20.84	0.009	07	0.18	37.253	15	38.4	20.95	18	38.4	A2	C9	4806	P -55 5227
53		7.3T	44 38.923	0.0024	13	-59 48 29.51	-0.016	10	0.59	38.807	21	05.3	28.73	19	00.6	A2	GC	17368	P -59 4448
54	9.1	9.4W	44 39.118	0.0023	05	-59 53 29.11	-0.009	05	0.14	39.110	14	46.4	29.08	14	46.4	B9	C0	3928	P -59 4449
55	9.6	9.4W	44 43.496	-0.0048	05	-58 8 34.13	-0.015	05	0.14	43.514	14	46.4	34.08	14	46.4	F8	C0	3929	P -57 5706
56	10.4		44 43.974	-0.0023	14	-52 1 16.40	-0.008	14	0.23	44.001	15	38.4	16.30	18	38.4	F0	C9	4807	P -51 5498
57		7.2T	44 44.259	-0.0030	12	-58 1 28.81	-0.027	11	0.58	44.408	19	02.9	27.44	18	99.8	A5	GC	17373	P -57 5707
58	8.2	8.1W	44 46.653	-0.0029	06	-58 9 33.33	-0.015	06	0.14	46.664	14	46.4	33.28	14	46.4	F0	C0	3931	P -57 5708
59		1.5H	44 47.037	-0.0050	02	-59 24 56.53	-0.017	02	0.08	47.223	04	13.0	55.88	04	11.9	B1	F4	481	P -59 4451
60	9.6	9.7W	44 51.191	-0.0058	14	-58 48 42.44	-0.017	14	0.15	51.211	14	46.4	42.38	14	46.4	B8	C0	3933	P -58 4510
61	10.0	8.4W	44 52.104	-0.0002	06	-56 46 18.36	-0.023	06	0.14	52.105	14	46.4	18.28	14	46.4	K5	C0	3934	P -56 5437
62	10.0	8.7W	44 57.438	-0.0028	12	-53 1 47.14	-0.011	12	0.22	57.470	15	38.4	47.00	18	38.4	K0	C9	4808	P -52 5894
63	9.7E	9.0Z	45 1.465	-0.0033	12	-51 10 15.06	-0.014	12	0.65	1.616	36	04.3	14.41	36	04.3	K0	CZ	10690	P -50 5475
64	9.4E	8.5Z	45 3.435	-0.0010	12	-50 14 21.48	-0.020	12	0.64	3.481	31	02.3	20.50	28	02.3	K0	CZ	10691	P -49 5462
65		7.5T	45 6.091	0.0015	14	-59 19 14.48	-0.015	10	0.58	6.020	19	06.5	13.78	18	01.8	B8	GC	17376	P -58 4514
66	10.2		45 9.024	-0.0059	07	-56 51 41.39	-0.032	07	0.14	9.045	14	46.4	41.28	14	46.4	A0	C0	3936	P*-56 5440
67	8.0	8.0W	45 13.175	-0.0014	06	-59 56 14.16	0.006	06	0.14	13.180	14	46.4	14.18	14	46.4	B0	C0	3937	P -59 4460
68	10.0	8.7W	45 14.213	0.0001	15	-55 55 45.11	-0.040	15	0.24	14.213	15	38.4	44.65	18	38.4	K2	C9	4809	P -55 5233
69	9.4	9.5W	45 21.746	-0.0040	07	-55 34 44.92	-0.024	07	0.18	21.793	15	38.4	44.65	18	38.4	B9	C9	4810	P -55 5234
70	9.8	8.6W	45 29.432	-0.0033	12	-53 19 19.04	0.006	12	0.22	29.470	15	38.4	19.10	18	38.4	K0	C9	4811	P -52 5900
71	10.0E	9.7Z	45 34.602	-0.0018	12	-50 11 32.71	-0.005	12	0.71	34.683	45	03.9	32.46	45	03.9	A5	CZ	10696	P -49 5466
72	8.9	8.0W	45 41.004	-0.0006	05	-52 21 6.37	0.003	05	0.18	41.011	15	38.4	6.40	18	38.4	K0	C9	4812	P -51 5508
73	9.3	9.5W	45 46.542	-0.0051	14	-53 24 33.22	-0.001	14	0.23	46.601	15	38.4	33.20	18	38.4	B9	C9	4813	P -53 5339
74	9.9	10.0W	45 54.327	-0.0048	13	-57 35 23.42	-0.012	13	0.15	54.344	14	46.4	23.38	14	46.4	A0	C0	3938	P -57 5721
75	8.9	8.7W	45 58.630	-0.0042	07	-56 34 53.79	0.053	07	0.14	58.645	14	46.4	53.98	14	46.4	F8	C0	3939	P -56 5448
76	9.8E	8.2Z	46 5.040	-0.0013	12	-51 7 8.68	-0.003	12	0.70	5.100	45	04.3	8.52	45	04.3	K5	CZ	10702	P -50 5484
77	9.6	9.1W	46 9.908	-0.0045	06	-58 10 33.56	0.006	06	0.14	9.923	14	46.4	33.58	14	46.4	F5	C0	3940	P -57 5724
78	9.6	9.2W	46 17.520	0.0034	07	-51 54 56.66	-0.022	07	0.18	17.481	15	38.4	56.40	18	38.4	F5	C9	4814	P -51 5519
79	9.6	9.1W	46 19.502	-0.0092	07	-56 48 33.47	-0.024	07	0.14	19.535	14	46.4	33.38	14	46.4	F8	C0	3941	P -56 5451
80	9.1	9.5W	46 21.829	-0.0036	08	-53 14 21.29	-0.016	08	0.19	21.871	15	38.4	21.10	18	38.4	B9	C9	4815	P -52 5919
81	8.6E	8.5Z	46 24.152	-0.0022	12	-51 9 16.51	-0.014	12	0.70	24.253	45	04.3	15.85	45	04.3	A5	CZ	10706	P -50 5491
82	9.4	9.2W	46 27.758	-0.0002	15	-54 27 0.08	0.002	15	0.24	27.760	15	38.4	0.11	18	38.4	A0	C9	4816	P -54 5340
83	9.6	10.0W	46 30.391	-0.0006	07	-58 23 52.08	-0.001	07	0.14	30.394	14	46.4	52.08	14	46.4	A0	C0	3942	P -57 5726
84	10.0	10.0W	46 30.661	-0.0035	12	-57 39 44.39	-0.002	12	0.15	30.674	14	46.4	44.38	14	46.4	A2	C0	3943	P -57 5727
85	9.2	8.1W	46 32.297	-0.0003	07	-52 17 46.42	-0.010	07	0.18	32.301	15	38.4	46.30	18	38.4	K0	C9	4817	P -51 5524
86	8.9	8.6W	46 38.974	-0.0049	06	-58 36 53.92	0.016	06	0.14	38.991	14	46.4	53.98	14	46.4	F0	C0	3944	P -58 4529
87	9.1	9.4W	46 39.710	0.0000	07	-59 50 30.09	-0.003	07	0.14	39.710	14	46.4	30.08	14	46.4	A0	C0	3945	P -59 4478
88		7.1T	46 48.039	-0.0020	14	-53 5 21.88	-0.011	10	0.62	48.130	21	04.0	21.34	19	98.3	G5	GC	17408	P -52 5927
89	9.3	9.0W	46 48.238	-0.0176	07	-58 43 23.94	0.067	07	0.14	48.302	14	46.4	24.18	14	46.4	G0	C0	3946	P -58 4531
90	10.1	8.6W	46 55.078	-0.0028	07	-53 40 52.27	-0.023	07	0.18	55.110	15	38.4	52.01	18	38.4	K2	C9	4819	P -53 5342
91	9.8E	9.5Z	46 56.795	-0.0021	12	-50 10 25.30	-0.006	12	0.71	56.890	45	03.9	25.00	45	03.9	A0	CZ	10716	P -49 5485
92	7.3E	8.0Z	46 57.749	-0.0040	12	-51 7 13.31	-0.013	12	0.64	57.940	31	02.7	12.68	28	02.7	A2	CZ	10717	P -50 5496
93	10.3E	9.2*	47 0.245	-0.0033	12	-50 34 24.55	-0.013	12	0.64	0.402	31	02.7	23.95	28	02.7	K0	CZ	10718	P -50 5497
94		7.0T	47 0.878	-0.0014	12	-50 3 5.24	-0.023	12	0.67	0.947	21	98.7	4.02	19	96.3	K2	GC	17412	C -49 7408
95	9.2	9.4W	47 2.393	-0.0029	06	-57 44 38.49	-0.003	06	0.14	2.404	14	46.4	38.48	14	46.4	A0	C0	3947	P -57 5735
96	9.9	9.7W	47 5.080	-0.0003	07	-55 16 4.16	-0.044	07	0.18	5.084	15	38.4	3.65	18	38.4	A0	C9	4820	P -54 5342
97	8.9	9.0W	47 5.116	-0.0020	06	-58 19 32.24	-0.016	06	0.14	5.123	14	46.4	32.18	14	46.4	A0	C0	3948	P -57 5736
98	9.7	9.0W	47 9.282	-0.0051	07	-59 43 56.17	0.002	07	0.14	9.300	14	46.4	56.18	14	46.4	G5	C0	3949	P -59 4482
99	9.3	9.2W	47 17.585	-0.0022	07	-52 32 36.28	-0.023	07	0.18	17.610	15	38.4	36.01	18	38.4	F0	C9	4821	P -52 5931
00	8.1	7.9W	47 23.561	-0.0008	06	-57 46 54.50	-0.006	06	0.14	23.564	14	46.4	54.48	14	46.4	F2	C0	3951	P -57 5738

more recent catalogue; and there is no good reason to single out the latter and call it 'original' epoch. A weighted average of the first and second epochs would have been more appropriate, and, moreover, would have permitted one to calculate the standard deviation of the position at any epoch from a formula of the type mentioned above ((10) on page xiv of the introduction). This, however, would have required that the position and its dispersion for this epoch were computed. Considering the form in which the information incorporated in the SAO Catalog was originally available, this would not always have been easily possible and was, therefore, in the interest of homogeneity not done.

Further information given in the catalogue is the spectral type, given for only 83 percent of the stars, the source catalogue (as described below) and the star's number in it, and the Durchmusterung number, the sequence of preference being Bonner Durchmusterung (BD), Cordoba Durchmusterung (CoD) and Cape Photographic Durchmusterung (CPD).

6.633 The Sources. The mean positions and the proper motions which, of course, form the most essential part of the information given in the catalogue were extracted from a variety of sources in the following order of precedence: FK3, GC, CPC (Cape Annals, Vols. 17-20), Yale Catalogues (Transactions 11-14, 16-22, 24-27), AGK2 (Vols. 1, 2, 5-8), Cape Astrographic Zone, and the Third and the Fourth Melbourne General Catalogues. Since the AGK2 contains no proper motions, they were derived by comparison with the AGK1 where available and the Greenwich Zone of the Astrographic Catalogue (see section 6.75). Those stars in the CPC listed without proper motion (about 15%) were omitted. Otherwise, completeness to the level of the Yale and Cape Catalogue and the AGK2 was maintained.

All data was rigorously precessed to the mean coordinate system 1950.0 and reduced to the FK4 system. It was first reduced to the GC system, then to the FK3 by means of the tables in [Kopff, 1939], and finally to the FK4 system by means of the FK4 minus FK3 tables published in [Fricke and Kopff, 1963]. The positional data was not reduced from the system of the GC to that of the FK4 directly because at the time when the work was carried out, the necessary reduction tables in [Brosche, Nowacki, and Strobel, 1964] had not yet appeared.

Table 6.10 shows the sources for the position material in the SAO Catalog for those stars which are not in the FK3 or the GC. It is quite obvious that the single most important contribution was that of the AGK1 and its southern extension, since the Yale and Cape proper motions were obtained by comparing the newly determined positions with the AGK1.

This gigantic enterprise could be carried out by a limited staff in a reasonable time only by deliberate restrictions. The SAO Catalog is invaluable for the geodesist or the nonspecialist catalogue user who needs reasonably accurate positions and information about their ac-

TABLE 6.10 Sources for SAOC Data

Declination		Source
From	To	
+90°	+85°	Yale
+85°	+80°	AGK2 & Greenwich AC
+80°	+60°	AGK2 & AGK1
+60°	+50°	Yale
+50°	+30°	AGK2 & AGK1
+30°	-30°	Yale
-30°	-40°	CPC
-40°	-52°	Cape Astrographic Zone
-52°	-64°	CPC
-64°	-90°	Melbourne 3 & 4

curacy and cannot or does not want to hunt for it in various different catalogues.

6.634 Using the SAOC. The more experienced and critical user will know how to improve on the accuracy of the SAO Catalog positions when this is necessary. Two examples will illustrate this point: In large areas of the sky, the data in the SAO Catalog is based on Yale material only although material of equivalent high accuracy, say from the AGK2, is also available. One could in this case get an improved position of a star for, say, the epoch 1950.0 by reducing the AGK2 position to the FK4 system, then bringing it to the epoch 1950.0 by applying the proper motions printed in the SAO Catalog and taking a weighted mean with the position printed there. This same method may be applied whenever new high precision material becomes available.

As stated above, the catalogue was compiled in such a fashion that available GC positions took precedence over any other except the FK3 positions. The individual accuracy of the GC at the epoch 1930.0, say, is usually inferior to an AGK2, Yale or Cape position although the proper motions (when suitably corrected to refer to the FK4 system) are of high quality. One will find this well documented by inspecting the dispersions of the epoch 1950.0 positions of the GC stars with that of the non-GC stars in the SAO Catalog. One will generally obtain positions for the GC stars which are considerably more accurate than those printed by applying the GC proper motion (given in the SAO Catalog reduced to the FK4 system) to the AGK2, Yale, or Cape positions, or a mean of them, whichever is available. Needless to say, when doing this, attention must be paid to the standard deviations of the positions obtained from different sources. Positions and proper motions of the non-FK4 stars in the N30 will generally also be more accurate than those in the SAO Catalog. Note that they must be reduced to the FK4 system before use.

6.7 The Astrographic Catalogue (AC)

6.71 History

The most numerous collection of data capable of yielding accurate star positions is the Astrographic Catalogue (French: Carte du Ciel (CdC); German: Photographische Himmelskarte (PHK)). It consists essentially of a list of measurements of rectangular coordinates of star images on plates, together with identifying numbers and measures of their intensity, i.e., magnitude. Also available is the data required for the conversion to standard coordinates, i.e., plate constants and positions of the tangential points. Most of the presently available plate constants are, however, not sufficiently accurate to yield the stars' spherical positions with even approaching the maximum accuracy which the measurements are capable of producing.

The average epoch of the plates is around 1900 although the actual exposure dates vary from the later 1880's to 1950.

The last volume of measurements was published only recently (1965), and since then the catalogue may in a sense be regarded as completely available. It represents an example of truly international cooperation and was in the making for three quarters of a century.

The original stimulus for the Astrographic Catalogue probably came from the successful construction and operation of the so-called Normal Astrograph by the brothers Paul and Prosper Henry at the Paris Observatory in the last third of the nineteenth century. A normal astrograph is essentially a double telescope (a camera and a visual telescope side by side in the same tube, the latter being used for guiding during the exposures). The focal length of the camera is 3437 mm, producing a plate scale of one minute of arc per millimeter in the focal plane, and its field is $2^\circ 10'$ by $2^\circ 10'$ (13 cm × 13 cm).

Encouraged by the good results which had been achieved with this type of telescope, the most prominent astrometrists of the world resolved in 1887 at a conference held at the Paris Observatory to start the work.

6.72 Photography

The sky was covered in such a way that the plate centers lie on the parallels corresponding to all the full degrees of declination (coordinate system 1900) and are spaced in right ascension in intervals of about $2^\circ/\cos\delta$ or slightly less. In the adjacent full degree parallel, the centers were offset by about $1^\circ/\cos\delta$ in right ascension. Ideally, the overlap would in this way produce a 'corner-in-center' pattern (with fairly generous 'edge-on-edge' overlap) except that the convergence of the meridians causes deviations from it.

The original aim was the derivation of mean positions of all stars brighter than eleventh (photographic) magnitude with a precision corresponding to a standard deviation of not more than $0''.5$ in either coordinate which was the typical accuracy of a good relative meridian circle observation at that time. On some plates, images of stars with a magnitude of thirteen and even fainter were measured and recorded. Outside the Milky Way where the star density is naturally lower, fainter stars were included in the measuring lists in order to lessen the discrepancy between the number of stars found on plates inside and outside the Milky Way.

The rectangular coordinates of about two million stars were recorded (usually at least in duplicate, but for some stars even in quintuplicate, depending on where a star is situated within the overlap pattern). On the average, there are about 200 measured coordinate pairs per plate, but the scatter around this figure is considerable. On some plates in the Milky Way, the coordinates of well over 1000 star images were measured.

6.73 Measurements

A rectangular réseau (grid) of five millimeters mesh width was copied on all plates before development. The stars were measured with respect to the lines forming the réseau square enclosing them. This was done in order to render innocuous the expected shifts of the photographic emulsion, which later turned out to be quite insignificant anyway. No precautions, however, were taken to guard against magnitude equation (i.e., a systematic error depending on magnitude, which is characteristic for an individual plate). In the meantime this has been recognized as one of the most dangerous errors affecting photographic positions. In criticizing the procedures applied to the construction of the Astrographic Catalogues one ought to consider, however, that there was hardly any previous experience to guide the investigators. The fact that the photographic astrometric procedures of today are more efficient and more accurate is in no small measure due to the experience gained during the work on the Astrographic Catalogue.

The measurements themselves were executed differently at almost every participating observatory and at some institutions, the mode of measuring was changed in the middle of the work. At some observatories only one image per star was measured (usually there were three slightly displaced exposures of different length on any one plate), at others, two. Sometimes the plates were measured in two positions differing by 180° with respect to the measuring machine ('direct and reverse'), but sometimes in only one position, and so forth. A detailed description of all measuring would be of interest to a specialist only. The project was initiated and a good deal of it carried out by French astronomers, and a number of observatories in the British Empire (led by H.H. Turner, then director of the Oxford University Observatory) also participated. Thus, the measuring methods fell naturally into two groups.

The French astronomers and those who followed their example measured the plates on short screw measuring machines, recording (and publishing) the measurements to one micron and sometimes even to the purely nominal figure of one-tenth of one micron. Turner and his disciples used measuring machines which had no screw at all. Their machines were equipped instead with a cross formed of two scales in the eyepiece, which fitted exactly into a réseau square when viewed through the eyepiece. The measurements were made by bringing the star into the center of the cross and reading its coordinates with respect to the réseau lines by noting where the latter intersected the scales. The last significant figure of the readings of this type corresponds to five microns and, therefore, they are frequently referred to as 'estimates' in some of the modern literature on the subject. Objective analysis has shown that the zones measured following the Turner system are only slightly inferior to those measured with a short screw micrometer: the mean standard deviation of a position from one plate is 0ʺ.30 for the Turner system [Eichhorn and Gatewood, 1968] as compared to 0ʺ.24 for the French system [Heckmann, Dieckvoss and Kox, 1954].

It would, of course, have been desirable to measure all plates on a screw type measuring machine, but the Turner eyepiece scale permitted one to measure faster, no mean consideration, since the measurements were the main bottleneck in getting the work done. Nowadays, fully digitized and even fully automatic machines are available, and both systems used for the measurement of the AC plates are completely out of date.

6.74 Reduction

As was pointed out in section 6.222, conventional photographic astrometry can yield spherical coordinates only if the necessary reduction constants (plate constants and tangential point coordinates) are available. They are computed by means of reference stars whose positions should ideally be known for the epoch at which the plates were exposed with an accuracy comparable to the intrinsic accuracy of the measured rectangular coordinates. Lastly, a sufficient number (say, at least about ten per plate) of reference stars must be available.

As stated before, most AC plates were exposed around 1900, and the positions of the prospective reference stars in the AGK1 were as a rule the best ones available. Only in rare cases were the epochs of these positions contemporaneous with those of the plate exposures, and, although the reference star density was generally satisfactory, the systematic and accidental accuracy of the positions of the reference stars were mostly not.

Consider, furthermore, that a rigorous least squares solution, necessary for obtaining the highest accuracy for the plate constants which the measured rectangular coordinates are capable of yielding, involved the calculation of several product sums, a more tedious task at a time when the most widely used arithmetic auxiliaries were logarithms, and when the desk calculator was still considered by some practitioners an outlandish novelty which 'would never replace logarithmic calculation.' Because of all this, all the original plate constants that were published with the measured rectangular coordinates are inaccurate although some are better than others. (The standard deviation of a spherical coordinate (α or δ) computed with the originally given plate constants is typically about 1".)

Definitive plate constants will still have to be computed, except for the zones Paris and Bordeaux where definitive constants were published in [Heckmann, Dieckvoss, and Kox, 1949 and 1954], and the northern Hyderabad zone, for which [Eichhorn and Gatewood, 1968] give new constants.

Before the full potential accuracy of the AC will be available, new and definitive constants will have to be calculated for all zones. Also, all available measurements of one and the same star will have to be reduced (usually at least two on two different plates). At the present time, the extraction of accurate positions from the AC is still a task for the specialist.

6.75 The Zones of the AC

Table 6.11 gives a survey of the AC zones. In the column 'System,' S indicates that the measurements were made with an eyepiece scale (Turner system), and M indicates that the measurements were made with a short screw eyepiece micrometer.

As a rule, the measurements were published plate by plate in volumes containing the measurements made on the plates centered on the same degree of declination except for the regions around the celestial poles.

Since the range of the measurements is always 130 millimeters in both coordinates, an inspection of the numbers will easily reveal the scale and the zero point of the coordinates. If they range from, say, 1.000 to 27.000 it is obvious that they are in units of 5 millimeters (one millimeter being essentially one minute of arc) and that the coordinates of the plate center are (14.000, 14.000).

The catalogues are published by the observatories listed in the table, in the zones indicated, and are (if not out of print) available from them, except that the French zones (Uccle, Paris, Bordeaux, Toulouse, Algiers) are distributed by the Paris Observatory, while the Cape Zone is distributed by the Royal Greenwich Observatory.

TABLE 6.11 The Zones of the Astrographic Catalogue

Declination of Plate Centers		Observatory	System	Plates Taken	
From	To			From	To
+90°	+65°	Greenwich	S	1892	1905
+64	+55	Vatican	S	1898	1922
+54	+47	Catania	M	1898	1926
+46	+40	Helsingfors	M	1892	1903
+39	+36	Hyderabad	S	1928	1936
+35	+34	Uccle	M	1940	1950
+33	+25	Oxford	S	+33° and +32°	
				1932	1936
				+24° to +31°	
				1892	1904
+24	+18	Paris	M	1895	1902
+17	+11	Bordeaux	M	1895	1902
+10	+ 5	Toulouse	M	1893	1905
+ 4	- 2	Algiers	M	1891	1903
- 3	- 9	San Fernando	M	1894	1903
-10	-16	Tacubaya	S	1890	1912
-17	-23	Hyderabad	S	1914	1929
-24	-31	Cordoba	12% M 88% S	1909	1913
-32	-40	Perth	S	1901	1915
-41	-51	Cape	M	1892	1910
-52	-64	Sydney	S	1891	1915
-65	-90	Melbourne	M	1891	1915
+39	+32	Potsdam (incomplete)	M	1893	1900

6.8 Future Work

It is difficult for anyone to attempt his luck as a prophet. Obviously, modern electronic techniques are being adopted at the meridian circle as well as at the measuring machine to increase the accuracy and efficiency of the measuring process. Computers have permitted the introduction of more sophisticated reduction methods.

The satisfactory situation regarding the availability of a sufficient number of sufficiently accurate star positions is still far removed, especially on the southern hemisphere. Programs currently planned or underway for remedying this state of affairs are mainly the following: Southern reference and fundamental stars are being observed by the expeditions of the Pulkovo Observatory in Cerro Calan, Chile, of the U.S. Naval Observatory in El Leoncito, Argentina, and of the Hamburg-

Bergedorf Observatory in Perth, Australia. The photographic work of the Cape Observatory is continued by photographing the declination zone from -30° to 40°. The zone from -40° to -52°, as mentioned earlier, has already been completed. The remaining zones from -52° to -90° will be taken up later at epochs which will provide good proper motions. No provision has been made for a new set of plates for the Yale zones between declination 0° and -30°. Important work is also being performed at the U. S. Army Map Service.

The task of establishing one gigantic General Catalogue, incorporating all the information contained in previous catalogues (except the AC) is under serious study at Heidelberg by the astronomers of the Astronomisches Rechen-Institut.

References

Astronomische Gesellschaft. (1890-1912). Catalog der Astronomischen Gesellschaft, (AGK1). Erste Abteilung: "Catalog der Sterne bis zur neunten Groesse zwischen 80° noerdlicher und 2° suedlicher Deklination fuer das Aequinoctium 1875." Zweite Abteilung: "Catalog . . . zwischen 2° und 23° suedlicher Deklination fuer das Aequinoctium 1900." Wilhelm Engelmann, Leipzig.

Astronomisches Rechen-Institut, Berlin-Dahlem (Kopernikus-Institut). (1943). "Katalog der Anhaltsterne fuer das Zonenunternehmen der Astronomischen Gesellschaft," (AGK2R). Veroeff. d. Kopernikus-Instituts (Astronomisches Rechen-Institut) zu Berlin-Dahlem, 55.

Astronomisches Rechen-Institut, Heidelberg. (1963). "Preliminary Supplement to the Fourth Fundamental Catalogue," (FK4 Sup). Veroeff. d. Astronomischen Rechen-Instituts, Heidelberg, 11.

Astronomisches Rechen-Institut, Heidelberg. (annual). Apparent Places of Fundamental Stars Containing the 1535 Stars in the Fourth Fundamental Catalogue, (APFS). Verlag G. Braun, Karlsruhe.

Auwers, A. (1879). "Fundamental-Catalog fuer die Zonen-Beobachtungen am Noerdlichen Himmel," (FC). Publikationen der Astronomischen Gesellschaft, XIV.

Auwers, A. (1883). "Mittlere Oerter von 83 suedlichen Sternen fuer 1875,0 zur Fortsetzung des Fundamental-Catalogs fuer die Zonen-Beobachtungen der Astronomischen Gesellschaft, nebst Untersuchungen ueber die Relationen zwischen einigen neueren Sterncatalogen, insbesondere fuer den in Europa sichtbaren Theil des suedlichen Himmels," (FC). Publikationen der Astronomischen Gesellschaft, XVII.

Auwers, A. (1889). "Vorlaeufiger Fundamental-Catalog fuer die suedlichen Zonen der Astronomischen Gesellschaft." Astronomische Nachrichten, 121, 145.

Auwers, A. (1897). "Fundamentalcatalog fuer Zonenbeobachtungen am Suedhimmel und suedlicher Polar-Catalog fuer die Epoche 1900." Astronomische Nachrichten, 143, 361.

Auwers, A. (1898). "Vorlaeufige Verbesserung des Fundamental-Catalogs fuer die Zonen Beobachtungen der Astronomischen Gesellschaft und seine suedliche Fortsetzung, (Publ. XIV und XVII der Astron. Ges.)." Astronomische Nachrichten, 147, 47.

Auwers, A. (1900). "Gewichtstafeln fuer Sterncataloge." Astronomische Nachrichten, 151, 225.

Auwers, A. (1903). "Nachtraege zu den Tafeln zur Reduktion von Sterncatalogen auf das System des Fundamental-Catalogs der AG und zu den Gewichtstafeln fuer Sterncataloge." Astronomische Nachrichten, 162, 357.

Auwers, A. (1904). "Ergebnisse der Beobachtungen 1750-1900 fuer die Verbesserung des Fundamentalcatalogs des Berliner Jahrbuchs, (Publ. d. AG XIV und XVII)." Astronomische Nachrichten, 164, 225.

Auwers, A. (1905). "Weitere Nachweise der Grundlagen fuer die neuen Stern-Ephemeriden des Berliner Jahrbuchs." Astronomische Nachrichten, 168, 161.

Barney, I. (1951). "Supplementary volume to the Yale Zone Catalogues -30° to +30°." Transactions of the Astronomical Observatory of Yale University, 23.

Belyavsky, S. I. (1947). "Astrograficheskii katalog 11322 zvezd mezhdu 70° severnovo skoneniya i severnym polyasom poftorenye," (Astrographic catalog of 11322 stars between +70° and the celestial north pole; repetition). Publications of the Central (Trudy Glavnoy) Astronomical Observatory, Pulkovo, Series II, 60.

Boss, B. (1937). "General Catalogue of 33342 stars for the epoch 1950," (GC). Publications of the Carnegie Institution of Washington, 468. Reissued by Johnson Reprint Co., New York.

Boss, B., A. J. Roy, and W. B. Varnum. (1931). "Albany Catalogue of 20811 stars for the epoch 1910." Publications of the Carnegie Institution of Washington, 419.

Boss, L. (1903). "Positions and proper motions of 627 standard stars." The Astronomical Journal, 23, 17.

Boss, L. (1910). "Preliminary General Catalogue of 6188 stars for the epoch 1900," (PGC). Publications of the Carnegie Institution of Washington, 115.

Boss, L. and B. Boss. (1928). "San Luis Catalogue of 15333 stars for the epoch 1910." Publications of the Carnegie Institution of Washington, 386.

Brosche, P. (1966). "Representation of systematic differences in positions and proper motions of stars by spherical harmonics."

Veroeffentlichungen des Astronomischen Rechen-Instituts, Heidelberg, 17.

Brosche, P., H. Nowacki, and W. Strobel. (1964). "Systematic differences FK4 minus GC and FK4 minus N30 for 1950.0." Veroeffentlichungen des Astronomischen Rechen-Instituts, Heidelberg, 15.

Clemence, G. M. (1966). "Inertial Frames of Reference." Quarterly Journal of the Royal Astronomical Society, 7, 10.

Cordoba Observatory. (1913-1954). "Catalogo de . . . estrellas entre . . . y . . . declinacion austral (1875) para el equinoccio medio de 1900.0 durante los anos . . ." Resultados del Observatorio Nacional Argentino. Volume, observations period, year of issue and number of stars contained are as follows:

Cat.	Vol.	Observed	Appeared	No. of Stars
A	22	1891-1900	1913	15975
B	23	1891-1900	1914	15200
C	24	1895-1900	1925	12757
D	38	- - -	1954	16610

Eichhorn, H. (1962). A Summary of a Study of Star Catalogues and Their Accuracies. RCA Service Co., Missile Test Project, Math. Services TM 62-4.

Eichhorn, H. and G. Gatewood. (1967). "New Plate Constants for the Northern Hyderabad Zone (+35° to +40°) of the Astrographic Catalogue, in Part Computed by the Plate Overlap Method." The Astronomical Journal, 72, 9.

Eichhorn, H., W. D. Googe, and G. Gatewood. (1967). "Concerning the Plate Overlap Method." The Astronomical Journal, 72, 5.

Fricke, W., D. Brouwer, J. Kovalevsky, A. A. Mikhailov, and G. A. Wilkins. (1965). "Report to the Executive Committee of the Working Group on the System of Astronomical Constants." Bulletin Géodésique, 75, pp. 59-67.

Fricke, W. and A. Kopff. (1963). "Fourth Fundamental Catalogue resulting from the revision of the Third Fundamental Catalogue (FK3) carried out under the supervision of W. Fricke and A. Kopff in collaboration with W. Gliese, F. Gondolatsch, T. Lederle, H. Nowacki, W. Strobel and P. Stumpff," (FK4). Veroeffentlichungen des Astronomischen Rechen-Instituts, Heidelberg, 10.

Gill, D. and S. S. Hough. (1923). Zone Catalogue of 20,843 Stars included between 40° and 52° of southern declination, referred to the equinox 1900. Royal Observatory at the Cape of Good Hope, H. M. Stationery Office, London.

Gliese, W. (1963). "The Right Ascension System of the Fourth Fundamental Catalogue (FK4)." Veroeffentlichungen des Astronomischen Rechen-Instituts, Heidelberg, 12.

Gonnesiat, F. (1924). "Catalogue des 9997 étoiles comprises entre -17°50' et -23° declinaisision 1855 pour l'équinoxe de 1900.0."

Publications de l'Observatorie d'Alger, VIII. Gauthier-Villars, Paris.

Gyllenberg, W. (1948). "Systematic Corrections and Weights of 108 Star Catalogues." Lunds Universitest Årsskrift, N. F. Avd. 2, 44, 2; also Kungl. fysiografiska sällskapets handlingar N. F. 59, 2; also Meddelande från Lunds Astronomiska Observatorium Ser. II, 122.

Haramundanis, K. (1967). "Experience of the Smithsonian Astrophysical Observatory in the Construction and Use of Star Catalogues," The Astronomical Journal, 72, 5.

Heckmann, O., W. Dieckvoss, and H. Kox. (1949). "Die Ableitung der Eigenbewegungen von Sternen der AG Katalog mit Benutzung der Photographischen Himmelskarte, insbesondere in der Zone +20° bis 25° (Berlin B)." Sitzungsberichte der Deutschen Akademie der Wissenschaften zu Berlin, mathematisch-naturwissenschaftliche Klasse, Jahrgang 1948, VII.

Heckmann, O., W. Dieckvoss, and H. Kox. (1954). "New Plate Constants in the System of the FK3 for the Declination Zones +21°, +22°, +23°, +24° of the Astrographic Catalogue Paris." The Astronomical Journal, 59, 143.

Heinemann, K. (1937). "Das Deklinationssystem des FK3." Appendix (Anhang) V to [Kopff, 1937].

Heinemann. (1964). "Verzeichnis von Sternkatalogen 1900-1962. Mit Nachtraegen zu AN 174 No. 4176 und AN 248 No. 5947." Veroeffentlichungen des Astronomischen Rechen-Instituts, Heidelberg, 16.

Hoffleit, D. (1964). Catalogue of Bright Stars, Containing Data Compiled Through 1962, 3rd revised edition. Yale University Observatory, New Haven, Connecticut.

Jackson, J. and R. H. Stoy. (1954-1966). "Cape Photographic Catalogue for 1950.0." Annals of the Cape Observatory, 17 (1954), 18 (1955), 19 (1955), 20 (1958), 21 (1966). H. M. Stationery Office, London.

Jenkins, L. F. (1952). General Catalogue of Trigonometric Stellar Parallaxes (edition of 1952, containing all data available in May, 1950. 1963 supplement, containing all data available in December, 1962). Yale University Observatory, New Haven, Connecticut.

Kahrstedt, A. (1937). "Das Rektascensionssystem des FK3." Appendix (Anhang) IV to [Kopff, 1937].

Kahrstedt, A. (1961-1966). Index der Sternoerter 1925-1960 (Index II). Akademie Verlag, Berlin.

Vol.	Zone	Appeared
I	anonymae (all)	1961
II	-0° to -22°	1962
III	-22° to -40°	1964
IV	-41° to -61°; -62° to -89°	1965

Vol.	Zone	Appeared
V	+0° to +10°	1963
VI	+11° to +20°	1963
VII	+21° to +30°	1964
VIII	+31° to +50°	1965
IX	+51° to +89°	1966

Knobel, E. B. (1877). "The Chronology of Star Catalogues." Memoirs of the Royal Astronomical Society, 43, 1.

Koenig, A. (1933). "Reduktion photographischer Himmelsaufnahmen." Handbuch der Astrophysik, I, pp. 502-559. Verlag von Julius Springer, Berlin.

Kopff, A. (1937). "Dritter Fundamentalkatalog des Berliner Astronomischen Jahrbuchs, I. Teil: Die Auwers-Sterne fuer die Epochen 1925 und 1950," (FK3). Veroeffentlichungen des Astronomischen Rechen-Instituts zu Berlin-Dahlem, 54.

Kopff, A. (1938). "Dritter Fundamentalkatalog des Berliner Astronomischen Jahrbuchs, II. Teil: Die Zusatzsterne." Abhandlungen der Preussischen Akademie der Wissenschaften, mathematischen-naturwissenschaftliche Klasse, 3.

Kopff, A. (1939). "Vergleich des FK3 mit dem General Catalogue von B. Boss." Astronomische Nachrichten, 269, pp. 160-167.

Kopff, A. (1949). "A Comparison of the System of the General Catalogue and the Dritter Fundamentalkatalog." Monthly Notices of the Royal Astronomical Society, 109, 580.

Kopff, A. (1954). "Remarks on the Revision of the FK3 and Its Relation to N30." Monthly Notices of the Royal Astronomical Society, 114, 478.

Kopff, A. and H. Nowacki. (1934). "Zusatzsterne des Dritten Fundamentalkatalogs des Berliner Astronomischen Jahrbuchs." Astronomische Nachrichten, 252, 185.

Kopff, A., H. Nowacki, and W. Strobel. (1964). "Individual Corrections to FK3 and the Declination System of the Fourth Fundamental Catalogue (FK4)." Veroeffentlichungen des Astronomischen Rechen-Instituts, Heidelberg, 14.

La Plata Observatory. (1919-1938). Publicaciones del observatorio astronomico, tomo . . . Generally the titles of the individual catalogues are: "Catalogo La Plata . . . de . . . estrellas de declinaciones comprendidas entre . . . y . . . (1875) para el equinoccio . . ."

Vol.	Appeared	Reference to Table 6.6
5	1919	A
7	1929	B
8	1924	C
9	1936	D
10	1947	E
13	1938	F

Moennichmeyer. (1909). "Verbesserte Oerter des AGK Bonn nebst gelegentlich bestimmten Oertern von weiteren 757 Sternen der Zone +40° bis +50°." Veroeffentlichungen der Koeniglichen Sternwarte zu Bonn, 9.

Morgan, H. R. (1952). "Catalog of 5268 Standard Stars, 1950.0 Based on the Normal System N30." Astronomical Papers Prepared for the use of the American Ephemeris and Nautical Almanac, 13, part 3. Government Printing Office, Washington, D.C.

Peters, J. (1907). "Neuer Fundamentalkatalog des Berliner Astronomischen Jahrbuchs nach den Grundlagen von A. Auwers fuer die Epochen 1875 und 1900." (NFK). Veroeffentlichungen des Astronomischen Rechen-Instituts, Berlin-Dahlem, 33.

Preussische Akademie der Wissenschaften. (1922-). Geschichte des Fixsternhimmels, enthaltend die Kataloge des 18. und 19. Jahrhunderts (GFH). Abteilung I: "Der noerdliche Sternhimmel." Verlag von G. Braun, Karlsruhe. Abteilung II: "Der Suedliche Sternhimmel." Verlag von G. Braun, Karlsruhe, and Akademie Verlag, Berlin.

Ristenpart, F. (1901 a). "Sterncataloge und -Karten." Handwoerterbuch der Astronomie (W. Valentiner, ed.), 3/2, pp. 455-513. Verlag von Eduard Trewendt, Breslau.

Ristenpart, F. (1901 b). Verzeichnis von 336 Sterncatalogen. Verlag von Eduard Trewendt, Breslau.

Ristenpart, F. (1909). "Fehlerverzeichnis zu den Sternkatalogen des 18. und 19. Jahrhunderts." Astronomische Abhandlungen, Ergaenzungs-Hefte zu den Astronomischen Nachrichten, 3, 16, pp. 1-500.

Schlesinger, F. and I. Barney. (1939). "On the Accuracy of the Proper Motions in the General Catalogue, Albany 1938." The Astronomical Journal, 48, 51.

Schlesinger, F., I. Barney, A. J. J. van Woerkom, D. Hoffleit, and R. Jones. (-). "Yale Zone Catalogues." Transactions of the Astronomical Observatory of Yale University, 4, 5, 9, 11-14, 16-22, 24-27. Yale University Observatory, New Haven, Connecticut.

Schlesinger, F. and C. J. Hudson. (1916). "The Determination of star positions by means of a wide angle camera." Publications of the Allegheny Observatory of the University of Pittsburgh, 3, 9.

Schorr, R. and A. Kohlschütter. (1951-1958). Zweiter Katalog der Astronomischen Gesellschaft, fuer das Aequinoktium 1950 (AGK2). Oerter der Sterne bis zur neunten Groesse zwischen dem Nordpol und 2° suedlicher Deklination, abgeleitet aus photographischen Himmelsaufnahmen der Sternwarten Bergedorf und Bonn, 1-15, Hamburger Sternwarte (1951) and Universitaetssternwarte Bonn (1958).

Schorr, R. and W. Kruse. (1928). Index der Sternoerter 1900-1925, (Index I). Band 1: "Der noerdliche Sternhimmel." Band 2: "Der suedliche Sternhimmel." Verlag der Sternwarte Bergedorf.

Scott, F. P. (1967). "Status of the International Efforts to Improve the Catalogs of Positions and Motions of Stars to the 9th Magnitude." The Astronomical Journal, 72, 5.

Smithsonian Astrophysical Observatory. (1966). Star Catalog. Positions and Proper Motions of 258,997 Stars for the Epoch and Equinox of 1950.0 (SAOC). In four parts. Smithsonian Institution, Washington, D. C.

Spencer Jones, H. and J. Jackson. (1936). Proper Motions of Stars in the Zone Catalogue of 20843 Stars, 1900. Royal Observatory, Cape of Good Hope, H. M. Stationery Office, London.

Woolard, E.W. (1953). "Theory of the Rotation of the Earth Around Its Center of Mass." Astronomical Papers prepared for the use of the American Ephemeris and Nautical Almanac, XV, part 1. Government Printing Office, Washington, D. C.

7 OPTICAL INSTRUMENTS FOR ASTRONOMIC POSITION, AZIMUTH, AND TIME DETERMINATIONS

The instruments used in geodetic astronomy for latitude, longitude (time), and azimuth determinations may be divided into the following categories:

(1) Observatory (high precision) instruments which are permanently installed in observatories for repeated determinations of time and/or latitude.

(2) First-order (precision) portable instruments which, when used in field conditions, can provide standard deviations less than 0".3 for the final mean latitude or time (longitude), or 0".4 for the azimuth.

(3) Second-order (geodetic) instruments giving latitude and time (longitude) with standard deviations less than 1".0, or azimuth with less than 1".5.

(4) Lower order (surveying) instruments.

High precision instruments and methods are required, for example, in the precise observatory determination of universal, ephemeris, and atomic times, as well as in the observation of variations in time and latitude.

First-order instruments and methods are needed to establish the astronomic coordinates and azimuth at the origin of independent triangulation systems; to observe longitude and azimuth at Laplace stations for controlling the azimuths of first-order triangulation, trilateration, or traverse; to determine astrogeodetic deflections for determining the size, shape, and orientation of the best fitting reference ellipsoid, etc. [Bomford, 1962, pp. 88-92, 453-456].

Second-order instruments and methods can be applied, for example, to determine astrogeodetic deflections for geoidal studies or for the reference ellipsoid; to determine control points and azimuths for topographic and aerial surveys for small scale mapping; to provide azimuth control for second-order triangulation, trilateration, and traverse [Bomford, 1962, pp. 60-61, 93-94, 320-327].

Lower order instruments and methods are used chiefly in engineering and surveying practice, normally consisting of observations on the sun or on Polaris using conventional surveying instrumentation [Bouchard and Moffitt, 1965, pp. 375-412].

According to the nature of the instruments, they may be divided into the following three groups:

(1) Optical instruments, generally employed to measure the direction of a celestial object with respect to the observer's astronomic meridian and/or horizon (its azimuth and/or altitude).

(2) Timing devices, used either to record the epoch of some observation or to facilitate interpolation of time during the interval between two time determinations.

(3) Auxiliary equipment, used in recording or aiding the determinations of quantities other than directions and local times. Such instruments are, for example, the radio, the chronograph, the thermometer, and the barometer.

In this section, first- and second-order optical instruments are discussed in some detail. Observatory techniques are treated briefly, and lower order instruments are not considered. Nonoptical instruments are described in Chapter 8.

7.1 First-Order (Precision) Instruments

The basic types of first-order instruments which may be used for the purposes of geodetic astronomy are the following:

(1) The transit (passage) instrument for determining the instant of a star's transit over a given vertical circle (usually either the astronomic meridian or the prime vertical). For this purpose, the motion of the telescope is confined to the plane of the vertical and no precisely graduated vertical reading circle is provided. If set in the meridian, it is usually employed to provide the local sidereal time; if in the prime vertical, it usually aims at determining the astronomic latitude.

(2) The meridian circle is a combination of a transit instrument and a precisely graduated vertical reading circle. It serves to measure zenith distances of stars directly, from which in geodesy the astronomic latitude of the observing station is calculated. The local sidereal time may also be determined from the transit. In astronomy, the meridian circle is the most important and most widely used instrument for the determination of absolute star positions and, before the introduction of photographic methods, was the most frequently used instrument for the determination of relative star positions as well (see section 6.2).

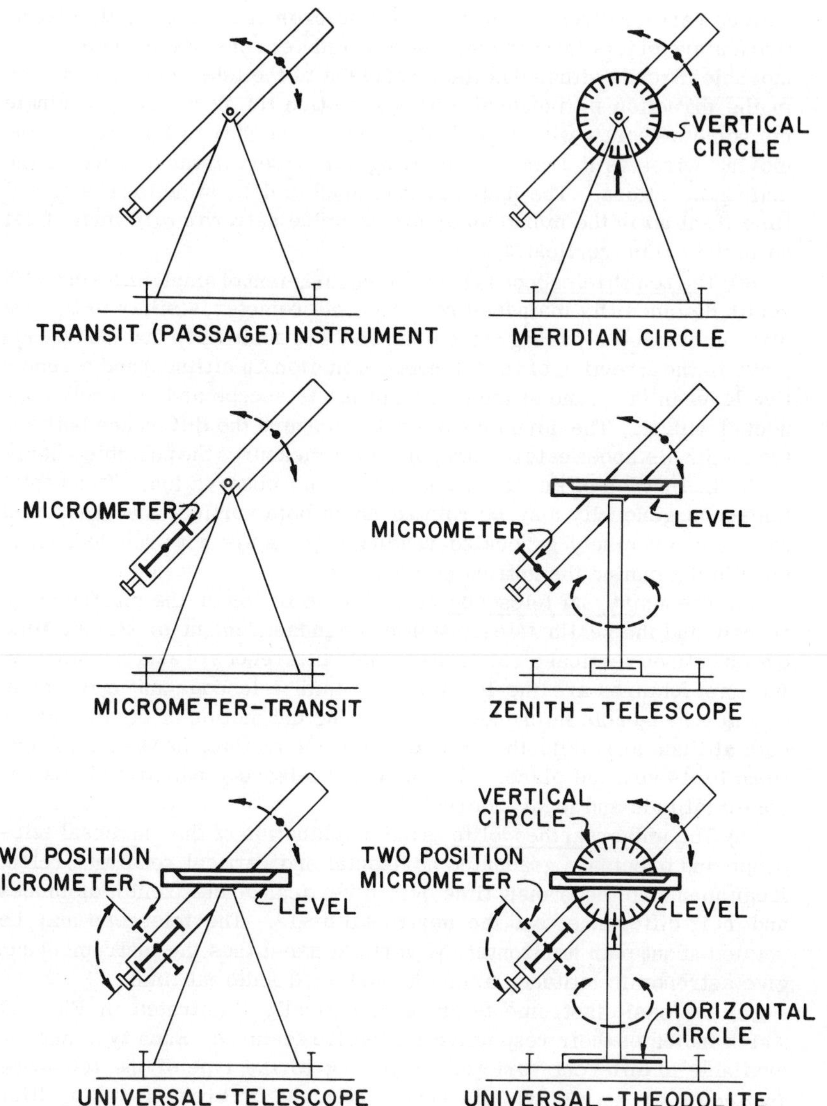

Fig. 7.1 Distinctive features of first-order optical instruments

(3) The micrometer-transit is a combination of the transit with a micrometer for the measurements of the instant of a star's transit over a given vertical circle by means of the micrometer which is the distintive feature of this instrument. The micrometer consists of one or more movable threads situated in the focal plane of the telescope and carried in the direction perpendicular to its motion (in azimuth) by a single micrometer screw with a graduated head. The star is followed by the moving wires as it transits, allowing for observations at each of the main graduations. The instrument is used to determine local sidereal time if set up in the meridian or to determine astronomic latitude if set up in the prime vertical.

(4) The zenith telescope is for the measurement of small differences of zenith distance. Its main features are a micrometer, similar to the one of the micrometer-passage instrument but in a position so that the threads move in the direction of the telescope's motion (in altitude) and a sensitive level in the plane of the motion of the telescope and securely connected with it. The former serves to measure the difference between the zenith distances of two stars, the latter measures the possible change in the inclination of the telescope during the observation. The zenith telescope generally may be rotated about both vertical and horizontal axes, but precisely graduated reading circles are not provided. It is used to determine the astronomic latitude.

(5) The universal telescope is the combination of the micrometer-transit and the zenith telescope for the measurement of transit time over a defined vertical circle and of small differences of zenith distances. Its main features are the level of the zenith telescope and a rotatable or interchangeable micrometer permitting the motion of the threads in both altitude and azimuth. The telescope's motion, however, is confined to its vertical plane. The universal telescope can provide astronomic latitude and local sidereal time.

(6) The universal theodolite is the combination of the universal telescope and precisely graduated horizontal and vertical reading circles. It can measure the transit time over a vertical circle, zenith distances and their difference, and the horizontal angle. The telescope may be rotated about both horizontal and vertical axes; thus the instrument can give astronomic latitude, azimuth, and local sidereal time.

These basic instruments are schematically illustrated in Fig. 7.1 which indicates their respective distinctive features. Each type may be available in different varieties according to the type of the telescope (reflector or refractor), its shape (straight or broken), its position (centric or eccentric), its plane of motion (horizontal or equatorial), etc.

Since the universal theodolites possess most, if not all, of the features of the other instruments and are more often used in modern geodetic as-

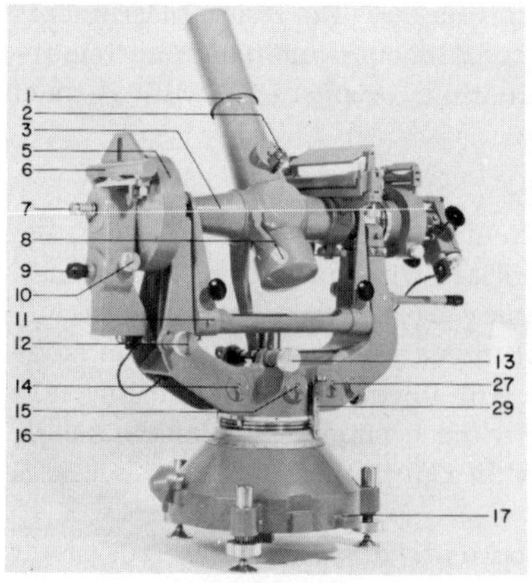

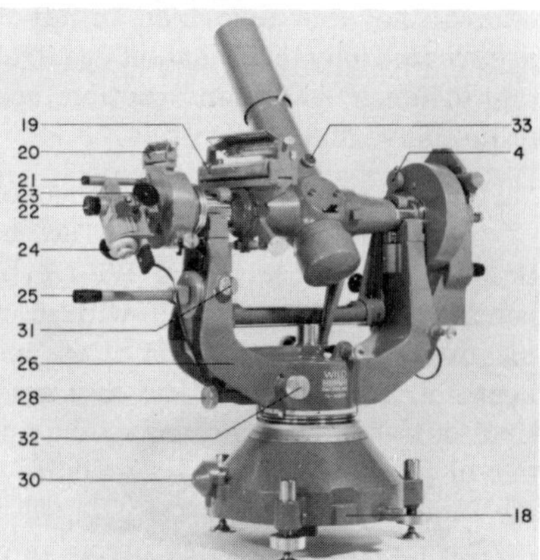

1 Telescope Tube
2 Lamp for Telescope Illumination
3 Horizontal Axis
4 Vertical Circle Drive
5 Vertical Circle
6 Vertical Circle Level
7 Vertical Circle Lamp
8 Vertical Clamp
9 Vertical Circle Eyepiece
10 Vertical Circle Micrometer
11 Suspension Level
12 Vertical Circle Level Fine-Adjustment Screw
13 Vertical Slow Motion Screw
14 Switch for Vertical Circle Lamp
15 Switch for Telescope Lamp
16 Horizontal Setting Circle
17 Circular Level
18 Fixing Cam
19 Horrebow Levels
20 Level for Vertical Setting Circle
21 Eyepiece for Vertical Setting Circle
22 Telescope Eyepiece
23 Eyepiece Micrometer Drive
24 Micrometer Drum
25 Horizontal Circle Eyepiece
26 Telescope Support
27 Switch for Setting Circle Lamps
28 Horizontal Clamp
29 Horizontal Setting Circle Index and Lamp
30 Horizontal Circle Drive
31 Horizontal Circle Micrometer Drive
32 Horizontal Slow Motion Screw
33 Telescope Illumination

Fig. 7.2 Wild T4 universal theodolite

tronomic work (not always with better results, though) than any of the others; they are described in full detail below. The other instruments are treated only to the extent required to point out some possible important differences in construction, application, or precision with respect to the universal theodolites.

7.11 The Wild T4 Universal Theodolite (Fig. 7.2)

7.111 Horizontal System. The vertical axis has a cylindric construction characteristic of the Wild theodolites with an upper conical ball bearing. The female axis with the upper conical widening is firmly screwed on the lower part of the instrument. The carrier of the horizontal circle rotates about this axis. The lower end of the hollow vertical axis fits exactly into the boring of the female axis, but the shoulder of its upper end rests on balls which run in the conical widening of the female axis.

The horizontal reading circle is of optical glass and can be set to any desired reading by turning the circle setting knob.

The optical micrometer of the horizontal circle is built into the support that carries the eyepiece-end of the telescope. The microscope eyepiece is horizontal in position and below the telescope eyepiece, so that aiming and circle reading can be done consecutively without changing position. The control knob for the optical micrometer is on the right of the horizontal circle eyepiece.

During reading, the lines of the upper and lower halves of the image are made to coincide by means of the micrometer knob (see Fig. 7.3).

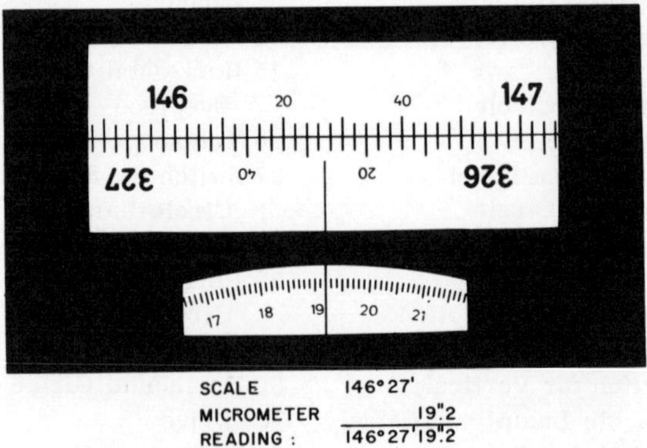

SCALE 146°27'
MICROMETER 19".2
READING : 146°27'19".2

Fig. 7.3 Reading the horizontal circle on the Wild T4

Then the upper figures are followed from the left to the last ten-minute line on the left of the index (in the figure, 146°20'), which is marked by the lower vertical line. Then the intervals from the index to the next diametrical ten-minute line of the right (to the 326°20' line) are counted and multiplied by the value of the interval, 2' each. This added to the first reading gives the circle reading to single minutes (in the figure, 146°27'). The seconds are read on the lower image of the optical micrometer which is graduated to 0".1, the numbers indicating the seconds of arc. The full reading in Fig. 7.3 thus is 146°27'19".2. The horizontal circle reading increases clockwise.

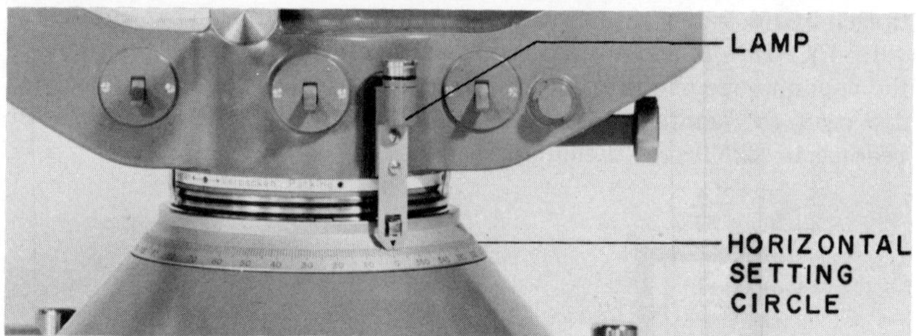

Fig. 7.4 Horizontal setting circle on the Wild T4

The horizontal setting circle (Fig. 7.4) on the outside of the base-cone allows for quick approximate orientation of the telescope to any desired direction. Orientation of the setting circle is done by hand after the telescope is directed to any point. For astronomic observations, the setting circle is normally set to zero if the telescope is pointing to the north.

7.112 Vertical System. The horizontal axis of the telescope rests on open bearings. The bearing on the eyepiece side is adjustable in height for correcting the tilt of the horizontal axis.

The vertical reading circle is made of optical glass and is placed on the horizontal axis opposite the telescope eyepiece. It is freely rotatable by turning a knob.

The optical micrometer is similar to the one of the horizontal circle, but the drum is graduated in units of 0".2.

The vertical circle level is developed as a coincidence level but is also graduated.

The eyepiece of the vertical circle is situated diametrically opposite the eyepiece of the horizontal circle; the micrometer knob is also on this side. Immediately after each vertical circle reading, the bubble deviation of the vertical circle level is read, transformed into seconds, and added to the circle reading. For example, in Fig. 7.5a the level correction in units of the bubble division (position of the left minus position of the right bubble end) is $2.4 - 0.9 = 1.5$. The value of a division in angular units is determined by means of the vertical circle by modifying the level readings with the level fine-adjustment screw and reading the vertical circle each time.

For reading the circle, the micrometer knob is turned until the lines of the upper and the lower half of the image of the circle coincide (Fig. 7.5b). The upper division is followed from the left to the last long line before the index line (in the figure, 34°20'). From here the intervals to the next diametric long line (214°20') are counted to the right and multiplied by the value of the interval, 4' each (the figure shows one interval =4'). This is added to the first reading (34°24' in the example). In the upper image of the optical micrometer, single minutes and seconds are read, the smallest interval being 0".2. In the figure the micrometer reading is 1'26".9; the complete reading is 34°25'26".9.

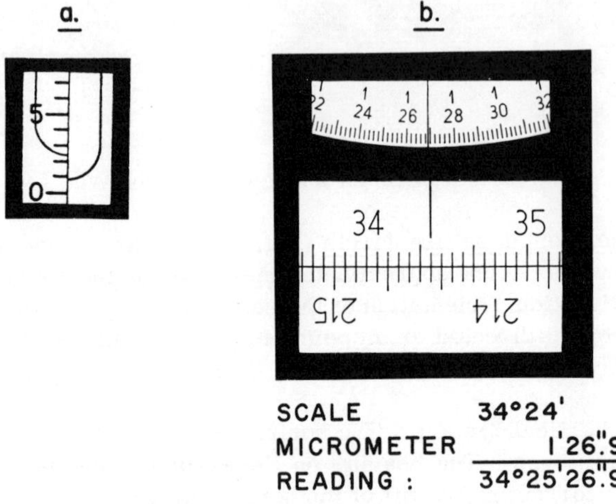

Fig. 7.5 Reading the vertical circle on the Wild T4

The vertical setting circle serves for the rapid setting of the telescope into a desired tilt. Its eyepiece is above the telescope eyepiece. The circle is divided into intervals of one degree of arc and correspondingly ciphered (Fig. 7.6). The reading scale is divided into six intervals, which allows reading to an estimated 1' (in the figure, 34°26'). The

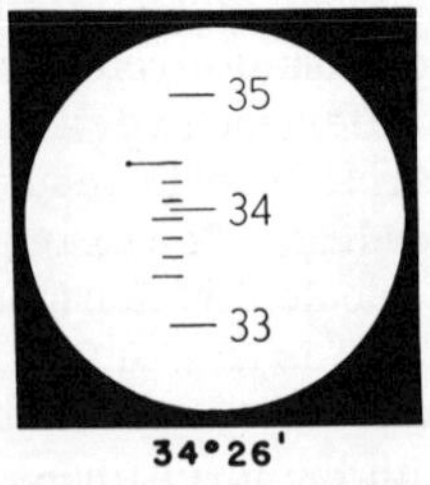

Fig. 7.6 Reading the vertical setting circle on the Wild T4

scale may be pre-set to any desired reading by means of the tangent screw at the lower part of the circle-housing. In order to set the telescope to a certain tilt, the reading of the setting circle is set to the desired zenith distance and the telescope is tilted until the small level on the top of the circle-housing is centered.

7.113 Telescope. The telescope is of the broken type; thus the eyepiece, situated at the end of the horizontal axis, always remains in the same position regardless of the tilt of the telescope. The quality of the telescope allows for the observation of stars brighter than about eighth magnitude.

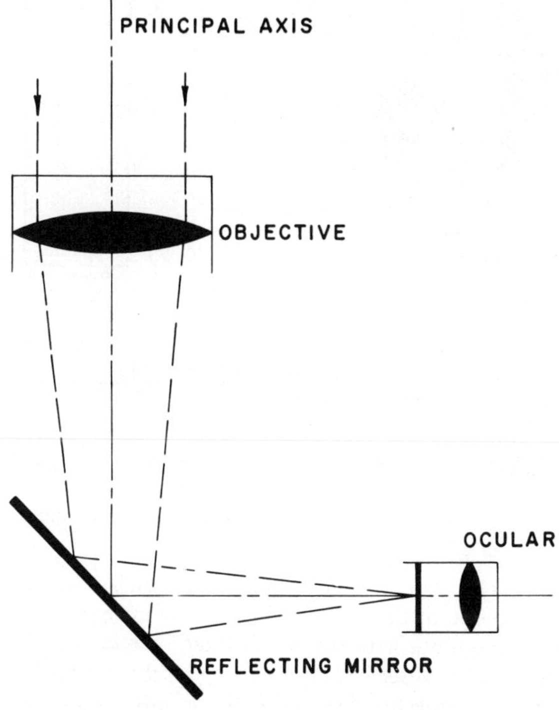

Fig. 7.7 Optical system of the Wild T4 telescope

Fig. 7.7 shows the optical system whose main features are the objective, the reflecting mirror, and the ocular or eyepiece (the objective and eyepiece actually are a combination of several lenses each). The rays parallel to the principal axis in the telescope are reflected by the mirror and join in the principal focus of the objective. The plane through the focus perpendicular to the principal axis is called the focal plane.

A glass reticle plate is in the focal plane in order that the telescope may be used for the measurement of angles. Cross lines (threads) are etched on this plate (Fig. 7.8b). The T4 has a fixed and a movable reticle, and, depending upon the circumstances, one or the other is used for observation. The line which connects the center point of the fixed cross marks with the center of the objective is called the line of sight of the telescope or the collimation axis.

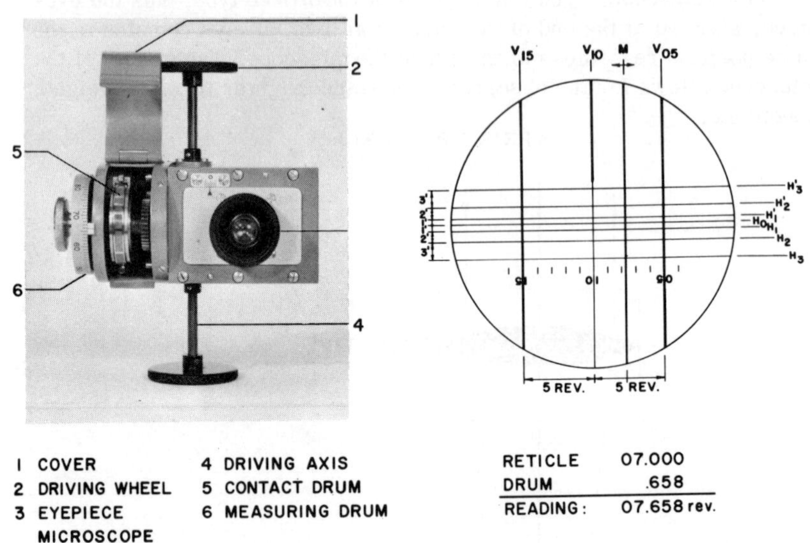

a. TELESCOPE EYEPIECE

b. RETICLE PLATE WITH MOVABLE THREAD

Fig. 7.8 Telescope eyepiece and impersonal micrometer of the Wild T4

7.114 Telescope Eyepiece and Impersonal Micrometer. The telescope eyepiece of the T4 consists of the eyepiece microscope and the glass reticle plate with the set of fixed threads and a movable line, the displacement of which can be measured with the impersonal micrometer. Fig. 7.8 shows the microscope with the micrometer drum as seen from the outside (a), and the view in the eyepiece with the fixed and moving threads (b). The central cross, defined by the lines V_{10} and H_0, is always used for measuring directions in azimuth or in altitude, for terrestrial targets as well as for stars. The additional lines V_{05}, V_{15}, and H_3, H_2, H_1, H_1', H_2', H_3' are used in connection with the micrometer and for certain types of star observations.

The movable line M is always parallel to V_{10}; the micrometer screw allows parallel movement of this line through the whole field. The movable line is also parallel to the axis of the micrometer drive connecting the wheels. The whole eyepiece is rotatable through ninety degrees between two stops. Therefore, the micrometer can be used for measuring small angles in the directions of azimuth and of altitude. When using the micrometer to measure horizontal angles, the driving-axis is placed at a right angle to the telescope tube; it is placed parallel to it for measuring zenith distances.

The micrometer drive acts on the measuring screw, which when it turns displaces a sledge, bearing a glass plate with the etched movable thread. The amount of displacement of this line is read on the measuring drum. One revolution of the drum displaces the line by approximately 155".

The range of the micrometer is twelve drum revolutions (≃30'). For the counting of full drum revolutions, the stationary reticle plate has an auxiliary graduation of ten divisions, every fifth numbered (05, 10, 15). The line 10 marks the central position. Thus, for example, in the figure the reading for the position of the moving thread is 07.658 revolutions (the full number is read from the plate, the fraction from the measuring drum).

The contact drum firmly connected with the measuring drum has ten contact strips distributed evenly around its circumference and two additional auxiliary contacts for marking the zero position. When a star is followed with the movable line (time determination), an electric circuit is closed with every one-tenth turn of the drum, which (simultaneously with the beat of a chronometer or radio) may be registered on a chronograph.

7.115 Levels. In addition to the levels already mentioned (those of the vertical and setting circles), the T4 has a small circular plate level on the alidade for the approximate leveling of the instrument and two precise levels: the suspension and the twin Horrebow levels.

The suspension (striding) level is a long level (approximately 180 mm) for measuring the tilts of the horizontal axis and is mounted free of strain in a protective tube (Fig. 7.9). The length of the bubble is adjustable. The vial has about ninety viewable divisions, every tenth numbered.

The two Horrebow levels can be firmly connected with the horizontal axis. They measure changes in the tilt of the telescope. The length of the bubbles is also adjustable. The divisions on the two levels are numbered differently (one is numbered from 0-80, the other from 100-180) so that the levels can be distinguished from the readings as recorded (Fig. 7.10).

7.116 Electric System. On the lower margin of the instrument there are two plug sockets for connecting the battery and a two-pole socket for connecting the eyepiece micrometer with a chronograph. In order to save the battery's energy as much as possible, all current consuming units on the instrument can be switched on and off separately.

The entire instrument is wired internally. Connections between rotating parts are provided by slip-rings, which need to be protected against dirt or weather as otherwise they often malfunction.

The specifications of the Wild T4 are summarized in Table 7.1.

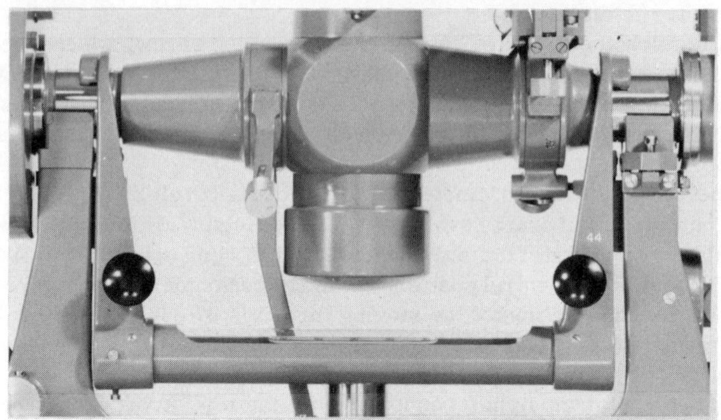

Fig. 7.9 Suspension level of the Wild T4

7.12 The Kern DKM3-A Universal Theodolite (Fig. 7.11)

The DKM3-A is a modified version of the DKM3 theodolite, the principal change being the addition of the telescope eyepiece with the impersonal micrometer for differential measurements. The Kern Company has improved the DKM3-A, since it was first made available, in an attempt to produce an instrument with first-order capabilities and has recently developed a promising new striding level and twin Horrebow level systems. According to tests the DKM3-A now yields observations which meet first-order requirements [Carter, 1965 and 1966].

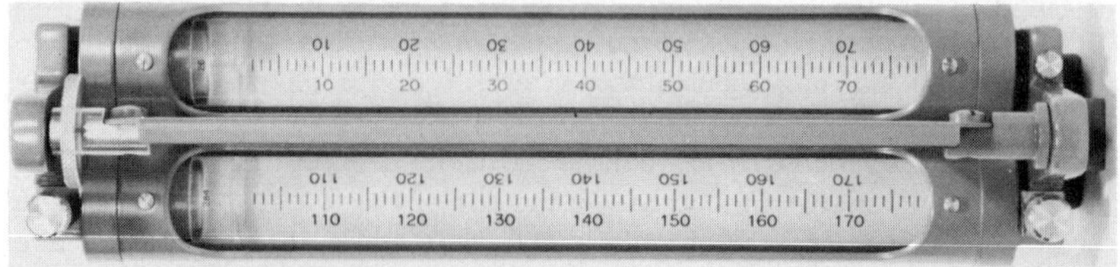

Fig. 7.10 Twin Horrebow levels of the Wild T4

7.121 Vertical Axis. The vertical axis system is a precision ball bearing type consisting of a short vertical axis that serves as a centering pivot only and does not support the weight of the alidade, and of a concentric ring of steel ball bearings that run between two optically flat races and support the alidade. The result is a highly stable, free running system that is particularly insensitive to temperature variations and lubricant viscosities.

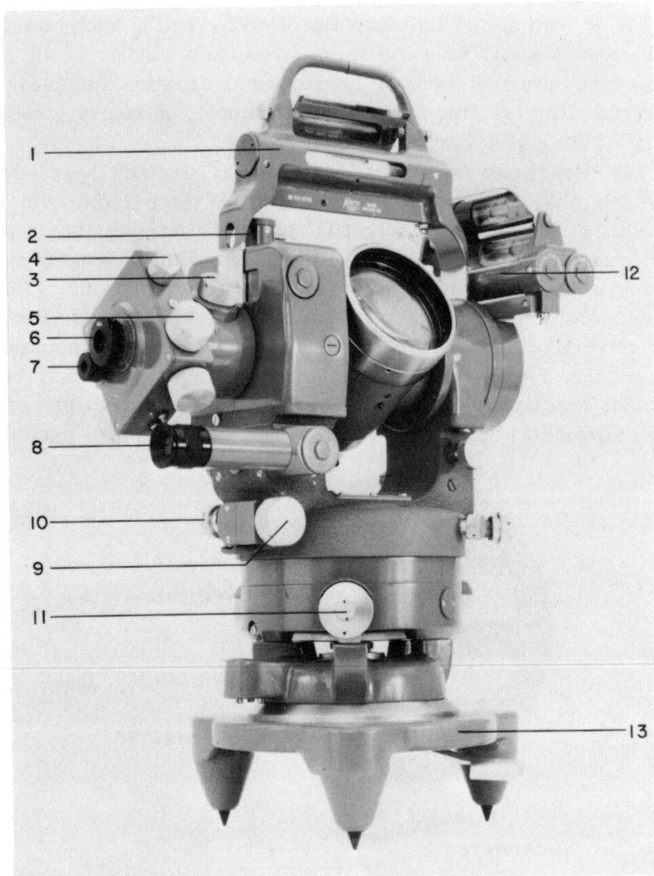

1 Striding Level
2 Viewing Prism for Collimation Level
3 Telescope Focus
4 Rheostat Control Knob
5 Tracking Knob
6 Telescope Eyepiece
7 Impersonal Micrometer Eyepiece
8 Circle Reading Eyepiece
9 Reading Micrometer Knob
10 Collimation Level Control Knob
11 Leveling Knob
12 Horrebow Levels
13 Trivet

Fig. 7.11 Kern DKM3-A universal theodolite

The DKM3-A eliminates many of the instabilities inherent in leveling with footscrews by the use of the well-known Kern system of leveling cams. A cam on the intersurface of each leveling knob bears on inclined supporting surfaces on the instrument's anchor plate. As the leveling knob is turned about its horizontal axis the instrument is raised or lowered. The leveling range is limited, and, therefore, preliminary leveling of the supporting base is necessary.

7.122 Precisely Graduated Reading Circles and Eyepiece. The DKM3-A is also a repeating theodolite. Both the horizontal and vertical reading circles can be displaced in either direction by any desired amount by means of the circle orienting gears. Both are made of optical glass and etched with two concentric sets of graduations. The outer set consists of single lines and the inner set of double lines. The images of diametrically opposed portions of the circles appear superimposed in the reading eyepiece, and coincidence is effected by placing the single graduations symmetrically within the double graduations by means of the micrometer knob (Fig. 7.12). The same knob is used to bring

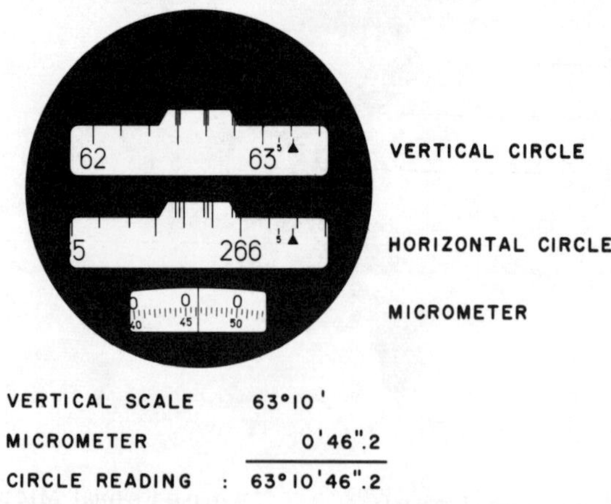

Fig. 7.12 Reading the vertical and horizontal circles on the Kern DKM3-A

either the vertical or the horizontal circle into coincidence. The vertical circle, horizontal circle, and micrometer readings appear from top to bottom in a single reading eyepiece located below the telescope eyepiece. A triangular index appears in each circle reading window and is used to read the circle to the multiple of five minutes appearing next to the index on the left (the vertical circle in the figure reads 63°10'). The remaining minutes, seconds, and decimal part of the second are then read from the micrometer (in the figure, 0'46".2) and added to the above value (the full reading of the vertical circle in the example is 63°10'46".2).

The horizontal reading circle graduations increase in a clockwise direction. The vertical circle readings start at 0° at the zenith and increase counterclockwise. Since the instrument is not equipped with any type of setting circle, the vertical circle must be used to set zenith distances. The triangular index may be used to estimate minutes directly for this purpose. For precise vertical circle readings, the ends of the vertical circle level must first be put in coincidence.

7.123 Telescope. The telescope is the broken type and consists of a system of convergent lenses and concave mirrors so designed and arranged as to provide a high degree of chromatic correction and excellent luminosity and resolving power, while eliminating undesirable stray light (Fig. 7.13).

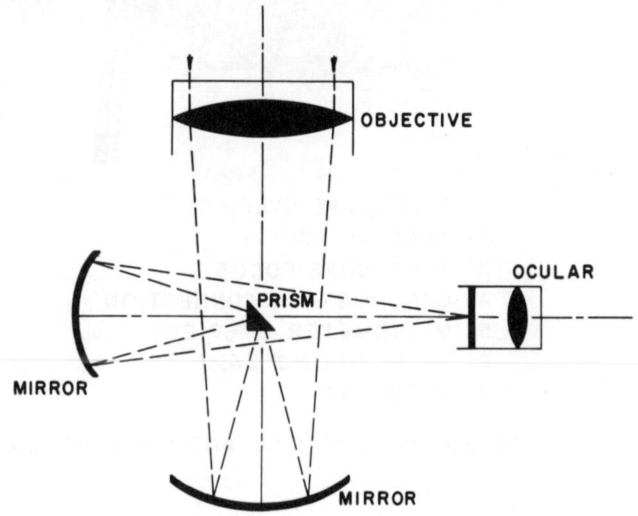

Fig. 7.13 Optical system of the Kern DKM3-A telescope

The system is adequate to permit the use of stars brighter than seventh magnitude. Diffraction screens are available for use during observations on bright stars.

7.124 Telescope Eyepiece and Impersonal Micrometer. The telescope eyepiece consists of the microscope, the movable and fixed reticles on which the threads are etched, and the impersonal micrometer (Fig. 7.14).

The movable reticle shown in Fig. 7.14b is of the bifilar type, but a single line type is also available. The fixed reticle pattern is a composite of lines that are used for various purposes. The five parallel lines, for example, may be used for observing star transits when the micrometer is not used. The ten short lines numbered from 05 to 15 corres-

I TELESCOPE EYEPIECE
2 TRACKING KNOBS
3 TELESCOPE FOCUS
4 CHRONOGRAPH CONNECTION
5 MICROMETER EYEPIECE
6 ILLUMINATING LAMP
7 RHEOSTAT

a. TELESCOPE EYEPIECE AND MICROMETER

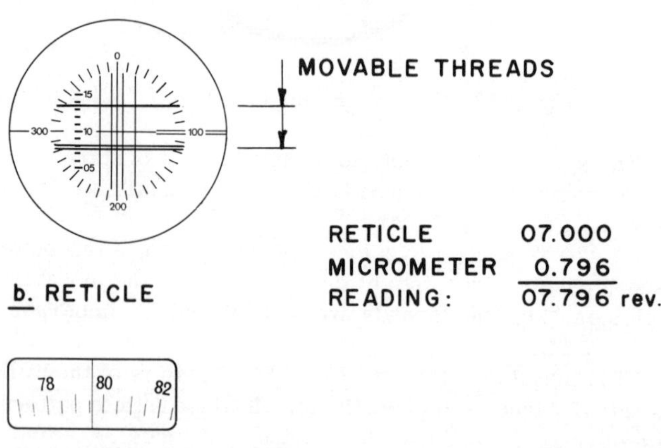

b. RETICLE

c. MICROMETER SCALE

Fig. 7.14 Telescope eyepiece and impersonal micrometer of the Kern DKM3-A

pond to whole turns of the micrometer as in the Wild T4, and they are used in conjunction with the movable reticle to measure azimuth or zenith distance differences. In addition, there is a system of azimuth graduations that may be used to orient the instrument in azimuth so that a particular star will pass through the center of the reticle—a condition necessary for certain observations.

The impersonal micrometer can be rotated ninety degrees about its axis to facilitate measurements in azimuth or in zenith distance. Although its basic functions are the same as those performed by that of the Wild T4 theodolite, several improvements and observer conveniences have been incorporated into its design. The micrometer is completely optically read. To read the micrometer, one first reads the number of whole turns from the position of the movable wire in relation to the fixed field of reticles (07 in Fig. 7.14b) and then reads the fractional part of the turn, or 'drum value,' through a special eyepiece located immediately adjacent to the telescope ocular (0.796 in Fig. 7.14c). The system is internally lighted, and the lighting intensity is controlled by a rheostat conveniently located on the micrometer housing.

The movable reticle is controlled by two sets of tracking knobs reciprocally linked and situated at right angles to one another. Thus, whether the micrometer is in the azimuth or zenith distance measuring position, and regardless of the zenith distance at which the measurements are being made, one set of tracking knobs is convenient.

A plug on the micrometer housing provides the connections for an exterior chronograph cable rotating with the alidade, thereby sometimes restricting the observer's freedom of movement (the Wild T4 is internally wired). In each full turn of the micrometer ten contact strips and one identification contact are registered. When the registering apparatus is not required, the commutator may be made inoperative by means of a switch located on the micrometer housing.

A factory installed autocollimating eyepiece is available upon request. The illuminator is permanently mounted on the micrometer housing and is connected to the instrument lighting system. The bright lines of the autocollimating diagram appear against a dark background. The diagram and the fixed reticle are on surfaces of a beam splitter cube. The lines of sight determined by the fixed reticle and bright-line pattern are aligned to each other within 0ʺ.5. Both the fixed reticle and the movable reticle can be used for autocollimation in combination with the bright-line pattern.

7.125 Levels. The instrument is equipped with a conventional plate level and a graduated vertical circle collimation level. The collimation bubble is viewed as a split image in a rotatable prism mounted at the top of the standard nearest the observer and is brought into coincidence by use of the collimation level slow motion screw. In addition, the DKM3-A also has precise twin Horrebow levels and a striding level which serves the same purpose as the suspension level of the Wild T4.

The striding level is not set directly on the horizontal axis, but is supported by two seats that are parallel to the axis and rigidly attached to the standards. The level is heavily weighted for stability and can be placed on the instrument in only one position, i. e., the level cannot be reversed on the seats. A centering adjustment screw is conveniently located at one end of the vial housing. The striding level interferes with the viewing of the Horrebow levels; thus, when the latter is used, either it has to be removed, or the observer has to go around the instrument to read the twin levels. Another disadvantage of the position of the striding level is that since one of its supporting legs is very close to the face of the observer, at low temperature (below 10°C) differential temperature effects on the level housing cause variations in the bubble position.

The vial is of the chamber type and has approximately forty viewable divisions. Every fifth division is numbered in a continuous scheme throughout the length of the vial, increasing toward the observing ocular.

The Horrebow levels are designed to be attached only after the instrument has been set up. A counterweight attached to the instrument during regular use must be removed before the Horrebow levels can be mounted. Three screws connect the Horrebow mount to the instrument directly opposite the observing ocular. The levels may be clamped to the horizontal axis with the telescope in any position and a slow motion tangent screw is provided for final adjustment. One mount is provided with a centering adjustment screw to facilitate alignment of the two bubbles. The entire mount must be removed in order to adjust the bubbles' lengths.

The vials are also of the chamber type and are numbered similar to the striding level. The scale of one Horrebow level is biased by 100 to eliminate any confusion in readings. The bubbles are all read as reflected images viewed in overhead mirrors. Lighting units are available for both the striding and Horrebow levels.

As can be seen from the foregoing, the DKM3-A has certain disadvantages due to its design and compactness, but these are mostly offset by its merits, such as its portability (it is about one-third the size and weight of the Wild T4) and its excellent optical system. The specifications of the instrument are summarized in Table 7.1.

7.13 The Zeiss (Jena) Theo-003 Universal Theodolite (Fig. 7.15)

The Theo-003 Universal Theodolite is similar in operation to the Wild T4 or to the Kern DKM3-A. Its major novelty is the elimination of the striding and the Horrebow levels by the incorporation of tilt compensators in the optical system of the telescope and in the index of the vertical circle. As today's technology does not allow the manufacture

TABLE 7.1 Universal Theodolite Specifications

Component	Unit	Universal Theodolites: Wild T4	Kern DKM3-A	Zeiss Theo-003
Telescope				
Aperture	mm	60	72	65
Magnification	X	65	27-45	32-75
Focal Length	mm	550	510	810-940
Horizontal Circle				
Diameter	mm	250	100	250
Graduation Interval	'	2	10	4
Micrometer Graduation Interval	"	0.1	0.5	0.2
Vertical Circle				
Diameter	mm	145	100	200
Graduation Interval	'	4	10	4
Micrometer Graduation Interval	"	0.2	0.5	0.2
Level Sensitivity				
Striding Level	'/2mm	1.0	1.5	automatic compensation
Horrebow Level	"/2mm	1.5	1.5	automatic compensation
Impersonal Micrometer				
Range	'	30	20	20
Equatorial Drum Value/ revolution	"	150	120	120
Graduation/revolution	no.	100	120	120
Weight (approximate)	kg	55	15	60
Price in the U. S., 1966 (approximate)	$	9,300	8,900	29,000

of precision level vials with the level sensitivity and stability desired (see section 7.171), their elimination is a welcome feature.

7.131 Precisely Graduated Reading Circles. The Zeiss Theo-003 has precision graduated horizontal and vertical reading circles. Both circles are read in a common reading microscope with the aid of a coincidence micrometer and can be employed as repeating circles. The vertical index is stabilized with a pendulum unit and thus the need for a vertical circle level is eliminated. Since there is no need to center or to read the bubble after the telescope is set in the proper vertical positions, the working speed is greatly increased in methods where direct readings on the vertical circle are required [Groedel, 1966]. A camera, together with a clock and a writing pad, may also be attached for recording the readings of the circles photographically.

Fig. 7.15 Zeiss (Jena) Theo-003 universal theodolite

7.132 Telescope and the Automatic Tilt Compensator. The broken telescope system consists basically of three telescopes inserted into one another [Roeder, 1966]:

(1) telescope for measuring horizontal angles,

(2) telescope for measuring vertical angles,

(3) view finder telescope.

In order to reduce systematic errors in horizontal angle measurements due to the tilt of the horizontal axis, and to its changes in tilt resulting from manufacturing defects of the vertical axis and the supporting bearings, the telescope is provided with an automatic compensation device consisting of a pendulum-mirror situated at the midpoint of the focal length of the objective (Fig. 7.16). A ray of light passes through the objective, the angular mirrors, the pendulum-mirror, the bore of the upper mirror, and an afocal group of lenses which compensate for nonaxial chromatic errors at the reticle in the focal plane of the eyepiece. The magnification can be varied in three stages by changing the eyepiece.

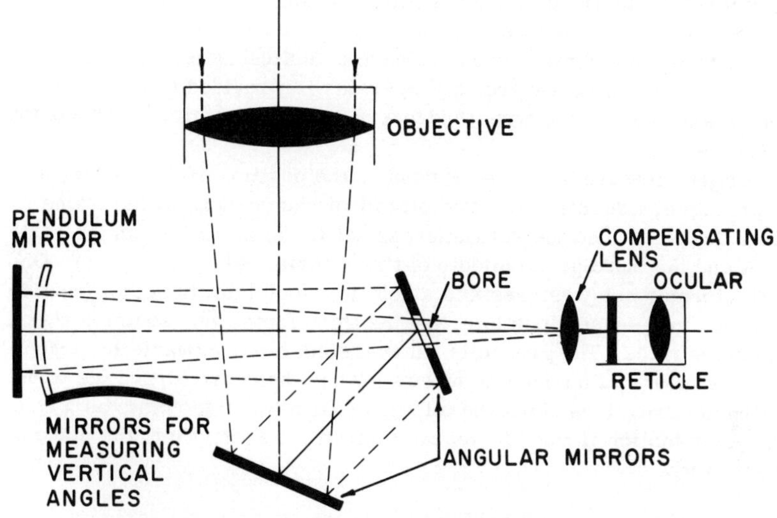

Fig. 7.16 Optical system of the Zeiss (Jena) Theo-003 telescope

When measuring vertical angles the effect of the pendulum-mirror is eliminated by inserting a switching mirror with a fixed axis on the front of the pendulum-mirror.

The view finder telescope and the main telescope have a common reticle and eyepiece. If a plane mirror is inserted, the light coming through the objective of the main telescope is blocked. The ray observed in this position comes through the view finder telescope objective which allows a much larger field of view ($2^{\circ}\!.5$ as opposed to 45').

The eyepiece may be interchanged with an impersonal micrometer eyepiece similar to that of the Wild T4 or the Kern DKM3-A. This is used when electrically recording star transits on a chronograph or when measuring zenith distance differences.

The specifications of the Zeiss (Jena) Theo-003 are summarized in Table 7.1.

7.14 Universal Telescopes

The universal telescopes differ from the universal theodolites in that they do not possess precisely graduated horizontal and vertical reading circles (thus the instrument can be used for differential measurements in azimuth and in zenith distance only), and their motion about a vertical axis is limited to about five degrees to provide means of setting the instrument in the plane of observation. The instruments are normally much heavier than their theodolite counterparts, which fact, together with their restricted motion, is considered an advantage by many experienced observers from the point of view of stability.

Since the instrument does not have a vertical axis about which its telescope could be reversed, an apparatus is provided to lift the horizontal axis out of the wyes and carry the weight while reversing the telescope.

Another feature of the instrument is the position of the footscrews. In principle, generally two are placed at the corners of the frame on one side (e.g., when the instrument is set in the meridian, on the north side) and the third at the middle of the opposite side (e.g., south). One of the former set (west) rests in a hole in a footplate, the other (east) in a groove in a second footplate, and the third footscrew (south) rests on a plane surface. The groove is cut in a steel block movable in the footplate by means of horizontal screws. When these are turned the whole instrument revolves about the hole in the first (west) footplate as a center. This motion is used to set the instrument exactly into the plane of observation.

Fig. 7.17 Askania AP70 universal telescope

Such universal instruments are the Askania AP 100 and AP 70 (Fig. 7.17) and their predecessor, the Bamberg broken telescope. Their specifications are listed in Table 7.2. The Askania instruments have as accessories rotatable circular base plates equipped with exterior setting circles permitting observations in any vertical plane, photographic attachments for recording both the hanging and the Horrebow levels, and mo-

TABLE 7.2 Universal Telescope Specifications

Component	Unit	Universal Telescope		
		Askania		Bamberg
		AP 70	AP 100	
Telescope				
Aperture	mm	70	100	70
Magnification	X	40-102	64-165	
Focal Length	mm	645	1031	670
Level Sensitivity				
Hanging Level	"	1	1	1.2
Horrebow Level	"	1	1	1.5
Impersonal Micrometer				
Range	'	30	30	30
Equatorial Drum Value/revolution*	"	80(160)	50(100)	80(160)
Graduation/revolution	no.	100	100	100
Weight (approximate)	kg	150	210	120
Price	$	no longer manufactured		

*First number for zenith distance differences, second (in parentheses) for azimuth (time).

tor driven impersonal micrometers. The AP100 can observe stars of a magnitude of ten or brighter. The limiting magnitude for the other two instruments is about seven. Observations on these instruments are faster and somewhat more accurate than those on the universal theodolites. Unfortunately, they are no longer manufactured.

7.15 Miscellaneous Instruments

In addition to the universal theodolites, there are also others such as the AY2/10 universal theodolite used in the Soviet Union [Grishin, 1962] or the Gigas-Askania TPR theodolite (Fig. 7.18) [Douglas, 1964]. Since these instruments have mechanical features, optical components, and reading systems essentially similar to those already described, they are not discussed here.

It should be mentioned that in addition to the hand and motor driven impersonal micrometers [Niethammer, 1947, pp. 40-45] other devices have also been developed to reduce personal systematic observational errors. For example, various universal and transit instruments have been coupled with photoelectric attachments to record star passages photoelectrically. In these instruments the impersonal micrometer (or the fixed reticle) is replaced by a diaphragm with slits symmetrical to the optical center. When the star passes over the slits, electric im-

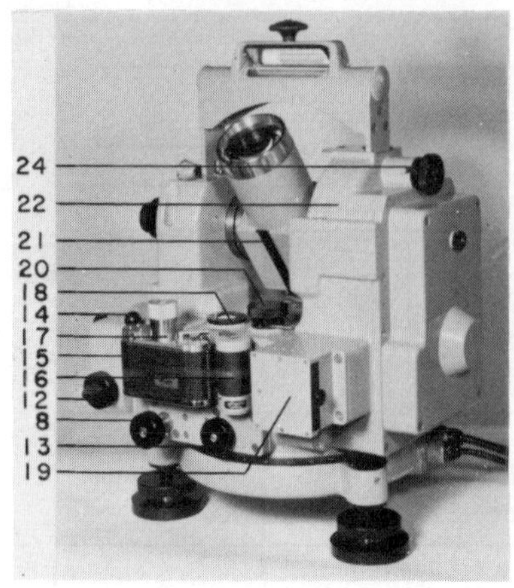

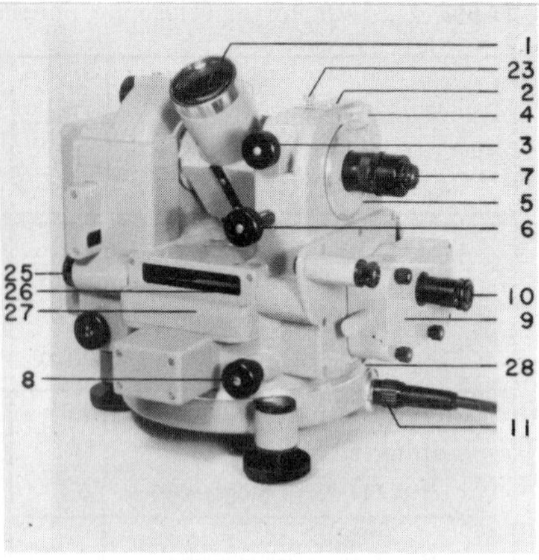

1 Telescope Objective
2 Slot for Mounting the Striding Level on the Horizontal Axis
3 Telescope Clamp
4 Magnifier
5 Vertical Setting Circle
6 Telescope Slow Motion Screw
7 Eyepiece of Main Telescope
8 Horizontal Slow Motion Screw
9 Optical Micrometer
10 Micrometer Eyepiece
11 Cable Connector
12 Micrometer
13 Clamp for Upper Plate
14 Camera Mounting Lever
15 Recording Camera
16 Shutter Release Solenoid
17 Camera Spring Housing
18 Circular Level
19 Time Piece Case
20 Horizontal Circle Setting Knob
21 Coarse Sights
22 Vertical Circle
23 Horizontal Axis Locking Screws for Transport
24 Vertical Circle Setting Knob
25 Plate Level Adjustment Screws
26 Plate Level
27 Plate Level Illumination
28 Horizontal Setting Circle

Fig. 7.18 Askania theodolite TPR

pulses are generated and recorded [Kuznetsov, 1961; Moreau, 1966]. In other experiments the slits have been replaced by the edge of a prism. The light is reflected in turn from both faces of the prism, intersecting along the edge, into two photocells whose output differences are registered. The star is over the edge when the difference of the outputs is zero [Tsubokawa, 1954 and 1957]. The prism edge provides a more abrupt record for start and finish than the slits, but it gives only a single reading for a pass.

A similar attempt to reduce the systematic human effect on the observations is the use of the Hunter shutter eyepiece [de Graaff-Hunter, 1938]. This device consists of a graduated reticle and a thin pivoted metal bar which in its ordinary position covers the moving star. Every third second the bar is drawn aside so that the star is visible for a few hundredths of a second when its position on the reticle is recorded.

Other developments, reaching for the same goal, strive to drive the telescope of a theodolite in azimuth and elevation by the output of an analog computer, which currently calculates altitude and azimuth and/or the rate of change of these quantities, and to register simultaneously the readings on the horizontal and vertical circles and the time when the star is at the center of the crosshairs. The observer's only duty is to keep the star at this point by means of fine adjustment levers. Tracking motors and registering apparatus which can be attached to the theodolite have been developed for the Askania TPR and prototypes for the DKM3-A [Ramsayer, 1966; Gigas, 1966, pp. 253-255].

A somewhat different approach is to keep the theodolite stationary, replace the standard reticle with a square grid, and apply a shutter mechanism to chop the star trail through the grid which is then photographed with a reflex camera. The times of the star trail-breaks are recorded, together with the horizontal and vertical circle readings [Schnädelbach, 1966; Kuntz and Schnädelbach, 1967].

As has been pointed out in the introduction of section 7.1, the transit, the meridian circle, the micrometer transit, and the zenith telescope are also first-order instruments. Their importance lies in positional astronomy and in the historical development of the universal instruments, since they possess one or more of their principal features. In modern geodetic work they are very seldom used, and thus for their description the reader is referred to the literature [Chauvenet, 1891, Vol. II; Bowie, 1917; Nassau, 1931, pp. 186-218].

7.16 Adjustment of the Instruments

Each optical instrument described in the preceding sections defines an instrumental coordinate system whose axes coincide with certain parts of the instrument in question: the first (z) is parallel to the vertical axis, the second (x) is parallel to the horizontal axis, and the third (y) is parallel to the geometric (physical) axis of the telescope. The purpose of the adjustment is twofold:

(1) To make the axes x and z perpendicular to each other and to assure that the plane yz forms a right angle with the plane xz.

(2) To assure that at the observation point the instrumental vertical axis z coincides with the axis z of the horizon system so that it points to the instantaneous local zenith.

The first requirement is achieved internally by adjustments performed on certain parts of the instrument; the second is achieved externally by setting up the instrument with appropriate procedures. Possible small residual effects after adjustment are eliminated or at least reduced by proper observing procedures and by measuring the imperfections and taking them into account in the calculations. For the latter, the measuring devices (levels, micrometer drum, etc.) need to be calibrated (see section 7.17).

The adjustment of an instrument by the observer should be restricted to those manipulations which he can perform in the field or at the office by means of relatively simple tools and operations. Adjustments which require the aid of a qualified instrument man (complete dismantling of the equipment or certain parts of it) should be done in the factory which is responsible for assuring that, for example, the following basic requirements are met:

(1) The graduated horizontal reading circle is perpendicular to the vertical axis.

(2) The graduated vertical reading circle is perpendicular to the horizontal axis.

(3) The division errors of the graduated reading circles are small.

(4) The micrometer used for reading the graduated circles are adapted to the accuracy of these circles and they have a small error of run.

(5) The rotation axes and the respective graduated reading circles are concentric.

(6) The rotation axes and the graduated reading circles do not wobble as the instrument is rotated.

(7) The level vials have an adequate radius of curvature which is constant, at least at the used portions of the vial.

(8) The materials used are such that no appreciable deformations arise on any part of the instrument when the temperature changes.

(9) The instrument when delivered is basically adjusted for all other systematic error sources to the extent that the observer should be required to do small conventional adjustments only.

The conventional adjustments to be performed by the observer are described in the following sections.

7.161 General Adjustment. After the instrument is set up, it should be inspected. The pivots and wyes of the instrument and levels should be carefully cleaned with watch oil, which must afterwards be wiped off to minimize the accumulation of dust. The lens should be examined occasionally to see that it is tight in its cell. It may be dusted off with a camel's-hair brush and, when necessary, may be cleaned by rubbing it gently with soft, clean tissue paper, first moistening the glass slightly by breathing on it.

The length of the level bubbles is adjusted to be about 1/4 to 1/3 of the lengths of the respective vials.

7.162 Vertical Axis Adjustment. The horizontal axis of the telescope and the striding (or similar) level are set parallel to the line joining two footscrews, and the settled position of the bubble is read. Although a reading at one end of the bubble is adequate, the mean center may be determined by summing the end readings and dividing by two. The instrument is then turned 180°, and the level is read in the same way as before. The bubble is then set, by means of the two footscrews, to the arithmetic mean of the direct and reverse readings. The instrument is then turned 90° and the level is read. Turning the instrument 180° and reading at the same end once again, the arithmetic mean of the two readings is set on the level by means of the third footscrew. This process is continued until the same mean reading is obtained in any quadrant, within one graduation. The vertical axis should now be pointing at the zenith.

7.163 Horizontal Axis Adjustment. After the vertical axis has been leveled to within one graduation, the horizontal axis must be made perpendicular to the vertical axis. The horizontal motion is locked and the striding level read. Then the level is reversed by taking it off the horizontal axis, turning it 180°, and reading it. The arithmetic mean is set on the level by turning the horizontal axis adjustment screw (on the Wild T4 the large capstan head screw at the top of the support on the eyepiece side). This process is repeated until the variation is no greater than two graduations.

7.164 Striding Level Adjustment. Since the level normally is not situated directly under or above the horizontal axis of the telescope it must be ascertained whether or not its axis is parallel to the horizontal axis (error of wynd).

This is determined by slowly rocking the level about the carefully leveled horizontal axis of the telescope. If the bubble moves more than two or three graduations, it must be recentered by use of the horizontal screws on the level frame. This is repeated until the movement is less than three graduations.

After the adjustment for wynd is perfected, the bubble is brought to the center again by the leveling footscrews. Then the striding level is reversed. If the bubble does not return to the center, half of the discrepancy is corrected for by the footscrews and half by the vertical adjusting screws on the level. This process of adjustment is repeated until the lack of adjustment does not exceed one or two divisions of the level.

7.165 Horizontal Collimation Adjustment. The middle thread of the fixed reticle of the telescope is set in the line of collimation by the customary method of direct and reverse pointings on a distant object and correction for half of the difference by adjusting the collimation screws. An alternate method is to record the readings of the micrometer head when the movable thread is set on an object near the center of the field with the instrument in both the direct and the reverse positions. Setting the micrometer head at the mean reading puts the movable thread in the line of collimation. The fixed middle thread should then be made to coincide with the movable one by adjusting the collimation screws. These screws either move the reticle horizontally or change the position of the reflecting mirror (see Fig. 7.7).

Wherever practicable, the adjustment for collimation should be made at infinite focus on a distant terrestrial object, or on the threads of a theodolite or collimator which has previously been adjusted to infinite focus. The theodolite or collimator is set up just in front of the telescope of the instrument. If necessary, the threads of the theodolite are artificially illuminated. Occasionally, if neither a distant object nor a theodolite is available for making the collimation adjustment, a near object may be used for the purpose. In this case, however, collimation error may exist when the telescope is in infinite focus. No attempt should be made to reduce the collimation error to zero. If it is already less than 2'', it should not be changed, for experience has shown that frequent adjustment of an instrument causes looseness in the screws and the moving parts.

7.166 Vertical Collimation Adjustment. Adjustment of the vertical collimation is made in the same way as on regular theodolites. The zenith distance of a well-defined object is measured with the instrument in the direct and the reversed positions (the deviation of the vertical circle (index) level is to be taken into consideration). The mean of the two readings is set on the vertical circle using the level actuating knob, and then the bubble of the index level is centered by use of the screw acting on the level holder (in some instruments depending on the graduation system of the vertical circle, 360° or 180° should be deducted from the reading in the reversed positions before taking the mean). During this operation the horizontal thread of the telescope must remain on the target. The collimation error should be less than 2''.

The same adjustment procedure should also be applied to the vertical setting circle. Care should be taken that when the initial readings are made, the bubble of the setting circle level is centered.

7.167 Thread Adjustment. The threads, usually etched on the glass reticle, should be adjusted for verticality and centering.

With the eyepiece locked in position, a sharp well-defined object is observed and the telescope is moved vertically. If the object tracks the thread through the field of view, the thread is vertical; if the object moves off the vertical thread, an adjustment must be made with the appropriate adjustment screw. Verticality must be ascertained in both eyepiece positions.

When the eyepiece is rotated, it is necessary that the center of the threads remain in the same relative position. This may be accomplished by placing the center of the threads on a sharp well-defined point and rotating the eyepiece 90°. If the center comes off the point, one half the correction to bring it back on the point is made by means of the appropriate screws. Excessive pressure must not be applied to these screws because this causes strain and new errors.

7.168 Focusing. Focusing is required for both the eyepiece microscope and the objective.

The eyepiece should be focused by turning the telescope to a light surface such as the sky and moving the eyepiece in and out until that position is found in which the most distinct vision of the micrometer threads is obtained.

The objective should then be focused by directing the telescope to some well-defined distant object and moving the micrometer draw tube in and out, thus changing the distance from the objective to the plane in which the micrometer threads move, until there is no apparent change of relative position (or parallax) of the micrometer threads or of the image of the object when the eye is shifted sideways or up and down in front of the eyepiece. The focus of the objective will need to be checked at night, using a star as the object, and corrected if necessary. Unless the focus is made nearly right in daylight, none but the brightest stars will be seen at all at night, and the observer may lose time trying to learn the cause of the trouble. If the objective is focused at night, a preliminary adjustment should be made on a bright star and the final adjustment on a faint star, as it is almost impossible to get a very sharp image from a bright star. A planet or the moon is an ideal object on which to make a preliminary focusing of the objective. After a satisfactory focus has been found, the drawtube is clamped in position with screws generally provided for that purpose.

7.169 Miscellaneous Comments. The preceding adjustments cannot and should not always be made. The actual procedure will depend on the type of instruments used and how often they are used. For example, the universal telescopes do not need vertical axis adjustment since there is no vertical axis. Similarly, on the Zeiss Theo-003, the tilt compensator makes the adjustment of the horizontal axis and of the striding level (there is none) unnecessary.

With the exception of the vertical axis adjustment, the adjustments need not be made at every station. The observer must examine and correct them often enough to make certain that the errors due to them are always within allowable limits.

There are other possible adjustments, peculiar to a certain type of equipment, which have not been mentioned here. For these the instruction manuals provided by the factory should be consulted. The procedures as outlined above should serve as a general guide only.

7.17 Instrument Calibration

As has been mentioned earlier, certain parts of the first-order instruments need to be calibrated to allow the determination of imperfections in the adjustment (e.g., inclination of the horizontal axis) and possible small instabilities or deformations in certain instrumental components during the observation (e.g., telescope tilt).

Instrument components which are used for measuring (reading circles, impersonal micrometer, etc.) have to be calibrated also. The most important calibrations, that of the reading circles, and initial calibrations of the other components are done by the factory. However, since calibration parameters may change with time, temperature, barometric pressure, transportation, etc., some components need to be recalibrated periodically by the observer. The number of such calibrations and their methods depend on the type of the instrument. The most important basic calibrations which should be performed periodically on modern instruments are the following:

(1) Level calibration (striding, Horrebow, vertical index).

(2) Impersonal micrometer calibration for (a) lost motion, (b) width of the contact strips, (c) equatorial value of one drum revolution.

(3) Equatorial value of thread-distances on the fixed reticle if the impersonal micrometer is not used (see section 7.44).

(4) Horizontal collimation if observations are not done in two telescope (direct and reversed) faces (see section 7.44).

7.171 Conventional Level Calibration. The level value in terms of sensitivity in seconds of arc per graduation must be determined for the level vials of the hanging level, Horrebow levels, and the vertical circle level. The sensitivity of the levels is determined on a level trier which is basically a very simple apparatus, consisting mainly of a beam pivoted on one end and driven vertically by a micrometer screw at the other. It is mounted on a very stable base and it is usually protected by a glass case (Fig. 7.19). The screw moves the beam a known number of seconds of arc (measured at the pivot) per revolution or fraction thereof.

The level vial should be securely fastened to the level trier without applying any strain on the vial. The length of the bubble should be made about 1/3 to 1/2 of the length of the vial.

For testing, the beam is lowered until the bubble is at one end of the vial. The beam is then moved, by means of a screw, a successive number of whole seconds, the bubble being read and recorded to one-tenth of a division each time, until the bubble is at the opposite end of the vial. The motion is then reversed and the bubble is brought successively back to the other end. The vial is then reversed on the trier and the process is repeated. This makes a total of four level determination sets. Sufficient time must be allowed between successive movements to allow the bubble to come to rest.

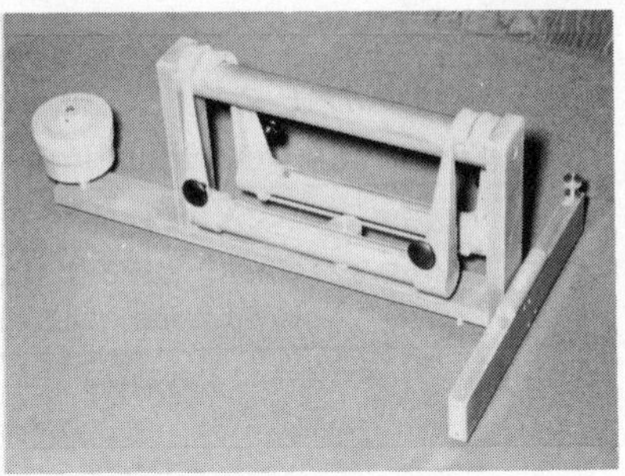

Fig. 7.19 Level trier

It is imperative that the vial be in the testing room at least twelve hours at a uniform temperature before testing. A unit of movement for the screw should be chosen that will cause a movement of more than one and less than three graduations of the bubble.

During the test the following data is recorded or computed (see Example 7.1):

(1) Reading of the micrometer screw (M).
(2) Reading of the right end of bubble (BR).
(3) Reading of the left end of bubble (BL).
(4) Center of the bubble in vial (BC).
(5) Displacement of the bubble (D).

Item (4) is computed by taking the mean of BL and BR; item (5) is computed by taking the difference of two consecutive BC's.

A least squares determination of the level sensitivity from each set is computed and the arithmetic mean is taken.

The sensitivity d in seconds of arc per division is

$$d = \frac{[xx] - \frac{[x]^2}{n}}{[xy] - \frac{[x][y]}{n}} \tag{7.1}$$

where x is the reading of the micrometer on the level trier (M), y is the mean reading of the ends of the bubble (BC), and n is the number of settings. The brackets denote summation of the enclosed values.

When the readings of the micrometer increase uniformly by a constant value D, the numerator of equation (7.1) is independent of x and equal to

$$\frac{n(n^2-1)}{12}D^2. \tag{7.2}$$

The test and level value computation of one position with this method are illustrated in Example 7.1.

EXAMPLE 7.1

Level Sensitivity Determination

Level Vial No. 64252 — Wild T4 theodolite
Date: October 27,1966 — Observer: J.L.Hammer
Temperature 23°.0 C — Columbus, Ohio

M(x)	BR	BL	BC(y)	D
0.00	21.6	60.6	41.1	
2.00	23.3	62.4	42.8	1.7
4.00	25.0	64.0	44.5	1.7
6.00	26.7	65.8	46.2	1.7
8.00	28.5	67.4	47.9	1.7
10.00	30.2	69.2	49.7	1.7
12.00	32.0	71.0	51.5	1.8
14.00	33.9	72.9	53.4	1.9
16.00	35.5	74.5	55.0	1.6
18.00	37.1	76.1	56.6	1.6
20.00	39.0	78.0	58.5	1.9
22.00	40.8	79.7	60.2	1.7
24.00	42.6	81.5	62.0	1.8
26.00	44.5	83.4	63.9	1.9
28.00	46.3	85.2	65.7	1.8
30.00	48.1	87.1	67.6	1.8
32.00	50.1	89.0	69.5	1.9

[x] = 272.00
[y] = 936.1
[xy] = 16420.80
[xx] = 5984.00

$$d = \frac{5984.0 - \frac{272^2}{17}}{16420.8 - \frac{272.0 \times 936.1}{17}} = 1''.1308/\text{division}$$

$m_d = 0''.0694$/division (standard deviation)

mean bubble length = $38.97\dot{6}$ divisions

Equation 7.1 is based on the observation equation

$$y = c + \frac{x}{d} \tag{7.3}$$

where c is a constant for the set. This model assumes that the sensitivity (the radius of curvature) is the same across the vial, and the solution yields the least square value of d based on this assumption.

Tests on level vials have shown that today's technology does not allow the manufacture of level vials with the constant radius of curvature desired; thus the sensitivity depends on the position of the bubble and also on its length [Barnes and Mueller, 1966]. According to these investigations, if the level sensitivity is determined the conventional way using equation (7.1), it should be established by running the bubble over the central portion of the vial which is used in the field. Further, the length of the bubble on the field should be kept the same as it was at the time of the sensitivity determination. If this procedure is not followed then the following is recommended:

(1) Each level should undergo tests to establish how its sensitivity varies with bubble length and position. As many bubble lengths should be tested as practical and each length's conventional sensitivity should be established as described above but with at least twelve sets of data. The results of such a test are shown in Fig. 7.20.

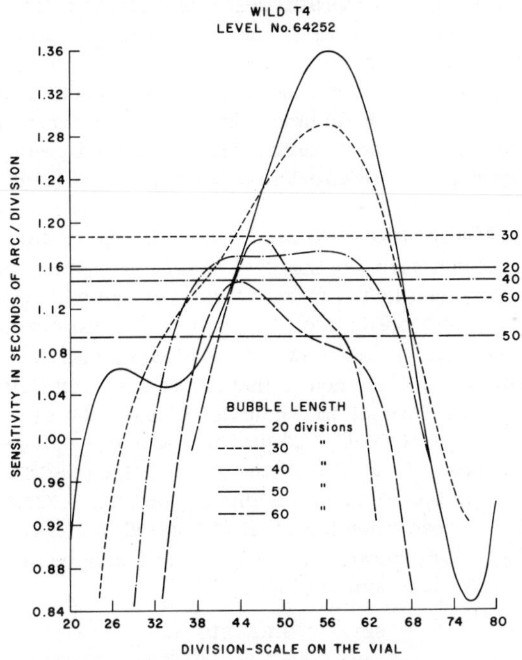

Fig. 7.20 Dependence of level sensitivity on bubble position and length

(2) The corrections for the length and position at the time of the observation should be applied to the conventional mean least square sen-

sitivity, computed from equation (7.1). The corrections can be determined from the diagrams obtained in item (1) above. The conventional mean least square sensitivity should be established periodically, but at least once in each season.

The corrections to the conventional mean sensitivity obtained in this way could be significant, especially for striding levels. For example, from Fig. 7.20 it can be seen that if level No. 64252 at the time of the observation is centered at the 55th division on the vial and has a bubble length of 50 divisions, and its conventional mean sensitivity in the laboratory was determined with a 30 division-long bubble, the correction is equal to $1.088 - 1.188 = -0\overset{''}{.}1$/division.

It is common knowledge that during one night of observation the changing temperature affects the length of the bubble (as the temperature drops the bubble length of a striding level may increase as much as ten to twenty divisions). The procedure as outlined above removes this effect from the observation results. It does not, however, remove the effect of temperature changing the radius of curvature of the vial, thus the sensitivity. Another temperature effect, due to the fact that the ends of the level may have different temperatures, also causes systematic errors. Since this 'gradient' effect moves the bubble towards the end of higher temperature but does not change the sensitivity (unless it causes deformations on the vial), it is not discussed here. In latitude determination from zenith distance differences, where the Horrebow levels are situated in a north-south direction with a generally sizable temperature gradient, this effect is especially significant and is to be considered.

7.172 Level Calibration with the Universal Theodolite. This method allows for the determination of the level sensitivity in the same environment where it is used in the field. The level trier is replaced by the carefully adjusted theodolite itself on which the level in question is placed with its axis parallel to the horizontal reading circle.

The principle of this method is that if the horizontal circle is tilted by the angle i, with respect to the horizon, about the axis CH containing the center of the circle C, and the level CL is rotated about the inclined vertical axis CZ′, the tilt of the level δ (the position of the bubble) will depend on the horizontal angle A (see Fig. 7.21). Thus if the bubble position is read when the level is rotated into several azimuths and the tilt angle i is known, the sensitivity d may be determined.

From the figure it is evident that

$$\sin \delta = \sin i \ \sin A,$$

or, since both i and δ are small angles,

$$\delta = i \sin A. \qquad (7.4)$$

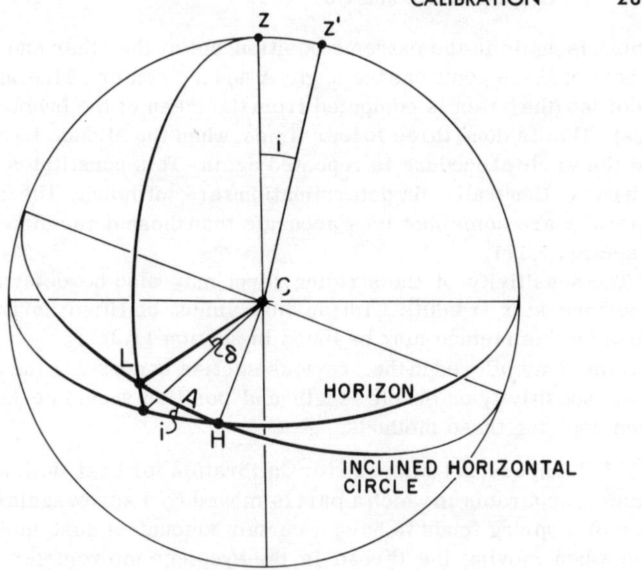

Fig. 7.21 Dependence of level tilt on the inclination of the alidade

Applying this equation for two level positions, 1 and 2, and taking the difference yields

$$\delta_2 - \delta_1 = i(\sin A_2 - \sin A_1). \tag{7.5}$$

If between these two positions the center of the bubble moved $y_2 - y_1$, the change in inclination may also be expressed as

$$\delta_2 - \delta_1 = d(y_2 - y_1) \tag{7.6}$$

where d is the level sensitivity. From equations (7.5) and (7.6),

$$d = i \frac{\sin A_2 - \sin A_1}{y_2 - y_1} \tag{7.7}$$

is the sensitivity of the level.

In the observational procedure the carefully adjusted instrument is pointed on a well-defined target situated in the plane of the telescope and a footscrew. The horizontal thread is then moved with the vertical slow motion screw through a small vertical angle depending on the range of the level to be calibrated. This angle should be determined from the vertical reading circle. The horizontal thread is moved back to the target with the footscrew mentioned above. The inclination of the horizontal circle i now is the same as the angle set previously on the vertical circle. The instrument with the level is rotated until the bubble reaches the extreme position of that portion of the vial which is to be calibrated. Then the alidade is rotated with the horizontal motion screw until the bubble is again in the extreme position but at the other end of the vial. In both of these positions the angle A and the center of the bubble y are recorded (the latter is computed from the mean of the bubble-end readings). This is done three to four times, when the alidade is rotated 180° and the whole procedure is repeated again. This constitutes one determination. Generally six determinations are sufficient. The final results generally are somewhat less accurate than those determined according to section 7.171.

The sensitivity of the striding level may also be determined from observing star transits with an instrument of tilted horizontal axis. More on this method may be found in section 11.311.

Comments offered in the previous section concerning the dependence of the sensitivity on bubble length and position should be kept in mind when applying these methods.

7.173 Impersonal Micrometer Calibration for Lost Motion. Any mechanical apparatus in which a part is moved by a screw against the pressure of a spring tends to have a certain amount of lost motion; this is true when moving the thread in the eyepiece micrometer. When the screw is turned in one direction, the coil springs in the micrometer are compressed. Conversely, when the motion of the screw is reversed, the springs are partially released.

To show the effect of lost motion, it is assumed that a given star moves from left to right and that the times are recorded with the star at the positions 1, 2, 3, and 4 consecutively, and that the instrument is then reversed and the times are recorded with the star in the corresponding positions 4′, 3′, 2′, and 1′ (Fig. 7.22). Owing to the lost motion in the mechanism, the micrometer indicates that the star is in the positions 4″, 3″, 2″, and 1″ respectively. It is evident that the times after reversal would all be recorded early by the amounts (4′ - 4″), (3′ - 3″), etc.

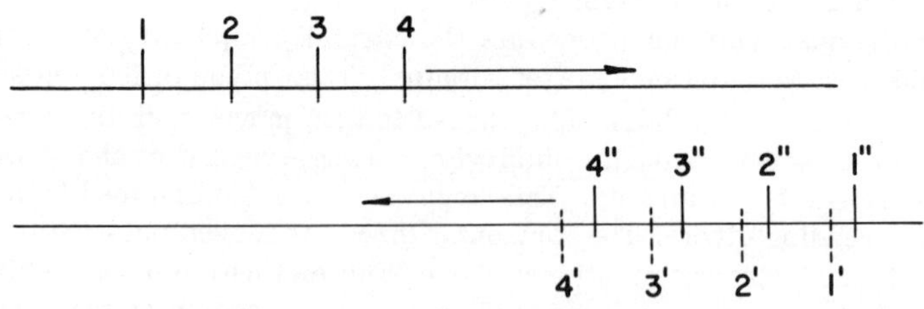

Fig. 7.22 Lost motion in the impersonal micrometer

The determination of lost motion usually is made in three positions of the telescope, at zenith distances of 45°, 0°, and 315°. The observing is performed upon the fixed threads and requires very careful bisection and close concentration on the part of the observer. The eyepiece must be so illuminated and focused that the threads appear as sharp black lines on an opaque background.

With the telescope set, the movable thread is placed outside the graduation. The handwheels are turned so as to increase the micrometer. readings until the movable thread bisects the first thread. The micrometer reading is recorded as an increasing value for that thread for that telescope setting.

In the same manner the values for the center and last threads are determined and recorded. This completes one set of 'increasing' values of a particular telescope zenith distance.

The same procedure is followed next but the micrometer wheels are turned in the opposite direction, and the drum is read and recorded for last, center, and first threads, in this order. These readings constitute the 'decreasing' values.

A subtraction, in the sense of increasing value minus decreasing value, gives the lost motion for each observation. A total of ten such observations in each of three telescope positions are performed. The mean of all observations is the lost motion m.

Example 7.2 illustrates the determination of lost motion m for one telescope setting. The value m should be determined at least once every year.

7.174 Impersonal Micrometer Calibration for Mean Width of Contact Strips. A star passage in the eyepiece micrometer is recorded electrically in ten to fifteen positions during every revolution of the micrometer. The contact must remain closed for a short time ($0^s.1$) to ensure the operation of the chronograph. The contacts have finite dimensions and the actual recording of a position occurs at the edge and not at the center of the strip (see Fig. 7.8a).

As soon as the index tip makes contact with the edge of each strip, the circuit is closed and a break occurs on the chronograph record. The instrument is reversed and the micrometer is turned in the opposite direction, causing the opposite edge of each strip to make contact first. It is evident that the contact is made too early in either direction by one-half the mean width of the contact strips.

The determination of the mean width of the strips is made by first connecting the micrometer to the chronograph so that when the circuit is closed between them an audible 'click' results as the micrometer is turned from strip to strip. A volt-ohm meter or a light bulb may also be used.

EXAMPLE 7.2

Lost Motion Determination

Date: July 31, 1965 — Wild T4 #48986

Telescope position: 45° — Observer: C. Friberg

Columbus, Ohio

No.	Thread								
	05			10			15		
	Decr.	Incr.	Diff.	Decr.	Incr.	Diff.	Decr.	Incr.	Diff.
1	98.6	98.7	0.1	98.5	98.4	-0.1	99.4	99.5	0.1
2	98.5	98.7	0.2	98.7	98.4	-0.3	99.6	99.5	-0.1
3	98.6	98.5	-0.1	98.4	98.4	0.0	99.6	99.4	-0.2
4	98.5	98.4	-0.1	98.4	98.5	0.1	99.4	99.6	0.2
5	98.4	98.4	0.0	98.5	98.6	0.1	99.6	99.4	-0.2
6	98.3	98.2	-0.1	98.5	98.4	-0.1	99.5	99.5	0.0
7	98.3	98.2	-0.1	98.4	98.4	0.0	99.4	99.6	0.2
8	98.5	98.2	-0.3	98.5	98.5	0.0	99.6	99.4	-0.2
9	98.2	98.3	0.1	98.5	98.5	0.0	99.5	99.5	0.0
10	98.3	98.3	0.0	98.6	98.5	-0.1	99.4	99.4	0.0
Sum			-0.3			-0.4			-0.2

m = (-0.3 - 0.4 - 0.2)/30 = -0.030 division (from one telescope position)

m = -0.032 division (from three telescope positions)

Starting at a point in front of the first strip, the micrometer is turned slowly in the direction of increasing value until the relay closes on the strip. At this point the micrometer is read. The turning is continued and the micrometer is read as the relay is closed on each strip. After the reading of the last strip the motion is reversed. This time the opposite edge of each strip is read, starting at the last strip and working toward the first.

A reading on both edges of each of the ten strips is considered a complete set. Subtracting the readings on each strip (decreasing reading minus increasing reading) gives the width of each strip in terms of micrometer divisions. The mean of the reading of all strips in one set gives the mean of that set. A mean of ten sets gives the mean width of the strips s; its determination from two sets is shown in Example 7.3. The value s should be determined at least once every year.

7.175 Impersonal Micrometer Calibration for Equatorial Value of One Drum Revolution. The values m and s (lost motion and mean width of the contact strips) as previously determined are in terms of micrometer graduations and as such may not be applied directly as corrections. Thus, the value of the micrometer graduations in seconds of time, or

EXAMPLE 7.3

Determination of the Mean Width of Contact Strips

Date: July 31, 1965				Wild T4 #48986 Observer: C. Friberg Columbus, Ohio		
No.	Set #1			Set #2		
	Incr.	Decr.	Diff.	Incr.	Decr.	Diff.
1	99.4	0.5	1.1	99.3	0.5	1.2
2	09.3	10.5	1.2	09.3	10.5	1.2
3	19.5	20.4	0.9	19.1	20.4	1.3
4	29.2	30.4	1.2	29.2	30.5	1.3
5	39.3	40.5	1.2	39.1	40.5	1.4
6	49.1	50.5	1.4	49.2	50.5	1.3
7	59.1	60.5	1.4	59.2	60.5	1.3
8	69.2	70.5	1.3	69.2	70.5	1.3
9	79.1	80.5	1.4	79.1	80.5	1.4
10	89.2	90.5	1.3	89.2	90.5	1.3
Mean (s)			1.24			1.30
s = 1.278 divisions (from 10 sets)						

the angle subtended by one graduation of the micrometer, must be determined by actually tracking stars with the micrometer.

After the instrument has been put into the meridian, a star within 20° of the equator ($+20° > \delta > -20°$) is tracked through the field, without reversing the instrument. The times of a selected single contact-strip are scaled from the chronograph record on which chronometer or radio beats are also recorded.

Since all stars are at an infinite distance in comparison to terrestrial measurements, all star observations are made with the eyepiece focused at infinity. An extremely clear night and a star of magnitude four to five are best.

If the time difference of the consecutive contacts on the selected strip are t_1, t_2, t_3, — t_n, and t is the arithmetic mean, the equatorial value R of one revolution is determined from the expression

$$R = t \cos \delta \tag{7.8}$$

where δ is the star's declination. The value t should be determined from tracking the star through ten revolutions of the drum. The procedure is repeated for four to six stars, and the arithmetic mean is taken as the equatorial value of one drum revolution. The corresponding value for one drum division is

$$r = \frac{R}{n} \tag{7.9}$$

where n is the number of graduations on the drum.

Example 7.4 illustrates the determination of R from tracking a single star.

EXAMPLE 7.4

Equatorial Drum Value Determination

Date: July 30,1966	Wild T4 #48986
Star: APFS # 1480	Observer: J.L.Hammer
$\delta = -2^{\circ}01'$	Columbus, Ohio

Turn	0 Strip Times	Diff.(t)
0	$18^h44^m59^s.14$	
		$10^s.15$
1	45 09.29	
		10.25
2	19.54	
		10.00
3	29.54	
		10.12
4	39.66	
		10.21
5	49.87	
		10.31
6	46 00.18	
		10.11
7	10.29	
		10.03
8	20.32	
		10.21
9	30.53	
		10.10
10	40.63	
Mean		$t=10^s.149$

$R = t \cos \delta = 10.149 \times 0.99938 = 10^s.143 = 152''.14$ (from one star)

$R = 152''.39$ (from six stars)

If latitude is determined at the station from measuring the zenith distance difference of two stars with the impersonal micrometer, the value R should be determined as part of the latitude computation (see sections 10.211 and 10.214).

A change in equatorial drum value will occur whenever there is a change in focus of the eyepiece; therefore, it is important that the focus remain constant. If the focus is changed, a new value must be determined. Extreme changes in temperature will necessitate a new determination.

7.2 Second-Order (Geodetic) Instruments

The basic types of second-order instruments which may be used for the purposes of geodetic astronomy are the following:

(1) Geodetic theodolites to measure horizontal and vertical angles or to observe star transits.

(2) Astrolabes to observe star crossings through a selected almucantar.

The geodetic theodolites are generally the types which are for angle measurements in first-order triangulation, with possible suitable attachments for night observations on stars (ocular prism and electric illumination). They are used in second-order latitude, sidereal time, or azimuth determinations.

The astrolabes are optical instruments built either directly as such or as attachments to fit the telescopes of geodetic theodolites or levels. The horizontality of the system is assured either by a pool of mercury in a small container attached to a prism (prismatic astrolabe) or by a pendulum compensator similar to those of automatic levels (pendulum astrolabe). Both of these systems assure that star crossings are observed on a selected almucantar defined by the construction of the instrument. The astrolabes normally are used for the simultaneous determination of latitude and sidereal time by observing stars at equal altitudes (see section 11.52).

Generally the observations with the astrolabes are simpler and faster than those with the geodetic theodolites. However, to achieve the same accuracy in astronomic coordinates, more stars and more observations on a single star are needed than when the position is determined with theodolites.

7.21 Geodetic Theodolites

The geodetic theodolites are instruments similar to the universal theodolites described earlier. The major differences are the following:

(1) The telescope is straight.

(2) The diameters of the graduated reading circles are generally smaller.

(3) The instrument is smaller and lighter.

(4) There is no impersonal eyepiece micrometer.

(5) There are no special levels (striding and Horrebow) except as special accessories.

General requirements are that the diameters of the circles should not be smaller than 100 mm, with a minimum graduation interval of 10', and a direct micrometer reading of 0".5 (estimated to 0".05). Instruments fulfilling these specifications are, for example, the Wild T3 and the Kern DKM3 geodetic theodolites. The specifications of these instruments are listed in Table 7.3 and they are shown in Figs. 7.23 and 7.24 respectively. The former figure shows the T3 with the astrolabe attachments to be explained later and the reading of the instrument (note the necessary double reading on the micrometer). The DKM3 is read the same way as the DKM3-A shown in Fig. 7.12.

It is possible that for second-order azimuth determinations, instruments with horizontal circles as small as 75 mm will be found to give errors no larger than those due to the horizontal refraction of light. Such so-called 'one-second' theodolites are, for example, the Wild T2, the Kern DKM2, and the Askania Tu 400. Their specifications are also listed in Table 7.3.

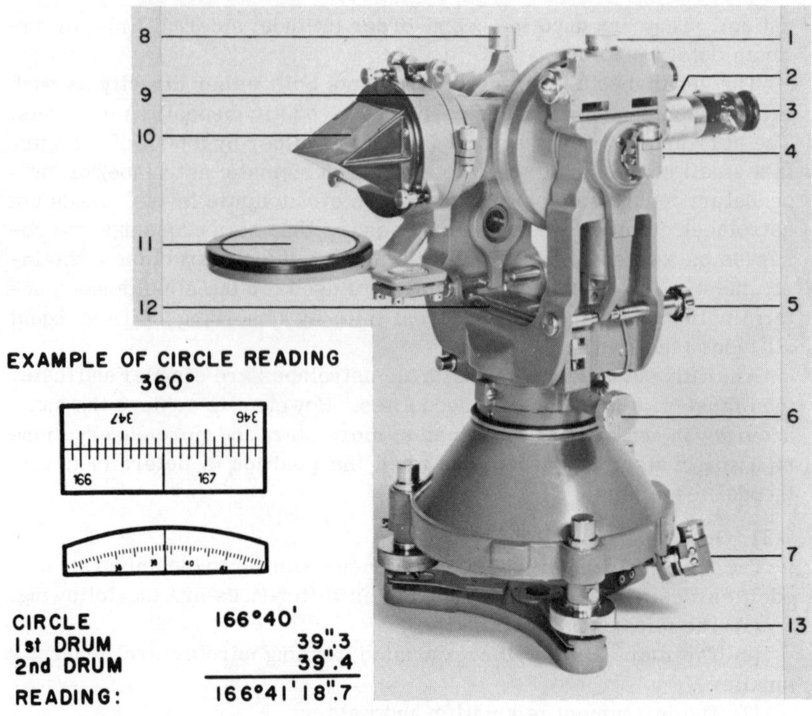

1 Vertical Circle Level Prism	7 Horizontal Circle Lamp
2 Focusing Ring	8 Vertical Clamp
3 Telescope Eyepiece	9 Astrolabe Attachment Ring
4 Vertical Circle Lamp	10 Astrolabe Prism
5 Vertical Circle Level Screw	11 Mercury Pool
	12 Alidade Level
6 Horizontal Clamp	13 Foot Screw

Fig. 7.23 Wild T3 geodetic theodolite with astrolabe attachment

The geodetic theodolites need to be adjusted for vertical axis, horizontal and vertical collimation, cross wire alignment, and focusing. The procedure is similar to that described in section 7.16.

For second-order observational purposes the calibration of the theodolite components as provided by the factory is satisfactory with the possible exception of the optional parts, such as the striding or Horrebow levels, which should be calibrated by the observer as described in section 7.17.

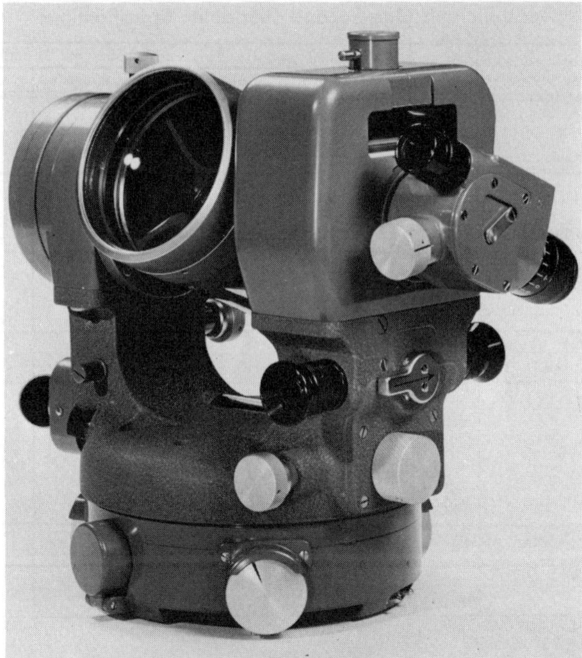

Fig. 7.24 Kern DKM3 geodetic theodolite

7.22 Prismatic Astrolabes

The prismatic astrolabe consists mainly of three parts (Fig. 7.25):

(1) A telescope with an appropriate reticle in the focal plane.

(2) A prism of the angle equal to the altitude of the almucantar, placed immediately in front of the objective with one face (BC) at right angles to the optical axis of the telescope.

(3) A mercury surface, placed just below the lower face of the prism, establishing the horizon.

These parts form a rigid unit, which is rotatable about the vertical axis to allow observations in any vertical plane. A horizontal setting circle permits the instrument to be set on a star of given azimuth.

Fig. 7.25 shows two parallel rays S and S′ of a star of approximate altitude a. Ray S′ enters the prism at a right angle to the surface AB, is reflected, and emerges from the surface BC. Ray S is reflected by the mercury surface, enters the prism, and also emerges from the surface BC. Thus the two emerging rays are parallel and produce a single image of the star at F, provided that the star at that instant is on the almucantar defined by the prism angle ($a = 45°$ or $60°$); otherwise, the observer will see two images moving opposite to each other. Only the time of coincidence is recorded. Better timing is obtained if the images

TABLE 7.3 Geodetic and One-Second Theodolite Specifications

Component	Unit	Geodetic Theodolite		One-Second Theodolite		
		Wild T3	Kern DKM3	Wild T2	Kern DKM2	Askania Tu 400
Telescope						
Aperture	mm	60	68	40	45	45
Magnification	X	24, 30, 45	27, 45	28	30	30
Horizontal Circle						
Diameter	mm	140	100	90	75	90
Graduation Interval		4	10	20	10	20
Vertical Circle						
Diameter	mm	96	100	70	70	70
Graduation Interval		8	10	20	10	20
Optical Micrometer						
Graduation Interval	"	0.2	0.5	1.0	1.0	1.0
Level Sensitivity						
Plate	"/2mm	7	10	20	20	20
Collimation	"/2mm	12	10	30	30	automatic compensator
Striding*	"/2mm		4	5	10	5
Horrebow*	"/2mm		4	5		
Weight (approximate)	kg	11	11.2	5.5	3.6	4.5
Price in the U.S., 1966 (approximate)	$	3,200	4,000	1,800	1,800	1,600

*Optional Accessory

move parallel to each other and do not come to coincidence but appear to pass each other when they are on the same level, as shown schematically in Fig. 7.26.

Prismatic astrolabes are basically of two types: (1) instruments originally built for this purpose, (2) astrolabe attachments, consisting of the prism and the mercury pool, which are fitted to the telescope of a geodetic theodolite.

The former type is illustrated in Fig. 7.27 which shows the 45° prismatic astrolabe of Cooke, Troughton and Simms, Ltd. [Clark, 1963, Vol.II, pp.68-69]; the latter is shown in Fig.7.23 where the Wild T3 precision theodolite is seen with its astrolabe attachment.

The prismatic astrolabe needs to be adjusted for the horizontality of the mercury surface to the extent that it will ensure that the mercury will not run to one side of its container. This is achieved by the customary adjustment of the vertical axis (section 7.162) and, in case of the astrolabe attachment, by turning the telescope into its horizontal position. Additional adjustments for the cross threads and focusing may also be necessary (sections 7.167 and 7.168), together with making the surface BC of the prism perpendicular to the optical axis of the telescope (Fig. 7.25). This latter adjustment is done by first viewing the brightly illuminated reticle through the eyepiece. If the surface BC is not in its proper position both the reticle and its reflection on the BC surface can be seen. If this surface is at right angles to the optical

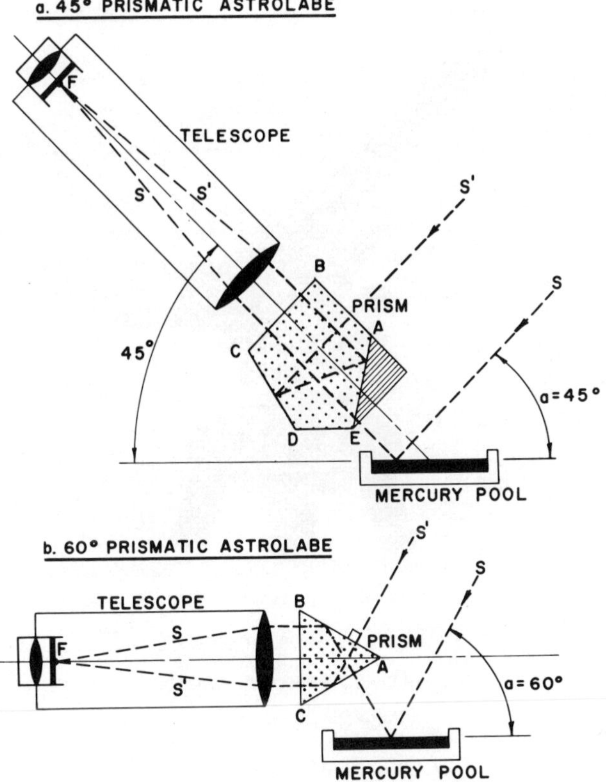

Fig. 7.25 Principle of prismatic astrolabes

axis, the two images coincide. The adjustment is made by moving the prism, i.e., by rotating the surface BC by means of appropriate screws. The horizontal distance between the two images of the star observed (SS′ in Fig. 7.26) can be adjusted by rotating the telescope about its optical axis, if the construction of the instrument permits such a motion.

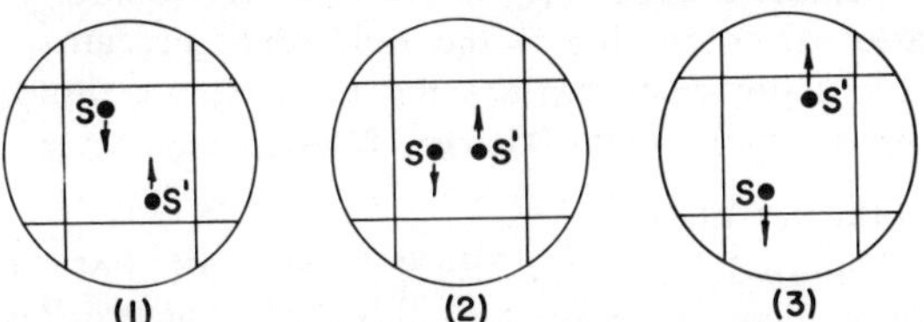

(1) BEFORE THE STAR REACHES THE ALMUCANTAR.
(2) THE INSTANT THE STAR IS ON THE ALMUCANTAR.
(3) SOME TIME LATER.

Fig. 7.26 View through the astrolabe eyepiece

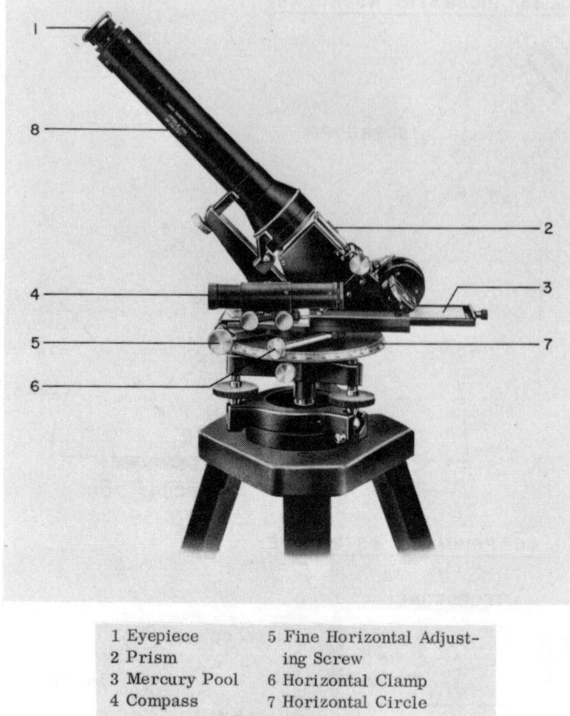

1 Eyepiece
2 Prism
3 Mercury Pool
4 Compass
5 Fine Horizontal Adjusting Screw
6 Horizontal Clamp
7 Horizontal Circle
8 Telescope

Fig. 7.27 Cooke 45° prismatic astrolabe

Calibration generally is not necessary, except for the angle of the prism. However, the deviation of this angle from its nominal value is determined from the adjustment of the observation results simultaneously with the latitude and longitude determinations (see section 11.523).

An instrument similar to the prismatic astrolabe is the circumzenithal, in which the prism is replaced by a system of mirrors inside the instrument, and the mercury surface is also inside. Fig. 7.28 shows the schematic cross section of the Nušl-Frič circumzenithal. When equipped with impersonal registering micrometer this instrument is capable of first-order work [Buchar, 1938].

7.23 Pendulum Astrolabes

The pendulum astrolabe permits the observation of a star at a constant altitude as does the prismatic astrolabe, but it differs in one important respect from the instrument just described. With the pendulum astrolabe, only one image of a star is observed as it crosses a fixed thread(s) of the reticle. The ray of light from the star either

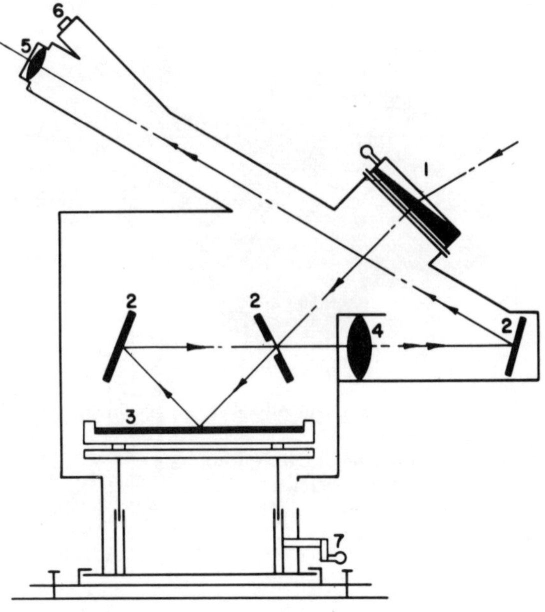

1 Reversable Prism to Change Zenith Distance
2 Reflecting Mirrors
3 Mercury Pool
4 Objective
5 Eyepiece
6 Finder Telescope
7 Mercury Container Release Lever

Fig. 7.28 Nušl-Frič circumzenithal

enters a prism at the front of the telescope or is reflected from a pendulum mirror situated inside the instrument. The angle of the prism or the position of the pendulum-mirror defines the altitude of the star when it appears on the center thread of the reticle. The horizontality of the system in the prismatic case is assured by a compensator pendulum which keeps the telescope horizontal, while in the pendulum-mirror case it is assured by this mirror itself which compensates for any tilt of the instrument from the horizon.

As an illustration, Figs. 7.29 and 7.30 show the Zeiss (Oberkochen) and the Zeiss (Jena) pendulum astrolabes respectively. The Zeiss (Oberkochen) instrument is basically the Ni2 automatic level with prism and illuminating attachments, seen at the objective and the eyepiece ends of the telescope, respectively. The Zeiss (Jena) instrument is equipped with an impersonal registering micrometer. The star image is kept in contact with the horizontal cross hair over an altitude range of 4!5 by the hand rotation of a parallel plate in the ray path, during which 27 elec-

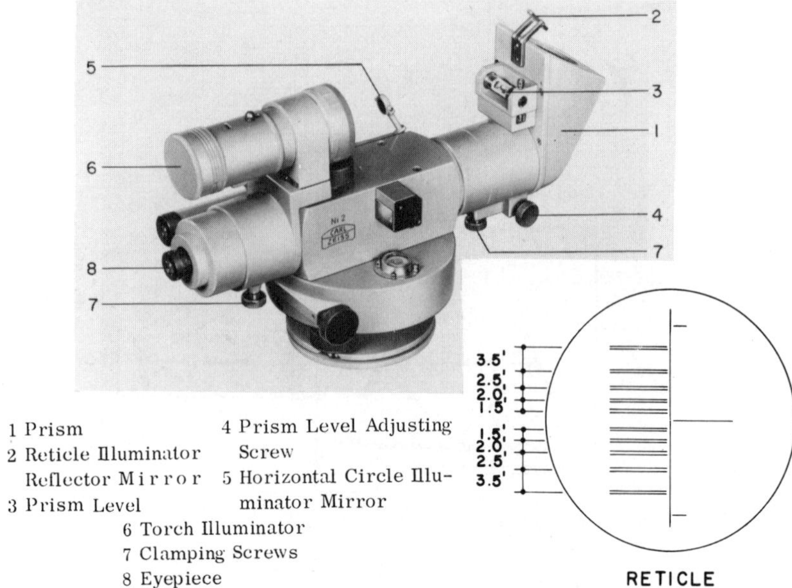

Fig. 7.29 Zeiss (Oberkochen) Ni2 pendulum astrolabe

trical contacts are made. Its size is about 30.5 ×30 ×25 cm and it weighs about 10 kg. Precisions of both of these instruments, as quoted by the manufacturers, approach first-order requirements [Drodofsky, 1963; Hemmleb, 1966].

Other similar instruments are the Willis pendulum astrolabe (David White Co. of Milwaukee, Wisconsin) and its more modern version with impersonal electric transit detector manufactured by the Perkin-Elmer Corporation of Norwalk, Connecticut.

The advantages of the pendulum astrolabes over the prismatic ones are that the mercury pool, inconvenient in the field, is eliminated, and that several readings on a single star are possible, provided that the reticle has a number of fixed threads (Fig. 7.29) or that the instrument is equipped with an impersonal micrometer.

7.3 Observatory (High Precision) Instruments

Optical instruments in this category are nonportable, generally rather large and heavy, and permanently installed. They are primarily used for repeated high precision observatory time and/or latitude determinations; thus they are considered astronomic rather than geodetic instruments. Their brief description is included because they point to the direction in which future instruments of geodetic astronomy are being developed (portable zenith cameras, etc.).

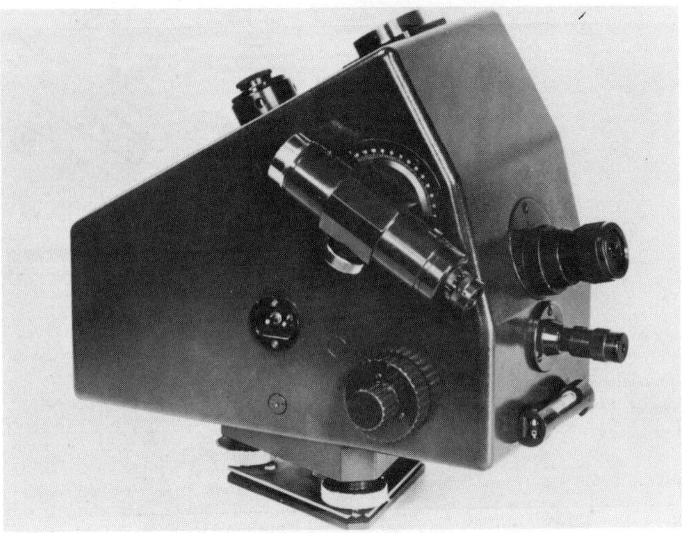

Fig. 7.30 Zeiss (Jena) pendulum astrolabe

In addition to the classical observatory instruments, such as the large meridian circles, micrometer-transits, or zenith telescopes which are similar but larger than those mentioned in section 7.15, there are some other more recent instruments also used (almost exclusively nowadays) in observatory time and/or latitude determinations. Such instruments, for example, are the following:

(1) Floating zenith telescope (FZT).
(2) Photographic zenith tube (PZT).
(3) Impersonal prismatic astrolabe (by Danjon).
(4) Moon camera (after Markowitz).

The floating zenith telescope is not essentially different from the ordinary zenith telescope, but instead of the Horrebow level, a source of many troublesome errors at observations of high precision, the telescope is floating in a large basin (containing about 100 kg of mercury). The leveling is considered to be secured by the free surface of mercury. In addition, the micrometer is replaced by photographic plates on which the stars are photographed and measured after the observation. The instrument may be used for the determination of both latitude and sidereal time [Hattori, 1951 and 1953].

The photographic zenith tube and the Danjon impersonal prismatic astrolabe are the most precise observatory instruments at present for latitude and sidereal time determinations (standard deviation = 0".075 from 14-16 star observations during one night or 0".2 from a single determination). The Danjon astrolabe is at the same time at least as accurate an instrument for the determination of absolute right ascensions

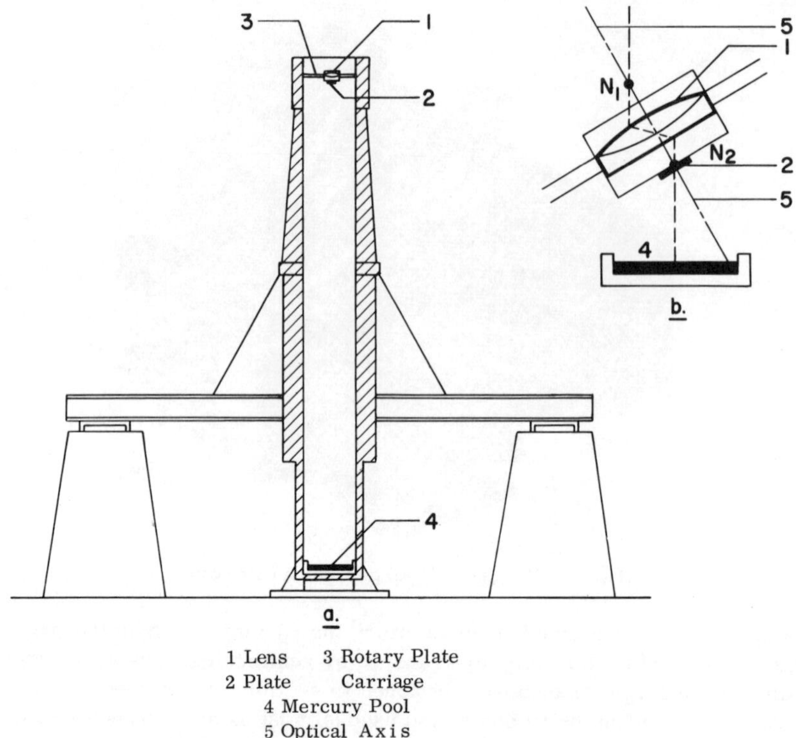

1 Lens 3 Rotary Plate
2 Plate Carriage
4 Mercury Pool
5 Optical Axis

Fig. 7.31 Principle of photographic zenith tube

and declinations. Its disadvantage in this respect over the meridian circle is that the faintest stars which can be effectively observed are of magnitudes of six. The Markowitz moon camera is used in ephemeris time determination. These latter instruments are discussed briefly below.

7.31 Photographic Zenith Tube (PZT)

The PZT is mounted in a vertical position; thus only stars which transit near the zenith can be observed (Figs. 7.31 and 7.32). The light rays from the star pass through the lens, are reflected by a basin of mercury at the lower end of the tube, and come to focus in the focal plane where the photographic plate is situated. The lens has two nodal points, N_1 and N_2 (Fig. 7.31b). The light ray which is directed to one nodal point leaves the second in a parallel direction; thus the path of the ray is $ZN_1N_2GN_2N_1Z$. The optical axis of the tube intersects the path of the ray in N_1 and N_2.

The underlying optical principle of the PZT is that its optical system is designed so that the inner nodal point N_2 lies in the focal plane, where the photographic plate rigidly connected to the lens holder is located.

1 Mercury Pool 3 Photographic Plate
2 Tube Carriage

Fig. 7.32 Photographic zenith tube of the U.S. Naval Observatory

After reflection from the mercury surface, a light ray from the zenith forms an image on the plate which coincides with the inner nodal point; thus tilt or horizontal translation on the lens does not alter the position of the zenith on the plate. The position of the image of a star which is not exactly at the zenith is not sensibly displaced by these motions; however, rotation of the lens about its axis by 180° results in a star image symmetrical about the zenith.

The photographic plate is mounted in a carriage which is driven by a motor synchronized to follow the stars by moving the plate carriage in an east-west direction. In addition, the carriage and lens cell can be rotated 180° by a motor driven rotary between exposures. The motion of the carriage during exposures triggers timing pulses which are recorded with respect to a crystal clock.

Four exposures of twenty seconds each are generally made of each star in alternating rotary positions, two before and two after transit. The interval between the mean exposure times is generally thirty seconds.

Each exposure is started with the center of the plate ten seconds west of the meridian. At the end of the exposure it is ten seconds east. Reversal of the rotary brings the center of the plate back to ten seconds west.

The images of a star as they appear on the PZT plate are shown in Fig. 7.33. The arrows indicate the direction of the star's motion with

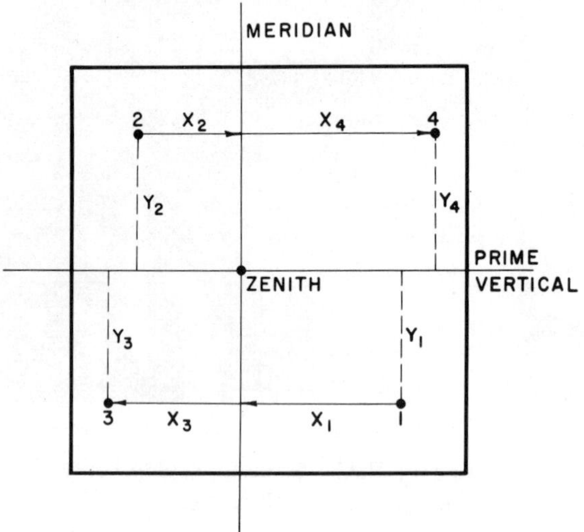

Fig. 7.33 The images of a star on the PZT plate

respect to the meridian. The numbers indicate the sequence of exposures. If the exposures were symmetrical with respect to transit time on the meridian, the images would form a rectangle.

The photographic zenith tubes used in the United States are the Ross type, operated between 1911 and 1955, and the Richmond type, developed by Sollenberg and Markowitz, in operation since 1949. These instruments have an aperture of 200 mm, effective focal length of 3780 mm to 5170 mm, and a field of 24' to 34'. There are other PZT's also on the market, such as the Askania FZT 250/3750 (aperture/focal length) and others. There were about nine PZT's participating in time and latitude services around the world as of 1965 (Greenwich, England; Hamburg, Germany; Mizusawa, Japan; Mount Stromlo, Canberra, Australia;

Neuchâtel, Switzerland; Ottawa, Canada; Richmond, Florida; Tokyo, Japan; and Washington, D. C.) [Yumi, 1967].

For further information see [Markowitz, 1960, pp. 88-107] on which the material in this section is mainly based, and section 11.532.

7.32 Danjon-OPL Impersonal Astrolabe

The Danjon astrolabe, built by the Société Optique et Précision de Levallois (OPL) differs from an ordinary 60° prismatic astrolabe with a broken telescope by its ingenious self-registering micrometer [Danjon, 1960, pp. 115-137]. The most important item of this device is a double Wollaston prism situated just in front of the focal plane (Fig. 7.34). The prism is preceded by an aberration compensator consisting of two plane-parallel quartz plates with their crystallographic axes inclined at 45° to the incident ray of light.

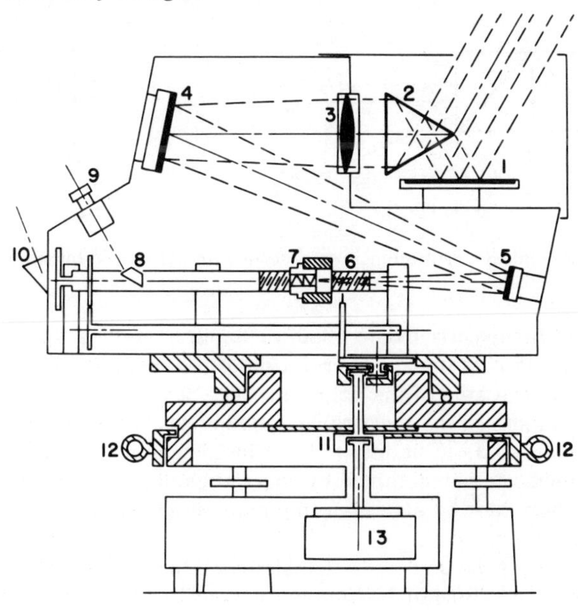

1 Mercury Pool
2 Prism
3 Objective
4 Large Mirror
5 Small Mirror
6 Micrometer Screw
7 Wollaston Prism and Compensator
8 Reflecting Prism
9 Eyepiece
10 Prism for Reading Drum
11 Differential Gear
12 Differential Correction Wheel
13 Motor

Fig. 7.34 Schematic cross section of the Danjon astrolabe

The Wollaston prism splits the light rays S and S′ of a star in such a manner that two parallel (S_1 and S_1') and two diverging (S_2 and S_2') rays emerge (Fig. 7.35). By a displacement of the Wollaston prism, parallel to the optical axis of the telescope, the parallel images S_1 and S_1' can be made coincident while they traverse the central part of the field of view. To maintain coincidence of the images, the prism has to be displaced at a speed in accord with that of the images S_1 and S_1', which is proportional to the sine of the azimuth and to the cosine of the latitude. The diverging images S_2 and S_2' are screened off. The prism is displaced by a motor-driven micrometer that carries electrical contacts connected to a chronograph. The speed of the motor can be set proportional to the cosine of the latitude of the observing station. When the

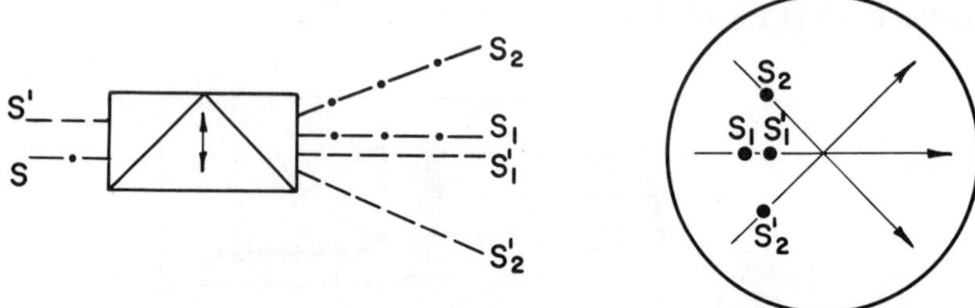

Fig. 7.35 Image formation in a double symmetrical Wollaston prism

astrolabe is turned around to observe another star, a speed reducer automatically varies the rate of rotation of the micrometer screw proportionally to the sine of the azimuth. The operator has to make only slight adjustments in the traversing speed of the micrometer by means of a handwheel so as to maintain the coincidence of the images. In fact this coincidence is not obtained by superimposing the two images but by placing them side by side along the same horizontal line between two illuminated wires.

Fig.7.36 shows the Danjon-OPL instrument in the U. S.Naval Observatory. It has an aperture of 100 mm and a focal length of 1000 mm. There were about eleven impersonal astrolabes (OPL and other types) participating in time and latitude services around the world as of 1965 (Algiers, Algeria; Besancon, France; Cape Town, South Africa; Paris, France; Poltava, U. S. S. R.; Potsdam, Germany; Quito, Equador; Santiago, Chile; Sao Paolo, Brazil; Uccle, Belgium; Zi-Ka-Wei, China) [Yumi, 1967].

Another astrolabe whose accuracy approaches that of the Danjon instrument is the one reported in [Tsubokawa, 1967]. Here the ray from a star is deflected by a constant angle using a mirror system of two reflecting planes and is incident upon the objective of the telescope whose collimation axis is kept vertical with a mercury horizon. The passage

Fig. 7.36 Danjon impersonal prismatic astrolabe at the U.S. Naval Observatory

of a star image on the almucantar is observed by photoelectronic means to eliminate personal error. The standard deviation of observation of a single star is reported as 0".15 to 0".30 for a set of 20 to 30 stars. Observations of 5 to 10 sets provide standard deviations, in both latitude and longitude, of better than 0".1. For further information see [Danjon, 1960, pp. 115-137], and section 11.531.

7.33 Markowitz Moon Camera

The precise position of the moon may be determined by photographing it in the background of stars. An ordinary camera attached to any telescope cannot be used with success for this purpose because the moon is much brighter and moves faster than the surrounding stars. On the resulting pictures either an overexposed trail of the moon would be seen in the background of stars (long exposure, camera driven at the sidereal rate) or a sharp image of the moon would be seen without any stars except the very bright ones (short exposure, camera driven at the lunar rate). There were several attempts in the past to solve this problem. Probably the best of these experiments, which during the International Geophysical Year found application on a worldwide scale, is the camera designed by W. Markowitz, former director of the Time Service Division, U. S. Naval Observatory. In his dual rate camera, the image of the moon is kept motionless relative to the stars, during ten or twenty seconds of exposure in which the moon and the stars are exposed simultaneously. The camera shown in Fig. 7.37 may be attached to visual or photographic refractors of about 200 or more millimeter aperture and of focal length preferably between 2000 to 6000 mm.

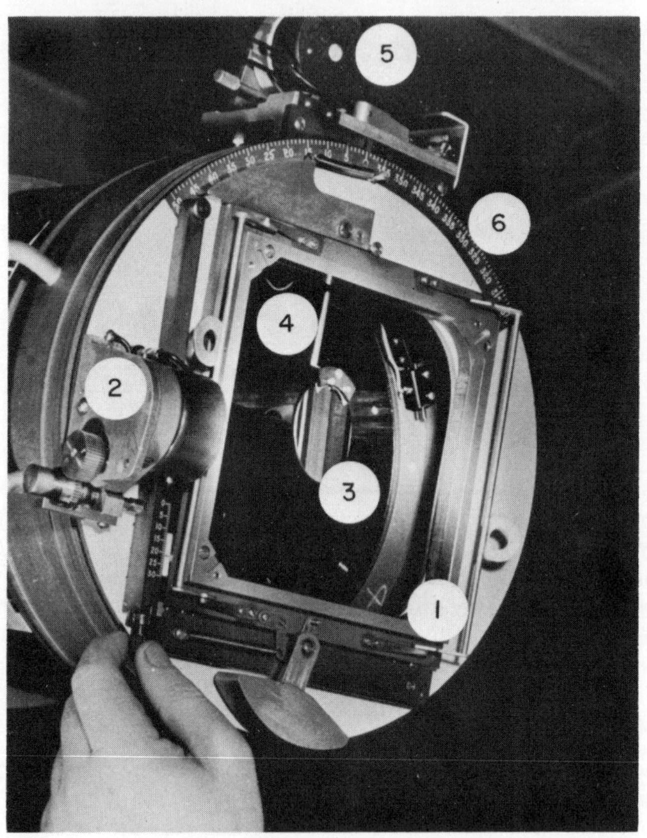

1 Plate Carriage	4 Tilt Arm
2 Carriage Drive and Micrometer	5 Filter Drive and Micrometer
3 Filter	6 Setting Circle

Fig. 7.37 Markowitz dual rate moon camera at the U.S. Naval Observatory

The principle of the operation is that while the photoplate within the stationary camera is driven at the sidereal rate in the direction of the star's motion, a 1.8 mm thick filter somewhat larger than the apparent size of the moon's image is rotated on a tilt-arm which keeps the moon's image fixed with respect to the stars, and at the same time reduces its light intensity by a factor of 1000 to permit long enough exposures for recording faint stars (to about a magnitude of nine) without overexposing the moon's image. The fictitious epoch of the observation is the instant when the filter is parallel to the plate, since at that time there is no displacement of the moon's image relative to the stars.

The plate carriage is driven by a synchronous motor and micrometer, seen on the left side of the figure. For operation at the correct rate, which depends on the declination of the moon, adjustment of the driving speed is possible. A second synchronous motor and micrometer, seen on the top of the camera, drives the tilt-arm which tilts the filter. The rate of tilt can be adjusted to match the speed of the moon relative to the stars. The platform carrying the tilt apparatus can also be moved so that the axis of tilt is made perpendicular to the apparent path of the moon with respect to the stars. The recording of the time is made by means of a chronograph tape on which time marks are made by a contact mounted on the tilt-arm. The camera is rotated 180° between successive exposures, and plates are measured in pairs by means of a measuring engine connected directly to a digitizer. The reduction of the plate is explained in section 11.7.

The camera has been primarily designed to measure the difference between ephemeris or atomic and universal times and to determine changes in the rate of rotation of the earth.

The first dual rate camera was put into operation in 1952 at the U.S. Naval Observatory. Twenty cameras were constructed during the International Geophysical Year. For further details see [Markowitz, 1954 and 1960, pp. 107-114].

7.4 Equations of the Optical Instruments

As pointed out in section 7.16, each optical instrument needs to be factory- and field-adjusted for the purposes of (a) making the horizontal and vertical axes perpendicular to each other, (b) making the principal axis of the telescope perpendicular to the horizontal axis, (c) making the principal axis and the line of sight, marked by the central thread, coincide, and (d) making the vertical axis of the instrument coincide with the true vertical at the station (placing the horizontal graduated circle into the plane of the observer's horizon). After these adjustments have been carefully performed, small residual effects are further reduced by measuring the imperfections and taking them into account in the calculations through the use of the 'instrument equations' to be described in this section.

The measurable imperfections in the case of vertical and horizontal angle measurements are generally given in the following components:

b = inclination or altitude of the horizontal axis with respect to the horizon at the observation point, positive when the circle end of the axis is above the horizon.

c = collimation or the distance of the thread from the principal axis, positive when the thread is on the opposite side of the principal axis from the graduated vertical circle.

i = inclination of the vertical axis to the vertical at the observation point.

If the time of a star's transit over a certain vertical plane (e.g., the meridian) is to be measured, then, in addition to the quantities b and c, the horizontal angle between the telescope's principal axis and the vertical plane in question must also be considered. This 'azimuth setting error' is denoted by 'a' and is positive when the axis is west from the vertical plane.

7.41 Equation for Horizontal Angle and Azimuth Measurement.

Let Fig. 7.38 illustrate the magnified effects of the residual adjustment imperfections b, c, and i on the azimuth and zenith distance of the object (e.g., star) S. The point is bisected by the horizontal and vertical threads of the telescope at O. The direction of the sight line OS is registered by making appropriate readings on the graduated horizontal and vertical circles of the instrument. In the figure, the circle A_0H rep-

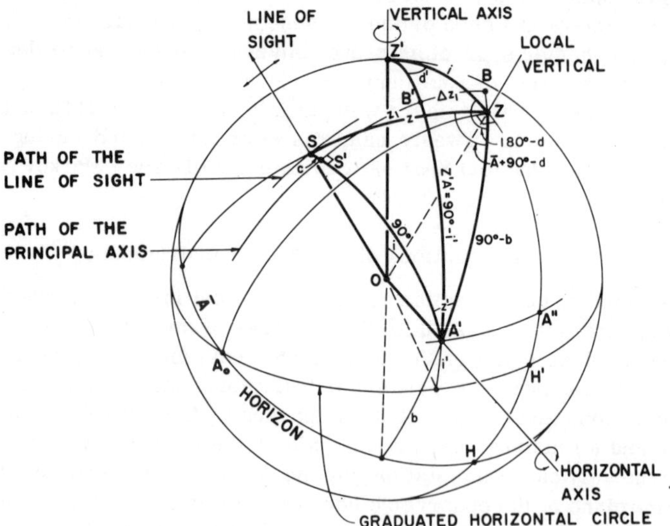

Fig. 7.38 The effect of residual instrumental adjustment errors on horizontal and vertical angle measurements

resents the horizon and Z represents the zenith. Let the vertical axis of the instrument be inclined to the vertical at O by the amount i, so when extended it intersects the celestial sphere in the instrumental zenith Z'. Let A_0H' be a great circle of which Z' is the pole; thus its plane contains the graduated horizontal reading circle of the instrument (provided that the vertical axis is perpendicular to the graduated circle). Further suppose that the horizontal axis of the instrument OA' makes the small angles b and i' with the horizon and the horizontal reading circle respectively. As the instrument revolves about its vertical axis, the extension of the horizontal axis to the celestial sphere describes a small circle $A'A''$ parallel to A_0H'. As the telescope revolves upon the horizontal axis, its principal axis describes the great circle $S'B'$ of which A' is the pole (provided that the principal axis is perpendicular to the horizontal axis), and the given thread describes a small circle parallel to $S'B'$. The angle $\bar{A}$ on the circle A_0H is the horizontal angle between the points S and A_0, and z is the zenith distance of S.

Applying spherical trigonometry, the following relations are produced from the triangle $A'ZZ'$:

$$\left.\begin{aligned} \cos b \cos d &= \cos d' \cos i' \cos i - \sin i' \sin i \\ \cos b \sin d &= \sin d' \cos i' \\ \sin b &= \cos d' \cos i' \sin i + \sin i' \cos i \end{aligned}\right\} . \tag{7.10}$$

Considering that i, i', and b are always so small that their cosines can be taken as unity and their sines may be substituted by their arcs,

$$\begin{aligned} d &= d', \\ b &= i \cos d' + i' = i \cos d + i'. \end{aligned} \tag{7.11}$$

From the triangle $A'ZS$,

$$-\sin c = \sin b \cos z - \cos b \sin z \sin(\bar{A} - d),$$

or, since b and c are small,

$$\sin(\bar{A} - d) = b \cot z + c \operatorname{cosec} z. \tag{7.12}$$

Hence, $\sin(\bar{A} - d)$ is also a small quantity, and the angle $\bar{A} - d$ is either nearly 0° or nearly 180°. When the vertical circle at the end of the horizontal axis is to the left of the observer [C. L.], as supposed in Fig. 7.38, it is evident that $\bar{A}$ and d are nearly equal and $\bar{A} - d$ is nearly 0°. If the instrument is revolved around its vertical axis and the telescope is directed to the same point S, the vertical circle will be on the right of the observer [C. R.] and the angle d will be increased by 180°. In this case, therefore, $180^\circ - (\bar{A} - d)$ will be a small quantity. Substituting then $\bar{A} - d$ or $180^\circ - (\bar{A} - d)$ for $\sin(\bar{A} - d)$ in equation (7.12),

$$\bar{A} = d + b \cot z + c \operatorname{cosec} z \qquad [\text{C. L.}],$$
$$\bar{A} = d + 180^\circ - b \cot z - c \operatorname{cosec} z \qquad [\text{C. R.}].$$

The angle d is not read directly from the graduated horizontal circle; but if R is the actual reading when the line of sight of the telescope is on the object S and R_0 is the fictitious reading when the point A' in the figure is at A'' (in this case, the horizontal axis is in the plane of the tilt of the vertical axis OZZ', and the telescope when horizontal is directed toward the point A_0), then

$$d = d' = R - R_0 \quad [C.L.],$$
$$d + 180^\circ = R - R_0 \quad [C.R.],$$

and, therefore,

$$\bar{A} = R - R_0 \pm b \cot z \pm c \operatorname{cosec} z \qquad (7.13)$$

where the upper sign corresponds to [C.L.] and the lower sign to [C.R.].

If the azimuth of the point A_0 is now denoted by A_0, then the azimuth of the star is

$$A = A_0 + \bar{A},$$

or

$$A = R + \Delta A \pm b \cot z \pm c \operatorname{cosec} z \qquad (7.14)$$

where $\Delta A = A_0 - R_0$ is called the index correction. This equation gives the azimuth of the point S, provided that the inclination of the horizontal axis b, the collimation c, and the index correction ΔA are known. Their determination is described in section 7.44. The circle reading R for circle right differs by 180° from that of circle left. In the above equation, it is assumed that the former [C.R.] have been increased or diminished by 180° when two observations in different positions are compared.

7.42 Equation for Vertical Angle Measurement

In Fig. 7.38 let the object S be observed on the horizontal thread SS' which is perpendicular to the great circle $S'B'$ (also representing the vertical circle of the instrument) and coincides with the arc $A'S'$ extended. The point B' in which $A'Z'$ meets the circle $S'B'$, represents the extremity of that diameter of the vertical reading circle which is in the plane of the vertical and horizontal axes of the instrument. The arc $B'S'$, or the angle $B'A'S'$ which it measures, is then the zenith distance as given directly by the circle from the readings for B' and S'. Let the fictitious reading of the circle, when the thread is at B', be denoted by V_0 and the reading on the object be denoted by V, and put $B'S'$ or $B'A'S' = z_1$; then, for circle left,

$$z_1 = V_0 - V, \qquad (7.15)$$

the graduation of the circle is assumed to increase from right to left.

The point B in which $A'Z$ meets the circle $S'B'$ represents the extremity of that diameter of the vertical reading circle which is in the plane defined by the vertical of the observer and the horizontal axis of the instrument. For different azimuths the relative position of B and B' is different, and they coincide only when the point A′ is in the plane of the circle ZZ′. Their relative position at any time is given by the vertical circle level. Let ℓ_0 be the fictitious reading of the level when B and B′ coincide and ℓ be the reading in any other case; then, denoting BB′ by Δz_1,

$$\Delta z_1 = \ell_0 - \ell \tag{7.16}$$

where the left-hand end of the level is taken as the positive end when the observer is facing the circle, and ℓ is half the algebraic sum of the readings of the ends of the bubble.

Denoting the observed zenith distance, i.e., the arc BS′ by $z' = z_1 + \Delta z_1$ and using spherical trigonometry in the triangle AS′Z, it is evident that the true zenith distance z may be computed from

$$\cos z = -\sin c \sin b + \cos c \cos b \cos z', \tag{7.17}$$

which is the basic relation to correct the observed zenith distance for collimation and the inclination of the horizontal axis. Substituting

$$\cos z' = \cos^2 \tfrac{1}{2} z' - \sin^2 \tfrac{1}{2} z',$$

and recognizing that the quantities b, c, and $z - z'$ are very small, equation (7.17) yields

$$\Delta z = z - z' = \left(\frac{c+b}{2}\right)^2 \sin 1'' \cot \tfrac{1}{2} z' - \left(\frac{c-b}{2}\right)^2 \sin 1'' \tan \tfrac{1}{2} z' \tag{7.18}$$

where Δz denotes the correction for collimation and the inclination of the horizontal axis, and c and b are in units of seconds of arc.

Using the above relations the zenith distance of the object S given by the observation [C. L.] is

$$z = z' + \Delta z = z_1 + \Delta z_1 + \Delta z,$$

or, after substituting equations (7.15) and (7.16),

$$z = V_0 - V + \ell_0 - \ell + \Delta z \quad [\text{C. L.}]. \tag{7.19}$$

In this equation, the constants V_0 and ℓ_0 are unknown. If the instrument is revolved 180° about its vertical axis and the zenith distance of S is observed in circle right, then, since

$$z_1 = V' - V_0, \text{ and } \Delta z_1 = -(\ell_0 - \ell'),$$

the zenith distance is computed from

$$z = V' - V_0 - \ell_0 + \ell' + \Delta z' \quad [\text{C. R.}] \tag{7.20}$$

where V' is the vertical circle reading, ℓ' is the vertical circle level reading, and $\Delta z'$ is the correction for collimation and inclination, all in the second [C. R.] observation.

The mean of equations (7.19) and (7.20) gives

$$z = \tfrac{1}{2}(V' - V) + \tfrac{1}{2}(\ell' - \ell) + \tfrac{1}{2}(\Delta z' + \Delta z), \qquad (7.21)$$

where when V' is numerically less than V, it should be increased by 360°.

The difference of equations (7.19) and (7.20) determines the unknown constant

$$V_0 + \ell_0 = \tfrac{1}{2}(V' + V) + \tfrac{1}{2}(\ell' + \ell) + \tfrac{1}{2}(\Delta z' - \Delta z) = Z \qquad (7.22)$$

which can then be determined by observing a distant nonmoving point or the cross thread of a collimating telescope in circle left and circle right positions. If these observations are made in the middle of the telescope's field, in most cases it may be assumed that $\Delta z' - \Delta z = 0$; thus

$$Z = \tfrac{1}{2}(V' + V) + \tfrac{1}{2}(\ell' + \ell). \qquad (7.23)$$

With this constant determined, a single observation of a moving point, in either position of the instrument, will suffice to determine its zenith distance, since then from equation (7.19), (7.20), and (7.22),

$$\begin{aligned} z &= Z - (V + \ell) + \Delta z \quad [\text{C.L.}], \\ z &= (V' + \ell') - Z + \Delta z' \quad [\text{C.R.}]. \end{aligned} \qquad (7.24)$$

The constant Z represents the 'zenith point of the instrument,' since, when the observed object is in the zenith, it is equal to the corrected circle reading. In the method of observation described above, a knowledge of the deviations b, c, etc. is generally not required; it is only necessary that they should be small. If their magnitude is appreciable, then Z should be determined from equation (7.22) rather than from (7.23).

7.43 Equation for Transit Time Measurement

Let Fig. 7.39 again represent the celestial sphere. The horizontal axis of the instrument lies in the vertical plane ZA', and A' is the point in which this axis extended to the west meets the sphere. Let $Z'P$ be the great circle described by the principal telescope axis (provided that it is perpendicular to the horizontal axis); A' is the pole of this circle. The parallel small circle through S is described by the line of sight. Its constant angular distance from the principal axis is the collimation c. Let b again denote the altitude of the point A', $270° - a$ its azimuth, $90° - h_A$ its hour angle, and δ_A its declination. Let Φ be the latitude of the observer O, and δ be the declination of the observed point S.

The spherical triangle NCP-A'-Z produces the trigonometric relations

$$\left.\begin{aligned} \cos\delta_A \sin h_A &= \sin b \cos\Phi + \cos b \sin a \sin\Phi \\ \cos\delta_A \cos h_A &= \cos b \cos a \\ \sin\delta_A &= \sin b \sin\Phi - \cos b \sin a \cos\Phi \end{aligned}\right\}, \qquad (7.25)$$

which determine h_A and δ_A when a and b are given. If τ denotes the negative hour angle of S, then from the spherical triangle A'-NCP-S,

$$-\sin c = \sin\delta_A \sin\delta - \cos\delta_A \cos\delta \sin(\tau - h_A);$$

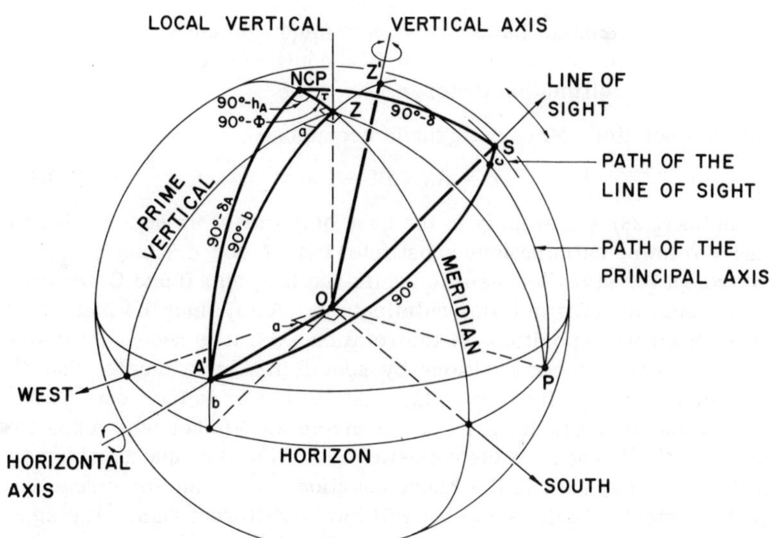

Fig. 7.39 The effect of residual instrumental adjustment errors on transit time measurements

thus

$$\sin(\tau - h_A) = \tan\delta_A \tan\delta + \sin c \sec\delta_A \sec\delta \qquad (7.26)$$

which determines τ. The general formulas (7.25) and (7.26) can be considerably simplified when the object is observed near the meridian, which is the common practice in geodetic work.

If the observation is made very close to the meridian (e.g., within 10" - 20"), then a, b, and c can easily be reduced to quantities so small that they can be substituted for their sines, and their cosines can be assumed equal to unity. Further, since h_A, δ_A, and τ will be quantities of the same order as a, b, and c, the general formulas (7.25) will become

$$\begin{aligned} h_A &= b\cos\Phi + a\sin\Phi, \\ \delta_A &= b\sin\Phi - a\cos\Phi, \end{aligned} \qquad (7.27)$$

and (7.26) gives

$$\tau = h_A + \delta_A \tan\delta + c\sec\delta \qquad (7.28)$$

which is Bessel's formula for computing the correction to be added to the observed sidereal clock time of transit of a star to obtain the clock time of the star's transit over the meridian. Substituting equation (7.27) into (7.28), the latter can be reduced to the following form:

$$\tau = a\sin(\Phi - \delta)\sec\delta + b\cos(\Phi - \delta)\sec\delta + c\sec\delta \qquad (7.29)$$

which is known as Mayer's formula. It is customary to identify the coefficients of this equation as follows:

$$\begin{aligned}\text{azimuth factor} &= A = \sin(\Phi - \delta)\sec\delta,\\ \text{level factor} &= B = \cos(\Phi - \delta)\sec\delta,\\ \text{collimation factor} &= C = \sec\delta.\end{aligned}$$

With this notation, Mayer's formula becomes

$$\tau = Aa + Bb + Cc. \tag{7.30}$$

Formulas (7.28)-(7.30) apply to the case of a star at its upper culmination. For lower culminations substitute $180^\circ - \delta$ for δ. Thus A is positive except for stars between the zenith and the pole; B and C are positive except for stars in lower culminations. Also, since the instrument may be used in two positions of the rotation axis, it is necessary to distinguish between these positions by specifying, for example, that the graduated vertical circle is on the east [C. E.] or in the west [C. W.]. If the value of c has been found for circle west (taken with a positive sign when the thread is on the opposite side of the principal axis from the circle) as is the case in the above equations, its value for circle east will be numerically the same but will have a different sign. The signs of the quantities a and b do not depend on the position of the vertical circle.

7.44 Determination of the Constants of the Instrumental Equations

7.441 Inclination of the Horizontal Axis (b). The residual inclination of the horizontal axis is usually determined by means of the striding level of the instrument, which is either mounted on two legs terminating on the axis or hangs from the axis by recurved arms (see sections 7.115 and 7.125).

The graduations of the level vials are numbered either consecutively from one end to the other (assume higher readings toward the vertical circle) or centrally when the zero graduation is near the center and the numbers increase toward both ends (assume positive readings on the side of the vertical circle). In the former case, the center of the bubble is at the graduation

$$\ell = \frac{w + e}{2} \quad [\text{consecutive numbering}];$$

in the latter case it is at

$$\ell = \frac{w - e}{2} \quad [\text{central numbering}],$$

where w and e denote the readings at the two ends of the bubble (assume that w is on the side of the vertical circle). Reversing the level on the horizontal axis (or revolving the instrument 180° about its vertical axis) will produce a new position for the center of the bubble, provided that the horizontal axis is inclined. The new readings for the two types of graduations mentioned will be

$$\ell' = \frac{w' + e'}{2} \quad \text{[consecutive numbering]},$$

$$\ell' = \frac{w' - e'}{2} \quad \text{[central numbering]},$$

where w' and e' denote the readings at the same ends of the bubble at which w and e were read previously (on the side of the vertical circle the reading is w' when the level is reversed, but e' when the instrument is revolved).

The inclination of the horizontal axis, in the case of consecutive numbering, in units of the vial graduation, is equal to the half difference of the readings corresponding to the bubble centers in the two level positions; in the case of the central numbering system, it is equal to the simple mean of the two readings. Multiplying this value with the sensitivity of the level d, determined as shown in sections 7.171 and 7.172, gives the inclination in angular units. Thus

$$\begin{aligned} b &= \frac{d}{2}(\ell - \ell') = \frac{d}{4}[(w - w') + (e - e')] \quad \text{[consec. numbering]}, \\ b &= \frac{d}{2}(\ell + \ell') = \frac{d}{4}[(w + w') + (e + e')] \quad \text{[central numbering]}. \end{aligned} \tag{7.31}$$

Using the numbering system and sign convention explained above, b will be positive when the vertical circle end of the horizontal axis is elevated.

7.442 Horizontal Collimation (c). The most convenient method of finding the residual collimation c is to employ a collimating telescope placed at the height of the horizontal axis. The cross thread of the collimator is observed at the point S in Fig. 7.38, with $z = 90°$. Observing in both circle left and circle right positions, let R and R' be the readings of the horizontal circle (the latter changed by 180°); then from equation (7.14)

$$A = R + \Delta A + c,$$
$$A = R' + \Delta A - c.$$

Thus

$$c = \tfrac{1}{2}(R' - R) \tag{7.32}$$

which relation gives c with its proper sign for circle left. If the collimator is not at the same level with the horizontal axis, then equation (7.14) gives

$$c = \tfrac{1}{2}(R' - R) \sin z - b \cos z \tag{7.33}$$

where z is the zenith distance of the collimator, and b is the inclination of the horizontal axis assumed to be constant during the observations and determined as described in the previous section.

If a collimator is not available or not practicable, observations on a known single star in both positions will suffice to determine c. If a star near the pole is taken it may be assumed that its zenith distance will have the same values for both observations in circle left and circle right positions; thus equation (7.14) will yield the following:

$$\begin{aligned} A &= R + \Delta A + b \cot z + c \operatorname{cosec} z, \\ A' &= R' + \Delta A - b \cot z - c \operatorname{cosec} z, \end{aligned} \tag{7.34}$$

from which

$$c = \tfrac{1}{2}[R' - R - (A' - A)] \sin z - b \cos z \tag{7.35}$$

where A' and A are the star's computed azimuths at the times when the readings R' and R were made in circle right and left positions (the former changed by 180°), respectively.

In conventional geodetic work, the observations are usually performed in the two positions of the instrument (i. e., [C. L.] and [C. R.]), which will cancel the residual collimation error of the central thread. For this reason generally the value of c need not be known.

7.443 Horizontal Index Correction (ΔA). The index correction may be determined from the observation of any known star in either position of the instrument by again utilizing equation (7.14) as follows:

$$\Delta A = A - (R + b \cot z + c \operatorname{cosec} z) \tag{7.36}$$

where A is the computed azimuth of the star at the time of the reading R. If observations are made on a polar star in the circle left and right positions, then from equations (7.34) the index correction is

$$\Delta A = \tfrac{1}{2}[(A + A') - (R + R')] \tag{7.37}$$

where the prime denotes quantities in the circle right position. In this case, the knowledge of b and c is not required.

7.444 Instrumental Zenith Point (Z). The instrumental zenith point Z may be determined by observing the cross threads of a collimating telescope or a distant nonmoving point and using equation (7.23), as described in section 7.42.

If Z is to be determined from circle left and circle right observations of a known star at the times T and T', the difference between the zenith distances for the interval $\Delta T = T' - T$ must be computed. If the interval is small this may be done through the use of the differential formula (3.32). The change in zenith distance Δz is

$$\Delta z = \Delta T \frac{dz}{dh} = -\Delta T \sin A \cos \Phi$$

where A is the azimuth of the star at the time $(T + T')/2$, and Φ is the latitude of the observer. The zenith point of the instrument is then computed after equation (7.23) as follows:

$$Z = \tfrac{1}{2}(V' + V) + \tfrac{1}{2}(\ell' + \ell) - \tfrac{1}{2}\Delta z \tag{7.38}$$

where the notation is the same as in section 7.42 (when V' is numerically less than V it should be increased by 360°).

7.445 Meridian (Azimuth) Setting Error (a). Observe the upper transits of two stars of different apparent declinations δ and δ' over the meridian. Let T and T' be the sidereal clock times of transit, ΔT_o be the clock correction (difference between the time shown on the clock and the apparent sidereal time) at any assumed time T_o, and δT be the hourly rate. The clock corrections at the times of observation, assuming a linear rate, then are

$$\begin{aligned} \Delta T &= \Delta T_o + \delta T(T - T_o)\,, \\ \Delta T' &= \Delta T_o + \delta T(T' - T_o)\,. \end{aligned} \tag{7.39}$$

If α and α' are the apparent right ascensions of the stars at the time of observation (corrected for diurnal aberration, if great accuracy is required), then in the case of an imperfectly adjusted instrument

$$\begin{aligned} \alpha &= T + \Delta T + \tau, \\ \alpha' &= T' + \Delta T' + \tau', \end{aligned}$$

or, using equation (7.30),

$$\begin{aligned} \alpha &= T + \Delta T + Aa + Bb + Cc\,, \\ \alpha' &= T' + \Delta T' + A'a + B'b + C'c \end{aligned} \tag{7.40}$$

where A, B, C are the respective azimuth, level and collimation factors for the star with the coordinates α and δ, and the primed factors for the star with the coordinates α' and δ'.

If the values of ΔT from equations (7.39) are substituted into (7.40) and the known quantities in the resulting equations are denoted by t and t', i.e.,

$$\begin{aligned} t &= T + \delta T(T - T_o) + Bb + Cc\,, \\ t' &= T' + \delta T(T' - T_o) + B'b + C'c\,, \end{aligned} \tag{7.41}$$

then equations (7.40) reduce to

$$\begin{aligned} \alpha &= t + \Delta T_o + Aa\,, \\ \alpha' &= t' + \Delta T_o + A'a \end{aligned} \tag{7.42}$$

where t and t' are the observed clock times reduced to the assumed epoch T_o; ΔT_o and a are unknowns.

Subtracting the two equations from each other yields

$$\alpha' - \alpha = t' - t + a(A' - A)$$

from where the azimuth setting error is

$$a = \frac{(\alpha' - t') - (\alpha - t)}{A' - A}. \tag{7.43}$$

If a is positive the telescope pointing north is too far to the west and if a is negative, too far to the east. For accurate determination select two stars, one transiting north of the zenith and the other south of the zenith within the shortest time interval possible so that t and t′ will not be affected seriously by the error in δT. Also the declination of the stars should be such that the factor $|A'-A|$ (the denominator of (7.43)) is as large as possible (in the northern hemisphere A is negative for north stars in upper culmination). This will be achieved, if both stars are observed at upper culmination, by selecting one near the pole and the other as far from it as possible. The value for any $|A|$ should always be larger than 0.5. Finally, since the right ascensions must be accurately known, only fundamental (FK4) stars should be used.

7.446 Inclinations of the Horizontal and Vertical Axes (i′ and i). In the methods of observation described above, the knowledge of the deviations i′ and i of the horizontal and vertical axes, shown in Fig. 7.38, from their normal positions is not required. It is only necessary that they should be small. Their values, however, for the purpose of checking the perfection of the instrumental adjustment, may be easily determined by means of the striding level. Let the graduated horizontal reading circle be set at any assumed reading R, and then also at $R+120°$ and $R+240°$, and let b_1, b_2, and b_3 be the inclination of the horizontal axis, determined as shown in section 7.441, in the three positions. Applying equation (7.11) in these positions

$$\left.\begin{aligned} i\cos d' + i' &= b_1 \\ i\cos(d'+120°) + i' &= b_2 \\ i\cos(d'+240°) + i' &= b_3 \end{aligned}\right\}, \tag{7.44}$$

the sum of which, since $\cos(d'+120°) + \cos(d'+240°) = -\cos d'$, gives

$$i' = \frac{1}{3}(b_1 + b_2 + b_3) \tag{7.45}$$

which determines i′.

Equation (7.45) subtracted from the first equation of (7.44) yields

$$i\cos d' = \frac{2b_1 - b_2 - b_3}{3}, \tag{7.46}$$

and the difference of the second and third equations of (7.44) gives

$$i\sin d' = \frac{b_3 - b_2}{\sqrt{3}}, \tag{7.47}$$

which determines i (and d′).

References

Barnes, G.L. and I. I. Mueller. (1966). "The Dependence of the Level-Sensitivity on the Position and Length of the Bubble on the Wild T4 Theodolite." Bulletin Géodésique, 81, pp. 277-286.

Bomford, G. (1962). Geodesy. Oxford University Press, London.

Bouchard, H. and F. H. Moffit. (1965). Surveying (fifth edition). International Textbook Company, Scranton, Pennsylvania.

Bowie, W. (1917). "Determination of Time, Longitude, Latitude, and Azimuth." U.S. Coast and Geodetic Survey Special Publication, 14, U.S. Government Printing Office, Washington, D.C.

Buchar, E. (1938). "Measures de l'equation personelle dans la méthode des hauteurs égales." Bulletin Géodésique, 60.

Carter, W. E. (1965). A Field Evaluation of the Kern DKM3-A Astronomical Theodolite for Precise Astronomic Position Determination. Thesis, The Ohio State University.

Carter, W. E. (1966). A Comparison of the Wild T4 and Kern DKM3-A Astronomical Theodolites for Precise Astronomic Longitude Determination. The 1381st Geodetic Survey Squadron, U.S. Air Force.

Chauvenet, W. (1891). A Manual of Spherical and Practical Astronomy (fifth edition). Philadelphia. Reprinted by Dover Publications, Inc., New York, 1960.

Clark, D. (1963). Plane and Geodetic Surveying (fifth edition). Constable and Company, Ltd., London.

Danjon, A. (1960). "The Impersonal Astrolabe." Stars and Stellar Systems (G. P. Kuiper and B. M. Middlehurst, eds.), I. University of Chicago Press.

de Graff-Hunter, J. (1938). "Some Experimental Observations for Longitude." Proceedings of the Royal Society, Series A, 166, 925, pp. 197-213.

Douglas, O. (1964). "Die Entwicklung des Gigas-Theodolites der Askania-Werke." Askania-Warte, 21, 64, pp. 5-8.

Drodofsky, M. (1963). "The Ni-2 Astrolabe." Zeiss Mitteilungen Ueber Fortschritte der Technischen Optik, 3, 2, pp. 79-100.

Gigas, E. (1966). Physikalisch-Geodätische Messverfahren. F. Dümmler Verlag, Bonn.

Grishin, B.S. (1962). Adjustment of Geodetic Instruments (in Russian). Geodesisdat, Moscow.

Groedel, E. (1966). "003 Theodolite, A New Jena Astronomical Geodetic Universal Instrument." Jena Review, Special Issue for the Spring Fair, Leipzig.

Hattori, T. (1951). "Latitude Observations with Floating Zenith Telescope at Mizusawa." Publications fo the International Latitude Observatory of Mizusawa, I, 1.

Hattori, T. (1953). "Latitude Observations with Floating Zenith Telescope at Mizsawa." Publications of the International Latitude Observatory of Mizusawa, I, 2.

Hemmleb, G. (1966). Beobachtungsergebnisse mit dem Astrolab des VEB Carl Zeiss Jena. Presented at the IAG Symposium on Geodetic Instruments and Measuring Techniques, Budapest, April 14-20.

Kuntz, E. and K. Schnädelbach. (1967). Simultaneous Determination of Latitude, Longitude and Azimuth with a Theodolite by Photographing the Star Passages. Presented at the XIVth General Assembly of the IUGG-IAG, Lucerne, September 25-October 7.

Kuznetsov, A. N. (1961). "Longitude Determination with the AY2/10 Universal Astronomic Instrument in which Star Passages are Photoelectrically Recorded" (in Russian). Trudy MIIGAiK, 48, pp. 149-154, Moscow.

Markowitz, W. (1954). "Photographic Determination of the Moon's Position and Applications to the Measure of Time, Rotation of the Earth, and Geodesy." The Astronomical Journal, 59, pp. 69-73.

Markowitz, W. (1960). "The Photographic Zenith Tube and the Dual Rate Moon Camera." Stars and Stellar Systems (G.P. Kuiper and and B. M. Middlehurst, eds.) I., University of Chicago Press.

Moreau, R.L. (1966). "Photoelectric Observations in Geodetic Astronomy." Canadian Surveyor, XX, 4.

Nassau, J.J. (1948). Practical Astronomy (second edition). McGraw-Hill Publishing Co., New York.

Niethammer, T. (1947). Die Genaunen Methoden der Astronomisch-Geographischen Ortsbestimmungen. Verlag Birkhäuser, Basel.

Ramsayer, K. (1966). Versuche zur Verbesserung der astronomischen Ortsbestimmung durch automatische Sternverfolgung. Presented at the IAG Symposium on Geodetic Instruments and Measuring Techniques, Budapest, April 14-20.

Roeder, R. (1966). "The Telescope System of the 'Theo 003' Universal Theodolite." Jena Review, Special Issue for the Spring Fair, Leipzig.

Schnädelbach, K. (1966). "Simultane Ortsbestimmung durch Photographie der Sternbahnen." Deutsche Geodaetische Kommission bei der Bayerischen Akademie der Wissenschaften, Reihe C: Dissertationen, 99.

Tsubokawa, I. (1954). "A Method of Photoelectric Observations of a Light Spot, and Its Application to Geodesy." Journal of the Geodetic Society of Japan, I, pp. 20-24.

Tsubokawa, I. (1957). "A Method of Photoelectric Observations of a Light Spot, and Its Application to Geodesy." Journal of the Geodetic Society of Japan, IV, pp. 6-14.

Tsubokawa, I. (1967). On Electronic Astrolabe. Presented at the XIVth General Assembly of the IUGG-IAG, Lucerne, September 25-October 7.

Yumi, S. (1967). Annual Report of the International Polar Motion Service for the Year 1965. Central Bureau of the International Polar Motion Service, Mizusawa.

8 TIMEKEEPING AND TIME DISSEMINATION

8.1 Timekeeping Instruments

Most observations in geodetic astronomy are either made to obtain time or are recorded with respect to time. The epoch of time generally is determined through astronomic observations and the interpolation between successive epochs is performed by man made timekeeping devices. These instruments thus provide the intermediate epochs and the time interval in the time scale (atomic, universal, sidereal, etc.) chosen. The geodesist, when observing celestial objects for position determinations, is chiefly concerned with the epoch of time. The uniformity of the time interval as kept by the timekeeping device is of greatest interest to the physicist, and also to the geodesist in certain specialized applications (e.g., in dynamic satellite geodesy).

A basic timekeeping instrument generally consists of the following components:

(1) An oscillator which assures the uniformity of time intervals.

(2) An auxiliary device regulated by the oscillator through a divider arrangement which displays a record of time either continuously on a clockface or chronograph, or upon command on other indicators such as a print out.

(3) A power source to provide continuous operation of the oscillator and related components.

If these three elements are present, the instrument is called a clock or (if small) a chronometer. If the auxiliary display-device is missing, the instrument is termed an oscillator or a frequency standard. Depending on the power source and the type of oscillator used, the timekeeper may be one of the following types.

Type	Power Source	Oscillator
mechanical	electricity, gravity or spring	pendulum or balance wheel
quartz	electricity	quartz crystal
atomic	electricity	atom or molecule

Regardless of the type of instrument in use, the local timekeepers must be calibrated against a recognized central standard in order to assure that the time scale is the same for the users. These central standards are based either on the earth's rotation for rotational times (as realized by repeated observatory sidereal time determinations), or, since ephemeris time is not immediately available, on the atomic clocks for (practically) uniform time.

Modern instruments used as central standards and also as high precision local timekeepers are generally atomic or quartz-crystal oscillators. Their operation as timekeepers is based on the fact that the elapsed time interval is equal to the count of the cycles of oscillations divided by the frequency. Thus the time is proportional to the reciprocal of frequency. Referring its frequency to a desired time scale (atomic, universal, etc.), such a frequency standard can serve as a basis for time measurements of that scale. In view of this relationship, it seems justified to discuss frequency standards at some length regardless of the fact that they are not considered to be primarily geodetic instruments.

A reasonable model of a good frequency standard can be approximated by the following relationship:

$$\varphi = \varphi_0 + f_0 t + \frac{A}{2} t^2 + B(t)$$

where φ is the phase in cycles, f_0 is the nominal frequency, $A/2$ is the drift-coefficient, $B(t)$ is the inherent noise function of the oscillator, and t is the time elapsed from the instant when $\varphi = \varphi_0$.

The output frequency f is then

$$f = \frac{d\varphi}{dt} = f_0 + At + \dot{B}(t),$$

and the output time is

$$t = \int \frac{1}{f} \, d\varphi .$$

Based on these expressions the frequency standards can be separated into two groups:

(1) Primary frequency standards, which have a defined nominal frequency f_0 and ideally a drift-coefficient, $A = 0$. These standards require no other reference and are used where long-term stability is of primary importance, e.g., as central (national) time or frequency standards.

(2) Secondary frequency standards, whose nominal frequency and stability are determined from measurements relative to a primary standard.

In either primary or secondary case, f_0 and A are assumed to be known; therefore, the main contribution to the uncertainty in t is the noise function B(t). Present day technology provides frequency standards or timekeepers within the following ranges of fractional standard deviation:

mechanical	1×10^{-5}	—	1×10^{-6} ,
quartz crystal	1×10^{-7}	—	2×10^{-13} ,
atomic	1×10^{-11}	—	1×10^{-14} .

Since there is no universal agreement on how to specify the performance of frequency standards, the term 'standard deviation' is used here as a representative number for precision, stability, or reproducibility. The term 'standard error' is representative of the accuracy, which in the case of a primary standard is estimated; in the case of a secondary standard it is relative to a primary standard. The precision of a clock generally is much higher than its accuracy.

The precision of a clock depends also on the time interval during which it is operated. As an example Fig. 8.1 illustrates typical stability variations of laboratory-type cesium beams, hydrogen masers, and quartz crystal oscillators [Halford, 1968]. Note, for example, that there is a region where quartz crystal oscillators are superior even to hydrogen masers (see section 8.11).

The estimated fractional standard error of the primary atomic frequency standards at present (1967) is between 1×10^{-11} and 1×10^{-12}, depending on the model. It is likely that this figure can be reduced to 2-3 parts in 10^{13} within the next few years. The accuracies of the secondary timekeepers relative to the primary frequency standards are determined from periodic comparisons. These measurements show great variations in accuracy between the different models. The geodetic criterion which distinguishes a good timekeeper from a poor one is not so much its accuracy but rather the stability of its oscillator which should provide a rate-of-change curve suitable for interpolation.

8.11 Primary Frequency Standards

An acceptable primary standard of time must satisfy the following requirements: continuity of operation, generation of an interval-unit which remains constant with respect to its adopted definition, accuracy

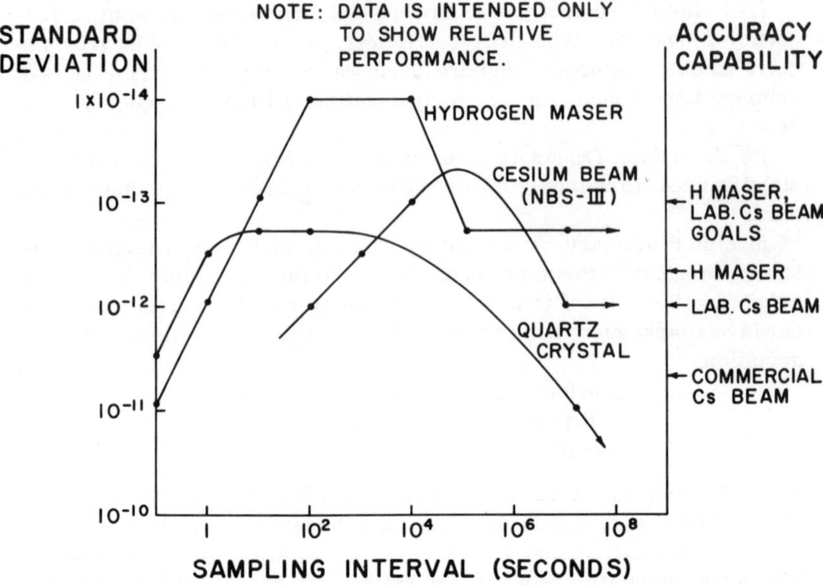

Fig. 8.1 Stability of atomic and quartz crystal oscillators

greater than or equal to that of secondary standards, capability for accumulating the interval-units to give an epoch, and accessibility to all who need it. Although astronomic standards based on the observed positions of celestial bodies meet these requirements, atomic standards have been developed since 1945 which possess a much higher accuracy and greatly improved precision, making it possible to provide better results in a much shorter averaging time. Various types of atomic standards have been developed using the following basic techniques [McCoubrey, 1966]: (a) atomic beam and magnetic reasonators; (b) gas-filled resonance cells; (c) maser oscillators.

The most widely used techniques at present are the cesium beam oscillators, both in the laboratory-type long beam and the commercial short beam versions; the hydrogen masers; and the rubidium gas cells using different buffer gases.

The <u>cesium atomic beam</u> standard was developed by Essen and Parry in 1955 at the National Physical Laboratory, Teddington [Essen and Parry, 1957]. Since then, laboratory cesium beams have been installed in about ten locations (National Bureau of Standards (NBS), Boulder, Colorado; National Research Council (NRC), Ottawa, Canada; etc.). The principle of the operation of a cesium atomic frequency standard is the following (Fig. 8.2). Neutral cesium atoms are emitted from an oven and are formed into a beam. The beam passes through two non-homogeneous magnetic fields. The first magnet 'A' sorts out those

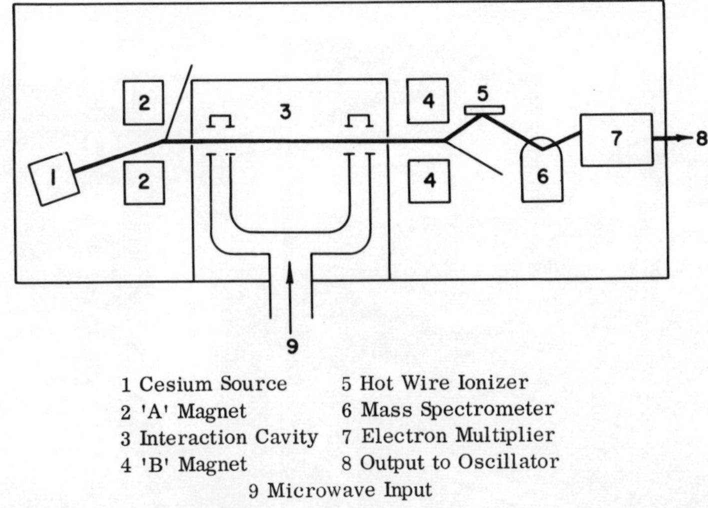

1 Cesium Source
2 'A' Magnet
3 Interaction Cavity
4 'B' Magnet
5 Hot Wire Ionizer
6 Mass Spectrometer
7 Electron Multiplier
8 Output to Oscillator
9 Microwave Input

Fig. 8.2 Schematic diagram of a cesium beam frequency standard

atoms that have antiparallel spin. Those with parallel spin, i.e., same direction of nuclei and electron spin, pass into a uniform magnetic field space (interaction cavity) and are excited by electromagnetic energy. At resonance (when the exciting microwave frequency equals the natural transition frequency of the atoms), a change of state occurs by absorption or stimulated emission, depending on the initial state of the atoms. Upon passing through the second nonhomogeneous magnetic field 'B' only those atoms that have undergone transition are focused on a hot wire ionizer where they are given a positive charge and sent back through the mass spectrometer to the electron multiplier. If the frequency of the applied electromagnetic energy is not exactly equal to the transition frequency of the cesium atoms, no transition occurs and no output signal is received from the ionizer detector. The frequency of the quartz crystal oscillator applying to the electromagnetic field is regulated until a signal from the detector is received. When this is the case, the frequency of the oscillator equals the transition frequency of the cesium atom. Due to the fact that the cesium beam works in conjunction with an external oscillator, it is called a passive atomic resonator. Fig. 8.3 is a picture of the cesium beam oscillator NBS-III in Boulder, Colorado. For further details and references, the reader is referred to [Mockler, 1964; Beehler et al., 1965].

In recent years, second generation transportable cesium standards have also been developed and used with great success both as central standards and as 'flying clocks' used in the comparisons of standards. Such instruments, for example, are manufactured by the Hewlett-Pack-

Fig. 8.3 The NBS-III primary cesium beam frequency standard at the National Bureau of Standards, Boulder, Colorado

ard Company, Palo Alto, California (Models 5060A and 5061A); Ebauches, S. A., Neuchâtel, Switzerland (Oscillatom B-5000); and others. Fig. 8.4 shows the Hewlett-Packard Model 5061A cesium beam frequency standard.

The hydrogen maser frequency standard was developed during 1960-1962 by Goldenberg, Kleppner, and Ramsey at Harvard [Goldenberg et al., 1960; Crampton et al., 1963; Kleppner, 1962; Kleppner et al., 1962]. It is an active atomic frequency standard which provides constant frequency without being coupled to an external oscillator. The frequency is derived from stimulated emission of electromagnetic energy, i.e., from the energy release associated with transitions of atoms from a higher to a lower energy level. By means of special arrangements the interaction time between hydrogen atoms in high energy states and a microwave radio frequency field is lengthened to about one second. The long interaction time stimulates the desired radiation of energy. The radiated energy is amplified by electronic devices to a useful power level. Hydrogen masers have shown an extremely high frequency stability over several months of operation [Vessot et al., 1964]. A portable hydrogen

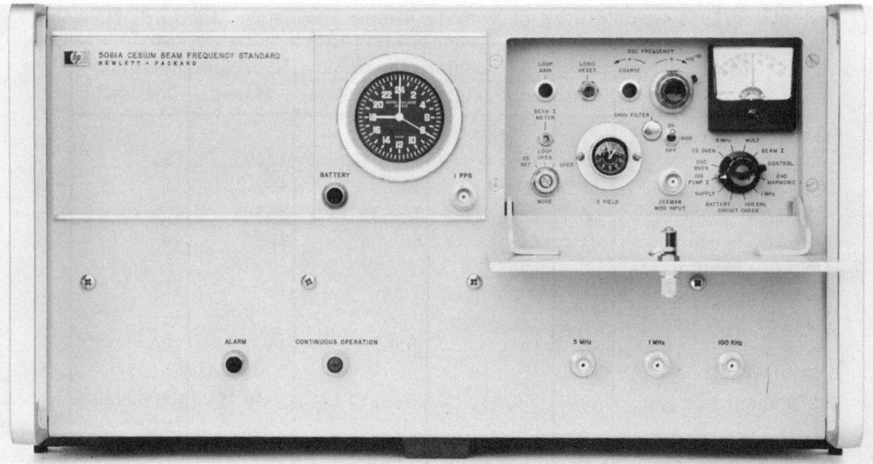

Fig. 8.4 Hewlett-Packard cesium beam frequency standard, model 5061A

maser has been developed by Varian Associates, Beverly, Massachusetts (Varian H10), which is now manufactured by Hewlett-Packard. For more details the reader is referred to [Ramsey, 1962; Vessot and Peters, 1962; and Beehler, 1967].

In the gas-filled rubidium resonance cells the resonant rubidium atoms diffuse slowly through a buffer gas during their interaction with the cavity fields. This mechanism makes it possible to realize a relatively long interaction period in a very compact apparatus. However, the diffusion process involves atomic collisions which produce a small (3×10^{-7}) collision offset in frequency, which must be established by a calibration procedure. The buffer gas filling must involve atomic particles which are nonmagnetic in order to avoid destruction of the energy states of interest during collisions. The gas-filled resonance cell has an outstanding degree of simplicity. It is also light in weight due to the lack of heavy magnets.

Table 8.1 summarizes the main characteristics of the frequency standards mentioned. It is evident that the comparison of the different types of atomic frequency standards is complicated by the many quantitative factors relating to performance and physical characteristics which differ in each case. However, on a qualitative basis it may be stated that rubidium gas cell frequency standards are, at the present time, the most compact, light weight atomic oscillators, and they have a high level of short-term stability. The long-term stability in this case exceeds that of the best quartz crystal oscillators by two or more orders of magnitude without the need for long warm-up or stabilization periods. The cost of rubidium frequency standards is less than that of other atomic standards. Cesium atomic beam frequency standards are somewhat

TABLE 8.1 Characteristics of Primary Atomic Frequency Standards (1967)

	Unit	Cesium Beam Laboratory	Cesium Beam Commercial	Hydrogen Maser	Rubidium Gas Cell
Nominal frequency (approximate)	MHz	9193	9193	1420	6834
Size	dm^3	710	40	465	17
Weight	kg	?	25 - 30	365	10 - 20
Power consumption	watt	?	60	200	40
No.of manufacturers	—	—	4[(2)]	1[(3)]	4[(4)]
No. of units produced	—	10	200	10	200
Estimated Standard Error		$1 - 2\times10^{-11}$	$3 - 6\times10^{-11}$	5×10^{-12}	depends on buffer gas
Standard Deviation - (1 sec)[(1)]		1×10^{-11}	5×10^{-11}	1×10^{-12}	6×10^{-12}
(100 sec)[(1)]		1×10^{-12}	5×10^{-12}	1×10^{-14}	6×10^{-13}
(1 day)[(1)]		1×10^{-13}	3×10^{-13}	1×10^{-13}	2×10^{-12}
Approximate cost	$	?	15,000	55,000	10,000

(1) See also Fig. 8.1

(2) Hewlett-Packard, National, Ebauches S.A., Continental Electronics (Pickard and Burns).

(3) Hewlett-Packard (Varian)

(4) Tracor, General Technology, Varian Associates, Rohde and Schwarz

larger and heavier than rubidium standards, and the short-term stability is limited. However, the long term stability is very high. The cesium standard also has a high degree of reproducibility which qualifies it for service as a primary standard without the need for calibration. The atomic hydrogen maser, at present, has the highest degree of short-term and long-term stability when compared with the other commerically produced atomic frequency standards. It is also the largest and the cost is influenced by the more advanced performance characteristics. For further details on atomic standards, see [Allan, 1966; McCoubrey, 1966 and 1967; Beehler, 1967; and Steiner, 1968].

8.12 Clocks and Chronometers

As mentioned previously, the term clock is reserved for an oscillator which is coupled with an auxiliary device to display either continuously or on command the record of time, and which is permanently installed; the term chronometer is used to designate a portable clock.

The history of clocks antedates the twentieth century by about 3500 years [Marrison, 1948, pp. 510-531]. Throughout the ages improvements in timekeeping centered around the power supply and the oscillatory device of the clocks. Due to the increasing demand for better precision, technology made mechanical (pendulum) clocks after some 400 years of use obsolete for most precise astronomic purposes. Mechanical (spring-driven) chronometers, however, are still widely used in geodetic astronomy, even for first-order position determinations, regardless of the fact that their price is not much less than that of a quartz crystal chronometer of medium quality. Since the middle of this century, observatories and geodesists in satellite work use quartz or atomic clocks and chronometers almost exclusively.

8.121 Mechanical. The first satisfactory oscillatory device was found to be the pendulum. Huygens, following an observation of Galileo, constructed the first pendulum clock in 1567. Since then, mechanical clocks employing a pendulum or a balance wheel as an oscillatory device have achieved a relatively high state of precision. The source of power for these clocks are weights or springs. Time is usually indicated through suitable mechanical gearing on a clock face, or through electric contacts on a chronograph.

Electric clocks utilize electric currents as a source of power or as a means to sustain the oscillatory motion of the pendulum. The most famous of the electric clocks is a so-called free-pendulum clock, the Shortt clock. It consists of two separate clocks which operate in synchronism. The timekeeping element is a free swinging pendulum that receives an impluse from a falling lever every half-minute. The lever is released by an electromagnetic signal from a slave clock. Upon delivery of the impulse, a synchronizing signal is transmitted back to the slave clock in order to assure that the following impulse to the free pendulum is given exactly one-half minute after the preceding impluse. The pendulum swings in a temperature- and pressure-controlled chamber.

Shortt clocks were used at several observatories, e.g., at the U. S. Naval Observatory and at the Royal Greenwich Observatory for keeping precise time until the mid-1940's, when they were replaced by quartz crystal clocks and later by atomic clocks. The precision of a Shortt clock is remarkably high, about 1×10^{-6}. Other mechanical clocks with very high precision are the Riefler and the Leroy clocks.

Mechanical clocks and chronometers, though generally compensated for changes in temperature and barometric pressure, should be used in a moderately controlled environment. It is also important that they be wound at regular intervals and moved with care. Chronometers should not be quickly horizontally rotated and should not be transported unless they have run out and their moving parts have been arrested. Fig. 8.5 illustrates a spring-driven chronometer with electric contacts by Ulisse Nardin. There are hundreds of similar chronometers on the market with prices ranging up from $500.

Fig. 8.5 Nardin mechanical sidereal chronometer with electric contacts

8.122 Quartz Crystal. The fundamental difference between pendulum clocks and quartz crystal (or atomic) clocks is that the latter's oscillatory device does not depend (practically) on gravity. Hence the rate of a quartz crystal clock or chronometer will not depend on the geographical location. Quartz crystal oscillators have come into almost universal use as reliable secondary standards of high short-term stability (see Fig. 8.1).

The heart of every quartz crystal frequency and time standard is a quartz vibrator that controls the frequency of an electronic oscillator. Two properties make quartz an ideal vibrator: its piezo-electric property and its high mechanical and chemical stability. Due to the latter, it requires only a very small amount of energy to sustain oscillation; this fact is important since the amount of disturbance of the rate of oscillation, i.e., the frequency, is proportional to the amount of this energy [Vigoureux, 1939, p. 1].

If quartz crystals are subjected to compression in certain directions relative to two crystal faces, negative electric charges are produced at the edge between those faces and positive electric charges at the opposite side of the crystal (the crystal becomes polarized). Conversely, if the crystal is placed between two electrodes of different electric potential, usually thin metallic coatings deposited on the crystal by evaporation, mechanical stresses are produced in certain directions within the crystal. The former phenomenon is called the direct piezo-electric effect, the latter is called the inverse piezo-electric effect.

If alternating electric current is applied to the electrodes, the crystal is set in mechanical vibration the frequency of which is equal to that of the applied electric field. Resonance occurs when the applied frequency coincides with the natural frequency of vibration of the crystal. In this case, the amplitude of vibration becomes considerably large and correspondingly large direct piezo-electric effects are produced, which react on the electric circuit employed for establishing the difference of electric potential. The frequency at which resonance occurs is primarily dependent on the elastic properties of quartz and the dimensions and cut of the quartz element (plates or rods) used.

It is also possible to connect the quartz element to an electronic tube in such a way that self-maintained oscillations are generated [Vigoureux and Booth, 1950, pp. 89-106]. One of the best circuits for this purpose is the so-called Pierce circuit. In this circuit the impulses generated by the piezo-electric effect are fed back to the quartz plate through the plate-grid capacitor of the tube. If the capacity of the oscillatory circuit is less than the value which would make the frequency of the oscillatory circuit equal to the natural frequency of vibration of the quartz element, the impulse is fed back in the right phase. If the damping of the quartz is small, self-maintained oscillations are produced. Usually the piezo-electric effects are amplified by electron tubes or transistors and a small amount of the amplified power is fed back to the crystal to sustain oscillation.

The resonant frequency of quartz crystals ages, i.e., tends to drift higher with time. The drift is greatest after the initial mounting and becomes more constant after a period of several years. The frequency of vibration depends also on the temperature and pressure of the ambient air; therefore, the crystal of an oscillator is thermostatically housed in a small oven.

Without going into further details it seems obvious from what has been said that the performance of a crystal oscillator depends to a large extent on the cut and mounting of the crystal, on the selection of the circuitry to sustain oscillation, on temperature and pressure control, and on the rate of aging. The theoretical limit of reliability of a quartz oscillator, provided all problems concerning mounting, etc. are solved, is given by the inherent stability of the quartz crystal itself.

From the early 1940's until about 1960, major services such as the U.S. National Bureau of Standards and the British Post Office used quartz crystal standards as the basis for precise time and frequency transmissions. The day-to-day precision of these devices was about 1×10^{-8}.

The heart of any quartz crystal clock is a secondary quartz standard as described above. However, since the frequency required for clock operation may be 1000 cycles/second (1 kHz) and the frequency of the quartz vibrator is generally of the order of 10^5 to 10^7 cycles/second (0.1-10 MHz), complex electronic frequency dividers are required to slow the frequency down to a useful level for clock driving. Thus a complete quartz crystal clock in principle comprises one or more precision crystal oscillators, a frequency divider and a time recorder in the form of either a synchronous motor driving the hands or a decade counter.

Large quartz crystal clock assemblies have been used at major observatories since about 1942 in connection with astronomic time observations. Until the advent of the atomic frequency standards, their drift rates were determined with respect to the earth's rotation, corrected for known variation in rotation speed. Quartz crystal clocks have attained a high state of reliability. Using the best available crystals, the day-to-day precision (stability) of a quartz clock based on a 2.5 MHz oscillator is about 0.2 μs (2×10^{-12}). A concise account of the development of quartz crystal clocks is given in [Marrison, 1948, pp. 523-588; Gerber and Sykes, 1966].

Portable precision timekeepers are of greatest interest to the geodesist. Portable compact quartz chronometer units or assemblies have appeared on the market in large numbers since about 1960. In fact, they have become indispensable in connection with the observations of artificial satellites, where timing accuracy is of utmost importance. Some of the commercially available crystal chronometers are listed in Table 8.2. The table is based mostly on information contained in [Robbins, 1960-1966]. Additions were made according to information provided directly by manufacturers. In all cases the listed data is based on manufacturer's description. Figs. 8.6 and 8.7 illustrate the Tracor (Sulzer A5) and the Newtek (Chronofax 102) quartz crystal clocks. The

Fig. 8.6 Tracor quartz crystal chronometer, model Sulzer-A5

TABLE 8.2 Characteristics of Secondary Quartz Crystal Frequency Standards or Chronometers (1967)

Manufacturer (description)	Advertised day-to-day precision (Temp. Range)	Dimensions in cm (weight in kg)	Power Supply M=Main B=Battery (days in power)	Output C=Clockface P=Printout H=Pulse or Wave	Price $
Ebauches, S.A. Neuchâtel, Switzerland (Model B-850)	5×10^{-7} (4°C-36°C)	8×8×13 (1)	M or B (5)	C and H	880
Ebauches, S.A. Neuchâtel, Switzerland (Model B-800)	1×10^{-9} (-20°C-60°C)	48×27×13	M or B (0.2)	C and H	2900
Littlemore Scientific Eng. Co., Oxford, G.B. (Chronocord, MK III)	1×10^{-9} (-10°C - 50°C)	36×28×25 (9)	B (0.6)	P to ms	4500
Newtek, Inc. Woodside, New York (Chronofax Model 103)	5×10^{-9}	30×33×13 (8)	B(1)	P to ms	12000; Time cor-relator, 3000; Power pack, 1000
E. Norrman Labs. Williams Bay, Wisconsin (Model 304)	1×10^{-7} (ovened)	48×30×27 (10)	M	H	950
Omega, S. A. Bienne, Switzerland (Omega Time Recorder 2)	1×10^{-7} (ovened)	41×35×17 (11)	B	P to 0.ˢ01	4200

TABLE 8.2 (continued)

Manufacturer (description)	Advertised day-to-day precision (Temp. Range)	Dimensions in cm (weight in kg)	Power Supply M=Main B=Battery (days in power)	Output C=Clockface P=Printout H=Pulse or Wave	Price $
Patek Philippe Geneva, Switzerland (Chronotome Model GP)	5×10^{-7} (4°C-36°C)	24×14×10 (3.8)	B(270)	H (also P to 0.ˢ1)	1050; Chronograph, printing to 0.ˢ1, 540
W.F. Sprengnether Instrument Co., Inc. St. Louis, Missouri (Model TS-100)	1×10^{-6} (0°C - 50° C)	33×25×11 (6.5)	M or B	C and H	900
Tracor Inc. Austin, Texas (Sulzer Models A5-5, A5-2.5)	Model A5-5 1×10^{-10} Model A5-2.5 2×10^{-11}	15×15×31 (19)	M or B (0.75)	C and H	A5-5, 4450; A5-2.5, 4950
VEB Funkwerk, Erfurt, German Democratic Republic (Model 2019)	5×10^{-8} (10°C-40°C)	53×24×35 (16)	M or B (0.1)	C and H	1830
Voumard Machines Co. Neuchâtel, Switzerland (Isatome)	1×10^{-8} (0°C - 40°C)	27×14×9 (5)	B(0.25)	H	1800

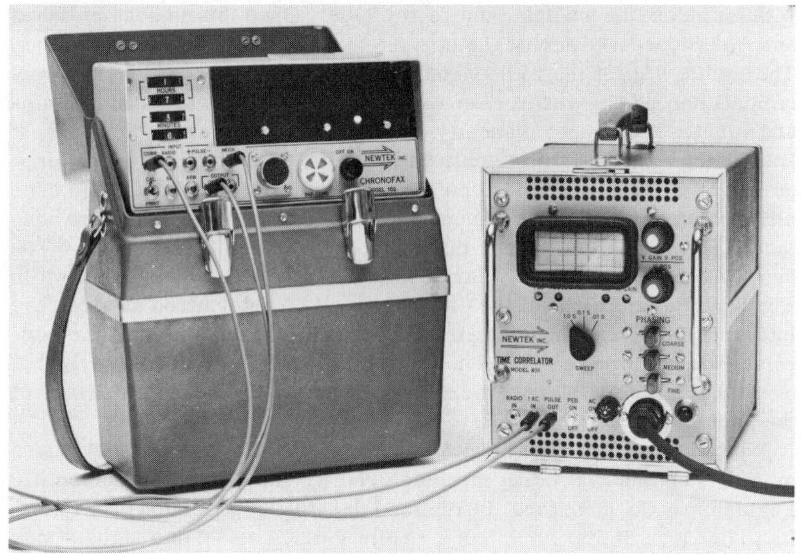

SAMPLE PRINTOUT

HOURS						MINUTES							SECONDS							0^{s}1				0^{s}01				0^{s}001			
TENS		UNITS				TENS			UNITS				TENS			UNITS				UNITS				UNITS				UNITS			
2	1	8	4	2	1	4	2	1	8	4	2	1	4	2	1	8	4	2	1	8	4	2	1	8	4	2	1	8	4	2	1
	●		●	●	●		●	●		●	●			●			●	●	●		●	●	●	●					●		●

TIME SHOWN: $17^h36^m27^s.785$

Fig. 8.7 Newtek quartz crystal chronometer, model Chronofax-102, with sample printout

former is the most accurate listed in the table. The latter is a unique (and expensive) system representing a chronometer with the most versatile geodetic applications.

The Newtek-Chronofax is a compact field chronometer that can operate twenty-four hours from internal batteries. It is usually set by synchronizing the one-per-second pulse derived from an internal 3 MHz oscillator with a received radio signal. Setting is accomplished by means of a special time correlator unit whose main features are a cathode ray tube, sweep speed adjustment, and continuous phasing control. A 1 kHz output from the Chronofax triggers the cathode ray tube sweep. The second pulses from the standard time station, received via a good receiver, are displayed on the tube as short flashes at the instant of triggering. By successive adjustment of the sweep speed and phasing control the flashes originating from the radio signal are brought into coincidence with an index line on the cathode ray tube. Once this is accomplished an arm is depressed during the next interval between two second pulses. The leading edge of the radio signal pulse sets the clock, which from now on beats seconds in synchronism with the radio signals to 0.5 ms. Hours and minutes are preset manually on the panel. Once the clock is set, it keeps step with the radio signals with a drift of less than 1ms/day for a period of a few days. The time of any event is printed on voltage sensitive paper to milliseconds upon command from, for example, the contact strip of an impersonal micrometer or radio time signals. The printed time record is in binary coded decimal. A sample is shown in Fig. 8.7. The drift rate of the Chronofax can be determined at arbitrary intervals simply by connecting it and the radio receiver to the time correlator and reading the difference between clock pulse and received pulse on the cathode ray tube, or by simply having the chronometer times of the radio signals printed out.

A similar printing crystal chronometer is the 'Chronocord' developed by A. R. Robbins and being manufactured by the Littlemore Scientific Engineering Co. at Oxford. Its claimed day-to-day precision is 1×10^{-9}. Its printout in digital form has a resolution of 1 ms. Time signals may be taken to about 1 ms without a separate unit such as the oscilloscope, and the chronometer may be set to this accuracy [Robbins, 1967].

8.123 Atomic. The term 'atomic clock' has been assigned to a system consisting of oscillators which are stabilized in frequency by one or more primary atomic frequency standards. Typical examples of atomic clocks are the A.1 and the NBS-A of the U. S. Naval Observatory in Washington, D. C. and of the National Bureau of Standards in Boulder, Colorado, respectively.

The A.1 scale is obtained by averaging the oscillation counts of 16-20 commercial cesium beam frequency standards (most of them Hewlett-Packard 5060-A). The frequencies of these resonators are compared periodically with primary standards located about the world (see section 5.6). The resonators are operated continuously. Through electronic divider circuitry, a series of highly accurate output frequencies are made available, including the one pulse per second (1 pps) output to which times in the United States are referenced. Another atomic clock in the Naval Observatory is the UTC (USNO), previously known as the 'Master-Clock.' This clock's frequency is presently (1966-68) offset by -300×10^{-10} relative to the frequency of A.1; thus the clock looses $86400000 \times 300 \times 10^{-10} = 2.592$ ms per day with respect to A.1 (A.1 is ahead of UTC (USNO)). The significance of this time scale is that the time kept is very close (within $0^s.1$) to UT2 (see section 5.42). More on this problem is explained in section 8.212. A portion of the clock system at the U. S. Naval Observatory in Washington, D. C. is shown in Fig. 8.8.

Fig. 8.8 The atomic clock system at the U.S. Naval Observatory, Washington, D.C.

The NBS-A system is obtained by averaging the oscillation counts of five portable (cesium beam, rubidium gas cell, and quartz crystal) oscillators, weighting each according to its inherent noise level. The oscillators are operated continuously and their frequencies are compared daily to the primary cesium beam frequency standard NBS-III (Fig. 8.3). The National Bureau of Standards also maintains atomic clocks keeping step adjusted (NBS(SA)) and also frequency offset (UTC(NBS), formerly

NBS-UA) atomic times, which control the transmissions of the NBS radio stations WWVB and WWV respectively (see section 8.222). The significance of these systems is the same as that of UTC(USNO) and is explained in section 8.212.

It is noteworthy that the clocks keeping the same time scale (i.e., AT or UTC) at the U.S. Naval Observatory and at the National Bureau of Standards are kept both in frequency and in epoch very close to each other and also with most major atomic clocks (systems) in the world. It has been already mentioned in section 5.61 that the difference A.1-(NBS-A) at present (end of 1967) is only a few milliseconds (10.9 ms). The difference UTC (NBS)-UTC (USNO) is even smaller (18 μs).

Portable atomic chronometers may be formed by associating suitable high-precision oscillators, frequency dividers, and time recorders (e.g., synchronous motor time displays, decade counters) with portable atomic frequency standards.

8.2 Time and Frequency Dissemination and Coordination

Since the primary standards are inaccessible to most geodesists, calibration of secondary standards or local timekeepers against the primary is often made through standard broadcasts from the national time services. (The term standard time used in the literature and adhered to here in connection with radio time broadcasts should not be confused with the time assigned to a time zone, e.g., 'Eastern Standard Time.' In the present context, standard time refers to a time kept by some central standard, such as the NBS-A.) Broadcasts of precise time and frequency referred to a certain standard usually provide frequency with or without time signals in some scale (universal or atomic).

Since geodesists at present are primarily concerned with the epoch of time, time-broadcasts are treated more extensively than frequency transmissions. The latter types are mainly for laboratory use, for worldwide stabilization of secondary frequency standards, for calibration of transmitters at radio stations, etc.

8.21 Bureau International de l'Heure

The present day radio communication network has reached the state in which time signals can be received practically anywhere on the surface of the earth from one or more radio stations. It is obvious that chaos would result if a high degree of coordination, based on international agreements, would not have been established. The coordinating agency, as it has been mentioned already (sections 4.131 and 5.4), is primarily the Bureau International de l'Heure (BIH) in Paris, France. The role of this office is four-fold:

(1) To provide predicted coordinates for the motion of the pole and seasonal variations in the rotation speed of the earth.

TABLE 8.3 Circular 'A' of the BIH (Sample)

BUREAU INTERNATIONAL DE L'HEURE

Tableau A

Corrections Δ Ts qu'il faut ajouter au temps universel pour l'affranchir des variations à courtes périodes de la rotation terrestre. Corrections données pour 0h TU, en $0^s\!,0001$.

Date 1967	J.J. 2439	Δ Ts en $0^s\!,0001$	Date 1967	J.J. 2439	Δ Ts en $0^s\!,0001$
Jan. 4	494,5	- 46	Juil. 23	694,5	+ 52
9	499,5	- 38	28	699,5	+ 15
14	504,5	- 31	Août 2	704,5	- 22
19	509,5	- 25	7	709,5	- 59
24	514,5	- 20	12	714,5	- 95
29	519,5	- 14	17	719,5	-129
Fév. 3	524,5	- 8	22	724,5	-160
8	529,5	- 1	27	729,5	-189
13	534,5	+ 6	Sept. 1	734,5	-215
18	539,5	+ 15	6	739,5	-237
23	544,5	+ 25	11	744,5	-256
28	549,5	+ 36	16	749,5	-270
Mars 5	554,5	+ 49	21	754,5	-281
10	559,5	+ 64	26	759,5	-287
15	564,5	+ 80	Oct. 1	764,5	-290
20	569,5	+ 98	6	769,5	-289
25	574,5	+116	11	774,5	-284
30	579,5	+136	16	779,5	-277
Avril 4	584,5	+157	21	784,5	-266
9	589,5	+178	26	789,5	-253
14	594,5	+198	31	794,5	-238
19	599,5	+219	Nov. 5	799,5	-222
24	604,5	+238	10	804,5	-204
29	609,5	+255	15	809,5	-186
Mai 4	614,5	+271	20	814,5	-168
9	619,5	+284	25	819,5	-150
14	624,5	+295	30	824,5	-133
19	629,5	+302	Déc. 5	829,5	-117
24	634,5	+305	10	834,5	-102
29	639,5	+304	15	839,5	- 88
Juin 3	644,5	+300	20	844,5	- 75
8	649,5	+291	25	849,5	- 64
13	654,5	+278	30	854,5	- 54
18	659,5	+261	35	859,5	- 46
23	664,5	+241			
28	669,5	+216			
Juil. 3	674,5	+188			
8	679,5	+157			
13	684,5	+124			
18	689,5	+ 89			

Note : Δ Ts a été calculé par la formule

$$\Delta Ts = +0^s\!,022 \sin 2\pi t - 0^s\!,012 \cos 2\pi t - 0^s\!,006 \sin 4\pi t + 0^s\!,007 \cos 4\pi t,$$

t en années tropiques :
pour 1967 jan. 4 à 0h TU, t = 1967,0081

TABLE 8.4 Circular 'B/C' of the BIH (Sample)

BUREAU INTERNATIONAL DE L'HEURE

CIRCULAIRE B/C NO 139 (30 OCTOBRE 1967)

COORDONNEES DU POLE INSTANTANE RAPPORTEES AU POLE MOYEN DE L'EPOQUE ET CORRECTIONS DE LONGITUDE TU1-TU0,A 0H TU.

VALEURS INTERPOLEES ET EXTRAPOLEES

DATE	J.J.	X	Y							
1967.68	2439	INTERPOLATION		AL	BA	BG	BL	BO	BS	BU
SEPT 16	749.5	0".002	-0".022	0".0011	-5	17	13	18	16	12
26	759.5	-0.005	-0.029	0.0015	-4	25	19	25	21	18
OCT. 6	769.5	-0.013	-0.035	0.0018	-2	32	25	32	26	24
		EXTRAPOLATION								
DEC. 15	839.5	-0.018	0.044	-0.0021	18	-29	-22	-31	-29	-20
25	849.5	-0.018	0.045	-0.0022	18	-30	-23	-32	-30	-20
JAN. 4	859.5	-0.017	0.046	-0.0022	18	-31	-24	-33	-31	-21

1967.68	CT	G	H	HP	IR	KH	L	LP	M	MI	MP
SEPT 16	-8	18	19	14	-5	13	21	-5	16	15	-4
26	-12	24	27	19	-1	21	32	-4	26	20	-3
OCT. 6	-16	29	33	23	3	29	42	-3	35	25	-1
DEC. 15	16	-35	-35	-26	24	-19	-32	18	-23	-26	15
25	17	-36	-36	-27	25	-19	-33	18	-24	-27	16
JAN. 4	17	-37	-37	-27	24	-20	-35	18	-25	-28	15

1967.68	MS	MZ	N	NK	NM	O	PA	PG	PR	PT	PU
SEPT 16	9	-9	15	13	1	5	17	-7	17	18	21
26	11	-10	21	19	8	2	22	-6	23	25	32
OCT. 6	11	-9	26	26	16	-2	27	-4	30	32	42
DEC. 15	-21	25	-29	-19	12	-18	-32	22	-29	-32	-32
25	-22	25	-29	-19	12	-18	-33	22	-30	-33	-33
JAN. 4	-22	25	-30	-20	11	-18	-34	22	-31	-34	-35

1967.68	RC	RG	RJ	SC	SF	TA	TO	U	VJ	W	ZI
SEPT 16	2	20	-4	-3	11	3	-8	19	17	4	-4
26	0	29	-4	-1	14	9	-8	24	25	1	-3
OCT. 6	-1	35	-4	0	16	14	-8	29	32	-2	-2
DEC. 15	-7	-33	13	14	-22	1	22	-34	-29	-14	15
25	-7	-34	13	14	-22	1	22	-35	-29	-14	16
JAN. 4	-7	-35	13	14	-22	0	22	-35	-31	-13	16

TABLE 8.5 Circular 'D' of the BIH (Sample)

BUREAU INTERNATIONAL DE L'HEURE

(B.I.H.)

61, Avenue de l'Observatoire

75 - PARIS (14ème)

Circulaire D 15

Paris, le 1 février 1968

HEURE DEFINITIVE ET COORDONNEES DU POLE A 0h TU

1. TEMPS COORDONNE TUC

Date (à 0h TU)	J.J. 2439	x	y	TU2 déf - TUC en 0ˢ0001	TU1 déf - TUC en 0ˢ0001	A3 - TUC
1967 déc 5	829,5	-0",007	+0",042	0825	0942	6ˢ1353
10	834,5	- 5	+ 40	0843	0945	1483
15	839,5	- 4	+ 39	0861	0949	1613
20	844,5	- 4	+ 38	0878	0953	1742
25	849,5	- 6	+ 37	0894	0958	1872
30	854,5	- 9	+ 37	0909	0963	2001
1968 jan 4	859,5	- 11	+ 37	0923	0969	2131
		- 4	+ 270			

x et y sont rapportés au pôle moyen de la date jusqu'au 1er janvier 1968 ; à l'Origine Conventionnelle Internationale ensuite.
A3 est l'échelle internationale de temps atomique.

TEMPS D'EMISSION DES SIGNAUX HORAIRES SUIVANT TUC, pour décembre 1967,

E = TUC - Signal en 0ˢ0001

Signal	E	Signal	E	Signal	E
CHU	+ 4	FTH42, FTK77, FTN87	+ 4	OMA, 50 kHz	+ 7
DAM	- 1	HBG	+ 1	OMA, 2,5 MHz	+ 1
DAN, DAO	+ 2	IAM	- 3	RWM (1)	- 909
DCF 77 (DHI)	+ 9	IBF	- 4	VHP	+ 9
DGI	+ 6	MSF, GBR, GPB, GIC	+ 4	VNG	+ 12
DIZ	- 3	NSS, 21,4 kHz	- 20	WWV, WWVH	+ 1
FFH	+ 5	NSS (O.C.)	+ 1	ZUO	+ 3
FTA 91	+ 10	OLB5, OLD2	+ 9		

(1) et autres émissions d'U.R.S.S.

TUC est aussi à 1 ms près environ, l'échelle de temps d'émission de :
JAS22 (JAQ56), JJY, LOL, NBA, PPE.

2. TEMPS ATOMIQUE A SAUTS TAS : A3 - TAS = + 6ˢ1546 en décembre 1967
(saut de 200 ms le 1er décembre 1967 à 0h TU).

TAS est, à moins de 1 ms près, le temps d'émission de : DCF77 (PTB), WWVB.

3. INFORMATIONS
Suivant les recommandations de l'U.A.I. et de l'U.G.G.I.(1967), les résultats du BIH sont rapportés à l'Origine Conventionnelle Internationale, à partir du 1er janvier 1968. Ceci affecte les coordonnées du pôle. Le système des longitudes du B.I.H. a été modifié de sorte que TU1 déf et TU2 déf ne subissent pas de sauts.

B. GUINOT

year [International Astronomical Union, 1964]. The amount of offset, i.e., the value of s, is announced by the BIH for a year in advance. (For the years 1966-68, the value of s, applied on January 1.0 UT, 1966 was -300×10^{-10}. During 1964 and 1965, the value of s was -150×10^{-10}.)

It follows that

$$\text{transmission (UTC) interval} = (1 - s) \text{ atomic (A) interval} \qquad (8.2)$$

thus, for example, during the years 1966-68

$$1^s\text{(UTC)} = (1 + 300 \times 10^{-10})^s\text{(A)},$$

meaning that the atomic clock daily gains $86400000 \times 300 \times 10^{-10} = 2.592$ms with respect to the clock which controls the transmission (the latter is late).

Owing to unpredictable changes in UT2, the adopted value of s may not suffice to keep the epoch of UTC in step with UT2. In order to retain a close agreement, periodic step adjustments in phase are made. The amount of adjustment and the manner in which it is applied is defined through international agreements, according to which, at present (1967), the step adjustment is exactly ±100 ms applied at 0^hUT of the first day of the month when required. The need of an adjustment is determined by the BIH upon consultation with the major time services. Adjustments are announced at least forty-five days in advance to all transmitting stations. The step adjustment in phase does not change the transmission frequency. It is equivalent to advancing or retarding the transmission clocks by 100 ms, depending on the sign of the adjustment. The total effect of the adjustments in phase and the frequency offset is that the maximum difference between UT2 and UTC is $0\overset{s}{.}1$. To illustrate the effect and importance of the step adjustments, Fig. 8.9 shows the variation of the difference A3-UTC for 1965-66, as provided in the Bulletin Horaire (UT2 and UTC are both referred to the 'Mean Observatory' described in section 2.213).

Those time services whose time signal transmissions fulfill equation (8.2) and at which step adjustments are applied so that

$$|\text{UT2} - \text{UTC(i)}| < 100 \text{ ms},$$

are called '<u>coordinated</u>' stations. All others are '<u>noncoordinated</u>.'

Since early 1967 some noncoordinated stations (WWVB, DCF77, etc.) emit time signals strictly on the atomic scale, e.g., no frequency offset is applied. In order to keep these signals near UT2, step adjustments of a multiple of 100 ms may be necessary. The common time scale for these signals is designated as 'Stepped Atomic Time'(SAT). The BIH-adopted epoch difference, referred to the 'Mean Observatory,' for January 1, 1967, is

$$\text{A3} - \text{SAT} = 5\overset{s}{.}354600.$$

This quantity naturally changes in steps of some multiple of 100 ms (usually 200 ms) whenever a step adjustment is made. During 1967, for example, step adjustments totaling 800 ms were applied, thus by January 1, 1968 the above difference was increased to $6^{s}1546$. The actual transmitted SAT(i)-s (e.g., NBS(SA) for WWVB, see section 8.222) do not differ generally from SAT by more than a few tenths of a millisecond.

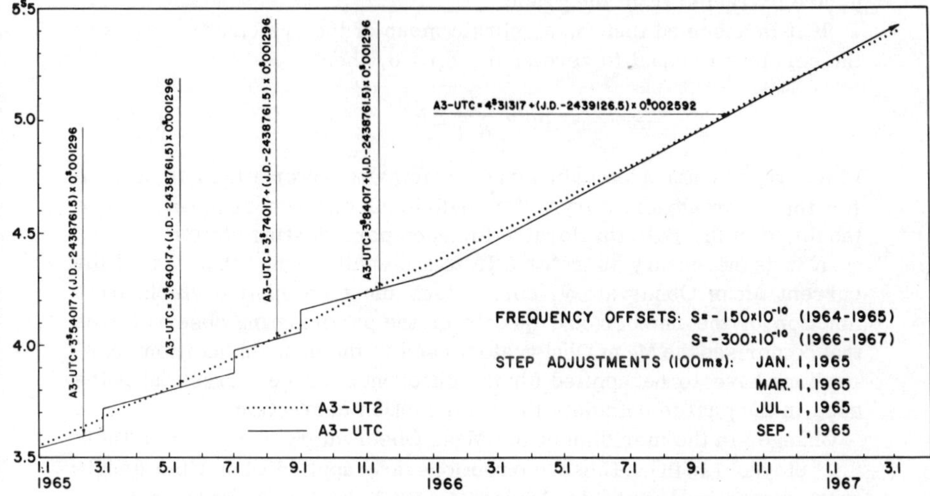

Fig. 8.9 Deviation of UTC from UT2, 1965–1966

8.213 The Mean Observatory (the Greenwich Mean Astronomic Meridian). The Greenwich Mean Astronomic Meridian or the astronomic meridian of the Mean Observatory is the reference (zero) meridian to which the final (definitive) UT2 epoch, thus all final astronomic longitudes, is referred. Its location is determined by the BIH from the observations at over forty time services listed in table 3 of the Bulletin Horaire [Stoyko, 1964a].

In principle, the calculation of the definitive UT2 of the Mean Observatory, or the corrections UT2-UT2(i) or UT2-UTC(i), is as follows.

Let 1, 2, 3, . . . , n be a group of time services that receive a signal emitted from one station. Then

$$h_1 - t_1 + r_1 + p_1 = h_2 - t_2 + r_2 + p_2 = \ldots = h_n - t_n + r_n + p_n = \text{UTC} \quad (8.3)$$

where h_i is the time of reception of the signal, t_i is the propagation delay (see sections 8.231 and 8.232), r_i is the accidental error, p_i is the systematic error (error in conventional longitude, clock error, etc.) at the time service i.

If one specific observatory is chosen as reference, then

$$h_i - h_1 - (t_i - t_1) = (p_1 - p_i) + (r_1 - r_i) \qquad (8.4)$$

where $i = 2, 3, \ldots, n$, and 1 refers to the reference observatory. The left side of equation (8.4) is known and can be represented graphically by a smoothed $(r_1 - r_i = 0)$ curve. For a particular epoch, values of $R_i = p_1 - p_i$ are extracted from the graph.

If it is assumed that the algebraic mean of the systematic errors of the services is equal to zero, i.e., $\Sigma p_i = 0$, then

$$p_1 = \frac{1}{n-1} \Sigma R_i \qquad (8.5)$$

which represents a certain mean systematic error which is adopted for the Mean Observatory. The individual differences $p_1 - R_i = p_i$ are tabulated in the Bulletin Horaire for each participating station.

If it is necessary to refer UT2 to a meridian other than that of the current Mean Observatory, corrections must be applied which are a function of the number and quality of the participating observatories that comprised the Mean Observatory used in the past. In addition, corrections have to be applied for the different average terrestrial poles used in the past to calculate the polar motion correction.

Changes in the meridian of the Mean Observatory are given in Table 8.6 [Stoyko, 1964b]. These corrections are applied when time (longitude) determined prior to January 1, 1962, has to be reduced to the system used since that date. Time corrections due to the changes in the average terrestrial pole of the epoch are given in Table 8.7.

TABLE 8.6 Corrections to Universal Time Due to the Changes in the Meridian of the Mean Observatory, 1931–1958

Year	$\Delta\Lambda_0$	Year	$\Delta\Lambda_0$	Year	$\Delta\Lambda_0$	Year	$\Delta\Lambda_0$
1931	$-0^s.0030$	1938	$-0^s.0022$	1945	$-0^s.0010$	1952	$-0^s.0049$
1932	-0.0029	1939	-0.0025	1946	-0.0035	1953	-0.0047
1933	-0.0016	1940	-0.0034	1947	-0.0047	1954	-0.0038
1934	-0.0024	1941	-0.0017	1948	-0.0045	1955	-0.0052
1935	-0.0023	1942	-0.0018	1949	-0.0045	1956	-0.0062
1936	-0.0023	1943	-0.0017	1950	-0.0040	1957	-0.0072
1937	-0.0025	1944	-0.0011	1951	-0.0050	1958	-0.0049

For the period 1931 to 1955 inclusive, the final UT2 time of emission in the present (1967) system is given by

$$\mathrm{UT2} = \mathrm{UT2(old\ tabulation)} + \Delta p + \Delta\Lambda_0 + \Delta\Lambda_s \qquad (8.6)$$

where Δp is taken from Table 8.7, $\Delta\Lambda_0$ is taken from Table 8.6 and $\Delta\Lambda_s$ is the seasonal variation applicable to the epoch in question (see section 5.43).

For the period 1956 to 1961 inclusive, the corresponding formula is

$$UT2 = UT2 \text{ (old tabulation)} + \Delta p \qquad (8.7)$$

where Δp is taken from Table 8.7.

TABLE 8.7 Corrections to Universal Time Due to the Changes in the Average Terrestrial Pole, 1931–1963

Year	Δp	Year	Δp	Year	Δp	Year	Δp
1931	$0^s.0025$	1940	$0^s.0037$	1948	$0^s.0067$	1956	$0^s.0017$
1932	0.0030	1941	0.0017	1949	0.0037	1957	0.0025
1933	0.0093	1942	0.0005	1950	0.0058	1958	0.0002
1934	0.0085	1943	0.0004	1951	0.0015	1959	0.0044
1935	0.0090	1944	0.0002	1952	0.0034	1960	0.0044
1936	0.0097	1945	0.0021	1953	0.0041	1961	0.0044
1937	0.0106	1946	0.0028	1954	0.0001	1962	0.0000
1938	0.0109	1947	0.0028	1955	0.0006	1963	0.0000
1939	0.0118						

For the period 1962 to 1967 inclusive, the values given in the Bulletin Horaire, applicable to the epoch in question, are used without correction (see section 8.215).

8.214 The Bulletin Horaire. The official publication of the BIH, the Bulletin Horaire (BH) contains all the information required for reduction of arbitrary epochs to UT0, UT1, UT2, or AT. In 1965, the BH was radically changed, the first issue in the new form being Series J, No. 1. The new series covers periods after January 5, 1964, and is published once every two months.

Prior to the change two series of the BH were in existence: a lettered series (e.g., Series H, No. 6) which contained corrections to time signals to arrive at the final (definitive) UT2, and a numbered series (e.g., Series 6, No. 10), which gave times of reception of signals from coordinated (and some noncoordinated) stations at the Paris Observatory. In addition, important information, e.g., coordinates of the pole, seasonal variation, definitions, were contained in one or the other of the series. The last issues of the old series are those mentioned as examples above.

The new form of the BH combines both the old lettered and numbered series. Sample pages are included in Tables 8.8 through 8.11 to facilitate understanding. The tables in the BH contain the following information for a period of two months, about one year in arrears of the date of publication:

(1) UT1 - UT0(i) for dates at ten-day intervals for each participating observatory (Table 8.8). The final coordinates of the true celestial pole are also given for a period of about sixty days, referred to the average terrestrial pole of the epoch (see section 4.131). The table is identical

TABLE 8.8 Bulletin Horaire: Table 1, Coordinates of the Pole and Time Corrections UT1-UT0 (Sample)

Nov. - Déc. 1967 2

1. TU1 - TU0 à 0h TU, en $0^{s}{,}0001$

Date 1967	J.J. 2439	x	y	Al	BA	BG	Bl	Bo	Bs	Bu
Nov. 5	799,5	-0″,022	+0″,012	$-0^{s}{,}0005$	+11	-2	-1	-3	-6	00
15	809,5	-18	+30	-14	+14	-18	-13	-19	-19	-11
25	819,5	-13	+40	-20	+15	-27	-21	-29	-27	-19
Déc. 5	829,5	-7	+42	-21	+13	-31	-23	-32	-29	-22
15	839,5	-4	+39	-19	+11	-29	-22	-30	-27	-21
25	849,5	-6	+37	-18	+11	-27	-21	-28	-25	-19
35	859,5	-11	+37	-18	+13	-26	-19	-27	-25	-18

1967	CT	G	H	HP	Ir	Kh	L	LP	M	Mi	MP
Nov. 5	+2	-9	-6	-5	+21	+3	+1	+12	+4	-5	+10
15	+10	-23	-23	-17	+21	-10	-18	+15	-12	-17	+13
25	+15	-32	-33	-24	+19	-19	-31	+15	-23	-24	+13
Déc. 5	+17	-33	-35	-25	+15	-23	-37	+13	-28	-26	+11
15	+16	-31	-33	-24	+12	-22	-35	+11	-27	-25	+9
25	+15	-29	-31	-22	+13	-20	-32	+12	-25	-23	+10
35	+14	-29	-30	-22	+17	-18	-29	+13	-22	-23	+11

1967	MS	Mz	N	Nk	Nm	O	Pa	PG	Pr	Pt	Pu
Nov. 5	-9	+13	-6	+1	+19	-15	-7	+13	-4	-5	+1
15	-16	+19	-19	-10	+13	-16	-21	+17	-19	-21	-18
25	-18	+21	-26	-18	+8	-14	-29	+18	-27	-30	-31
Déc. 5	-18	+20	-28	-22	+2	-11	-31	+16	-30	-33	-36
15	-16	+18	-26	-21	0	-8	-29	+14	-28	-31	-35
25	-15	+18	-25	-19	+1	-9	-27	+14	-26	-29	-32
35	-17	+19	-24	-17	+6	-12	-27	+17	-25	-28	-29

1967	Q	Rc	Rg	RJ	Sc	SF	Ta	To	U	VJ	W
Nov. 5	0	-7	-1	+7	+11	-6	+10	+11	-7	-2	-12
15	0	-6	-20	+10	+12	-15	+4	+17	-22	-17	-12
25	0	-5	-31	+11	+11	-19	0	+19	-31	-27	-11
Déc. 5	0	-3	-35	+10	+9	-20	-4	17	-33	-30	-8
15	0	-2	-34	+9	+7	-18	-5	+15	-31	-29	-6
25	0	-3	-31	+9	+8	-17	-3	+15	-29	-27	-7
35	0	-4	-29	+10	+10	-18	-1	+17	-28	-25	-9

1967	Zi
Nov. 5	+10
15	+13
25	+13
Déc. 5	+11
15	+10
25	+10
35	+12

2. TU2 - TU1

Voir Bulletin Horaire J19.

TABLE 8.9 Bulletin Horaire: Table 3, List of Participating Stations (Sample)

3 Nov. - Déc. 1967

3. SERVICES HORAIRES participant à la formation de l'heure définitive et résidus (obs.-calc.).

Observatoire	Inst.	Abr.	Longitude	Latitude	obs.-calc. en $0^s,0001$ 1967-68 90	95	00
ALGER (BOUZAREAH)	1	Al	$-0^h 12^m 8^s,463$	+36°48,1	+138	+ 90	+115
BELGRADE	2	Bl	-1 22 3,233	+44 48,2	-285	-233	-108
BESANCON	1	Bs	-0 23 57,025	+47 14,9	- 62	- 21	+ 62
BOROWA GORA	3	BG	-1 24 8,913	+52 28,6	+254	+245	+239
BOROWIEC	2	Bo	-1 8 18,437	+52 16,6	+425	+446	+329
BRATISLAVA	2	Br	-1 8 28,765	+48 10	-105	- 83	- 29
BUCAREST	2	Bu	-1 44 23,115	+44 24,8	+524	+406	+451
BUENOS-AIRES Géog.	2	BAg	+3 54 4,471	-34 34,4	- 19	-264	-330
BUENOS-AIRES Nav.	2	BAn	+3 53 25,194	-34 37,3	-109	-164	-200
GREENWICH	4	G	-0 1 21,102	+50 52,3	- 71	-110	- 79
HAMBOURG Hydrogr.	4	H	-0 40 3,679	+53 35,8	-183	-123	- 61
HAUTE PROVENCE	1	HP	-0 22 52.009	+43 55,9	-112	- 71	- 77
IRKOUTSK Astr.	3	Ira	-6 57 22,748	+52 16,7	+103	+209	+196
IRKOUTSK Mes. (2 inst.)	1	Irma	-6 57 11,843	+52 16,4	+408	+469	+446
IRKOUTSK Mes.	3	Irmf	-6 57 11,843	+52 16,4	+233	+249	+336
LA PLATA	2	LP	+3 51 43,639	-34 54,5	- 79	-114	-140
LE CAP	1	CT	-1 13 54,671	-33 56,1	- 31	- 75	- 33
LENINGRAD Astr.	3	La	-2 1 10,800	+59 56,5	+ 90	+140	+144
LENINGRAD Mes.	2	Lmi	-22 1 15,930	+59 55,1	-260	-390	-346
LENINGRAD Mes.	3	Lmf	-2 1 15,930	+59 55,1	-220	-200	-176
MILAN	2	Mi	-0 36 45,831	+45 28,0	+187	+269	+342
MIZUSAWA	1	Mza	-9 24 31,602	+39 8,1	-129	-153	-117
MIZUSAWA	4	Mzp	-9 24 31,443	+39 8,1	- 29	- 73	- 7
MONT STROMLO	4	MS	-9 56 1,406	-35 19,3	+ 72	+ 75	-
MOSCOU Astr.	3	Ma	-2 30 10,695	+55 42,0	+ 59	+ 21	+ 17
MOSCOU Mes.	1	Mma	-2 28 55,597	+55 58,7	-221	-219	-204
MOSCOU Mes.	3	Mmf	-2 28 55,597	+55 58,7	+139	+ 71	- 4
NEUCHATEL	4	N	-0 27 49,779	+46 59,8	- 3	+ 89	+ 22
NIKOLAIEV	3	Nk	-2 7 53,817	+46 58,3	+222	+124	+ 60
NOVOSSIBIRSK (2 inst.)	1	Nma	-5 31 38,193	+55 2	+ 2	- 13	+ 1
NOVOSSIBIRSK (2 instr.)	2	Nmi	-5 31 38,193	+55 2	+ 42	+ 77	+111
OTTAWA	4	O	+5 2 51,940	+45 23,6	- 1	- 28	- 25
PARIS	1	Pa	-0 9 20,921	+48 50,2	+ 58	+ 10	+ 2
PECNY	2	Pyi	-0 59 9,363	+49 54,9	+ 6	- 93	-100
PECNY	5	Pyc	-0 59 9,363	+49 54,9	-174	- 73	:0
POTSDAM	1	Pta	-0 52 16,069	+52 22,9	+ 56	- 13	- 71
POTSDAM n° 2	2	Pti	-0 52 16,069	+52 22,9	-	-	-
POTSDAM n° 3	2	Ptj	-0 52 16,069	+52 22,9	+ 56	+ 67	+ 59
POULKOVO	3	Pu	-2 1 18,572	+59 46,3	+ 40	+ 31	+ 24
PRAGUE	2	Pri	-0 57 34,886	+50 4,6	+266	+287	+330
PRAGUE	5	Prc	-0 57 40,808	+50 4,6	+176	+217	+260
QUITO	1	Q	+5 13 59,734	-00 14,6	-112	-128	-149
RICHMOND	1	Rca	+5 21 31,719	+25 36,8	+353	+336	+383
RICHMOND	4	Rcp	+5 21 31,718	+25 36,8	-197	- 64	-117
RIGA	3	Rg	-1 36 27,716	+56 57,1	+ 12	+ 43	+ 56
RIO-DE-JANEIRO	2	RJ	+2 52 53,467	-22 53,7	+ 84	+ 69	+ 72
SAN-FERNANDO	2	SF	+0 24 49,241	+36 27,7	-121	-139	-
SANTIAGO-DU-CHILI	1	SC	+4 42 11,855	-33 23,8	- 50	- 84	- 61
SAO PAULO	1	SP	+3 6 29,440	-23 39,2	+224	+189	+171
TACHKENT	2	Tai	-4 37 10,488	+41 19,5	+ 39	-137	-196
TACHKENT	3	Taf	-4 37 10,488	+41 19,5	+219	+263	+244
TOKYO (Mitaka)	4	To	-9 18 9,930	+35 40,3	- 18	- 42	-107
UCCLE	1	U	-0 17 25,937	+50 47,9	+388	+379	+391
WASHINGTON	4	W	+5 8 15,729	+38 55,3	- 63	0	+ 24

1 : astrolabe, 2 : instr. des passages, 3 : instr. des passages photoélectrique
4 : PZT, 5 : circumzénithal A = -0,011 B = -0,033

TABLE 8.10 Bulletin Horaire: Table 6, Atomic Time (Sample)

5 Nov. - Déc. 1967

HEURE DEFINITIVE

6. TEMPS ATOMIQUE

Date (0h TU)	J.J. 2439	TU2 déf - A3	TU1 déf - A3
1967 nov. 5	799,5	$-5^s,9816$	$-5^s,9594$
10	804,5	9937	9733
15	809,5	-6,0060	9874
20	814,5	0183	-6,0015
25	819,5	0302	0152
30	824,5	0417	0284
déc. 5	829,5	0528	0411
10	834,5	0640	0538
15	839,5	0752	0664
20	844,5	0864	0789
25	849,5	0978	0914
30	854,5	1092	1038
1968 janv.4	859,5	1208	1162

Ecarts des temps atomiques individuels.

Etalon i		A3-TAi en $0^s,0001$ nov.-déc. 1967	Origine
Boulder	Bld	- 5	1 janv. 1961
Teddington	ET	+ 26	"
Tokyo	To	- 1	1 nov. 1964
Ottawa	NRC	0	1 mars 1965
Stockholm (FOA)	Sto	- 1	"
Washington (Obs.Nav.)	WNO	- 2	1 juin 1965
Paris	Pa	+ 7	1 fév. 1966
Johannesbourg	J	+ 2	1 déc. 1966
Braunschweig (PTB)*	Bk	0	1 nov. 1967
Mount Stromlo**	MS	0	"

* Etalon à césium HP 5060A, n° 274 ; Physikalisch-Technische Bundesanstalt Braunschweig, Allemagne, RF.

** Etalon à césium HP 5060, N° 205 ; Mount Stromlo Observatory, Canberra, ACT, Australie.

TABLE 8.11 Bulletin Horaire: Table 7, Coordinated Time UTC (Sample)

Nov. - Déc. 1967 6

HEURE DEFINITIVE

7. TEMPS COORDONNE TUC.

Heure définitive de l'émission des signaux horaires coordonnés dans le système TUC.

Date (0h TU)	J.J. 2439	TU2 déf - TUC en $0^s_,0001$	TU1 déf - TUC en $0^s_,0001$	A3 - TUC
1967 nov. 5	799,5	0760	0982	$6^s_,0576$
10	804,5	0768	0972	0705
15	809,5	0775	0961	0835
20	814,5	0782	0950	0965
25	819,5	0792	0942	1094
30	824,5	0807	0940	1224
déc. 5	829,5	0825	0942	1353
10	834,5	0843	0945	1483
15	839,5	0861	0949	1613
20	844,5	0878	0953	1742
25	849,5	0894	0958	1872
30	854,5	0909	0963	2001
1968 jan.4	859,5	0923	0969	2131

Ecarts individuels E : E = TUC - Signal émis.

Signal	fréquence en kHz	E en $0^s_,0001$ nov.	déc.	Signal	fréquence en kHz	E en $0^s_,0001$ nov.	déc.
CHU	toutes fr.	+ 4	+ 4	MSF(3)	toutes fr.	+ 4	+ 4
DAM	toutes fr.	- 1	- 1	NBA	24	(4)	(4)
DAN (1)	2614	+ 2	+ 2	NBA	autres fr.	+ 15	+ 28
DCF77(DHI)	77,5	+ 8	+ 9	NSS	21,4	- 28	- 20
DGI	185	+ 5	+ 6	NSS	autres fr.	+ 1	+ 1
DIZ	4525	- 5	- 3	OLB5	3170	+ 7	+ 9
FFH	2500	+ 6	+ 5	OMA	50	+ 8	+ 7
FTA91	91,15	+ 9	+10	OMA	2500	0	+ 1
FTH42(2)	7428	+ 4	+ 4	PPE	toutes fr.	(4)	(4)
HBG	75	0	+ 1	RWM(5)	toutes fr.	-908	-909
IAM	5000	- 6	- 3	VHP	toutes fr.	+ 9	+ 9
IBF	5000	- 7	- 4	VNG	toutes fr.	+ 11	+ 12
JAQ56	16170	-12	-12	WWV	toutes fr.	+ 1	+ 1
JJY	toutes fr.	- 8	-10	WWVH	toutes fr.	+ 1	+ 1
LOL	toutes fr.	+ 2	+ 1	ZUO	toutes fr.	+ 4	+ 3

(1) ainsi que DAO (2775 kHz)

(2) et FTK77, FTN87.

(3) et GBR, GPB30 B, GIC27, GIC33, GIC37.

(4) pas d'information disponible.

(5) et autres signaux d'URSS.

to the interpolated portion of the BIH Circular 'B/C' for the same time period (see Table 8.4).

(2) UT2 - UT1 = $\Delta\Lambda_s$ for the tabulated dates as calculated from equation (5.47). It appears only in the first issue of each year and is identical to the BIH Circular 'A' (see Table 8.3).

(3) List of the participating observatories with their coordinates. In addition it gives for each one-twentieth of the year the residual 'observation-computation' (i.e., UT0(i) - UT0, the latter representing the value used by the BIH to determine UT1). A sample page is shown in Table 8.9.

(4) Information on transmission stations, e.g., changes in schedules, frequencies, new stations. The first issue of the year lists all coordinated stations and some important noncoordinated stations.

(5) Lists of frequency offsets and step adjustments together with small adjustments for the improvement of time signal coordination.

(6) UT2 - A3 and UT1 - A3 at 0^hUT for the tabulated dates (Table 8.10). A3 is the atomic time adopted by the BIH (see section 5.6). Also given are the monthly mean systematic deviations A3 - AT(i), where AT(i) denotes other individual atomic time scales.

(7) UT2 - UTC, UT1 - UTC, and A3 - UTC all referred to the Mean Observatory for the tabulated dates at five-day intervals. Also listed are the monthly mean differences E = UTC - UTC(i) (Table 8.11). (It should be noted that UTC as used in the BH refers to a fictitious signal emission time of the Mean Observatory and that 'signal emis' stands for UTC of a specific station, i.e., UTC(i).)

(8) UT2 - SAT, UT1 - SAT, A3 - SAT for the tabulated dates at five-day intervals, are referred to the Mean Observatory. Also listed are the monthly mean differences E = SAT - SAT(i), where SAT(i) is the equivalent of UTC(i) for stations transmitting stepped atomic times (e.g., NBS(SA), see sections 8.212 and 8.222).

(9) UT2 - UTC(i) for coordinated stations excluded from table 7 and for certain noncoordinated stations, at the time indicated at the head of the respective columns.

In order to speed up the availability of UT2 and UT1, certain quantities contained in tables 7 and 8 (UT2 - UTC, UT1 - UTC, A3 - UTC, UTC - UTC(i) and SAT - SAT(i)) are also distributed by the BIH only one month in arrears in its Circular 'D' (Table 8.5). Since the quantity UTC generally differs from the signals (UTC(i)) of the better coordinated stations only by a maximum of a few milliseconds, this publication allows the determination of the final UT2 or UT1 to that accuracy. In this publication, as in the BH tables 7-9, the corrections are given in units of 0.1 ms without signs, thus they should be used as explained at the end of section 8.211.

8.215 Recent Changes in the Procedures of the BIH. Commencing on January 1, 1968, the following changes were introduced in the procedures of the BIH [Guinot, 1967]:

(1) The reference pole was changed from that of the 'epoch' to the average terrestrial pole of 1900-05 described in section 4.131 and was designated as the 'Conventional International Origin' (CIO). The differences between the coordinates of the true celestial (instantaneous) pole referred to the CIO and to the various BIH poles in use during the years 1955-1967, are listed in Table 8.12 and are illustrated in Fig. 8.10.

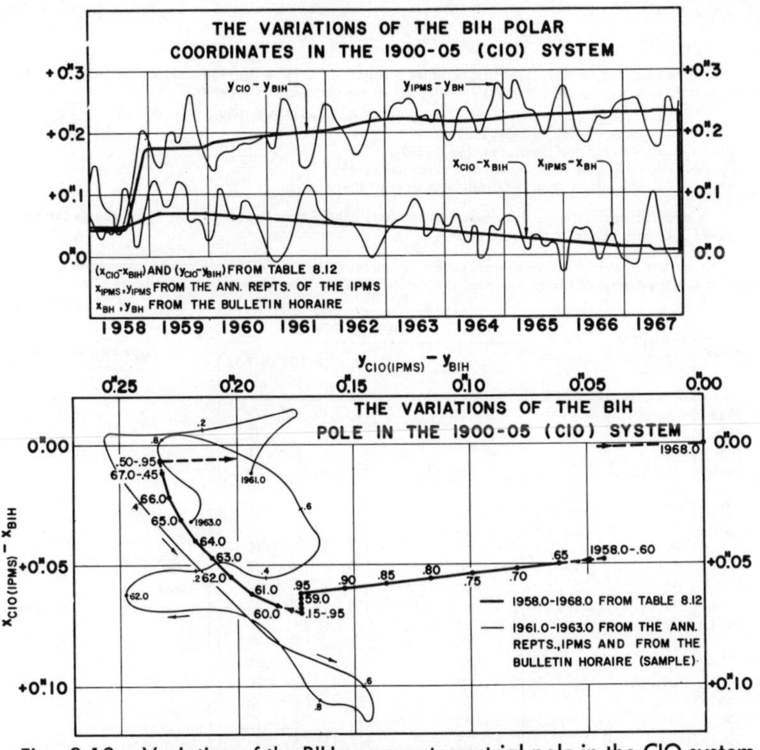

Fig. 8.10 Variation of the BIH average terrestrial pole in the CIO system, 1958–1967

In the figure the differences between the coordinates of the true celestial pole as actually published by the BIH (referred to its various reference poles) and by the International Polar Motion Service (referred to the CIO) are drawn with thin lines; the values of Table 8.12 are shown in heavy lines.

TABLE 8.12 Coordinates of the Average Terrestrial Pole Used by the BIH in the System of the Conventional International Origin, and the Resulting Corrections to Universal Time, 1955-1967

BUREAU INTERNATIONAL DE L'HEURE

CORRECTIONS A AJOUTER AUX RESULTATS PUBLIES PAR LE BIH
pour les rapporter au système de 1968

Depuis 1968, 1er janvier, le BIH rapporte les résultats à l'Origine Conventionnelle Internationale. Les résultats antérieurs, depuis 1955, avaient été affectés par les changements suivants :

- changements d'origine pour les coordonnées du pôle (1959 et 1968),
- passage du FK3 au FK4 (1962),
- corrections des longitudes (1962, 1968),
- correction de la constante de l'aberration (1968),
- modifications de la constitution de l'observatoire moyen.

On donne ici les corrections à ajouter aux résultats publiés pour la période 1955-1967, afin de les rendre homogènes avec ceux publiés à partir de 1968.

Etant donné la petitesse de ces corrections, elles peuvent être considérées comme invariables durant chaque année, sauf pendant l'intervalle août 1958-février 1959, où elles devront être interpolées.

Année	Centièmes d'année	Date commune	Corrections à ajouter			Explications
			$x_{OCI} - y_{BIH}$	$y_{OCI} - y_{BIH}$	à TU1 ou TU2 publiés	
1955			+0",048	+0",043	+0ˢ0025	
1956			"	"	+ 24	
1957			"	"	+ 24	
1958	00 à 60		"	"	+ 24	
	65	août 26,3	+ 50	+ 62	+ 15	Passage du pôle de Cecchini au pôle moyen de la date.
	70	sept.13,5	+ 52	+ 80	+ 07	
	75	oct. 1,8	+ 54	+ 99	- 02	
	80	20,1	+ 56	+ 117	- 10	
	85	nov. 7,3	+ 58	+ 136	- 18	
	90	25,6	+ 60	+ 154	- 27	
	95	déc. 13,9	+ 62	+ 173	- 35	
1959	00	jan 1,1	+ 64	"	- 27	
	05	19,4	+ 66	"	- 28	
	10	fév 6,6	+ 68	"	- 28	
	15 à 95		+ 70	"	- 28	
1960			+ 67	+ 183	- 32	Adoption du FK4 et de nouvelles longitudes.
1961			+ 62	+ 194	- 37	
1962			+ 55	+ 203	+ 04	
1963			+ 47	+ 211	+ 03	
1964			+ 40	+ 218	+ 04	
1965			+ 31	+ 224	+ 05	
1966			+ 22	+ 229	+ 34	Adoption de l'Origine Conventionnelle Internationale, de nouvelles longitudes, d'une nouvelle valeur de la constante de l'aberration.
1967	00 à 45		+ 12	+ 232	+ 20	
	50 à 95		+ 7	+ 233	+ 20 *	
1968			0	0	0	

* Les longitudes utilisées par le BIH ont été modifiées de sorte que le saut de TU1 et TU2 au début de 1968 est entièrement dû au changement de valeur de la constante de l'aberration.

B. GUINOT
1 mars 1968

The BIH reference poles were defined as follows:

— 1955.0 - 1958.60: the average terrestrial pole of 1949-58 as derived by Cecchini,

— 1958.65 - 1959.15: smooth transition from Cecchini's 1949-58 pole to the 'average terrestrial pole of the epoch (see section 4.131),

— 1959.20 - 1967.5: the average terrestrial pole of the epoch.

Since from January 1, 1968, the BIH will use the CIO as its reference pole, there will be a discontinuity in the coordinates of the instantaneous pole of the amounts

$$x_{CIO} - x_{BIH} = 0''.007,$$
$$y_{CIO} - y_{BIH} = 0''.233.$$

If the continuity of the quantity $\Delta\Lambda_P = UT1 - UT0$ computed from equation (4.40), thus the continuity of UT1, is going to be perserved then it will be necessary to change the adopted longitudes of the time services or the longitude of the Mean Observatory. Another possibility is to step adjust UT1 appropriately and keep the adopted longitudes. In any case, the necessary revisions due to this transition of the reference pole must take place in the years following 1968.

(2) The measurements of time and latitude are to be used simultaneously for computing the coordinates of the pole (in the past only latitude determinations were utilized).

(3) The computations are to be carried out every fifth day using unsmoothed data directly from the observations. The tabulated values are to be slightly smoothed.

(4) Circulars 'A,' 'B/C,' and 'D' are to be essentially unchanged. Circulars 'B/C' and 'D' are to appear about one month in arrears.

(5) The Bulletin Horaire will be discontinued after the J24 issue (corrections for November-December, 1967). In a sense it is to be replaced by the monthly Circular 'D' and by the new Annual Reports of the BIH which will contain the following information:

(a) the description of computational methods,

(b) the contents already published in Circulars 'A' and 'D,'

(c) the differences UTC - UTC(i) for observatories not covered in Circular 'D,'

(d) the residuals 'observations-computation' in time (e.g., UT0(i) - UT0) and in latitude for the participating observatories,

(e) the differences A3 - AT(i),

(f) the characteristics of the main time services,

(g) other information regarding observatories, timekeeping, and distribution.

The annual reports are to be published six months in arrears.

(6) The atomic time A3 used by the BIH will be replaced by the International Atomic Time Scale (IATS).

8.22 Time Services

The purpose of the time services is twofold:

(1) To establish local standards and methods for measuring time and frequency and study their properties and limitations.

(2) To define, establish, and operate widespread time scales and services for synchronization and timekeeping purposes.

Means to accomplish the first goal have been discussed in some detail in sections 8.11 and 8.123. This section deals essentially with the problems associated with the second goal.

8.221 Radio Time Signals. The major types of radio time signals in use are the following: (1) English system, (2) modified rhythmic system, (3) international ONOGO system, (4) technical broadcast system. These systems are briefly described below. Additional information on these and on other systems (USSR, Japanese-NDB, etc.) may be found in [Hydrographic Office, 1962-1964].

In the English system, time signals are radiated for five minutes preceding the hour, i.e., from 55^{m} to 60^{m}. Each second is marked by a $0^{s}.1$ tick. At the minute, the tick is lengthened to $0^{s}.6$. The commencement of each tick is the reference point. The last five seconds of each minute thus are graphically represented by

55 56 57 58 59 60 1
· · · · · — ·

The modified rhythmic system consists of 306 signals emitted in 300 seconds of mean time. Transmission is usually made for five minutes only, either before or after the hour.

In each five minute period signals Nos. 1, 62, 123, 184, 245, and 306 are single dashes of $0^{s}.4$ duration and commence at exact minutes. Each dash is followed by sixty ticks of $0^{s}.1$ duration. The instants of commencement of tick or dash are evenly spaced at intervals of 60/61 parts of one second of mean time. The signals produce a vernier effect with the second-breaks of a mean time or sidereal chronometer. With the latter, coincidences occur at intervals of 72 seconds of sidereal time.

The name of the international ONOGO system is derived from the sequence of Morse code letters transmitted during three minutes preceding the hours.

The sequence of transmission is as indicated in Fig. 8.11. Each dash is of one second duration followed by one second of silence; each dot is of $0^{s}.25$ duration. The preparatory signals from $57^{m}00^{s}$ to $57^{m}49^{s}$ are not time signals. In the modified ONOGO system, the letter O is replaced by a dot marking each second.

The technical broadcast system is fully described in the next section.

It is typically represented by the stations WWV and WWVH, and it is the system that satisfies the requirements of geodetic astronomy best because time signals are emitted continuously.

TIME	SIGNAL REPRESENTATION	LETTER
$57^m00^s - 57^m40^s$	—— • • —— —— • • etc.	X
57 55 – 58 00	55^s 56^s 57^s 58^s 59^s 60^s	O
58 08 – 58 10	08 09 10	N
58 18 – 58 20	18 19 20	N
⋮	⋮	⋮
58 48 – 58 50	48 49 50	N
58 55 – 59 00	55 56 57 58 59 60	O
59 06 – 59 10	06 07 08 09 10	G
59 16 – 59 20	16 17 18 19 20	G
⋮	⋮	⋮
59 46 – 59 50	46 47 48 49 50	G
59 55 – 00 00	55 56 57 58 59 60	O

Fig. 8.11 The international ONOGO system

8.222 Transmissions Controlled by the U. S. National Bureau of Standards. Every major country has its own service broadcasting standard time and/or frequency primarily for the regulation of the ever-expanding radio communication and telegraph networks and for purposes of safe air and sea navigation. The transmissions are either on high frequency (HF = 3-300 MHz), on low frequency (LF = 30-300 kHz), or on very low frequency (VLF = 10-30 kHz). The time and frequency services in the United States will serve as a representative example since their reliability and precision is above average. In this country, the National Bureau of Standards (NBS) and the Department of the Navy broadcast standard times and frequencies from several stations. Transmissions from the NBS stations are essentially based on the NBS-A clock and those from the Navy stations are based on the A.1 system (see section 8.123).

As described previously, to ensure reliability five portable oscillators run continuously at the National Bureau of Standards in Boulder, Colorado. The cycles of each oscillator are counted to get the indicated time; the frequency of each is checked daily against NBS-III. A weighted average of the five results is used to form the atomic clock (scale) NBS-A and its frequency offset and step adjusted version, the UTC(NBS). Both are shown in Fig. 8.12 together with the associated time and frequency measurement console consisting of the cycle counters. The time NBS-A is transmitted on 60 kHz to the transmitting station WWVB at Fort Collins, Colorado, which is phase locked to the reference

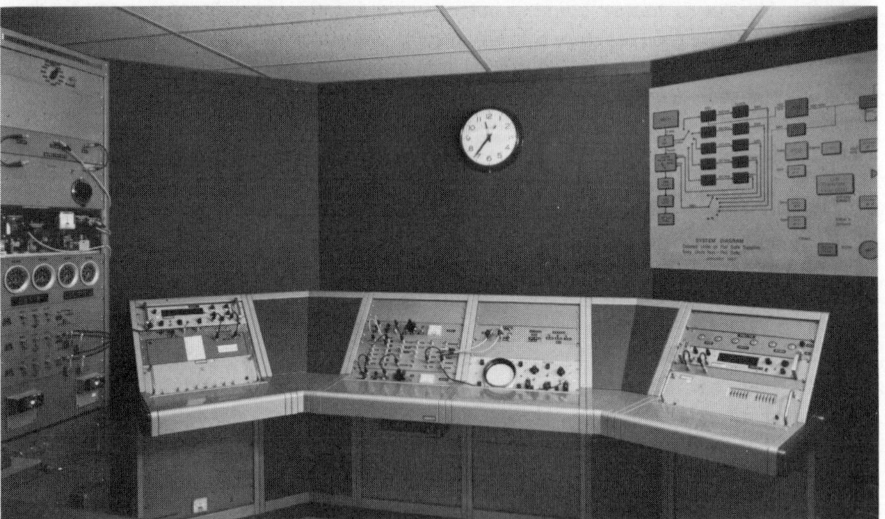

Fig. 8.12 Time and frequency measurement console, and the NBS-A and UTC (NBS) clocks at the National Bureau of Standards in Boulder, Colorado

signals. Oscillators at stations WWVL and WWV (also at Fort Collins) are directly connected to the WWVB system with coaxial cables. The transmission at WWVB is periodically step adjusted to provide the stepped atomic time scale NBS(SA). The transmissions at WWV and WWVH are frequency offset and step adjusted and within a few microseconds are the same as UTC(NBS). Station WWVL is also offset by the same amount. The cesium beam oscillators at the transmission stations periodically are compared to UTC(NBS) by means of portable clocks. Transmissions at WWV and WWVH are on high frequencies, at WWVB on low frequency, and at WWVL on very low frequencies. The NBS frequency and time facilities are illustrated in Fig. 8.13.

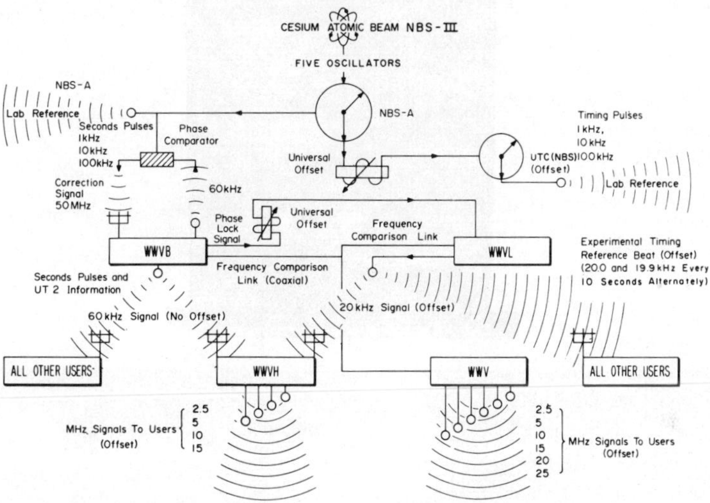

Fig. 8.13 NBS time and frequency facilities (1967)

Transmissions from stations WWV and WWVH are utilized most often in geodetic work. They provide the following information [National Bureau of Standards, 1967]: (1) standard radio frequencies, (2) standard audio frequencies, (3) standard musical pitch, (4) standard time intervals, (5) time signals, (6) corrections to arrive at a close approximation to UT2, (7) propagation forecasts (WWV only), (8) geophysical alerts.

Standard radio frequenices are broadcast on 2.5, 5, 10, 15, and 25 MHz from WWV and on 2.5, 5, 10, and 15 MHz from WWVH. (The 2.5 MHz transmitter and the station WWV are shown in Fig. 8.14. The 2.5 MHz antenna is left off the building.) The broadcasts are continuous through twenty-four hours, except for silent periods. The silent period of WWV commences at 45^m15^s after each hour and ends at 49^m15^s; that of WWVH commences at fifteen mintues and ends approximately at nineteen minutes after each hour.

As mentioned the oscillators generating the HF transmission frequencies are offset to provide UTC(NBS) scale. The offset is transmitted in international Morse code. (The code for 1966-1968 is M300, indicating the quantity 's' in equation (8.2).)

Fig. 8.14 Station WWV and the 2.5 MHz transmitter

Corrections to the transmitted frequencies are continuously determined with respect to NBS-A and are published monthly in 'Proceedings of the Institute of Electrical and Electronic Engineers' under the heading 'Standard Time and Frequency Notices,' and also in the 'NBS Time and Frequency Services Bulletin'(s) available from the Radio Standards Laboratory, Institute for Basic Standards, NBS, Boulder, Colorado.

Standard time intervals are given by second-pulses at precise intervals. Due to the frequency offset mentioned above and in section 8.212 the transmitted time intervals in 1966-1968 are exactly $(1+300\times 10^{-10})$ times the atomic interval (NBS-A). Intervals of one minute are marked by omission of the 59th second-pulse of every minute and by commencing each minute with two pulses spaced $0^s.1$ apart. The first pulse marks the minute. The two-, three- and five-minute intervals are synchronized with the second-pulses and are marked by the beginning or ending of the periods when audio frequencies are not transmitted. The pulse duration is five milliseconds. At WWV each pulse consists of five cycles of a 1000 Hz frequency, at WWVH each pulse contains six cycles of a 1200 Hz frequency. The audio frequencies are interrupted for forty milliseconds for each second-pulse. The pulse begins ten milliseconds after commencement of the interruption.

Station identification and universal time is announced in international Morse code during the first half (WWV) and second half (WWVH) of the fifth, tenth, fifteenth, etc. minute during the hour, i. e., twelve times per hour. The first two figures give the hour (in the 24-hour system); the last two figures give the minute past the hour when the tone returns.

Station identification and announcement of universal time in voice is made at WWV during the second half of each fifth minute during the hour, i.e., twelve times per hour; at WWVH a similar voice announcement is made during the first half of each fifth minute during the hour. For example, the voice announcement at WWV at 5^h35^mUTC in English is: 'National Bureau of Standards, WWV, Fort Collins, Colorado; next tone begins at five hours, thirty-five minutes Greenwich mean time.'

In addition to the above information, WWV broadcasts a timing code for one minute commencing at 7^m, 12^m, 17^m, etc. of each hour (except during the silent period). It is of value when simultaneous observations are made at widely separated stations. The code format broadcasted is generally known as the NASA 36-Bit Time Code. The code is produced at a 100 pps rate and is carried on a 1000 Hz modulation. It contains the epoch of UTC in seconds, minutes, hours, and day of the year as shown with the code key in Fig. 8.15. The binary-coded decimal (BCD) system is used. The 1000 Hz modulation is synchronous with the code pulses so that millisecond resolution is possible. A 'binary zero' consists of two cycles of 1000 Hz amplitude modulation, and a 'binary one' pulse consists of six cycles of 1000 Hz amplitude modulation (see enlargement of 100 pps time code in Fig. 8.15).

Corrections to the epoch of the signals to arrive at a close approximation to UT2 are given in international Morse code during the second half of the nineteenth minute of each hour on WWV and during the second

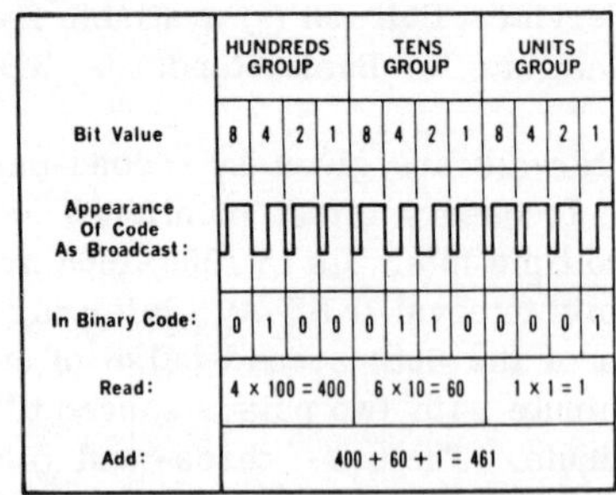

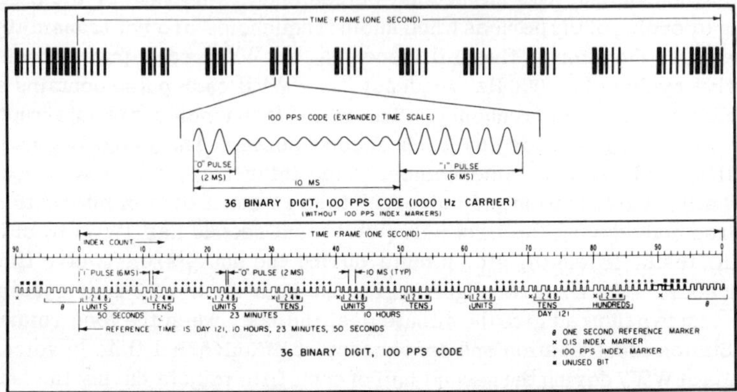

Fig. 8.15 Chart of time code transmissions from WWV

half of the 49th minute on WWVH. The symbols broadcasted are UT2, then either AD or SU, followed by a three-digit number which represents the correction in milliseconds. The correction is added to the time of the signal if AD is broadcast and subtracted when SU is broadcast. The corrections are derived from extrapolated data supplied by the USNO. New values appear during the hour following 0^hUT and remain unchanged for the following twenty four hours. Final corrections to UTC are issued periodically by the USNO in its 'Time Signals Bulletin' (see section 8.233).

Propagation forecasts for radio waves (at WWV only) and geophysical alerts are broadcast in international Morse code on all frequencies. For further details the reader is referred to [National Bureau of Standards, 1967, p. 5].

Stations WWVB and WWVL provide standard frequencies (WWVB also standard time intervals, time signals, timing codes, and corrections to the signals to arrive at predicted UT2) continuously broadcasted on 60 kHz and 20 kHz respectively. WWVL also transmits at 19.9 kHz alternating with the 20 kHz every ten seconds. The purpose of this transmission is to allow experimentation on time synchronization using dual frequencies (see section 8.225).

The carrier frequency of WWVL is offset from NBS-A by the same amount as WWV and WWVH. Since January 1, 1965, the carrier frequency of WWVB is maintained without offset with respect to the NBS-A; thus the interval between two second-signals is exactly one atomic second. For this reason, the time signals from WWVB depart from UT2 at a different rate than those of WWV and WWVH. To keep the time transmitted close to UT2, 200 ms step adjustments in phase are made at 0^{h}UT on the first day of a month at the discretion of the NBS upon the advice of the BIH. The time signals are coded so that they resemble the code of WWV (see [National Bureau of Standards, 1967, p. 8]).

Station identification is given at WWVL by international Morse code during the 1st, 21st, and 41st minute of each hour. WWVB identifies by a unique timing code and by advancing the carrier phase 45° at 10^{m} after each hour and returning to normal phase at 15^{m} after each hour.

A summary of the hourly broadcast schedules of the NBS stations are shown in Fig. 8.16.

8.223 Standard Time and Frequency Broadcasts. Radio stations located about the world which broadcast standard time and frequency are listed in Tables 8.13 and 8.14 and are shown in Fig. 8.17. Stations listed in Table 8.13 are coordinated stations; those listed in Table 8.14 are noncoordinated (see section 8.212). Station coordinates are given in general to the nearest minute of arc (longitudes positive to the east).

The data provided in both tables is based on information given in the Bulletin Horaire, Series J, No. 19, and it reflects the situation as of September, 1967. It should be understood that the given broadcast schedules and station locations are subjects of unpredictable changes and the current situation should be investigated when the broadcast of a particular station is required.

8.224 Loran-C and Omega Transmissions. Loran (LOng RAnge Navigation)-C is a pulsed, hyperbolic radio-navigation system developed at the Massachusetts Institute of Technology during World War II, which operates on LF carrier frequency of 100 kHz. Owing to favorable propagation characteristics, it can be used to disseminate precise time to about 20 μs on a worldwide basis over moderate (1500-3000 km) distances using manual receiving techniques. The necessary receiver is manufactured by the Aerospace Research Inc. in Brighton, Massachusetts. Special automatic equipment (Lorchron) capable of better than 1 μs synchronization is also available.

A Loran-C chain consists of a master (M) and several slave stations (W, X, Y, Z). At present there are 25 active stations in eight chains in the general area of Hawaii, the Aleutians, the U. S. East Coast, the

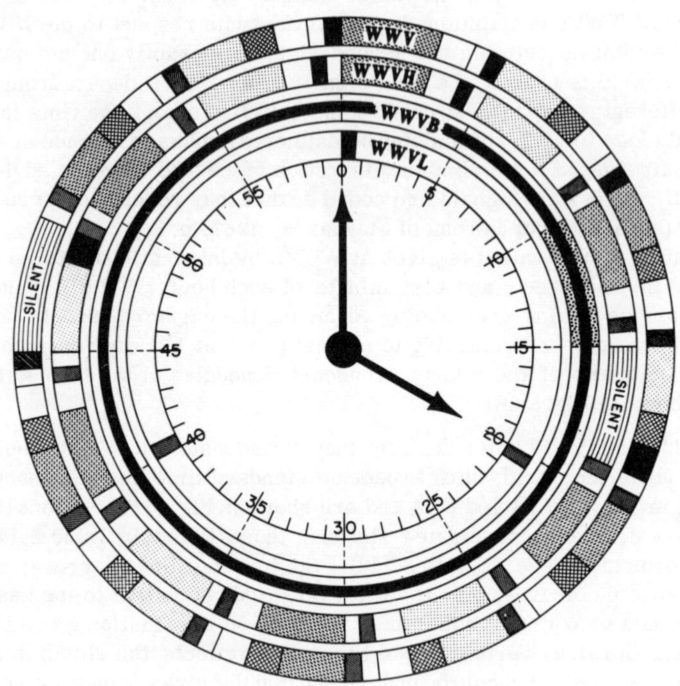

SECONDS PULSES - WWV, WWVH - CONTINUOUS EXCEPT FOR 59th SECOND OF EACH MINUTE AND DURING SILENT PERIODS

WWVB - SPECIAL TIME CODE

WWVL - NONE

STATION ANNOUNCEMENT	100 PPS 1000 Hz MODULATION WWV TIMING CODE
WWV - MORSE CODE - CALL LETTERS, UNIVERSAL TIME, PROPAGATION FORECAST	TONE MODULATION 600 Hz
VOICE - GREENWICH MEAN TIME	TONE MODULATION 440 Hz
MORSE CODE - FREQUENCY OFFSET (ON THE HOUR ONLY)	GEOALERTS
WWVH - MORSE CODE - CALL LETTERS, UNIVERSAL TIME,	IDENTIFICATION PHASE SHIFT
VOICE - GREENWICH MEAN TIME	UT-2 TIME CORRECTION
MORSE CODE - FREQUENCY OFFSET (ON THE HOUR ONLY)	SPECIAL TIME CODE
WWVL - MORSE CODE - CALL LETTERS, FREQUENCY OFFSET	

Fig. 8.16 Hourly broadcast schedule of the NBS radio stations (1967)

TABLE 8.13 Coordinated Standard Time and Frequency Broadcasts (1967)

No.	Call Sign	Location	Longitude Latitude	Carrier Frequency (kHz)	Operating Schedule (UT)	Accuracy
1	CHU	Ottawa, Canada	-75°45' +45°18'	3330; 7335; 14670	Continuous on all frequencies	5×10^{-9}
2	DAM	Elmshorn, Germany	+9°40' +53°46'	8638.5; 16980 6475.5; 12763.5 4265; 8638.5	$11^h55^m - 12^h06^m$ $23^h55^m - 0^h06^m$ Mar 20-Sept 21 $23^h55^m - 0^h06^m$ Sept 20-Mar 21	
3	DAN	Norddeich, Germany	+7°08' +53°36'	2614	$11^h55^m - 12^h06^m$ $23^h55^m - 0^h06^m$	
4	DCF-77	Mainflingen, Germany	+9°01' +50°01'	77.5	$1^h, 2^h, 7^h, 10^h, 19^h, 19^h30^m, 20^h, 21^h, 22^h, 23^h, 0^h$ no transmission from 19^h on Sat until 10^h Sun	
5	DIZ	Nauen, Germany, E.	+12°55' +52°39'	4525	Continuous	
6	FFH	Chevannes, France	+2°27' +48°32'	2500	$9^m45^s-20^m$; 30^m-40^m; $49^m45^s-60^m$ each hour between $8^h-16^h25^m$ daily except Sat and Sun	2×10^{-10}
7	FTA-91	St. André de Corcy, France	+4°55' +45°55'	91.15	$8^h, 9^h, 9^h30^m, 13^h, 20^h, 21^h, 22^h30^m$	
8	FTH-42	Pontoise, France	+2°07' +49°04'	7428	$9^h, 21^h$	
	FTK-77			10775	$8^h, 20^h$	
	FTN-87			13873	$9^h30^m, 13^h, 22^h30^m$	
9	GBR	Rugby, Great Britain	-1°11' +52°22'	16	$3^h, 9^h, 15^h, 21^h$	1×10^{-10}
	GIC-27			7397.5	$9^h, 21^h$	
	GIC-33			13555	$9^h, 21^h$	
	GIC-37			17685	$9^h, 21^h$	
	GPB-30B			10331.5	$9^h, 21^h$	
10	HBG	Prangins, Switzerland	+6°15' +46°24'	75	Continuous	2×10^{-11}

TABLE 8.13 (continued)

No.	Call Sign	Location	Longitude Latitude	Carrier Frequency (kHz)	Operating Schedule (UT)	Accuracy
11	HBN	Neuchâtel, Switzerland	$+6°57'$ $+46°58'$	5000	5^m- 10^m, 15^m- 20^m, 25^m- 30^m, 35^m- 40^m, each hour	
12	IAM	Rome, Italy	$+12°27'$ $+41°52'$	5000	From 7^h30^m to 8^h30^m for 10 mins every 15 mins, daily, except Sun	1×10^{-10}
13	IBF	Turin, Italy	$+7°46'$ $+45°02'$	5000	From 6^h50^m to 7^h30^m and 10^h 50^m to 11^h30^m daily, except Sun	1×10^{-10}
14	JAS-22	Oyama, Japan	$+139°48'$ $+36°16'$	16170	From 12^h25^m to 12^h30^m	
15	JJY	Koganei, Japan	$+139°31'$ $+35°42'$	2000; 5000; 10000; 15000	Continuous, except 25^m- 34^m each hour	1×10^{-10}
	JG2 AR			20	From 5^h30^m - 7^h30^m	1×10^{-10}
16	LOL-1	Buenos Aires, Argentina	$-58°21'$ $-34°27'$	5000; 10000; 15000	0^h- 1^h, 12^h- 13^h, 15^h- 16^h, 18^h- 19^h, 21^h- 22^h	1×10^{-9}
	LOL-2			8030	1^h, 13^h, 21^h	
	LOL-3			17180	1^h, 13^h, 21^h	
17	MSF	Rugby, Great Britain	$-1°11'$ $+52°22'$	60	Continuous from 13^h to 16^h	1×10^{-10}
				2500; 5000; 10000	0^m- 5^m, 10^m- 16^m, 20^m- 25^m, 30^m- 35^m, 40^m- 45^m, 50^m- 55^m (alternating with HBN) each hour	
18	NBA	Balboa, Canal Zone, USA	$-79°39'$ $+9°04'$	24	55^m- 60^m each hour except between 23^h55^m and 24^h	5×10^{-11}
				147.85; 5448.5; 11080; 17697.5; 22515	5^h, 10^h, 17^h, 23^h starts 5^m before the hour	
19	NPG	Mare Island, USA	$-122°16'$ $+38°06'$	114.95; 4010; 6428.5; 9277.5; 12966; 17055.2; 22635	6^h, 12^h, 18^h, 24^h starts 5^m before the hour	
20	NPN	Guam, USA	$+144°43'$ $+13°27'$	484; 4955; 8150; 13530; 17530; 21760;	6^h, 12^h, 18^h, 24^h starts 5^m before the hour	
				21.4	55^m- 60^m each hour	5×10^{-11}

TABLE 8.13 (continued)

No.	Call Sign	Location	Longitude Latitude	Carrier Frequency (kHz)	Operating Schedule (UT)	Accuracy
21	NSS	Annapolis, USA	-76°27' +38°59'	162; 5870; 9425; 13575; 17050.4; 23650	$2^h, 6^h, 8^h, 12^h, 14^h, 18^h, 24^h$ starts 5^m before the hour	
22	OLB-5	Podebrady, Czechoslovakia	+15°08' +50°09'	3170	Continuous from 10^h to 11^h	
	OLD-2			18985	12^h30^m to 13^h on Wed and Fri	
23	OMA	Liblice, Czechoslovakia	+14°53' +50°04'	50	Continuous except 10^h to 11^h each day	1×10^{-9}
				2500	5^m- 15^m, 25^m- 30^m, 35^m-40^m, 50^m- 60^m, each hour	1×10^{-9}
24	PPE	Rio de Janeiro, Brazil	-43°11' -22°54'	8720	0^h30^m; 13^h30^m; 20^h30^m	
	PPR			1305; 4244; 6421; 8634; 17194	1^h30^m, 14^h30^m, 21^h30^m	
25	VHP	Belconnen, Australia	+149°08' -35°15'	44	3^h(except Tues and Wed) and 8^h	
				4286; 6428.5; 8478; 12907.5; 17256.8	$3^h, 8^h, 14^h, 20^h$	
				22485	$3^h, 8^h$	
26	VNG	Lyndhurst, Australia	+145°02' -38°00'	5425	12^h15^m to 22^h	
				7515 12005	Continuous 22^h15^m to 12^h	1×10^{-9}
27	WWV	Ft. Collins, Colorado, USA	-105°02' +40°41'	2500; 5000; 10000; 15000; 20000; 25000;	Continuous except 45^m- 49^m each hour (see Fig. 8.14)	5×10^{-11}
28	WWVH	Maui, Hawaii, USA	-156°28' +20°46'	2500; 5000; 10000; 15000	Continuous except 15^m- 19^m each hour (See Fig. 8.14)	1×10^{-10}
29	ZUO	Olifantsfontain, S. Africa	+28°14' -25°58'	5000	Continuous	1×10^{-10}
30		Johannesburg, S. Africa	+28°04' -26°11	10000	Continuous	1×10^{-10}

TABLE 8.14 Noncoordinated Standard Time and Frequency Broadcasts (1967)

No.	Call Sign	Location	Longitude Latitude	Carrier Frequency (kHz)	Operating Schedule (UT)	Accuracy
a	BVP	Shanghai, China	+121°26' +31°12'	5430; 9351 5000; 10000; 15000	11^h, 13^h, 15^h, 17^h, 19^h, 21^h Continuous	
b	DCF-77[1]	Mainflingen, Germany	+9°00' +50°01	77.5	7^h28^m; 10^h28^m; 19^h11^m; 19^h41^m	5×10^{-10}
c	LQB-9 LQC-28	Monte Grande, Argentina	-58°31' +34°45'	8167 17551.5	20^h50^m 10^h5^m; 11^h50^m; 22^h5^m	
d	RES[2] RWM[2] RAT[2]	Moscow, USSR	+37°18' +55°45'	100 100000; 15000 2500; 5000		
e	RID[2] RKM[2]	Irkoutsk, USSR	+104°18' +52°18'	10004; 15004 5004; 10004; 15004		
f	RIM[2] RCH[2]	Tashkent, USSR	+69°15' +41°19'	5000; 10000; 15000 2500	 5^m-10^m; 15^m-20^m; 34^m-40^m; 45^m-50^m every hour	
g	RTA[2]	Novossi-birsk, USSR	+82°58' +55°04'	4996; 9996; 14996		
h	XSG	Shanghai, China	+121°26' +31°12'	4586; 414.5; 8502; 12871.5	3^h; 9^h	
i	WWVB[1]	Fort Collins, USA	-105°03' +40°40	60	Continuous	2×10^{-11}

[1]These stations transmit stepped atomic time (SAT)

[2]These stations are synchronized with respect to each other to 0.1ms

Mediterranean, the North Atlantic, the Norwegian Sea, the Northwestern Pacific, and Southeast Asia (see Table 8.15 and Fig. 8.15) [Shapiro, 1968].

As an example, in the U. S. on the East Coast the Naval Observatory controls the time pulse broadcast by the master station located at Cape Fear, North Carolina. The frequency of the cesium beam oscillator at Cape Fear is synchronized with the UTC(USNO). The other stations comprising the chain (Table 8.15) monitor the Cape Fear signals. Thus,

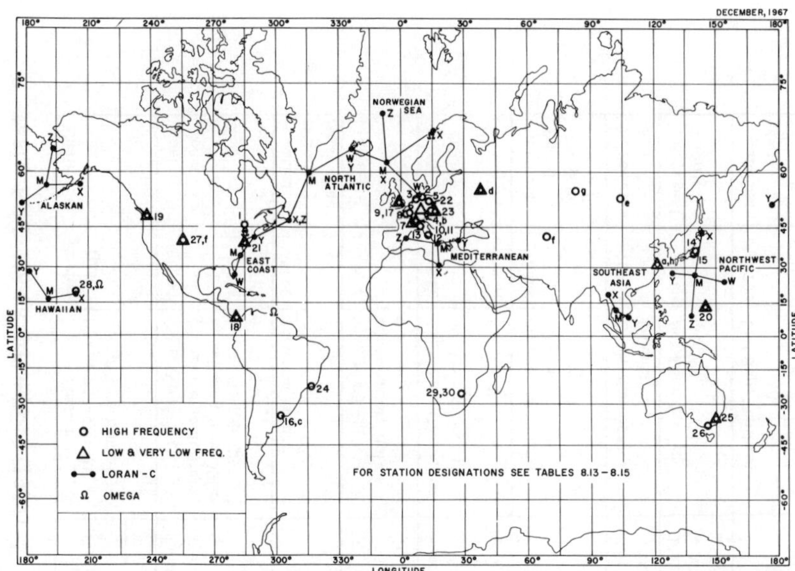

Fig. 8.17 Standard frequency/time services, Loran-C and Omega

the emission of time pulses from any one of the Loran-C stations is synchronized in the time scale UTC(USNO). To prevent interference of the pulses, the slave stations have fixed emission delays (varying from about 14 ms to 84 ms) with respect to Cape Fear which is held constant to about 0.1 μs. The mode of time pulse transmission at present is as follows: The nominal transmission consists of a group of eight pulses spaced one millisecond apart. A ninth pulse, not one millisecond from the eighth, identifies the master station. (Blinking of this pulse indicates that the Loran-C system is not operating.) The groups of eight pulses have a repetition of eighty milliseconds; thus, there are 12.5 repetitions per second.

A once-per-second pulse, two milliseconds before each 0^s of UTC, is transmitted by Cape Fear only. Thus, the sequence of pulses emitted from Cape Fear is

Once-per-second pulse		
1st pulse of first cycle	at	$59^s.998$ UTC
1st pulse of 1st group of 8	at	0.000
1st pulse of 2nd group of 8	at	0.080
⋮		⋮
1st pulse of 12th group of 8	at	0.960

TABLE 8.15 Loran-C Networks (1967)

Chain	Period (μs)	Master Power 250–400 kW, except where noted.	Slaves Power 250–400 KW, except where noted. Emission delay given in μs with reference to Master.			
			W	X	Y	Z
East Coast	100000	Cape Fear, N.C. ('T' slave sometimes broadcasts from Wildwood, N.J. at ED 84028.7 μs)	Jupiter, Fla. 13695.5	Cape Race, Newfoundland 4 MW 36389.6	Nantucket Island Mass. 52542.5	Dana, Indiana 68564.2
Central Pacific (Hawaiian)	59600	Johnston Island		Upolo Point, Hawaii 15971.8	Kure Island 35252.4	
Mediterranean	79600	Simeri Chichi (Catanzaro). Italy		Matratin, Libya 14107.6	Targabarun, Turkey 32273.3	Estartit, Spain 50999.7
North Atlantic	79300	Angissoq, Greenland 500 kW	Sandur, Iceland 4 MW 15068.2	Ejde, Faeroe Islands 27803.9		Cape Race, Newfoundland 4 MW 48212.3
Northern Pacific (Alaskan)	79800	St. Paul, Pribiloff Islands		Sitkinak, Alaska 14284.4	Attu, Aleutian Islands 31875.3	Port Clarence Alaska 1850 kW 53069.1
Norwegian Sea	79700	Ejde, Faeroe Islands	Sylt, Germany 30065.6	Bø, Norway 15048.1	Sandur, Iceland 48944.5	Jan Mayen 63216.3
Northwest Pacific	99700	Iwo Jima 4 MW	Marcus Island 4 MW 15283.3	Hokkaido, Japan 36685.2	Gesashi Okinawa 59463.0	Yap 4 MW 80746.5
Southeast Asia	49700	Sattahip, Thailand		Lampang, Thailand 13182.8 (comp)	Con Son Island, South Viet-Nam 29522.1 (comp)	

Once-per-second pulse	at	0.998
1st pulse of 1st group of 8	at	1.040
1st pulse of 2nd group of 8	at	1.120
⋮		⋮
1st pulse of 12th group of 8	at	1.920
Once-per-second pulse	at	1.998
1st pulse of new cycle	at	2.000
⋮		⋮
etc.		etc.

It is therefore possible to measure the signal 100 times per second.

As of 1967, only the U. S. East Coast Loran-C chain has been synchronized (to about 25 μs) in time and frequency with the U. S. Naval Observatory. There are plans to synchronize the North Atlantic chain in the near future.

The Omega navigation system, under development by the U. S. Navy and Tracor Inc. in Austin, Texas, in many respects is similar to Loran-C. It differs, however, in three important respects: the longer range (8000-10000 km instead of 1500-3000 km), the use of radio waves in the VLF(10-14 kHz) instead of in the LF (100 kHz) area, and the use of phase difference instead of time difference measurement [Chi and Witt, 1966]. Omega transmitters are located at present in Hawaii (Haiku), New York (Forestport), Norway (Aldra), and in the West Indies (Trinidad).

The transmissions of the Loran-C and Omega systems are coordinated by the U. S. Naval Observatory. Changes in transmissions, in station locations, information on synchronization, etc. are published periodically in the 'Time Service Announcement'(s) available upon request from the Time Service Division of the USNO.

8.225 International Coordination of Time and Frequency Services. Coordination between time or frequency services was achieved during recent years to the extent that transmissions from most stations can now be used to maintain synchronism in epoch to about one millisecond using HF or to 50-100 μs with LF/VLF signals. Frequency can be maintained to about a few parts in 10^{11} utilizing LF/VLF transmissions.

The coordination to maintain the necessary quality of the broadcasts from widely separated stations is achieved principally through <u>monitoring signals</u> emitted from several stations at one receiving station. For example, the U. S. Naval Observatory regularly monitors time broadcasts from stations WWV, NBA, NSS, NPG, VHP, CHU, GBR, LOL3, and others. The National Bureau of Standards monitors WWVB, WWVL, NSS, GBR. On the other side of the Atlantic Ocean, The Royal Greenwhich Observatory monitors signals emitted from stations WWV, NSS, GBR, and others mainly located about Europe. The net result of this monitoring system is that corrections can be assigned to the times of emission of the signals with respect to a primary standard, e.g., A.1 or NBS-A. Thus, no matter where time-dependent observations are made (provided they are referred to time signals emitted from one of the coordinated stations), the epoch of observations or the frequency of the clock may be referred to an internationally consistent system.

It should be noted that, although time signals from coordinated stations are emitted in close synchronism, the reception governs the accuracy of the observations. Propagation anomalies, which may not be rigorously determined, usually prevent full realization of the precision inherent in the signal as transmitted. In order to reduce this problem, experiments are being conducted using VLF transmissions on alternating multiple frequencies [Fey and Looney, 1966]. Results using the dual frequencies of WWVL indicate a present synchronization precision of 20-50 μs. It is expected that with better receivers and three alternating frequencies, this figure can be reduced to 1-5 μs. It is obvious that such a system makes stringent demands on the stability of the transmitted signal and on the receiving system as well as that of the propagation medium.

Experiments to synchronize master clocks by physically transporting quartz crystal or atomic clocks via airlines from station to station were started in 1960. Details and results of the first experiments are given in [Reder et al., 1961]. The conclusion was that clocks anywhere on earth could be synchronized to 5 μs or better. Between 1964 and 1966 other experiments followed which not only correlated time between the USA, Europe, and Asia but also provided data on traveling times of HF and VLF radio signals. Results indicated that time could be correlated to about 1 μs [Bodily, 1965; Bodily et al., 1966]. In September-October, 1967, time was correlated with Hewlett-Packard portable cesium standards (Model 5061A) between 53 places in 18 countries to about 0.1 μs. Two clocks exhibited time differentials of 1.7 and 3.5 μs over 41 days, corresponding to average frequency differences of 5×10^{-13} and 10×10^{-13} respectively [Bodily et al., 1967]. This mode of coordination seems to be the best today.

In addition to signal monitoring and transporting clocks, it is also possible to achieve time coordination via artificial satellites. In 1962 the U. S. Naval Observatory and the National Physical Laboratory in England conducted an experiment using the active satellite Telstar I to synchronize the USNO master clock and that of the Royal Greenwich Observatory. In this experiment, time pulses of 5 ms in length were transmitted simultaneously over the satellite circuit from the satellite ground stations at Andover, Maine and Goonhilly Downs, Cornwall. Each station measured the time difference between emitted and received pulse. From these differences the relative setting of the station clocks was determined. It was found that the station clock at Goonhilly was 72.6 μs (standard deviation = 0.8 μs) ahead of the Andover clock. The satellite ground station clocks were related to the observatory clocks through VLF transmissions from MSF (60 kHz) in Great Britain and by Loran-C in the United States respectively. During the experiment the MSF time signals were monitored at the RGO. Thus, the Goonhilly clock could be directly related to the master clock at the RGO, and the Andover clock was directly related to the USNO master clock. The results of the comparison showed that on August 27, 1962, the RGO time standard was ahead of the USNO clock by 2234 μs (standard deviation = 20 μs). The overall results established that time synchronization to 1 μs is possible between satellite ground stations using active satellites such as Telstar. The limiting factors in the Telstar experiment were chiefly the VLF and the Loran-C links between the satellite ground stations and observatory clocks and also the uncertainty in the height of the satellite which was computed from Minitrack observations. For further details, especially on equipment used for the experiment, the reader is referred to [Steele et al., 1964].

Similar experiments were carried out to synchronize clocks at Mojave, California, and Haskima, Japan, more recently via the Relay II satellite. The results indicated that synchronization to 0.1 μs is a possibility [Frequency Control Symposium, 1965; Markowitz et al., 1966].

The National Bureau of Standards is also experimenting with time synchronization via the ATS satellite using two-way VFH (137-149 MHz) transmissions. Results indicate a precision of 1-10 μs. Microwave transmissions could reduce this figure to about 0.1 μs. Other radar experiments, bouncing signals from the moon, suggest possible precisions also in the microsecond range [Gatterer, 1968].

8.23 Corrections to Time Signals

When a time signal is received via radio, it is affected by the characteristics of the propagation medium and is referenced to the system of the time service, UTC(i). Corrections need to be applied for the propagation delay of the time signal between the transmitting and receiving antennas and also for reducing the epoch of the emission to the desired time system (to UT2, UT1, UT0, AT, etc.). Depending on the type of equipment used and on the accuracy desired, possible transmitter and receiver delays may also need to be accounted for.

8.231 Corrections for the Propagation Delay of HF Signals. High frequency radio signals propagate in a complex manner between the ionosphere and the earth's surface. Variations in the height of the ionosphere and its profile cause variations in travel time of the HF signals between the transmitter and receiver. The principal factors that affect the propagation time of HF signals are the great circle distance between transmitter and receiver antennas, the number of reflections between the ionosphere and the earth's surface, and the actual height of the reflection layers of the ionosphere. Several methods to calculate the transmission delay are in use: For distances up to about 160 km, a ground-wave propagation path can be assumed. Thus, the distance traveled is approximately equal to the length of the great circle, which can be calculated by a variety of methods. For distances up to 2400 km, one-reflection propagation path can be assumed. During the daytime the reflecting layer is assumed to be the E layer, at about 125 km above the earth's surface (during the night the E layer does not exist). Long distance HF transmissions are usually reflected by the F2 layer which has an average height of 350 km above the earth's surface. A one-reflection path may be assumed for distances up to 4000 km. Considering now that reflections from both E and F2 layers, or even others, may be received and that the number of reflections is uncertain, it is obvious that the calculation of the transmission delay is at best a close approximation. Furthermore, it is possible that the signal was not received via the shortest route but had come around the earth the longer way.

Once the length of the propagation path D is determined, the transmission delay Δt_{HF} is usually computed with sufficient accuracy from

$$\Delta t_{HF} = \frac{D}{V_{HF}} \tag{8.8}$$

where V_{HF} =285km/ms is the recommended mean propagation velocity of HF radio waves. Recent 'flying clock' experiments (see section 8.225) indicated that this value might vary between 280-290km/ms. Special graphs for the estimation of transmission delays with an argument of distance per reflection are in existence. One such graph with explanations is given in [Hewlett-Packard, 1965, p. 4-5].

The BIH uses an empirical formula to determine the transmission delay for HF waves over distances between 1000 km and 40000 km [Stoyko, 1964a, p. 2]. The propagation velocity V_{HF} is calculated from

$$V_{HF} = (290 - \frac{a}{d+b})\ \text{km/ms} \tag{8.9}$$

where d=Dkm/1000, a=139.41, b=2.9) (empirical constants). The transmission delay is computed from (8.8) using the generally too small values of V_{HF} from (8.9).

The transmission delay has to be subtracted from the clock time obtained from a time comparison.

8.232 Correction for the Propagation Delay of LF and VLF Signals. LF or VLF radio waves can be assumed to propagate parallel to the earth's surface. Irregularities in the ionosphere height have much less influence since the ionosphere acts more as a boundary than as a reflector. For this reason the determination of the transmission delay is much simpler. It may be calculated directly from equation (8.8), substituting the mean propagation velocity for LF or VLF radio waves for V_{HF}. A reasonable value for $V_{LF/VLF}$ seems to be 290 km/ms, though the BIH uses 252 km/ms [Stoyko, 1964a, p. 2]. The former value is based on direct measurements of the time difference between VLF signals received directly and from around the world [Brown, 1949]. The BIH value was obtained in 1933 from the mean reception times at observatories with individual results varying between 160 and 398 km/ms.

8.233 Corrections to the Emitted Time Signals for Epoch Reference. In addition to the propagation delay, further corrections are needed to make the emitted UTC(i) refer to the desired time system. These systems are either UT2, UT1, UT0 (at the time service) or AT, or occasionally ET. If the selected AT is that of the time service in question, the correction AT(i) - UTC(i) is available immediately. The differences between UTC(i) and UT2, UT1, or UT0 are available generally after some months, while for a good ET reference the observer has to wait several years.

The final time of emission of the signal in any one of these systems (except in ET) may be deduced from the information contained in the publications of the BIH (UT2, UT1, UT0, and A3) or in the various time service bulletins published by the major observatories. In these latter bulletins preliminary and final corrections are given to UTC(i) transmitted from or coordinated by the time service in question and from closely associated stations, which enables the user to reference the emitted epoch to the adopted reference time system UT2(i). (E.g., in the case of the U. S. Naval Observatory, the reference system is either A.1 in atomic time or the adopted UT2 (USNO), as determined from the weighted means of universal time determinations at the USNO in Washington, D. C. and at its substation in Richmond, Florida. The adopted UT2(USNO) is not exactly equal to the UT2 of the BIH Mean Observatory since the latter is based on observations at all the participating observatories. See Examples 8.1 and 8.2.) The systematic deviations UT2-UT2(i) are published in the Bulletin Horaire in its table 3 for all participating observatories (Table 8.9). (From 1968 the differences UT0-UT0(i) will be published in the Annual Reports of the BIH.)

Examples for the various time service bulletins are the following documents published by the USNO:

(1) <u>List of VLF and HF transmissions suitable for precise time measurements</u>. Includes call sign, geographic location, frequencies, radiated power, etc.

(2) <u>Schedule of U.S. Navy time signal transmissions in VLF and HF bands</u>. Indicates times of broadcast, frequencies, etc.

(3) <u>Schedule of U.S. Naval VLF transmissions, including the OMEGA system</u>. Includes location, frequencies, power radiated, maintenance periods, type of emission, etc. and is updated as required by 'Time Service Announcement'(s).

(4) Daily relative phase values, issued weekly. Includes observed phase and time differences between VLF, LF, OMEGA, and Loran-C stations and the UTC(USNO) clock. The supplement to daily relative phase values contains instructions for use of phase value bulletins and messages.

(5) Teletype messages, dispatched daily. Contain daily relative phase and time differences between VLF, LF, OMEGA, and Loran-C stations and the UTC(USNO) clock.

(6) Daily VLF frequency values, issued weekly. Indicates the observed frequency offset of VLF stations relative to the UTC (USNO) clock, averaged over a 24-hour interval.

(7) Extrapolated corrections (Table 8.16), issued weekly. Contains A.1-UTC (USNO) and UT2(USNO)-UTC(USNO) predicted two weeks in advance.

(8) Time service announcements of BIH-adopted offset from nominal frequency and phase adjustments.

(9) Time service announcements pertaining to Loran-C. Includes change in transmissions, repetition rates, general information, null ephemeris, etc.

(10) Preliminary times of emissions (Table 8.17), issued weekly. Indicates preliminary time differences between UT0, UT1, UT2, A.1, and UTC(USNO). Also includes provisional coordinates of the pole.

(11) Time signals bulletin (Table 8.18). Contains final corrections to UTC(i) (for certain coordinated stations) in the form UT2(USNO)-UTC(i); the seasonal variations $\Delta\Lambda_s$; the polar motion corrections $\Delta\Lambda_p$ for Washington and Richmond; the differences UT2(USNO)-A.1; the results of daily observations at Washington and Richmond in the form UT2 (observed)-UT2(USNO); the differences A.1-UTC(i) (for NBA, NSS, WWV, etc.); and the daily and monthly values of frequency offsets ('s' in equation (8.1)) for NBA and NSS. This bulletin is published periodically about six to nine months in arrears.

(12) Photographic zenith tube (PZT) information. Contains results, catalogs, papers, etc.

(13) Dual-rate moon-position camera program. Includes information on programs, results, etc.

(14) U.S. Naval Observatory circulars and other general information pertaining to time determination, measurement, and dissemination.

Comparing these documents with those of the BIH, it may be seen that the information in items (8) and (11) above corresponds to that contained in the Bulletin Horaire; the information in item (10) corresponds to that contained in the BIH Circular 'B/C.' The corrections, however, are in reference to the adopted time reference systems (UT2(USNO) vs. UT2; A.1 vs. A3; etc.). For further important information on the use of these documents, see U.S. Naval Observatory Time Service Notices Nos. 6 and 9 through 11.

TABLE 8.16 U.S. Naval Observatory Extrapolated UT2—UTC(USNO) Corrections (Sample)

U. S. NAVAL OBSERVATORY
WASHINGTON, D.C. 20390

18 January 1968

PRELIMINARY TIMES AND COORDINATES OF THE POLE, SERIES 7 NO. 3

I. EXTRAPOLATED CORRECTIONS

Extrapolated corrections of coordinated time signals are issued by the U. S. Naval Observatory weekly with the predictions two weeks in advance.

These predictions are based on observations of UT2 made at Washington and Richmond. Linear curves are fitted to the unweighted, nightly results of 1, 2, and 3 months observations at each station separately. The curves are extrapolated and combined into a single value giving Richmond a weight of two and Washington a weight of one. Experience and judgment are used if the results from the 1, 2, or 3 months curve fitting differ extensively, the last observed month being given the largest weight.

Thus the predictions are the difference UT2-UTC(USNO). However, they are equally accurate for all coordinated time signals. The estimated accuracy is about .005 seconds.

		Extrapolated UT2-UTC(USNO)
1968 Jan	25	+ .085
	26	+ .085
	27	+ .085
	28	+ .086
	29	+ .086
	30	+ .086
	31	+ .086

Rate + 0.2 ms/day

TABLE 8.17 U.S. Naval Observatory Preliminary Emission Times and Provisional Coordinates of the Pole (Sample)

II. PRELIMINARY EMISSION TIMES for Signals from NSS, NBA, GBR, WWV, CHU, and Other Coordinated Stations

For 17 January 1968

UT0 - UTC,	085
UT1 - UTC,	081
UT2 - UTC,	078
A.1 - UTC,	$6^{s}.283$

UT0 is the reading of a clock which indicates time UT0. Similarly for UT1, UT2, and A.1. "UTC" is the reading of the transmitting clock. The quantities tabulated above are the amounts, in milliseconds, by which the signals, UTC, are emitted late with respect to clocks which indicate UT0, UT1, UT2, and A.1, respectively.

III. PROVISIONAL COORDINATES OF THE POLE

For 17 January 1968

B.I.H.	x	y
Mean Pole	$-0''.014$	$+0''.046$
Conventional Pole	$-0''.007$	$+0''.279$

TABLE 8.18 U.S. Naval Observatory Time Signals Bulletin (Sample)

Bulletin 215 — 20 June 1968

I. FINAL TIMES OF EMISSION, UT2 — UTC

UT2 − UTC

Signal:	NSS L.F.(1)*	NBA 24kc	WWV All Freq	CHU 7335kc	GBR 16kc	LOL 17183kc	NBA H.F.(2)*	NSS H.F.(3)*	NPG 17055kc	FTN87 13873kc
K	0007	0114	0080	0026	0199	0299	0118	0002	0139	0220
UT	2h	13h	3h	13h	15h	21h	17h	2h	18h	13h
1967 Jul 8	0514	0521	0510	0530	0550	0524	0536	0516	0408	0524
18	0539	0530	0535	0542	0556		0543	0514	0430	0538
28	0558	0542	0553	0566	0574		0561	0539	0460	0560
Aug 7	0568	0567	0564	0577	0593	0558	0576	0555	0477	0575
17	0578	0577	0574	0588	0607		0586	0566	0501	0584
27	0589	0587	0585	0598	0618		0599	0580	0486	0592
Sep 6	0601	0598	0597	0604	0631	0596	0608	0593	0499	0605
16	0620	0616	0616 (4)*	0623	0647	0616	0629	0617	0505	0626
26	0645	0643	0639	0647	0672	0638	0653	0640	0515	0647
Oct 6	0670	0669	0664	0671	0698		0678	0658	0530	0670
16	0682	0682	0676	0682	0708		0691	0667	0538	0680
26	0674		0668	0672	0701		0681	0658	0528	0670
Nov 5	0665		0659	0663	0694	0661	0678	0652	0523	0665
15	0664		0657	0663	0694	0662	0678	0644	0520	0662
25	0677		0671	0680	0709	0678	0697	0650	0536	0680
Dec 5	0724		0718	0726	0759	0711	0746	0694	0600	0729
15	0752		0746	0754	0784	0746	0780	0727	0670	0753
25	0780		0774	0782	0809	0772	0814	0755	0720	0781
1968 Jan 4	0804		0797	0805	0830	0790	0817	0772	0726	0802

*NOTES.

(1) Low frequencies 21.4, 121.5, and 162.0 kc/s.

(2) High frequencies 5448.5, 11080, and 17697.5 kc/s.

(3) High frequencies 5870, 9425, 13575, 17050, and 23650 kc/s.

(4) WWV was advanced 200 μs on 20 September 1967 at 0000 UT.

7. The seasonal variation, S, which is independent of location, and the observed value for the polar variation, P, are obtained from the Bureau International de l'Heure (B.I.H.). To convert UT2 to UT1 and UT0 use the formulas:

$$UT1 = UT2 - S$$
$$UT0 = UT2 - V$$

where V = S + P.

	S	V(W)	V(R)
1967 Jul 8	+0.016	+0.013	+0.014
18	+0.009	+0.006	+0.008
28	+0.002	0.000	0.000
Aug 7	-0.006	-0.007	-0.007
17	-0.013	-0.014	-0.013
27	-0.019	-0.019	-0.019
Sep 6	-0.024	-0.024	-0.024
16	-0.027	-0.027	-0.027
26	-0.029	-0.029	-0.029
Oct 6	-0.029	-0.029	-0.029
16	-0.028	-0.028	-0.028
26	-0.025	-0.026	-0.026
Nov 5	-0.022	-0.023	-0.023
15	-0.019	-0.020	-0.019
25	-0.015	-0.016	-0.016
Dec 5	-0.012	-0.012	-0.012
15	-0.009	-0.009	-0.009
25	-0.006	-0.007	-0.007
1968 Jan 4	-0.005	-0.006	-0.005

2

TABLE 8.18 (continued)

II. ADOPTED UT2 – A.1

8. The following are the adopted differences, UT2 – A.1, for every tenth day, used to derive the final times of emission. A.1 is a system of atomic time based on cesium resonators of the Naval Observatory and other laboratories. The values are based on PZT observations of Washington and Richmond, Florida, and are smoothed over an interval of about two months. The quantity, UT2 – A.1, is the difference between a clock indicating UT2 and one indicating A.1.

UT2 – A.1

Date	UT2 – A.1	Date	UT2 – A.1
1967 Jul 8.0	$-5^{s}.7303$	1967 Oct 6.0	$-5^{s}.9480$
18.0	-5.7537	16.0	-5.9727
28.0	-5.7778	26.0	-5.9994
Aug 7.0	-5.8027	Nov 5.0	-6.0262
17.0	-5.8276	15.0	-6.0523
27.0	-5.8524	25.0	-6.0769
Sep 6.0	-5.8771	Dec 5.0	-6.0981
16.0	-5.9011	15.0	-6.1212
26.0	-5.9246	25.0	-6.1443
		1968 Jan 4.0	-6.1679

III. OBSERVATIONS

9. The quantities marked O-A give the difference, UT2 (Observed) – UT2 (Adopted). τ, the fraction of a tropical year, is measured from the nearest beginning of a Besselian year.

Date 1967	Julian Date 2439000+	τ	WASHINGTON Stars	WASHINGTON O-A	RICHMOND Stars	RICHMOND O-A
Jul 1.3	672.8	-.5037	15	-.014	--	---
2		-.5010	17	+.002	18	+.004
3	674.8	-.4982	--	---	--	---
4		-.4955	--	---	--	---
5		-.4928	18	-.010	13	-.008
6		-.4900	--	---	5	+.020
7		-.4873	--	---	6	+.012
8	679.8	-.4845	9	+.005	19	+.012
9		-.4818	--	---	17	+.012
10		-.4791	17	+.002	16	+.009
11		-.4763	--	---	--	---
12		-.4736	9	-.013	23	.000
13	684.8	-.4708	--	---	24	-.001
14		-.4681	--	---	28	-.003
15		-.4654	--	---	15	+.007
16		-.4626	--	---	--	---
17		-.4599	16	-.014	--	---
18	689.8	-.4572	--	---	7	+.019
19		-.4544	17	-.016	15	+.012
20		-.4517	12	-.009	28	+.011
21		-.4489	--	---	17	+.015
22		-.4462	17	-.003	--	---
23	694.8	-.4435	9	-.017	--	---
24		-.4407	14	+.014	--	---
25		-.4380	15	+.001	19	+.001
26		-.4353	--	---	27	+.002
27		-.4325	--	---	12	+.004
28	699.8	-.4298	--	---	13	+.005
29		-.4270	--	---	18	+.011
30		-.4243	--	---	20	-.010
31		-.4216	11	-.014	9	+.001
Aug 1		-.4188	--	---	7	-.008
2	704.8	-.4161	12	-.012	14	+.002
3		-.4134	--	---	6	-.001
4		-.4106	--	---	16	+.010
5		-.4079	--	---	--	---
6.3		-.4051	--	---	18	+.002

TABLE 8.18 (continued)

Date 1967	Julian Date 2439000+	τ	WASHINGTON Stars	WASHINGTON O-A	RICHMOND Stars	RICHMOND O-A
Dec 15.3	839.8	-.0465	--	---	25	+.002
16		-.0437	11	+.009	7	-.005
17		-.0410	24	+.009	16	-.025
18		-.0383	--	---	19	-.013
19		-.0355	--	---	23	-.017
20	844.8	-.0328	28	+.005	17	-.013
21		-.0300	9	+.003	17	+.013
22		-.0273	--	---	14	+.005
23		-.0246	--	---	25	-.001
24		-.0218	28	+.011	29	+.001
25	849.8	-.0191	--	---	--	---
26		-.0164	--	---	11	-.023
27		-.0136	16	+.015	25	-.006
28		-.0109	--	---	9	+.015
29		-.0081	--	---	--	---
30		-.0054	29	+.012	20	-.003
31.3	855.8	-.0027	10	+.022	19	-.013

IV. TIMES OF EMISSION, A.1 – UTC, AND DEVIATIONS IN FREQUENCY ON A.1

10. The system of atomic time, A.1, has as point of origin 0000 UT2 (USNO) on 1 January 1958. At that moment A.1 coincided with UT2.

The scale is based on the atomic definition of the second as the duration of exactly 9,192,631,770 periods of the microwave transition between the hyperfine levels F = 4, m_F = 0 and F = 3, m_F = 0 of the $^2S_{\frac{1}{2}}$ ground state of the unperturbed Cs^{133} atom.

The standard frequency was realized initially by averaging the measured frequencies of VLF transmitters relative to the cesium beam standards of up to nine widely distributed laboratories and observatories. The uniformity of the time scale has been gradually improved by the use of the latest technology available. Currently the scale is based on the operation of a number of independently operating cesium standards of the U.S. Naval Observatory at Washington and Richmond. In addition, two laboratory hydrogen masers are being used as very high precision interpolation oscillators. The masers are operated at NRL, Washington, D.C. The system of time so derived is currently uniform in rate to within approximately 1 part in 10^{12}.

11. A.1 – UTC is the difference between a clock which indicates atomic time, A.1, and a clock synchronized with the emitted time signal.

12. ΔF/F is the deviation in frequency of the carrier wave of a VLF station with respect to the standard A.1 frequency. It is given by the formula:

$$\frac{\Delta F}{F} = \frac{\text{Carrier} - f(A.1)}{f(A.1)}$$

ΔF/F is expressed in units of 1 part in 10^{10}.

13. The above frequency of cesium was determined with respect to ephemeris time. (See W. Markowitz, R.G. Hall, L. Essen, and J.V.L. Parry, Physical Review Letters, 1, 105,1958.) The observed time difference between the ephemeris and A.1 time scales is

$$E.T. = A.1 + 32^s\!.15$$

and the available astronomical observations indicate that this difference is constant. (See American Ephemeris and Nautical Almanac, 1968, pg. xv.) The system A.1 may, therefore, be used to extrapolate E.T. forward with greater precision than is possible using current astronomical observations.

14. Values of A.1 – UTC for intermediate dates may be obtained by interpolation, except when indicated.

TABLE 8.18 (continued)

A.1 − UTC

0^h UT

		UTC(WWV)	UTC(NSS)	UTC(USNO)
1967	Jul 8	$5^{s}.7813$	$5^{s}.7817$	$5^{s}.7811$
	18	5.8072	5.8076	5.8070
	28	5.8331	5.8336	5.8329
	Aug 7	5.8591	5.8595	5.8589
	17	5.8850	5.8854	5.8848
	27	5.9109	5.9113	5.9107
	Sep 6	5.9368	5.9372	5.9366
	16	5.9627	5.9631	5.9625
	26	5.9885	5.9891	5.9885
	Oct 6	6.0144	6.0150	6.0144
	16	6.0403	6.0409	6.0403
	26	6.0662	6.0668	6.0662
	Nov 5	6.0921	6.0927	6.0921
	15	6.1180	6.1187	6.1180
	25	6.1440	6.1446	6.1440
	Dec 5	6.1699	6.1705	6.1699
	15	6.1958	6.1964	6.1958
	25	6.2217	6.2223	6.2217
1968	Jan 4	6.2476	6.2483	6.2476

Daily Values of $\Delta F/F$ for NSS

1967	JUL	AUG	SEP	OCT	NOV	DEC
1	-300.1	-300.0	-300.0	-299.9	-300.1	-300.0
2	-300.1	-300.1	-300.0	-300.0	-300.0	-300.0
3	-300.0	-300.1	-300.0	-300.0	-300.0	-300.0
4	-300.1	-300.0	-300.0	-300.0	-300.0	-300.0
5	-300.1	-299.8	-300.0	-300.0	-300.0	-300.0
6	-300.0	-299.8	-300.0	-299.9	-300.0	-300.0
7	-300.1	-299.8	-300.0	-300.0	-300.1	-299.9
8	-300.0	-299.9	-300.0	-300.0	-300.1	-300.0
9	-300.0	-299.8	-299.9	-300.0	-300.1	-300.0
10	-300.1	-299.8	-299.9	-300.0	-299.9	-300.0
11	-300.0	-299.9	-300.0	-299.9	-300.0	-300.0
12	-300.1	-299.9	-300.0	-300.0	-300.1	-299.9
13	-300.1	-299.8	-300.0	-300.0	-300.0	-300.0
14	-299.9	-299.9	-300.0	-300.0	-300.0	-300.0
15	-300.2	-299.8	-300.0	-300.0	-300.0	-300.0
16	-300.2	-300.0	-299.9	-300.0	-300.0	-300.0
17	-300.0	-300.0	-300.0	-300.1	-300.0	-300.0
18	-300.2	-300.0	-300.0	-300.0	-299.9	-299.9
19	-300.0	-300.0	-300.0	-300.0	-300.0	-300.1
20	-300.2	-300.0	-299.9	-299.9	-300.0	-300.0
21	-300.1	-299.9	-300.0	-300.1	-300.0	-299.9
22	-300.1	-299.9	-300.0	-300.0	-300.0	-300.0
23	-300.1	-300.0	-300.0	-299.9	-300.0	-300.0
24	-300.1	-299.9	-299.9	-300.1	-300.0	-300.0
25	-300.1	-300.1	-300.0	-300.0	-300.0	-300.0
26	-300.1	-300.0	-300.1	-299.9	-300.0	-300.0
27	-300.1	-300.0	-299.9	-300.0	-299.9	-299.9
28	-300.1	-299.9	-300.0	-300.0	-300.0	-300.0
29	-300.2	-300.0	-299.9	-300.0	-300.0	-300.0
30	-300.1	-300.0	-300.0	-300.0	-300.0	-300.0
31	-300.1	-300.0		-299.9		-300.0
Monthly Values of $\Delta F/F$	-300.09	-299.94	-299.98	-299.99	-300.01	-299.99

J. M. MCDOWELL
Captain, U. S. Navy
Superintendent

7

Examples 8.1 and 8.2 illustrate the procedures to be followed when reducing an emitted time signal to the desired time system using the Bulletin Horaire and the USNO Time Signals Bulletin, i.e., to the Greenwich Mean Astronomic Meridian and to the adopted meridian of the USNO ($\Lambda = -5^h08^m15^s.729$) respectively. From the two examples it can be seen that at the specified epoch, UT2-UT2 (USNO) = 1.7 ms and A.1-A3 = 35.9 ms. The BH, Series J, No. 4, table 3, gives the following systematic monthly differences for August: UT2-UT2(WASH) = 3.3 ms and UT2-UT2(RICH) = 0.2 ms. The adopted UT2(USNO) is determined by averaging UT2(WASH) and UT2(RICH) giving Richmond a weight of two. Using the same averaging for the monthly systematic differences published in the BH, UT2-UT2(USNO) = 1.23 ms, which agrees fairly well with the difference as computed in the numerical examples.

EXAMPLE 8.1

Referencing an Emitted Time Signal to the BIH System

Station: WWV — Epoch: $3^h23^m14^s.2020$ UTC(WWV)
Observatory: USNO, Washington, D. C., Richmond, Florida — Date: August 19, 1964

No.	Source	Type of Correction	Correction 0.1 ms
1	BH Ser.J No. 4, t.1	UT1-UT0(WASH)	+ 131
2	No. 4, t.1	UT1-UT0(RICH)	+ 80
3	No. 1, t.2	UT2-UT1	- 140
4	No. 4, t.6	UT2-A3	-32898
5	No. 4, t.7	UT2-UTC	- 1255
6	No. 4, t.7	UTC-UTC(WWV)	2

Type of Time	Equation	Time: August 19,1964
UT2	UTC(WWV) +(5)+(6)	$3^h23^m14^s.0763$
UT1	UT2-(3)	3 23 14.0903
UT0(WASH)	UT1-(1)	3 23 14.0772
UT0(RICH)	UT1-(2)	3 23 14.0823
A3	UT2-(4)	3 23 17.3661

8.3 Time Receiving and Comparison

The previous section contains information of the dissemination and coordination of precise time and frequency on a worldwide scale. The geodesist wants to use this time in the various geodetic operations requiring different timing accuracies with a minimum loss in precision.

EXAMPLE 8.2

Referencing an Emitted Time Signal to the USNO System

Station: WWV Observatory: USNO, Washington, D.C., Richmond, Florida		Epoch: $3^h23^m14\overset{s}{.}2020$ UTC(WWV) Date: August 19, 1964	
No.	Source	Type of Correction	Correction 0.1 ms
1	Time Signals Bulletin 207-5	UT2(USNO) - UTC(WWV)	- 1274
2	6	UT2(USNO) - UT1(USNO)	- 143
3	6	UT2(USNO) - UT0(WASH)	- 13
4	6	UT2(USNO) - UT0(RICH)	- 63
5	8	UT2(USNO) - A.1	-33274

Type of Time	Equation	Time - August 19, 1964
UT2(USNO)	UTC(WWV) + (1)	$3^h23^m14\overset{s}{.}0746$
UT1(USNO)	UT2(USNO) - (2)	3 23 14.0889
UT0(WASH)	UT2(USNO) - (3)	3 23 14.0759
UT0(RICH)	UT2(USNO) - (4)	3 23 14.0809
A.1	UT2(USNO) - (5)	3 23 17.4020

Time or frequency receiving via transmitted signals does not require too much of a discussion. The main instrument required is a good commercial receiver of the HF or LF/VLF type with a direct or indirect (amplified) output to a device which allows the signals to be displayed on a screen (oscilloscope), on a tape (chronograph), or on a numerical decoder, etc.

The comparison of the received time signals with a local timekeeping device must yield one or both of the following quantities:

(1) Corrections to the epoch of the time kept by the local timekeeper.

(2) Variations in the rate of the local timekeeper.

The epoch correction may be obtained without the use of radio signals through the use of one of the methods in section 8.225 (transporting clocks, via artificial satellites, etc.). The stability of the clocks, on the other hand, is determined almost exclusively by means of the time or frequency broadcasts.

This section outlines mostly those methods which require broadcasts as a standard of comparison. The achievable accuracy is naturally limited to the precision by which the signals can be received and the length of time during which the comparisons are made. At present the precision of HF frequency comparisons, using relatively simple equipment, may vary from a few parts in 10^8 to a few parts in 10^7 when short-term (one second to one hour) measurements are made, but a few parts in 10^{10} have also been achieved when signals were averaged over a 30-day period. The precision of HF time comparisons varies between 0.5 ms and 1 ms, depending on the distance between the transmitter and the receiver [Morgan, 1967]. The reception of LF/VLF signals requires more complex equipment compared to HF receivers. The maximum precision of the received frequency at present varies from 1×10^{-10} to 1×10^{-12}, depending on the frequency of the signal, on the distance traveled, on the averaging period, etc. Maximum timing precision is around 100 μs using standard LF/VLF frequency transmissions and in the order of 0.1 μs with Loran-C or Omega.

The discussion is divided into three groups:

(1) High precision comparisons, yielding local clock times precise to a few milliseconds or fractional frequency offsets ($\Delta f/f$) to within a few parts in 10^{10} or better.

(2) Precise (first-order) comparisons, yielding local clock times to about $0^s.01$ or somewhat better.

(3) Geodetic (second-order) comparisons.

In geodesy high precision methods are applied in satellite and occultation work, first-order comparisons are used in first-order astronomic longitude determinations. Second-order methods are applied in other lower order applications when time is required. It is obvious that the comparison method selected should correspond to the quality of the local timekeeper, e.g., high precision comparison should be made only if the timekeeper is at least a stable quartz crystal chronometer.

8.31 High Precision Time Comparisons Using HF Transmissions

The general characteristics of HF signal propagation have been discussed in section 8.231. For best results, the time comparisons should be scheduled for either an all-daylight or all-night propagation path.

The most frequently used high precision time comparison methods utilizing HF transmissions are the following:

(1) Tick phasing adjustment.

(2) Stroboscopic comparison.

(3) Comparison with delay counters.

In these three methods the basic equipment consists of the local timekeeper, the HF radio receiver and the comparison device, which is, depending on the method, either an oscilloscope, a stroboscope, or a delay counter.

8.311 Comparison by Tick Phasing Adjustment. When using this method the timekeeper at the station must be such that the phase of the generated second-pulses should be manually shiftable either continuously or in small increments.

The oscilloscope may be of the multiple trace or single trace type. It must permit sweep-speed adjustment from about 0.2s/cm to about 1ms/cm (1 cm on the oscilloscope is swept by the signal pulse in one millisecond).

The second-pulse generated by the local clock triggers the sweep of the oscilloscope. The received second-pulses (e.g., for WWV) are also fed to the oscilloscope where they are displayed during the sweep. Initially, the station clock pulse and the received pulse may be apart by as much as a half second. By changing the sweep speed, the leading edge of the received pulse may be located with respect to the station clock pulse. By means of the phasing control and successive oscilloscope sweep-speed adjustment, the two pulses are brought into near coincidence so that higher resolution is possible, and then their distance expressed in time units is recorded either visually or photographically from the oscilloscope scale. This is the correction to the initial setting of the station clock.

The oscilloscope reading normally establishes only the fractional-second part of the difference between the station clock and the received ticks. The total correction to the station clock may comprise additional full hours, minutes, and seconds. Usually hour-, minute-, and second-counters (or the hands of a clock face) are preset to a desired value and manually released at an appropriate instant, e.g., just before a minute tick (the double tick of the WWV) is received.

The drift rate of the station clock may be determined by repeating the comparisons at frequent intervals. The method may be used with certain variations depending on the type of instrumentation. In the case of continuous phasing control, for instance, the pulses from the station clock and the radio signal are brought initially into full coincidence. The station clock then beats the seconds in synchronism with the received second-pulses. Successive checks at known intervals on the synchronism may show progressive deviations of the station clock pulses from the received pulses from which the rate of the station clock may be determined. In this connection see the description of the Newtek-Chronofax at the end of section 8.122 .

Time comparators can also be used in connection with this method. If hooked up to the oscilloscope, the time comparator will show the exact time difference between the clock pulse and the leading edge of the received pulse. The comparator may also produce modulated time markers on the oscilloscope at small intervals which facilitate readings to about ten microseconds. For further details see [Hewlett-Packard, 1965, pp. 2-4 to 2-7] and [Puckle, 1951].

8.312 Comparison with Stroboscopic Flashes. Contrary to oscilloscopes which display the wave form of an impulse, stroboscopes exhibit flashes at the instant of triggering caused by an impulse. Mechanical and electronic stroboscopes are used for time comparisons.

The main parts of the mechanical stroboscope are a cathode ray tube and a synchronous motor driven by the output of the station clock. The motor drives a disk which rotates once per second carrying a scale of (say) 100 units. During the time comparison, each received radio pulse triggers a stroboscopic flash used as an index marker. At the instant of the flashes the scale reading on the disk is recorded. If the rotation speed of the disk is properly calibrated, the readings give the time difference between station clock and received pulse to better than 10 ms.

For higher accuracy, a combination of oscillograph (polar) and electronic stroboscope is used. Instead of the rotating disk, a rotating circle with a superimposed adjustable graduated scale is produced on the oscilloscope. The pulses of the station clock or of the radio signal trigger the stroboscopic flashes which appear in the circle. With an adjustable scale, difference between the flashes can be read to 0.5 ms.

A more modern stroboscope consists of four cathode ray tubes with horizontal and vertical electrostatic deflection plates. The horizontal sweep of each of the tubes is derived from a linear time base which is triggered by the station clock pulse. Superimposed on the first tube there is a time scale in units of $0\overset{s}{.}1$, and on the fourth tube there is one in units of 0.1 ms. The received pulse is fed to the tubes via the vertical deflection plates and is superimposed on the scales. This permits reading of the difference between the time of horizontal sweep triggering (derived from the station clock) and the time of reception of the pulse from the radio source, to 0.1 ms. Photographic registration, auxiliary counters and other modifications, similar to those mentioned in connection with the tick phasing adjustment, could also be made part of this comparison system.

8.313 Comparison with Delay Counter. Time comparison of high accuracy may also be made using time comparators independent of oscilloscopes.

In this case the delay between the pulses of the station clock and the radio signal is counted directly, in units of milliseconds or in smaller units, by a decade counter. The decade counter is started by the station clock pulse and stopped by the received pulse. A drawback is that the station clock starts the counter on time, but the cutoff by the received signal may be less precise due to fading and jitter. Nevertheless, millisecond accuracy may be obtained under favorable reception conditions counting only the large portion of the delay (to about 20 ms) by a decade counter. Displaying the residual portion on an oscilloscope and measuring it by means of a linear scale, improves the results tenfold.

8.32 High Precision Frequency Comparisons Using LF/VLF Transmissions.

The importance of frequency comparisons stems from the fact that while the capabilities of HF propagation remain at precisions of around 1 ms and are limited by ionospheric fluctuation, synchronizations with precisions in the order of 1 μs are now being made by the other methods (portable clock, etc.) described in section 8.225. The difficulty in applying these methods is that extreme reliability is required of the remote clock to maintain the synchronization for long periods. The obvious approach to the solution of this problem is to make use of the extreme stability of LV/VLF propagation and its extremely low attenuation rate, permitting worldwide reception from a few transmitters (the VLF coverage of WWVL, for example, extends over an area of 130° in latitude and 180° in longitude, centered around Fort Collins; the LF coverage is much smaller, e.g., WWVB covers an area of about 30° in latitude and 60° in longitude). The simplest and most accurate way to make use of these advantages is to synchronize a remote clock by use of a portable clock, then maintain the synchronization by manual adjustment, obtaining the necessary drift corrections from VLF reception.

The basic equipment needed for VLF comparisons is much more complex than that for HF work, and it consists of a station clock with oscillator output, a VLF receiver with proper antenna, a VLF phase comparator, and possibly an oscilloscope. The complexity and the high cost of this equipment are the reasons why HF, and not LF or VLF, is mostly used in ordinary geodetic time synchronizations, even at the cost of some loss in accuracy (1 ms vs. 0.1 ms).

Once the initial setting of the station clock is determined the local oscillator frequency is compared with a VLF frequency by means of a phase comparator. While comparisons over several hours may be made, it is preferable to extend the periods to more than 24 hours to minimize the variations associated with diurnal phase shifts caused by ionospheric changes. The phase comparators are either decade counters or self-recording. The latter produce a plotted record of the phase differences in time units; the former are electronic counters that record elapsed time intervals. In the case of the decade counter, the interval count is started and stopped by the frequency output of the station clock (e.g., 1 kHz) and the received frequency (e.g., 20 kHz from WWVL) respectively.

The drift rate of the local oscillator is calculated then either from the time interval counts or from the recorded phase differences (the fractional frequency difference between the local standard and the received frequency is equivalent to the rate of change in phase difference measured over a known time interval).

Example 8.3 illustrates the calculations to obtain the correct time at a certain observation epoch. The station clock at the U.S. Coast and

Geodetic Survey satellite tracking station at Christmas Island is synchronized to the U. S. Naval Observatory time substation USNONC in California via a Sulzer A5 quartz crystal oscillator, while its stability is maintained through VLF signals. (It should be mentioned here that in order to facilitate direct comparisons with the primary standard in Washington, D.C., the USNO has set up a system of time reference and monitor substations at the following locations(1967): Boulder, Colorado; Fuchu-Shi, Tokyo-To, Japan; Geneva, Switzerland; Miami (Perrine), Florida; Palo Alto, California; Wahiawa, Oahu, Hawaii. The times kept by the clocks at these stations are known to an accuracy of better than 2.5 μs with respect to UTC(USNO)).

EXAMPLE 8.3

High Precision Time Comparison with Portable Quartz Crystal Oscillators and VLF Signal Monitoring

Synchronized Clock: Sulzer Model A2.5 Quartz Crystal Oscillator No. (6059-2) at the USCGS Satellite Tracking Station No. 6059 at Christmas Island.

Portable Clocks: Sulzer Model A5 Quartz Crystal Oscillator No. (M5A-8) and No. (M5A-20).

Reference Clock: UTC (USNONC) at the U. S. Naval Observatory Time Reference Substation, USNONC, in Palo Alto, California. This clock is a cesium beam oscillator whose drift is considered negligible. During the period of the comparison, UTC(USNONC) - UTC(USNO) = 0; thus the synchronization may be considered to be with respect to UTC(USNO).

VLF Signals: Stations NLK (18.6 kHz) and NPM (23.4 kHz).

The drift rates of the quartz crystal oscillators are assumed to be a linear function of time, i.e.,

$$\frac{d\Delta T}{dt} = a + bt$$

where ΔT is the clock error, and t is the time from some initial epoch when $a = d\Delta T/dt$. The drift parameters a and b are determined by a least squares fit of a straight line to the daily drift data, obtained from VLF phase comparisons. The equation for the clock error, from the integration of the above expression, is

$$\Delta T = \Delta T_0 + at + \frac{b}{2}t^2 = \Delta T_0 + (a + \frac{b}{2}t)\, t$$

where ΔT_0 is the integration constant (the vertical intercept of the quadratic error curve when $t = 0$), determined first when the oscillator in question is synchronized with the standard.

The drift rate $d\Delta T/dt$ is expressed either in units of time (μs/day) or in units of fractional frequency offset (shift) $\Delta f/f$. The conversion from one set of units to the other is

$$8.64\,\mu s/day = 1 \times 10^{-10}\,(\Delta f/f).$$

The clock error ΔT_0 is always given in units of time (μs).

The aim of the time synchronization and stabilization procedure is to keep the station oscillator (6059-2) always within a certain error limit (a few tenths of a millisecond) with respect to UTC(USNO). This is achieved by periodically synchronizing the station clock via the portable clock to the UTC(USNO) to a few microseconds, and by keeping the fractional frequency offsets within $\pm 10 \times 10^{-10}$ by resetting the frequency of the oscillator whenever its drift approaches this limit. The drift rate of the portable oscillator is determined also from VLF phase comparisons made before the trip (precalibration) and after the trip (postcalibration).

In this example the events took place as shown below:

1. Sequence of Events

	1967 UT Date	Event
Synchronization Trip No. 1	June 1.75 - 7.75	Precalibration of (M5A-8)
	8.05	Synchronization: (USNONC) - (M5A-8)
	8.99 - 9.00	Synchronization: (M5A-8) - (6059-2)
	9.85	Synchronization: (USNONC) - (M5A-8)
	10.00 - 13.75	Postcalibration of (M5A-8)
Maintain Stability	12.75	Reset (6059-2) oscillator frequency by 40 dial divisions (2 dial division = 1×10^{-10} in $\Delta f/f$)
	20.75	Reset (6059-2) oscillator frequency by 28 dial divisions
	28.75	Reset (6059-2) oscillator frequency by 28 dial divisions
	July 7.75	Reset (6059-2) oscillator frequency by 24 dial divisions
	20.75	Reset (6059-2) oscillator frequency by 16 dial divisions
Synchronization Trip No. 2	15.75 - 20.00	Precalibration of (M5A-20)
	20.22	Synchronization: (USNONC) - (M5A-20)
	20.99	Synchronization: (M5A-20) - (6059-2)
	21.88	Synchronization: (USNONC) - (M5A-20)
	22.00 - 27.75	Postcalibration of (M5A-20)

The next table illustrates the calibration of the portable oscillator (M5A-8). The variation of its drift rate is shown in Fig. 8.18.

2. Variations in the Drift Rate of Portable Oscillator No.(M5A-8) from VLF Phase Comparisons at the Base Station

	Julian Date[1] 2439600+	ΔT[2] Drift μs	VLF[3] Correction μs	Mean Epoch 2439600+	$\frac{d\Delta T}{dt}$ Drift Rate μs/day	$\frac{\Delta f}{f}$[4] Drift Rate 10^{-10}	Residual from Linear Drift Rate μs/day
Precalibration at Base	43.25	84	0	43.75	84	9.7	3.0
	44.25	99	0	44.75	99	11.4	-2.9
	45.25	104	0	45.75	104	12.0	1.0
	46.25	115	0	46.75	115	13.3	- .9
	47.25	122	0	47.75	122	14.1	.9
	48.25	133	0	48.75	133	15.4	-1.0
	49.25						
Clock in Transit							
Post Calibration at Base	51.50	119	0	51.875	158	18.3	1.3
	52.25	172	0	52.75	172	19.9	-4.1
	53.25	179	0	53.75	179	20.7	-2.1
	54.25	181	0	54.75	181	20.9	4.8
	55.25						

Equation of Drift Rate determined from least squares fit:

$$\frac{d\Delta T}{dt} = 66.88755 + 8.97574\ (\text{JD} - 2439641.50)\ \mu s/\text{day}$$

Notes:
(1) JD 2439643 = 1967 June $1^{d}5$ UT
(2) From VLF phase comparator record
(3) From USNO or NBS
(4) $1 \times 10^{-10} = 8.64\ \mu s$/day

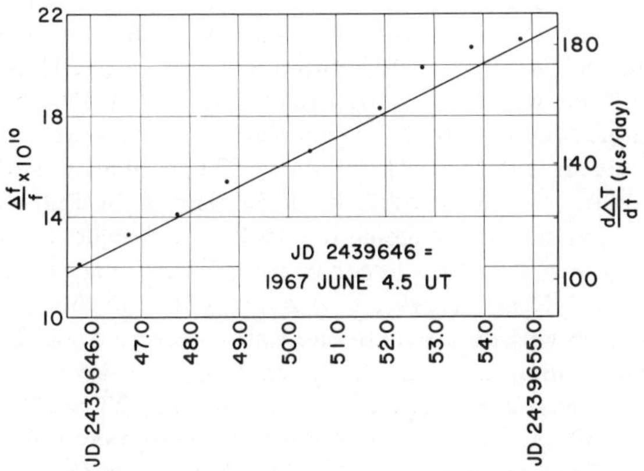

Fig. 8.18 Drift rate of the Sulzer-A5 quartz crystal oscillator M5A-8

The next table illustrates the synchronization computations. The result of this synchronization is that on 1967 June $9^{d}.005$ UT (JD 2439650.505), UTC(6059-2)-UTC(USNONC)=1 μs. The second trip on 1967 July $20^{d}.993$UT (JD 2439692.493) using the oscillator (M5A-20) showed that this difference increased to 244.4 μs.

3. Synchronizations with the Portable Oscillator No.(M5A-8) While in Transit

Julian Date	Station Oscillator	Dial Reading[1] to Synchronize	Mean[2] Epoch	$(\frac{d\Delta T}{dt})_m$[3] Mean Drift Rate	ΔT[4] Drift	ΔT_0[5] Clock Error
2439600+		μs	2439600+	μs/day	μs	μs
49.550	USNONC	494				
50.499	6059-2	4759	50.024	143.40	136.1	4401
Station Oscillator (6059-2) Retarded 4400 μs						
50.505	6059-2	358	50.027	143.42	137.0	1
51.350	USNONC	237	50.450	147.22	265.0	8[6]

Notes:
(1) The unit of dial division on the oscillator clock (M5A-8) corresponds to 1 μs.
(2) Mean Epoch = $\frac{1}{2}$(JD + 2439649.55)
(3) $(\frac{d\Delta T}{dt})_m$ = 66.88755 + 8.97574 (Mean Epoch - 2439641.50) μs/day
(4) $\Delta T = (\frac{d\Delta T}{dt})_m$ (JD - 2439649.55) μs
(5) ΔT_0 = Dial reading at station - Dial reading at USNONC at JD 2439649.55 + ΔT
(6) Trip closure

Fig. 8.19 is a sample of the record of the (NLK, 18.6 kHz) VLF phase comparator for June 1-15, 1967. From the record it can be read that, for example, on the oscillator (6059-2) the phase between 18^{h}UT on June 15 and 18^{h}UT on June 16 shifted 31 - 40 = -9 μs. Thus at the mean epoch 6^{h}UT on June 16, 1967 = 1967 June $16^{d}.25$ UT, the drift rate $d\Delta T/dt$ = -9 μs/day = -1.0×10^{-10}. Similar records are available for the other station (NPM, 23.4 kHz) also. These quantities plotted daily form the drift rate curve $d\Delta T/dt = a + bt$, shown in Fig. 8.20 as $\Delta f/f$, for each period between successive oscillator frequency resets. The figure also shows the clock error curve $\Delta T = \Delta T_0 + at + (b/2)t^2$ for each period. For the first interval, t = 0 corresponds to the epoch when the synchronization with the portable oscillator occurred and therefore ΔT_0 = 1 μs; for any other interval the epoch t = 0 corresponds to the epoch when the oscillator frequency was last reset and $\Delta T_0 = \Delta T$ at the end of the previous interval. In Fig. 8.20 the drift curve shows clearly the frequency resets employed to keep the rate $\Delta f/f$ within the predetermined limits. Note how well the second time check on June 20 fits the error curve.

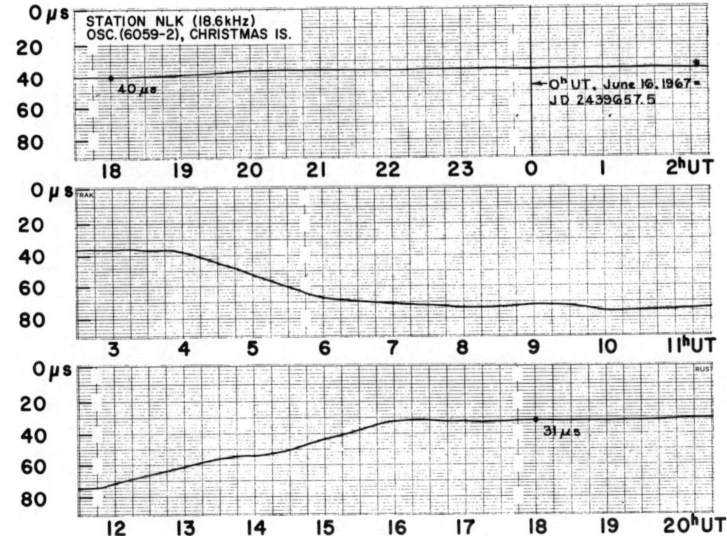

Fig. 8.19 Record of VLF phase comparisons, VLF (18.6 kHz): station oscillator (6059-2). Sample

As a check it is also useful to plot the drift rate curves from both the 18.6 kHz and the 23.4 kHz comparisons against each other. This kind of comparison is shown in Fig. 8.21.

The use of the clock error curves is self explanatory. If the error read from the curve is removed from the time registered by the oscillator, the result is UTC(USNONC), which in this particular example is equal to UTC(USNO). For example, from Fig. 8.20 at 10^h29^mUT on June 30, 1967, $\Delta T = 196\,\mu s$; thus the correction to be applied to the oscillator time to get UTC(USNO) is $-196\,\mu s$.

8.33 First-Order Time Comparisons

The necessary instruments used in connection with these methods almost invariably consist of a chronometer, either mean or sidereal, preferably of the quartz crystal type, an HF receiver, a chronograph, and an amplifier.

Chronographs are devices that produce a graphical record of the beats of a chronometer and/or a radio signal. They are usually either a drum or tape type (Figs. 8.22 and 8.23). Both types may have one or two pens and are driven by either electric current, or clockwork, or weights at adjustable but constant speeds (convenient speed for time comparisons is 1 cm/s).

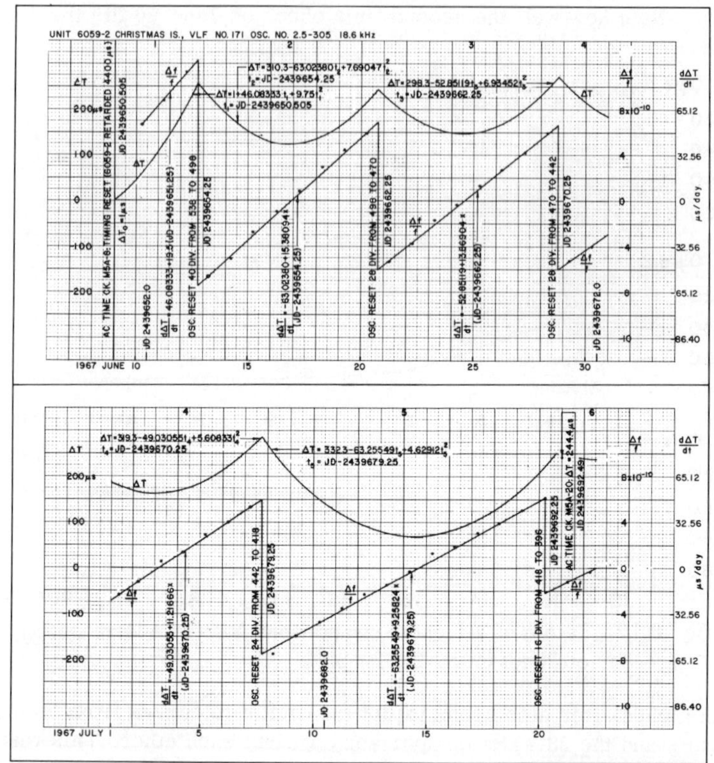

Fig. 8.20 Drift rate and clock error curves for the station oscillator (6059-2) using 18.6 kHz signals

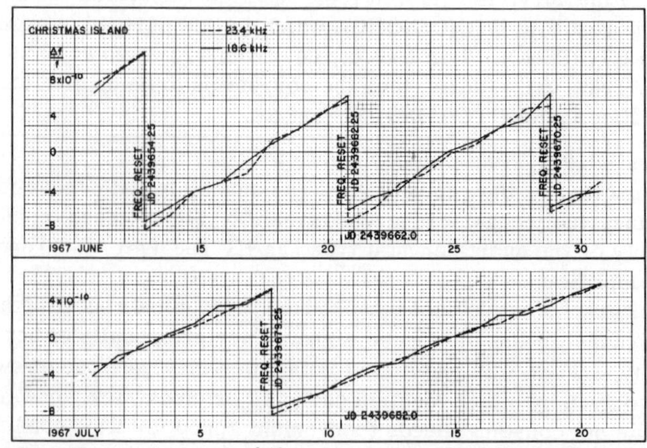

Fig. 8.21 Comparison of drift rate curves for the station oscillator (6059-2) using 18.6 kHz and 23.4 kHz signals

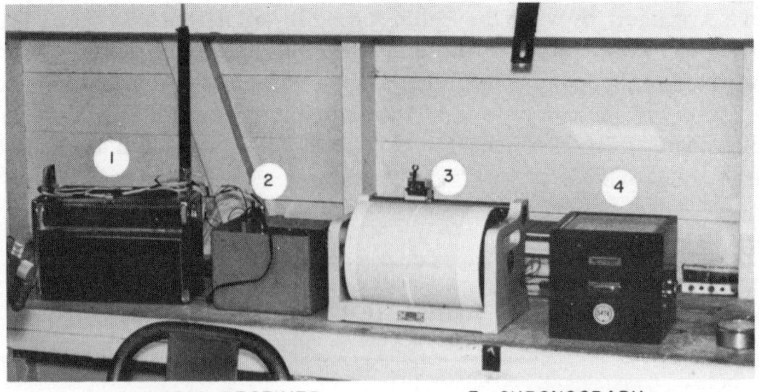

1 RADIO RECEIVER 3 CHRONOGRAPH
2 AMPLIFIER 4 CHRONOMETER

Fig. 8.22 First-order time comparison equipment with one-pen drum chronograph

The chronometers should have electric break circuits by which the pen circuit of the chronograph can be broken for about $0^{s}.2$. Mechanical chronometers generally break either each second or every other second only, depending on the type used.

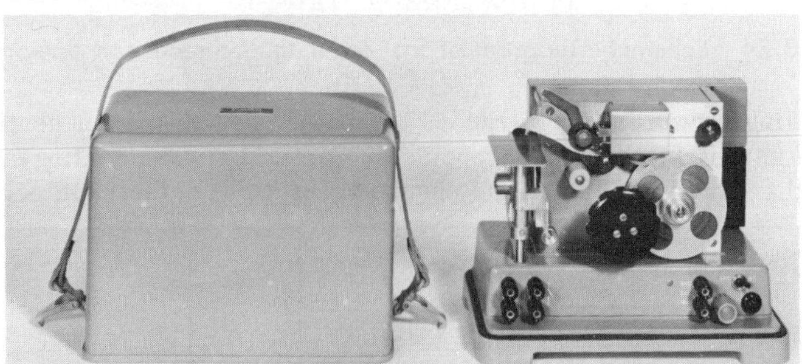

Fig. 8.23 Favag two-pen tape chronograph

The small current which can be passed through the clock often needs amplification to work the chronograph pen. The necessary amplifiers may contain magnetic or electronic relays, the latter being preferable. An audio filter, which could be part of the amplifier, filters out noise and audio frequencies during the time when the second-pulses are recorded.

During the time comparison the chronometer ticks and the radio pulses are fed to the chronograph via the amplifier. In the case of a two pen chronograph each signal may actuate one pen; with a single pen chronograph both signals actuate the same pen. In the former case a 'pen equation' (due to parallax) should be determined by switching both pens to the chronometer output. Similar reasoning applies to the case when two amplifiers are used, or when the radio signal is amplified only. For higher accuracy the relay delay of the amplifier has to be determined also.

The graphical record of the comparison will show a continuous record of the received second-pulses and the chronometer ticks in the form of a broken line. The leading edges of the breaks are taken at the instant of the second-breaks (Fig. 8.24).

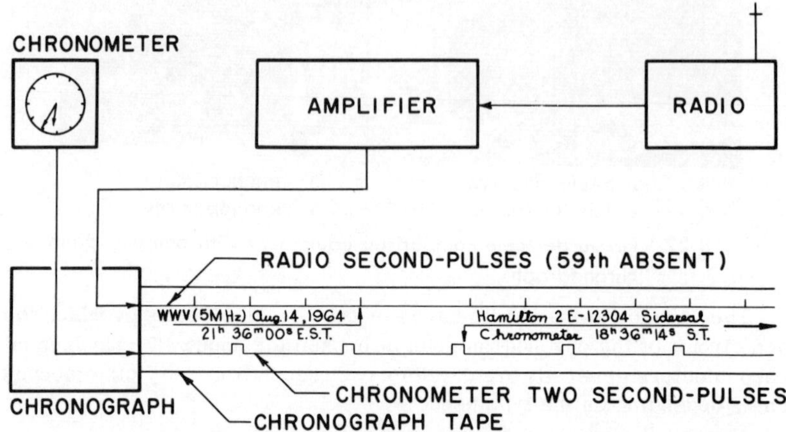

Fig. 8.24 Schemetic diagram of first-order time comparison equipment

Scaling the chronograph record by means of a transparent scaling fan (Fig. 8.25) establishes the chronometer correction. The scaling fan permits the determination of the chronometer time of the radio signal to

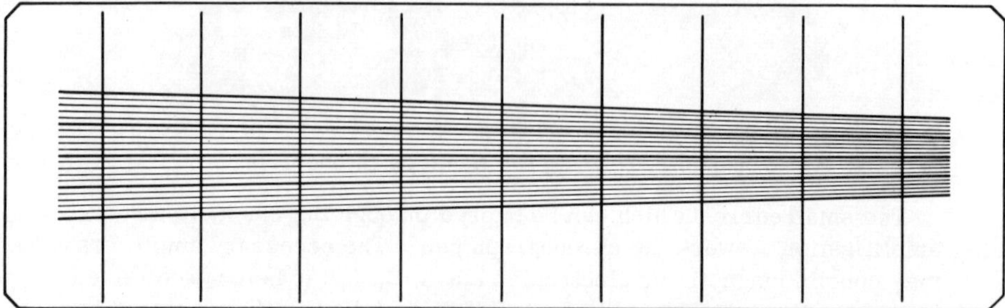

Fig. 8.25 Chronograph record scaling fan

$0^{s}.1$, and $0^{s}.01$ can be estimated. One comparison is extended over about 21 unambiguous radio breaks, and a mean chronometer and mean radio time for the interval is calculated. Hours, minutes, and seconds must be identified on the chronograph record for both chronometer and radio time.

Depending on whether a mean time or a sidereal chronometer is used,

the correction to the chronometer time is found directly from the comparison, or through further computations (see Example 8.4).

Repetition of the comparison will establish another chronometer correction. From the difference between the corrections and the known time interval between comparisons, the rate of the chronometer can be calculated by generally assuming that the chronometer correction changes either linearly or parabolically with time [Milasovszky, 1960].

For further details the reader is referred to [Hoskinson and Duerksen, 1947, pp. 5-14] and [Thorson, 1965, pp. 5-14].

Example 8.4 illustrates how the chronometer time of an HF signal is determined from the mean of 21 scalings and the computations necessary to determine the rate of a sidereal chronometer.

EXAMPLE 8.4

First-Order Time Comparison with HF Signal Monitoring

1. Scaling Record of Radio-Chronometer Comparison

Station: Lanum, Ohio
Local Date: August 14, 1964

Transmitting Station: WWV 5 MHz
Chronometer: Hamilton 2E-12304 sidereal

No.	EST of Signal	Chronometer Time of Signal	EST of Signal	Chronometer Time of Signal	EST of Signal	Chronometer Time of Signal
1	$19^h56^m34^s$	$16^h56^m35\overset{s}{.}60$	$21^h01^m27^s$	$18^h01^m39\overset{s}{.}23$	$21^h35^m41^s$	$18^h35^m58\overset{s}{.}82$
2	35	36.60	28	40.24	42	59.82
3	36	37.61	29	41.24	44	61.83
4	37	38.61	30	42.25	45	62.83
5	38	39.61	31	43.25	46	63.84
6	39	40.61	32	44.25	48	65.84
7	40	41.62	33	45.26	49	66.85
8	41	42.62	34	46.27	50	67.85
9	42	43.62	35	47.27	51	68.86
10	43	44.62	36	48.27	52	69.86
11	44	45.63	37	49.28	53	70.87
12	45	46.63	38	50.28	54	71.88
13	46	47.63	39	51.28	55	72.88
14	47	48.64	40	52.28	56	73.88
15	48	49.64	41	53.28	57	74.88
16	49	50.64	42	54.29	58	75.89
17	50	51.65	43	55.29	60	77.90
18	51	52.65	44	56.29	61	78.90
19	52	53.66	45	57.30	62	79.90
20	53	54.66	46	58.30	64	81.90
21	54	55.67	47	59.30	65	82.90
Mean	$19^h56^m44\overset{s}{.}0$	$16^h56^m45\overset{s}{.}630$	$21^h01^m37\overset{s}{.}0$	$18^h01^m49\overset{s}{.}271$	$21^h35^m53\overset{s}{.}0$	$18^h36^m10\overset{s}{.}866$

2. Chronometer Rate Determination

Station: Lanum, Ohio — Assumed longitude: $-5^h34^m10^s.609$
Chronometer: Hamilton 2E-12304 sidereal — Assumed latitude: $39°39'39''.65$

Local date	August 14, 1964	August 14, 1964	August 14, 1964
Transmitting station	WWV	WWV	WWV
Frequency	5 MHz	5 MHz	5 MHz
EST of signal[1]	$19^h56^m44^s.000$	$21^h01^m37^s.000$	$21^h35^m53^s.000$
Zonal correction[2]	+5	+5	+5
Greenwich date	August 15, 1964	August 15, 1964	August 15, 1964
UTC(WWV) of signal	$0^h56^m44^s.000$	$2^h01^m37^s.000$	$2^h35^m53^s.000$
GMST at 0^hUT	$21^h33^m43^s.771$	$21^h33^m43^s.771$	$21^h33^m43^s.771$
Solar to sidereal interval correction	$9^s.319$	$19^s.978$	$25^s.608$
GMST of signal	$22^h30^m37^s.090$	$23^h35^m40^s.749$	$24^h10^m02^s.379$
Λ	$-5^h34^m10^s.609$	$-5^h34^m10^s.609$	$-5^h34^m10^s.609$
MST of signal = GMST + Λ	$16^h56^m26^s.481$	$18^h01^m30^s.140$	$18^h35^m51^s.770$
Chronometer time of signal[1]	$16^h56^m45^s.630$	$18^h01^m49^s.271$	$18^h36^m10^s.866$
Chronometer correction	-19.149	-19.131	-19.096
Rate per minute (assuming linear chronometer correction)	0.277 ms		1.021 ms

[1] From the previous table.
[2] See Examples 5.1 and 5.2 for conversion of solar time to sidereal time.

8.34 Second-Order Time Comparisons

This group of comparisons comprises aural and aural-visual methods, and the so-called extinction method. These methods are used mainly in connection with rhythmic signals (61 ticks per minute and a prolonged dash at the minute).

In the aural comparison the operator listens to the radio and chronometer ticks by means of earphones and records the chronometer times of coincident ticks.

In the aural-visual method one listens to the radio ticks and observes the second-hand of the chronometer, noting the chronometer time (of the dashes).

In the extinction method the chronometer is wired to the radio in such a way that the radio is silenced when coincident seconds occur. These coincidences are noted.

Rather elaborate techniques have been worked out in the past to obtain reliable results. For further details the reader is referred to [Biddle, 1958, pp. 192-197; Roelofs, 1950, pp. 84-95].

References

Allan, D.W. (1966). "Statistics of Atomic Frequency Standards." Proceedings of the Institute of Electrical and Electronics Engineers, 54, 2.

Beehler, R.E. (1967). "A Historical Review of Atomic Frequency Standards." Proceedings of the Institute of Electrical and Electronics Engineers, 55, 6.

Beehler, R. E. and D. J. Glaze. (1966). "The Performance and Capability of Cesium Beam Frequency Standards at the National Bureau of Standards." Institute of Electrical and Electronics Engineers, Transactions on Instrumentation and Measurement, IM-15, 1-2.

Beehler, R. E., R. C. Mockler, and J. M. Richardson. (1965). "Cesium Beam Atomic Time and Frequency Standards." Metrologia, 1, 3.

Biddle, C. A. (1958). Textbook of Field Astronomy. H. M. Stationery Office, London.

Blair, B. B. and A. H. Morgan (1965). "Control of WWV and WWVH Standard Frequency Broadcasts by VLF and LF Signals." Journal of Research, National Bureau of Standards (Radio Science), 69D, 7.

Bodily, L.N. (1965). "Correlating Time from Europe to Asia with Flying Clocks." Hewlett-Packard Journal, 16, 8.

Bodily, L. N., D. Hartke, and R.C. Hyatt. (1966). "World-Wide Time Synchronization, 1966," Hewlett-Packard Journal, 17, 12.

Bodily, L.N. and R. C. Hyatt. (1967). "Flying Clock Comparisons Extended to East Europe, Africa and Australia." Hewlett-Packard Journal, 19, 4.

Brown, J.N. (1949). "Round-the-World Signals at Very Low Frequency." Journal of Geophysical Research, 54

Chi, A.R. and S.N. Witt. (1966). "Time Synchronization of Remote Clocks Using Dual VLF Transmissions." Proceedings of the 20th Annual Symposium on Frequency Control, April 19, Atlantic City, New Jersey.

Crampton, S. B., D. Kleppner, and N. F. Ramsey. (1963). "Hyperfine Separation of Ground State Atomic Hydrogen." Physical Review Letters, 11, 7.

Essen, L. and J. Parry. (1957). "The Cesium Resonator as a Standard of Frequency and Time." Philosophical Transactions of the Royal Society of London, 250, Series A, pp. 45-69.

Fey, L. and C.H. Looney, Jr. (1966). "A Dual Frequency VLF Timing System." Institute of Electrical and Electronics Engineers, Transactions on Instrumentation and Measurement, IM-15, 4.

Frequency Control Symposium. (1965). "From Tuning Forks to Flying Clocks." Frequency, May-June.

Gatterer, L. (1968). Satellite Timing Techniques and Moonbounce Radar Timing. Proceedings of the Frequency and Time Seminar, National Bureau of Standards, Boulder, Colorado, February 28-March 1.

Gerber, E.A. and R. A. Sykes. (1966). "State of the Art—Quartz Crystal Units and Oscillators." Proceedings of the Institute of Electrical and Electronics Engineers, 54, 2.

Goldenberg, H. M., D. Kleppner, and N. F. Ramsey. (1960). "Atomic Hydrogen Maser." Physical Review Letters, 5, 8.

Guinot, B. (1967). Rapport sur les travaux du Bureau International de l'Heure. Presented at the XIVth General Assembly of the International Union of Geodesy and Geophysics, Luzerne, September 25-October 7.

Halford, D. (1968). Measurement of Frequency Stability. Proceedings of the Frequency and Time Seminar, National Bureau of Standards, Boulder, Colorado, February 28-March 1.

Hewlett-Packard. (1965). "Frequency and Time Standards." Hewlett-Packard Co., Application Note, 52, Palo Alto, California.

Hoskinson, A. J. and J. A. Duerksen. (1947). "Manual of Geodetic Astronomy." USCGS Special Publication, 237, U.S. Government Printing Office, Washington, D. C.

Hudson, G. E. (1967). "Some Characteristics of Commonly Used Time Scales." Proceedings of the Institute of Electrical and Electronics Engineers, 55, 6.

Hydrographic Office, U.S. Navy. (1962-64). "Radio Navigational Aids." Hydrographic Office Publication, 117A and 117B, U.S. Government Printing Office, Washington, D. C.

International Astronomical Union XIIth General Assembly, Hamburg. (1964). "Clarifications and Resolutions." Bulletin Géodésique, 74, pp. 315-316.

Kleppner, D. (1962). "Properties of the Hydrogen Maser." Applied Optics, 1, p. 55.

Kloppner, D., H. M. Goldenberg, and N. F. Ramsey. (1962). "Theory of the Hydrogen Maser." Physical Review, 126, 1.

Markowitz, W. (1964a). "Time Determination and Distribution: Current Developments and Problems." Proceedings of the International Conference on Chronometry, Lausanne, 1, pp. 157-177.

Markowitz, W. (1964b). "International Frequency and Clock Synchronization." Frequency, July-August.

Markowitz, W. and C. A. Lidback. (1966). "Clock Synchronization via Relay II Satellite." Proceedings of the 1966 Conference on Precision Electromagnetic Measurements, National Bureau of Standards, Boulder, Colorado, June 21-24.

Marrison, W. A. (1948). "The Evolution of the Quartz Crystal Clock." Bell Systems Technical Journal, 27, pp. 510-558.

McCoubrey, A. O. (1966). "A Survey of Atomic Frequency Standards." Proceedings of the Institute of Electrical and Electronics Engineers, 54, 2.

McCoubrey, A. O. (1967). "The Relative Merits of Atomic Frequency Standards." Proceedings of the Institute of Electrical and Electronics Engineers, 55, 6.

Milasovszky, B. (1960). "Astronomical Longitude Determination Under Special Conditions." Publications of the Technical University for Heavy Industry, XXI, pp. 177-202, Miskolc, Hungary.

Mockler, R.C. (1964). "Atomic Frequency and Time Interval Standards." Journal of Research, National Bureau of Standards (Radio Science), 68D, 5.

Morgan, A. H. (1967). "Distribution of Standard Frequency and Time Signals." Proceedings of the Institute of Electrical and Electronics Engineers, 55, 6.

National Bureau of Standards, U. S. (1967). "NBS Standard Frequency and Time Services." National Bureau of Standards Miscellaneous Publication, 236, U.S. Government Printing Office, Washington, D.C.

Puckle, O.S. (1951). Time Bases. John Wiley & Sons, Inc., New York.

Ramsey, N.F. (1962). "The Atomic Hydrogen Maser." Institute of Radio Engineers, Transactions, I-11, pp. 177-182.

Reder, F., P. Brown, G. Winkler, and G. Bickard. (1961). "Final Results of a World-Wide Clock Synchronization Experiment (Project WOSAC)." Proceedings of the 15th Annual Symposium on Frequency Control, U. S. Army Research and Development Laboratory, Fort Monmouth, New Jersey, p. 226.

Robbins, A. R. (1960-66). "Commercially Available Portable Crystal Chronometers." Circular Nos. 1, 1A, 1B, 1C, 1D distributed to members of Special Study Group 3.04 of the International Association of Geodesy.

Robbins, A.R. (1967). Report of the IAG Special Study Group 3.04. Presented at the XIVth General Assembly of the International Union of Geodesy and Geophysics, Luzerne, September 25-October 7.

Roelofs, R. (1950). Astronomy Applied to Land Surveying. N. V. Wed. J. Ahrend & Zoon, Amsterdam.

Shapiro, L. D. (1968). "Loran-C Timing and Frequency Comparison." Frequency, 6, 3.

Steele, McA. J., W. Markowitz, and C.A. Lidback. (1964). "Telstar Time Synchronization." Institute of Electrical and Electronics Engineers, Transactions on Instrumentation and Measurement, IM-13, 4.

Steiner, C. (1968). "Atomic Clocks." Frequency, 6, 3-5.

Stoyko, A. (1964a). "Heure Definitive des Signaux Horaires et le Temps

Atomique. Historique Definitions, Method du Calcul, Utilisation." Bulletin Horaire, Series H, 1, pp. 1-10.

Stoyko, A. (1964b). "La Reduction des Heures Definitives dans une Systeme Uniforme et un Pole Moyen de l'Epoque." Bulletin Horaire, Series H, 2, pp. 41-44.

Thorson, C. W. (1965). "Second Order Astronomical Position Determination Manual." U.S. Coast and Geodetic Survey Publication, 64-1, U. S. Government Printing Office, Washington, D. C.

Vessot, R. F. C. and H. E. Peters. (1962). "Design and Performance of an Atomic Hydrogen Maser." Institute of Radio Engineers, Transactions, I-11, 3 and 4.

Vessot, R.F.C., H. E.Peters, and J. Vanier. (1964). "Recent Developments in Hydrogen Masers." Frequency, July-August.

Vigoureux, P. (1939). Quartz Oscillators and Their Application. H. M. Stationery Office, London.

Vigoureux, P. and C.F. Booth. (1950). Quartz Vibrators. H.M. Stationery Office, London.

9 DETERMINATION OF ASTRONOMIC AZIMUTH

In this and in the following two chapters certain selected methods of astronomic position (latitude, longitude) and azimuth determinations are given. The discussion in each chapter is divided according to the required precision which generally lies between the following limits:

Type of Work	Required Standard Deviation	
	Latitude or Longitude	Azimuth
observatory	0".015 - 0".09	—
first-order	0".1 - 0".3	0".2 - 0".4
second-order	0".4 - 1".0	0".5 - 1".5

The precision of the final position or azimuth depends on many factors: the quality of the instruments, the method of position determination and that of the observation, the extent to which the optimum observational conditions for the given method are observed, the number of repetitions, the recognition and successful removal of systematic errors, the experience of the observer, and the latitude of the observer. These considerations guide the choice of the program and the method of determination and the instruments to be used to obtain the various degrees of precision.

The choice and use of the instruments and the elimination of the systematic errors arising from their imperfections have been described in Chapters 7 and 8. In the reduction computations, stars are generally updated only to their apparent positions; thus astronomic refraction and diurnal aberration are viewed as systematic errors affecting the observations. Compensating corrections for their effects are directly applied to the observed quantities as outlined in sections 4.214 and 4.232.

Where practicable, those aspects of the above factors affecting the final results which have not been discussed in the previous chapters will be taken into account in the summary of the position determination methods given here and in the next two chapters, the purpose being to provide a basis for the selection and evaluation of a method for a particular purpose and to determine the number of repetitions that might be required to attain a desired precision.

The astronomic azimuth of a line, connecting two points on the earth's surface, may be determined by measuring the horizontal angle between the line and the vertical plane of a star and by adding or subtracting this angle to or from the azimuth of the star. If the observer's position and the star's identity are known, only a single quantity need be determined to solve the astronomic triangle for azimuth. This may be either the star's hour angle or its zenith distance. In either case, the only major instruments required are a geodetic theodolite and a chronometer. In addition to these, for the hour angle determinations, an HF radio receiver with an amplifier and a chronograph with a manual key could prove useful; to determine the refraction angle when measuring zenith distances, a thermometer and a barometer are necessary.

9.1 Second-Order Azimuth Determination

9.11 Azimuth by Star Hour Angles

The hour angle of the star is not directly observed but is determined by the reading of a sidereal chronometer at the instant of the star's transit across the vertical thread. The star's azimuth may then be computed from equations (3.1) and (3.3) which by division yield

$$\tan A = \frac{\sin h}{\sin \Phi \cos h - \cos \Phi \tan \delta} \tag{9.1}$$

where the azimuth A is measured clockwise from north through 360°, and the hour angle $h = ST - \alpha$ is measured westward from upper culmination through 24^h in accordance with the sign convention established earlier.

It is desirable to select stars and a method which will minimize the effect of systematic errors in the results. The sources of possible systematic errors include the error $d\Phi$ in the assumed latitude and dh in the hour angle, the latter arising from an error in the chronometer correction. Differentiation of equation (9.1) or equations (3.28) and

(3.31) yields the effect of these errors on the computed azimuth:

$$dA = \sin A \cot z \, d\Phi + \cos \Phi (\tan \Phi - \cos A \cot z) dh. \quad (9.2)$$

The error in the latitude is thus eliminated when the star's azimuth is either 0° or 180° (at culmination), and the error in the hour angle is eliminated when the star is observed at elongation when, from equations (3.5) and (3.51),

$$\tan \Phi = \cos A \cot z.$$

Both conditions cannot be satisfied simultaneously, but the effects will be eliminated if the azimuth is determined from the mean of two stars selected so that

$$\sin A_1 \cot z_1 = -\sin A_2 \cot z_2, \quad (9.3)$$

$$2 \tan \Phi - \cos A_1 \cot z_1 = \cos A_2 \cot z_2. \quad (9.4)$$

Charts designed for the purpose of aiding in the selection of stars in accordance with these expressions are available in the literature, e.g., in [Stoch, 1963]. By investigating the conditions for which (9.3) and (9.4) will concurrently have the minimum effect, taking into account the systematic instrumental errors mentioned in section 7.41, and restricting z to a maximum of 75° due to refraction, the following general conclusions may be arrived at:

(1) At stations of latitude less than 15°, two stars should be observed, one south at $z_2 = 75^\circ$ and one north at z_1 complying with

$$\cot z_1 = 2 \tan \Phi + \cot 75^\circ. \quad (9.5)$$

This method of azimuth determination (by the hour angles of stars near culmination) is described in section 9.113.

(2) At latitudes greater than 15°, Polaris should be observed at any hour angle (see section 9.111). If this is not possible, two north stars should be observed near the meridian, above and below the pole and complying with

$$\cot z_1 + \cot z_2 = 2 \tan \Phi. \quad (9.6)$$

This method of azimuth determination (by the hour angles of stars near culmination) is described in section 9.113.

(3) If the latitude is accurately known, then the only systematic error is that of the hour angle, which as shown before may be eliminated by observations at elongation, keeping z as large as possible to minimize the effects of the inclination of the horizontal axis and collimation as expressed by equation (7.14). This method of azimuth determination (by the hour angles of stars near elongation) is treated in section 9.112.

The above methods can also be used for first-order azimuth determinations with appropriately chosen instruments, number of repetitions, and other small modifications as described in section 9.2.

9.111 Azimuth by the Hour Angle of Polaris. The horizontal angle between Polaris and the reference mark is measured and the ST of pointing at the star is recorded. The hour angle of the star is deduced from the observed time through its right ascension, and its azimuth is computed from equation (9.1) or from the 'Pole Star Table' of the 'American Ephemeris and Nautical Almanac.' The azimuth of the line is obtained by adding or subtracting the measured horizontal angle.

A suggested sequence of observation is the following:

(1) Direct on mark, record horizontal circle reading.

(2) Direct on star, record time and horizontal circle reading.

(3) Repeat (2).

(4) Repeat (1).

(5) Reverse telescope and repeat (1) - (4).

This constitutes one determination. The sequence of reduction is:

(1) If the time kept by the chronometer is not sidereal, compute the AST corresponding to each time of observation.

(2) Compute the apparent hour angle corresponding to each observation:

$$h = AST - \alpha.$$

(3) Compute the azimuth of Polaris and the corresponding azimuth of the reference mark by adding or subtracting the horizontal angle deduced from the circle readings.

If the theodolite is equipped with a striding level, a correction for the inclination of the horizontal axis ($\pm b \cot z$) could be applied in accordance with equation (7.14); thus appropriate readings on both ends of the level should be made after the pointings on the star. Correction for collimation is unnecessary since the observations are made in alternate positions of the telescope.

The final azimuth will be the mean obtained from all the observations. Generally eight determinations provide satisfactory second-order results.

Computations may be shortened by using the mean direction and time for each pair of observations, or for all observations in one determination, as a single observation. The azimuth so computed will be the azimuth A_0 corresponding to the average sidereal time (hour angle) of the n observations

$$T_0 = \frac{T_1 + T_2 + \ldots + T_n}{n} \tag{9.7}$$

but will not be equal to the required mean of the azimuths A_1, A_2, . . . ,

A_n corresponding to the times (hour angles) of the n pointings since the rate of change of the azimuth is constantly varying on account of the curvature (nonlinearity) of the apparent path of the star. The difference

$$\Delta A_C = \frac{A_1 + A_2 + \ldots + A_n}{n} - A_O \qquad (9.8)$$

is small but not always negligible. It is compensated by computing a second-order 'curvature correction' ΔA_C, which is to be added to the azimuth A_O. Since the azimuth of a star is a function of the sidereal time (hour angle) T_i of the observation, it may be expressed as

$$A_i = f(T_i) = f(T_O + \tau_i) = A_O + \frac{dA}{dh}\tau_i + \tfrac{1}{2}\frac{d^2A}{dh^2}\tau_i^2. \qquad (9.9)$$

Adding the n equations of the above form, dividing the sum by n, and making use of equations (9.7) and (9.8), it becomes evident that

$$\Delta A_C = \tfrac{1}{2}\frac{d^2A}{dh^2}\left(\frac{\tau_1^2 + \tau_2^2 + \ldots + \tau_n^2}{n}\right). \qquad (9.10)$$

In order that ΔA_C be expressed in seconds of arc, it is necessary to multiply the right-hand side of this equation by sin 1"; and since τ_i (in seconds of arc) is small, it is possible to make the following approximation:

$$\tfrac{1}{2}\tau_i^2 \sin 1'' = \frac{2\sin^2\frac{1}{2}\tau_i}{\sin 1''} = m_i. \qquad (9.11)$$

With this understanding, equation (9.10) may be written in the form

$$\Delta A_C'' = \frac{d^2A}{dh^2}\frac{1}{n}\sum_{i=1}^{n} m_i. \qquad (9.12)$$

From equation (3.4)

$$\frac{dA}{dh} = -\frac{\cos\delta\cos h}{\sin z\cos A}; \qquad (9.13)$$

therefore,

$$\frac{d^2A}{dh^2} = \frac{\sin z\cos A\sin h\cos\delta - \sin z\sin A\cos h\cos\delta\,(dA/dh)}{\sin^2 z\cos^2 A}. \qquad (9.14)$$

The substitution of sin z from equation (3.4) and dA/dh from (9.13) into (9.14) yields

$$\frac{d^2A}{dh^2} = \frac{\cos^2 h \tan A - \cot A \sin^2 h}{\sin^2 h \cot^2 A}$$

$$= \frac{\tan A}{\sin^2 h}\left(\frac{\cos^2 h \sin^2 A - \sin^2 h \cos^2 A}{\cos^2 A}\right)$$

$$= \frac{\tan A}{\sin^2 h}\left(\frac{\cos^2 h - \cos^2 A}{\cos^2 A}\right) = C_A . \qquad (9.15)$$

With this expression, equation (9.12) yields the curvature correction for azimuth observations:

$$\Delta A_C'' = \frac{C_A}{n} \sum_{i=1}^{n} m_i . \qquad (9.16)$$

For a star within a few degrees of the meridian and for the purpose of this correction, $\cos A \approx 1$; thus equation (9.15) yields

$$C_A = -\tan A,$$

and, therefore,

$$\Delta A_C'' = -\frac{\tan A}{n} \sum_{i=1}^{n} m_i . \qquad (9.17)$$

The quantity m_i may frequently be found tabulated in various publications, e.g., in table 9 in [Roelofs, 1950] or in table XII in [Hoskinson and Duerksen, 1952]. The latter publication, in table XIV, also provides values for the entire curvature correction, based on equation (9.17) and $n = 2$, for azimuths between 0° and $\pm 3^\circ 10'$.

The curvature correction for Polaris is negligibly small if the mean azimuth A_0 is computed from the times of a single pair of direct and reverse observations ($n = 2$) less than $1^m 30^s$ apart from each other.

The method of determining azimuth by the hour angle of Polaris is illustrated in Example 9.1.

EXAMPLE 9.1

Azimuth by the Hour Angle of Polaris

Observer: A. Gonzalez-Fletcher
Computer: A. Gonzalez-Fletcher
Local Date: 3-31-1965
Location: OSU old astro pillar
$\Phi \cong 40°00'00''$
$\Lambda \cong -5^h32^m10^s$
Star Observed: FK4 No. 907
(α Ursae Minoris [Polaris])

Instruments: Kern DKM3-A theodolite (No. 82514); Hamilton sidereal chronometer (No. 2E12304); Favag chronograph with manual key; Zenith transoceanic radio.
Azimuth Mark: West Stadium

1. Level Corrections (Sample)

$$\Delta A = \frac{d}{4}\,[(w+w') - (e+e')]\cot z$$

	Determination No.		
	1	2	3
d/4 cot z	0.″347	0.″347	0.″347
w + w′	41.8	41.7	41.9
e + e′	43.3	43.3	42.8
ΔA	-0.″52	-0.″59	-0.″31

d = 1.″6/division

2. Azimuth Computation (Sample)

$$\tan A = \sin h/(\sin\Phi \cos h - \cos\Phi \tan\delta)$$

		Determination No.		
		1	2	3
1	Mean chronometer reading of direct and reverse pointings	$9^h22^m02.^s35$	$9^h36^m13.^s85$	$10^h11^m38.^s16$
2	Chronometer correction (computed similar to Example 8.4)	$-5^m51.^s25$	$-5^m51.^s28$	$-5^m51.^s33$
3	AST	$9^h16^m11.^s10$	$9^h30^m22.^s57$	$10^h05^m46.^s83$
4	α (apparent, see Example 10.2)	$1^h57^m53.^s46$	$1^h57^m53.^s46$	$1^h57^m53.^s46$
5	h = AST - α	$7^h18^m17.^s64$	$7^h32^m29.^s11$	$8^h07^m53.^s37$

		Determination No.		
		1	2	3
6	h (arc)	109°34'24".60	113°07'16".65	121°58'20".55
7	δ (apparent, see Example 10.2)	89°06'12".92	89°06'12".92	89°06'12".92
8	cos h	-0.33501582	-0.39267890	-0.52951032
9	tan δ	63.91162817	63.91162817	63.91162812
10	sin Φ cos h	-0.21534402	-0.25240913	-0.34036267
11	cos Φ tan δ	48.95914742	48.95914742	48.95914742
12	(10) - (11)	-49.17449144	-49.21155655	-49.29951009
13	sin h	0.94221251	0.91967535	0.84830350
14	tan A = (13)/(12)	-0.01916059	-0.01868820	-0.01720714
15	A (at the average h)	358°54'08".33	358°55'45".73	359°00'51".12
16	Time difference between direct and reversed pointings (2τ)	$7^m12^s.4$	$2^m44^s.5$	$2^m36^s.5$
17	Curvature correction (equation (9.17))	0".5	0".1	0".1
18	A (Polaris) = (15) + (17)	358°54'08".8	358°55'45".8	359°00'51".2
19	Circle (horizontal) reading on Polaris	258°25'48".9	258°27'28".2	258°32'39".0
20	Correction for dislevelment (ΔA)	-0".5	-0".6	-0".3
21	Corrected circle reading = (19) + (20)	258°25'48".4	258°27'27".6	258°32'38".7
22	Circle reading on Mark	00°01'13".7	00°01'14".7	00°01'17".5
23	Angle between Mark and Polaris = (22) - (21)	101°35'25".3	101°33'47".1	101°28'38".8
24	Azimuth of Mark = (18) + (23)	100°29'34".1	100°29'32".9	100°29'30".0

3. Final Observed Azimuths from Eight Determinations

Determination No.	Azimuth of Mark	v	vv
1	100°29'34".1	-2".39	5.71
2	32 .9	-1.19	1.42
3	30 .0	1.71	2.92
4	31 .3	0.41	0.17
5	32 .0	-0.29	0.08
6	30 .3	1.41	1.99
7	31 .6	0.11	0.01
8	31 .5	0.21	0.04
Final (mean)	100°29'31".71	[v] = -0".02	[vv] = 12.35

Standard deviation of an azimuth determination:

$$m_A = \sqrt{\frac{[vv]}{n-1}} = \sqrt{\frac{12.35}{8-1}} = 1''.33$$

Standard deviation of the mean azimuth:

$$M_A = \frac{m_A}{\sqrt{n}} = \frac{1''.33}{\sqrt{8}} = 0''.47$$

Result:

$$A_T\,(\text{West Stadium}) = 100^\circ 29' 31''.7 \pm 0''.5$$

(In these calculations the effect of diurnal aberration on the azimuth has been neglected. If this is not considered permissible, the procedure outlined in section 9.21 should be followed.)

9.112 Azimuth by the Hour Angles of Stars Near Elongation. The star's practically constant azimuth near elongation is obtained from the simple relation

$$\sin A\ (\text{elongation}) = \cos\delta \sec\Phi. \qquad (3.50)$$

Of the two solutions, the smaller ($0^\circ < A < 90^\circ$) is for the eastern elongation. The azimuth of the western elongation is 360° minus the smaller solution. The azimuth of the line is obtained by applying the measured horizontal angle between the reference mark and the star to the azimuth of the star.

Stars must be selected prior to observation for which purpose the time of elongation is computed from

$$ST = \alpha + h$$

where the hour angle of the elongation h is given by

$$\cos h\ (\text{elongation}) = \cot\delta \tan\Phi. \qquad (3.52)$$

Of the two solutions, the smaller ($0^h < h < 6^h$) is for the western elongation, and the larger ($18^h < h < 24^h$) is for the eastern elongation. Factors to keep in mind are the following: For a star to elongate, δ must be greater than Φ; the lower the altitude of observation, the less will be the effect of the inclination of the horizontal axis and of collimation; and the greater the declination, the lower will be the altitude of elongation. One star will suffice, but the usual procedure is to use two stars (star groups) paired east and west. Polaris may be observed as a single star, but the time of elongation is often inconvenient.

A suggested sequence of observation is:

(1) Carefully level instrument and record horizontal circle reading on the reference mark.

(2) Five to fifteen minutes before elongation, intersect the star and record sidereal time and horizontal circle reading.

(3) Repeat (2).

(4) Repeat (1).

(5) Reverse telescope and repeat (1) - (4).

(6) Repeat sequence for the other star of the pair.

This constitutes two determinations (one on each star). Normally, eight determinations are used to compute the final azimuth.

The reduction procedure is:

(1) Compute the star's apparent azimuth at elongation from (3.50).

(2) Apply curvature correction to the computed azimuth of elongation to obtain the azimuth of one determination corresponding to the mean epoch of the four pointings:

$$A = A\ (\text{elongation}) + \Delta A_C. \tag{9.18}$$

Similar to equation (9.16),

$$\Delta A_C = \frac{C_A}{4} \sum_{i=1}^{4} m_i. \tag{9.19}$$

It may be shown that in the above equation

$$C_A = \frac{\sin\delta \cos\delta}{\sin z\ (\text{elongation})} \tag{9.20}$$

in which

$$\sin z\ (\text{elongation}) = \cos\Phi \sinh\ (\text{elongation}). \tag{9.21}$$

The term m_i is as in equation (9.11), with $\tau_i = T\ (\text{pointing})_i - T\ (\text{elongation})$.

(3) To obtain the azimuth of the mark, compute the average observed angle between the star and the reference mark for each determination and add to the azimuth of the star computed from equation (9.18).

(4) The mean of all determinations is the final azimuth.

If the theodolite is equipped with a striding level, a correction for the inclination of the horizontal axis ($\pm b \cot z$) could be applied in accordance with equation (7.14); thus appropriate readings on both ends of the level should be made after the pointings on the star. Correction for collimation is unnecessary since the observations are made in alternate positions of the telescope.

Example 9.2 illustrates the procedure. Since in this example the latitude of the station was only approximately known, the result as expected is somewhat inferior to that of Example 9.1.

EXAMPLE 9.2

Azimuth by Hour Angles of Stars Near Elongation

Observer: A. Gonzalez-Fletcher
Computer: A. Gonzalez-Fletcher
Local Date: 4-22-1965
Location: OSU old astro pillar
$\Phi \cong 40^\circ 00' 00''$
$\Lambda \cong -5^h 32^m 10^s$
Stars Observed:
FK4 No. 909 (51H Cephei) near west elongation
FK4 No. 913 (δ Ursae Minoris) near east elongation

Instruments: Kern DKM3-A theodolite (No. 82514); Hamilton sidereal chronometer (No. 2E12304); Favag chronograph with manual key, Zenith trans-oceanic radio.
Azimuth Mark: West Stadium

1. Computation of Auxiliary Quantities (Sample)

For equations see section 3.45

	Star No. 909 (West elongation)
δ (apparent, see Example 10.2)	$87^\circ 06' 10''.47$
α (apparent, see Example 10.2)	$7^h 24^m 42^s.33$
$\cos \delta$	.05054225
$\sec \Phi$	1.30540729
$\sin A = \cos \delta \sec \Phi$	.06597822
A (elongation)	$356^\circ 13' 01''.12$
$\cot \delta$	.05060693
$\tan \Phi$	
$\cos h = \cot \delta \tan \Phi$	.04246426
h (arc)	$87^\circ 33' 58''.48$
h (time)	$5^h 50^m 15^s.90$
AST $= \alpha + h$	$13^h 14^m 58^s.23$
$\sin \delta$	.99872
$\sin \delta \cos \delta$	.05048
$\sin h$	.99910
$\sin z = \cos \Phi \sin h$	.76535
$\sin \delta \cos \delta / \sin z$	.06596

2. Computation of $\Sigma m/n$ (Sample)

Star No.	Determination No.	Telescope Positions: D(irect) or R(eversed)	AST (Corrected Chronometer Reading)	τ = Time of Pointing - Time of Elongation = AST - $13^h14^m58^s.2$	$m = 2\rho'' \sin^2 \frac{\tau}{2}$	$\frac{\Sigma m}{n}$ (n = 4)
909	1	D	$13^h06^m40^s.2$	$-8^m18^s.0$	135".25	95".55
		D	13 07 14.7	-7 43.5	117 .16	
		R	13 09 01.7	-5 56.5	69 .31	
		R	13 09 25.2	-5 33.0	60 .48	
	2	D	13 10 43.7	-4 14.5	35 .33	17 .14
		D	13 11 14.2	-3 44.0	27 .37	
		R	13 13 35.7	-1 22.5	3 .71	
		R	13 13 55.7	-1 02.5	2 .13	
	3	D	13 15 18.2	0 20.0	0 .22	5 .69
		D	13 15 43.2	0 45.0	1 .10	
		R	13 17 03.7	2 05.5	8 .59	
		R	13 17 31.7	2 33.5	12 .85	

3. Azimuth Computation (Sample)

$$A = A\,(\text{elongation}) + (\sin\delta\cos\delta/\sin z)\,\frac{\Sigma m}{n}$$

	Star No. 909	Determination No. 1	2	3
1	$\Sigma m/n$ (see above)	95".55	17".14	5".69
2	$\frac{\sin\delta\cos\delta}{\sin z}\frac{\Sigma m}{n}$	6".30	1".13	0".37
3	A (elongation)	356°13'01".12	356°13'01".12	356°13'01".12
4	A = (3) + (2)	356°13'07".42	356°13'02".25	356°13'01".49
5	Circle reading (corrected, as in Example 9.1, for dislevelment) on star	356°13'12".07	356°13'07".62	356°13'08".08
6	Circle reading on mark	100°29'34".85	100°29'31".20	100°29'33".40
7	Angle between star and mark = (6) - (5)	104°16'22".78	104°16'23".58	104°16'25".32
8	Azimuth of mark = (7) + (4)	100°29'30".20	100°29'25".83	100°29'26".81

4. Final Observed Azimuth from Eight Determinations

Star No.		Determination No.	Azimuth of Mark	v	vv
	909	1	100°29'30".20	-0".01	0.00
		2	25.83	4.36	19.00
		3	26.81	3.38	11.42
		4	26.53	3.66	13.40
	913	5	33.25	-3.06	9.36
		6	30.58	-0.39	0.15
		7	32.39	-2.20	4.84
		8	35.91	-5.72	32.72
Final (Mean)			100°29'30".19	[v]=0".02	[vv]=90.89

Standard deviation of an azimuth determination:

$$m_A = \sqrt{\frac{[vv]}{n-1}} = \sqrt{\frac{90.89}{8-1}} = 3''.60$$

Standard deviation of the mean azimuth:

$$M_A = \frac{m_A}{\sqrt{n}} = \frac{3''\!.60}{\sqrt{8}} = 1''\!.27$$

Result:

$$A_T \text{ (West Stadium)} = 100^\circ 29' 30''\!.2 \pm 1''\!.3$$

(In this example the effect of diurnal aberration on the final azimuth has been considered to be negligible.)

9.113 Azimuth by the Hour Angles of Stars Near Culmination. The horizontal angle between a reference mark and the vertical plane of the instrument is measured and the azimuth of this vertical plane is obtained, in a manner similar to that for transit time determination, by observing the transit time of stars as explained in section 7.445. Two stars near upper culmination should be observed in rapid succession, one in the south at a low altitude and the other in the north as close to the pole as possible. Noting the chronometer times of the upper transits of the pair, the error of the setting of the instrument in azimuth is computed through equations (7.43) and (7.41). This, applied to the horizontal angle between the reference mark and the vertical plane of the instrument, provides the azimuth of the mark. Eight such determinations will provide a final azimuth of sufficient second-order precision. For further details see [Johns, 1955 and 1957; Bhattacharji, 1958].

9.114 Effect of Random Observational Errors. The methods described represent the optimum conditions for the elimination of systematic errors on account of the assumed latitude and of the measured hour angle and which are due to the inclination of the horizontal axis of the instrument and its collimation. The analysis of the effect of the random observational errors, which can be reduced mainly by repeating the observations, is a more difficult task primarily because this effect is dependent on the instrument used. It is worthy of consideration, however, since too many observations result in a waste of time while too few may give unsatisfactory results. The discussion here is limited to the instruments categorized as geodetic theodolites (e.g., Wild T-3) or as one-second theodolites (e.g., Wild T-2). Random errors representative for these instruments have been determined by special investigations together with formulas to estimate the expected (apriori) accuracy attainable by the main methods just described.

The random errors affecting the azimuth determinations by the hour angle of stars are the following [Roelofs, 1950, pp. 102-103]:

m_{hp} Standard error of one pointing with the vertical thread (horizontal pointing) on a star. So far as is known, this has been investigated only in the case of stars near the meridian. Several investigators have arrived at the rather surprising conclusion that the accuracy of pointing is fairly independent of the velocity of the star's apparent motion. The estimated standard errors are

$$m_{hp} = 1''.8 \quad [\text{geodetic theodolite}],$$
$$m_{hp} = 2''.5 \quad [\text{one-second theodolite}].$$

m_{ha} Standard error of measuring the horizontal angle between a star and an azimuth mark, disregarding the error of pointing on the star. This includes the standard errors of both circle readings involved and the error of pointing on an azimuth mark. These values are estimated to be

$$m_{ha} = 0''.8 \quad [\text{geodetic theodolite}],$$
$$m_{ha} = 2''.5 \quad [\text{one-second theodolite}].$$

m_{ℓ} Standard error of leveling the theodolite by means of the plate level:

$$m_{\ell} = 0''.9 \quad [\text{geodetic theodolite}],$$
$$m_{\ell} = 1''.5 \quad [\text{one-second theodolite}].$$

m_t Standard error of timing on account of variations in the personal equation of the observer, errors in reading the chronometer or chronograph record, and accidental errors in the chronometer or chronograph:

$$m_t = 0''.5 = 0^s.03 \quad [\text{chronograph}],$$
$$m_t = 1''.5 = 0^s.10 \quad [\text{without chronograph}].$$

Roelofs also derived the basic relationship between these random errors and the a priori standard error of the final mean azimuth determined by the hour angles of stars [Roelofs, 1950, p. 110]:

$$m_A^2 = \frac{1}{n} F m_t^2 + \frac{1}{n} (m_{hp}^2 + m_{ha}^2) \tag{9.22}$$

where n is the number of pointings on the star, and

$$F = \cos^2\Phi\,(\tan\Phi - \cos A \cot z)^2 + m(2\tan^2\Phi + \cot^2 z - 2\tan\Phi \cos A \cot z),$$
$$m = (m_{hp}^2 + m_{\ell}^2)/m_t^2$$

As an example, for the case of azimuth determination from the hour angle of Polaris, when $\tan\Phi \simeq \cot z$ and $A \cong 0°$ (thus $F = m \tan^2\Phi$), equation (9.22) yields

$$m_A^2 = \frac{1}{n}(m_{hp}^2 + m_{\ell}^2)\tan^2\Phi + \frac{1}{n}(m_{hp}^2 + m_{ha}^2).$$

Using the estimated observational standard errors listed above for a geodetic theodolite and n=16, the expected standard error for the final azimuth is

$$m_A = \left[\frac{1}{16}(1.8^2 + 0.9^2)\tan^2 40° + \frac{1}{16}(1.8^2 + 0.8^2)\right]^{\frac{1}{2}} = 0''.65$$

which value is in fair agreement with the final result of Example 9.1.

9.12 Azimuth by Star Altitudes

This method has the advantage that no precise knowledge of time is required for the observations. Both horizontal and vertical circles are read on the star, and the star's azimuth is computed from equation (3.5) using the measured altitude corrected for refraction:

$$\cos A = \frac{\sin\delta - \sin\Phi \sin a}{\cos\Phi \cos a}. \tag{9.23}$$

As before, it is desirable to select a star program which will minimize the effects of systematic and random errors on the final result. The main sources of systematic error include the error $d\Phi$ in the assumed latitude and da in the reduced altitude arising from an error in the residual refraction. Differentiating equation (9.23), their effects on the azimuth are

$$\begin{aligned} dA \sin A = & (\tan a - \cos A \tan\Phi)d\Phi + \\ & + (\tan\Phi - \cos A \tan a)da. \end{aligned} \tag{9.24}$$

The effect of $d\Phi$ is zero when

$$\tan a = \cos A \tan\Phi,$$

which, according to equations (3.2) and (3.3), occurs when the hour angle is either 90° or 270°. The effect of da is zero when

$$\tan\Phi = \cos A \tan a,$$

which, according to equations (3.5) and (3.51), occurs at elongation. To eliminate the influence of both errors, two stars must be observed such that the errors neutralize each other. This will occur when

$$a_1 = a_2 \text{ and } A_1 = 360^\circ - A_2.$$

Therefore, azimuth should be derived from two stars at about the same altitude and symmetrical with respect to the meridian.

The random observational errors affecting the azimuth determination by star altitudes, in addition to those pertaining to horizontal angle measurements as described in section 9.115, are the following [Roelofs, 1950, p. 122]:

m_{vp} Standard error of pointing the horizontal thread (vertical pointing) on a star:

$$m_{vp} = 1''.8 \quad \text{[geodetic theodolite]},$$
$$m_{vp} = 2''.5 \quad \text{[one-second theodolite]}.$$

m_{va} Standard error of measuring the altitude of a star, disregarding the error of pointing the telescope on the star. It includes the error of circle reading and of centering the bubble of the vertical circle level:

$$m_{va} = 0''.4 \quad \text{[geodetic theodolite]},$$
$$m_{va} = 1''.8 \quad \text{[one-second theodolite]}.$$

m_{tr} Standard error of tracking (ceasing to turn the telescope at the instant the star's image arrives at the intersection of the cross hairs) in the case where simultaneous horizontal and vertical pointings are required on a star:

$$m_{tr} = 1''.0.$$

The effect of random errors on the determination of azimuth from star altitudes is [Roelofs, 1950, p. 123]:

$$m_A^2 = \frac{1}{n}\Big\{(m_{hp}^2 + m_\ell^2)\tan^2 a + [(m_{va}^2 + m_{vp}^2)\operatorname{cosec}^2 A + m_{tr}^2\cos^2\Phi]\,(\tan\Phi - \cos A\tan a)^2 + (m_{hp}^2 + m_{ha}^2)\Big\}. \qquad (9.25)$$

The conditions under which (9.25) and the effect of systematic errors will reach a minimum will be met when stars are selected as follows.

(1) At latitudes less than 15°, east and west stars should be observed near elongation at an altitude of approximately 15°.

(2) At latitudes greater than 15°, Polaris should be observed near elongation, or stars at hour angles of approximately 90° and 270° and at approximately 15° altitude.

Computations necessary for the selection of stars to be observed at elongation have already been covered in section 9.112; to select stars at hour angles of 90° and 270°, first compute their approximate declinations using expression (3.2) which (when $h = 90^\circ$ or 270°) yields

$$\sin \delta = \frac{\sin a}{\sin \Phi}.$$

Then the approximate right ascensions may be determined by adding 6^h to the local sidereal times when the observations are to be made for east stars, or by subtracting 6^h for west stars. As in the case of azimuth by hour angles, there are star selection charts available which will aid in programming the star observations, conforming with the conditions above [Stoch, 1963].

A suggested observing sequence is the following:

(1) Set up and level the instrument carefully; record horizontal circle reading on the reference mark.

(2) Set horizontal thread in advance of the star, and track with vertical thread using the slow motion screw until the star reaches the horizontal thread. Check (and read, if necessary) the vertical circle level and record vertical and horizontal circles and the chronometer.

(3) Repeat (2).

(4) Repeat (1).

(5) Reverse telescope and repeat (1) - (4).

(6) Repeat sequence for other star of the pair.

(7) Periodically record pressure and temperature for refraction correction. This constitutes two determinations (one on each star). Eight determinations should yield satisfactory second-order results.

The reduction sequence is:

(1) Correct observed altitudes for refraction.

(2) Compute the apparent azimuth of the star for each observation from equation (9.23).

(3) Compute the azimuth of the reference mark for each observation by applying (adding or subtracting) the respective horizontal angle to the azimuth of the star. The final azimuth is the mean of the individual azimuths.

Since the apparent motions of stars in zenith distance is not linear, the zenith distance of a star obtained from the mean of the measured zenith distances of each pointing is not equal to the required zenith distance corresponding to the mean times of the pointings. For this reason a curvature correction, similar to that discussed in section 9.111 for azimuth, but of opposite sign, needs to be applied to the average zenith distance of each determination corrected for refraction. The interpretation of this correction is

$$\Delta z_C = z_0 - \frac{z_1 + z_2 + \ldots + z_n}{n} \tag{9.26}$$

where z_1, z_2, ..., z_n are the zenith distances corresponding to the times of the n pointings, and z_0 is the zenith distance corresponding to the average sidereal time of the n observations. The curvature correction may be computed from an equation similar to equation (9.16):

$$\Delta z_C'' = \frac{C_z}{n} \sum_{i=1}^{n} m_i \tag{9.27}$$

where m_i is from expression (9.11) and

$$C_z = \cos^2\Phi \cos A\,(\tan\Phi - \sec A \cot z). \tag{9.28}$$

Example 9.3 illustrates the computational procedure when determining azimuth by star altitudes (the effect of diurnal aberration has been neglected).

EXAMPLE 9.3

Azimuth by Star Altitudes

Observer: A. Gonzalez-Fletcher
Computer: A. Gonzalez-Fletcher
Local Date: 4-22-1965
Location: OSU old astro pillar
$\Phi \simeq 40°00'00''$
$\Lambda \simeq -5^h32^m10^s$
Temperature: 9°.3 C
Barometric Pressure: 765 mm

Instruments: Kern DKM3-A theodolite (No. 82514); Hamilton sidereal chronometer (No. 2E12304); Zenith transoceanic radio.
Azimuth Mark: West Stadium
Stars Observed:
FK4 No. 1534 (41 Cygni) near $h = 18^h$.
FK4 No. 305 (χ Geminorum) near $h = 6^h$.

1. Correction of Observed Zenith Distances for Refraction and Curvature (Sample)

$$z = z_m' + \Delta z_R + \Delta z_C$$

z_m' = mean observed zenith distance from direct and reversed pointings
Δz_R = refraction angle (see section 4.232)
Δz_C = curvature correction = $C_z(\Sigma m)/n$, where m is computed from equation (9.11)
C_z = $\cos^2\Phi \cos A\,(\tan\Phi - \sec A \cot z)$

	Star No.	1534		
	Determination No.	5	6	7
1	z'_m (from 2 direct and 2 reversed pointings)	72°27'30"0	71°35'00"0	70°47'30"0
2	Δz_R	183"8	174"6	166"7
3	$z_m = z'_m + \Delta z_R$	72°30'33"8	71°37'54"6	70°50'16"7
4	$\Sigma m/n$ (as in Example 9.1 or 9.2)	5"54	3"92	2"42
5	$\cos A \cong (\sin\delta - \sin\Phi\cos z_m)/\cos\Phi\sin z_m$*	0.425	0.414	0.405
6	cot z	0.315	0.332	0.348
7	sec A cot z	0.741	0.802	0.859
8	$\tan\Phi$ - sec A cot z	0.098	0.037	- 0.020
9	$\cos^2\Phi\cos A$	0.249	0.243	0.238
10	C_z = (9) × (8)	0.024	0.009	- 0.005
11	$\Delta z_C = C_z\Sigma m/n$ = (10) × (4)	0"13	0"04	- 0"01
12	$z = z_m + \Delta z_C$ = (3) + (11)	72°30'33"9	71°37'54"6	70°50'16"7

* In this equation $\sin\delta \cong \cos z_m/\sin\Phi$

2. Azimuth Computation (Sample)

$$\cos A = (\sin\delta - \sin\Phi\sin a)/\cos\Phi\cos a$$

	Star No.	1534		
	Determination No.	5	6	7
1	a = 90° - z	17°29'26"1	18°22'05"4	19°09'43"3
2	δ (apparent at $h=18^h$, see Example 10.2)	30°14'44"88	30°14'44"88	30°14'44"88
3	sin δ	.50371066	.50371066	.50371066
4	sin a	.30054905	.31512180	.32824069
5	sin Φ sin a	.19318921	.20255639	.21098905
6	sin δ - sin Φ sin a	.31052145	.30115427	.29272161
7	cos a	.95376636	.94905124	.94459411
8	cos Φ cos a	.73062742	.72701543	.72360107
9	cos A = (6)/(8)	.42500656	.41423367	.40453452
10	A (since star is near $h=18^h$)	64°50'56"1	65°31'44"2	66°08'17"1

	Star No.	1534		
	Determination No.	5	6	7
11	Horizontal circle reading (corrected, as in Example 9.1, for dislevelment) on star	64°50'51".3	65°31'32".6	66°08'18".4
12	Circle reading on Mark	100°29'22".8	100°29'22".1	100°29'25".8
13	Angle between star and Mark = (12) - (11)	35°38'31".5	34°57'49".5	34°21'07".4
14	Azimuth of Mark = (13) + (10)	100°29'27".6	100°29'33".7	100°29'24".5

3. Final Observed Azimuth from Eight Determinations

Star No.		Determination No.	Azimuth of Mark	v	vv
Star No.	305	1	100°29'31".5	-1".81	3.28
		2	31.8	-2.11	4.45
		3	34.5	-4.81	23.14
		4	28.8	+0.89	0.79
	1534	5	27.6	+2.09	4.37
		6	33.7	-4.01	16.08
		7	24.5	+5.19	26.94
		8	25.1	+4.59	21.07
		Final (Mean)	100°29'29".19	[v]=+0.02	[vv]=100.12

Standard deviation of an azimuth determination:

$$m_A = \sqrt{\frac{[vv]}{n-1}} = \sqrt{\frac{100.12}{8-1}} = 3''.78$$

Standard deviation of the mean azimuth:

$$M_A = \frac{m_A}{\sqrt{n}} = \frac{3''.78}{\sqrt{8}} = 1''.34$$

Result:

$$A_T(\text{West Stadium}) = 100°29'29''.2 \pm 1''.3.$$

9.13 Miscellaneous Methods of Azimuth Determination

The methods previously covered may be considered classical, generally capable of achieving second-order, or sometimes even better, precisions. In the following, certain methods of limited accuracy, each designed for some special purpose, are described briefly.

9.131 Azimuth by Equal Altitudes. Measurement of the horizontal angles between a reference mark and the two positions of a star at equal altitudes will give the angle between the meridian and the mark by taking the mean of the two observed angles. The method is not precise but has the advantage of being independent of knowledge of the star's coordinates and the observer's position. Unless observations are made soon before and after culmination, several hours are required for the determination. On the other hand, as may be seen from equation (9.24), the effect of a small error in altitude on the azimuth theoretically approaches infinity near the meridian. The optimum position for observations would be near the prime vertical plane, but this would require an extremely long waiting period between the two measurements unless the observations are near the zenith, in which case the azimuth again becomes indeterminate. The practical application of this method is for approximate meridian settings when time is not available.

A bright star is selected east of the meridian which may be easily identified some time later. The reference mark is sighted, and the horizontal circle reading is recorded. The star is then intersected, and both the horizontal and vertical circle readings are recorded. An equal number of direct and reverse pointings are made. With the star west of the meridian, the same altitudes are set on the vertical circle using the same face for each individual altitude as before. The star is tracked in azimuth until reaching the horizontal wire, and the horizontal circle reading is recorded. The mean of all the observed angles will give the angle between the reference mark and the meridian.

9.132 Azimuth by the Hour Angles of Almucantar Crossings. This method, described in [Thornton-Smith, 1954-1959], was devised as an expedient method for obtaining an azimuth of moderate precision in the southern hemisphere where there is a dearth of circumpolar stars close to the pole and observations at elongation result in uncomfortably high altitudes. It is most convenient for latitudes of $30^\circ - 50^\circ$ and consists of a series of pointings with the vertical thread at known instants of time to a star as it crosses the almucantar whose altitude is equal to the latitude of the observer. The computed azimuth of almucantar passage, corrected for curvature to the mean times of observations, is added to the mean horizontal angle between the star and reference mark to give the azimuth of the mark.

The author suggests selection and visual identification of a star with polar distance of $20° - 40°$, east or west of the meridian that appears to be approaching the appropriate almucantar. A series of direct and reverse observations are taken between the star and the reference mark, recording the time and the horizontal circle readings of each pointing on the star. Observations are continued until the vertical circle reading shows that the star has passed the almucantar.

9.133 Azimuth by the Rate of Change of Zenith Distance. If the horizontal angle between a reference mark and a star is observed midway between two timed observations of altitude on the star, an approximate value of the azimuth may be obtained without having to identify the star. Taking the difference in altitude and the elapsed time between two moderately spaced altitude observations, the approximate rate of change in zenith distance may be calculated from

$$\frac{dz}{dh} = \frac{a_1 - a_2}{T_2 - T_1} .$$

The azimuth of the star at the mean time of the two observations is obtained from equation (3.32) which yields

$$\sin A = - \frac{dz}{dh} \sec \Phi . \qquad (9.29)$$

The azimuth of the mark is calculated by adding the horizontal angle measured to the computed azimuth of the star. Though the quantity dz/dh is maximum near the prime vertical plane, $\sin A$ changes very slowly there. For this reason, a better defined azimuth will usually result from observing stars near the azimuths 45°, 135°, 225°, or 315°.

9.2 First-Order Azimuth Determination

The main purpose of first-order astronomic azimuth determination is to control the geodetic azimuth in first-order triangulation, trilateration, or traverse networks, by defining it through the Laplace equation [Bomford, 1962, p. 90]:

$$\alpha = A - \Delta\Lambda \sin \varphi$$

where α and A are the respective geodetic and astronomic azimuths, $\Delta\Lambda$ = astronomic longitude - geodetic longitude; and φ is the latitude of the station. Since in the computations of geodetic positions the calculation of the azimuth involves a greater accumulation of errors than the longitude, it is necessary at frequent intervals (at least near the intersections of primary chains and at the end of any chain which does not form part of a closed circuit) to observe an astronomic azimuth. With properly spaced astronomic control, the accumulation of errors in azimuth between consecutive Laplace stations (where both astronomic azimuth and longitude are known) will not be large and, when distributed properly, will have a minimum effect on the geodetic azimuths of the intermediate stations.

The specifications of a first-order azimuth for use in geodetic control in the U. S. are the following [Hoskinson and Duerksen, 1952]:

(1) In the northern hemisphere between latitudes of 15° and 55°, Polaris preferably should be used, though any circumpolar star will do, especially near elongation. With time known to better than one second, Polaris may be used at any hour angle. The method of observation (including the use of the geodetic theodolite) and reduction is essentially the same as in section 9.111. The differences are pointed out below and in section 9.21.

In high latitudes it is difficult to observe on Polaris with a geodetic theodolite. For this reason it is more satisfactory to use a universal theodolite and its micrometer (see section 9.22).

(2) To reduce the possible effects of horizontal refraction, observations should be made on at least two nights.

(3) No determination which gives a residual of 5" or greater from the mean should be accepted.

(4) The azimuth should depend upon at least 24 acceptable determinations, not less than 12 of them being on any one night.

(5) The standard deviation of the final mean azimuth should not exceed 0".45.

At low latitudes ($\Phi < 15^\circ$), if the latitude of the observer is well known, the azimuth may be determined from the hour angles of stars near elongation as described in section 9.112, except that the azimuth should be based on the mean of 48 acceptable determinations not less than 24 being made on each of two nights. At least two star pairs should be observed each with elongations east and west of the observer at an altitude of about 15°. When the observer is within 2° of the equator he may also select those stars that are on his prime vertical plane at an altitude of 30°. These are to be observed at the same altitudes as above. Other specifications are identical to those for the Polaris observations described here and in the following section. For more details see [Smith, 1961; Opie, 1965; Szpunar, 1966 and 1967].

In addition to these classical methods, a rather new method attributed to A.N. Black should also be mentioned. The main feature of this method is that the direct result of horizontal angle measurements, between at least three stars and the reference mark by means of the micrometer of a universal theodolite, is the geodetic azimuth without the need for determining $\Delta\Lambda$ separately [Black, 1951 and 1953; Bomford, 1962, pp. 333-335]. Additional features are that, though the method is most accurate and convenient at high latitudes, it can be used in all latitudes and that, with some additional effort, it can provide both $\Delta\Lambda$ and $\Delta\Phi$ (= astronomic latitude - geodetic latitude) together with the azimuth. This latter feature makes it particularly appealing when controlling primary traverses, where the azimuth and both $\Delta\Lambda$ and $\Delta\Phi$ (to determine the profile of the geoid along the traverse) may all be wanted.

The principal instruments needed are essentially the same as for second-order work, i. e., a geodetic theodolite equipped with a striding level, sidereal chronometer possibly coupled with a chronograph equipped with a manual key, and an HF receiver with an amplifier. When the 'micrometer' method is used, a universal theodolite or a universal telescope must be substituted for the geodetic theodolite; a thermometer and a barometer are also needed.

The principal corrections for the elimination of systematic errors include the following:

(1) Corrections for the inclination of the horizontal axis and collimation as given by equations (7.14) and (7.30). The latter needs to be applied only when the observations are made in one position of the telescope. The correction for the inclination should be applied to the horizontal circle readings on both the star and the reference mark. The zenith distances for this purpose need to be known only approximately. Appropriate values may be obtained from readings on the vertical circle and through the application of equations (7.23) and (7.24).

(2) Correction for diurnal aberration which, as applied to the azimuth, from equations (4.55) and (3.31) is

$$\Delta A_D'' = \pm 0''.32 \cos A \cos\Phi \operatorname{cosec} z. \qquad (9.30)$$

The negative sign is to be applied when $90^\circ < A < 270^\circ$, i.e., the correction is always positive.

(3) Correction for the curvature of the star's path using equation (9.16) or (9.17).

9.21 Azimuth by the Hour Angle of Polaris: Direction Method

In latitudes $15^\circ < \Phi < 55^\circ$ the observation and reduction procedures are identical to those described in section 9.111 with the following exceptions (in addition to the general specifications mentioned in the previous section):

(1) One night's azimuth program usually consists of 16 determinations, each in a different position of the horizontal circle. For Laplace stations, of the combined determinations on the first and on the second nights, a minimum of 24 must be acceptable. For other stations, one night's azimuth of 16 determinations is considered sufficient.

(2) While in the second-order case the availability of a striding level is an option, here it is a requirement. In northern latitudes where the altitude of Polaris is high, the sensitivity of the level should be at least 4''/division (2 mm).

(3) In the computations the azimuth of Polaris should always be determined from equation (9.1).

(4) The azimuth should be corrected for diurnal aberration using equation (9.30). Since for Polaris and for latitudes smaller than 74° the greatest variation of the correction from its mean value of 0''.32 is 0''.02, it is generally satisfactory to simply add 0''.32 to the final mean azimuth of the reference mark.

9.22 Azimuth by the Hour Angle of Polaris: Micrometer Method

Precise azimuth determinations by the direction method described above become increasingly more difficult with increasing latitude. Beyond latitudes of 50° - 55°, the standard method of Polaris observations with a geodetic theodolite cannot be used at all satisfactorily.

One method which has been used in the past is the observation of hour angles of stars near culmination as described in section 9.113, but by means of a universal theodolite or universal telescope. This method, in practice actually a by-product of the determination of astronomic longitude (see section 11.31), does not lend itself to convenient repetition of observations, thus to high precision.

The micrometer method on Polaris also requires a universal theodolite or universal telescope but the observations can be made very rapidly and with any desired number of repetitions. The method actually is no different from the one described in the previous section except that the horizontal angle between Polaris and the reference mark is measured with the eyepiece micrometer instead of the horizontal circle. It follows that since the range of the micrometer in measuring angles is not greater than about 25', an auxiliary reference mark may be needed in the horizontal direction of Polaris. The azimuth of the auxiliary mark thus determined and the appropriately measured horizontal angle between this mark and the actual reference mark define the azimuth of the latter.

It is desirable to make the observations when Polaris is near elongation in order that its movement in a horizontal direction will be slow and many observations may be made before Polaris moves beyond the range of the micrometer. The azimuth of the auxiliary mark (or its supplement if the mark is south of the station) should be 5' - 8' less (more) than the azimuth of Polaris at eastern (western) elongation. The distance of the auxiliary mark from the station should be such that the infinite focusing of the telescope could remain unchanged. Sometimes in practice it is difficult to meet these requirements, which is a distinct disadvantage of the micrometer method.

The suggested sequence of observations is the following.

(1) Record temperature and barometric pressure and measure the altitude of the auxiliary mark to the nearest minute of arc.

(2) Set micrometer thread on mark. Record micrometer reading and the position of the instrument (e.g., circle or eyepiece east).

(3) Without turning the instrument about its vertical axis, raise the telescope until it is in line with Polaris. Set micrometer thread on Polaris. Record time and micrometer reading.

(4) Record both ends of striding level.

(5) Reverse instrument.

(6) Repeat (3) and (4).

(7) Repeat (2).

This completes one azimuth determination. Ordinarily 32 such determinations should be made for a Laplace azimuth. After the last determination, repeat (1). A minimum of 24 determinations with a standard error of 0".45 for the final mean azimuth will be acceptable.

The observations may be made on a single night unless there is reason to believe that the horizontal refraction is not negligible (due to weather or terrain conditions). Otherwise, they should be made on two different nights with a minimum of eight determinations on either night.

The impersonal micrometer's equatorial value of one drum revolution must also be known. It is desirable to determine it, in accordance with section 7.175, either immediately before or after the azimuth observations at each station.

In the observations at high latitudes considerable care should be used to guard against variable personal equation between the pointings on the star and on the mark [Robbins, 1960]. Since Polaris is at high elevation, the moving thread during the pointing will appear to be inclined to the vertical; while when pointing on the mark, it will be nearly vertical. Personal equation may differ considerably for these two conditions. The observer should try to keep it constant by turning his head sidewise for the pointing on the star.

The reduction of the observations is the same as in section 9.21 or 9.111 illustrated in Example 9.1 except in one respect: The horizontal angle (corrected for the inclination of the horizontal axis) between Polaris and the reference mark in one azimuth determination is not computed from the mean horizontal circle and level readings corresponding to the direct and reversed pointings on each, but from the readings on the impersonal micrometer. Let M and P denote the horizontal directions of the mark and of Polaris, respectively, measured in a clockwise direction from the north with respect to the collimation axis. Either of them can be computed from

$$\overline{A} = (\pm)\frac{R}{2}(M_E - M_W) \operatorname{cosec} z + b \operatorname{cotan} z \qquad (9.31)$$

where R is the equatorial drum value of the micrometer, M_E and M_W are the micrometer readings in eyepiece east and west positions respectively, z is the observed zenith distance of the object of observation (including refraction), and b is the inclination of the horizontal axis after equation (7.30). The difference of the micrometer readings is multiplied by the cosecant of the zenith distance for the same reason as is the collimation error in equation (7.14): Only the principal (collimation) axis of the telescope moves along a great circle of the celestial sphere when the telescope is turned on its horizontal axis; the pointing thread, if away from the principal axis, will move on a small circle parallel to the great circle. The sign in parenthesis at the front of the equation depends on the type of instrument used, i.e., on the numbering system of the micrometer. For the Bamberg universal telescope, for example, the positive sign should be used; the Wild T-4 universal theodolite requires the opposite sign convention. If the mark is north of the station, the correction from equation (9.31), when positive, represents the angle M ($M=\bar{A}$); when the correction is negative, its explement is equal to M ($M=360^\circ-\bar{A}$). If the mark is south of the station, then the supplement of the correction is equal to M ($M=180^\circ-\bar{A}$). In the case of Polaris, the explement of the absolute value of the correction (9.31) (regardless of whether it is positive or negative) represents the angle P ($P=360^\circ-|\bar{A}|$). With this sign convention the horizontal angle (corrected for the inclination of the horizontal axis) between Polaris and the auxiliary mark is

$$M - P. \tag{9.32}$$

Example 9.4 illustrates the computation of this angle, which, when added to the azimuth of Polaris as computed from equation (9.1) for the mean time of the two pointings (as in Example 9.1), will determine the azimuth of the mark from one determination.

9.3 Corrections to the Observed Azimuth

As pointed out in section 2.33, the 'observed' or 'true' astronomic azimuth A_T is the angle between the instantaneous astronomic meridian plane of the observer at P and his vertical plane containing the reference mark Q, measured in a plane perpendicular to the vertical at P (Fig. 9.1). The instantaneous astronomic meridian plane contains the observer's vertical and is parallel to the instantaneous rotation axis of the earth. In certain applications a 'reduced' astronomic azimuth A is required which, for example, may be defined as the angle between the average astronomic meridian plane of the observer's projection (along his plumb line) on the geoid (P_0) and his geoidal normal plane contain-

EXAMPLE 9.4

Horizontal Angle by the Micrometer Method (Sample)

Observer:	H. J. Seaborg	Instrument:	Bamberg #21
Date:	8-31-1944	d	= 1".366/div
Location:	Redoubt, Alaska	$\frac{1}{2}$R	= 79".0515
Azimuth Mark:	Muth	Chronometer:	#3479

	Polaris		Mark (south)	
Determination	1	2	1	2
Micrometer: $M_E - M_W$	-0.388	-0.581	-0.731	-0.686
Level: (w - w') + (e - e')	-0.9	2.0		
z	29°15'39"	29°14'43"	89°37'.5	89°37'.5
cot z	1.7848	1.7859		
(d/4) cot z	0.610	0.610		
sin z	0.488786	0.488549	0.999980	0.999980
$\frac{1}{2}R\ (M_E - M_W)$	-30".6720	-45".9289	-57".79	-54".23
$\frac{1}{2}R\ (M_E - M_W)$ cosec z	-62".75	-94".01	-57".79	-54".23
b cot z	-0".55	1".22		
$\overline{A}$	-63".30	-92".79	-57".79	-54".23
P(M)	359°58'56".70	359°58'27".21	(180°00'57".79)	(180°00'54".23)
M - P	180°02'01".1	180°02'27".0		

ing the reference mark projected along the ellipsoidal normal onto the reference ellipsoid (Q_0), measured in a plane tangent to the geoid at P_0. The difference between the true azimuth, determined as described in sections 9.1 and 9.2, and this reduced azimuth may be divided into the following components (corrections):

(1) Correction for the motion of the pole.

(2) Correction for the curvature of the plumb line between P and P_0.

(3) Correction for the skew of the normal of Q.

Depending on the type of the geodetic application and on its required accuracy, some or all of the above corrections will have to be applied to the observed azimuth. If for some practical reason (blocked line of sight, etc.) the observer and/or the reference mark are situated off their respective true positions, further corrections are needed to compensate for these eccentricities.

9.31 Correction for the Motion of the Pole

This correction when applied to the observed (true) azimuth will refer it to the average astronomic meridian thus making it independent from the varying position of the instantaneous terrestrial pole. The correction as derived in section 4.133 is

$$\Delta A_P = A - A_T = -(x_P \sin\Lambda + y_P \cos\Lambda)\sec\Phi. \qquad (4.41)$$

where x_P and y_P are the coordinates of the true pole in units of seconds of arc. In most cases this correction is barely significant (0".1 - 0".3).

9.32 Correction for the Curvature of the Plumb Line

This correction when applied to the observed azimuth will in effect transfer the observation station from the ground (geop) to the geoid. It may be visualized as the difference between the true azimuth of the line PQ observed from P (Fig. 9.1) and the imaginary azimuth of the line P_OQ as

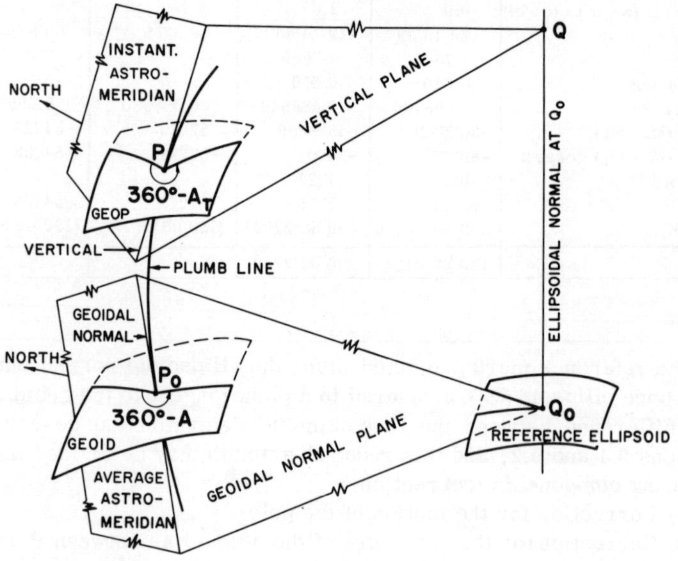

Fig. 9.1 Corrections to the observed azimuth for the curvature of the plumb line and the skew of the normal

determined by the observer at P_O. If $\delta\Phi$ and $\delta\Lambda$ denote the respective differences in the astronomic latitudes and longitudes of the points P and P_O,

$$\delta\Phi = \Phi_P - \Phi_{P_O},$$
$$\delta\Lambda = \Lambda_P - \Lambda_{P_O},$$

then the curvature correction may be calculated from the customary equation for 'direction correction for the deflection of the vertical,' well known in geodesy when reducing observed directions to the reference surface in connection with first-order triangulations [Bomford, 1962, pp. 93-94] adopted for this purpose:

$$\Delta A_{PC} = A - A_T = -(\delta\Phi \sin A - \delta\Lambda \cos\Phi \cos A) \tan a \tag{9.33}$$

where a is the altitude of the reference point. The quantities $\delta\Phi$ and $\delta\Lambda$ depend on the curvature properties of the plumb line between the points P and P_0 (Fig. 9.1) which in turn are functions of the variation of the gravity field in the vicinity of the station. The change in the coordinates may be calculated from [Heiskanen and Moritz, 1967, p. 194]:

$$\begin{aligned} \delta\Phi &= -\int_{P_0}^{P} \frac{1}{g}\frac{\partial g}{\partial x} \operatorname{cosec} 1''\, dh, \\ \delta\Lambda \cos\Phi &= -\int_{P_0}^{P} \frac{1}{g}\frac{\partial g}{\partial y} \operatorname{cosec} 1''\, dh \end{aligned} \tag{9.34}$$

where g is the gravity, $\partial g/\partial x$ and $\partial g/\partial y$ are the respective north and east components of the horizontal gradients of gravity at some point on the plumb line between P and P_0. Since the variations of the horizontal gravity gradient components along the plumb line are generally not known, the above integrals must be solved through certain approximations. If, for example, it is assumed that the gravity and its gradients are constants (which is the equivalent of assuming that this section of the plumb line is a curve of constant curvature), the above equations yield

$$\begin{aligned} \delta\Phi &= -\frac{H}{\bar{g}}\frac{\partial \bar{g}}{\partial x} \operatorname{cosec} 1'', \\ \delta\Lambda \cos\Phi &= -\frac{H}{\bar{g}}\frac{\partial \bar{g}}{\partial y} \operatorname{cosec} 1'' \end{aligned} \tag{9.35}$$

where H is the orthometric height of P (see section 2.4) and the overbars denote mean values along the plumb line of P between the geoid and the ground. As a further approximation the quantities $\bar{g}$, $\partial\bar{g}/\partial x$, and $\partial\bar{g}/\partial y$ may be substituted either by their values at the station (determined from surface gravity or torsion balance observations [Mueller, 1963]) or, worse, by their 'normal' values based on equation (2.15). In the latter case, after substituting

$$\left.\begin{aligned} \bar{g} &\cong \gamma \\ \frac{\partial \bar{g}}{\partial x} &\cong \frac{1}{R}\frac{\partial \gamma}{\partial \varphi} \cong \frac{2\gamma\beta \sin\varphi \cos\varphi}{R} \\ \frac{\partial \bar{g}}{\partial y} &\cong \frac{1}{R\cos\varphi}\frac{\partial \gamma}{\partial \lambda} = 0 \end{aligned}\right\}, \tag{9.36}$$

(R is the radius of the earth), equations (9.35), with current constants, yield

$$\delta\Phi = -\frac{\beta H \sin 2\varphi}{R \sin 1''} = -0''.00017\ H \sin 2\varphi, \qquad \delta\Lambda \cos\varphi = 0 \tag{9.37}$$

where H is to be substituted in meters.

In most cases, owing to the uncertainties involved in the calculations of $\delta\Phi$ and $\delta\Lambda$, the curvature correction ΔA_{pc} is neglected. If, however, the reference mark Q is at considerable altitude above (or below) the horizon, due to the factor tan a in equation (9.33), neglecting the correction could lead to a sizable systematic error when the azimuth is to be referred to the geoid.

9.33 Correction for the Skew of the Normal

This correction when applied to the true azimuth will in effect transfer the reference mark from the ground to the reference ellipsoid. The correction may be visualized by setting the vertical thread of a perfectly adjusted theodolite at P on the mark Q and realizing that when the telescope is rotated about its horizontal axis the thread will not stay on the ellipsoidal normal unless the latter is in the vertical plane of the instrument. The fact that the normal 'skews' will make it necessary to rotate the instrument somewhat about its vertical axis when the (imaginary) setting on Q_0 is attempted. The amount of the necessary rotation is equal to the correction sought. This 'skew normal correction' is also well known to geodesists and is conventionally applied to every direction observed in first-order triangulation nets as part of the necessary reductions of observed quantities to the reference ellipsoid. The correction is [Bomford, 1962, p. 94]:

$$\Delta A_S = A - A_T = \frac{e^2 h \cos^2\varphi \sin 2A}{2R \sin 1''} \tag{9.38}$$

where e is the first eccentricity of the reference ellipsoid, R is the radius of the earth, and h is the height of Q above the reference ellipsoid. If the latter is expressed in meters, the above equation with current constants yields

$$\Delta A_S = 0''.00011\ h \cos^2\varphi \sin 2A. \tag{9.39}$$

This is a small correction, less than 0".1/1000 m, which can reasonably be ignored except in mountainous country.

9.34 Correction for the Eccentricity of the Station

It may happen in practice that for some reason (blocked line of sight, etc.), the azimuth of the reference mark cannot be observed directly from the station. In this case a nearby (few hundred meters maximum) eccentric station needs to be set up from where the azimuth is determined as explained previously and then its final value is to be reduced to the station. In Fig. 9.2 let P denote the station, Q denote the reference mark, and P′ denote the eccentric station. The observed azi-

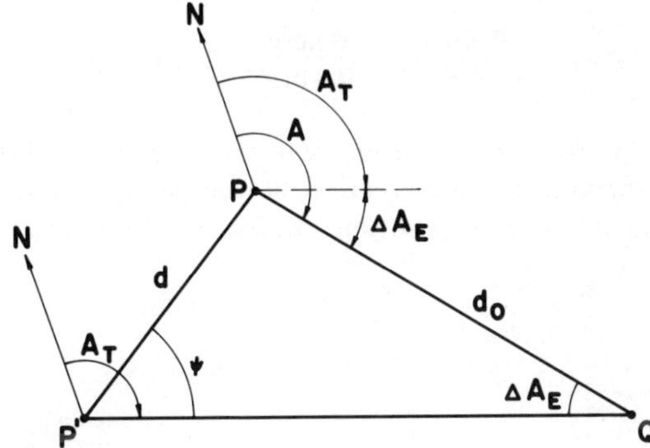

Fig. 9.2 Correction to the observed azimuth for the eccentricity of the station

muth at P′ is A_T, while its value reduced to P is A. Applying the law of sines in the triangle PP′Q and realizing that the small angle ΔA_E may be substituted by its sine, the eccentricity correction is

$$\Delta A_E = A - A_T = \frac{d \sin\psi}{d_o \sin 1''} \tag{9.40}$$

where d and d_o are the respective distances PP′ and PQ, ψ is the angle QP′P, all measured in the field. If the heights of the points P′, P, and Q are considerably different from each other, the measured distances d_o and d may need to be projected to the horizon of P, and the angle ψ be corrected for the inclination of the horizontal axis.

In the derivation of the above equation it has been assumed that the stations P and P′ are close enough in longitude that the convergence of their meridians can be considered negligible. This not being the case, equation (9.40) should be supplemented by the convergence term

$$\mu = \Delta\lambda \sin\varphi \tag{9.41}$$

where $\Delta\lambda$ is the longitude difference between P and P′ to be calculated from the identity for the length of a short parallel arc:

$$R \cos\varphi \, \Delta\lambda \sin 1'' = d \sin(A_T - \psi);$$

thus,

$$\Delta\lambda = \frac{d \sec \varphi}{R \sin 1''} \sin (A_T - \psi) \qquad (9.42)$$

where R is approximated by the mean radius of the earth.

Substituting (9.42) into (9.41) and adding the result to (9.40), the correction for station eccentricity becomes

$$\Delta A_E = \frac{d \sin\psi}{d_0 \sin 1''} + \frac{d \tan\varphi}{R \sin 1''} \sin (A_T - \psi). \qquad (9.43)$$

9.35 Correction for the Eccentricity of the Reference Mark

It may also happen that for some reason the reference mark is slightly set off from its desired position. In Fig. 9.3 let Q represent the de-

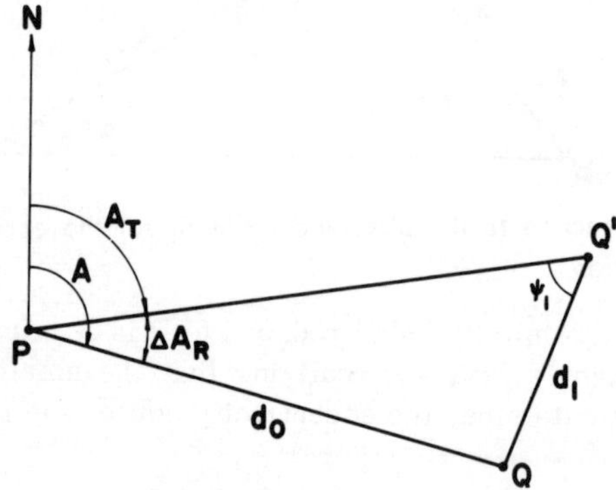

Fig. 9.3 Correction to the observed azimuth for the eccentricity of the reference mark

sired location and Q′ represent the eccentric reference mark as observed. If d_1 and ψ_1 are the respective measured distance and angle defining the relative positions of Q and Q′, and considering that the angle ΔA_R is small, the correction to be applied to the measured azimuth is

$$\Delta A_R = A - A_T = \frac{d_1}{d_0 \sin 1''} \sin\psi_1 \qquad (9.44)$$

where d_0 is the distance PQ. Again, if the heights of the points P, Q, and Q′ are significantly different from each other, the measured distances d_0 and d_1 may need to be projected to the horizon of P and the angle ψ_1 be corrected for the inclination of the horizontal axis.

References

Bhattacharji, J.C. (1958). "Comments on 'Azimuth Determination Without Circle Readings' by R.K.C. Johns." Bulletin Géodésique, 50.

Black, A. N. (1951). "Laplace Points in Moderate and High Latitudes." Empire Survey Review, 11, 82.

Black, A. N. (1953). "A Note on Azimuth Determination." Empire Survey Review, 12, 89.

Bomford, B. G. (1962). Geodesy. Oxford University Press, London.

Forrester, W. D. (1951). "A Discussion on the Accuracy Required in Astronomic Determinations for Laplace Azimuth Control." Transactions, American Geophysical Union, 32, 6.

Heiskanen, W.A. and H. Moritz. (1967). Physical Geodesy. W.H. Freeman and Company, San Francisco and London.

Hoskinson, A. J. and J. A. Duerksen. (1952). "Manual of Geodetic Astronomy." U. S. Coast and Geodetic Survey Special Publication, 237, U. S. Government Printing Office, Washington, D. C.

Johns, R. K. C. (1955). "Azimuth Determination by Equatorial Stars." Bulletin Géodésique, 37.

Johns, R. K. C. (1957). "Azimuth Determination Without Circle Readings." Bulletin Géodésique, 45.

Mueller, I. I. (1963). "Geodesy and the Torsion Balance." Journal of the Surveying and Mapping Division, Proceedings of the American Society of Civil Engineers, 89, SU3.

Opie, B.P. (1965). "The Accuracy of Circum-Elongation Observations for Azimuth." Survey Review, 18, 137-138.

Robbins, A.R. (1960). "Personal Equation in Determination of Geodetic Azimuth." Bulletin Géodésique, 57.

Roelofs, R. (1950). Astronomy Applied to Land Surveying. Amsterdam.

Smith, A. F. (1961). "Criteria for the Determination of Azimuth by Observing Stars at Elongation." Empire Survey Review, 16, 121.

Stoch, L. (1963). "Selecting Stars for Azimuth Determination." Bulletin Géodésique, 69.

Szpunar, W. (1966). "Method of Determining Azimuth on the Basis of Observations of Pairs of Stars in Symmetrical Elongation." Bulletin Géodésique, 82.

Szpunar, W. (1967). "Determination of the Azimuth by Observation of Groups of Stars in Elongation." Bulletin Géodésique, 84.

Thornton-Smith, G. J. (1951). "Curvature Correction in Precise Azimuth Observations." Empire Survey Review, 11, 80 and 82.

Thornton-Smith, G. J. (1954). "An Azimuth Observation in the Almucantar." Empire Survey Review, 12, 94.

Thornton-Smith, G. J. (1956). "Semi-Graphic Almucantar Azimuths." Empire Survey Review, 13, 102.

Thornton-Smith, G. J. (1958). "Azimuth Controlled Almucantar Observations." Empire Survey Review, 14, 109.

Thornton-Smith, G. J. (1959). "A Circum-Almucantar Observation for Azimuth." Empire Survey Review, 15, 112.

10 DETERMINATION OF ASTRONOMIC LATITUDE

Astronomic latitude is defined as the angle between the vertical at the observation point and the plane of the instantaneous equator measured in the plane of the meridian. It may also be considered as the true altitude of the celestial pole.

In the astronomic triangle (Fig. 3.8), assuming that the identity of the star to be used in the latitude determination is known, the declination is the only known quantity. To solve for the latitude, two additional quantities must be measured, normally the altitude (zenith distance) and the hour angle of the star. The basic relation between these quantities is expressed by

$$\cos z = \sin\delta \sin\Phi + \cos\delta \cos h \cos\Phi. \qquad (3.2)$$

The effect of small systematic errors in the measurements of zenith distance and hour angle on the determination of latitude from the differentiation of the above equation is

$$d\Phi = -\sec A\, dz - \cos\Phi \tan A\, dh. \qquad (10.1)$$

It is apparent that a small error in h (i.e., in timing) will have no effect, and a small error in z will have minimal effect, when measurements are made at transit on the meridian. For the foregoing reason most latitude determinations are based on meridian zenith distances determined either from direct measurements at culmination or indirectly by reducing zenith distances measured near the meridian ('circum-meridian') to the meridian. The major source of systematic error in

z, the effect of astronomic refraction, is further reduced by generally basing the latitude on stars paired north and south of the zenith and transiting approximately at the same zenith distance. If the zenith distances are observed ex-meridian, the hour angle (i. e., the sidereal time) of the observations must be known accurately.

In certain applications it may be practical to determine the latitude from hour angle and horizontal angle (rather than vertical angle) measurements. The basic relations between these quantities are equations (9.1) and (9.2). In this case in order to reduce the effect of timing errors the stars should be observed at elongation.

The instruments used for latitude determinations are normally the same as those used for azimuth: a geodetic theodolite (universal theodolite or telescope for first-order work), a chronometer (possibly coupled with a chronograph equipped with a manual key), a radio receiver, a thermometer, and a barometer. When the latitude is determined from horizontal angle measurements, the geodetic theodolite should have a striding level to measure the inclination of the horizontal axis.

10.1 Second-Order Latitude Determination

10.11 Latitude by Meridian Zenith Distances

The measured zenith distance or altitude of a star at meridian transit, corrected for refraction, enables the latitude to be derived from relations (3.42) - (3.44) as follows:

$$\Phi = \delta_N - z_N = \delta_N + a_N - 90^\circ \qquad \text{(upper transit, north)} \qquad (10.2)$$
$$\Phi = \delta_S + z_S = 90^\circ - a_S + \delta_S \qquad \text{(upper transit, south)} \qquad (10.3)$$
$$\Phi = 180^\circ - \delta_N - z_N = 90^\circ + a_N - \delta_N \qquad \text{(lower transit, north)} \qquad (10.4)$$

where z is the zenith distance determined in accordance with equation (7.24) and $a = 90^\circ - z$ is the altitude. The subscripts N and S denote north and south stars respectively.

The largest systematic errors affecting such a determination include the refraction and the error of setting the instrument in the meridian (diurnal aberration has no effect on the declination when the star is on the meridian). The first, together with the need to know the zenith point of the instrument, may be effectively eliminated by observing stars in north-south pairs at about the same altitude, as then the latitude may be computed from altitude differences and will be unaffected by constant errors of altitude measurement. The latitude may be computed either from the sum of equations (10.2) and (10.3) (north star at upper culmination):

$$\Phi = \tfrac{1}{2}(\delta_S + \delta_N) + \tfrac{1}{2}(z_S - z_N) = \tfrac{1}{2}(\delta_S + \delta_N) + \tfrac{1}{2}(a_N - a_S) \qquad (10.5)$$

or from the sum of equations (10.3) and (10.4) (north star at lower culmination):

$$\Phi = \tfrac{1}{2}(\delta_S - \delta_N) + \tfrac{1}{2}(z_S - z_N) + 90^\circ = \tfrac{1}{2}(\delta_S - \delta_N) + \tfrac{1}{2}(a_N - a_S) + 90^\circ. \quad (10.6)$$

Random errors of observation include those of vertical pointing m_{vp} and of measuring the altitude m_{va} (see section 9.12). Ignoring possible random errors of refraction, the expected accuracy may be obtained from [Roelofs, 1950, p. 142]:

$$m_\Phi^2 = \frac{1}{2n}(m_{va}^2 + m_{vp}^2) \quad (10.7)$$

where n is the number of star pairs.

The star program for the observations is simple to compile. Stars should be listed in the order to be observed, including name, number, magnitude, chronometer time of transit and altitude of transit. Since there are generally fewer north stars in the catalogues than south stars, they should be selected first and paired with a south star. For a north star at upper transit, α=ST at which the observation is to begin; at lower transit, $\alpha = ST + 12^h$. The declination should be selected so that the altitude at transit is not less than 15° nor greater than can be comfortably observed with the available equipment. The approximate declination limits may be determined from equations (10.2) or (10.4). A south star is then selected so that its right ascension is approximately equal to that of the north star at upper transit or 12^h less than that of the north star at lower transit. The difference in the right ascensions should be large enough to allow the recording of the first observation and the preparation for the second (about four minutes). The approximate declination required may be determined from equations (10.5) and (10.6) after setting $z_N = z_S$:

$$\begin{aligned} \delta_S &= 2\Phi - \delta_N \quad &\text{(north star at upper culmination)},\\ \delta_S &= 2\Phi + \delta_N - 180^\circ \quad &\text{(north star at lower culmination)}. \end{aligned} \quad (10.8)$$

Prior to beginning observations, the vertical plane of the instrument must be set in the meridian within 1' to 5' depending on the final accuracy desired. This may be accomplished in several ways:

(1) If the longitude is accurately known, a control star may be selected culminating at a low altitude in the south (see section 7.445). The chronometer time of culmination is computed and the star is tracked in azimuth up to the computed time of culmination.

(2) If an approximate value of the longitude is known, the meridian setting may be obtained by pointing the vertical thread of the telescope on Polaris at any hour angle and recording time and the horizontal circle reading. The azimuth of Polaris at this instant as computed from equation (9.1) may be taken as the azimuth of the instrument, which then is rotated into the meridian.

(3) If the longitude is unknown, meridian setting may be obtained by the method outlined in section 7.445.

With the instrument set in the meridian, the horizontal circle may be zeroed. The altitude of the telescope is set for the first star, and when the star enters the field it is tracked with the horizontal thread until it reaches the vertical thread. The vertical circle reading is then recorded after centering its level. The other star of the pair is observed in the same face, so the telescope is turned 180°. Pressure and temperature are recorded for refraction correction.

The reduction procedure starts with the correction of the observed altitudes for refraction and continues with the computation of the latitude from each pair of stars from equation (10.5) or (10.6). This constitutes one determination. The final latitude will be the mean of latitudes obtained from four to eight determinations. The precision does not increase significantly above that obtainable from six pairs. Example 10.1 illustrates the most important aspects of the computational procedure. For the first-order version of this method see section 10.222.

10.12 Latitude by Circum-Meridian Zenith Distances

This method is similar in principle to the previous one except that several zenith distance (or altitude) observations are made on each star before and after meridian passage. Each observed zenith distance is reduced to the meridian by computation and the mean of these reduced zenith distances is used for the latitude computation in equations (10.5) or (10.6). The number of stars to be observed may be reduced to one pair and the same precision obtained, provided that the total number of observations remains the same as in the previous method. The advantage of having to use fewer stars is offset by the more complicated observation and reduction procedure.

The procedure for selection of stars remains the same as outlined in section 10.11 except that a period of at least twenty minutes should be allowed between star transits to permit time for observation.

EXAMPLE 10.1

Latitude by Meridian Zenith Distances

Observer: A. Gonzalez-Fletcher	Instruments: Kern DKM3-A
Computer: A. Gonzalez-Fletcher	theodolite (No. 82514);
Local Date: 4-9-1965	Hamilton sidereal chro-
Location: OSU old astro pillar	meter (No. 2E12304)
$\Phi \cong 40°00'00''$	
$\Lambda \cong -5^h32^m10^s$	

1. Refraction Correction (Sample)

FK4 Star No.	z(observed)	Δz_R^m (Table 4.8)	Pressure (mm)	C_B (Table 4.9)	Temp. (C°)	C_T (Table 4.10)	$\Delta z_R = \Delta z_R^m C_B C_T$	$z = z(\text{obs.}) + \Delta z_R$
335 (N)	08°10'20".9	8".3	763.6	1.005	7.0	1.011	8".4	08°10'29".3
352 (S)	05 27 17 .8	5 .6	763.6	1.005	6.1	1.014	5 .7	05 27 23 .5
1244 (S)	13 39 38 .8	14 .1	763.6	1.005	5.8	1.015	14 .4	13 39 53 .2
358 (N)	11 49 43 .6	12 .1	763.6	1.005	5.6	1.016	12 .4	11 49 56 .0
368 (N)	19 11 36 .6	20 .1	764.2	1.006	5.6	1.016	20 .5	19 11 57 .1
384 (S)	16 24 07 .0	17 .1	764.2	1.006	4.4	1.020	17 .5	16 24 24 .5
394 (N)	16 08 59 .7	16 .8	765.0	1.007	4.4	1.020	17 .3	16 09 17 .0
405 (S)	16 37 21 .9	17 .3	765.0	1.007	3.9	1.022	17 .8	16 37 39 .7

2. Latitude Computation (Sample)

Determination	Star Pair (N/S)	δ_N (apparent, see Ex. 10.2)	δ_S (apparent, see Ex. 10.2)	$\frac{1}{2}(\delta_N + \delta_S)$	z_N	z_S	$(z_S - z_N)$	$\Phi = \frac{1}{2}(\delta_N + \delta_S) + \frac{1}{2}(z_S - z_N)$
1	335/352	48°11'00".12	34°32'35".67	41°21'47".90	08°10'29".3	05°27'23".5	-1°21'32".89	40°00'15".00
2	358/1244	51 50 27 .15	26 20 05 .63	39 05 16 .39	11 49 56 .0	13 39 53 .2	0 54 58 .60	40 00 14 .99
3	368/384	59 12 27 .75	23 35 33 .52	41 24 00 .64	19 11 57 .1	16 24 24 .5	-1 23 46 .30	40 00 14 .34
4	394/405	56 09 47 .55	23 22 19 .71	39 46 03 .63	16 09 17 .0	16 37 39 .7	0 14 11 .35	40 00 14 .90

3. Final Observed Latitude from Eight Determinations

Determination No.	Φ	v	vv
1	40°00'15".00	-0".16	0.0256
2	14.99	-0.15	0.0225
3	14.34	0.50	0.2500
4	14.90	-0.06	0.0036
5	14.51	0.33	0.1089
6	15.85	-1.01	1.0201
7	14.13	0.71	0.5041
8	14.99	-0.15	0.0225
Final (mean)	40°00'14".84	[v] = 0".01	[vv] = 1.9573

Standard deviation of a latitude determination:

$$m_\Phi = \sqrt{\frac{[vv]}{n-1}} = \sqrt{\frac{1.9573}{8-1}} = 0''.53$$

Standard deviation of the mean latitude:

$$M_\Phi = \frac{m_\Phi}{\sqrt{n}} = \frac{0''.53}{\sqrt{8}} = 0''.19$$

Result:

$$\Phi_T = 40^\circ 00' 14''.84 \pm 0''.19 \,.$$

The suggested observation procedure is the following:

(1) Approximately ten minutes prior to culmination begin a series of zenith distance observations, recording the time and vertical circle reading of each. Continue observations until about ten minutes after culmination, making a total of approximately twenty observations as symmetrical as possible with respect to the meridian. The telescope may be reversed at the meridian, but it is not necessary if the star is adequately paired with one on the opposite side of the zenith.

(2) Repeat the procedure for the other star of the pair.

The reduction sequence is:

(1) Determine the chronometer time of culmination T_0. This may be determined from the right ascension of the star if the chronometer correction is known. Otherwise, altitudes may be plotted against time,

and the time corresponding to the greatest altitude on the plot may be used.

(2) Subtract T_0 from each of the observed chronometer times T_i, which results in either h or $24^h - h$ of the star.

(3) Correct each observed zenith distance or altitude for refraction as shown in Example 10.1.

(4) Compute the reduction to the meridian for each observation. The necessary correction

$$\Delta z_m = z(\text{meridian}) - z(\text{observed})$$

may be obtained through the utilization of the curvature correction

$$\Delta z''_m = C_z m \tag{10.9}$$

where from equation (9.11) with h from step (2) above

$$m = \frac{2 \sin^2 \frac{1}{2} h}{\sin 1''}, \tag{10.10}$$

and

$$C_z = \pm \cos \Phi \cos \delta \operatorname{cosec} z \tag{10.11}$$

where the positive sign corresponds to lower culminations and the negative sign to upper culminations.

In the exceptional case where observations are made more than $8^m - 10^m$ off on either side of the meridian or where the zenith distances are unusually small, the correction to the meridian (10.9) needs to be supplemented by an additional term. The complete (Delambre) correction in this case is as follows [Chauvenet, 1891, p. 239]:

$$\Delta z''_m = C_z m + C_z^2 n \cot z \tag{10.12}$$

where

$$n = \frac{2 \sin^4 \frac{1}{2} h}{\sin 1''}. \tag{10.13}$$

It should be noted that the quantities z and Φ used to compute C_z are approximate values. If these differ substantially from the final values, an iteration may be necessary.

(5) Reduce observed zenith distances (z_i) or altitudes (a_i), already corrected for refraction to the meridian as follows:

$$\begin{aligned} z_m &= z_i + \Delta z_{mi}, \\ a_m &= a_i - \Delta z_{mi}. \end{aligned} \tag{10.14}$$

(6) Compute the latitude from equation (10.5) or (10.6).

For a computational example, see the south-star portion of Example 10.2 following section 10.132.

10.13 Latitude by the Zenith Distance of Polaris and of a South Star

Due to its small polar distance, latitude determination may be made by zenith distance or altitude observations on Polaris at any hour angle (rather than at culmination as in the previous methods). For the elimination of systematic errors, however, it should be paired with a south star of the same altitude observed in the circum-meridian mode, essentially as explained in the previous section.

10.131 Latitude by the Zenith Distance of Polaris. The observations consist of a series (approximately twenty) of alternate direct and reversed zenith distance or altitude measurements on Polaris at any hour angle, recording sidereal time, periodic temperature, and barometric pressure. After applying corrections for refraction and curvature (if the time between a pair of direct and reversed pointings is longer than 1^m30^s), the latitude may be computed from

$$\Phi = a - x \qquad (10.15)$$

where x is a combination of the polar distance of Polaris and the correction to the meridian. It is a quantity of the same order of magnitude as the polar distance.

Denoting the polar distance by P, equation (3.2) may be written in the following form:

$$\sin a = \cos P \sin \Phi + \sin P \cos h \cos \Phi$$

where the altitude a and the hour angle h are derived from the observation and Φ is the required latitude. Since P is small (at present less than 1°), it is possible to develop Φ in a series of powers of P and retain as many terms as needed to acquire any given degree of precision. The following relations are utilized for this purpose:

$$\begin{aligned}
\sin \Phi &= \sin(a - x) = \sin a - x \cos a - \tfrac{1}{2}x^2 \sin a + \tfrac{1}{6}x^3 \cos a + \ldots \\
\cos \Phi &= \cos(a - x) = \cos a + x \sin a - \tfrac{1}{2}x^2 \cos a - \tfrac{1}{6}x^3 \sin a + \ldots \\
\sin P &= P - \tfrac{1}{6}P^3 + \ldots \\
\cos P &= 1 - \tfrac{1}{2}P^2 + \ldots
\end{aligned}$$

which, when substituted in the above equation, give

$$\begin{aligned}
\sin a = \sin a - x \cos a + P \cos h \cos a - \\
- \tfrac{1}{2}(x^2 - 2x P \cos h + P^2) \sin a + \ldots . \qquad (10.16)
\end{aligned}$$

From this, the general expression for the correction x is the following:

$$x = P\cos h - \tfrac{1}{2}(x^2 - 2xP\cos h + P^2)\tan a + \\ + \tfrac{1}{6}(x^3 - 3x^2P\cos h - 3xP^2 - P^3\cos h) + \\ + \tfrac{1}{24}(x^4 - 4x^3P\cos h + 6x^2P^2 - 4xP^3\cos h + P^4)\tan a - \ldots . \tag{10.17}$$

As a first approximation

$$x = P\cos h. \tag{10.18}$$

The second approximation is obtained by substituting (10.18) into the second term of (10.17) and neglecting the third powers of P and x:

$$x = P\cos h - \tfrac{1}{2}P^2\sin^2 h\tan a. \tag{10.19}$$

Substituting this value in the second and third terms of (10.17), a third approximation is obtained which, if substituted in the second and subsequent terms of (10.17), results in the following fourth approximation:

$$x = P\cos h - \tfrac{1}{2}P^2\sin^2 h\tan a + \tfrac{1}{3}P^3\cos h\sin^2 h - \\ - \tfrac{1}{8}P^4\sin^4 h\tan^3 a + \tfrac{1}{24}P^4(4-9\sin^2 h)\sin^2 h\tan a. \tag{10.20}$$

The last term of this equation in latitudes where Polaris is used never amounts to more than 0".01. The maximum combined value of the third and fourth terms does not exceed 0".1 in any practical case. Hence, for most second-order purposes equation (10.19) will yield satisfactory results. In special cases when the neglected terms reach their maximum values (third term at $\cos h = 1/\sqrt{3}$; fourth term at $\sin h = 1$), their possible inclusion in the calculations should be evaluated.

Another factor to be considered is the fact that when Polaris is observed off the meridian, any error in h (e. g., erroneous longitude) will affect the latitude by the amount expressed by the second term of equation (10.1). For this reason it is imperative that the sidereal time of the observation be known with sufficient accuracy.

For a computational example see the Polaris portion of Example 10.2 following the next section.

10.132 Latitude by the Combined Zenith Distances of Polaris and the South Star. The observational and computational procedures for both the south star and Polaris are essentially the same as described in sections 10.12 and 10.131. Exceptions or additions are the following:

The observation sequence begins with ten pointings on Polaris in alternating telescope positions (five determinations). These are followed by the twenty pointings on the selected south star (of an altitude approximately that of Polaris) of which ten alternating pointings are made before transit and ten alternating pointings after transit (total of ten determinations). The sequence is concluded by a second series of ten alternating pointings (five determinations) on Polaris. In order to be able to follow this procedure, observations on Polaris should begin about 25^m to 30^m before the transit of the south star. For optimum results the pointings on the south star should be as symmetrical as possible with respect to the meridian, and the times of the vertical circle readings should be recorded by means of a chronograph equipped with a manual key.

The computations on Polaris follow equations (10.15) and (10.19); thus for each determination

$$\Phi_P = a - P \cos h + \tfrac{1}{2}P^2 \sin 1'' \tan a \sin^2 h \qquad (10.21)$$

where a is the mean altitude from a pair of direct and reversed pointings corrected for curvature (when necessary) and refraction, P is the apparent polar distance ($90^\circ - \delta$) in seconds of arc, and h is the apparent hour angle (AST $-\alpha$).

The computation on the south star follows equations (10.3) and (10.14). Curvature correction for each pair of direct and reversed pointings as given by equations (9.27) and (10.11) also needs to be applied. Thus the latitude from each determination is

$$\Phi_S = 90^\circ - a + \delta \qquad (10.3)$$

where

$$a = a_i - C_z (m_1 + m_2 + C_z n \tan a) \qquad (10.22)$$

and

$$C_z = -\cos\Phi \cos\delta \sec a, \qquad (10.11)$$

$$m_1 = \frac{2\sin^2\frac{1}{2}\tau}{\sin 1''} = \frac{(15\tau^s)^2}{2} \sin 1'' = 0''\!.00054541(\tau^s)^2, \qquad (9.11)$$

$$m_2 = \frac{2\sin^2\frac{1}{2}h}{\sin 1''}, \qquad (10.10)$$

$$n = \frac{2\sin^4\frac{1}{2}h}{\sin 1''} \quad (n = 0 \text{ when } h < \pm 8^m). \qquad (10.13)$$

In these equations a_i is the mean altitude from one determination (one pair of direct and reversed pointings) corrected for refraction, τ is half of the time interval between the direct and reversed pointings, and h is the average hour angle of the star corresponding to the two pointings. If the quantities a and Φ used to compute C_z differ substantially from the final values, an iteration may be necessary.

The final latitude of the station is the mean of those determined from the Polaris and the south star observations, i.e.,

$$\Phi = \tfrac{1}{2}(\Phi_P + \Phi_S). \qquad (10.23)$$

This method of latitude determination is the standard adopted by the U. S. Coast and Geodetic Survey for second-order work. Their specification calls for eight determinations on each star with a spread of less than 7".0. The rejection of three determinations on either star is cause for reobservation of the latitude. The final (mean) latitude must have a standard deviation of 0".6 or better [Thorson, 1965].

Example 10.2 illustrates the entire procedure including the preparation of the observing list, the field record, the calculation of the apparent places of the stars from the 'Apparent Places of Fundamental Stars' (APFS), and the latitude calculations. As may be seen from the result, the observed latitude is in fact one of first order. As far as the use of the APFS is concerned, item 3 in the example serves as a guide for the other examples in this book regarding the general interpolation procedures when using this ephemeris.

EXAMPLE 10.2

Latitude by the Combined Zenith Distances of Polaris and a South Star

Observer: J. D. Bossler
Recorder: S. F. Cushman
Computer: H. D. Preuss
Local Date: 8-14-1964
Location: Lanum, Ohio (near USCGS triangulation station)

$\Phi \cong 39°39'39''.65$

$\Lambda \cong -5^h34^m10^s.609$

Instruments: Wild T3 theodolite (No. 26588); Hamilton sidereal chronometer (No.2E12304); Favag chronograph with manual key; Zenith transoceanic radio.

Stars Observed:
FK4 No. 907 (α Ursae Minoris [Polaris])
FK4 No. 1461 (-11° 4411 B. D. [Serpentis])

1. Observing List for (South) Stars

Calculation of right ascension and declination limits

Observation	α		δ	
Period	EST	ST $\cong \alpha$	Upper limit	Lower limit
Beginning	20^h	$16^h56^m.8$	-6°20'.5	-16°20'.5
End	22	18 56.1	-5 55.6	-15 55.6

ST = EST - zonal correction + GST at 0^hUT + solar to sidereal interval correction + Λ (west longitude negative)

$\delta = \Phi$ + (altitude of Polaris (a) + 5°) - 90°, where

$$a = \Phi - (a_0 + a_1 + a_2)$$

where the quantities a_0, a_1, a_2 are taken from the Pole Star Table of the 'Ephemeris.'

Stars selected from APFS, 1964, as possible candidates
(magnitude < 6.0)

FK4 Star No.	Magnitude	ST of transit	δ	T3 vertical circle setting	
				D(irect)	R(eversed)
1438	4.73	$16^h47^m.9$	$-10°43'$	$109°49'$	$70°11'$
1450	5.58	17 07.8	-10 29	109 56	70 04
1461	5.68	17 32.8	-11 13	109 34	70 26
658	3.64	17 35.5	-15 23	107 29	72 31
673	3.50	17 57.0	- 9 46	110 17	69 43
696	4.73	18 27.2	-14 35	107 53	72 07
1482	4.06	18 33.3	- 8 16	111 02	65 58
1486	4.74	18 40.3	- 9 05	110 38	69 22
702	5.09	18 41.6	- 8 19	111 01	68 59

$D = 135° - \frac{z}{2}$; $R = 45° + \frac{z}{2}$; $z = \Phi - \delta$

2. Latitude Field Record

Star No.	Set No.	Approximate chronometer time	*Scaled seconds	Telescope position	Vertical circle reading ° ' "	"	*"	*Observed altitude, a_1	*Chronometer time, (T_1)	Remarks
907 (Polaris)	1	$17^h 20^m 06^s$	$06^s.71$	D	109 32 22.6	22.6	45.2	39° 05' 50".2	$17^h 20^m 29^s.86$	T_s = 13°.1 C. B_s = 734.3 mm
		20 53	53.00	R	70 26 27.5	27.5	55.0			
		22 03	03.69	D	109 32 28.3	28.4	56.7	39 06 04.6	17 21 47.24	
		21 30	30.79	R	70 26 26.0	26.1	22.1			
		22 48	48.49	D	109 32 30.3	30.4	60.7	39 06 19.7	17 23 05.34	
		23 22	22.20	R	70 26 20.5	20.5	41.0			
		24 23	23.50	D	109 32 34.6	34.7	69.3	39 06 31.1	17 24 07.55	
		23 51	51.60	R	70 26 19.1	19.1	38.2			
		25 05	05.10	D	109 32 36.5	36.6	73.1	39 06 45.0	17 25 24.06	T_e = 13°.4 C.
		25 43	43.02	R	70 26 14.0	14.1	28.1			B_e = 733.9 mm
	2	17 40 30	30.26	D	109 34 19.5	19.5	39.0	39 09 40.1	17 41 00.85	T_s = 12°.8 C. B_s = 733.1 mm
		41 31	31.44	R	70 24 29.4	29.5	58.9			
		42 52	52.50	D	109 34 27.8	27.9	55.7	39 10 00.1	17 42 34.10	
		42 15	15.71	R	70 24 27.8	27.8	55.6			
		43 20	20.49	D	109 34 27.8	27.9	55.7	39 10 11.9	17 43 39.88	
		43 59	59.27	R	70 24 21.9	21.9	43.8			
		45 31	31.80	D	109 34 35.3	35.3	70.6	39 10 31.2	17 45 19.40	
		44 47	47.00	R	70 24 19.7	19.7	39.4			
		46 13	12.83	D	109 34 36.7	36.6	73.3	39 10 44.1	17 46 27.83	T_e = 12°.5 C.
		46 43	42.83	R	70 24 14.6	14.6	29.2			B_e = 733.1 mm

Quantities * are determined when observed data is reduced.

D, R = direct and reverse telescope position, respectively.

$T_{s,e}$; $B_{s,e}$ = temperature and barometric pressure reading, respectively (s = start, e = end).

Latitude Field Record (continued)

Star No.	Set No.	Approximate chronometer time	*Scaled seconds	Telescope position	Vertical circle reading ° ′ ″	″	*″	*Observed altitude, a_1	*Chronometer time (T_1)	*2τ	Remarks
1461	1	$17^h27^m38^s$	38.^{s}55	D	109 32 46.6	46.7	93.3	39°07′ 26.″8	$17^h28^m00.^s26$	43.^{s}43	T_s = 13.°5 C. B_s = 733.8 mm
		28 22	21.98	R	70 26 03.2	03.3	06.5				
		29 44	44.04	D	109 32 55.1	55.1	110.2	39 07 50.8	17 29 26.68	34.73	
		29 09	09.31	R	70 24 59.7	59.7	119.4				
		30 17	17.30	D	109 32 56.7	56.8	113.5	39 08 04.9	17 30 33.75	32.90	
		30 50	50.20	R	70 24 54.3	54.3	108.6				
		31 51	50.99	D	109 32 60.0	60.0	120.0	39 08 13.9	17 31 36.84	28.30	
		31 22	22.69	R	70 24 53.0	53.1	106.1				
		32 37	37.42	D	109 34 00.5	00.5	01.0	39 08 17.2	17 32 53.56	32.29	
		33 09	09.71	R	70 24 51.9	51.9	103.8				
	2	$17^h34^m20^s$	20.^{s}36	D	109 32 59.4	59.5	118.9	39°08′ 15.″1	$17^h34^m03.^s62$	33.^{s}49	
		33 47	46.87	R	70 24 51.9	51.9	103.8				
		35 07	07.19	D	109 32 58.8	58.8	117.6	39 08 07.9	17 35 24.10	33.81	
		35 41	41.00	R	70 24 54.8	54.9	109.7				
		37 09	09.49	D	109 32 53.1	53.1	106.2	39 07 52.0	17 36 44.92	49.15	
		36 20	20.34	R	70 24 57.1	57.1	114.2				
		37 42	42.54	D	109 32 50.7	50.7	101.4	39 07 32.3	17 38 01.42	37.76	
		38 20	20.30	R	70 26 04.5	04.6	09.1				
		39 29	29.60	D	109 32 42.0	42.1	84.1	39 07 07.2	17 39 15.80	27.60	T_e = 12.°8 C. B_e = 733.1 mm
		39 02	02.00	R	70 26 08.4	08.5	16.9				

Quantities * are determined when observed data is reduced.

D, R = direct and reverse telescope position, respectively.

$T_{s,e}$; $B_{s,e}$ = temperature and barometric pressure reading, respectively (s = start, e = end).

2τ = time interval between direct and reverse pointing.

3. Calculation of Star Position

Greenwich Date: August 15, 1964

A. Coordinates of Star No. 907 (Polaris)

Mean ST of observation	$17^h33^m.0$
Assumed longitude	-5 34.2
GST = ST - Λ	23 07.2
GST at 0^hUT August 15, 1964	21 33.7
Sidereal interval	1 33.5
Correction sidereal to solar interval	-0.3
Approximate UT of observation	$1^h33^m.2$
Approximate α of Polaris	$1^h59^m.3$
"UT of Transit, First Point of Aries" from the 'Ephemeris'	2 25.9
Approximate UT of transit of Polaris	$4^h25^m.2$

From comparing the approximate UT of transit of Polaris with the approximate UT of observation it follows that transit has not yet occurred, hence the table for Polaris in the APFS has to be entered on August 14, 1964.

Interpolation factor:

$$n = \frac{h^G\,(= GST - \alpha)}{24^h} = 0.8805$$

Coordinate	α	δ
Tabulated values for August 14, 1964	$1^h59^m18^s.30$	$89°05'37''.52$
First difference $\times$ n	+1.24	+0.15
Apparent position	$1^h59^m19^s.54$	$89°05'37''.67$

B. Coordinates of Star No. 1461

Determination of interpolation factor, n:

Mean ST of observation	17^h33^m
Assumed longitude (Λ)	-5 34
GST = ST - Λ	23 07
α (approximate)	17 33
h^G = GST - α	5^h34^m

$$n = \frac{1}{10}\left[\frac{h^G}{24^h} + \text{number of transits since last tabulated date}\right] = \frac{1}{10}\left[\frac{5^h34^m}{24^h} + 9\right] = 0.9232$$

Note: From the calculations for Polaris it is known that the mean epoch of observation is approximately August 15.1. The nearest tabulated dates in the APFS for Star No. 1461 are August 5.9 and August 15.8. It follows that August 14.8 < 15.1 < 15.8, hence 9 transits have occurred since last tabulated date.

Determination of short period terms of nutation:

$$\Delta\alpha^s = d\alpha(\psi)\,d\psi + d\alpha(\epsilon)\,d\epsilon$$

$$\Delta\delta'' = d\delta(\psi)\,d\psi + d\delta(\epsilon)\,d\epsilon$$

where $d\alpha(\psi)$, $d\alpha(\epsilon)$, $d\delta(\psi)$, and $d\delta(\epsilon)$ are tabulated under each star and $d\psi$ and $d\epsilon$ are interpolated from table I in APFS.

The interpolation factor, m, for $d\psi$ and $d\epsilon$ is:

$$m = \frac{\text{GST - GST at } 0^h\text{UT}}{24^h} = \frac{23^h07^m - 21^h34^m}{24^h} = 0.0646$$

From table I, APFS

Date	$d\psi$	$d\epsilon$	Date	$d\psi$	$d\epsilon$
Aug. 15 Aug. 16	-0.155 +0.008 -0.147	+0.007 -0.037 -0.030	Aug. 15.1	-0.154	+0.0046

$$\Delta\alpha^s = (+0.066)(-0.154) + (-0.002)(+0.0046) = -0^s.010$$

$$\Delta\delta'' = (-0.05)(-0.154) + (-0.99)(+0.0046) = +0''.003$$

Apparent right ascension and declination of Star No. 1461:

Coordinate	α	δ
Tabulated values for Aug. 5.9	$17^h32^m47^s.897$	$-11°13'\ 06''.95$
First differences	-0.095	+0.203
*Second differences	+0.005	+0.010
Short periodic nutation	-0.010	+0.003
Apparent position	$17^h32^m47^s.797$	$-11°13'\ 06''.734$

*Second differences are taken from table VI of APFS.

4. Latitude Computation

Greenwich Date: August 15, 1964
Barometer: 734.4 mm
Temperature: 12°.98 C

Star No. 907 (Polaris)	$\alpha = 1^h59^m19^s.54$				$\delta = 89°05'37''.67$			
	Set No. 1				Set No. 2			
Chron. time	$17^h21^m47^s.24$	$17^h23^m05^s.34$	$17^h24^m07^s.55$	$17^h25^m24^s.06$	$17^h41^m00^s.85$	$17^h42^m34^s.10$	$17^h43^m39^s.88$	$17^h45^m19^s.40$
Chron. correction	-20.14	-20.14	-20.14	-20.14	-20.13	-20.13	-20.13	-20.13
AST	17 21 27.10	17 22 45.20	17 23 47.41	17 25 03.92	17 40 40.72	17 42 13.97	17 43 19.75	17 44 59.27
α	1 59 19.54	1 59 19.54	1 59 19.54	1 59 19.54	1 59 19.54	1 59 19.54	1 59 19.54	1 59 19.54
$h = AST - \alpha$	15 22 07.56	15 23 25.66	15 24 27.87	15 25 44.38	15 41 21.18	15 42 54.43	15 44 00.21	15 45 39.73
Observed altitude	39°06′04″.6	39°06′19″.7	39°06′31″.1	39°06′45″.0	39°09′40″.1	39°10′00″.1	39°10′11″.9	39°10′31″.2
Refraction angle	-1 08.0	-1 08.0	-1 08.0	-1 08.0	-1 07.8	-1 07.8	-1 07.8	-1 07.8
Altitude (a)	39 04 56.6	39 05 11.7	39 05 23.1	39 05 37.0	39 08 32.3	39 08 52.3	39 09 04.1	39 09 23.4
$a_m = \Sigma a/n$, (n=4)			39°05′.28				39°08′.97	
$\tan a_m$			0.812340				0.814113	
$\cos h$	-0.63565390	-0.63125918	-0.62774420	-0.62340322	-0.56873054	-0.56313990	-0.55918005	-0.55316539
$\sin^2 h$	0.59594386	0.60151193	0.60593770	0.61136917	0.67654573	0.68287416	0.68731727	0.69400729
$P = 90° - \delta$			3262″.33		3262″.33			
$P^2/412530$			25″.799		25″.799			
$P\cos h$	-2073″.713	-2059″.376	-2047″.909	-2033″.747	-1855″.387	-1837″.148	-1824″.230	-1804″.608
B*	12″.490	12″.606	12″.699	12″.813	14″.210	14″.343	14″.436	14″.576
$\Phi = a - P\cos h + B$	39°39′42″.80	39°39′43″.68	39°39′43″.71	39°39′43″.56	39°39′41″.90	39°39′43″.79	39°39′42″.77	39°39′42″.58

$$* B = \frac{P^2}{2} \sin 1'' \tan a_m \sin^2 h = \frac{P^2}{412530} \tan a_m \sin^2 h$$

Latitude Computation (continued)

Greenwich Date: August 15, 1964
Barometer: 734.2 mm
Temperature: 13°.02C

Star No. 1461	$\alpha = 17^h32^m47^s.80$				$\delta = -11°13'06''.73$			
	Set No. 1				Set No. 2			
Chron. time	$17^h28^m00^s.26$	$17^h29^m26^s.68$	$17^h30^m33^s.75$	$17^h31^m36^s.84$	$17^h34^m03^s.62$	$17^h35^m24^s.10$	$17^h36^m44^s.92$	$17^h38^m01^s.42$
Chron. Correction	-20.14	-20.13	-20.13	-20.13	-20.13	-20.13	-20.13	-20.13
AST	17 27 40.12	17 29 06.55	17 30 13.62	17 31 16.71	17 33 43.49	17 35 03.97	17 36 24.79	17 37 41.29
α	17 32 47.80	17 32 47.80	17 32 47.80	17 32 47.80	17 32 47.80	17 32 47.80	17 32 47.80	17 32 47.80
$h = AST - \alpha$	-05 07.68	-03 41.25	-02 34.18	-01 31.09	00 55.69	02 16.17	03 36.99	04 53.49
2τ	43.43	34.73	32.90	28.30	33.49	33.81	49.15	37.76
Observed altitude	39°07'26''.8	39°07'50''.8	39°08'04''.9	39°08'13''.9	39°08'15''.1	39°08'07''.9	39°07'52''.0	39°07'32''.3
Refraction angle	-1 07.9	-1 07.8	-1 07.8	-1 07.8	-1 07.8	-1 07.8	-1 07.8	-1 07.9
Altitude (a_1)	39 06 18.9	39 06 43.0	39 06 57.1	39 07 06.1	39 07 07.3	39 07 00.1	39 06 44.2	39 06 24.4
$m_1=(15\tau)^2/2\rho''$	0.26	0.16	0.15	0.11	0.15	0.16	0.33	0.19
$m_2=2\rho''\sin^2(h/2)$	51.63	26.70	12.96	4.53	1.69	10.13	25.68	46.98
$m=m_1+m_2$	51.89	26.86	13.11	4.64	1.84	10.29	26.01	47.17
C_zm*	-50.50	-26.14	-12.76	-4.52	-1.79	-10.01	-25.31	-45.91
$a=a_1-C_zm$	39 07 09.40	39 07 09.14	39 07 09.86	39 07 10.62	39 07 09.09	39 07 10.11	39 07 09.51	39 07 10.31
$\Phi=\delta-a+90°$	39°39'43''.87	39°39'44''.13	39°39'43''.41	39°39'42''.65	39°39'44''.18	39°39'43''.16	39°39'43''.76	39°39'42''.96

*$a_m = \Sigma a_i/n$ = 39°06'.79

$\cos a_m$ = 0.775902

$\cos\Phi$ = 0.769834

$\cos\delta$ = 0.980892

$C_z = -\cos\delta\cos\Phi\cos a_m$ = -0.973222

5. Final Observed Latitude from Sixteen Determinations

Star No.	Set No.	Latitude	v	vv
907	1	39°39'42".80	0".51	0.26
		43.68	-0.37	0.14
		43.71	-0.40	0.16
		43.56	-0.25	0.62
	2	41.90	1.41	1.99
		43.79	-0.48	0.23
		42.77	0.54	0.29
		42.58	0.73	0.53
1461	1	43.87	-0.56	0.31
		44.13	-0.82	0.67
		43.41	-0.10	0.01
		42.65	0.66	0.44
	2	44.18	-0.87	0.76
		43.16	0.15	0.22
		43.76	-0.45	0.20
		42.96	0.35	0.12
Final (mean)		39°39'43".31	[v] = 0".05	[vv] = 6.95

Standard deviation of a latitude determination:

$$m_\Phi = \sqrt{\frac{[vv]}{n-1}} = \sqrt{\frac{6.95}{16-1}} = 0''.68$$

Standard deviation of the mean latitude:

$$M_\Phi = \frac{m_\Phi}{\sqrt{n}} = \frac{0''.68}{\sqrt{16}} = 0''.17$$

Result:

$$\Phi_T = 39°39'43''.3 \pm 0''.2$$

10.14 Latitude by Equal Zenith Distances of Two Stars

From the declinations of two stars and their hour angles, deduced from the times of observation, the latitude may be determined by observing two stars at the same altitude without the use of the altitude itself. From equation (3.2) applied to both stars, the latitude is given by

$$\tan \Phi = \frac{\cos \delta_1 \cos h_1 - \cos \delta_2 \cos h_2}{\sin \delta_2 - \sin \delta_1}. \tag{10.24}$$

Stars of a pair should be north and south of the zenith and symmetrical to the prime vertical plane (i.e., on the same side of the meridian as close to it as is convenient). The method for practical purposes is the same as latitude by meridian altitudes but requires a more careful selection of stars, and the accuracy of the results is largely dependent on maintaining a constant altitude with the theodolite. The method is primarily applicable with an astrolabe, but it offers no advantage over the more conventional second-order methods where geodetic theodolites are used. For the first-order version of this method, see section 10.221.

10.15 Latitude by the Azimuths of a Star Pair Near Elongation

This method, suggested in [Bhattacharji, 1959] and one on the same principle in [Bowie, 1956], departs from the usual convention in that horizontal rather than vertical angles are measured.

From equation (9.2) it is apparent that small errors in time will have no effect on the determination of the azimuth of a star at elongation, but azimuth will be affected by an error in the assumed latitude. Therefore, if the horizontal angle between an east and a west star at elongation can be accurately measured, the difference between the measured angle and that computed from the azimuths of the stars will be due to an error in the assumed latitude. Based on the first term of equation (9.2), the correction to the assumed latitude may be computed from

$$d\Phi = \frac{dH}{\sin A_E \cot z_E - \sin A_W \cot z_W} \tag{10.25}$$

in which

$$dH = (A_W - A_E) - (H_W - H_E) \tag{10.26}$$

where the quantity $(A_W - A_E)$ is the horizontal angle between the two stars at elongation (the difference of their azimuths) as computed using the assumed latitude; the quantity $(H_W - H_E)$ represents the measured value of the same (the difference of the horizontal circle readings).

An observing list of east-west pairs reaching elongation near the same time and altitude should be prepared giving the name, number, magnitude, azimuth, altitude, and time of observation. These may be computed from formulas in section 9.112.

The accuracy of the result is directly affected by any error in measuring horizontal angles and therefore is sensitive to the inclination of the horizontal axis. Great care must be taken in the elimination or determination of this source of systematic error, the best results being obtained by the use of a striding level. It is also advisable to refer the horizontal circle readings to an auxiliary reference mark to insure the orientation of the horizontal circle.

The suggested observation procedure is the following:

(1) Read horizontal circle on the reference mark.

(2) As the star approaches elongation, make two to four pointings, recording time and horizontal circle readings.

(3) Reverse telescope and reobserve reference mark.

(4) Repeat (2).

(5) Repeat (1) - (4) for the other star of the pair.

A recommended sequence of reductions is the following:

(1) Compute the stars' azimuths from equation (9.1) using the assumed latitude and their hour angles from equation (3.52).

(2) Compute the corrections for the inclination of the horizontal axis and for the curvature (as in Example 9.2) and apply them to the horizontal circle readings.

(3) Compute dH from equation (10.26).

(4) Compute the zenith distances of the stars at elongation from equation (3.51).

(5) Compute $d\Phi$ from equation (10.25).

(6) Compute the final latitude by adding $d\Phi$ to the assumed latitude.

The random errors affecting the determination are the same as in section 9.114. The expected standard error derived from a single pair may be determined from

$$m_{\Phi}^2 = (m_{HW}^2 + m_{HE}^2)/(\sin A_E \cot z_E - \sin A_W \cot z_W)^2 \qquad (10.27)$$

where

$$m_H^2 = \frac{1}{n}[(m_{hp}^2 + m_{\ell}^2)\cot^2 z + m_{hp}^2 + m_{ha}^2].$$

10.16 Miscellaneous Methods of Second-Order Latitude Determination

The preceding second-order methods include the more common ones and those capable of producing the required precision. There are numerous other methods, some of them mentioned briefly below, which either are of limited accuracy or are seldom used in the field for various reasons.

10.161 Latitude by Ex-Meridian Zenith Distances.

The zenith distance and sidereal time of observation of any star of known declination allows the astronomic triangle to be solved for the latitude. With the zenith distance corrected for refraction and the hour angle deduced from the time of observation, the latitude may be computed from

$$\sin(F + \Phi) = \cos F \cos z \operatorname{cosec} \delta \tag{10.28}$$

where

$$\tan F = \cot \delta \cos h.$$

The method is inferior in accuracy to those involving meridian observations as the effect of systematic errors, in the zenith distance and time measurements, on the derived latitude increases rapidly off the meridian.

10.162 Latitude by Equal Zenith Distances of One Star. Observation of the chronometer times of the instants when a star reaches the same altitude east and west of the meridian enables the determination of the latitude without knowing the chronometer correction. One-half of the elapsed time, corrected for the chronometer rate, gives the hour angle for the computation of latitude by the formulas of the preceding method. It also provides an approximate chronometer correction since the mean of the chronometer times should correspond to the transit time of the star. Again, the expected accuracy is inferior for the same reasons as in the previous method.

10.163 Latitude by the Hour Angles of Stars in the Prime Vertical Plane. When a star is in the prime vertical plane ($A = 90^\circ$ or 270°) and its hour angle is determined, the latitude of the observer may be computed from equation (3.48) which yields

$$\cot \Phi = \frac{\cos h}{\tan \delta}. \tag{10.29}$$

Generally six to ten stars are observed paired east-west. The result will be free of refraction error, but, since according to equation (9.2) errors in timing are eliminated only when the stars elongate and not when they are in the prime vertical plane, the accurate knowledge of time is necessary.

10.164 Approximate Latitude by the Rate of Change of the Zenith Distance Near the Prime Vertical. An approximate determination of the latitude may be obtained by the rapid observations of two zenith distances and the corresponding times on a star near and symmetrical to the prime vertical plane. Using the differences of the times and of the zenith distances to determine the approximate rate of change of the zenith distance dz/dh, as in section 9.133, the latitude from equation (3.32) is obtained by

$$\cos \Phi = \pm \frac{dz}{dh} \tag{10.30}$$

where the minus sign is for a star in the east, and the plus sign is for a star in the west.

10.2 First-Order Latitude Determination

10.21 The Horrebow-Talcott Method

The method most widely used for the determination of first-order astronomic latitude with portable instruments (e.g., universal theodolites) is the Horrebow-Talcott method originally announced by Peter Horrebow in 1732 and rediscovered by Captain Andrew Talcott of the U.S. Engineers in 1834. It was adopted as the standard method for first-order latitude determinations by the U.S. Coast and Geodetic Survey in 1851. The method is based on the measurement of the difference between the zenith distances of two stars which are on the meridian, on opposite sides of the zenith.

The advantage of the method is that it does not require the measurement of the zenith distance itself and that it reduces the uncertainties of refraction corrections and other systematic errors due to the reading of the vertical circle. The high degree of precision obtainable is mostly due to the fact that by measuring small zenith distance differences only, most of the systematic errors are cancelled.

The stars are observed in pairs whose zenith distance differences are so small that they may be measured in the field of the telescope by means of the impersonal micrometer without changing the inclination (vertical angle) of the telescope. Any incidental change in the inclination between the observations on the north and south stars is measured by using the two Horrebow levels clamped to the horizontal axis (in effect, to the telescope) of the instrument. The observations consist of centering the movable wire on the first star as it crosses the meridian and reading the turns and divisions of the micrometer. The turns are read from the numbers in the field of view; the divisions (part of a turn) are read from the micrometer head (see, e.g., Fig. 7.8). After reading the ends of the Horrebow levels, the instrument is reversed and the procedure is repeated for the second star. The difference in zenith distance between the stars is the difference of the micrometer readings corrected as discussed later.

The latitude is determined either from equation (10.5) or from (10.6) in which the zenith distance difference in terms of the micrometer readings is calculated from

$$z_S - z_N = (\pm)(M_W - M_E)R \qquad (10.31)$$

where M_E and M_W are the respective micrometer readings (in units of turns of the micrometer) with eyepiece east and west, and R is the value of one turn of the micrometer in seconds of arc. The sign in parenthesis depends on the type of instrument used, i.e., on the system of graduations of the micrometer. For example, the positive sign should be used for the Wild T4 theodolite, the instrument assumed to be used in this section. The scale of the micrometer of the T4 is marked in such a way that for eyepiece west and a south star the graduations increase with increasing zenith distance, while for a north star graduations decrease with increasing zenith distance; for eyepiece east and a south star graduations decrease with increasing zenith distance, while for a north star graduations increase with increasing zenith distance. In case of another type of instrument, certain small modifications to the procedures described below might be necessary.

10.211 Latitude Equation: Corrections to the Observed Zenith Distance Differences. The corrections to be applied are the following:

(1) Correction for the inclination (tilt) of the telescope.

(2) Correction for differential refraction.

(3) Reduction to the meridian (if the observations were made slightly off the meridian).

The tilt of the telescope during the observations on a star pair is determined from the readings of both ends of the Horrebow levels in both positions of the instrument. In accordance with equation (7.31) for the case of consecutive numbering, the tilt from the mean of the readings on the two levels is

$$b' = \frac{1}{2}\left\{\frac{d_1}{4}\left[(N_1+S_1)_W - (N_1+S_1)_E\right] + \frac{d_2}{4}\left[(N_2+S_2)_W - (N_2+S_2)_E\right]\right\}$$

or, assuming that

$$d_1 = d_2 = \frac{d_1 + d_2}{2},$$

$$b' = \frac{d_1 + d_2}{16}\left[(N_1 + N_2 + S_1 + S_2)_W - (N_1 + N_2 + S_1 + S_2)_E\right] \qquad (10.32)$$

where N_1 and S_1 are the respective readings on the north and south ends of one of the Horrebow levels whose sensitivity per division is d_1. The subscripts E and W refer to the position of the eyepiece (east and west).

The correction for the difference in refraction of the stars in a pair may be computed from equation (4.76); using the first term only,

$$r = \Delta z_{RS} - \Delta z_{RN} = A_1(\tan z_S - \tan z_N) = A_1 \frac{\sin(z_S - z_N)}{\cos z_S \cos z_N}.$$

Since z_S and z_N are nearly equal, they may be substituted for this purpose by their mean value z_m, in which case the above equation with $A_1 = 57''.9$ (see Table 4.8 at $z = 45^\circ$) yields the standard correction for differential refraction:

$$r = 57''.9 \sin(z_S - z_N) \sec^2 z_m. \qquad (10.33)$$

One-half of this correction is the latitude correction tabulated, for example, in table VIII in [Hoskinson and Duerksen, 1952]. Its value at the extreme situation of $z_m = 45^\circ$ and $z_S - z_N = 20'$ is $r/2 = 0''.67$. The correction has the same sign as the micrometer difference.

If a star is observed on the telescope's middle vertical thread slightly off the meridian, its zenith distance should be reduced to the meridian according to equations (10.9) - (10.11). In practice it may also happen that the instrument is in the meridian but the star is observed at a certain distance from the middle vertical thread. In this case, the reduction to the meridian should be computed from the triangle shown in Fig. 10.1. In this figure Q is a star south of the zenith on the meridian. The

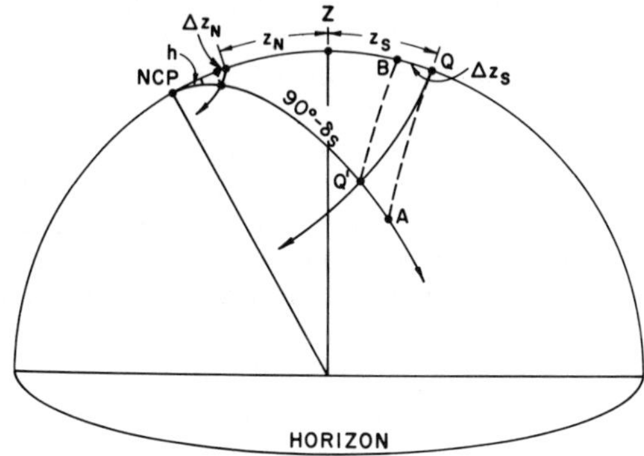

Fig. 10.1 Reduction of zenith distances to the meridian

line AQ is the projection of the micrometer thread if the star is observed on the meridian and BQ' is the projection of the micrometer thread if the star is observed off the meridian at Q'. The distance $QB = \Delta z_S$ is the correction to be applied to the observed zenith distance to reduce it to the meridian. From the spherical triangle NCP-B-Q', disregarding the subscripts,

$$\cos h = \tan\delta \tan(90^\circ - \delta - \Delta z)$$

from which

$$\tan\delta = \cos h \tan(\delta + \Delta z). \qquad (10.34)$$

From trigonometry

$$\tan(\delta + \Delta z) = \frac{\tan\delta + \tan\Delta z}{1 - \tan\delta \tan\Delta z}$$

and

$$\cos h = (1 - 2\sin^2\tfrac{1}{2}h).$$

After substituting the last two relations into equation (10.34) and clearing the fraction, expanding and neglecting the term $(2\sin^2\frac{1}{2}h \tan\Delta z)$, it yields

$$\tan\Delta z = \sin\delta \cos\delta\,(2\sin^2\tfrac{1}{2}h) \qquad (10.35)$$

or, since Δz is a small angle $\tan \Delta z \simeq \Delta z$, and also $\sin\delta \cos\delta = \frac{1}{2}\sin 2\delta$, the correction to the meridian in seconds of arc becomes

$$\Delta z = \sin 2\delta \, \frac{\sin^2 \frac{1}{2} h}{\sin 1''} . \tag{10.36}$$

The reduction for the north star is the same. The zenith distances of south stars and north stars in lower culmination are decreased, while the zenith distances of north stars in upper culmination are increased when observing off the meridian. Therefore, the correction to be added to the zenith distance difference $z_S - z_N$ is:

$$\text{reduction to the meridian} = \Delta z_S \pm \Delta z_N \tag{10.37}$$

where the positive sign applies to north stars near upper culmination and the negative sign to those near lower culmination. When the south star's declination is negative, the sign of the correction Δz_S also becomes negative. One-half of the correction Δz is the latitude correction, positive values of which may be obtained directly from table X of [Hoskinson and Duerksen, 1952] using δ and h as arguments.

Utilizing equations (10.5), (10.6), (10.31) - (10.33), and (10.37), the latitude from one star pair, when the north star is in upper culmination, is computed from

$$\Phi = \tfrac{1}{2}(\delta_S + \delta_N) + \tfrac{1}{2}(M_W - M_E)R + b' + \tfrac{1}{2} r + \\ + \tfrac{1}{2}(\Delta z_S + \Delta z_N) . \tag{10.38}$$

When the north star is in lower culmination, the latitude equation becomes

$$\Phi = \tfrac{1}{2}(\delta_S - \delta_N) + \tfrac{1}{2}(M_W - M_E)R + b' + \tfrac{1}{2} r + \\ + \tfrac{1}{2}(\Delta z_S - \Delta z_N) + 90^\circ . \tag{10.39}$$

In these equations the equatorial drum value R usually is entered with its approximate value only. A correction to R is computed together with Φ from the adjustment of the available latitude determinations (see section 10.214).

10.212 Observing List. The generally accepted program for a first-order astronomic latitude determination consists of the observation of sixteen pairs of stars. The entire latitude may be observed on a single night. The observation can be accomplished so rapidly that the number of pairs is often increased to 24.

Considerable effort goes into preparation of the observing list. Pairs of stars must be selected whose zenith distances are nearly equal. It is therefore necessary to select the stars from a catalogue which has a great many stars listed. The GC or the SAO catalogue is generally used. Stars to be included in the observing list should satisfy the following requirements:

—stars brighter than the seventh magnitude should be used.
—stars within 30° of the zenith should be used; therefore, the limits in declination are $\Phi + 30^\circ \geq \delta \geq \Phi - 30^\circ$.
—the limits in right ascension are the local sidereal times of sunset and sunrise.
—the zenith distance difference between the two stars in a pair should be within the range of the micrometer (20' in the case of the Wild T4).
—the difference in the right ascensions of two stars of a pair should not be less than 1^m and not more than 10^m.
—the difference in the right ascensions of the last star of a pair and of the first star of the next pair should be at least 2^m.
—for a night's work, the algebraic sum of the micrometer differences $(M_W - M_E)$ should be less than the number of pairs observed to minimize the effect of error in the micrometer value.

In selecting the star pairs, if δ_1 is the declination of the first star, the declination of the second star (both in upper culmination) may be approximated from

$$\delta_2 \approx (2\Phi - \delta_1) \pm 20'.$$

Since there are fewer bright stars in the northern sky than in the southern sky, in northern latitudes it is more convenient to select the northern star first and then try to find a pair for it. The first table in Example 10.3 (following section 10.214) is a sample observing list. The column $(2\Phi - \Sigma\delta)$ represents the differences between the zenith distances in minutes of arc $(\Sigma\delta = \delta_N + \delta_S)$. In the column $a(2\Phi - \Sigma\delta)$ are the zenith distance differences W - E expressed in turns of the micrometer ($a = 60''/R$, R being the equatorial value of one turn of the micrometer in seconds of arc). The mean zenith distance of the pair (both stars in upper culmination) is equal to $\frac{1}{2}(\delta_N - \delta_S)$, which value serves to compute the vertical circle setting for the observation of the pair. The vertical setting circle of the Wild T4 is constructed in such a way that the setting is $\frac{1}{2}(\delta_N - \delta_S)$ for star N(S) and eyepiece position E(W) or $360^\circ - \frac{1}{2}(\delta_N - \delta_S)$ for star N(S) and eyepiece W(E). The approximate micrometer settings are computed in the last column as an aid in identifying the proper star. The micrometer setting for the Wild T4 is

$$10 \pm a(2\Phi - \Sigma\delta)/2$$

where the +(-) sign is for the eyepiece position W(E). The micrometer set to this value will bring the star near the movable thread in the field of view.

10.213 Observations. It is assumed that the instrument is set in the meridian and a sidereal chronometer with the approximate local sidereal time is running. The suggested sequence of observations is the following:

(1) At the appropriate time point the telescope to the proper zenith distance through the use of the setting circle. Turn the micrometer to its precomputed setting.

(2) Clamp the Horrebow levels to the horizontal axis and center the bubbles.

(3) Bisect the star with the movable thread as it crosses the vertical thread.

(4) Read the micrometer turns in the field and the divisions from the micrometer head.

(5) Record both ends of both Horrebow levels.

(6) Rotate the instrument by 180° and reset in the meridian as in (1).

(7) Repeat steps (3) - (5) for the second star.

This constitutes one latitude determination. Generally 16-24 determinations will give a first-order latitude of sufficient precision. The time of observation need not be recorded unless the star is observed off the meridian. Temperature and barometric pressure readings are taken at the beginning and at the end of the night's work.

It is extremely important that the vertical angle of the telescope and the Horrebow levels not be disturbed between observations on the stars of a pair. If after the turning of the instrument the bubble of a Horrebow level reaches its extreme position, it may be recentered by means of the vertical fine motion screw of the telescope.

A sample field record is shown in the second table of Example 10.3.

10.214 Computations. The latitude for each star pair is computed from equation (10.38) or (10.39). The observation equation can be obtained by taking its partial derivative with respect to the variables Φ and R/2:

$$d\Phi - M\,dR + \Delta\Phi = v \qquad (10.40)$$

where $d\Phi$ is the correction to the assumed latitude Φ_0 (usually taken as the average latitude from all the determinations), $M = M_W - M_E$, dR is the correction to $\frac{1}{2}R$, and $\Delta\Phi = \Phi_0 - \Phi$ where Φ is the latitude obtained from the pair.

Example 10.3 illustrates the entire computation procedure.

EXAMPLE 10.3

Latitude by the Horrebow-Talcott Method

Observer: L. I. Sukman
Recorder: S. F. Cushman
Computer: H. D. Preuss
Local Date: 5-14-1962
Location: OSU Old Astro Pillar
$\Phi \cong 40°00'00''.00$
$\Lambda \cong -5^h32^m10^s$

Instruments: Wild T4 theodolite (No. 64252); Hamilton sidereal chronometer (No.2E12304); Zenith transoceanic radio

1. <u>Observing List for Latitude</u> (Sample)

R = 152''.65 $a = \frac{60''}{R}$ = .393 revolution/minute of arc

Pair No.	G.C. No.	Magnitude	α	$\delta_{N,S}$	$(\delta_N - \delta_S)$	$\Sigma\delta = \delta_N + \delta_S$	$(2\Phi - \Sigma\delta)$	$a(2\Phi - \Sigma\delta)$ revolutions	Star Position	Eyepiece	Setting: Vertical Circle[1]	Setting: Micrometer[2]
20	19519	6.2	$14^h26^m33^s$	36°22'	07°11'	79°55'	+ 5'	+2.0	S	W	03°36'	11.0
	19593	6.7	14 29 54	43 33					N	E		9.0
21	19726	6.0	14 36 29	18 28	42 57	79 53	+ 7	+2.8	S	E	338 32	8.6
	19825	6.2	14 41 07	61 25					N	W		11.4
22	20060	6.8	14 53 00	30 13	19 34	80 00	0	0	S	W	09 47	10.0
	20119	5.7	14 55 08	49 47					N	E		10.0
23	20308	5.6	15 04 11	48 18	16 22	80 14	-14	-5.5	N	E	08 11	12.8
	20489	6.2	15 12 34	31 56					S	W		7.2
24	20544	6.5	15 14 37	69 05	57 57	80 13	-13	-5.1	N	W	331 02	7.4
	20722	7.0	15 23 01	11 08					S	E		12.6
25	20817	6.4	15 27 00	62 24	44 37	80 11	-11	-4.3	N	E	22 18	12.2
	20962	6.1	15 33 50	17 47					S	W		7.8
26	21164	5.9	15 42 59	17 23	45 20	80 06	- 6	-2.4	S	W	22 40	8.8
	21246	5.1	15 46 06	62 43					N	E		11.2
27	21383	6.9	15 53 07	65 24	50 53	79 55	+ 5	+2.0	N	E	25 26	9.0
	21428	5.7	15 55 30	14 31					S	W		11.0
28	21569	6.2	16 01 08	53 01	26 11	79 51	+ 9	+3.5	N	W	346 54	11.8
	21751	6.7	16 08 31	26 50					S	E		8.2

[1]Vertical circle setting is: $\frac{1}{2}(\delta_N - \delta_S)$ for star N (S) and eyepiece E (W); 360° − $\frac{1}{2}(\delta_N - \delta_S)$ for star N (S) and eyepiece W(E).

[2]Micrometer setting is: $10 \pm a(2\Phi - \Sigma\delta)/2$, +(−) for eyepiece W(E).

2. Latitude Field Record (Sample)

Temperature: 17°.5 C
Barometer: 765 mm

Pair No.	G.C. No.	Star Position	Eyepiece	Micrometer		Σ of Turns W - E	Level		Chronometer Time of Observations
				Turns	Divisions		North	South	
20	19519	S	W	10	87.1		133.0	116.0	$14^h26^m33^s$
							49.2	22.8	
	19593	N	E	8	70.1	+2	117.8	134.7	14 28 55
							24.6	51.2	
21	19726	S	E	8	20.4		119.9	139.9	14 36 30
							24.8	48.0	
	19825	N	W	11	06.3	+5	138.1	117.8	14 41 07
							45.0	21.9	
22	20060	S	W	9	85.9		146.5	125.8	14 53 00
							48.9	25.6	
	20119	N	E	9	68.2	+5	128.1	149.0	14 55 08
							26.8	50.0	
23	20308	N	E	11	99.2		119.3	140.2	15 04 12
							22.2	45.9	
	20489	S	W	6	84.2	0	139.0	117.9	15 12 34
							44.1	20.7	
24	20544	N	W	6	99.9		148.1	127.0	15 14 37
							49.0	25.4	
	20722	S	E	11	83.2	-5	128.6	150.0	15 23 00
							26.0	49.9	
25	20817	N	E	11	81.6		128.9	150.1	15 27 00
							26.8	50.2	
	20962	S	W	7	58.4	-9	147.1	125.8	15 33 50
							48.9	25.1	
26	21164	S	W	8	18.1		146.2	122.8	15 42 59
							46.8	22.9	
	21246	N	E	10	32.0	-11	125.9	147.3	15 46 06
							24.7	48.4	
27	21383	N	E	8	48.7		126.7	148.2	15 53 07
							25.4	49.2	
	21428	S	W	10	54.1	-9	145.5	123.9	15 55 31
							48.0	24.0	
28	21569	N	W	11	66.7		139.5	118.0	16 01 08
							44.4	20.7	
	21751	S	E	8	19.2	-6	121.8	143.3	16 08 31
							22.7	46.4	

3. Computation of Latitude (Sample)

Horrebow Level No. 2182: d_1 = 0".918/div.
No. 2616: d_2 = 1".045/div.
$(d_1 + d_2)/16$ = 0".122687/div.
R/2 = 76".328

Pair No.	Star: G. C. No.	Star: Position	Eyepiece	Micrometer: Reading	Micrometer: Difference	Level: N	Level: S	Level: Difference	δ(FK4)	$\frac{1}{2}(\delta_N + \delta_S)$	Corrections: Micrometer	Corrections: Level	Corrections: Refraction	Φ 40°00' +
	19519	S	W	10.871		133.0	116.0		36°22'00".39					
20					+2.170	49.2	22.8	-7.3		39°57'31".68	+2'45".63	-0".90	.+".05	16".46
	19593	N	E	8.701		117.8	134.7		43 33 02.97					
						24.6	51.2							
	19726	S	E	8.204		119.9	139.9		18 27 45.08					
21					+2.859	24.8	48.0	-9.8		39 56 37.20	+3 38.22	-1.20	+.07	14.29
	19825	N	W	11.063		138.1	117.8		61 25 29.31					
						45.0	21.9							
	20060	S	W	9.859		146.5	125.8		30 13 03.21					
22					+0.177	48.9	25.6	-7.1		40 00 02.93	+0 13.51	-0.87	0	15.57
	20119	N	E	9.682		128.1	149.0		49 47 02.65					
						26.8	50.0							
	20308	N	E	11.992		119.3	140.2		48 17 52.10					
23					-5.150	22.2	45.9	-5.9		40 06 48.48	-6 33.09	-0.72	-.11	14.56
	20489	S	W	6.842		139.0	117.9		31 55 44.86					
						44.1	20.7							
	20544	N	W	6.999		148.1	127.0		69 05 10.83					
24					-4.833	49.0	25.4	-5.0		40 06 24.50	-6 08.89	-0.61	-.13	14.87
	20722	S	E	11.832		128.6	150.0		11 07 38.16					
						26.0	49.9							
	20817	N	E	11.816		128.9	150.1		62 24 26.69					
25					-4.232	26.8	50.2	-9.1		40 05 39.02	-5 23.02	-1.12	-.10	14.78
	20962	S	W	7.584		147.1	125.8		17 46 51.36					
						48.9	25.1							
	21164	S	W	8.181		146.2	122.8		17 22 56.70					
26					-2.139	46.8	22.9	-7.6		40 02 59.13	-2 43.27	-0.93	-.05	14.88
	21246	N	E	10.320		125.9	147.3		62 43 01.56					
						24.7	48.4							

4. Summary of Latitude Computation

Pair No.	G C Nos.		Φ 40°00′ +	$\Delta\Phi_1 = \Phi_1 - \Phi$	$\Delta\Phi_2 = \Phi_2 - \Phi$	Micrometer Differences (M)	*Corrections (Mr)	*$\Phi = \Phi_2 + Mr$ 40°00′ +	*v
20	19519	19593	16″.46	-1″.542					
21	19726	19825	14.29	0.628	0″.547	2.859	-0″.118	14″.172	0″.716
22	20060	20119	15.57	-0.652	-0.733	0.177	-0.007	15.563	-0.675
23	20308	20489	14.56	0.358	0.277	-5.150	0.213	14.773	0.115
24	20544	20722	14.87	0.048	-0.033	-4.883	0.202	15.072	-0.184
25	20817	20962	14.78	0.138	0.057	-4.232	0.175	14.955	-0.067
26	21164	21246	14.88	0.038	-0.043	-2.139	0.088	14.968	-0.080
27	21383	21428	14.44	0.478	0.397	2.054	-0.085	14.355	0.533
28	21569	21751	14.93	-0.012	-0.093	3.479	-0.144	14.786	0.102
29	21983	22112	14.61	0.308	0.227	-3.032	0.125	14.735	0.153
30	22235	22344	13.91	1.008	0.927	-6.467	0.267	14.177	0.711
31	22412	22471	15.10	-0.182	-0.263	1.422	-0.059	15.041	-0.153
33	22807	22975	14.88	0.038	-0.043	-5.000	0.207	15.087	-0.199
34	23071	23127	14.82	0.098	0.017	-2.359	0.098	14.918	-0.030
35	23223	23315	14.52	0.398	0.317	-2.546	0.105	14.625	0.263
36	23433	23647	15.53	-0.612	-0.693	-1.433	0.059	15.589	-0.701
37	23741	23872	15.07	-0.152	-0.233	2.165	-0.090	14.980	-0.092
38	23993	24183	15.03	-0.112	-0.193	-1.042	0.043	15.073	-0.185
39	24346	24510	15.09	-0.172	-0.253	2.404	-0.099	14.991	-0.103
40	24714	24903	15.02	-0.102	-0.183	-0.021	0.001	15.021	-0.133
Sum (e.g., [$\Delta\Phi$])			1: 298″.36 2: 281.90	0″.00	0″.003	-23.744		282″.881	-0″.009
Mean (e.g., $\frac{[\Phi]}{n}$)			1: 14″.918 2: 14.837			- 1.2497		14″.888	
Square Sum (e.g., [$\Delta\Phi\Delta\Phi$])				5.3405	2.8376	201.0692			2.5451

NOTE: Pair No. 20 rejected. Subscripts 1 and 2 refer to quantities calculated from 20 pairs and to quantities calculated after rejections respectively.

* Computed after the adjustment.

5. Final Observed Latitude from Nineteen Determinations for the Latitude and for the Equatorial Drum Value of the Micrometer

The normal equations are

$$n d\Phi - [M]dR + [\Delta\Phi_2] = 19.000 d\Phi + 23.744 dR + 0.003 = 0$$

$$-[M]d\Phi + [MM]dR - [M\Delta\Phi_2] = 23.744 d\Phi + 201.0692 dR + 7.0926 = 0$$

The solution of the normal equations yields the following corrections and standard deviations:

$$d\Phi = 0''.052 \qquad M_\Phi = 0''.096$$
$$dR = -0''.041 \qquad M_R = 0''.029$$

Thus the observed values are

$$\tfrac{1}{2}R\ (\text{assumed}) = 76''.328$$
$$dR = -0.041 \quad \pm 0''.029$$
$$\tfrac{1}{2}R\ (\text{final}) = 76''.287 \quad \pm 0''.029 \;\|$$

and

$$\Phi_2\ (\text{assumed}) = 40^\circ 00' 14''.837$$
$$d\Phi = 0.052 \ \pm 0''.096$$
$$\Phi_T = 40^\circ 00' 14''.889 \ \pm 0''.096 \;\|$$

10.22 Alternative Methods for First-Order Latitude Determination

The primary method of first-order latitude determination is that using zenith distance differences as described in section 10.21. There are other less frequently used, but still satisfactory methods producing first-order results which are described below.

10.221 The Pevtsov Method [Pewcow, 1888]. The method is a refined version of the latitude determination by equal altitudes of two stars described in section 10.14. The instrument used is a universal theodolite where the change in the inclination of the telescope during the observation is determined by means of the Horrebow levels exactly as described in section 10.211. The correction to the latitude (computed from equation (10.24)) due to the inclination of the telescope is [Mühlig, 1960, p. 92]

$$\pm \frac{2 \sin z \cos\Phi}{\sin\delta_2 - \sin\delta_1} b' \tag{10.41}$$

where b' is computed from equation (10.32), and the positive sign is applied when the lower numbers on the vial are toward the stars.

In order to eliminate the systematic errors in timing, i.e., in the hour angles of the stars, the pair should theoretically be observed on the meridian (see equation (10.12)). Since it is very difficult, if not impossible, to find stars which culminate north and south of the zenith at exactly the same zenith distances, they must be observed off the meridian. It may be shown that if the azimuths of stars in a pair are such that $A_1 + A_2 = 180°$, i.e., if they are observed in symmetrical positions about either the east or the west side of the prime vertical, the latitude will again be free of systematic errors in h and it will not be affected by diurnal aberration either [Niethammer, 1947, pp. 64-65]. Refraction should have no effect because of the equal altitudes of the stars.

The preparation of an observing list is even more tedious than that of the Horrebow-Talcott method. The observations and the calculations, on the other hand, are very simple. No readings on the circles or on the micrometer need to be recorded and the only correction to be applied is that of (10.41). The zenith distances should be limited to 15°-60°. The horizontal angle between either side of the prime vertical plane and the star should be 50°-85°. The number of star pairs providing first-order precision is 16-24.

10.222 The Sterneck Method. If the purpose is to simplify the problem of preparing the observation list and pay the price by making the observations dependent on circle readings and possibly burden the final results by small residual systematic errors (due mostly to refraction), then the method to be used is the latitude by meridian zenith distances described in section 10.11. First-order results may be achieved by a more carefully selected observation program than is described there and by the use of universal instruments instead of geodetic theodolites. The name of the method is traced back to General Sterneck who apparently proposed to use it extensively the first time in connection with the Austrian grade measurements.

Since the main source of systematic error is due to refraction, stars should be observed in groups of four to six pairs, half of them culminating north of the zenith, the other half culminating south. Each pair within the group should have a difference in zenith distance as close to zero as possible or at least, in case of upper culminations, the stars should fulfill the equivalent condition of

$$\frac{\Sigma\delta}{n} \cong \Phi$$

where n is the number of stars observed. This condition together with the maximum advised zenith distance of 60° determine the declination range when selecting the stars. The right ascension difference between any two stars of a pair should be $3^m - 20^m$ to insure adequate time to make the necessary readings and then turn the instrument but not so long a time as to disturb its stability. The observation procedure is the same as in section 10.11 except that the ends of the vertical circle index level should also be recorded (immediately before the vertical circle is read) to be able to determine the precise zenith point of the instrument. The zenith distance is computed as explained in section 7.42 from the readings on the vertical circle and on its level. In order to minimize the effect of the systematic graduation errors of the vertical circle, it should be advanced (rotated) after each group by $180^\circ/m$, m being the number of groups observed one night. If the star is observed off the meridian, its zenith distance should be reduced to the meridian either by equations (10.9) - (10.11) or by (10.36) as explained in section 10.211. Observations on six groups (48-72 stars) during one night should produce satisfactory first-order results.

10.3 Corrections to the Observed Latitude

As pointed out in section 2.33, the 'observed' or 'true' astronomic latitude Φ_T is the angle between the vertical at the observer and the plane of the instantaneous equator (which is perpendicular to the instantaneous axis of rotation). Evidently, latitudes observed at different locations at different times are not directly comparable since they refer to different positions of the instantaneous equator and to the normals of different geops. In order to resolve this problem, it is customary to refer the latitude to the average terrestrial equator and to the geoid, a common reference surface to all observers. The difference between the true latitude Φ_T and its reduced value Φ may be divided into the following components (corrections):

(1) Correction for the motion of the pole.

(2) Correction for the curvature of the plumb line.

If for some special reason the observation is made from a place located off the center of the reference station, a further correction is needed to compensate for the eccentricity.

10.31 Correction for the Motion of the Pole

This correction when applied to the observed latitude will refer it to the average terrestrial equator, thus making it independent from the varying position of the instantaneous terrestrial pole. The correction as derived in section 4.133 is

$$\Delta\Phi_P = \Phi - \Phi_T = y_P \sin\Lambda - x_P \cos\Lambda \qquad (4.39)$$

where x_P and y_P are the coordinates of the true pole in units of seconds of arc. In most cases this correction is barely significant.

10.32 Correction for the Curvature of the Plumb Line

This correction when applied to the observed latitude will in effect transfer the observation station from the ground (geop) to the geoid.

The correction as described in section 9.32 is computed from

$$\Delta\Phi_{PC} = \Phi - \Phi_T = \delta\Phi = -\frac{H}{\bar{g}}\frac{\partial\bar{g}}{\partial x}\operatorname{cosec} 1'' \qquad (9.35)$$

where H is the orthometric height of the observation station, $\bar{g}$ and $\partial\bar{g}/\partial x$ are the gravity and the north component of its horizontal gradient, respectively, at the station. If the latter values are not available, the correction is approximated by its 'normal' value:

$$\Delta\Phi_{PC} = \delta\Phi = -0''.00017\ H \sin 2\varphi \qquad (9.37)$$

where H is to be substituted in meters and φ denotes either the geodetic or the observed astronomic latitude.

10.33 Correction for the Eccentricity of the Station

This correction is to be applied when the latitude is observed off the reference station, and it will reduce the observed latitude to the reference station. In Fig. 10.2 let P be the reference station and P′ be the eccentric station where the latitude Φ_T is observed. If the distance PP′ is d and the azimuth of the line is A, the identity for the length of a short meridian arc is

$$R\Delta\Phi_E \sin 1'' = d \cos A;$$

thus the correction

$$\Delta\Phi_E = \Phi - \Phi_T = \frac{d}{R \sin 1''}\cos A \qquad (10.42)$$

where R is approximated by the mean radius of the earth.

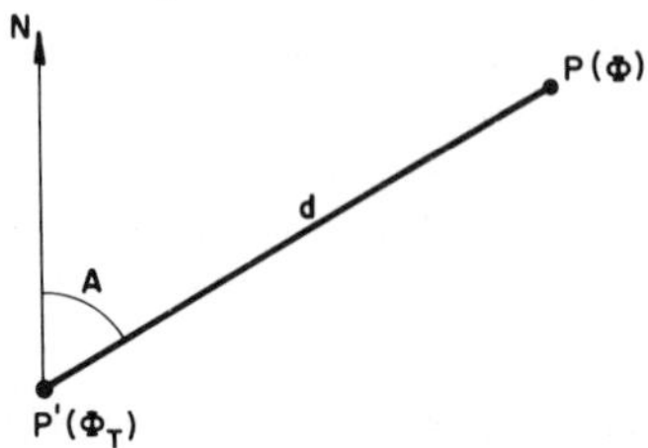

Fig. 10.2 Correction to the observed latitude for the eccentricity of the station

10.4 Simultaneous Determination of Latitude and Azimuth

This method proposed in [Ghosh, 1953] enables the determination of the latitude of a station and the azimuth of a reference mark from the station by three observations on a single star whose coordinates need not be known. Measurement of the altitude of the star and the corresponding horizontal angle from the reference mark, in three positions of the star, provide the data required for solution. The more widely spaced the observations are, the stronger the resulting solution will be.

The long time period required is the major disadvantage of the method, but it has conceivable field application in cases where a star catalogue may not be available. The solution outlined in the reference is a unique method utilizing Cartesian coordinates. A supplementary article outlines a solution for handling redundant observations through an approximate graphical solution [Gougenheim, 1954]. Still another author presents a direct solution based on equation (3.5) which in the interest of brevity is outlined here without derivations [Tárczy-Hornoch, 1956].

The star selected should be bright enough to be readily identifiable some hours later and in a position to be of sufficient altitude above the horizon throughout the observation period so that it will not be affected by anomalous refraction.

The observation procedure consists of an azimuth pointing on the reference mark, recording the horizontal circle reading. Direct and reverse pointings are made on the star in rapid sequence, recording both horizontal and vertical circles. A reverse pointing is then made on the azimuth mark. Temperature and pressure are recorded for refraction correction. Two more such sets are required at approximately equal and as extended as possible time intervals.

The reduction procedure starts with computing the zenith distances of the three positions z_1, z_2, and z_3 by taking the mean of the values obtained from the direct and reversed pointings and correcting each for refraction. Let L_1, L_2, and L_3 denote the horizontal circle readings corresponding to the three positions. Also let $H_1 (H_2)$ denote the horizontal angles between the first (second) and third positions of the star, and H_3 denote the horizontal angle between the third position and the north side of the meridian; thus

$$\begin{aligned} H_1 &= \pm(L_1 - L_3) \\ H_2 &= \pm(L_2 - L_3) \end{aligned} \quad [+(-) \text{ before (after) upper culmination}],$$

$$\begin{aligned} H_3 &= A_3 && [\text{star east of meridian}]\ , \\ &= 360^\circ - A_3 && [\text{star west of meridian}]\ . \end{aligned}$$

With this notation

$$\tan H_3 = \frac{(\sin z_2 \cos H_2 - \sin z_3) - K(\sin z_1 \cos H_1 - \sin z_3)}{\sin z_2 \sin H_2 - K \sin z_1 \sin H_1} \tag{10.43}$$

where

$$K = \frac{\cos z_3 - \cos z_2}{\cos z_3 - \cos z_1}\ .$$

The azimuth of the reference mark A is obtained by adding or subtracting the horizontal angle between the star in its third position and the reference mark H_R to the azimuth of the star A_3:

$$A = A_3 \pm H_R \tag{10.44}$$

where $A_3 = H_3$ or $360° - H_3$ depending on whether the star in its third position is east or west of the meridian respectively.

The latitude is computed from either of the equations below:

$$\tan\Phi = \frac{\sin z_2 \cos(H_3 + H_2) - \sin z_3 \cos H_3}{\cos z_3 - \cos z_2} = \frac{K[\sin z_1 \cos(H_3 + H_1) - \sin z_3 \cos H_3]}{\cos z_3 - \cos z_2} \tag{10.45}$$

This constitutes one latitude and azimuth determination. Second-order precision may be obtained through several (3-6) determinations either by observing the same star in more than three positions or by using several stars and applying a least squares solution in both cases.

A similar method is described in [Baldini, 1966] where the solution for azimuth and latitude is based on the recording of the horizontal circle readings when three or more different stars reach the same but unknown zenith distance.

References

Baldini, A. A. (1966). "New Method for Determining Azimuth and Latitude Independent of Time and Zenith Distance." GIMRADA Research Notes, 17, U. S. Army Engineer, GIMRADA, Fort Belvoir, Virginia.

Bhattacharji, J.C. (1959). "A Method of Determining Astronomic Latitude from Observations of a Star Pair Near Times of Elongation." Empire Survey Review, 15, 114.

Bowie, I. G. (1956). "Latitude Determination Without Vertical Circle Readings." Empire Survey Review, 13, 102.

Chauvenet, W. (1891). A Manual of Spherical and Practical Astronomy, 1, (fifth edition), Philadelphia. Reprinted by Dover Publications, Inc., New York, 1960.

Ghosh, S. K. (1953). "Determination of Azimuth and Latitude from Observations of a Single Unknown Star by a New Method." Empire Survey Review, 12, 87.

Gougenheim, A. (1954). "Determination of Latitude and Azimuth from Observations of an Unknown Star." Empire Survey Review, 12, 94.

Hoskinson, A. J. and J.A. Duerksen. (1952). "Manual of Geodetic Astronomy." U.S Coast and Geodetic Survey Special Publication, 237, U. S. Government Printing Office, Washington, D. C.

Mühlig, F. (1960). Grundlagen und Beobachtungs-Verfahren der Astronomisch-geodätischen Ortsbestimmung. Herbert Wichmann Verlag, Berlin.

Niethammer, Th. (1947). Die Genauen Methoden der Astronomisch-Geographischen Ortsbestimmung. Verlag Birkäuser, Basel.

Pewcow, M. (1888). Über die Bestimmung der geographischen Breite durch korrespondierende Höhen." Schriften der Käiserlichen russischen geographischen Gessellschaft, 17, 5 and also 32, 2, 1899, St. Petersburg.

Tárczy-Hornoch, A. (1956). "Determination of Azimuth and Latitude from Observations of a Single Unknown Star by a New Method." Empire Survey Review, 13, 99.

Thorson, C.W. (1965). "Second-Order Astronomical Position Determination Manual." U.S. Coast and Geodetic Survey Publication 64-1. U.S. Government Printing Office, Washington, D.C.

11 DETERMINATION OF ASTRONOMIC LONGITUDE AND TIME

11.1 Time and Longitude

11.11 Longitude Equations

Astronomic longitude is defined as the angle between the plane of the observer's astronomic meridian and the Greenwich Mean Astronomic Meridian plane (or the meridian plane of the Mean Observatory as defined by the BIH; see section 8.213) measured in a plane parallel to the equator. It is reckoned positive to the east from the meridian of the Mean Observatory to $360^\circ = 24^h$. The determination of longitude is based on the equation

$$\Lambda = AST - GAST \tag{5.4}$$

where the local apparent sidereal time, AST, is determined by means of astronomic observations; the Greenwich apparent sidereal time, GAST, is obtained through the synchronization of the local timekeeper with a central time standard. In first- and second-order work the synchronization is generally performed via HF radio time signals as explained in section 8.3, and the local timekeeper is a mechanical or a quartz crystal chronometer.

The actual solution for the longitude depends on the type of timekeeper used. When using a chronometer keeping mean solar time, the solution may be based on the following general relations:

$$AST = \alpha + h, \tag{3.7}$$
$$GAST = T_M + \Delta T_M . \tag{11.1}$$

Thus

$$\Lambda = \alpha + h - (T_M + \Delta T_M) \tag{11.2}$$

where α is the apparent right ascension of the star corrected for diurnal aberration, h is its hour angle determined from the astronomic observations, T_M is the time shown on the local timekeeper at the instant of the observation, and ΔT_M is the chronometer correction which, in the ordinary case when the synchronization is done with respect to a UTC standard, includes the following components:

(1) The epoch difference $\Delta T_M^o = UTC - T_M^o$, at the time of the synchronization T_M^o.

(2) The drift of the chronometer $(d\Delta T_M/dt)(T_M - T_M^o)$ which, during the generally short time interval between the synchronization and the observation (1-2 hours), may be assumed to be linear. This not being the case, the procedure in Example 8.3 should be followed.

(3) The UT1 - UTC correction at the epoch of the observation (see section 8.233).

(4) The difference between UT1 and GAST at the epoch of the observation (see section 5.3).

Note that in item (3) the correction is to UT1 which represents the actual rotation of the earth; thus GAST determined in item (4) will be free of the effect of the motion of the instantaneous terrestrial pole. For this reason, in order to keep the two terms on the right-hand side of equation (5.4) consistent, AST must also be freed from the effect of polar motion. This is done by applying the polar motion correction $\Delta\Lambda_P$ as given by equation (4.40). Considering all these, the longitude equation becomes

$$\Lambda = \alpha + h + \Delta\Lambda_P - \left[T_M + \Delta T_M^o + \frac{d\Delta T_M}{dt}(T_M - T_M^o) + \right.$$
$$\left. + (UT1 - UTC) + (GAST - UT1) \right]. \qquad (11.3)$$

The correction (UT1 - UTC) may be substituted by its predicted (extrapolated) value to be found in the Circulars 'B/C' of the BIH (see Table 8.4) or, for example, in the 'Preliminary Times and Coordinates of the Pole' bulletins of the USNO (see Table 8.17). The quantity (GAST - UT1) is computed as shown in Example 5.1 (steps 9-17) or in Example 5.2 (steps 9-13). In practice, due to the relatively short duration of the observations (1-2 nights), the polar motion correction $\Delta\Lambda_P$ is added to the final observed longitude and is not entered with each observation (see also section 11.41). This same procedure may also be satisfactory for the correction (UT1 - UTC) provided that the same UTC system is used during all the observations or that the UTC systems employed are sufficiently coordinated.

Another frequent approach is to use a sidereal chronometer which is set to keep the approximate local mean sidereal time T_S, based on an assumed value of longitude Λ_O. In this case the difference between the observed AST and the time shown on the chronometer is equal to the difference between the actual and assumed longitude $\Delta\Lambda$; thus

$$\Lambda = \Lambda_o + \Delta\Lambda \tag{11.4}$$

where

$$\Delta\Lambda = AST - (T_s + \Delta T_s) = \alpha + h + \Delta\Lambda_P - (T_s + \Delta T_s). \tag{11.5}$$

In this equation T_s is the time shown on the sidereal chronometer at the instant of the observation. The chronometer is set at a time T_s^o by synchronizing it with UTC_o and converting the synchronization epoch in UTC to approximate mean sidereal time (MST) as follows:

$$T_s^o = UTC_o + (UT1_o - UTC_o) + (GMST_o - UT1_o) + \Lambda_o \tag{11.6}$$

where $(UT1_o - UTC_o)$ is determined as explained after equation (11.3); $(GMST_o - UT1_o)$ is computed in accordance with section 5.3. Both corrections refer to the epoch of synchronization. The quantity ΔT_s in (11.5) is the chronometer correction which consists of the drift of the chronometer during the time interval $T_s - T_s^o$, and the equation of the equinox at the time of the observation; thus

$$\Delta T_s = \frac{d\Delta T_s}{dt}(T_s - T_s^o) + Eq.E\,. \tag{11.7}$$

The correction Eq. E may be eliminated from the chronometer correction by applying it to the term AST in equation (11.5) which in this case becomes

$$\Delta\Lambda = MST - (T_s + \Delta T_s). \tag{11.8}$$

If the time interval $T_M - T_M^o$ or $T_s - T_s^o$ in equations (11.3) and (11.7), respectively, is short (1-2 hours), the drift rate $d\Delta T/dt$ may generally be assumed to be a constant. If the interval is long, the drift rate should be determined as illustrated in Example 8.3 except that HF receptions rather than VLF signals are utilized.

11.12 Principles of Determining Local Sidereal Time

Regardless of whether the longitude is computed from equation (11.3) or from (11.5), it will depend, as far as the astronomic observations are concerned, on the determination of the hour angle of the star which together with its right ascension defines AST. In practice, the observations are made either ex-meridian, in which case the hour angle is calculated from zenith distance measurements, or in the meridian (or reduced to the meridian) when the hour angle is 0^h or 12^h. In the latter case the chronometer time of the star's transit is actually measured, and the longitude is determined from equations (4.3) or (4.5) with $h = 0^h$ or 12^h. In the ex-meridian case, the hour angle is computed from equation (3.2) which yields

$$\cos h = \frac{\cos z - \sin\delta \sin\Phi}{\cos\delta \cos\Phi}. \tag{11.9}$$

Systematic errors affecting the longitude determination in this case include those of the assumed latitude $d\Phi$, the observed zenith distance dz, the observed chronometer time (e.g., personal equation) dT, and the chronometer correction $d\Delta T$. The total effect, based on equations (3.32), (10.1), and (11.2), is expressed by

$$d\Lambda = -(\sec\Phi \cot A\, d\Phi + \sec\Phi \operatorname{cosec} A\, dz + dT + d\Delta T). \quad (11.10)$$

The effect of $d\Phi$ is eliminated by observing in the prime vertical plane, while that of dz is eliminated by observing east and west stars in pairs which are symmetrical to the meridian and are at the same altitude. The effects of dT and $d\Delta T$ cannot be eliminated regardless of star selection and are, therefore, the most serious source of errors in longitude determinations.

The primary instrument required for longitude work is a universal theodolite or telescope (or micrometer transit) for first-order determinations and a geodetic theodolite for second-order work. The auxiliary equipment consists of a good mechanical or quartz chronometer, an HF radio receiver with an appropriate amplifier, and a chronograph (equipped with a hand tappet for second-order work). Fig. 11.1 is a scheme of the hook-up of the auxiliary equipment.

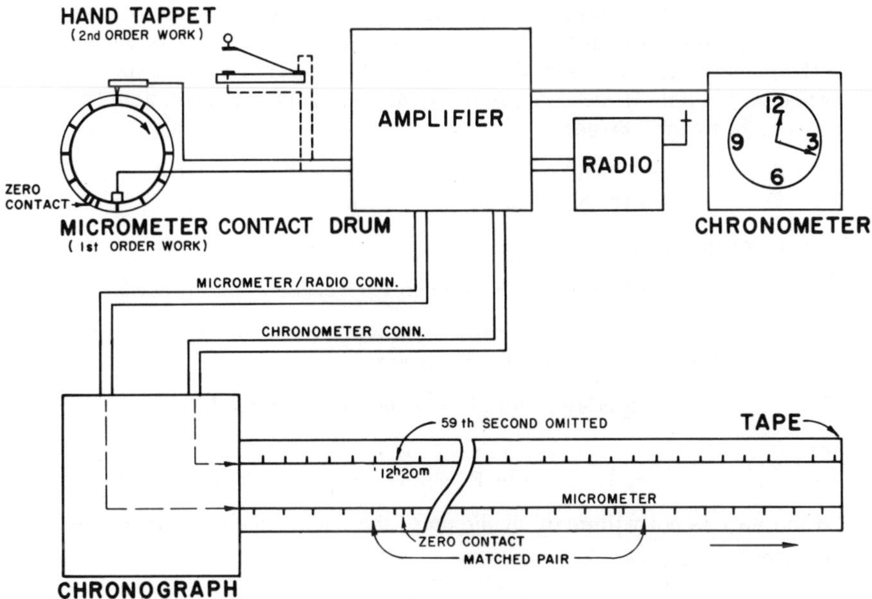

Fig. 11.1 Hookup diagram of auxiliary equipment for longitude work

In second-order work when zenith distances are utilized the observation essentially follows previously described procedures when the star's position on the horizontal thread is recorded on the reading circles of the theodolite. The time of the event should be recorded on the tape of the chronograph by pressing the hand tappet at the appropriate instant. The chronograph also registers the beats of the chronometer on the tape. The time of the event is then determined by interpolating the signal of the hand tappet between the signals of the chronometer beats. When transit time is observed through the vertical thread of the instrument, the same timing procedure should be followed. If the telescope of the theodolite is equipped with several vertical threads, the effect of random observational errors may be reduced by observing the transit over all of them and reducing the times recorded to the middle thread (near the collimation axis). An odd number of threads is always used, and they are placed as nearly equidistant as possible. If the threads were exactly equidistant, the mean of the observed times of transit over all of them could be taken as the time of transit over the middle one. Since it rarely happens that the threads are perfectly equidistant or symmetrical, it becomes necessary to determine their distances. Let f denote the angular interval of any thread from the middle one and F denote the time required by a star whose declination is δ to pass over this interval. If c is the collimation of the middle thread, the distance of the side thread from the collimation axis is $f+c$; and if τ is the (negative) hour angle of a star when on the middle thread, $F+\tau$ is its value when on the side thread. Applying equation (7.26) for transit time measurement to each thread

$$\sin(F+\tau-h_A) = \tan\delta_A \tan\delta + \sin(f+c) \sec\delta_A \sec\delta,$$
$$\sin(\tau-h_A) = \tan\delta_A \tan\delta + \sin c \sec\delta_A \sec\delta,$$

the difference of which is

$$2\cos(\tfrac{1}{2}F+\tau-h_A)\sin\tfrac{1}{2}F = 2\cos(\tfrac{1}{2}f+c)\sin\tfrac{1}{2}f\sec\delta_A\sec\delta.$$

At meridian transits the quantities $(\tau-h_A)$, c, and δ_A are very small; thus the above equation may be reduced to

$$2\cos\tfrac{1}{2}F\sin\tfrac{1}{2}F = 2\cos\tfrac{1}{2}f\sin\tfrac{1}{2}f\sec\delta$$

or

$$\sin F = \sin f\sec\delta. \qquad (11.11)$$

If the star is not within 10° of the pole, the above equation may be simplified to

$$F = f\sec\delta.$$

Either of these relations may be used to find f (the equatorial interval of a thread from the middle thread) by determining F from the observation of the time interval required for a star of declination δ and near the meridian to pass between the middle thread and the side thread in question. With the known equatorial intervals for each thread, the time of meridian transit over the middle thread will be

$$T_m = \frac{1}{n}\left[\sum_{i=1}^{n} T_i + \sec\delta \sum_{i=1}^{n} f_i\right] \tag{11.12}$$

where T_i is the observed transit time over the 'i' thread of equatorial interval f_i, and n is the number of threads (including the middle thread) over which the transit times are observed. Note that the f_i values have different signs on the opposite sides of the middle thread.

In first-order work the side threads are, in effect, replaced by the contact strips of the impersonal micrometer of the universal theodolite (or micrometer transit) used to track the star when it transits the field of view. As the micrometer (see Fig. 7.8) is rotated, the contacts are registered on the chronograph tape together with the beats of the chronometer. The times of the contacts (i. e., the transit times over imaginary threads) can then be interpolated between the chronometer signals on the tape and the transit time over the middle thread computed essentially as shown above and as described in section 11.31.

11.2 Second-Order Longitude Determination

11.21 Longitude by Zenith Distances Measured Near the Prime Vertical

In this method, complying with the requirements expressed by equation (11.10), a star pair consisting of one east star and one west star is observed near the prime vertical plane. The method is the standard used by the U. S. Coast and Geodetic Survey for second-order work [Thorson, 1965]. Their specifications call for observations within 5° of the prime vertical plane on a pair of stars whose zenith distances should not differ by more than 1° and whose altitudes range between 30° and 40°.

11.211 Observing List. The preparation of an observing list for longitude is more elaborate than that described for second-order latitude. The method limits the observations to a small portion of the sky in the east and west. The limits are given by the azimuth range of 85°- 95° and of 265°-275°, by an altitude range of 30°-40°, and by the usable nighttime hours. Using the specified altitude and azimuth limits, four border values are computed for declination and hour angle by the respective equations (3.5) and (11.9). These border values represent the terminal points of the star's apparent path through the usable area of the sky. Additional values of δ and h are calculated (or interpolated between the border values) for every degree of altitude and azimuth. (See the first table in Example 11.1 at the end of section 11.213.) If it is not desired to calculate a 24-hour observing list, the local sidereal time of the beginning and end of an anticipated observing period imposes further limitations.

The procedure then for setting up an observing list for a station is as follows:

(1) Calculate the ST for the beginning and end of the observing period for a desired date and approximate longitude.

(2) Compute the approximate range in right ascension from

$$\alpha = ST - h$$

where h may be interpreted as the approximate mean hour angle of stars in the prime vertical and computed from equation (11.9).

(3) With this information and the calculated declination limits, select suitable stars from the FK4 catalogue or from the APFS.

(4) Using the declinations of selected stars, interpolate from the declination table their zenith distances and azimuths for entry and exit from the usable area.

(5) Using the interpolated values of the zenith distance and azimuth, interpolate the hour angles of entry and exit from the hour angle table. (See the second table in Example 11.1 at the end of section 11.213.)

For field use, star charts may be prepared by plotting the stars' paths at a suitable scale with ST and with the vertical circle setting as coordinate axes (see Fig. 11.2 in Example 11.1). The points of intersections of east and west stars are the ideal positions for selecting a star pair because at those points their zenith distances are equal. It should be noted that a star which is in the usable area less than ten minutes cannot be used in general. When a star's path intersects the paths of a number of corresponding stars, it may be paired with more than one star. For example, in Fig. 11.2, east star No. 823 may be successfully paired with west star No. 1358 and also later with west star No. 522. In such a case considerable computation time is saved later on.

11.212 Observations. Zenith distance observations should begin on either the east or the west star of a pair as follows:

(1) Set the theodolite to the proper azimuth and altitude, both quantities corresponding to ST about ten minutes prior to the time of intersection of the east and west star paths as determined from the star chart.

(2) Follow four pointings in direct telescope position with four pointings in reversed position. Record the chronometer time of the star's transit across the horizontal thread (e.g., on the chronograph tape by means of the hand tappet).

(3) Immediately after the last direct and first reversed pointings, record the horizontal circle to assure that the same star is picked up after reversing the instrument.

(4) After completion of the observations on the first star, observe the second star of the pair in the same manner. The time interval between the last pointing on the first star of the pair and the first pointing on the second star of the pair should not exceed fifteen minutes to reduce the effects of changes in the state of the atmosphere.

(5) Record temperature and barometric pressure before and after the observations on each star. Steps (1)—(5) constitute one determination (2–4 determinations are suggested).

(6) Make radio-chronometer time comparisons, as described in section 8.33, at intervals dependent on the stability of the chronometer. (In case of a mechanical chronometer, the interval should be one hour or less.)

A sample field record is shown in the third table of Example 11.1 after section 11.213.

Random observational errors affecting the determination are m_t, m_{va}, and m_{vp} (see sections 9.115 and 9.12). Their effect on the AST obtained from an east-west pair with n observations per star may be expressed as [Roelofs, 1950, pp. 74-76, 163]

$$m^2_{\Lambda \cos\Phi} = \frac{1}{2n} m_t^2 \cos^2\Phi + \frac{1}{2n}(m_{va}^2 + m_{vp}^2).$$

To obtain the estimated standard error of the longitude, it is necessary to add to this the error of the chronometer synchronization (i. e., that of GAST).

11.213 Computations. The solution is based on equation (11.5), where the hour angle is calculated from equation (11.9) using the mean zenith distance (altitude) for a pair of direct and reversed pointings on the star. Thus one determination results in two corrections to the assumed longitude (one from each star), each being the average of four $\Delta\Lambda$ values (one from each pair of direct and reversed pointings). The longitude corrections resulting from each star need to be corrected for the curvature of the star path, for the effect of unsymmetrical refraction, for diurnal aberration (if significant), for personal equation, and for the propagation time of the radio signals (if used in the chronometer synchronization). The last three corrections, however, normally do not have significant variations during the time needed for the determination of the station longitude; thus generally they are applied to the average longitude computed from all determinations rather than to the individual ones based on a single star.

The curvature correction, due to the star's nonlinear motion during the interval between direct and reversed pointings, is given by equation (9.27) which is to be applied to the average zenith distance of the two pointings. Using the second term in equation (11.10), this correction may be transformed into a correction Δh_C, directly applicable to the hour angle (longitude). The transformation results in the following expression for the correction in seconds of time:

$$\Delta h_C = \frac{15\,\tau^2 \sin 1''}{2} \cos\Phi \cos A\,(\cot z \cot A - \tan\Phi \operatorname{cosec} A) \quad (11.13)$$

where τ is one-half the time interval between direct and reversed pointings in seconds of time. In practice, it may be satisfactory to calculate only one Δh_C for each star using the mean values of τ, z, and A which is then applied directly with the proper algebraic sign to $\Delta\Lambda$.

To reduce the effect of any uncorrected unsymmetrical refraction (arising from the possibility that the average altitudes of the east stars do not agree exactly with those of the west stars), the curvature corrected longitude corrections $\Delta\Lambda_m$ (east and west) may be averaged and one-half the difference of these mean values subtracted from each $\Delta\Lambda_m$ (east) and added to each $\Delta\Lambda_m$ (west). Thus the refraction corrected value for a star is

$$\Delta\Lambda = \Delta\Lambda_m \pm \frac{1}{2n}\left(\sum_E \Delta\Lambda_m - \sum_W \Delta\Lambda_m\right)$$

where the plus sign is for west stars, and n is the number of determinations (see item 5A in Example 11.1).

As mentioned in section 4.214, the diurnal aberration is usually subtracted from the observed parameters (e.g., AST) rather than being added to the apparent star coordinates. Thus the correction to be added directly to the average hour angle of each star (longitude), after equation (4.55), is

$$\Delta h_D = 0^{s}.02132 \cos\Phi \cos h \sec\delta.$$

As pointed out earlier, no appreciable error is introduced when a single diurnal aberration correction is added to the longitude of the station computed from all the determinations. In this case the quantities h and δ in the above equation will be substituted by their average values from all the observations (see item 5B in Example 11.1).

Second-order longitude determinations must also be corrected for the personal equation of the observer. To determine this systematic error, observations are made at a reference station of known first-order astronomic longitude. If the second-order longitude is reduced to the coordinate system of the first-order longitude, the difference between the two longitudes is a direct indication of the personal error (equation) of the observer. Thus when determining the personal equation care must be taken that the first-order longitude refers to the same time (UT1) and fundamental catalogue systems (FK4) as the second-order longitude. Normally personal equations are determined several times during an observational season, and the correction applicable to the second-order longitude is calculated through interpolation. The criterion which distinguishes a good observer from a poor one is not the magnitude of his personal equation but its constancy in time. Since the personal equation is determined from observations, its precision must be taken into account when calculating the precision of the station longitude (see item 5C in Example 11.1).

Since in ordinary circumstances the reference station is in the vicinity of the observation station, the corrections for the propagation times of the radio signals, used in the chronometer synchronizations, need not be applied but can be considered part of the personal equation.

Example 11.1 illustrates the entire computational procedure as outlined above. For alternative procedures see [Freislich, 1955].

11.22 Longitude by Transit Times

At upper transit, a star's right ascension is equal to the local sidereal time; thus the observation of the chronometer time of upper transit enables the longitude to be determined from equation (11.3) or (11.5). The accuracy attainable is inferior to the previous method mostly due to errors in setting the instrument in the meridian Additional serious systematic errors are those due to the possible inclination of the horizontal axis of the instrument and due to collimation. The former may be reduced by careful leveling, possibly by the use of a striding level if available; the latter could be eliminated by observing in both direct

EXAMPLE 11.1

Longitude by Zenith Distances Measured Near the Prime Vertical

Observer:	J. D. Bossler	Instruments: Wild T3 theodolite (No. 26588); Hamilton sidereal chronometer (No. 2E12304); Favag chronograph with hand tappet; Zenith transoceanic radio with amplifier.
Recorder:	S. F. Cushman	
Computer:	H. D. Preuss	
Local Date:	8-14-1964	
Location:	Lanum, Ohio (near USCGS triangulation station) $\Phi_O = 39°39'43''.30$ $\Lambda_O = -5^h34^m10^s.609$	

1. Interpolation Tables for a, A, and h

Interpolation table for $a = 90° - z$ and A with argument δ

A / a	95° 265°	94° 266°	93° 267°	92° 268°	91° 269°	90° 270°	89° 271°	88° 272°	87° 273°	86° 274°	85° 275°	E W
30°	15°08'	15°49'	16°31'	17°13'	17°54'	18°37'	19°19'	20°01'	20°44'	21°27'	22°10'	Declination
31	15 44										22 43	
32	16 20										23 16	
33	16 56										23 49	
34	17 32										24 22	
35	18 08										24 54	
36	18 43										25 26	
37	19 18										25 57	
38	19 53										26 29	
39	20 27										26 59	
40	21°02'	21°40'	22°18'	22°56'	23°35'	24°13'	24°52'	25°31'	26°11'	26°50'	27°30'	
	Declination											

East Stars: entering: top row or right column; leaving: bottom row or left column

West Stars: entering: bottom row or left column; leaving: top row or right column

Interpolation table for h, with argument a and A

A / a	95° 265°	94° 266°	93° 267°	92° 268°	91° 269°	90° 270°	89° 271°	88° 272°	87° 273°	86° 274°	85° 275°	E W
30°	4^h13^m	4^h15^m	4^h17^m	4^h20^m	4^h22^m	4^h24^m	4^h26^m	4^h28^m	4^h31^m	4^h33^m	4^h35^m	Hour Angle
31	4 10										4 31	
32	4 06										4 27	
33	4 03										4 24	
34	3 59										4 20	
35	3 56										4 16	
36	3 53										4 12	
37	3 49										4 08	
38	3 46										4 05	
39	3 42										4 01	
40	3^h39^m	3^h41^m	3^h43^m	3^h44^m	3^h46^m	3^h48^m	3^h50^m	3^h52^m	3^h53^m	3^h55^m	3^h57^m	
	Hour Angle											

East Stars: entering: top row or right column; leaving: bottom row or left column

West Stars: entering: bottom row or left column; leaving: top row or right column

2. Stars Selected from APFS, 1964

Star No.	Magnitude	Position	α	δ	Entering				Leaving			
					h	ST	a	A	h	ST	a	A
1565	4.8	E	21^h28^m	23°29′	4^h26^m	17^h02^m	32°24′	85°00′	3^h46^m	17^h42^m	40°00′	91°10′
1570	5.3	E	21 36	19 09	4 26	17 10	30 00	89 14	3 50	17 46	36 45	95 00
823	5.0	E	21 51	25 45	4 10	17 41	36 37	85 00	3 52	17 59	40 00	87 37
1579	6.6	E	21 55	21 04	4 32	17 23	30 00	86 32	3 39	18 16	40 00	94 57
1586	6.4	E	22 24	18 16	4 23	18 01	30 00	90 29	3 55	18 29	35 14	95 00
1589	6.0	E	22 27	26 35	4 04	18 23	38 12	85 00	3 54	18 33	40 00	86 18
1596	6.4	E	22 44	19 11	4 26	18 18	30 00	89 11	3 50	18 54	36 48	95 00
859	4.1	E	22 45	23 23	4 26	18 19	32 13	85 00	3 45	19 00	40 00	91 18
880	4.6	E	23 19	23 33	4 25	18 54	32 31	85 00	3 46	19 33	40 00	91 03
881	4.6	E	23 24	23 12	4 26	18 58	31 51	85 00	3 47	19 37	40 00	91 35
1339	6.0	W	13 05	21 22	3 40	16 45	40 00	265 32	4 33	17 38	30 00	273 53
1358	5.9	W	13 45	25 53	3 53	17 38	40 00	272 33	4 09	17 54	36 52	275 00
507	4.5	W	13 46	17 38	3 58	17 44	34 10	265 00	4 21	18 07	30 00	268 37
513	2.8	W	13 53	18 35	3 54	17 47	35 46	265 00	4 24	18 17	30 00	269 57
522	4.8	W	14 09	25 16	3 51	18 00	40 00	271 37	4 13	18 22	35 41	275 00
526	0.2	W	14 14	19 22	3 49	18 03	37 07	265 00	4 26	18 40	30 00	271 04
1378	5.4	W	14 25	19 23	3 49	18 14	37 08	265 00	4 26	18 51	30 00	271 06
1392	6.2	W	14 55	21 42	3 41	18 36	40 00	266 03	4 34	19 29	30 00	274 21
1396	5.0	W	15 06	25 01	3 50	18 56	40 00	271 14	4 15	19 21	35 13	275 00

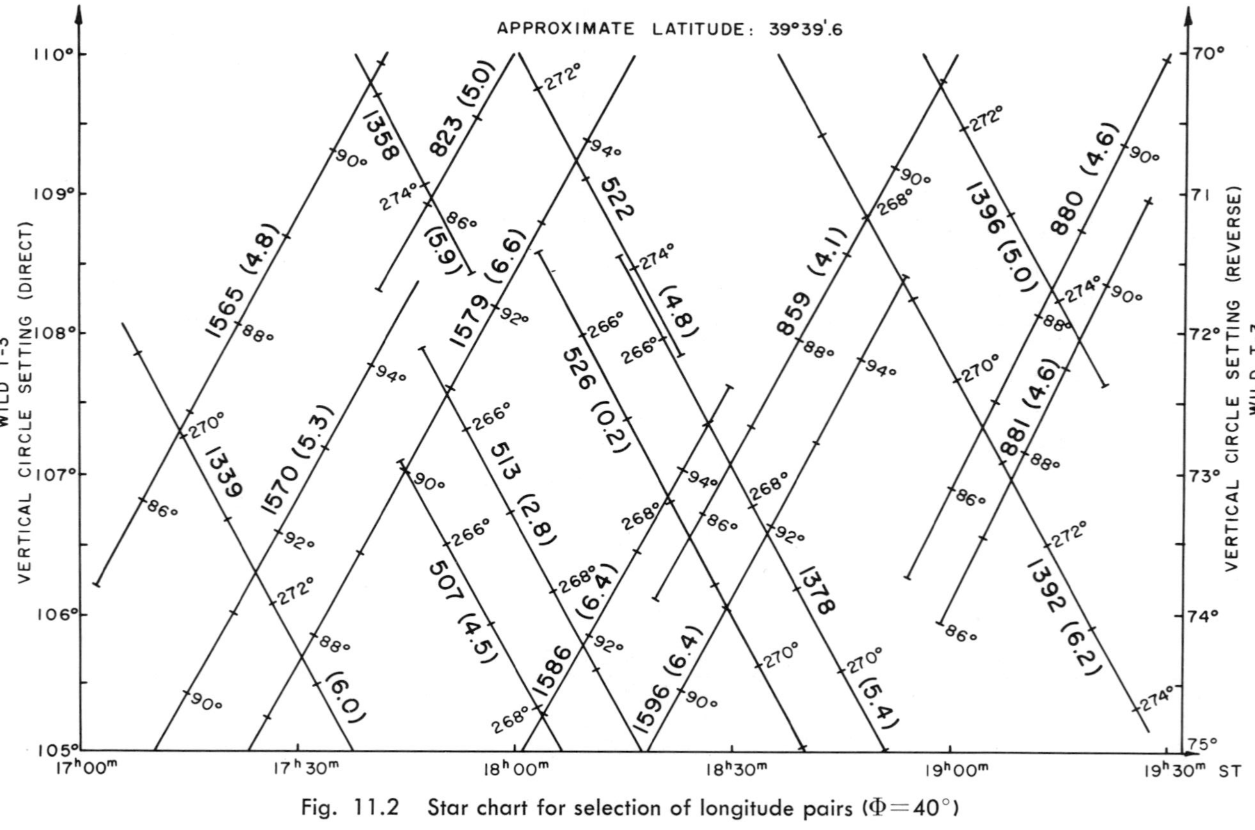

Fig. 11.2 Star chart for selection of longitude pairs ($\Phi = 40°$)

3. Longitude Field Record (Sample)

Pair No.	Star No.	Position	Approximate Chronometer time	*Scaled seconds	Telescope position	Vertical circle reading ° ′	″	″	*″	*Observed altitude, a′	*Chronometer time, (T_S)	*2τ	Azimuth	Remarks
1	1358	West	17ʰ53ᵐ09ˢ	09ˢ.11	D	108 36	04.9	04.9	09.8	37°02′46″.2	17ʰ54ᵐ00ˢ.24	1ᵐ.7	95°03′10″	T_s = 12°.2 C.
			54 51	51.37	R	71 32	41.8	41.8	83.6					B_s = 733.1 mm
			53 26	26.84	D	108 34	13.6	13.7	27.3	36 57 15.8	17 54 28.94	2.1		
			55 31	31.05	R	71 36	35.7	35.8	71.5					
			53 41	41.23	D	108 32	32.3	32.3	64.6	36 54 01.2	17 54 45.78	2.2		
			55 50	50.34	R	71 38	31.7	31.7	63.4					
			53 57	57.50	D	108 30	45.4	45.5	90.9	36 51 01.1	17 55 01.45	2.1		
			56 05	05.40	R	71 40	14.9	14.9	29.8				274 52 30	
	823	East	17 59 09	09.10	D	108 30	52.8	52.9	105.7	36 51 11.9	17 58 03.70	2.2	86 12 38	T_s = 12°.4 C.
			56 58	58.30	R	71 40	16.9	16.9	33.8					B_s = 733.1 mm
			59 37	37.32	D	108 34	14.1	14.1	28.2	36 56 37.0	17 58 31.93	2.2		
			57 26	26.54	R	71 36	55.6	55.6	111.2					
			59 55	55.26	D	108 36	05.5	05.6	11.1	37 00 17.1	17 58 51.13	2.1		
			57 47	47.00	R	71 34	57.0	57.0	114.0					
			60 15	15.02	D	108 38	02.1	02.1	04.2	37 06 08.7	17 59 21.60	1.8		
			58 28	28.19	R	71 30	57.7	57.8	115.5				266 06 36	

*Quantities are determined when observed data is reduced.

D, R = direct and reversed telescope positions, respectively.

$T_{s,e}$; $B_{s,e}$ = temperature and barometric pressure readings, respectively (s = start, e = end).

2τ = time interval between direct and reversed pointings in minutes of time.

4. Longitude Computation (Sample)

Barometer: 733.9 mm
Temperature: 12°26 C
C_B = 0.966
C_T = 0.992
$C_B \times C_T$ = 0.958272

Greenwich Date: August 15, 1964

Pair No. 1	East Star No. 823		δ = 25°10' 16".43		West Star No. 1358		δ = 25°53' 03".81	
Observed altitude	36°51' 11".9	36°56' 37".0	37°00' 17".1	37°06' 08".7	37°02' 46".2	36°57' 15".8	36°54' 01".2	36°51' 01".1
Refraction	-1 13.9	-1 13.7	-1 13.5	-1 13.2	-1 13.4	-1 13.6	-1 13.8	-1 13.9
Altitude	36 49 58.0	36 55 23.3	36 59 03.6	37 04 55.5	37 01 32.8	36 56 02.2	36 52 47.4	36 49 47.2
sin a	0.59948158	0.60074313	0.60159662	0.60295856	0.60217428	0.60089388	0.60013872	0.59943968
sin Φ sin δ	0.27146693				0.27863571			
cos Φ cos δ	0.69672092				0.69259159			
cos h	0.47079776	0.47260846	0.47383347	0.47578825	0.46714192	0.46529322	0.46420288	0.46319357
h	$-4^h07^m39^s.34$	$-4^h07^m11^s.10$	$-4^h06^m51^s.98$	$-4^h06^m21^s.44$	$4^h08^m36^s.26$	$4^h09^m04^s.99$	$4^h09^m21^s.92$	$4^h09^m37^s.59$
α	$22^h05^m22^s.18$				$13^h45^m03^s.25$			
AST = α + h	17 57 42.84	17 58 11.08	17 58 30.20	17 59 00.74	17 53 39.51	17 54 08.24	17 54 25.17	17 54 40.84
Chron. time (T_S)	17 58 03.70	17 58 31.93	17 58 51.13	17 59 21.60	17 54 00.24	17 54 28.94	17 54 45.78	17 55 01.45
Chron. corr. (ΔT_S)	-20.13	-20.13	-20.13	-20.13	-20.13	-20.13	-20.13	-20.13
ΔΛ'	-0.73	-0.72	-0.80	-0.73	-0.60	-0.57	-0.48	-0.48
$\Delta\Lambda'_m = \Sigma\Delta\Lambda'/n$	$-0^s.745$				$-0^s.532$			

Curvature Correction (Δh_C)

Star No.	2τ	A	Δh_C	$\Delta\Lambda_m = \Delta\Lambda'_m + \Delta h_C$
823	$2^m.1$	86°.2	$-0^s.01$	$-0^s.755$
1358	2.0	275.0	0.01	-0.522

Formulas:

$$\cos h = \frac{\sin a - \sin \Phi \sin \delta}{\cos \Phi \cos \delta}$$

$$\Delta\Lambda' = AST - (T_S + \Delta T_S)$$

5. Corrections for Refraction, Diurnal Aberration, and Personal Equation

A. Reduction of the Effect of Unsymmetrical Refraction

Star No.	$\Delta\Lambda_m$		$\Delta\Lambda$		v	vv
	East	West	East	West		
823	$-0\overset{s}{.}755$		$-0\overset{s}{.}655$		$-0\overset{s}{.}0445$	0.00198
823	-0.690		-0.590		0.0205	0.00042
859	-0.672		-0.572		0.0385	0.00148
859	-0.725		-0.625		-0.0145	0.00021
1358		$-0\overset{s}{.}522$		$-0\overset{s}{.}622$	-0.0113	0.00013
522		-0.480		-0.580	0.0305	0.00093
526		-0.538		-0.638	-0.0275	0.00076
1378		-0.502		-0.602	0.0085	0.00007
Mean	$-0\overset{s}{.}7105$	$-0\overset{s}{.}5105$	$-0\overset{s}{.}6105$	$-0\overset{s}{.}6105$	$[v]=0\overset{s}{.}0002$	$[vv]=0.00598$

$$\Delta\Lambda_E = \sum_{\text{EAST}} \frac{\Delta\Lambda_m}{4}; \quad \Delta\Lambda_W = \sum_{\text{WEST}} \frac{\Delta\Lambda_m}{4}$$

$\Delta\Lambda = \Delta\Lambda_m \pm \frac{1}{2}(\Delta\Lambda_E - \Delta\Lambda_W)$ where in this case

$$\tfrac{1}{2}(\Delta\Lambda_E - \Delta\Lambda_W) = -0\overset{s}{.}100$$

Standard deviation of a longitude determination:

$$m_{\Delta\Lambda} = \sqrt{\frac{[vv]}{n-1}} = \sqrt{\frac{0.00598}{8-1}} = 0\overset{s}{.}0292$$

Standard deviation of the mean longitude:

$$M_{\Delta\Lambda} = \frac{m_{\Delta\Lambda}}{\sqrt{n}} = \frac{0\overset{s}{.}0292}{\sqrt{8}} = 0\overset{s}{.}0103$$

B. Correction for Diurnal Aberration

$$\delta_m = \Sigma\,\delta/n = 22^\circ 22' 33''$$
$$h_m = \Sigma\,h/n = 5^h 08^m 37^s$$
$$\Delta h_D = 0''\!.3198 \cos\Phi \cos h_m \sec\delta_m = 0''\!.125 = 0\overset{s}{.}0083$$

C. Personal Equation

First-Order Standard Station: OSU Old Astro Pillar

$\Lambda_1 = -5^h32^m09\overset{s}{.}267 \pm 0\overset{s}{.}008$ (see Example 11.4)

No.	Second-Order Astronomic Longitude (Λ_2)	Local Date of Observation	$PE=\Lambda_1-\Lambda_2$	M_{PE}
1	$-5^h32^m09\overset{s}{.}384 \pm 0\overset{s}{.}012$	June 25, 1964	$0\overset{s}{.}117$	$0\overset{s}{.}015$
2	-5 32 09.308 ± 0.028	July 29, 1964	0.041	0.030
3	-5 32 09.245 ± 0.016	August 18, 1964	-0.022	0.018

Notes: Λ_1 and Λ_2 are corrected for the motion of the pole and are referenced to FK4. Λ_1 is further corrected for transmission time, thus the personal equation includes the correction for the propagation time of the radio signals.

$$M_{PE} = \sqrt{M^2_{\Lambda_1} + M^2_{\Lambda_2}}$$

Personal equation interpolated from the above table for August 14, 1964 is

$$PE = -0\overset{s}{.}009 \pm 0\overset{s}{.}021$$

6. Final Observed Longitude from Eight Determinations (32 Observations)

Assumed longitude	Λ_O	$= -5^h34^m10\overset{s}{.}609$
Longitude correction (from 5A)	$\Delta\Lambda$	$= -\ 0.610 \pm 0\overset{s}{.}010$
Diurnal aberration (from 5B)	Δh_D	$= 0.008$
Personal equation (from 5C)	PE	$= -\ 0.009 \pm 0.021$
Final observed longitude	Λ_T	$= -5^h34^m11\overset{s}{.}22 \pm 0\overset{s}{.}02$

and reverse faces of the telescope. The effect of random errors may be reduced by observing more than the minimum number of stars required when they transit over each vertical thread in the field of view, as explained in section 11.12.

11.221 Longitude by Meridian Transit Times. In this method the instrument is set into the meridian by one of the methods described in section 10.11. The chronometer times of transit of a north and a south star of about the same altitude are observed in the same face of the instrument and the longitude is computed from equation (11.3) or (11.5) with $h = 0^h$ (or 12^h). It is assumed that the primary source of error is that the star's transit is observed off the meridian, partly due to the inaccurate meridian setting and partly due to collimation. This error causes a timing error which is greater for slow moving stars than for fast ones. A correction to reduce this effect may be calculated from the rate at which a star crosses the meridian [Fallon, 1957]. From equation (9.13), applied for meridian transits and disregarding signs,

$$\frac{dA}{dh} = \cos\delta \operatorname{cosec} z. \tag{11.14}$$

A table may be worked out giving the time in seconds for a star to travel one minute of arc across the meridian. Assuming other errors are negligibly small, the difference of the longitudes computed from the observations on the north and south stars is apportioned according to the rates of the stars. For example, at $\Phi = 40^\circ$ for stars of $\delta_N = 70^\circ$ and $\delta_S = 30^\circ$, the respective rates for $dA = 1' = 4^s$ are $dh_N = 5\overset{s}{.}85$ and $dh_S = 0\overset{s}{.}80$. Thus, if the longitude difference from the two observations is $2\overset{s}{.}0$, the longitude as determined from the north star would be corrected by $2.0 \times 5.85/(5.85 + 0.80) = 1\overset{s}{.}76$, and the one from the south star would be corrected by $2.0 \times 0.80/(5.85 + 0.80) = 0\overset{s}{.}24$.

The selection of stars is quite simple since the ST equals the right ascension at meridian transit. The APFS can be scanned for suitable observation times, and the zenith distances for identification can be found from equations (3.42)—(3.44).

The field observations for this method may be accomplished as follows:

(1) Make a radio-chronometer time comparison as described previously.

(2) Set the instrument into meridian for either the north or south star of the pair, level it carefully, and set the zenith distance for the star. When the star crosses the vertical thread(s), record the chronometer time(s) (hand tappet is pressed).

(3) Turn the instrument to the other star of the pair without changing its face, and repeat the same procedure.

(4) Repeat the entire procedure for the other pairs of stars (12–16 are suggested when observing the transits over the middle thread only).

(5) Make radio-chronometer time checks as often as required.

The reduction for this method may be completed as follows:

(1) Compute the corrected chronometer time (e.g., $T_S + \Delta T_S$) for each observed (reduced) transit over the middle thread.

(2) Compute the apparent right ascension as shown in Example 10.2.

(3) Compute the longitude (or the correction to the assumed longitude) from equation (11.3) (or (11.5)) for each meridian transit.

(4) Compute the time for each star to travel 1' of arc in azimuth from equation (11.14).

(5) Determine the difference in computed longitude for each pair of stars.

(6) Calculate the correction to be applied to the computed longitude of each star by distributing the difference obtained in step (5) in proportion to the rates of motion from step (4).

(7) The mean of the longitudes computed by the other pairs, corrected again for personal equation (including propagation time of radio signals) and diurnal aberration, as explained in section 11.213, provides the final 'observed' longitude.

Field tests indicate that results obtainable with this method are of comparable precision to those from zenith distances measured near the prime vertical provided that the same number of longitude observations are made (e.g., 32, as in Example 11.1). The advantage of the method lies in the ease with which the stars can be selected and the longitude calculated. Example 11.2 illustrates the computational procedure for eight longitude observations (four star pairs). Note that the final precision, due to the insufficient number of observations, does not quite meet second-order standards. An increase in the number of observations to 32 (16 pairs) would bring the precision of the mean longitude in the neighborhood of the precision obtained in Example 11.1. For a first-order version of this method, see section 11.31.

EXAMPLE 11.2

Longitude by Meridian Transit Times

Observer: E.R. Therkelsen Recorder: A. Gonzalez-Fletcher Computer: E.R. Therkelsen Local Date: 4-9-1965 Location: OSU old astro pillar $\Phi_0 = 40^\circ 00' 00''.00$ $\Lambda_0 = -5^h 32^m 10^s.000$	Instruments: Kern DKM-3 theodolite (No. 82514); Hamilton sidereal chronometer (No. 2E12304); Favag chronograph with hand tappet; Zenith transoceanic radio with amplifier.

1. Longitude Correction from One Star Pair (Two Observations)
Solution based on equation (11.5)

		North Star (FK4-358)	South Star (FK4-1244)
1	T_S (chronometer time)	$9^h 30^m 38^s.22$	$9^h 22^m 45^s.78$
2	ΔT_S (chronometer correction)	-7.64	-7.63

1. Longitude Correction from One Star Pair (Two Observations) (continued)

		North Star (FK4-358)	South Star (FK4-1244)
3	$T_S + \Delta T_S$	$9^h30^m30^s.58$	$9^h22^m38^s.15$
4	α = AST (computed as in Ex. 10.2)	9 30 32.50	9 22 37.67
5	$\Delta\Lambda' = AST - (T_S + \Delta T_S)$	1.92	0.48
6	$\Delta\Lambda'_N - \Delta\Lambda'_S$	$2^s.40$	
7	δ	$51°49'$	$26°20'$
8	z	$11°49'$	$13°40'$
9	dh(eq. (11.14)) for $dA = 1' = 4^s$	$1^s.32$	$1^s.05$
10	$(6) \times (9) / (dh_N + dh_S)$	$1^s.33$	$1^s.07$
11	$\Delta\Lambda = (5) \pm (10)$	$0^s.59$	$0^s.59$

2. Mean Longitude from Eight Observations

Star Pair	$\Delta\Lambda$	v	vv
358/1244	$0^s.59$	$-0^s.350$	0.122
335/352	0.02	0.220	0.048
368/384	0.10	0.140	0.019
394/405	0.25	-0.010	0.000
	Mean $= 0^s.240$	$[v] = 0^s.000$	$[vv] = 0.189$

Standard deviation of the longitude from a single pair:

$$m_{\Delta\Lambda} = \sqrt{\frac{[vv]}{n-1}} = \sqrt{\frac{0.189}{4-1}} = 0^s.25$$

Standard deviation of the mean longitude:

$$M_{\Delta\Lambda} = \frac{m_{\Delta\Lambda}}{\sqrt{n}} = \frac{0.25}{\sqrt{4}} = 0^s.125$$

Mean longitude (uncorrected for diurnal aberration and personal equation):

$$\Lambda_T = \Lambda_O + \Delta\Lambda = -5^h32^m09^s.76 \pm 0^s.12.$$

11.222 Longitude by Transit Times Through the Vertical Plane of Polaris. The basic principle of this method is to observe the transit of a south star across the vertical plane of Polaris. The chronometer time of the observation is then corrected to obtain the time of the star's transit over the meridian by the method outlined in section 7.43. From the reduced chronometer time of meridian transit, the longitude is determined through the use of equation (11.3) or (11.5) as in the previous section.

The relations to be used in the calculations may be easily derived from equations (7.25) and (7.26) for transit time measurement. Assuming that by careful leveling and instrument adjustment the inclination of the horizontal axis b and the collimation error c are too small to seriously affect second-order work, and recognizing that the azimuth of Polaris A_P is equal to the negative value of the angle a, equation (7.25) becomes

$$\left.\begin{aligned} \cos\delta_A \sin h_A &= -\sin A_P \sin\Phi \\ \cos\delta_A \cos h_A &= \cos A_P \\ \sin\delta_A &= \sin A_P \cos\Phi \end{aligned}\right\}, \qquad (11.15)$$

which determine h_A and δ_A when A_P and Φ are given. The azimuth of Polaris A_P may be calculated from equation (9.1) using the approximate longitude of the station to allow the computation of its required hour angle ($h_P = AST - \alpha_P$). If the assumed longitude is significantly different from the final one, the calculations may have to be repeated. The correction τ to be added to the observed sidereal chronometer time of transit of the south star to obtain its time of transit over the meridian is given by equation (7.26) which, in this case, yields

$$\sin(\tau - h_A) = \tan\delta_A \tan\delta \qquad (11.16)$$

where h_A and δ_A are computed from (11.15) and δ is the declination of the south star.

The observing list for the south stars (about eight to ten minutes apart) may be prepared as described in the previous section, with zenith distances found from equation (3.43). The azimuth of a south star is computed by adding 180° to the azimuth of Polaris at the time of observation. The fact that the south star is observed slightly off the meridian requires a correction to the assumed time of meridian observation which can be computed from equation (11.14) where dA is taken as the differences in the azimuths of the stars and 180°. This correction converted to time is applied algebraically with the sign of dA, which is positive for a star with azimuth greater than 180° and negative for a star with azimuth less than 180°. The meridian zenith distances obtained are not exact either since the star is not on the meridian, but they are sufficiently close for star identification.

The observing procedure for this method may be as follows:

(1) Make a radio-chronometer time comparison prior to the observations as described before.

(2) Level the theodolite carefully.

(3) About four minutes before the south star is expected, direct the instrument on Polaris and record the chronometer time (e.g., with the hand tappet). Record also the position of the horizontal circle to facilitate the pointing on Polaris in the reversed position of the instrument in step (5).

(4) Turn the telescope about its horizontal axis to point to the south (the horizontal circle reading is unchanged!). Set the altitude for the star to be observed. When the star crosses the vertical thread(s), record the chronometer time(s).

(5) Reverse the instrument and repeat steps (3)–(4) for the second south star. This constitutes one longitude determination.

(6) Repeat steps (3)–(5) for other star pairs (six to eight determinations are suggested only when the middle vertical thread is utilized).

(7) Make radio-chronometer time comparisons as often as required. The steps of the calculations are self-explanatory from what has been said above and in the previous section. Again correction for personal equation (including propagation time of radio signals), as explained in section 11.213, needs to be applied to the mean longitude. Since the stars are ex-meridian diurnal aberration will affect both their right ascensions and declinations; thus their apparent coordinates should be corrected by means of equation (4.55) or (11.32). Field tests indicate that the method is capable of producing precisions comparable to those of the methods described previously, provided that the same number of observations are made. The method is limited to latitudes where Polaris is visible and at low enough altitudes to be observable by geodetic theodolites. For the first-order version of this method see section 11.321.

11.23 Longitude by Equal Zenith Distances

The mean of two chronometer times, corrected for rate, at which a star reaches the same altitude east and west of the meridian, represents the chronometer time of meridian transit. The longitude may then be computed from the previous basic relations. The determination is independent of the altitude itself but depends on the variation of refraction between the two observations. Observations should be made where the systematic errors have the least effect, i.e., near the prime vertical (see equation (11.10)). At altitudes convenient for observations this would necessitate a wait of several hours between observations. For these reasons, the method is seldom used for field determinations of longitude.

The disadvantage of the long waiting period for observations and the resulting errors can be reduced by observing two stars at the same altitude east and west of the meridian. If the east and west stars were of equal declination as well as of equal altitude, the mean of their right ascensions would give the ST corresponding to the mean of the chronometer times of observation. Since the probability of finding two stars fulfilling these conditions is quite small, a different approach is used based on equation (11.5), which yields the following:

$$\mathrm{AST} = (T_S + \Delta T_S) + \Delta\Lambda = \alpha + h. \tag{11.17}$$

Denoting the corrected chronometer time $T_S + \Delta T_S$ by T, applying the above equation for an east (E) and a west (W) star, and taking the mean, give the following relation for the correction to the assumed longitude:

$$\Delta\Lambda = \frac{\alpha_E + \alpha_W}{2} - \frac{T_E + T_W}{2} + h_m \tag{11.18}$$

where

$$h_m = \frac{h_E + h_W}{2}. \tag{11.19}$$

It is obvious that $h_E = -h_W$, thus the correction h_m is zero in the unlikely case when $\delta_E = \delta_W$ and $z_E = z_W$. This not being a practical situation, the longitude determination is reduced to the problem of determining the mean hour angle h_m of the star pair at the instant when their zenith distances are equal. The basis of this solution lies in equation (3.2) which when applied to the two stars (realizing that $z_E = z_W$) yields

$$\sin\Phi\,(\sin\delta_W - \sin\delta_E) + \cos\Phi\,(\cos\delta_W \cos h_W - \cos\delta_E \cos h_E) = 0. \tag{11.20}$$

Using the notation

$$\left.\begin{aligned} \delta_m &= \frac{\delta_E + \delta_W}{2} \\ \Delta\delta &= \frac{\delta_W - \delta_E}{2} \\ \Delta h &= \frac{h_W - h_E}{2} = \frac{T_W - T_E}{2} - \frac{\alpha_W - \alpha_E}{2} \end{aligned}\right\}, \tag{11.21}$$

and the subsequent substitutions

$$\delta_E = \delta_m - \Delta\delta, \quad \delta_W = \delta_m + \Delta\delta, \quad h_E = h_m - \Delta h, \text{ and } \quad h_W = h_m + \Delta h,$$

the parenthetical expressions in equation (11.20) become

$$\begin{aligned} \sin\delta_W - \sin\delta_E &= 2\sin\Delta\delta\cos\delta_m, \\ \cos\delta_W\cos h_W - \cos\delta_E\cos h_E &= -2(\cos\delta_m\sin\Delta h\cos\Delta\delta\sin h_m + \\ &\quad + \sin\delta_m\cos\Delta h\sin\Delta\delta\cos h_m). \end{aligned}$$

Substitution of these into equation (11.20) yields

$$\sin h_m + \frac{\tan \delta_m \tan \Delta\delta}{\tan \Delta h} \cos h_m = \frac{\tan \Delta\delta \tan \Phi}{\sin \Delta h}$$

where h_m is the only unknown. Manipulation of the above equation produces the following convenient solution:

$$h_m = n - m \tag{11.22}$$

where

$$\tan m = \tan \delta_m \tan \Delta\delta \operatorname{cotan} \Delta h,$$
$$\sin n = \tan \Phi \tan \Delta\delta \operatorname{cosec} \Delta h \cos m.$$

Using the correction h_m, equation (11.18) gives the desired $\Delta\Lambda$ to be added to the assumed longitude of the observer.

As mentioned earlier, best results are expected when the star pairs are observed near the prime vertical and symmetrical to the meridian where the effect of systematic errors, according to equation (11.10), is at its minimum. This being the case, the method of preparing an observing list is the same as described in section 11.21 with the additional restriction that an attempt is made to match the declinations even closer than before to keep the correction h_m small. It is difficult to find many stars in the APFS which will fulfill these requirements over a short observational period. For this reason, either the observations must be extended over quite a few hours or stars are to be selected from the SAO Catalog (in this connection see also section 11.322).

Once the star list is compiled the observations are accomplished as follows:

(1) Make a radio-chronometer time comparison prior to the observations, as in the previous methods.

(2) Prior to the observation time, set the azimuth and zenith distance of the star into the theodolite. Locate the star and place it near the vertical thread. When the star crosses the horizontal thread(s), record the chronometer time(s) (i. e., press the hand tappet). Bring the vertical collimation bubble into coincidence and read and record the vertical angle.

(3) With the same zenith distance still set into the theodolite, set the instrument to the azimuth of the other star of the pair and await the time of observation. When the west star is located, follow the same procedure as above.

(4) Repeat steps (2) and (3) for the other star pairs (12–16 are suggested). If the theodolite is equipped with several horizontal threads and an observation is made on each of them, the number of necessary star pairs may be reduced.

(5) Make thermometer and barometer readings before and after the observation of each star pair.

(6) Make radio-chronometer time checks as often as required.

The calculations are made in the following manner:

(1) Compute the corrected chronometer time T for each observation.

(2) If a significant change in temperature and/or barometric pressure is noted between the east and west observations, correct the chronometer time of the second observation utilizing the second term in the right-hand side of equation (11.10).

(3) From equation (11.22) compute the correction h_m for each pair of stars.

(4) Compute the longitude correction $\Delta\Lambda$ from (11.18) for each pair.

(5) The mean of the $\Delta\Lambda$'s added to the assumed longitude is the final result which still should be corrected for personal equation (including propagation time of radio signals) and diurnal aberration (either by adding it to the apparent coordinates of the stars using equations (4.55) or by equation (11.35)).

The accuracy of the results is largely dependent on maintaining a constant altitude with the theodolite. The method is primarily applicable with an astrolabe in which case, however, the stars are not observed near the prime vertical since very few stars will cross the almucantar defined by the angle of the prism at this location.

Example 11.3 illustrates the procedure for eight observations (four star pairs). Note again that the final precision due to the insufficient number of observations does not quite meet second-order standards. On the other hand, the result agrees very well with that of Example 11.2 where the same number of observations were used. For the first-order version of this method, see section 11.322.

EXAMPLE 11.3

Longitude by the Prime Vertical Transit Times of Equal Altitude Stars

Observer:	E. R. Therkelsen	Instruments: Kern DKM-3
Recorder:	A. Gonzalez-Fletcher	theodolite (No. 82514); Ham-
Computer:	E. R. Therkelsen	ilton sidereal chronometer
Local Date:	4-16-1965	(No. 2E12304); Favag chro-
Location:	OSU old astro pillar	nograph with hand tappet;
	$\Phi_0 = 40^\circ 00' 00\overset{''}{.}00$	Zenith transoceanic radio
	$\Lambda_0 = -5^h 32^m 10\overset{s}{.}000$	with amplifier.

1. Longitude Correction from One Star Pair (Two Observations)
Solution based on equations (11.18) and (11.22)

		East Star (FK4-526)	West Star (FK4-241)
1	$\alpha_{E,W}$ (computed as in Ex. 10.2)	$14^h14^m05^s.01$	$6^h20^m49^s.72$
2	$\delta_{E,W}$ (computed as in Ex. 10.2)	$19^\circ21'41''.77$	$22^\circ32'02''.25$
3	$T_{E,W}$ (corrected chronometer time)	$10^h19^m22^s.10$	$10^h25^m01^s.18$
4	$\delta_m = (\delta_E + \delta_W)/2$	$20^\circ56'52''.01$	
5	$\Delta\delta = (\delta_W - \delta_E)/2$	$1^\circ35'10''.24$	
6	$(T_W - T_E)/2$	$2^m49^s.54$	
7	$(\alpha_W - \alpha_E)/2$	$-3^h56^m37^s.65$	
8	$\Delta h = (6) - (7)$	$3^h59^m27^s.19 = 59^\circ51'47''.85$	
9	$\tan\delta_m$	0.38281870	
10	$\tan\Delta\delta$	0.02769109	
11	$\cot\Delta h$	0.58053601	
12	$\sin\Delta h$	0.86482994	
13	$\tan m = (9) \times (10) \times (11)$	0.00615407	
14	m	$0^\circ21'09''.35$	
15	cos m	0.99998106	
16	$\tan\Phi$	0.83909963	
17	$\sin n = (10) \times (15) \times (16)/(12)$	0.02686672	
18	n	$1^\circ32'22''.33$	
19	$h_m = n - m$	$1^\circ11'12''.97 = 4^m44^s.86$	
20	$(\alpha_E + \alpha_W)/2$	$10^h17^m27^s.37$	
21	$(T_E + T_W)/2$	$10^h22^m11^s.64$	
22	$\Delta\Lambda = (19) + (20) - (21)$	$0^s.59$	

2. Mean Longitude from Eight Observations

Star Pair	$\Delta\Lambda$	v	vv
526/241	$0^s.59$	$-0^s.352$	0.124
513/211	0.11	0.127	0.016
584/1200	0.34	-0.103	0.011
609/1220	-0.09	0.327	0.107
	Mean = $0^s.238$	[v] = $-0^s.001$	[vv] = 0.258

Standard deviation of the longitude from a single pair:

$$m_{\Delta\Lambda} = \sqrt{\frac{[vv]}{n-1}} = \sqrt{\frac{0.258}{4-1}} = 0^s.29$$

Standard deviation of the mean longitude:

$$M_{\Delta\Lambda} = \frac{m_{\Delta\Lambda}}{\sqrt{n}} = \frac{0.29}{\sqrt{4}} = 0\overset{s}{.}15$$

Mean longitude (uncorrected for diurnal aberration and personal equation):

$$\Lambda_T = \Lambda_O + \Delta\Lambda = -5^h32^m09\overset{s}{.}76 \pm 0\overset{s}{.}15$$

11.24 Longitude by Horizontal Angles

This method described in [Bhattacharji, 1958] is based on the same principle as the latitude determination by the same author, described in section 10.15. If the angle between a north and a south star is computed from their azimuths at the instant of observation, the difference between it and the measured angle will be due to an error in the AST (assumed longitude) and assumed latitude. For stars near culmination, according to equation (9.2), a small error in the assumed latitude has negligible effect on the computed azimuths, so a relation may be established between the error in the computed azimuths and the error in the assumed longitude. The method is sensitive to the dislevelment of the horizontal axis and good results will be difficult to obtain without the use of a striding level.

Based on the second term of equation (9.2), the correction to the assumed longitude, for upper culminations, may be computed from

$$\Delta\Lambda = \frac{dH}{\cos\delta_S \operatorname{cosec}(\Phi - \delta_S) - \cos\delta_N \operatorname{cosec}(\Phi - \delta_N)} \tag{11.23}$$

in which

$$dH = (A_N - A_S) - (H_N - H_S)$$

where the quantity $(A_N - A_S)$ is the horizontal angle between the two stars (the difference of their azimuths) as computed based on the assumed longitude from equation (9.1); the quantity $(H_N - H_S)$ represents the measured value of the same (the difference of the horizontal circle readings), corrected for the inclination of the horizontal axis and for the curvature of the star path (if multiple observations are made on each star).

In accordance with section 7.445, stars for each pair selected should be near their upper culminations and on opposite sides of the zenith, with the north star at a high declination and the south star at a low declination. An observing list will be simple to prepare as right ascensions correspond to the ST of transit and approximate zenith distances may be computed from equations (3.42)–(3.44).

The observation procedure may be accomplished as follows:

(1) Make a radio-chronometer comparison prior to the observations, as described in the previous methods.

(2) Set the azimuth and zenith distance of the first star of a pair into the instrument. When the star crosses the vertical thread, record the chronometer time and the reading on the horizontal circle. Also read and record the ends of the striding level. Make four observations in direct position and four observations in reverse in rapid succession.

(3) Locate the second star of the pair, and repeat the procedure.

(4) Repeat steps (2)–(3) for other pairs of stars (six to eight pairs are suggested).

(5) Make radio-chronometer time comparisons as often as required.

The reduction procedure may be the following:

(1) Correct the chronometer time of each observation to $AST = T_S + \Delta T_S$ using the assumed longitude of the station, and compute the average time of all the observations AST_O.

(2) Compute the azimuth A of each star corresponding to the time AST_O obtained in step (1), from equation (9.1), using $h = AST_O - \alpha$.

(3) Compute the horizontal angles $(A_N - A_S)$ for each star pair.

(4) Correct the horizontal circle readings on each star for the inclination of the horizontal axis (see equations (7.14) and (7.30)) and for curvature (see equation (9.16) and also Example 9.2) to obtain a single mean reading H on each star corresponding to AST_O.

(5) Compute the observed horizontal angles $(H_N - H_S)$ for each star pair.

(6) Compute the quantity $dH = (A_N - A_S) - (H_N - H_S)$ for each star pair.

(7) Compute the longitude correction $\Delta\Lambda$ from equation (11.23) for each star pair.

(8) The mean of the $\Delta\Lambda$ corrections obtained from all pairs added to the assumed longitude is the mean longitude of the station.

(9) Correct the mean longitude for diurnal aberration and personal equation (including transmission time) to obtain the final 'observed' longitude of the station.

The accuracy of the results is largely dependent on the successful elimination of the errors introduced by the inclination of the horizontal axis. This can hardly be done without a striding level which is not a standard accessory of geodetic theodolites. Field tests with instruments equipped with striding levels indicate that the precision of the method is comparable to that of previously described methods, provided that the same number of observations are used. In this connection, note that an observation is the horizontal angle between the positions corresponding to the mean times of one direct and one reversed pointing (to eliminate the collimation error) on each star of the pair. Thus one observation actually consists of two (one direct and one reversed) pointings on both stars of the pair, a total of four pointings. When observing as outlined above, each star pair will provide four observations giving one longitude determination. Six to eight such determinations are adequate for second-order results.

11.3 First-Order Longitude Determination

11.31 The Mayer Method

The method most widely used for the determination of first-order astronomic longitude with portable instruments (e.g., universal theodolites or telescopes) is the observation of the chronometer time of the meridian transits of stars when $\alpha = ST$; thus in principle it is identical to the method described in section 11.221. The differences are in the type of instrumentation, in the method of observation, and in the data reduction. The method was named after Mayer whose equation for transit time measurement, explained in section 7.43, is used in the data reduction. The Mayer method is the standard first-order longitude determination method used in the U.S. It was employed until 1914 utilizing transit instruments, which were then replaced by Bamberg universal telescopes, and later by Wild T4 universal theodolites [Hoskinson and Duerksen, 1952]. Both the method of observation and the data reduction are somewhat dependent on the type of instrument used. The discussion below assumes the use of the Wild T4 universal theodolite.

The Wild T4 universal theodolite (described in section 7.11) is equipped with an impersonal micrometer suitable for the observation of meridian transits of stars. The field of view consists of a series of fixed threads and a movable thread for tracking the star (Fig. 7.8). A contact drum (5 in the figure) with ten electrical contact strips spaced equidistantly around the drum is connected to the movable thread (M) by a system of gears. As a star is tracked with the movable thread, the contact drum rotates, and the electrical contacts are recorded on the tape of a chronograph. The chronograph simultaneously records the chronometer time. One contact has one additional contact strip on both sides and is called the zero contact. This is an important aid in ascertaining the proper contacts in a pair (Fig. 11.1). The star is tracked for about two and one-half revolutions of the contact drum as the star approaches the instrument's meridian. The tracking starts when the star reaches a point just outside the double threads V_{05} or V_{15}, depending on the position of the telescope (Fig. 7.8). When the star reaches the vicinity of division 7.5 (or 12.5), the observer stops tracking to allow sufficient time to reverse the instrument. The instrument is then reversed; and when the star reaches the movable thread (left in the vicinity of division 7.5 or 12.5), the star is tracked away from the meridian over the same part of the field where it was followed in.

The time of transit across the instrument's meridian can then be determined by scaling the chronometer time of a pair of symmetrical (matching) contacts from the chronograph tape (Fig. 11.1), one before and one after meridian transit, and taking the mean time. This procedure is, in effect, the same as observing the transit over a set of vertical threads symmetrical about the middle thread.

Let

T_b = contact time before meridian transit,

T_a = contact time after meridian transit,

$$T_m = \text{chronometer transit time} = \frac{T_b + T_a}{2}. \qquad (11.24)$$

In practice about ten such pairs of contacts are used, and the mean transit time of the ten pairs $T_S = \Sigma\, T_m/10$ is taken as the observed chronometer transit time. The scaling of the tape is performed by means of a scaling fan shown in Fig. 8.25.

11.311 Longitude Equation: Corrections to the Observed Meridian Transit Time. The mean chronometer time of meridian transit T_S, obtained by averaging the transit times from the symmetrical contacts, still needs to be corrected for the following systematic errors:

(1) Chronometer correction ΔT_S for synchronization and drift, as explained in section 11.11.

(2) Corrections according to Mayer's equation

$$\tau = Aa + Bb + Cc \qquad (7.29)$$

for errors due to the instrument's vertical plane being slightly off the meridian (Aa), due to the inclination of the horizontal axis (Bb), and due to collimation (Cc). In practice, the instrument's vertical plane is initially placed within one second of time in the meridian by means of the method described in section 7.445, and the residual azimuth setting error is generally determined together with the longitude from the transit observations. The collimation error is eliminated by observing in both faces of the telescope. Thus the only correction actually applied from Mayer's equation is that for the inclination of the horizontal axis Bb. The quantity b is determined as explained in section 7.441 from readings on the striding level, which is calibrated either conventionally by means of level triers (see section 7.171) or in the field using star observations [Dulian, 1967].

(3) Correction for the mean width of the contact strips and lost motion for reasons explained in section 7.173 and 7.174. The correction is computed from

$$\ell = \tfrac{1}{2} r(m + s) \sec \delta \qquad (11.25)$$

where m and s are the lost motion and the mean width of the contact strips, respectively, expressed in units of measuring-drum divisions and determined as shown in Examples 7.2 and 7.3; r is the equatorial value of one division of the measuring drum in units of seconds of time, determined as explained in sections 7.175 and 10.214; the factor $\sec\delta$ reduces the equatorial value of r to its value at the declination δ.

(4) Correction for diurnal aberration which, for first-order methods, is applied to each star separately. This correction at meridian transit after equation (4.55) is

$$\Delta h_D = \pm 0\overset{s}{.}021 \cos\Phi \sec\delta \tag{11.26}$$

where the positive sign is for lower culminations ($h = 12^h$), and the negative sign is for upper culmination ($h = 0^h$).

Using the above corrections and equation (11.5), the longitude correction, to be added to the assumed longitude Λ_0 as determined from the upper meridian transit of one star, is

$$\Delta\Lambda = \alpha - (T_S + \Delta T_S + Aa + Bb + Cc + \ell + \Delta h_D). \tag{11.27}$$

For lower transits add 12^h to the right-hand side of the equation. Separating the unknowns from the known quantities, the above yields the longitude equation applicable for each star:

$$\Delta\Lambda + Aa - (\alpha - t) = 0 \tag{11.28}$$

where

$$t = T_S + \Delta T_S + Bb + Cc + \ell + \Delta h_D$$

is the corrected chronometer time of meridian transit for upper culmination; for lower culminations subtract 12^h (or add 12^h to α in equation (11.28)). As mentioned earlier, since the observations are made in both faces of the telescope, the collimation correction Cc is generally neglected.

The disadvantage of using the longitude equation in the above form lies in the contradictory requirement for star selection to obtain accurate longitude ($\Delta\Lambda$) and azimuth (a) corrections. The best result for $\Delta\Lambda$ is expected from the observations of the relatively fast moving 'time stars' which can be well tracked and whose azimuth factors are small ($A < 0.75$); an accurate azimuth a is obtainable from the transit of the relatively slow 'azimuth stars' whose azimuth factors are large ($A > 0.75$). For this reason, especially in high latitudes, it is advisable to determine the azimuth, as explained in section 7.445, from azimuth stars distributed between the time stars. The correction Aa should be calculated separately (either once for the duration of the observations or as many times as required by the stability of the instrument) and used as a known quantity in the determination of $\Delta\Lambda$. In this case, the longitude equation, applicable for each of the time stars, becomes

$$\Delta\Lambda - (\alpha - t') = 0 \tag{11.29}$$

where

$$t' = T_S + \Delta T_S + Aa + Bb + Cc + \ell + \Delta h_D.$$

An alternative to this solution is the use of equation (11.28) for both time and azimuth stars and the weighting of the stars in proportion to their declinations and zenith distances in the least squares solution. The weighting functions generally have the form of $G = g\cos^2\delta$ or $G = 1/(1+g'\tan^2\delta)$ where the coefficients are determined empirically. The U.S. Coast and Geodetic Survey uses $g' = 0.32653$ when weighting observations at high latitudes [Hoskinson and Duerksen, 1952, p. 61]. The coefficient g is a function of the declination and zenith distance, and its value changes from 1 at $\delta = 0°$ to 1.5 at $\delta = 90°$. The normal equations in this type of solution will be the following:

$$\begin{aligned} [G_1]\ \Delta\Lambda + [G_1 A_1]a - [G_1\ (\alpha_1 - t_1)] &= 0, \\ [G_1 A_1]\ \Delta\Lambda + [G_1 A_1^2]a - [G_1 A_1 (\alpha_1 - t_1)] &= 0. \end{aligned} \tag{11.30}$$

The brackets denote summation. At moderate latitudes good results may be obtained by taking $G_1 = 1$ [Milasovszky, 1956 and 1964].

11.312 Observing List. The preparation of the observing program is relatively simple. All the stars must be selected from the FK4 catalogue or the APFS. A single longitude determination normally consists of observations of two sets of stars, six time stars in each, about equally divided between north and south. Two nights of three determinations, with at least one determination in a single night, provide the final observed longitude. The maximum number of stars per hour should be selected subject, in the U.S., to the following requirements [Hoskinson and Duerksen, 1952]:

—stars brighter than seventh magnitude should be used
—the azimuth factor A should be smaller than 0.6
—for any set: $\Sigma A < 1$, as near to zero as possible
—for any determination: $\Sigma A < 1$, as near to zero as possible
—each set should have about an equal number of north and south stars (3 and 3 or 3 and 4)
—about four to five minutes of time should be allowed between successive stars in a set to allow for the observations

In selecting the stars, the declination limits are computed from the relation for the azimuth factor in equation (7.29); with $A = 0.6$,

$$\tan\delta = (\sin\Phi - 0.6)\sec\Phi.$$

The right ascensions are limited by the sidereal times of sunset and sunrise.

Stars selected as outlined above will produce satisfactory first-order results at ordinary latitudes. At high latitudes or if the instrumental azimuth is required with great precision, the time stars should be supplemented by pairs of azimuth stars before and after each longitude determination, selected as outlined at the end of section 7.445. Diagrams aiding in the proper selection of stars, indicating the expected standard errors in longitude and azimuth, may be found in [Roelofs, 1966].

The observing list should also contain the appropriate vertical circle settings for both eyepiece (east and west) positions of the telescope. A sample observing list is shown in the first table of Example 11.4 following section 11.314.

11.313 Observations. The suggested sequence of observations is the following:

(1) Set the instrument within 15" in the meridian as outlined in section 7.445.

(2) Set the eyepiece micrometer in the azimuth (transit) position (see section 7.114).

(3) Set the chronometer to keep approximate local sidereal (or mean solar) time.

(4) Make a radio-chronometer time comparison.

(5) Set the vertical setting circle for the direct observation of the first star (see section 7.112).

(6) Rotate the telescope into the appropriate altitude until the bubble of the vertical setting circle settles at the center of its vial.

(7) Set the vertical setting circle for the reverse observation of the first star.

(8) As the star approaches the outer double threads in the field of view, turn on the chronograph.

(9) Track in the star with the movable thread through about 2.5 turns (25 contacts).

(10) Read and record the striding level.

(11) Reverse the telescope, utilizing the setting accomplished in step (7).

(12) When the star reaches the movable thread, track it out through the same number of contacts through which it was tracked in.

(13) Read the striding level.

(14) Repeat steps (5)–(13) for other stars until a set is obtained.

(15) Make radio-chronometer time comparisons after each set or as often as required.

(16) Observe six sets during two nights (at least two sets during any one night).

The recorder should record on the tape for each star any useful information such as the star's identification number, the time of any special chronometer contact (e.g., full minute or missing second as in Fig. 11.1), and readings on the measuring drum indicating the positions of the movable thread immediately before and after tracking.

The same procedure should be used for azimuth stars when used. A sample field record is shown in the second table of Example 11.4 following the next section.

11.314 Longitude Computations. The calculations start with abstracting the field records for the readings on the striding level and scaling the chronograph tape for the chronometer times of the symmetrical contacts. This abstract (the third table of Example 11.4 following this section) thus provides the basic information needed for setting up the observation equation (11.28) for each star. A least squares solution is then performed for each set to obtain the longitude correction $\Delta\Lambda$ and the azimuth setting error a. In this solution, if azimuth stars have also been observed and if an accurate value for the azimuth is sought, or if the determination is at high latitudes, the stars should be weighted as described in section 11.311. An alternative is the use of equation (11.29) to obtain $\Delta\Lambda$ only. The calculations (without azimuth stars) are illustrated in the fifth table of Example 11.4 which is followed by the summary of longitude computations containing the results of all sets and the corrections for the transmission time of the UTC radio signals and for UT1 - UTC. It is practical to apply the latter in two steps: (UT2 - UTC) is applied to the results of the individual sets, and then (UT1 - UT2) is applied to their mean value which is then further corrected for the motion of the pole by the quantity $\Delta\Lambda_P$, all in accordance with equations (11.5) and (11.6) (with $h = 0^h$ or 12^h). The corrected mean $\Delta\Lambda$ added to the assumed longitude is the final longitude of the station.

An additional feature of Example 11.4, normally unnecessary, is the reduction of the right ascensions of the stars to the FK4 system in the fifth table. The observations and the original calculations were performed in 1962 when the FK3 system was the one adopted for time determinations. Since today the reference system is that of the FK4, it seemed appropriate to illustrate the application of this correction which refers the old results to the new reference system. The corrections were taken from the tables in [Astronomisches Rechen-Institut, Heidelberg, 1961].

EXAMPLE 11.4

Longitude by Mayer's Method

Observer:	L. I. Sukman	Instruments: Wild T4 theodolite (No. 64252); Nardin sidereal chronograph (No. 127973); Favag chronograph; National radio with amplifier
Recorder:	S. F. Cushman	
Computer:	L. I. Sukman and H. D. Preuss	
Local Date:	5-24/25-1962	
Location:	OSU old astro pillar $\Phi_0 = 40^\circ 00' 00''.000$ $\Lambda_0 = -5^h 32^m 10^s.000$	

1. Observing List for Longitude (Sample)

FK4 Star No.	Magnitude	Right Ascension	Declination	Vertical Circle Setting East	Vertical Circle Setting West	Azimuth Factor A
1318	4.8	$12^h20^m37^s$	26°03'	346°03'	13°57'	0.27
461	5.2	12 24 00	39 14	359 14	00 46	0.02
466	5.7	12 27 50	21 06	341 06	18 54	0.35
467	5.4	12 28 13	58 37	18 37	341 23	-0.61
1322	5.4	12 31 48	33 27	353 27	06 33	0.14
470	4.3	12 31 58	41 34	01 34	358 26	-0.04
1323	4.8	12 32 59	22 50	342 50	17 10	0.32
473	5.2	12 33 15	18 35	338 35	21 25	0.37
1326	5.0	12 39 59	10 27	330 27	29 33	0.50
1327	5.4	12 43 22	45 39	05 39	354 21	-0.14
1328	5.2	12 43 43	07 53	327 53	32 07	0.54
1332	5.1	12 49 52	27 45	347 45	12 15	0.24
483	1.7	12 52 23	56 10	16 10	343 50	-0.50
484	3.7	12 53 43	03 36	323 36	36 24	0.59
485	2.9	12 54 16	38 31	358 31	01 29	0.03
488	3.0	13 00 18	11 10	331 10	28 50	0.48
1337	5.1	13 03 59	36 00	356 00	04 00	0.08
1338	5.7	13 04 11	45 28	05 28	354 32	-0.14
1339	6.0	13 04 31	21 21	341 21	18 39	0.34
491	6.0	13 08 20	38 42	358 42	01 18	0.03
492	4.3	13 10 07	28 04	348 04	11 56	0.24
1344	5.0	13 15 42	05 40	325 40	34 20	0.56
494	4.7	13 15 52	40 46	00 46	359 14	-0.01
1346	5.7	13 18 38	40 21	00 21	359 39	0.00
497	2.4	13 22 26	55 07	15 07	344 53	-0.45
1349	5.2	13 26 36	13 59	333 59	26 01	0.45
500	5.4	13 27 05	60 08	20 08	339 52	-0.68

2. Longitude Field Record (Sample)

Local Date: May 24, 1962

Set 1					Set 2				
A	ΣA	Star No.	Levels		A	ΣA	Star No.	Levels	
			W	E				W	E
	RCC[1]	CHU 7.336 MHz EST $20^h40^m00^s$ Chr. $12^h16^m37^s$				RCC	WWV 5.0 MHz EST $22^h10^m00^s$ Chr. $13^h46^m52^s$		
0.35		466	24.3 73.0 48.7	72.6 24.4 48.2	0.38		513	72.6 24.4 48.2	21.9 75.0 53.1
-0.04	0.31	470	73.0 24.8 48.2	24.3 73.2 48.9	0.62	1.00	516	24.0 72.9 48.9	74.8 21.9 52.9
0.50	0.81	1326	23.8 73.0 49.2	72.5 24.1 48.4	-0.10	0.90	1368	73.2 24.2 49.0	22.1 75.2 53.1
-0.14	0.67	1327	73.2 24.1 49.1	24.2 73.0 48.8		RCC	WWV 5.0 MHz EST $22^h32^m00^s$ Chr. $14^h08^m54^s$		
-0.50	0.17	483	23.3 72.6 49.3	72.1 23.7 48.4	-0.32	0.58	528	22.4 75.2 52.8	73.4 24.2 49.2
	RCC	CHU 7.336 MHz EST $21^h20^m00^s$ Chr. $12^h56^m43^s$			-0.34	0.24	531 (releveled)	74.2 23.7 50.5	23.1 74.6 51.5
-0.14	0.03	1338	73.0 23.7 49.3	23.6 73.0 49.4	0.03	0.27	535	23.7 75.0 51.3	74.6 24.0 50.6
	RCC	CHU 7.336 MHz EST $22^h00^m00^s$ Chr. $13^h36^m50^s$				RCC	WWV 5.0 MHz EST $22^h55^m00^s$ Chr. $14^h32^m00^s$		

[1] RCC = Radio-Chronometer time comparison
EST = Eastern Standard Time
Chr. = Chronometer Time

3. Abstract of Longitude Record

	Set 1																	
Star	466			470			1326			1327			483			1338		
Level	W 48.7	E 48.2		W 48.2	E 48.9		W 49.2	E 48.4		W 49.1	E 48.8		W 49.3	E 48.4		W 49.3	E 49.4	
	0.5			-0.7			0.8			0.3			0.9			-0.1		
Chronometer Time of Matching Contact Pairs	12^h		Sums	12^h		Sums	12^h		Sums	12^h		Sums	12^h		Sums	13^h		Sums
	27^m	28^m	55^m	30^m	32^m	62^m	39^m	40^m	79^m	42^m	44^m	86^m	51^m	53^m	104^m	03^m	04^m	07^m
	s	s	s	s	s	s	s	s	s	s	s	s	s	s	s	s	s	s
	6.7	44.3	51.0	58.1	69.1	127.2	13.9	54.8	68.7	23.9	32.5	56.4	12.9	45.9	58.8	27.6	66.8	94.4
	7.6	43.2	50.8	59.3	67.6	126.9	14.9	53.9	68.8	25.2	31.0	56.2	14.8	44.1	58.9	29.0	65.4	94.4
	8.9	42.1	51.0	60.7	66.3	127.0	15.8	52.9	68.7	26.6	29.5	56.1	16.4	42.2	58.6	30.5	63.8	94.3
	9.9	41.0	50.9	62.1	64.9	127.0	16.9	51.7	68.6	28.0	28.2	56.2	18.4	40.3	58.7	31.9	62.3	94.2
	10.9	40.0	50.9	63.3	63.6	126.9	18.0	50.8	68.8	29.6	26.7	56.3	20.3	38.5	58.8	33.5	60.8	94.3
	12.1	38.8	50.9	64.7	62.3	127.0	18.9	49.7	68.6	31.0	25.2	56.2	22.1	36.6	58.7	34.8	59.6	94.4
	13.2	37.8	51.0	66.1	60.9	127.0	20.1	48.7	68.8	32.5	23.7	56.2	24.0	34.8	58.9	36.2	58.1	94.3
	14.2	36.6	50.8	67.4	59.6	127.0	21.0	47.7	68.7	34.1	22.2	56.3	25.6	33.0	58.6	37.7	56.6	94.3
	15.4	35.5	50.9	68.8	58.1	126.9	22.1	46.6	68.7	35.4	20.8	56.2	27.4	31.2	58.6	39.2	55.2	94.4
	16.5	34.4	50.9	70.3	57.0	127.3	23.1	45.6	68.7	36.9	19.2	56.1	29.5	29.4	58.9	40.6	53.8	94.4
	Mean	50.91		Mean	127.02		Mean	68.71		Mean	56.22		Mean	58.74		Mean	94.34	
	Mean/2	25.455		Mean/2	63.510		Mean/2	34.355		Mean/2	28.110		Mean/2	29.370		Mean/2	47.170	
T_S	$12^h 27^m 55.^s455$			$12^h 32^m 03.^s510$			$12^h 40^m 04.^s355$			$12^h 43^m 28.^s110$			$12^h 52^m 29.^s370$			$13^h 04^m 17.^s170$		

4. Radio-Chronometer Comparison (Sample)

Local Date	May 24, 1962	May 24, 1962	May 24, 1962	May 24, 1962	May 24, 1962	May 24, 1962
Transmitting Station	CHU	CHU	CHU	WWV	WWV	WWV
Frequency	7.336 MHz	7.336 MHz	7.336 MHz	5.0 MHz	5.0 MHz	5.0 MHz
EST of Signal	$20^h 45^m 11^s.000$	$21^h 20^m 29^s.000$	$22^h 03^m 29^s.000$	$22^h 11^m 43^s.000$	$22^h 34^m 10^s.000$	$22^h 55^m 21^s.000$
Zonal Correction	5 00 00.000	5 00 00.000	5 00 00.000	5 00 00 000	5 00 00.000	5 00 00.000
Greenwich Date	May 25, 1962	May 25, 1962	May 25, 1962	May 25, 1962	May 25, 1962	May 25, 1962
UTC	01 45 11.000	02 20 29.000	03 03 29.000	03 11 43.000	03 34 10.000	03 55 21.000
GMST at 0^h UT	16 08 24.260	16 08 24.260	16 08 24.260	16 08 24.260	16 08 .24.260	16 08 24.260
Solar to Sidereal Interval Correction	0 17.279	0 23.077	0 30.141	0 31.494	0 35.282	0 38.662
GMST of Signal	17 53 52.539	18 29 16.337	19 12 23.401	19 20 38.754	19 43 09.442	20 04 23.922
Λ_0	-5 32 10.000	-5 32 10.000	-5 32 10.000	-5 32 10.000	-5 32 10.000	-5 32 10.000
MST of Signal	12 21 42.539	12 57 06.337	13 40 13.401	13 48 28.754	14 10 59.442	14 32 13.922
Chronometer Time of Signal	12 21 48.596	12 57 12.364	13 40 19.402	13 48 34.754	14 11 05.410	14 32 19.885
Chronometer Correction	$-6^s.057$	$-6^s.027$	$-6^s.001$	$-6^s.000$	$-5^s.968$	$-5^s.963$
Rate per Minute	0.85 ms		0.60 ms	1.42 ms		0.24 ms

5. Computation of a Longitude Set (Sample, based on equation (11.28))

Level No: 616

Level Value: 0".897/div.

Star No.	Set 1					
	466	470	1326	1327	483	1338
Level	0.5	-0.7	0.8	0.3	0.9	-0.1
b	$0^s.007$	$-0^s.010$	$0^s.012$	$0^s.004$	$0^s.013$	$-0^s.001$
B	1.104	1.336	0.885	1.424	1.725	1.419
α	$12^h 27^m 49^s.688$	$12^h 31^m 57^s.364$	$12^h 39^m 58^s.690$	$12^h 43^m 21^s.935$	$12^h 52^m 22^s.907$	$13^h 04^m 10^s.967$
T_s	12 27 55.455	12 32 03.510	12 40 04.355	12 43 28.110	12 52 29.370	13 04 17.170
ℓ	0.067	0.084	0.064	0.090	0.113	0.090
Δh_D	-0.017	-0.022	-0.016	-0.023	-0.029	-0.023
Bb	0.007	-0.013	0.011	0.006	0.022	-0.001
ΔT_s	-6.052	-6.048	-6.041	-6.039	-6.031	-6.023
Eq. E.	-0.840	-0.840	-0.840	-0.840	-0.840	-0.840
t	12 27 48.620	12 31 56.671	12 39 57.533	12 43 21.304	12 52 22.605	13 04 10.373
FK4-FK3	-0.004	-0.002	-0.001	-0.019	-0.014	0.019
α - t	$1^s.064$	$-0^s.691$	$1^s.156$	$0^s.612$	$0^s.288$	$0^s.613$

Star \ x	A	α - t	AA	A(α - t)	Aa	$\Delta\Lambda$	v
466	.347	$1^s.064$	.1204	$^s.3692$	$^s.3064$	$-^s.7576$	$^s.0256$
470	-.036	0.691	.0013	-.0249	-.0318	-.7228	-.0092
1326	.502	1.156	.2520	.5803	.4433	-.7127	-.0193
1327	-.141	0.612	.0199	-.0863	-.1245	-.7365	.0045
483	-.500	0.288	.2500	-.1440	-.4415	-.7295	-.0025
1338	-.136	0.613	.0185	-.0833	-.1201	-.7331	.0011
[x]	.036	$4^s.424$	.6621	$^s.6110$	Mean:	$-^s.7320$	$^s.0002$
[xx]		3.7792					.00114

Solution of Normal Equations Yields:

$\Delta\Lambda = 0^s.732035$ $\qquad$ $a = 0^s.882994$

$M_{\Delta\Lambda} = 0^s.0070$ $\qquad$ $M_a = 0^s.0207$

6. Summary of Longitude Computation

Set No.	Greenwich Date	Radio Station	a	M_a	$\Delta\Lambda_{1,2}$	$M_{\Delta\Lambda}$	v_1	v_2
1	May 25.10, 1962	CHU	0.ˢ8830	0.ˢ0207	0.ˢ7320	0.ˢ0070	-0.ˢ0039	-0.ˢ0222
2	May 25.15, 1962	WWV	0.8583	0.0778	0.7026	0.0277	0.0255	0.0072
3	May 26.10, 1962	WWV	1.0673	0.0342	0.7011	0.0148	0.0270	0.0087
4	May 26.14, 1962	CHU	1.0332	0.0308	0.6864	0.0114	0.0417	0.0234
5	May 26.16, 1962	CHU	0.4569	0.1755	0.8195	0.0620	-0.0914	
6	May 26.19, 1962	CHU	1.0669	0.0252	0.7268	0.0110	-0.0013	-0.0170
			Mean $\Delta\Lambda_1$ =		0.ˢ7281	$[v_{1,2}]$	0.ˢ0002	0.ˢ0001
			Mean $\Delta\Lambda_2$ =		0.ˢ7098	$[v^2_{1,2}]$	0.01149	0.00146

$$M_{\Delta\Lambda_{1,2,3}} = \sqrt{\frac{[vv]_{1,2,3}}{n(n-1)}}$$

$M_{\Delta\Lambda_1}$ = 0.ˢ0196

$M_{\Delta\Lambda_2}$ = 0.ˢ0084

Set No. 5 rejected since $v_1 > 3.5 \times 0.6745\, M_{\Delta\Lambda_1}$

Corrections

Set No.	Transmission time	-(UT2-UTC)	$\Delta\Lambda_3$	v_3
1	-0.ˢ0029	-0.ˢ0104	0.ˢ7187	0.ˢ0216
2	-0.0018	-0.0102	0.6906	-0.0065
3	-0.0018	-0.0101	0.6892	-0.0079
4	-0.0029	-0.0103	0.6732	-0.0239
6	-0.0029	-0.0102	0.7137	-0.0166
		Mean =	0.6971	$[v_3]$ = -0.ˢ0001
		$M_{\Delta\Lambda_3}$ =	0.ˢ0084	$[v_3^2]$ = 0.00142

Final longitude from five longitude sets (30 observations):

Assumed longitude	Λ_0 =	$-5^h 32^m 10.^s000$
Longitude correction	$\Delta\Lambda_3$ =	$0.697 \pm 0.^s008$
Correction to UT1	-(UT1-UT2) =	0.030
Polar motion	$\Delta\Lambda_p$ =	0.006
Final station longitude	Λ =	$-5^h 32^m 09.^s267 \pm 0.^s008$

Note: Subscripts 1, 2 and 3 refer to quantities computed from the complete sets (Nos. 1-6), from the sets remaining after rejection limits have been applied, and to quantities corrected for transmission time and signal correction respectively. Corrections (UT2-UTC), (UT1-UT2), and for polar motion are taken from U. S. Naval Observatory Bulletin No. 198.

11.32 Alternative Methods for First-Order Longitude Determination

The primary method of first-order longitude determination is based on the observations of meridian transit times as described in section 11.31. There are other less frequently used but still satisfactory first-order methods which are described below.

11.321 The Doellen Method. This method is a refined version of the longitude determination by transit times through the vertical plane of Polaris described in section 11.222 [Doellen, 1863]. The improvements lie in the method of observation and in the data reduction, the purpose being to reduce the effect of systematic and random errors on the results as much as the available instrumentation permits. As far as the observations are concerned, the technique is the same as in section 11.222 except for the following:

(1) Only time stars ($A < 0.6$) are observed in the south.

(2) The inclination of the horizontal axis is determined every time after the instrument is placed into the vertical plane of Polaris (i. e., after Polaris is sighted). For this purpose the ends of the striding level are read, then the level is reversed on the horizontal axis, and the ends are read again. The inclination is calculated from the four readings as shown in section 7.441.

(3) After rotating the telescope about its horizontal axis to the south and setting it into the appropriate zenith distance, the time of the south star's transit over the vertical plane of Polaris is recorded by means of the impersonal micrometer, similar to the method in section 11.31.

In the data reduction, the observed transit time ($T_S + \Delta T_S$) of each time star reduced to the meridian by the quantity τ from equation (11.16) needs to be further corrected for the following effects [Niethammer, 1947]:

(1) Inclination of the horizontal axis: $B'b$, where

$$B' = \sin(z + z_P) \sec\delta \operatorname{cosec} z_P \approx \sec\Phi. \tag{11.31}$$

(2) Collimation: $C'c$, where

$$C' = \pm \sec\Phi \cos\tfrac{1}{2}(z_P - z) \sec\tfrac{1}{2}(z_P + z), \tag{11.32}$$

the sign depending on the position of the instrument (see section 7.43).

(3) Diurnal aberration:

$$\Delta h_D = 0^s\!.0215 (\sin z + \sin z_P) \sec\delta. \tag{11.33}$$

In the above equations the quantities without subscripts refer to the south time star, while those with the subscript P refer to Polaris.

The longitude equation applicable for each time star, similar to equation (11.29), is

$$\Delta\Lambda - (\alpha - t'') = 0 \tag{11.34}$$

where

$$t'' = T_S + \Delta T_S + \tau + B'b + C'c + \Delta h_D.$$

Since the longitude is determined from pairs of south stars observed in the two faces of the instrument, the mean of their transit times will be practically free from the collimation error, provided that their declinations are approximately equal. This being the case, the term $C'c$ may be neglected. The mean longitude correction obtained from 15–18 star pairs will be of first-order precision which after being corrected (as shown in Example 11.4) for (UT1 - UTC), for transmission time of radio signals, and for polar motion, and which after being added to the assumed longitude, provides the final longitude of the station. Recent extensive tests showed that precisions obtained from the methods of Doellen and Mayer are practically equivalent [Proverbio, 1967]. An advantage of the former method is in its independence from the necessity of finding the azimuth correction a and from the stability of the instrument in azimuth for the entire duration of the observations. The disadvantage of the method lies in the long computational procedure.

11.322 The Tsinger Method. This method is a refined version of the longitude determination by equal zenith distances described in section 11.23 [Zinger, 1877]. The instrument used is a universal theodolite or a zenith telescope where any change in the inclination of the telescope during the observation (b') is determined by means of the Horrebow levels exactly as described in section 10.211. The effect of the inclination on the mean of the observed chronometer times $(T_E + T_W)/2$, from equation (3.32) with $dz = b'/2$, is

$$\pm \frac{b'}{2 \cos \Phi \sin A} \tag{11.35}$$

where b' is computed from equation (10.32), and the upper sign is applied when the lower numbers on the vial are toward the star.

The only additional correction to be added to each star pair is for diurnal aberration which, if not applied to the apparent coordinates of the stars, may be added directly to the mean of the observed chronometer times and which may be computed from [Niethammer, 1947, p. 54]:

$$-0^s\!.0215 \cos z. \tag{11.36}$$

With the above corrections the longitude equation (11.18) for the Tsinger method becomes

$$\Delta\Lambda = \frac{\alpha_E + \alpha_W}{2} - \frac{T_W - T_E}{2} + h_m \mp \frac{b'}{2 \cos \Phi \sin A} + 0^s\!.0215 \cos z \tag{11.37}$$

where the notation corresponds to that in section 11.23, and the upper sign of the inclination correction term is applied when the lower numbers on the vial are toward the star. The mean longitude correction obtained from 15–18 star pairs will be of first-order precision which, after being corrected (as shown in Example 11.4) for (UT1 - UTC), for transmission time of radio signals, and for polar motion, and which after being added to the assumed longitude, provides the final longitude of the station.

As mentioned previously, the stars within each pair should be as symmetrical to the meridian and as near the prime vertical plane as possible. The former condition will reduce systematic errors in the assumed latitude and in the measured quantity ($T_W - T_E$); the latter will reduce the systematic errors in z (e.g., refraction) and also in δ [Niethammer, 1947, p. 55]. These conditions create great difficulties in the preparation of an observing list, especially for first-order work, since the number of pairs required is in the order of 15–20. Tsinger himself in 1874 published a list for 160 pairs for the latitudes 30°–70°; in 1932 the list was extended to 500 pairs for the latitudes 34°–70° and was published in four volumes by the Institute for Theoretical Astronomy in the Soviet Union. The coordinates in these volumes are referred to 1950.0 but naturally are not in the FK4 system.

The main advantage of this method lies in the fact that relative instrument stability is required during the observations of only one pair. The complex star program and the lengthy calculations required are the main disadvantages. Precision of the final results may be increased by observing on several horizontal threads or by means of the impersonal micrometer. Naturally the method is also applicable with good astrolabes (e.g., the Danjon) in which case, however, observations cannot be restricted to the vicinity of the prime vertical plane for reasons already explained at the end of section 11.23.

11.4 Corrections to the Observed Longitude

As pointed out in section 2.33 the 'observed' or 'true' astronomic longitude Λ_T is the angle between the planes of the observer's instantaneous astronomic meridian and the astronomic meridian of the Mean Observatory (as defined by the BIH), measured in the plane of the instantaneous equator. Evidently true longitudes observed at different locations at different times are not directly comparable since they refer to different positions of the instantaneous rotation axis of the earth (which defines the direction of the instantaneous meridian plane) and to the normals of different geops. In order to resolve this problem it is customary to refer the longitude to the average terrestrial pole and to the geoid, a common reference surface to all observers. The difference between the true longitude Λ_T and its reduced value Λ may be divided into the following components (corrections):

(1) Correction(s) for the motion of the pole,

(2) Correction for the curvature of the plumb line.

If for some reason the observation is made from a place located off the reference mark, a further correction is needed to compensate for this eccentricity.

11.41 Corrections for the Motion of the Pole

The correction which eliminates the effect of the oscillation of the local astronomic meridian, due to the motion of the instantaneous terrestrial pole, on the longitude (actually on ST), as derived in section 4.133, is

$$\Delta\Lambda_P = \Lambda - \Lambda_T = -(x_P \sin\Lambda + y_P \cos\Lambda)\tan\Phi \qquad (4.40)$$

where it is understood that Λ_T has already been referenced to UT1 and corrected for the propagation time of radio signals, for personal equation, and for diurnal aberration. The application of the correction $\Delta\Lambda_P$ is explained in section 11.1 (also see Example 11.4)

The effect of polar motion on the meridian of the Mean Observatory, which is also part of the longitude definition, is taken care of by the fact that in the basic equation

$$\Lambda = AST - GAST = MST - GMST \qquad (5.4)$$

the only quantity referring to the Mean Observatory is GMST which is computed from UT1 already corrected for polar motion by the BIH. The important consideration here is that the origin to which the coordinates of the instantaneous pole x_P and y_P refer in equation (4.40) should always be the same as the one used by the BIH in computing UT1. If then, for example, the x_P and y_P coordinates are referenced to the CIO (see section 8.215), it is imperative that the BIH uses the same origin when calculating UT1. This being the case, the only polar motion correction in either equation (11.3) or (11.5) is the one given in (4.40). If, however, in the UT1 computations the BIH utilized its own average pole 'of the epoch' (as was the case until January 1, 1968), an additional polar motion correction is needed to take care of this difference in origins. This correction is of the following form

$$d\Delta\Lambda_P = \frac{1}{\Sigma G_i} \sum_{i=1}^{n} G_i\,(\Delta x \sin\Lambda_i + \Delta y \cos\Lambda_i)\tan\Phi_i \qquad (11.38)$$

where the subscript i denotes the observatory 'i' participating in the work of the BIH defining UT1 by astronomic time observations, G_i is the weight of the observatory assigned by the BIH, n is the number of observatories, $\Delta x = x_{CIO} - x_{BIH}$ and $\Delta y = y_{CIO} - y_{BIH}$ are as shown in Fig. 8.10. For the 46 observatories (used by the BIH in 1966), this correction is [Bomford and Robbins, 1968]

$$d\Delta\Lambda_P = 0.29\ \Delta x + 0.42\ \Delta y$$

where Δx and Δy are in seconds of arc. On January 1, 1968, when the BIH average pole of the epoch was changed to the CIO, $\Delta x = 0\rlap{.}''007$ and $\Delta y = 0\rlap{.}''233$; thus the 'origin' correction using the above equation amounted to about $0\rlap{.}''1 = 6.7$ ms.

11.42 Correction for the Curvature of the Plumb Line

This correction when applied to the station longitude will in effect transfer the station from the ground (geop) to the geoid. The correction as described in section 9.32 is based on equation (9.35) and is computed from

$$\Delta\Lambda_{PC} = \Lambda - \Lambda_T = -\frac{H}{\bar{g}}\frac{\partial \bar{g}}{\partial y} \sec\Phi \operatorname{cosec} 1'' \tag{11.39}$$

where H is the orthometric height of the observation station, $\bar{g}$ and $\partial\bar{g}/\partial y$ are the gravity and the east component of its horizontal gradient, respectively, at the station.

11.43 Correction for the Eccentricity of the Station

This correction is to be applied when the longitude is observed off the reference station, and it will reduce the observed longitude to the reference station. Using the notation of Fig. 10.2, the identity for the length of a short parallel arc is

$$\Delta\Lambda_E R \cos\Phi \sin 1'' = d \sin A;$$

thus the correction for the eccentricity is

$$\Delta\Lambda_E = \frac{d}{R\cos\Phi} \sin A \operatorname{cosec} 1'' \tag{11.40}$$

where R is approximated by the mean radius of the earth.

11.5 Simultaneous Determination of Longitude and Latitude

Methods described in this and the next (11.6) sections are designed to obtain the maximum amount of information from a single set of observations, economizing time spent in observation and computation where azimuth and/or more than one element of position are required. With proper regard to the selection of stars and balancing observations, many of the described methods are capable of achieving generally second-order, and in exceptional cases, first-order precisions comparable to those obtained from the independent methods described previously. The principles of certain methods used in observatory time and/or latitude determinations and requiring special equipment are also included for completeness.

11.51 Latitude and Longitude by Ex-Meridian Zenith Distances

The zenith distance of a star at a given time is a function of the observer's latitude and longitude and of the apparent right ascension and declination of the star corrected for diurnal aberration. According to equations (3.2), (3.7), and (5.4), the basic relationship between these quantities is

$$\cos z = \sin\delta \sin\Phi + \cos\delta \cos\Phi \cos(\text{GAST} + \Lambda - \alpha) \tag{11.41}$$

where z is the observed zenith distance corrected for refraction. It follows that by observing the zenith distances of at least two stars and the corresponding chronometer time (GMST), the coordinates of the station may be determined. The results may often be improved by observing more than two stars and by making several observations on each star either by pointing the telescope on each star several times in succession (and reading the zenith distance and chronometer time at each pointing) or, if the telescope is provided with several horizontal threads, by pointing the telescope on each star and reading the zenith distance once only, but observing the times when the star's image crosses the threads.

11.511 Latitude-Longitude Equation. A direct solution of a set of equations of the form of (11.41) being impracticable, the equation is converted into a linear form. Let Φ_0 and Λ_0 denote the respective approximate latitude and longitude of the observer. Also let

$$\Phi = \Phi_0 + \Delta\Phi,$$
$$\Lambda = \Lambda_0 + \Delta\Lambda$$

where $\Delta\Phi$ and $\Delta\Lambda$ may be considered as a new set of unknowns. Substituting the above expressions into (11.41) and expanding it in a series, omitting all terms of second- and higher-order, the following relation is obtained:

$$\begin{aligned}\cos z &= \cos z_0 + \Delta\Phi \sin z_0 \cos A_0 \\ &\quad + \Delta\Lambda \cos\Phi_0 \sin z_0 \sin A_0\end{aligned} \tag{11.42}$$

where z_0 and A_0 are the respective 'predicted' zenith distance and azimuth of the star computed from equations (3.1)–(3.3) with $\Phi = \Phi_0$ and $h = \text{GAST} + \Lambda_0 - \alpha$. Recognizing that the quantity $\Delta z = z_0 - z$ is small,

$$\cos z = \cos(z_0 - \Delta z) \cong \cos z_0 + \Delta z \sin z_0 .$$

Introducing this in equation (11.42), the fundamental latitude-longitude equation is obtained:

$$F\,\Delta\Phi + L\,\Delta\Lambda \cos\Phi_0 = \Delta z = z_0 - z \tag{11.43}$$

where

$$F = \cos A_0 ,$$
$$L = \sin A_0 .$$

The unknowns in this equation are the corrections $\Delta\Phi$ and $\Delta\Lambda$, while Δz is the difference between the predicted and observed zenith distances. Two such equations provide a solution. If redundant observations are available, a least squares solution may be applied in which each observation is weighted in inverse proportion to the square of its standard error which can be estimated from [Roelofs, 1950, p. 170]:

$$m^2_{\Delta z} = \frac{1}{nN} m_t^2 \cos^2 \Phi \sin^2 A + \frac{1}{n}\left(m^2_{va} + \frac{1}{N} m^2_{vp}\right) \tag{11.44}$$

where n is the number of observations on the star in question, N is the number of horizontal threads on which observations were made (generally, when $N > 1$, $n = 1$; and when $n > 1$, $N = 1$); the other symbols correspond to those in sections 9.114 and 9.12. The least squares solution will be from the following normal equations:

$$\begin{aligned} [G_i F_i^2]\,\Delta\Phi + [G_i F_i L_i]\,\Delta\Lambda \cos\Phi_0 - [G_i \Delta z_i F_i] &= 0\,, \\ [G_i F_i L_i]\,\Delta\Phi + [G_i L_i^2]\,\Delta\Lambda \cos\Phi_0 - [G_i \Delta z_i L_i] &= 0\,, \end{aligned} \tag{11.45}$$

where the subscript 'i' denotes quantities related to the observations of the star 'i';

$$G_i = m^2/m^2_{\Delta z_i}$$

denotes the weight of the observation (m^2 is an arbitrary factor, $m^2_{\Delta z_i}$ is from equation (11.44)); and the brackets indicate summation.

11.512 Star Selection. The main sources of systematic errors are the error in zenith distance dz (due mostly to anomalous refraction) and the constant error in the chronometer time dT (due mostly to personal equation). Their effect on Δz, from the differentiation of equation (11.41), is

$$d\Delta z = dz_0 - dz = -dT \cos\Phi_0 \sin A_0 - dz. \tag{11.46}$$

Stars should be selected in sets of fours, one star in each quadrant symmetrical to the meridian and the prime vertical; thus their azimuths should be A_i, $180° - A_i$, $180° + A_i$, and $360° - A_i$. If this condition is fulfilled, for each set

$$[G_i F_i L_i] = 0, \tag{11.47}$$

$$[G_i F_i] = [G_i L_i] = 0; \tag{11.48}$$

thus the solutions for the normal equations (11.45) are

$$\begin{aligned} \Delta\Phi &= \frac{[G_i \Delta z_i F_i]}{[G_i F_i^2]}\,, \\ \Delta\Lambda \cos\Phi_0 &= \frac{[G_i \Delta z_i L_i]}{[G_i L_i^2]}\,. \end{aligned} \tag{11.49}$$

Differentiating the above equations with respect to Δz_i, substituting $d\Delta z$ from (11.46), and using (11.48) yields

$$d\Delta\Phi = \frac{[G_1\, d\Delta z\; F_1]}{[G_1 F_1^2]} = 0,$$
$$d\Delta\Lambda \cos\Phi_0 = \frac{[G_1 d\Delta z_1 L_1]}{[G_1 L_1^2]} = -dT \cos\Phi_0, \tag{11.50}$$

indicating that if the stars are grouped as prescribed above the constant systematic errors in zenith distance will affect neither the latitude nor the longitude; and the constant timing errors will have no effect on the latitude, but they influence the longitude with their total amount. Experiments show that the systematic error in zenith distance is a function of the altitude. Since the relationship between these quantities is not known, it is also advisable to keep the altitudes of the stars within each set as near to each other as possible (within 5°). The lower limit of the altitudes should be 30° due to refraction problems; the upper limit is restricted only by the equipment used. For optimum results up to latitude 60°, the first azimuth should be $45^\circ \pm 5^\circ$; and when selecting the others for second-order work ($1''.0 < M_{\Phi,\Lambda} < 0''.4$), a total of 24–80 observations will produce satisfactory results. This means that either 6–20 sets should be observed with one pointing on each star or, for example, one set is sufficient with 6–20 pointings on each of the four stars. The number of observations required for first-order work ($0''.3 < M_{\Phi,\Lambda} < 0''.1$) is in the order of 300–350. This number can be achieved only by instruments having many horizontal threads or an impersonal micrometer. The fact that satisfactory results may be achieved without the necessity of observing a great number of stars and of requiring a rigid star program is the reputed advantage of the method. Stars could be identified at the time of the observation or may be identified later from the observational data.

If the stars are not well balanced in azimuth and altitude, it may be advisable to supplement equation (11.43) with an additional constant unknown K to absorb constant errors in zenith distance (e.g., collimation, vertical index error). In this case the normal equations will have the following form:

$$\left.\begin{aligned} [G_1F_1^2]\,\Delta\Phi + [G_1F_1L_1]\,\Delta\Lambda\cos\Phi_0 - [G_1F_1]K - [G_1\Delta z_1F_1] &= 0 \\ [G_1F_1L_1]\,\Delta\Phi + [G_1L_1^2]\,\Delta\Lambda\cos\Phi_0 - [G_1L_1]K - [G_1\Delta z_1L_1] &= 0 \\ -[G_1F_1]\,\Delta\Phi - [G_1L_1]\,\Delta\Lambda\cos\Phi_0 + [G_1]\,K + [G_1\Delta z_1] &= 0 \end{aligned}\right\} \tag{11.51}$$

where the notation is the same as in (11.43). The solution requires at least three observations.

11.513 Observations. The routine of observations differs slightly depending on whether an observing list is used. The basic sequence of observations, however, is the same in either case. For arbitrary selection it is best to scan the quadrants to see if there are four bright stars at approximately the same zenith distance. If not, it will be acceptable to start observations on any bright star visible at a convenient zenith distance as follows:

(1) Bring the star into the field of view, and level the vertical circle bubble.

(2) Observe a series of zenith distances in rapid succession, and record the chronometer time, recentering the vertical circle bubble if necessary, prior to each reading of the vertical circle. The number of intersections on each star will be previously determined.

(3) If identification of a star is in doubt, also read the horizontal circle.

(4) Repeat the procedure for the other stars of the set. Since constant errors in zenith distance do not affect the final result, all observations can be made in the same face of the instrument.

(5) Make radio-chronometer time comparisons and temperature and barometric pressure readings periodically, as often as required.

11.514 Least Squares Adjustment. The computational steps may be as follows:

(1) Compute the apparent coordinates of the stars, and correct for diurnal aberration.

(2) Correct observed zenith distances for refraction.

(3) Compute the mean zenith distance z and mean chronometer time of observation T for each star. Curvature correction, using equation (9.27), should be applied to make the mean zenith distance correspond to the mean time of observations.

(4) Compute the GAST for each star from the mean chronometer time, for example, from

$$\text{GAST} = T_S + \Delta T_S$$

where T_S is the sidereal time shown on the chronometer which was synchronized with UTC at the epoch T_S^O from

$$T_S^O = \text{UTC}_O + (\text{UT1}_O - \text{UTC}_O) + (\text{GMST}_O - \text{UT1}_O);$$

and ΔT_S is from equation (11.7). The notation corresponds to that in section 11.11.

(5) For each star, compute the zenith distance z_O from (11.41) using the assumed values of latitude (Φ_O) and longitude (Λ_O) and the quantity $\Delta z = z_O - z$.

(6) Compute the azimuth of each star from equation (3.1) using $h = \text{GAST} + \Lambda_O - \alpha$.

(7) Compute the coefficients of the observation equations of the form (11.43).

(8) Compute the weights of the observations using the reciprocal of equation (11.44). If the azimuth symmetry and the equal altitude rules

in the selections of stars are strictly observed, it will be satisfactory to use $G_1 = 1$ for all stars.

(9) Form and solve the normal equations (11.45).

(10) Final values are

$$\Phi = \Phi_0 + \Delta\Phi,$$
$$\Lambda = \Lambda_0 + \Delta\Lambda,$$

the latter still to be corrected for personal equation, etc. (see section 11.4).

Example 11.5 illustrates the computations based on a set of four stars with eight pointings on each star. The unknown K has been added to the observation equations (see item 3) to absorb constant errors in zenith distance (e.g., collimation and refraction). This procedure seemed to be justified by the fact that the symmetry condition which would have eliminated the effect of such errors was not perfect (the angles from the meridian for the four stars are $39\overset{\circ}{.}5$, $48\overset{\circ}{.}0$, $41\overset{\circ}{.}5$, and $39\overset{\circ}{.}0$ respectively).

11.515 Graphical Adjustment: Position Lines. The principle of a graphical solution is that when zenith distance is measured, the observation point lies on the circumference of a 'position circle' with its center at the substellar point of the star and an angular radius equal to the measured zenith distance. Observation to a second star produces another such circle which will intersect the first at the position of the observer. If more than two stars are observed, due to errors of observation, the position circles will no longer intersect at a point, and the problem becomes that of determining the most probable position of the observer. This position may be found graphically by means of 'position lines,' actually very short arcs of the position circles in the vicinity of their intersections, the curvature of the arcs being negligible.

EXAMPLE 11.5

Longitude and Latitude by Ex-Meridian Zenith Distances

Observer:	J. L. Passauer	Instruments: Wild T3 theodolite (No. 26794); Negus sidereal chronometer (No. 1304); USCGS drum chronograph with hand tappet; Zenith transoceanic radio with amplifier.
Computer:	J. L. Passauer	
Local Date:	9-12-1964	
Location:	OSU old astro pillar $\Phi_0 = 40^\circ 00' 00''.00$ $\Lambda_0 = -5^h 32^m 10^s.000$	

1. Computation of the Observed Mean Zenith Distance (Sample)

	Star: FK4-598	Barometric Pressure: 743 mm Temperature: 10° C			
		Pointings			
		1	2	3	4
1	Observed zenith distance	$53^\circ 34' 48''.20$	$53^\circ 53' 36''.60$	$53^\circ 58' 52''.60$	$54^\circ 02' 39''.00$
2	Refraction	$76''.70$	$77''.60$	$77''.70$	$77''.80$
3	$z = (1) + (2)$	$53^\circ 36' 04''.90$	$53^\circ 54' 54''.20$	$54^\circ 00' 10''.30$	$54^\circ 03' 56''.80$
4	Observed chr. time T_s	$22^h 28^m 56^s.98$	$22^h 41^m 29^s.47$	$22^h 42^m 12^s.71$	$22^h 42^m 43^s.19$
5	Mean $\Sigma T_s/n$	22 42 52.79	22 42 52.79	22 42 52.79	22 42 52.79
6	$\tau = (4) - (5)$	-3 55.81	-1 23.32	-0 23.32	-0 09.60
7	$m = \rho'' 2 \sin^2 \frac{1}{2}\tau$	$30''.30$	$3''.79$	$0''.87$	$0''.05$

1. Computation of the Observed Mean Zenith Distance (continued)

Star: FK4-598		Barometric Pressure: 743 mm Temperature: 10° C			
		Pointings			
		5	6	7	8
1	Observed zenith distance	54°06'51".80	54°13'02".80	54°17'52".00	54°22'38".90
2	Refraction	78".00	78".40	78".70	78".90
3	$z = (1) + (2)$	54°08'09".80	54°14'21".20	54°19'10".70	54°23'57".30
4	Observed chr. time T_s	$22^h43^m17^s.83$	$22^h44^m08^s.53$	$22^h44^m47^s.61$	$22^h45^m25^s.99$
5	Mean $\Sigma T_s/n$	22 42 52.79	22 42 52.79	22 42 52.79	22 42 52.79
6	$\tau = (4) - (5)$	0 25.04	1 15.74	1 54.82	2 33.20
7	$m = \rho'' 2 \sin^2 \frac{1}{2} \tau$	0".34	3".13	7".19	12".79

Mean Time of Obs:* $T_s = 22^h 42^m 52^s.79$

Average Zenith Distance:* $z' = 54^\circ 05' 05''.6$

Curvature Corr: $\Delta z_C = C_z/n \Sigma m = 0''.9$ (equation 9.27)

Mean Zenith Distance * $z = 54^\circ 05' 06''.5$

* from pointings Nos. 1–8

2. Reduction of Observations

	Star No.	598	103	717	1607
1	T_s	$22^h42^m52^s.79$	$22^h33^m01^s.76$	$22^h20^m49^s.18$	$22^h10^m08^s.33$
2	ΔT_s	-1 03 16.59	-1 03 18.32	-1 03 20.24	-1 03 22.01
3	AST	21 39 36.20	21 29 43.44	21 17 28.94	21 06 46.32
4	α	16 01 12.22	2 51 44.22	19 04 21.96	23 12 29.86
5	$h = AST - \alpha$	5 38 23.98	-5 22 00.78	02 13 06.98	-2 05 43.54
6	h	84°35'59".70	-80°30'11".70	33°16'44".70	-31°25'53".10
7	δ	58 39 54.40	52 36 58.60	-4 56 09.00	-6 14 21.40
8	sin h	0.996	-0.986	0.549	-0.521
9	cos h	0.0941098	0.1649920	0.8360077	0.8532650
10	sin δ	0.8541423	0.7945871	-0.0860400	-0.1086808
11	cos δ	0.5200393	0.6071502	0.9962917	0.9940767
12	sin Φ_o	0.6427876	0.6427876	0.6427876	0.6427876
13	cos Φ_o	0.7660444	0.7660444	0.7660444	0.7660444
14	(10 × 12)	0.5490321	0.5107507	-0.0553054	-0.0698587
15	(9 × 13 × 11)	0.0374908	0.0767384	0.6380441	0.6497672
16	cos z_o = (14 + 15)	0.5865229	0.5874891	0.5827387	0.5799085
17	z_o	54°05'21".70	54°01'15".50	54°21'23".80	54°33'21".20
18	sin z_o	0.8099	0.8092	0.8126	0.8147
19	sin A_o = -(11 × 8/18)	-0.6390	0.740	-0.673	0.636
20	A_o	320°25'	47°74'	222°30'	140°50'
21	cos A_o	0.769	0.672	-0.739	-0.771
22	z	54°05'06".50	54°00'49".40	54°21'33".10	54°33'20".50
23	Δz = (17 - 22)	15".20	26".10	-9".30	0".70

3. Least Squares Solution for Observations on a Set of Four Stars (Eight Pointings on Each)

A. Coefficients of the observation equations,

$$F_i\Delta\Phi + L_i\Delta\Lambda\cos\Phi_0 + K = \Delta z_i:$$

Star	$\Delta\Phi$	$\Delta\Lambda\cos\Phi_0$	K	$-\Delta z_i$
598	0.77	-0.64	1	-15.2
103	-0.67	0.74	1	-26.1
717	-0.74	-0.67	1	9.3
1607	0.77	0.64	1	-0.7

B. Normal equations and solution:

$\Delta\Phi$	$\Delta\Lambda\cos\Phi$	K	$-\Delta z$	
2.1823	0.0060	-0.0700	-35.5340	-33.4157
	1.8157	0.0700	-16.2650	-14.3733
		4.0000	-32.7000	-28.7000
			999.2300	914.7310
2.1823	0.0060	-0.0700	-25.5340	-33.4157
-0.0027	1.8157	0.0702	-16.1691	-14.2831
0.0321	-0.0387	3.9950	-33.2149	-29.2231
16.2828	8.9052	8.3141	0.4959	0.4722
16".5	8".6	8".3		

C. Inverse weight matrix:

0.4585		
-0.0018	0.5512	
0.0081	-0.0097	0.250

D. Solutions:

$\Delta\Phi = 16''.5$ $\Delta\Lambda = 11''.2$ $K = 8''.3$

Variance of unit weight

$$m_0 = \sqrt{\frac{[Gvv]}{n-\mu}} = \sqrt{\frac{0.4959}{4-3}} = 0.704$$

Standard deviation of the mean latitude:

$$M_\Phi = \sqrt{Q_{11}}\ m_0 = \sqrt{0.4585} \times 0.704 = 0''.48$$

Standard deviation of the Mean $\Delta\Lambda \cos\Phi$:

$$M_{\Delta\Lambda \cos\Phi} = \sqrt{Q_{22}}\ m_o = \sqrt{0.5512} \times 0.704 = 0\overset{''}{.}52$$

$$M_{\Delta\Lambda} = 0\overset{''}{.}69$$

Standard deviation of the mean K

$$M_K = \sqrt{Q_{33}}\ m_o = \sqrt{0.250} \times 0.704 = 0\overset{''}{.}35$$

E. Final Station Coordinates

$\Phi_o = 40°00'00\overset{''}{.}0$ $\qquad \Lambda_o = -83°02'30\overset{''}{.}0$

$\Delta\Phi = 16\overset{''}{.}5 \pm 0\overset{''}{.}5$ $\qquad \Delta\Lambda = 11\overset{''}{.}2 \pm 0\overset{''}{.}7$

$\Phi = 40°00'16\overset{''}{.}5 \pm 0\overset{''}{.}5$ $\qquad \Lambda = -83°02'18\overset{''}{.}8 \pm 0\overset{''}{.}7$

$\qquad\qquad = -5^h 32^m 09\overset{s}{.}25 \pm 0\overset{s}{.}05$

The computations are completed as in section 11.514, up to and including step (6). A rectangular coordinate system is constructed (see Fig. 11.3), and an 'azimuth line' is drawn for each star corresponding

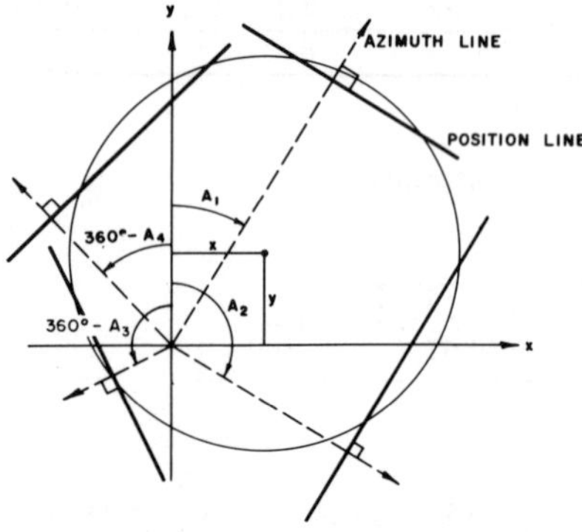

Fig. 11.3 Longitude and latitude by position lines

to the azimuths computed in step (6) above. Perpendicular to each azimuth line, the 'position line' is drawn at a distance Δz from the origin (toward the star if positive and away from the star if negative). A circle is inscribed with the best average tangential fit to the position lines. It follows from the form of equation (11.43) that the coordinates of the center of the circle represent the corrections to be applied to the assumed position, thus

$$\Delta\Phi = y,$$
$$\Delta\Lambda \cos\Phi_0 = x.$$

The accuracy of the graphical method is limited by the plotting errors. They have been found to be of about the same order as observational errors at a scale of 1 mm = 1". A favorable aspect of the graphical solution is that gross errors of observation or computation are shown immediately, while they may not be revealed in the analytic solution until computations are complete.

For further details and for other analytical and graphical solutions, the reader is referred to [Pring, 1952; Tait, 1958a,b; Roelofs, 1950, pp. 165-176].

11.52 Latitude and Longitude by Equal Zenith Distances

The method is similar in principle to the previous one with the exception that all the stars are observed at the same predetermined zenith distance. If all observations are at the same zenith distance, its value need not be known and it may be treated as an unknown in addition to the latitude and longitude. Thus, at least three observations are required. In practice, a large number of observations are made, and the unknowns are determined by least squares adjustment or graphically as in the previous method.

Special instruments, the astrolabes, which enable the observations of stars when they cross an almucantar of zenith distance generally 30° or 45°, have been designed for this method (see sections 7.22, 7.23, and 7.32). Theodolites may also be used either with astrolabe attachments or in their ordinary configuration if strict precautions are taken to preserve the constant zenith distance of the telescope. Such a precaution, for example, is that once observations are begun the vertical circle level adjustment screw must not be touched to preclude changing the zenith distance setting; any vertical level adjustment must be done with the footscrews. If Horrebow levels are available, they should be utilized as in section 10.221 or in 11.322.

11.521 Star Selection. The flexibility of the previous method is lost since observations must all be at the same zenith distance. Due to the large number of stars required, the preparation of an observing program is a laborious process unless special aids are available. A list of suitable stars observable at a zenith distance of 30° may be found in [Institute of Geographical Exploration, 1925] in which the approximate local sidereal times, azimuths, and right ascensions of stars are given when they cross the almucantar. The list covers each degree of latitude between +60° and -60° and is referred to the mean equator and equinox at 1925.0. Though the coordinates are outdated, the list is still an excellent aid for star selection. In the absence of such a list, the following process may be used.

For a star to reach a desired zenith distance z, its declination must be between the limits $(\Phi - z)$ and $(\Phi + z)$. With the assumed latitude and the desired zenith distance, declinations are computed for each 5° of azimuth between 0° and 180° from

$$\sin\delta = \cos z \sin\Phi_0 + \sin z \cos A \cos\Phi_0. \tag{3.5}$$

The corresponding hour angles are then computed from

$$\cos\delta \sin h = -\sin z \sin A. \tag{3.4}$$

A curve, hour angle versus declination, is thenplotted, and the corresponding azimuths may be noted on the curve (Fig. 11.4). Entering with

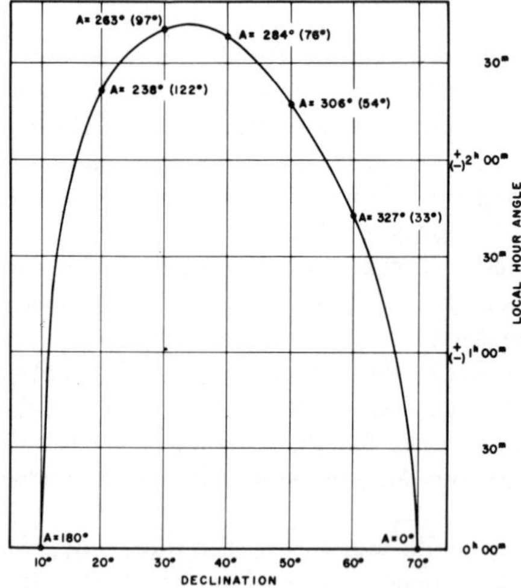

Fig. 11.4 Local hour angles and azimuths for 30° zenith distance ($\Phi = 40^\circ$)

the star's declination, the hour angle and azimuth at the time it reaches the desired zenith distance is determined by the curve. For convenience, the azimuth may also be plotted separately against the declination (Fig. 11.5). Such a graph is more accurate near the extreme declinations of $\Phi - z$ and $\Phi + z$.

The required right ascensions may be determined by subtracting the angles from the approximate sidereal times.

Stars are to be selected as before in sets of four, symmetrical to the meridian and to the prime vertical according to A_1, $180^\circ - A_1$, $180^\circ + A_1$, and $360^\circ - A_1$. The collective sets should be well distributed in azimuth, none closer than about 20° to the meridian and 20° to theprime vertical. Stars near the meridian will give strength to the latitude; those near the prime vertical will reinforce the longitude.

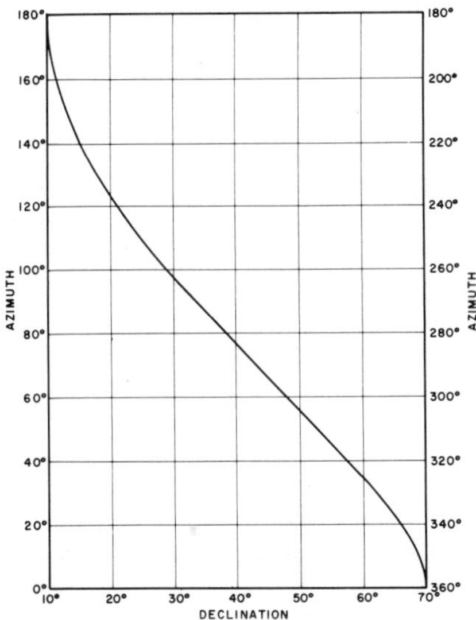

Fig. 11.5 Azimuths for 30° zenith distance ($\Phi = 40°$)

The stars can be selected from the catalogue directly; but if the desired list is to cover only a limited time period, the construction of another graph is recommended. Suppose that the sidereal times when the observations are to be started and ended are decided. The values of h for each ten degrees of declination is extracted from a chart like Fig. 11.4, and the right ascensions corresponding to these declinations and to the starting and ending epochs are computed from $\alpha = ST - h$. The declinations of the stars which will cross the almucantar during this time period can be plotted versus the right ascensions (Fig. 11.6). Then

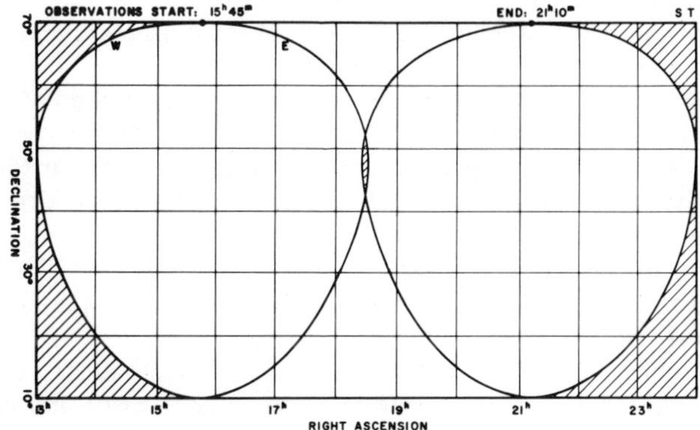

Fig. 11.6 Declination limits for stars at 30° zenith distance during an observation period ($\Phi = 40°$)

the star catalogue can be screened, and all stars that fall within the nonshaded area (the overlap at the center is shaded) can be observed in the selected time period. Right ascension and declinations of all such stars are listed. The local sidereal time of observation for each star can then be computed by adding the hour angle to the right ascension. If the range of the available time is long enough, some stars may be observed both in the east and in the west. The final list should be arranged in order of sidereal time of each observation giving the star number, magnitude, and azimuth. Each star should be at least two minutes from the next. It is possible to complete observations of this method without an observing list, but it is not recommended due to the large number of stars. This also makes it more difficult to plan the observations for proper azimuth balance.

The number of observations required is the same as in the previous method (24–80 on the average for second-order work, 300–350 for first-order). The number of stars may be reduced by observing on more than one horizontal thread. For example, in the field of view of the Zeiss Ni2 astrolabe (Fig. 7.29), there are twenty horizontal threads on which observations can be made; thus observations on four sets (of four stars each) yields 320 observations and a precision of about $M_{\Phi,\Lambda}=0''.2$. For further details on star selection, see [Roelofs, 1950, pp. 182-193].

11.522 Observations. Observations in this method may be accomplished with an ordinary theodolite in the following manner:

(1) Set up the instrument in the meridian, and level the instrument. Center the vertical circle bubble.

(2) Enter the zenith distance to be used on the vertical circle. Check the vertical circle bubble.

(3) Make sure the vertical circle clamp is tight, and rotate the instrument through the 360° checking the stability of the vertical circle and plate bubbles. Make any required adjustments.

(4) Point the instrument in the azimuth of the first star, and wait for the star to cross the horizontal thread(s).

The main difference between this procedure up to this point and that required when an astrolabe attachment (e.g., Fig. 7.23) is used is that the astrolabe attachment needs certain adjustments and the telescope of the instrument is kept horizontal. The following procedure is recommended:

(1) Level the theodolite carefully.

(2) Center the vertical circle bubble, make the telescope horizontal, and clamp it firmly.

(3) Make the rear face of the prism normal to the optical axis of the telescope. The telescope's eyepiece must be unscrewed and replaced with a special collimating attachment. This allows the cross hairs to be illuminated and to be seen reflected on the rear face of the prism.

Adjust the prism with the screws provided until the reflected image and cross hairs coincide (see section 7.22).

(4) Check that the telescope is still horizontal by reading the vertical circle. If not, bring it back with tangent screws.

(5) Put enough mercury in the receptacle to cover the surface to at least 1 mm depth. This will allow the mercury to present a level surface without interference from defects in the surface of the receptacle. Each time the mercury surface is cleaned, it should be checked and more mercury added if necessary.

(6) Clean the surface of the mercury periodically by rotating and drawing a glass rod across the surface. The impurities will be collected on the rod and can be wiped off with a piece of tissue.

(7) If the reflected image vibrates, place a wind-shade on top of the mercury receptacle.

(8) Point the instrument in the azimuth of the first star, and wait for the star to cross the almucantar (this happens when the direct and reflected images of the star coincide).

When a regular (prismatic or pendulum) astrolabe is used, the necessary preparations depend on the type of instrument used, but basically they are similar to what is described above.

In all cases when the star crosses the almucantar(s) (crosses the horizontal thread(s) or the images of the star coincide), register the chronometer time of the event by means of the chronograph and the hand tappet (impersonal micrometer, if available). Also make radio-chronometer time comparisons, and record temperature and barometric pressure periodically as often as required.

11.523 Computations. The required calculations are exactly the same as those described in the previous method (in section 11.511 and 11.514) with the following exceptions:

(1) In the latitude-longitude equation (11.43), z is the nominal zenith distance of the selected almucantar (30° or 45°).

(2) The inclusion of K as an unknown in the normal equations (11.51) is mandatory, and it represents any deviation from the nominal zenith distance z, due to imperfections in the prism angle, etc.

Another method of reducing equal zenith distance observations for longitude and latitude may be found in [Baldini, 1963].

11.53 Observatory Longitude (Time) and Latitude Determinations

As mentioned earlier several astronomic observatories participate in continuous programs of time and/or latitude observations for such purposes as contributing observations to the work of the BIH or the IPMS. The determination of UT at the BIH is based on (daily) sidereal time observations at some 45 observatories; the motion of the pole is determined from periodic latitude observations at the IPMS and other observatories (though either repeated longitude or azimuth observations could also provide the same information [Bhattacharji, 1966]). In some observatories time and latitude are observed independently using the methods described before (most often Mayer's method for time and the Horrebow-Talcott method for latitude); but most of them employ either the Danjon-OPL astrolabe or the Photographic Zenith Tube (PZT) (see sections 7.31 and 7.32) for combined time and latitude determinations, though the resulting latitude is seldom utilized or even computed. Theoretically, both of these instruments are suitable for first-order field observations, but their required permanent installation (especially the PZT's) together with their weight and size prevent their general use outside the observatories. The principles of their use, however, may be utilized in the field when suitable instrumentation is available.

11.531 Observation with the Danjon-OPL Astrolabe. The OPL astrolabe is described in section 7.32, and its full view is shown in Fig. 7.36. In the astrolabe the coincidence of the two star images depends on the zenith distance of the star, on the effective angle of the main prism, and on the position of the Wollaston prism with respect to the focal plane of the instrument (see Fig. 7.34). The position of the Wollaston prism at which the star crosses the almucantar is given when the primary optical plane of the prism coincides with the focal plane. Let this position be designated as zero position and the corresponding micrometer reading as V. The value of V varies slightly during a night's observation due to temperature changes within the instrument. It is usually determined before and after observations of a group of stars by means of a special autocollimator built into the instrument. From these readings, the value of V corresponding to the time of observation is usually determined by linear interpolation. Each observed star thus has three sets of readings associated with it: the micrometer readings, the corresponding time pulses, and V. In addition, the azimuth of the star is also recorded to determine certain asymmetries of the path with respect to the prism to be discussed later.

Let the micrometer readings be $x_1, x_2, \ldots, x_n$ and the corresponding times, recorded with respect to a sidereal clock, be $T_1, T_2, \ldots, T_n$. The mean micrometer reading then is

$$x_0 = \frac{1}{n} \sum_{i=1}^{n} x_i$$

which corresponds to the average time of observation

$$T_0 = \frac{1}{n} \sum_{i=1}^{n} T_i.$$

Let the effective angle of the main prism (the equilateral prism in front of the objective) at T_0 be $90^\circ - z'$. Since the reading V for the zero position of the prism corresponds to the apparent zenith distance of the almucantar, the observed zenith distance corresponding to x_0 is

$$z = z' - S(x_0 - V)$$

where S is a scale factor which may be determined from the obvious relation based on equation (3.32):

$$S\frac{x_{i+1} - x_i}{T_{i+1} - T_i} = S\frac{\Delta x_i}{\Delta T_i} = -\frac{dz}{dh} = \cos\Phi \sin A.$$

A good average value for S thus may be obtained from

$$S = \frac{\sum_{i=1}^{n} \Delta T_i \cos\Phi \sin A}{\sum_{i=1}^{n} \Delta x_i}$$

for each star observation.

Let

$$\zeta = z - z' = S(V - x_0) \tag{11.52}$$

represent the excess of the observed zenith distance over the zenith distance of the apparent almucantar at the time of the observation, still to be corrected for refraction (Δz_R) and, when the star is not observed symmetrically about the center of the field of view, for asymmetry (Δz_S) as follows [Thomas, 1965, p. B290]:

$$\Delta z_S = \frac{1}{4} \sin 2z' (\Delta A)^2 \operatorname{cosec} 1'' \tag{11.53}$$

where ΔA in radians is the difference between the mean azimuth of the star during the observation and the azimuth of the center of the field. In equation (11.52), second-order terms, due to the fact that the zenith distance does not change linearly, are neglected, since their contribution during the observation is too small to be considered. The term Δz, used in section 11.51 and 11.52 as the observed quantity, in this case thus becomes

$$\Delta z = z_0 - z = z_0 - (z' + \zeta + \Delta z_R + \Delta z_S) \tag{11.54}$$

where z_0 is the predicted zenith distance for the epoch T_0, using the assumed longitude and latitude of the station in the calculations. The solution will be from equations (11.51) in which K represents a correction to the nominal zenith distance of the almucantar (of the prism) z'.

The accuracy of time determinations with the Danjon astrolabe is affected chiefly by observational errors and by errors in the adopted star positions. The latter is reduced by internally adjusting the coordinates of the stars together with the unknown parameters. The result is an internally consistent 'astrolabe catalogue' containing the coordinates of the stars used in the observations [Danjon, 1960]. One or two sets of stars are generally observed during one night, each set containing about 28 stars (seven stars in each quadrant). Each set is observed within two hours and is well balanced in azimuth as described before. (An astrolabe catalogue contains sufficient information for twelve sets of stars.) Counting on 10–20 contacts on each star, a set provides 280–560 observations; thus during one night 280-1120 observations may be made. The precision (standard deviation) obtainable from one night's work is in the order of 3–7 ms (0".04–0".1).

11.532 Observation with the PZT. The PZT is described in section 7.31, and its full view is shown in Fig. 7.32. The four images of a star as they appear on the photographic plate are shown in Fig. 7.33. In principle, longitude and latitude are calculated from previously derived equations using the star's hour angle and zenith distance as determined from measurements on the plate.

Let the sidereal epochs of the four exposures be T_1, T_2, T_3, T_4. An interval timer measures the interval between the impulse from the camera shutter to the nearest second of the clock. Let these intervals be p and q for the motor positions west and east respectively. If the clock-second following T is denoted by t, the exposure times starting with motor west may be expressed as

$$T_1 = t - p, \qquad T_3 = t + 60^s - p,$$
$$T_2 = t + 30^s - q, \qquad T_4 = t + 90^s - q. \tag{11.55}$$

The mean time of the sequence is

$$T_O = \tfrac{1}{2}(T_1 + T_4) = t + 45^s - \tfrac{1}{2}(p + q). \tag{11.56}$$

Subtracting equations (11.55) from (11.56), the exposure times are

$$T_1 = T_O - 45^s - s, \qquad T_3 = T_O + 15^s - s,$$
$$T_2 = T_O - 15^s + s, \qquad T_4 = T_O + 45^s + s, \tag{11.57}$$

where $s = \frac{1}{2}(p - q)$. Starting with motor east, s has the opposite sign. If the time kept by the clock is approximate sidereal time and the (unknown) clock correction at the epoch T_O is ΔT_S, the hour angle of the star corresponding to the mean epoch T_O is

$$h_O = T_O + \Delta T_S - \alpha.$$

The unknown clock correction (i.e., the longitude correction) may then be computed from

$$\Delta T_S = h_O - T_O + \alpha \tag{11.58}$$

provided that h_O can be determined from the plate.

Several reduction techniques are in use for determining h_O from measurements on the photographic plate [Markowitz, 1960; Takagi, 1961;

Tanner, 1955; Thomas, 1964]. Imagine a plane tangent to the celestial sphere at the zenith. In this plane a rectangular coordinate system with origin at the zenith is positioned. The positive axis x is directed to the east; the positive axis y is directed toward the south. It can be shown that for plates of such small angular size the rectangular coordinates of a north star in seconds of arc are given by

$$\begin{aligned} x &= -15\,h\cos\delta, \\ y &= -\,(z + \Delta z) \end{aligned} \qquad (11.59)$$

where h is the hour angle in seconds of time (on the east side of the meridian, use $h - 24^h$), δ is the star's declination, z is its meridian zenith distance in seconds of arc, and Δz, given by equation (10.36), is the correction for reducing z to the ex-meridian position of the star at the instant of the observation. The objective projects this coordinate system into the focal plane. A second coordinate system is assumed in the photographic plate to coincide with the xy system at time T_1 or T_3, i.e., when the motor is west. If the plate coordinates of the star's images at time T_1 through T_4 are X_1, X_2, X_3, X_4 and Y_1, Y_2, Y_3, Y_4 (Fig. 7.33), then the following relations exist:

$$\left.\begin{aligned} SX_1 &= x_1 & SY_1 &= y_1 \\ SX_2 &= -x_2 & SY_2 &= -y_2 \\ SX_3 &= x_3 & SY_3 &= y_3 \\ SX_4 &= -x_4 & SY_4 &= -y_4 \end{aligned}\right\}, \qquad (11.60)$$

where S is a scale factor to be found from the X distances between images 1 and 3 or 2 and 4 on the plate and the corresponding 60^s time interval. Substituting hour angles for times in equation (11.57), applying the resulting $h_1 \ldots h_4$ hour angles in (11.59), and using the resulting coordinates $x_1 \ldots x_4$ in (11.60), the following relation may be obtained:

$$S(X_1 - X_2 - X_3 + X_4) = -(h_1 + h_2 - h_3 - h_4)\cos\delta = 120^s\cos\delta. \quad (11.61)$$

When δ is known, this gives the scale factor in seconds of time per revolution of the micrometer head (of the engine measuring the XY coordinates).

From the quantity

$$S(X_1 - X_4 + X_3 - X_2) = x_1 + x_2 + x_3 + x_4$$

which, with equation (11.59), gives

$$S(X_1 - X_4 + X_3 - X_2) = -4\,h_o\cos\delta,$$

the hour angle corresponding to the time T_o may be expressed as

$$h_o = \frac{S(-X_1 + X_2 - X_3 + X_4)}{4\cos\delta}. \qquad (11.62)$$

With h_0 known, the clock correction is determined from equation (11.58), the longitude correction is calculated from equation (11.5); also AST = $T_0 + \Delta T_s$.

PZT observations also yield the latitude of the observing station. The observed double zenith distance is the distance between images 1 and 4, or 2 and 3, measured in the plate perpendicular to the prime vertical. From equation (11.60)

$$S(Y_1 - Y_2 + Y_3 - Y_4) = y_1 + y_2 + y_3 + y_4.$$

Substitutions from equation (11.59), through the use of (11.57) with $T_1 = h_1$, yields

$$15\,S(Y_1 - Y_2 + Y_3 - Y_4) = -4\,(z + \Delta z_0) \tag{11.63}$$

where z is the meridian zenith distance in seconds of arc and Δz_0 is the reduction from the meridian computed from equation (10.36) with h_0. From the above equation the meridian zenith distance of the star is

$$z = -\left(\frac{15\,S\,(Y_1 - Y_2 + Y_3 - Y_4)}{4} + \Delta z_0\right), \tag{11.64}$$

from which the latitude may be calculated by means of equation (3.42).

The above is the principle of the reduction method. In practice, further considerations must be given to the coordinate system of the measuring engine, to the focal length of the PZT, to diurnal aberration, etc. Correction for refraction need not be applied because it is taken care of in the determination of the scale factor. Usually two scale factors are determined each night: One determined from equation (11.61) is used in the time determination; the other, computed from (11.63) when applied to a north and a south star (both should give the same latitude), is used in the latitude determination.

The PZT stars observed are generally divided into groups of ten stars each. Each night two groups of stars are observed which result in standard deviations of 3–5 ms (0".04–0".08). The catalogue of star positions used in the PZT time determinations is based as a whole on a fundamental system, such as the FK4. Internally, however, the positions are determined from the PZT observations themselves by imposing the condition that all the stars should give the same latitude. The PZT catalogue is therefore free of accidental and periodic errors in the fundamental catalogue. On the other hand, due to the condition imposed on the adjustment, the catalogue is suitable for determining the variation of latitude; but it is not suitable for determining the latitude itself. For further details the reader is referred to [Markowitz, 1960].

11.6 Simultaneous Determination of Longitude, Latitude, and Azimuth

11.61 Longitude, Latitude, and Azimuth from Horizontal Directions

This method involves the simultaneous determination of longitude, latitude, and azimuth based on the motion of stars in azimuth. Since most theodolites provide more accurate horizontal circles than vertical ones, and since vertical refraction does not affect the observations, the method is very attractive, as evidenced by the relatively great number of recent studies written on the subject [Gougenheim, 1954; Ney, 1954; Thornton-Smith, 1955; Roelofs, 1956 and 1961; Merritt, 1964; White, 1966; Adams, 1968].

11.611 The Fundamental Equation. Most methods are based on equation (9.2) which in a slightly modified form is

$$dA = \sin A_0 \cot z_0 \, \Delta\Phi + (\tan\Phi_0 - \cos A_0 \cot z_0)\, \Delta\Lambda \cos\Phi_0 \qquad (9.2)$$

with

$$\Delta\Phi = \Phi - \Phi_0,$$
$$\Delta\Lambda = \Lambda - \Lambda_0,$$
$$dA = A - A_0,$$

where Φ_0, Λ_0 are the closely approximated respective latitude and longitude of the station; A is the observed azimuth; A_0 and z_0 are the respective predicted azimuth and zenith distance based on Φ_0 and Λ_0 at the epoch of the observation. The observed azimuth, according to equation (7.14) (assuming that the collimation error is eliminated by observing in both faces of the instrument), is

$$A = R + \Delta A \pm b \cot z \qquad (7.14)$$

where R is the reading on the horizontal circle, ΔA is the (unknown) index correction, and b is the inclination of the horizontal axis which should be determined from the readings on a striding level if available. Assuming that $\Delta A = \Delta A_0 + d\Delta A$ where ΔA_0 is an approximate index correction, substitution of (7.14) in equation (9.2) yields the fundamental relation

$$-d\Delta A + \Delta\Phi \sin A_0 \cot z_0 + \Delta\Lambda \cos\Phi_0 (\tan\Phi_0 - \cos A_0 \cot z_0) =$$
$$= R + \Delta A_0 \pm b \cot z_0 - A_0. \qquad (11.65)$$

Using the notation

$$F = \sin A_0 \cot z_0$$
$$L = \tan\Phi_0 - \cos A_0 \cot z_0,$$
$$\Delta a = R + \Delta A_0 \pm b \cot z_0 - A_0,$$

the above equation has the following convenient form:

$$-d\Delta A + F\Delta\Phi + L\Delta\Lambda \cos \Phi_0 = \Delta a. \qquad (11.66)$$

The 'quasi-' observation on the right-hand side contains the known or measured quantities; the left-hand side contains the three unknowns representing directly the corrections to the assumed elements. A minimum of three well-separated horizontal direction observations provide a unique solution.

If more than three stars are observed, a least squares adjustment will provide the solution. The observation equations will be based on (11.66). The normal equations will be the following:

$$\left.\begin{array}{l} [G_i]\ d\Delta A - [G_iF_i]\ \Delta\Phi - [G_iL_i]\ \Delta\Lambda\cos\Phi_0 + [G_i\Delta a_i] = 0 \\ -[G_iF_i]\ d\Delta A + [G_iF_i^2]\ \Delta\Phi + [G_iF_iL_i]\ \Delta\Lambda\cos\Phi_0 - [G_iF_i\Delta a_i] = 0 \\ -[G_iL_i]\ d\Delta A + [G_iF_iL_i]\ \Delta\Phi + [G_iL_i^2]\ \Delta\Lambda\cos\Phi_0 - [G_iL_i\Delta a_i] = 0 \end{array}\right\} (11.67)$$

where G_i denotes the weight of the quasi-observation Δa_i of the star i. It is computed from

$$G_i = \frac{m^2}{m_{Ai}^2} \qquad (11.68)$$

where m_{Ai} may be estimated for the star 'i' from equation (9.22), and m^2 is an arbitrary factor. The brackets denote summation. Note that the weight depends on the position of the star (A_0, z_0) and on the number of observations on the star (n).

11.612 Star Selection. The proper selection of stars theoretically is a straightforward process. Similar to section 11.511, it may be shown that if the common factors in the normal equations are zero, the errors in $\Delta\Phi$, $\Delta\Lambda$, and $d\Delta A$ are independent. In the case of equation (11.67), this would impose the following conditions:

$$[G_iF_i] = [G_iL_i] = [G_iF_iL_i] = 0.$$

In addition to these, the systematic errors should also be eliminated—in this case, those of timing (personal equation, etc.) and of the horizontal direction (inclination of the horizontal axis, collimation, etc.). For the best determination of the three unknowns, the coefficients $[G_i]$, $[G_iF_i^2]$, and $[G_iL_i^2]$ should be at maximum. It is obvious that finding stars which fulfill all these conditions is very difficult. For this reason, the most practical solution is to select stars in all four quadrants as in section 11.512 and set up the normal equations with the appropriate weights. Solving the normal equations and differentiating the results, similar to section 11.512, it may be shown that

$$\left.\begin{aligned} d\Delta\Phi &= 0 \\ d\Delta\Lambda \cos\Phi_0 &= -dT \cos\Phi_0 \\ dd\Delta A &= d\Delta a \end{aligned}\right\} . \qquad (11.69)$$

Thus constant errors in timing affect only the longitude, and constant errors in the horizontal direction affect only the azimuth.

General guidelines for the star selection, based on equation (9.2), are the following:

(1) Good latitude stars are those near elongation.

(2) Good longitude stars are those near the meridian.

(3) Good azimuth stars are those near the pole.

Thus if only three stars are observed, there should be one from each group. Further stars should be observed in as symmetrical as possible positions with respect to the meridian and to the prime vertical. The stars should be observed as near as possible to a selected almucantar. A graphical star selection method using charts is presented in [White, 1966]. These charts also allow the estimation of the final precision of the results.

11.613 Observation. The following procedure is recommended:

(1) Point to the reference mark, and record the horizontal circle.

(2) Point the instrument in the direction of the selected star. Clamp the instrument when the star is one to two minutes away from the vertical thread.

(3) Record the chronometer time when the star crosses the vertical thread (preferably near the horizontal thread).

(4) Record the horizontal circle reading.

(5) Repeat (2)–(4).

(6) Reverse the instrument, and repeat (2)–(5).

(7) Repeat (1).

(8) If a striding level is available, read the ends of the bubble each time after step (3).

(9) Make radio-chronometer time comparisons at appropriate intervals.

This constitutes two observations (the mean of two direct and two reverse pointings) to be used in the reduction. Increased precision may be obtained by increasing the number of stars or increasing the number of observations on each star. The number of observations should be about the same as in section 11.51 (about 24–80 observations for second-order work using a geodetic theodolite, 300–350 observations for first-order work using universal theodolites).

11.614 Computations. The computational steps may be as follows:

(1) Assume values for latitude, longitude, and index-correction (Φ_0, Λ_0, ΔA_0).

(2) Compute the mean chronometer time of observation for each star.

(3) Correct the horizontal circle readings for the inclination of the horizontal axis ($\pm$b cotan z; see equation (7.14)), and compute the mean horizontal circle reading for each star corresponding to the time obtained in step (2) by taking the mean of the corrected individual readings, and by applying the correction for curvature (equation (9.16)).

(4) Compute the azimuth A_0 (from equation (9.1)) and the zenith distance z_0 (from equation (11.41)) for each star using the approximate values Φ_0 and Λ_0. In these equations, $h = \text{GAST} + \Lambda_0 - \alpha$ where GAST is determined from the mean chronometer time obtained in step (2) as in step (4) of section 11.514.

(5) Compute the coefficients F, L, and Δa of the observation equations similar to (11.65).

(6) Compute the weights G_i of the quasi-observations Δa_i (equation (11.68)).

(7) Form and solve the normal equations (11.67).

(8) The final values are

$$\Phi = \Phi_0 + \Delta\Phi,$$
$$\Lambda = \Lambda_0 + \Delta\Lambda,$$
$$\Delta A = \Delta A_0 + d\Delta A.$$

If the mean horizontal circle reading on the azimuth mark from a direct and reversed pointing is R_M, its azimuth is given by

$$A_M = R_M + \Delta A \pm b \cot z. \qquad (7.14)$$

The longitude still needs to be corrected for personal equation. The apparent star coordinates used in the calculations should be corrected for diurnal aberration and preferably referred to the FK4 fundamental system.

11.62 Alternative Methods of Determining Longitude, Latitude, and Azimuth Simultaneously

The methods described here consist of combinations of observations on stars in special configurations by which it is possible to obtain position and azimuth with a minimum of time and computation. They are not designed for the complete elimination of systematic errors but with care will provide reasonable accuracy. The computations can be done by means of simplified versions of previously derived equations.

11.621 Determination from Two Stars, at Altitude Equal to Declination. In this method latitude, longitude, and azimuth are determined from observation of chronometer time and horizontal circle readings on two stars when their altitudes equal the declinations of the stars [Thornton-Smith, 1955]. Both stars are observed in the same face of the instrument, and the reduction includes a correction to be applied to the derived quantities due to a possible constant error in zenith distance. Two stars of about equal declination and in an altitude range of 30°–50° on opposite sides of the meridian are required. The major difficulty of the method is that the altitude cannot be set until the star is selected, and the star must be about to attain the altitude equal to its declination at the time the observation is desired. A preplanned observing list will expedite field observations and will not require much time in preparation.

11.622 Determination from Two Ex-Meridian Stars. In this method devised by de Graaff-Hunter, latitude, longitude, and azimuth are determined by observations of two stars, differing in azimuth by about 90°, noting horizontal and vertical circle readings as well as the chronometer time [Clark, 1961]. Stars may be selected in the field, near the center of adjacent quadrants at zenith distances convenient for observation, but less than 60° due to refraction. Brighter stars should be used; and if not visually identified, the information from observations is sufficient to compute the right ascension and declination for identification in a star catalog. Each star in turn is intersected on both threads and the times are recorded, the observations being repeated on the same face a few minutes later. Face is then changed on the instrument, and another couple of observations are taken on the same pair of stars. It is claimed that the observations on one pair of stars and the calculation can be completed in the field in less than one hour.

11.623 Determination from Meridian Transits. This method, also devised by de Graaff-Hunter, is designed to provide a rapid determination of longitude and azimuth as well as a latitude of moderate accuracy. It consists of setting the line of collimation approximately in the meridian and observing the times and altitudes of transit of two stars differing in declination by approximately 90° with a small time interval between transits [Clark, 1961]. The advantages of the method are that it is simple to use in the field; identification of stars is easy since the star is observed in the meridian, and the calculations can be done very quickly. Precision may be increased by observing several pairs with change in face between pairs.

11.7 Determination of Ephemeris Time

As mentioned in section 5.5, the definition of Ephemeris Time (ET) is based on the motion of the earth around the sun, i.e., on Newcomb's theory of the sun. In practice, ET may be obtained by finding, in the gravitational ephemeris (with ET argument) of a heavenly body in the solar system, the time corresponding to an observed position (e.g., right ascension and declination) of that body. Since the sun itself is not suitable for the rapid determination of ET, the IAU recommended that ET be determined from observations of the moon whose theory of motion is considered sufficiently known at present. Most often the position of the moon is obtained by photographing it in the background of stars with the dual-rate moon camera, the principles of design and operation of which have already been described in section 7.33. In this section the general method of determining ET from such lunar photography is briefly explained.

11.71 Determination of the Direction of the Moon

As is seen in Fig. 11.7, the image of the moon is a large, partly bright disk surrounded by the reference stars. The method of deter-

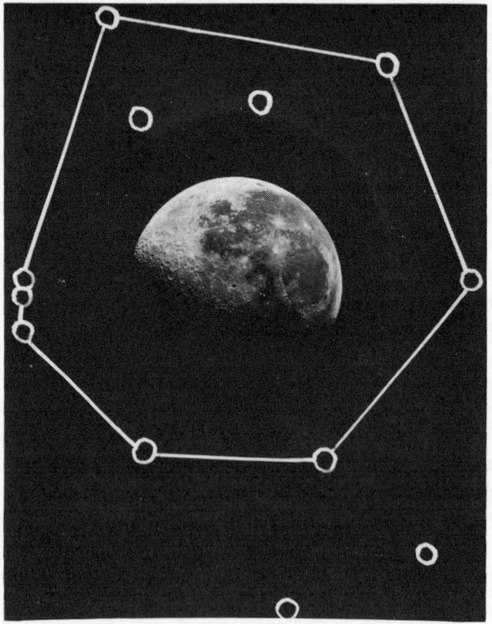

Fig. 11.7 Photograph taken by the Markowitz dual rate moon camera

mining the direction of its geometrical center is essentially the same as the one described in section 6.222 for relative star positions. The difficulty, of course, is that since the moon's center cannot be found on the plate, its machine coordinates cannot be measured directly but must be calculated from measurements made on the limb. A method used by the U. S. Naval Observatory utilizes some ten reference stars and about 30–40 points on the moon's limb, whose X and Y coordinates are measured on the measuring engine. A few preliminary measures are made to determine the center and radius of a provisional circle which fits the bright limb. The coordinates of points six degrees apart on this circle are then computed. In measuring the moon, one screw is set on a computed coordinate, say Y, and the other screw moves the cross threads up to the limb in X. Thereby, a differential correction to the radius is obtained which is of the form $dr = (X/r)dX$ since $dY = 0$. The dr's are combined in groups to give about ten or twelve 'normal points.' If a coordinate measuring machine which has an accurate rotating stage is used, the dr's could be measured without preliminary computation. The determination of the plate constants allows correction to be made for all first-order terms, including first-order differential refraction. These corrections are applied to the normal points, and the position of the center of the circle of best fit is found by least squares. The direction of the moon as determined by this method is comparable with the direction computed from the lunar ephemerides after parallax corrections. For further details on this, the reader is referred to [Markowitz, 1960, pp. 112-113].

The results of the above-described reduction method are apparent geocentric right ascension and declination of the moon for a certain observation epoch UT2. The coordinates are referred to the true equator and equinox of the date by either updating the star coordinates from the epoch of the catalogue to the epoch of observation before the determination of the plate constants or by updating the derived coordinates of the moon, if mean star coordinates are used as given in the catalogue. The star positions are taken from the Yale Zone catalogues, and a correction to a fundamental catalogue (FK4) is applied.

11.72 Interpolation of Ephemeris Time

Since 1960 computed values of the apparent right ascension and declination of the moon, published in national ephemerides, e.g., in the 'American Ephemeris and Nautical Almanac,' are calculated from Brown's lunar theory. The coordinates are tabulated as a function of ephemeris time at hourly intervals to $0^s.001$ in right ascension and $0''.01$ in declination. These tables are entered with the observed coordinates of the moon and two values of ET are taken out by interpolation with the respective arguments right ascension and declination, the former being more reliable. The two values are combined with suitable weights and the difference

$$\Delta T = ET - UT2$$

is obtained. An example is given in the 'Improved Lunar Ephemeris, 1952–1959' in which the calculation of the lunar ephemeris is also described. As mentioned in section 5.51, a distinction should be made between the theoretical value of ET and $ET0 = UT2 + \Delta T$ when ΔT is obtained from observations of the moon. The practical determination of ΔT is affected by observational errors and by possible defects in the lunar (and solar) theories. The latter would obviously introduce systematic errors. Observational errors may be of systematic and random nature. Systematic effects arise from the not well known topography of the moon. Due to libration, the topography of the moon's limb varies and introduces systematic errors in the measurement of the radii. The use of lunar profiles, e.g., from the Watts charts, and the determination of limb corrections would reduce this effect (see section 12.71).

References

Adams, L.P. (1968). "Astronomical Position and Azimuth By Horizontal Directions." Survey Review, 19, 148.

Angus-Leppan, P.V. (1955). "A Note on Calculation of Position Lines." Empire Survey Review, 13, 98.

Astronomisches Rechen-Institut, Heidelberg. (1961). "Definitive Corrections FK4–FK3 for the Fundamental Stars for the Year 1962, Supplement to the 'Apparent Places of Fundamental Stars 1962'." Mitteilungen d. Astronomischen Rechen-Instituts, Heidelberg, Serie B, 5.

Baldini, A.A. (1963). Method of Independent Equations for Determining Latitude, Clock Correction and Zenith Distance. Presented at the XIIIth General Assembly of the International Union of Geodesy and Geophysics, Berkeley, California, September.

Bhattacharji, J.C. (1958). "A Method of Determination of Astronomical Latitude and Longitude When Time and Horizontal Angles Are Observed." Empire Survey Review, 14, 110.

Bhattacharji, J. C. (1966). "Azimuth Variation from Observations of Close Circumpolar Stars." Survey Review, 18, 139.

Bomford, G. and A. R. Robbins. (1968). "The Direction of the Minor Axis of Geodetic Reference Spheroids." Continental Drift, Secular Motion of the Pole, and Rotation of the Earth (Wm. Markowitz and B. Guinot, eds.). D. Reidel Publishing Co., Dordrecht, Holland.

Capon, L. B. (1954). "Latitude, Longitude and Azimuth." Australian Surveyor, 2.

Clark, D. (1961). Plane and Geodetic Surveying, Vol. II, 4th edition. Constable and Company, London.

Danjon, A. (1960). "The Impersonal Astrolabe." Stars and Stellar Systems (G.P. Kuiper and B. M. Middlehurst, eds.), I. University of Chicago Press.

Doellen, W. (1863). Die Zeitbestimmung vermittelst des tragbaren Durchgangsinstrumentes im Verticale des Polarsternes, St. Petersburg.

Dulian, B. (1967). "Determination of the Angular Value of a Graduation Interval of Hanging Level Tube on a Transit Instrument by Observations of Stars." Bulletin Géodésique, 83.

Fallon, N.R. (1957). "Azimuth Setting and Longitude by Transits." Empire Survey Review, 14, 104.

Freislich, J. G. (1955). "A Comparison of Three Methods Used for Calculating the Longitude from a Pair of Well-Balanced Stars." Empire Survey Review, 13, 96.

Gougenheim, A. (1954). Théorie et pratique de la méthode des droites d'azimut. Bulletin Géodésique, 32.

Hickerson, T. F. (1949). "Determination of Position and Azimuth by Single and Accurate Methods." Transactions of American Society of Civil Engineers, 114.

Hoskinson, A. J. and J. A. Duerksen. (1952). "Manual of Geodetic Astronomy." U.S. Coast and Geodetic Survey Special Publication, 237.

Institute of Geographical Exploration. (1925). Complete 60° Star Lists for Position Fixing by the Equal Altitude Method, Harvard University, Cambridge, Massachusetts.

Markowitz, W. (1960). "The Photographic Zenith Tube and the Dual-Rate Moon Camera." Stars and Stellar Systems (G. P. Kuiper and B. M. Middlehurst, eds.), I. University of Chicago Press.

Merritt, E. L. (1964). The Determination of Astronomic Position and Azimuth by Random Observations of Stars with a Theodolite. Presented at the American Congress on Surveying and Mapping Regional Convention, Kansas City, Missouri, September.

Milasovszky, 'B. (1956). "Clock Correction and Azimuth Constant in Astronomical Time Determination." Publications of the Faculties of Mining and Geotechnics, 29. Technical University, Sopron, Hungary.

Milasovszky, B. (1964). "Analysis of the Meridian Method of Time Determination." Acta Technical Academiae Scientiarum Hungaricae 47, 3-4.

Murthy, V.N.S. (1959). "Simultaneous Determination of Latitude, Azimuth and Time by Observations to a Pair of Stars." Empire Survey Review, 15, 111.

Ney, C. H. (1954). "Geographical Positions from Stellar Azimuths." Transactions, American Geophysical Union, 35, 3.

Niethammer, T. (1947). Die Genauen Methoden der Astronomisch-Geographischen Ortsbestimmung. Verlag Birkhäuser, Basel.

Pring, R.W. (1952). "Some Notes on Astronomy as Applied to Surveying." Empire Survey Review, 11, 85.

Proverbio, E. (1967). "Time and Longitude Determination with the Döllen Method." Bulletin Géodésique, 83.

Roelofs, R. (1950). Astronomy Applied to Land Surveying. N.V. Wed. J. Ahrend and Zoon, Amsterdam.

Roelofs, R. (1956). "Latitude, Longitude and Azimuth." Australian Surveyor, 16, 1.

Roelofs, R. (1961). "Simultaneous Determination of Latitude, Longitude, and Azimuth." Annales Academiae Scientiarum Fennicae, A. III.

Roelofs, R. (1966). "Selection of Stars for the Determination of Time, Azimuth and Laplace Quantity by Meridian Transits." Netherlands Geodetic Commission, Publications on Geodesy, New Series, 2, 2.

Smith, W.P. (1949). "Astronomic Fix by Two Ex-Meridian Stars." Empire Survey Review, 10, 71.

Tait, G. B. (1958a). "A Proposed Method of Solution of the Position Line Problem." Empire Survey Review, 14, 107.

Tait, G.B. (1958b). "Solution of the Position Line Problem: A Further Development of the Rate of Change of Latitude Method." Empire Survey Review, 14, 109.

Takagi, S. (1961). "On the Reduction Method of the Mizusawa Photographic Zenith Tube." Publications of the International Latitude Observatory of Mizusawa, III, 2.

Tanner, R. W. (1955). "Method and Formulae Used in PZT Plate Measurements." Publications of the Dominion Observatory, XV, 4.

Thomas, D. V. (1964). "Photographic Zenith Tube: Instrument and Methods of Reduction." Royal Observatory Bulletins, 81. H. M. Stationery Office, London.

Thomas, D. V. (1965). "Results Obtained with a Danjon Astrolabe at Herstmonceux." Royal Observatory Bulletins, 92. H. M. Stationery Office, London.

Thornton-Smith, G. J. (1955a). "Latitude, Longitude and Azimuth from Two Stars." Empire Survey Review, 13, 97.

Thornton-Smith, G. J. (1955b). "Latitude, Longitude and Azimuth." Australian Surveyor, 5.

Thornton-Smith, G. J. (1956). "Latitude, Longitude and Azimuth." Australian Surveyor, 2.

Thornton-Smith, G.J. (1959a). "Simultaneous Determination of Latitude, Azimuth and Time by Observations to a Pair of Stars." Empire Survey Review, 15, 114.

Thornton-Smith, G.J. (1959b). "Comments on 'Solution of Position Line Problem, A Further Development of the Rate of Change of Latitude Method' by G. B. Tait." Empire Survey Review, 15, 112.

Thorson, C. W. (1965). "Second-Order Astronomical Position Determination Manual." U.S. Coast and Geodetic Survey Publication 64-1, U. S. Government Printing Office, Washington, D. C.

White, L. A. (1966). "General Theory for Horizontal Angle Observations in Astronomy." Survey Review, 18, 141-142.

Zinger, N. (1877). Die Zeitbestimmung aus correspondierenden Höhen verschiedener Sterne. Translation by H. Kelchner of the Russian edition published in Moscow in 1874, Leipzig.

12 SOLAR ECLIPSES AND OCCULTATIONS

12.1 Introduction

12.11 Definitions

The term eclipse may be applied to any obscuration of the light of one celestial body by another. The eclipse of the sun by the moon is called a solar eclipse, while an eclipse of the sun by one of the inferior planets is called the transit of the planet. The occultation of a star or of a planet is an eclipse of the star or of the planet by the moon. A lunar eclipse is an eclipse of the moon by the earth. As far as geodetic applications are concerned, only two of these phenomena, the solar eclipses and the occultation of a star, will be investigated here.

The same general principles may be used in the computation of all the different eclipses. The investigation of solar eclipses, with which this discussion begins, will involve nearly everything required in the other cases.

12.12 Two Different Points of View

An eclipse may be calculated from two different points of view:

(1) That of an observer watching the sky from a definite point on the earth. The position of the moon in the sky and of the sun also, if it is an eclipse, is affected by parallax. The observer sees the moon advancing eastward with respect to the stars till the moon's limb touches the sun's limb externally or covers the star. Then in the case of total eclipse there are three other contacts, the last being an external one as the moon leaves the sun. In an occultation there is only one further contact, an emersion, at which time the advancing moon leaves the star behind. In the computations an equatorial Cartesian coordinate system

is used and the parallax must be calculated accurately. The computations are rather tedious.

(2) The other point of view is that of an observer from a point somewhere outside the earth. He sees the shadow cast by the moon fall upon the earth, cutting off the light of the sun or star and advancing from west to east while the earth turns on its axis, thus presenting a varying aspect to the observer in external space. This point of view, usually credited to Bessel, has the advantage of generality and relative simplicity of computation and, therefore, will be used in the discussions. Chauvenet's exposition of Bessel's method will be the main reliance in this chapter. In some cases Chauvenet has simplified Bessel's analysis without sacrificing the required accuracy [Chauvenet, 1863; Bessel, 1876].

In the computations, the so-called Besselian type of Cartesian coordinate system is used which will be discussed in more detail later.

Another important factor should be pointed out here. The observation of the eclipses and occultations consists of the recording of the universal time of the beginning or ending of the phenomena. The universal time would depend on the geodetic coordinates of the observer, i. e., his position with reference to the body of the earth and not on the direction of the vertical of the observer. This fact is the key to the applications and usefulness of eclipses in geodesy.

12.2 Fundamentals of the Theory of Solar Eclipses

12.21 General Prediction

A solar eclipse is due to the obscuration of the sun's light by the moon. The moon must be at or near conjunction with the sun, i. e., it must be new moon. In addition, it is evident that the eclipse cannot take place unless the moon is on or near the ecliptic. These two conditions are illustrated in Fig. 12.1. The symbols in the figure have the following meanings:

M and S — the positions of the moon and the sun respectively on the celestial sphere at conjunction in longitude,

M′ and S′ — the positions of the moon and the sun on the celestial sphere when their apparent distance is the least,

N — the nodal point of the moon's orbit, i.e., the intersection of the ecliptic and the orbit of the moon,

i_M — the inclination of the moon's orbit to the ecliptic,

β_M — the latitude of the moon at conjunction,

Σ — the distance M′S′.

General prediction is the determination of the limits which deter-

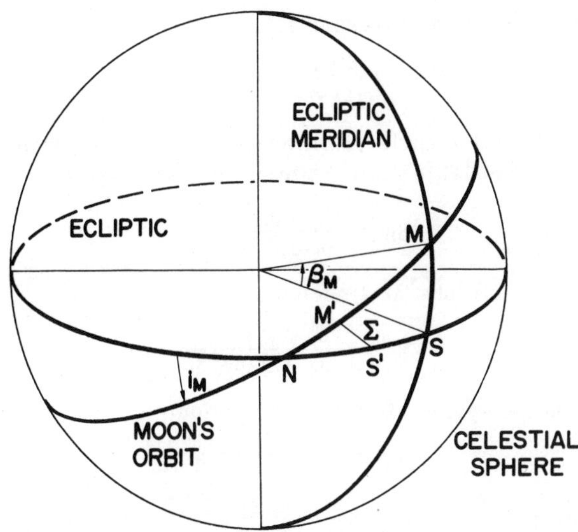

Fig. 12.1 The moon and the sun in conjunction in longitude

mine the possibility of the occurrence of a solar eclipse for some part of the earth. For this purpose the spherical triangle SMN will be regarded as a plane triangle (Fig. 12.2).

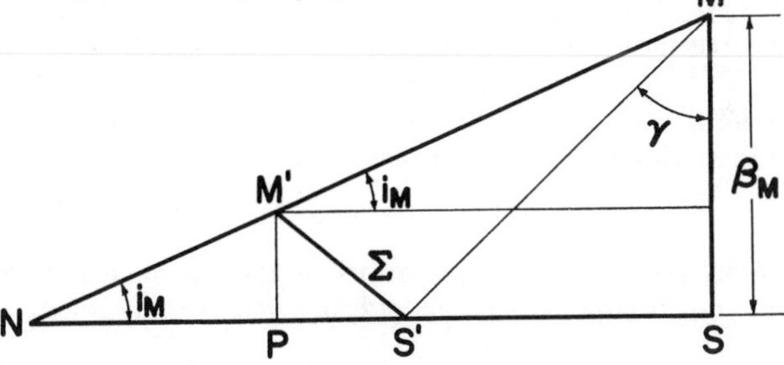

Fig. 12.2 Simplified diagram of the moon and the sun at conjunction

It can be seen from the figure that

$$SS' = \beta_M \tan \gamma.$$

Introducing the symbol λ as the quotient of the motion of the moon in longitude (SP) divided by that of the sun (SS′) yields

$$SP = \lambda \times SS' = \lambda \beta_M \tan \gamma.$$

Hence,

$$S'P = SP - SS' = \beta_M (\lambda - 1) \tan \gamma,$$

and

$$M'P = MS - SP \tan i_M = \beta_M - \lambda \beta_M \tan \gamma \tan i_M.$$

Therefore from these expressions,

$$\Sigma^2 = (M'P)^2 + (S'P)^2 = \beta_M^2 [(\lambda - 1)^2 \tan^2\gamma + \\ + (1-\lambda \tan i_M \tan\gamma)^2]. \tag{12.1}$$

In order to find the angle γ, differentiate this expression with respect to γ and set it equal to zero. After some manipulation,

$$\tan\gamma = \frac{\lambda \tan i_M}{(\lambda-1)^2 + \lambda^2 \tan^2 i_M} . \tag{12.2}$$

Substituting (12.2) into (12.1) yields

$$\Sigma^2 = \frac{\beta_M^2 (\lambda-1)^2}{(\lambda-1)^2 + \lambda^2 \tan^2 i_M}$$

or, defining the symbol I by the expression

$$\tan I = \frac{\lambda}{\lambda - 1} \tan i_M , \tag{12.3}$$

results in

$$\Sigma = \beta_M \cos I. \tag{12.4}$$

The apparent distance of the centers of the sun and the moon as seen from the observer's position may be less than Σ by the difference of the horizontal parallaxes of the bodies, i. e.,

$$\text{minimum apparent distance} = \Sigma - (\pi_M - \pi_S)$$

where π_M, π_S are the horizontal parallaxes of the moon and of the sun respectively.

It is obvious that an eclipse will occur when this distance is less than the sum of the semi-diameters of the bodies; thus when

$$\Sigma - (\pi_M - \pi_S) < k_M + k_S$$

where k_M and k_S are the semi-diameters of the moon and of the sun respectively. Or, using expression (12.4)

$$\beta_M \cos I < \pi_M - \pi_S + k_M + k_S . \tag{12.5}$$

In order to get an approximate limit for β_M, some approximate calculations are done. Equation (12.3) using the values of $i = 5^\circ\!.1453964$ and $\lambda = 13.5$ yields

$$\sec I = 1.00472 ,$$

and with this from (12.5)

$$\beta_M < (\pi_M - \pi_S + k_M + k_S) + 0.00472\, (\pi_M - \pi_S + k_M + k_S).$$

The mean value of the second term is about 25'', thus

$$\beta_M < \pi_M - \pi_S + k_M + k_S + 25'' . \tag{12.6}$$

Using the greatest values for π_M, k_M, and k_S and the least for π_S

($\pi_M = 61'30''$; $k_M = 16'45''$; $k_S = 16'18''$; $\pi_S = 8''.65$) from the 'American Ephemeris and Nautical Almanac,' the above equation yields

$$\beta_M < 1^\circ 34' 49''.3 \,;$$

using the least values for π_M, k_M, and k_S and the greatest for π_S ($\pi_M = 53'53''$; $k_M = 14'41''$; $k_S = 15'46''$; $\pi_S = 8''.96$),

$$\beta_M < 1^\circ 24' 36''.0.$$

This means as an approximation that solar eclipse is certain if at new moon $\beta_M < 1^\circ 24' 36''$, impossible if $\beta_M > 1^\circ 34' 49''.3$, and doubtful between.

Expression (12.5) may be used for more precise general prediction. This type of general prediction is not necessary if, for instance, the 'American Ephemeris and Nautical Almanac' is at hand, because it contains the predictions of solar eclipses.

EXAMPLE 12.1

General Prediction of a Solar Eclipse

The 'American Ephemeris and Nautical Almanac' for the year 1963, on pages 26, 60, and 314 indicates the following:

	Sun		Moon		
ET	Longitude Mean Equinox of 1963.0		Apparent Longitude	Apparent Latitude	Age
July 18 0^h	114°40'17".4	3435".8	77°04'13".27	-2°54'39".05	26.5 days
19 0	115 37 33 .2	3436 .4	91 19 38 .10	-1 44 02 .12	27.5
20 0	116 34 49 .6	3437 .0	105 25 08 .64	-0 27 51 .40	28.5
21 0	117 32 06 .6	3437 .5	119 16 50 .88	0 48 44 .25	0.1
22 0	118 29 24 .1		132 51 40 .09	2 00 58 .60	1.1

By simply comparing this data it is apparent that the sun and moon will be in conjunction in longitude on July 20, 1963. Closer interpolation shows that the conjunction will happen at $20^h 29^m 11^s$ ET. Since at this instance the apparent latitude of the new moon $\beta_M < 1^\circ 24' 36''$, the solar eclipse is certain.

12.22 Condition of the Beginning or Ending of a Solar Eclipse at a Given Place on Earth

The moon during a solar eclipse will cast two shadow cones (Fig. 12.3), one which is obtained by imagining a cone externally tangent to the surfaces of the moon and the sun, and one internally tangent to them.

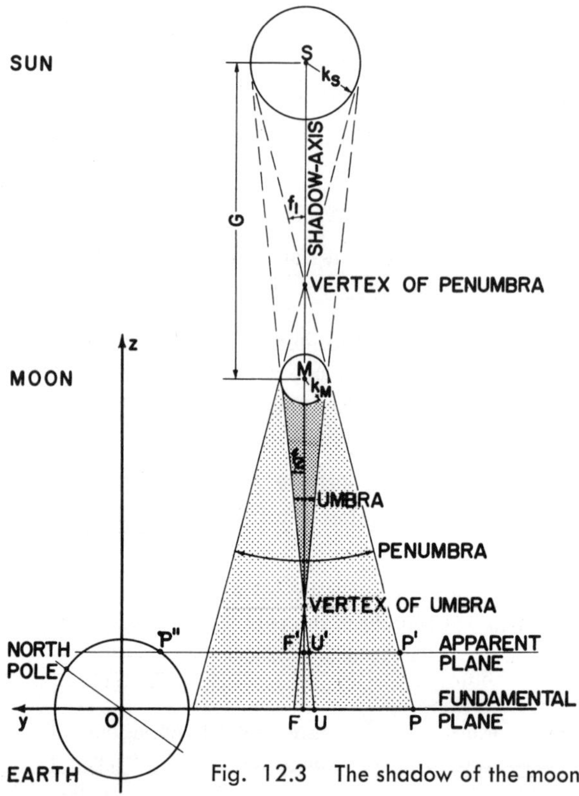

Fig. 12.3 The shadow of the moon

The first shadow cone is called the umbra (shaded in the figure), and the second is called the penumbra. The axis of these cones is called the shadow axis. When the penumbra covers or uncovers a given point on the earth, the observer there sees the external contacts between the limbs of the sun and the moon (first and fourth contacts). Covering or uncovering by the umbra corresponds to the two internal contacts (second and third contacts). If the observer is situated within the umbra above its vertex, the eclipse seen will be a total eclipse, i.e., the moon presents a complete barrier to the light from the sun. If he is below the vertex he will notice that the moon's apparent diameter is smaller than that of the sun and, this way, the moon will not completely cover the solar disk. This type of eclipse is called an annular eclipse. Finally, an observer situated within the penumbral cone will find that the moon conceals only part of the solar disk; the eclipse is then said to be a partial eclipse.

As mentioned in section 12.12, the Besselian type of Cartesian coordinates will be used in the discussion. As is seen in Fig. 12.3, the origin of this system is at the center of the earth, O. The axis z is taken parallel to the shadow axis, and it is positive towards the sun. The xy plane is called the fundamental plane, and it is perpendicular to the shadow axis. The plane yz contains the north pole, and y is taken positive northward. The axis x is perpendicular to the page and it is positive to the east. Actually, it is the intersection of the fundamental plane with the equatorial plane. The plane containing the observer and parallel to the fundamental plane is said to be the apparent plane.

The following notations will be used:

r_M	earth-moon distance, OM,
r_S	earth-sun distance, OS,
ℓ_1	radius of the penumbral shadow in the fundamental plane, FP,
ℓ_2	radius of the umbral shadow in the fundamental plane, FU,
L_1	radius of the penumbral shadow in the apparent plane, $F'P'$,
L_2	radius of the umbral shadow in the apparent plane, $F'U'$,
ξ, η, ζ	the Besselian coordinates of the observer,
Δ	the distance between the observer and the shadow axis in the apparent plane, $P''F'$.

It is evident from what has been said that the condition of the occurrence of the exterior contacts of the limbs of the sun and of the moon is that the distance between the observer and the center of the shadow should be equal to the apparent penumbral radius, i.e.,

$$\Delta = L_1.$$

It is also obvious that the condition for the occurrence of the interior contacts is that the distance between the observer and the center of the shadow should be equal to the apparent umbral radius, i.e.,

$$\Delta = L_2.$$

From these conditions it can be seen that in order to be able to predict the time of the occurrence of any contact at some place on the earth, the following must be computed:

—the position of the shadow axis at any given time,

—the distance of the observer from the shadow axis in the apparent plane,

—the radius of the shadow cone in the apparent plane.

Having all these it will be possible to write an analytical expression for the condition of the occurrence of the contacts.

12.221 Position of the Shadow Axis at Any Given Time. The shadow axis intersects the celestial sphere at a point. The right ascension (a) and declination (d) of this point define the direction of the shadow axis, i. e., its position in space. In this section this direction will be determined at any given time.

In accomplishing this goal an equatorial coordinate system of Cartesian type s, p, t will be used (Fig. 12.4). The origin of the system is the center of the earth; the axis p coincides with the rotation axis and is positive to the south; the axis s is the projection of the earth-sun direction (OS) onto the equatorial plane, and the axis t forms a right-hand system with s and p. The position of the sun (S) and of the moon (M) is defined by their right ascensions, declinations, and distances from the origin, α_S, δ_S, r_S, and α_M, δ_M, r_M respectively.

Projecting the moon-sun distance, G in Fig. 12.4, into the p, s, and t direction, three equations are obtained:

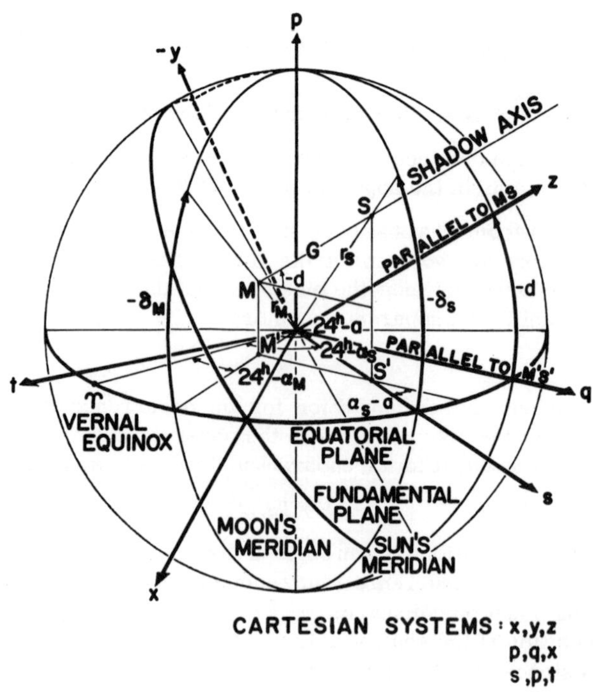

Fig. 12.4 Relation between the different Cartesian coordinate systems

$$\left.\begin{aligned} G \sin d &= r_S \sin \delta_S - r_M \sin \delta_M \\ G \cos d \cos (a-\alpha_S) &= r_S \cos \delta_S - r_M \cos \delta_M \cos (\alpha_M - \alpha_S) \\ G \cos d \sin (a-\alpha_S) &= \qquad - r_M \cos \delta_M \sin (\alpha_M - \alpha_S) \end{aligned}\right\}, \qquad (12.7)$$

or, by substituting

$$g = \frac{G}{r_S} \quad \text{and} \quad b = \frac{r_M}{r_S},$$

$$\left.\begin{aligned} g \sin d &= \sin \delta_S - b \sin \delta_M \\ g \cos d \cos (a-\alpha_S) &= \cos \delta_S - b \cos \delta_M \cos (\alpha_M - \alpha_S) \\ g \cos d \sin (a-\alpha_S) &= \qquad - b \cos \delta_M \sin (\alpha_M - \alpha_S) \end{aligned}\right\}. \qquad (12.8)$$

In these equations α_M, δ_M, α_S, δ_S, and r_S may be found in the 'American Ephemeris and Nautical Almanac.' If r_M is not given, b must be computed as follows:

$$b = \frac{r_M}{r_S} = \frac{\sin \pi_S}{\sin \pi_M}$$

where π_M and π_S are the equatorial horizontal parallaxes of the moon and of the sun respectively. Using the mean earth-sun distance as a unit, it is also true that

$$\sin \pi_S = \frac{\sin \pi_S^M}{r_S}$$

where π_S^M is the mean equatorial horizontal parallax of the sun. Hence

$$b = \frac{\sin \pi_S^M}{r_S \sin \pi_M}. \qquad (12.9)$$

Expression (12.9) is a very convenient form for computing b, because π_M and r_S are given in the 'American Ephemeris and Nautical Almanac' and π_S^M is a constant.

From equations (12.8) the position of the shadow axis (a, d) and also G may be computed with any accuracy desired.

An approximate solution of equations (12.8), given by Chauvenet, which could be used with great accuracy if $\alpha_M - \alpha_S < 1^\circ 43'$ and $a - \alpha_S < 17'$, is the following:

$$a = \alpha_S - \frac{b}{1-b} \cos \delta_M \sec \delta_S (\alpha_M - \alpha_S),$$

$$d = \delta_S - \frac{b}{1-b} (\delta_M - \delta_S),$$

$$g = 1 - b \quad \text{and} \quad G = r_S g,$$

which in many cases will yield

$$a = \alpha_S - b(\alpha_M - \alpha_S),$$
$$d = \delta_S - b(\delta_M - \delta_S).$$

12.222 Distance of a Given Observation Place from the Shadow Axis at a Given Time. In order to compute the distance between a given observation place and the shadow axis measured in the apparent plane, first the Besselian coordinates x and y of the shadow axis and of the observer are determined. If these coordinates are known, the distance is given by simple geometrical formulas. In the computation of the Besselian coordinates of the shadow axis, a new equatorial Cartesian coordinate system p, q, x will be used, with origin at the center of the earth (Fig. 12.4). The axis p is the same as the one used in the previous section; the axis q is parallel to the projection of the line MS into the equatorial plane; the axis x is perpendicular to both and, as can be seen from the figure, coincides with the axis x of the Besselian coordinate system.

In this system the coordinates of the moon are as follows:

$$\left.\begin{aligned} p &= -r_M \sin \delta_M \\ q &= r_M \cos \delta_M \cos (\alpha_M - a) \\ x &= r_M \cos \delta_M \sin (\alpha_M - a) \end{aligned}\right\}. \tag{12.10}$$

The Besselian coordinates of the moon may be computed from these coordinates by rotating the system about its axis x into the Besselian system xyz. The transformation equations are

$$\left.\begin{aligned} x &= x \\ y &= -p \cos d - q \sin d \\ z &= -p \sin d + q \cos d \end{aligned}\right\}. \tag{12.11}$$

Substituting the p, q, and x values from (12.10) yields

$$\left.\begin{aligned} x &= r_M \cos \delta_M \sin (\alpha_M - a) \\ y &= r_M \sin \delta_M \cos d - r_M \cos \delta_M \sin d \cos (\alpha_M - a) \\ z &= r_M \sin \delta_M \sin d + r_M \cos \delta_M \cos d \cos (\alpha_M - a) \end{aligned}\right\}. \tag{12.12}$$

As can be seen from Fig. 12.3, the x and y coordinates of the shadow axis and of the sun are the same as of the moon. The z coordinate of the sun is

$$z_S = z + G$$

where z is calculated from (12.12) and G from (12.8).

The next step is the computation of the Besselian coordinates of the observer. From Fig. 12.5 it is evident that the p, q, and x coordinates of the observer are

$$\begin{aligned} p &= -\rho \sin \varphi', \\ q &= \rho \cos \varphi' \cos (\theta - a), \\ x &= \rho \cos \varphi' \sin (\theta - a), \end{aligned}$$

where ρ is the <u>true</u> distance of the observer from the geocenter, φ' is the geocentric latitude, and θ is the right ascension of the observer's meridian, i.e., the negative local sidereal time.

Transforming these to Besselian coordinates by means of equations (12.11) the following relations are obtained:

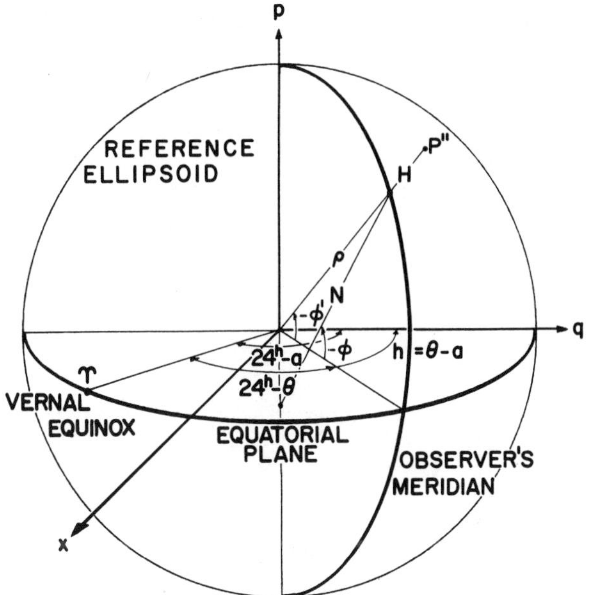

Fig. 12.5 The position of the observer

$$\left.\begin{aligned}\xi &= \rho \cos\varphi' \sin(\theta - a)\\ \eta &= \rho \sin\varphi' \cos d - \rho \cos\varphi' \sin d \cos(\theta - a)\\ \zeta &= \rho \sin\varphi' \sin d + \rho \cos\varphi' \cos d \cos(\theta - a)\end{aligned}\right\}, \qquad (12.13)$$

or, remembering that in the case of a rotational ellipsoid

$$\begin{aligned}\rho \cos\varphi' &= (N + H) \cos\varphi,\\ \rho \sin\varphi' &= [N(1 - e^2) + H] \sin\varphi,\end{aligned}$$

thus the coordinates of the observer, with sufficient accuracy, are

$$\left.\begin{aligned}\xi &= (N + H) \cos\varphi \sin(\theta - a)\\ \eta &= (N + H)[(1 - e^2) \sin\varphi \cos d - \cos\varphi \sin d \cos(\theta - a)]\\ \zeta &= (N + H)[(1 - e^2) \sin\varphi \sin d + \cos\varphi \cos d \cos(\theta - a)]\end{aligned}\right\} \qquad (12.14)$$

where N is the radius of curvature in the prime vertical plane, H is the elevation of the observer above the ellipsoid, φ is the geodetic latitude, and e is the eccentricity of the ellipsoid.

The term $(\theta - a)$, the hour angle of the shadow axis, will be denoted by the letter h. The hour angle may be expressed in several ways, such as

$$h = \theta - a = h^G + \lambda = h^E + \lambda - 1.0027\,\Delta T,$$

where h^G is the Greenwich hour angle of the shadow axis, h^E is its ephemeris hour angle, λ is the longitude of the observer, and ΔT is the difference between ephemeris time and universal time as published in the 'Ephemeris.'

Fig. 12.6 shows the apparent plane with the position of the observer P'' and the center of the shadow F'. It is evident that the distance Δ may be calculated as follows:

$$\Delta^2 = (x - \xi)^2 + (y - \eta)^2. \tag{12.15}$$

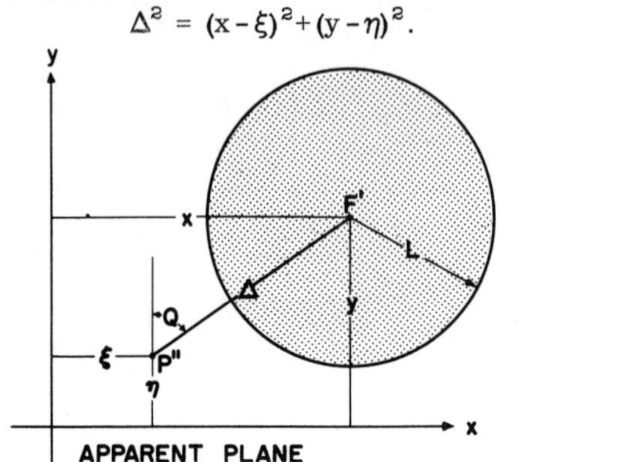

Fig. 12.6 The projection of the observer and of the shadow on the apparent plane

Introducing the angle Q, two more equations may be written for later use:

$$\begin{aligned} \Delta \sin Q &= x - \xi, \\ \Delta \cos Q &= y - \eta. \end{aligned} \tag{12.16}$$

In these equations x and y should be calculated from (12.12), and ξ and η from (12.14).

12.223 Radius of the Shadow. Fig. 12.7 shows the geometry of the shadow. It is evident from the figure that the angles of the two shadow cones may be computed as follows:

$$\sin f_1 = \frac{k_S + k_M}{G},$$

$$\sin f_2 = \frac{k_S - k_M}{G},$$

or, in a general form,

$$\sin f_{1,2} = \frac{k_S \pm k_M}{G} \tag{12.17}$$

where plus should be used for the penumbral cone and minus for the umbral cone.

The following expressions may be justified by observing the figure:

$$\begin{aligned} \ell_1 \cos f_1 &= z \sin f_1 + k_M, \\ \ell_2 \cos f_2 &= z \sin f_2 - k_M. \end{aligned}$$

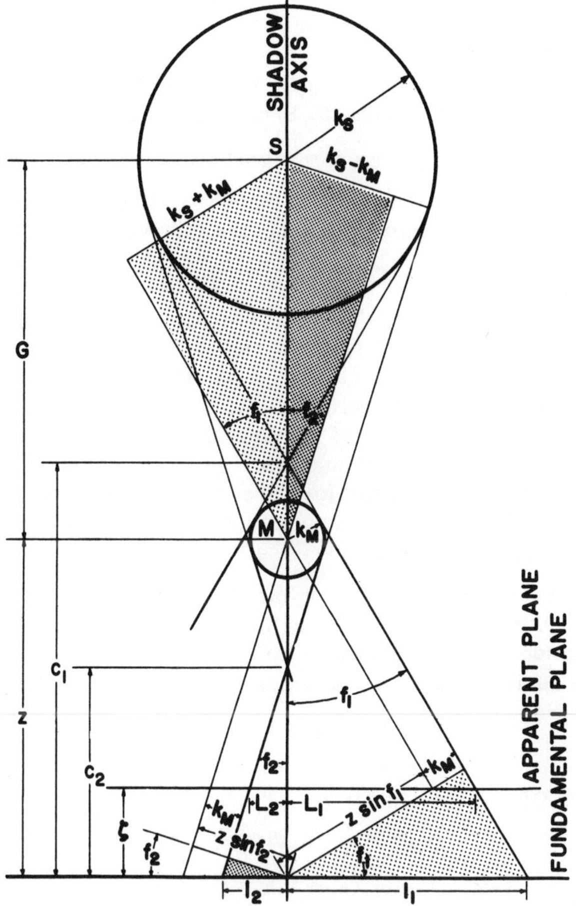

Fig. 12.7 The geometry of the shadow

From here the radii of the shadows in the fundamental plane are

$$\ell_{1,2} = \frac{z \sin f_{1,2} \pm k_M}{\cos f_{1,2}},$$

or, in a different form,

$$\ell_{1,2} = z \tan f_{1,2} \pm k_M \sec f_{1,2} \tag{12.18}$$

where the subscript 1 and plus sign correspond to the penumbra, and the subscript 2 and minus sign correspond to the umbra.

The distance of the vertices of the cones from the fundamental plane are

$$c_{1,2} = z \pm \frac{k_M}{\sin f_{1,2}} \tag{12.19}$$

where the indices and signs should be used with the same meanings as before. Using this, expression (12.18) may be written in a more general form

$$\ell_{1,2} = c_{1,2} \tan f_{1,2}. \tag{12.20}$$

From the figure it is obvious also that the radii of the shadows in the apparent plane may be calculated from the following formula:

$$L_{1,2} = \ell_{1,2} - \zeta \tan f_{1,2} = (c_{1,2} - \zeta) \tan f_{1,2}. \tag{12.21}$$

For the penumbral cone, i. e., for partial eclipse, $c_1 - \zeta$ is always positive and, therefore, L_1 is positive also. For the umbral cone $c_2 - \zeta$ is negative when the vertex of the cone is situated below the apparent plane, i. e., in case of total eclipse L_2 will be negative. In the case of an annular eclipse the vertex of the umbral cone will be above the apparent plane, therefore $c_2 - \zeta$ and also L_2 will be positive. The same sign convention holds naturally for the radii of the shadows in the fundamental plane, i.e., ℓ_1 is always positive, ℓ_2 is positive for an annular eclipse and negative for a total eclipse.

12.224 Fundamental Equation of the Theory of Eclipses. The fundamental equation of the theory of eclipses is the analytical expression of the condition of the occurrence of a contact, i.e., of the beginning or ending of a solar eclipse. As is already known, disregarding the indices, this condition is

$$\Delta = L.$$

Using the symbol $i = \tan f$, this expression may be expanded by means of equations (12.15) and (12.21) to

$$(x - \xi)^2 + (y - \eta)^2 = (\ell - i\zeta)^2, \tag{12.22}$$

or, using expression (12.16) rather than (12.15),

$$\begin{aligned} (\ell - i\zeta) \sin Q &= x - \xi, \\ (\ell - i\zeta) \cos Q &= y - \eta. \end{aligned} \tag{12.23}$$

Either of these two sets of equations may be regarded as the fundamental equation of the theory of eclipses.

12.225 Besselian Elements of the Eclipse. Those parameters which define the position and the geometry of the shadow at a given time are said to be the Besselian elements of an eclipse. They are

—the coordinates of the shadow axis in the fundamental plane, x and y,

—the direction of the shadow axis at a given time, d and $h^G = \theta^G - a$, or d and $h^E = h^G + 1.0027\ \Delta T$,

—the radii of the shadows in the fundamental plane, ℓ_1 and ℓ_2,
—the angles of the cones f_1 and f_2.

All these elements are listed in the 'Ephemeris' for ten-minute intervals for the duration of the eclipse. Note that they are all independent of the location of the observer.

12.3 Prediction of a Solar Eclipse for a Given Place

12.31 Time of the Contact

The prediction of a solar eclipse for a given place consists of finding an instant T when the fundamental equations (12.23) are satisfied. Let T_0 be an assumed time of the contact, let τ be the difference between this time and the contact time, $\tau = T - T_0$. Let x, y, d, a, ℓ, and i be the Besselian elements of the eclipse; let ξ, η, ζ be the Besselian coordinates of the observer calculated for the time T_0; and x', y', ξ', and η' denote the hourly variations in time of the quantities x, y, ξ, and η respectively. Then at the contact time T the Besselian coordinates of the shadow axis and of the observer will be

$$\left.\begin{array}{l} x + x'\tau \\ y + y'\tau \\ \xi + \xi'\tau \\ \eta + \eta'\tau \end{array}\right\} \qquad (12.24)$$

The quantities x and y do not vary uniformly. In order to obtain their values with accuracy from available or possibly precomputed tables, second and even third differences should be employed in the interpolation, depending on the interval used in the tabulation. The variations x' and y' may be calculated by means of the same tables by taking the difference between the consecutive tabulated values and multiplying them properly with the ratio: hour/interval. For instance, if the quantities x,y are tabulated for every ten minutes, then the difference between two consecutive values multiplied by six would give the hourly variations x', y'. However, if the tabulation is for every hour only, then the variation should be computed as follows:

$$x' = \frac{1}{2}(x_1 - x_{-1}) - \frac{1}{6} d_x^3,$$

$$y' = \frac{1}{2}(y_1 - y_{-1}) - \frac{1}{6} d_y^3,$$

where x_1, y_1 are the quantities computed for the time $T_0 + 1^h$, and the values x_{-1}, y_{-1} correspond to $T_0 - 1^h$; d^3 denotes the third difference. Example 12.2 illustrates the procedure for x.

EXAMPLE 12.2

Variations in the Besselian Coordinates

	x	d_x^1	d_x^2	d_x^3
$T_0 - 2^h$	-1.171856			
		+0.545297		
$T_0 - 1^h$	-0.626559		+18	
		+0.545315		-45
T_0	-0.081244		-27	
		+0.545288		-60
$T_0 + 1^h$	+0.464044		-87	
		+0.545201		
$T_0 + 2^h$	+1.009245			

From here, for the time T_0,

$$x' = \frac{1}{2}(0.464044 + 0.626559) + \frac{1}{6}\left[\frac{1}{2}(0.000045 + 0.000060)\right] =$$
$$= 0.545310 \text{ earth radii/hour.}$$

The variations of the Besselian coordinates of the observer ξ, η, and ζ are more uniform and, therefore, may be obtained by differentiating equations (12.13) in which the latitude, longitude, and geocentric radius of the observer are to be taken as constants. After differentiation, the results are

$$\left.\begin{aligned}
\xi' &= \frac{d\xi}{dT} = \frac{\partial\xi}{\partial h}\frac{\partial h}{\partial T} = h'\rho\cos\varphi'\cos h = \\
&= h'(-\eta\sin d + \zeta\cos d) \\
\eta' &= \frac{d\eta}{dT} = \frac{\partial\eta}{\partial h}\frac{\partial h}{\partial T} + \frac{\partial\eta}{\partial d}\frac{\partial d}{\partial T} = h'\xi\sin d - d'\zeta \\
\zeta' &= \frac{d\zeta}{dT} = \frac{\partial\zeta}{\partial h}\frac{\partial h}{\partial T} + \frac{\partial\zeta}{\partial d}\frac{\partial d}{\partial T} = -h'\xi\cos d + d'\eta
\end{aligned}\right\} \qquad (12.25)$$

In these expressions, $d' = \partial d/\partial T$ is the variation of the declination of the shadow axis and $h' = \partial h/\partial T$ is the variation of the hour angle which may be substituted for by the variation of the ephemeris hour angle.

Using the x', y', ξ', η' values computed as described above, the Besselian coordinates x, y, ξ, and η for the contact time T may be calculated by (12.24). The fundamental equations (12.23) for the time T, neglecting the variations of L during the short period τ, will take the following form:

$$\begin{aligned} L\sin Q &= x - \xi + (x' - \xi')\,\tau, \\ L\cos Q &= y - \eta + (y' - \eta')\,\tau. \end{aligned} \qquad (12.26)$$

Reserving Δ and Q for points inside the shadow and using m and M auxiliaries for the time T_0, equations (12.26) will read as follows:

$$\begin{aligned} m\sin M &= x - \xi, \\ m\cos M &= y - \eta. \end{aligned} \qquad (12.27)$$

Since x, y, ξ, and η are known quantities, m and M may be calculated from here. They are the relative polar coordinates of the center of the shadow with respect to the observer.

Similar equations may be written for the variations also. In Fig. 12.8 the hourly variations are shown, in a coordinate system in the apparent plane, in such a way that the shadow is kept motionless while the observer moves. The distance $P_1''P_2''$ = n shown is the magnitude of the relative velocity, and N is the direction of the relative motion. In other words, the quantities n and N give the polar coordinates of the relative motion of shadow and observer. From the figure it is evident that for the variations the following two equations hold:

$$\begin{aligned} n \sin N &= x' - \xi', \\ n \cos N &= y' - \eta'. \end{aligned} \qquad (12.28)$$

Since x', y', ξ', and η' are known quantities, n and N may be calculated from here.

Substituting equations (12.27) and (12.28) into (12.26) yields

$$\begin{aligned} L \sin Q &= m \sin M + \tau n \sin N, \\ L \cos Q &= m \cos M + \tau n \cos N. \end{aligned} \qquad (12.29)$$

Subtracting the product of sin N and the second equation in (12.29) from the product of cos N and the first equation yields

$$L \sin (Q - N) = m \sin (M - N).$$

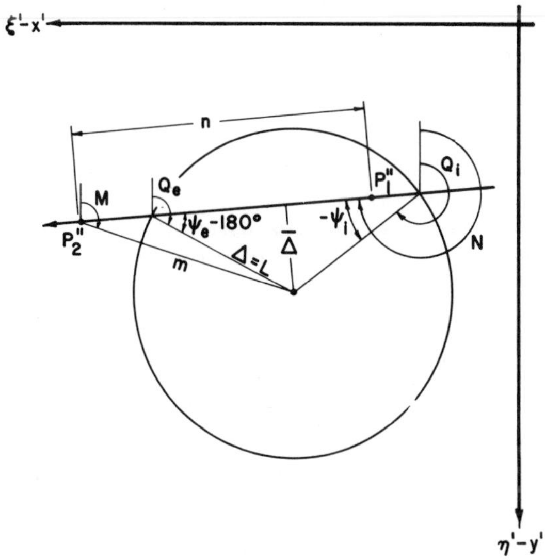

Fig. 12.8 The relative polar coordinates of the center of the shadow with respect to the observer

Adding the product of $\sin N$ and the first equation in (12.29) to the product of $\cos N$ and the second equation yields

$$L \cos (Q - N) = m \cos (M - N) + n\tau.$$

Using the symbol $\psi = Q - N$, these equations yield

$$\begin{aligned} \sin \psi &= \frac{m}{L} \sin (M - N), \\ \tau &= \frac{L}{n} \cos \psi - \frac{m}{n} \cos (M - N). \end{aligned} \tag{12.30}$$

The second equation in (12.30) is the solution of the problem. It gives τ, which when added to the assumed time T_0 is the contact time T. In the equation, m, M are computed from (12.27); n, N are computed from (12.28); L with proper sign is computed from (12.21); ψ is computed from the first equation in (12.30). It should be noted that the first equation does not determine the sign of $\cos\psi$ in the second equation. The following rule will take care of this ambiguity: $\cos\psi$ must be negative for the beginning of a partial or annular solar eclipse and for the end of a total eclipse. In all other cases it should be positive. This convention together with the sign of $\sin\psi$ in equation (12.30) will determine the quadrant of ψ.

The following times can then be computed:

$$\begin{aligned} &\text{immersion:} \quad T_i = T_{0i} \mp \frac{L_i}{n_i} \left| \cos \psi_i \right| - \frac{m_i}{n_i} \cos (M_i - N_i), \\ &\text{middle:} \quad T_m = T_{0m} - \frac{m_m}{n_m} \cos (M_m - N_m), \\ &\text{emersion:} \quad T_e = T_{0e} \pm \frac{L_e}{n_e} \left| \cos \psi_e \right| - \frac{m_e}{n_e} \cos (M_e - N_e), \end{aligned}$$

where the lower signs correspond to a total eclipse and the upper signs correspond to all other cases. The indices i, m, and e refer to immersion, middle, and emersion respectively.

For greater accuracy, the times resulting from the calculation outlined above should be taken in place of the original approximate time and a second approximation performed. For the second approximation, the following formula may be more convenient:

$$\tau = \frac{m}{n} \frac{\sin (M - N - \psi)}{\sin \psi}. \tag{12.30a}$$

This formula, however, is not very accurate if ψ is small.

12.32 The Position Angle

The prediction of an eclipse for a given place includes the determination of the point on the sun's limb at which the first contact is to take place. This <u>position angle</u> is given by

$$P = N + \psi, \qquad (12.31)$$

which is the angular distance of the point of contact from the north point of the sun's limb. It is reckoned positive to the east. A second approximation is commonly needed because the assumed time for which ψ was computed is not exactly the time of contact.

An observer who wishes to know the angle from the vertex of the sun's limb (the point on the limb nearest the zenith) must involve himself in another computation. This involves the calculation of the parallactic angle (the distance of the vertex from the north point) in an astronomic triangle having the vertices, pole, zenith and the shadow axis (Fig. 12.9).

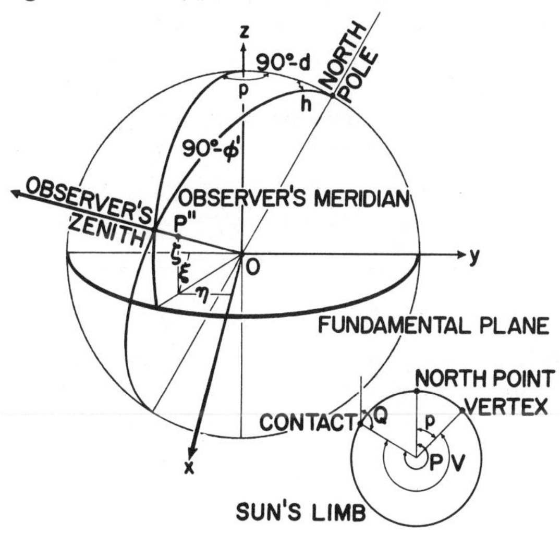

Fig. 12.9 The position angle

Applying equations (3.27) and (3.28), the following relation is obtained:

$$\tan p = \frac{\cos \varphi' \sin h}{\sin \varphi' \cos d - \cos \varphi' \sin d \cos h}.$$

Comparing this equation with (12.13), it may be seen that

$$\tan p = \frac{\xi}{\eta}, \qquad (12.32)$$

sin p having the same algebraic sign as ξ.

The angle V between the contact point and the vertex, reckoned positive towards the east, is given by

$$V = P - p = N + \psi - p. \qquad (12.33)$$

It will not be difficult to recompute ξ and η for the required instants. If we wish to make use of the values ξ and η already found, we can pass to their values at the required instant by means of two additional terms of MacLaurin's series for each of them. Using values already computed yields

$$\tan p = \frac{\xi + \tau\xi' - \tau^2\xi}{\eta + \tau\eta' + \tau^2\eta_2}$$

where the quantity τ is the interval between the times of which ξ and η were computed and of the required contact, and

$$\eta_2 = \rho \cos\varphi' \sin d \cos h.$$

12.33 Time and Degree of Maximum Obscuration

At the instant of greatest obscuration, the observer is the farthest from the edge of the shadow, i. e., the distance $L - \Delta$ is maximum. Assuming that the variation of L may be ignored, it can be said that the greatest obscuration will occur when Δ is minimum. Denoting this time by $T_1 = T_0 + \tau_1$, the quantity τ_1 may be calculated from the two equations preceding (12.30). Substituting ψ_1 for Q - N, and Δ for L yields

$$\Delta \sin\psi_1 = m \sin(M - N),$$
$$\Delta \cos\psi_1 = m \cos(M - N) + n\tau_1.$$

The sum of their squares gives

$$\Delta^2 = m^2 \sin^2(M - N) + [m \cos(M - N) + n\tau_1]^2.$$

Since m and M are computed for the time T_0 and N is nearly constant, the first term in this expression is approximately constant. Therefore, Δ is minimum when the second term is zero, and this happens if

$$\tau_1 = -\frac{m \cos(M - N)}{n}, \tag{12.34}$$

which is the second term in (12.30) already calculated.

In this case the minimum distance $\bar{\Delta}$ is (Fig. 12.8)

$$\bar{\Delta} = m \sin(M - N) = L \sin\psi.$$

The degree of obscuration $(\bar{M})$ is usually expressed by the fraction of the sun's apparent diameter which is covered by the moon's disc. When the observer is so far in the penumbra as to be on the edge of the umbra, the obscuration is total. In this case, the distance of the observer from the edge of the shadow is equal to the algebraic sum $L_1 + L_2$. In any other case the distance of the observer from the edge of the penumbra is $L_1 - \bar{\Delta}$; therefore, approximately,

$$\bar{M} = \frac{L_1 - \bar{\Delta}}{L_1 + L_2}. \tag{12.35}$$

12.34 Correction for Refraction

Since in the prediction of a solar eclipse absolute accuracy is not required, the effect of the astronomic refraction has not been treated in the preceding sections. The effect, though very small, must be treated before the geodetic applications of observed eclipses, in which the greatest possible accuracy is required, are discussed.

Let $S'M'DP''$ be the path of the ray of light from the limb of the sun S', passing through the limb of the moon at the contact point M', entering the atmosphere at D and observed at P'' (Fig. 12.10). It is evident that the observer at P'' sees the apparent contact of the limbs at the same time that an observer at P_1'' would see the true contact if there were no atmosphere, i.e., refraction. This means that if the point P_1'' is substituted for P'' in the computation, i.e., if $\rho + \Delta H$ is used instead of ρ, the effect of refraction is taken fully into account. The problem now is the determination of the fictitious elevation difference ΔH.

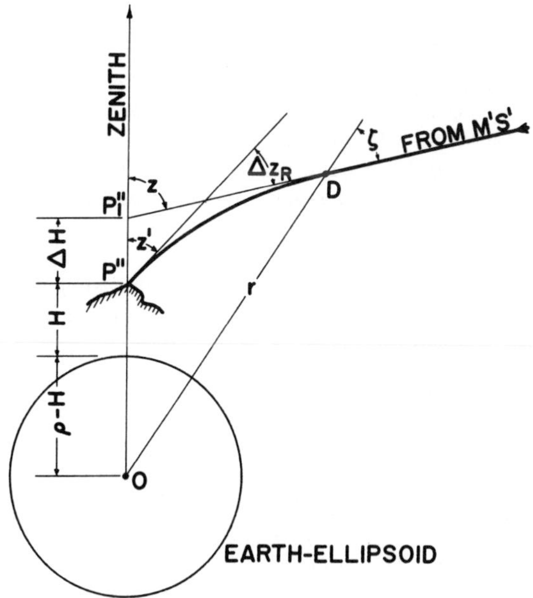

Fig. 12.10 Effect of refraction

Applying the law of refraction expressed by equation (4.68) at the points D ($\mu = 1$) and P'' ($\mu = \mu_0$, $r = \rho$, $\zeta = z'$), the following equation is obtained:

$$r \sin \zeta = \rho \mu_0 \sin z'.$$

From the triangle $OP_1''D$,

$$r \sin \zeta = (\rho + \Delta H) \sin z.$$

Substituting this into the first equation yields,

$$1 + \frac{\Delta H}{\rho} = \mu_0 \frac{\sin z'}{\sin z} .$$

Introducing the symbol $\Delta z_R = z - z'$,

$$1 + \frac{\Delta H}{\rho} = \mu_0 \frac{\sin (z - \Delta z_R)}{\sin z} = \mu_0 (\cos \Delta z_R - \cot z \sin \Delta z_R) .$$

Since Δz_R, the correction to the zenith distance due to refraction, is a very small quantity, it may be assumed that $\cos \Delta z_R = 1$ and $\sin \Delta z_R = \Delta z_R$. Hence,

$$1 + \frac{\Delta H}{\rho} = \mu_0 - \mu_0 \Delta z_R \cot z,$$

or, as the final result, the correction ΔH may be computed from

$$\frac{\Delta H}{\rho} = (\mu_0 - 1) - \mu_0 \Delta z_R \cot z$$

where Δz_R may be taken from any refraction correction table available and is to be substituted in radians.

The formulas for the refractive height ΔH must be used with great caution. In most cases the usually unknown atmospheric variables at the observer, e.g., humidity of the air, are of a magnitude to vitiate any correction. Further, the formulas are very sensitive to round-off errors for certain ranges of z, requiring the computation of z to an accuracy not warranted by the uncertainites in local atmospheric conditions.

12.4 Prediction of an Occultation

12.41 General Method

Lunar occultation will occur when the moon obscures the light of a star. We assume that it is already known that the star will be occulted at the given place. The limits where the occultation will be observable may be determined as described in section 12.43. In certain respects, occultation is a simplified solar eclipse where the star replaces the sun. This point of view will allow us to use most of the equations derived for solar eclipses by simply replacing the parameters of the sun with those of the star. Since the star can be considered as at an infinite distance from the earth, the former moon-sun distance G will be infinite, and the diameter of the sun k_S will be zero. Substituting these quantities in equation (12.17), $f_1 = f_2 = 0°$ is obtained, which means that the two shadow cones become a shadow cylinder the diameter of which is the apparent diameter of the moon $2k_M$. The axis of this shadow cylinder passes through the star; therefore, its direction is determined by the direction of the star.

In other words, making the following replacements in the previous equations, the formulas needed for the predictions of lunar occultations are obtained:

—the apparent right ascension of the shadow axis, $a = \alpha$ (the right ascension of the star),

—the apparent declination of the shadow axis, $d = \delta$ (the declination of the star),

—the moon-earth distance, $r_M = \operatorname{cosec} \pi_M$,

—the apparent diameter of the sun, $k_S = 0$ (the apparent diameter of the star),

—the moon-sun distance, $G = \infty$ (the moon-star distance),

—the shadow angles, $f_1 = f_2 = 0$,

—the radii of the shadow cones, $L_{1,2} = \ell_{1,2} = k_M$ (apparent semidiameter of the moon),

where π_M is the horizontal parallax, and k_M is the apparent semidiameter of the moon whose sine is taken as $0.2725026 \sin \pi_M$ [Nautical Almanac Offices, 1966]. It should be noted here again that in all calculations the unit of distance is the semidiameter of the earth. Thus, for example, when k_M is needed in this unit (e.g., equations (12.40)), its angular value in seconds of arc obtained from the previous relation needs to be divided by the quantity $\sin \pi_M/\sin 1''$. Using these substitutions, the calculations are as follows:

(1) The Besselian coordinates of the moon (and of the shadow axis) from (12.12) will be

$$\left.\begin{aligned} x &= \cos \delta_M \sin (\alpha_M - \alpha) \operatorname{cosec} \pi_M \\ y &= [\sin \delta_M \cos \delta - \\ &\quad - \cos \delta_M \sin \delta \cos (\alpha_M - \alpha)] \operatorname{cosec} \pi_M \\ &= [\sin (\delta_M - \delta) \cos^2 \tfrac{1}{2} (\alpha_M - \alpha) + \\ &\quad + \sin (\delta_M + \delta) \sin^2 \tfrac{1}{2} (\alpha_M - \alpha)] \operatorname{cosec} \pi_M \\ z &= [\sin \delta_M \sin \delta + \\ &\quad + \cos \delta_M \cos \delta \cos (\alpha_M - \alpha)] \operatorname{cosec} \pi_M \\ &= [\cos (\delta_M - \delta) \cos^2 \tfrac{1}{2} (\alpha_M - \alpha) - \\ &\quad - \cos (\delta_M + \delta) \sin^2 \tfrac{1}{2} (\alpha_M - \alpha)] \operatorname{cosec} \pi_M \end{aligned}\right\} . \qquad (12.36)$$

The second expressions in y and z are more convenient for the time of occultations because the first equations are the differences of two large but nearly equal numbers.

From equations (12.13) the Besselian coordinates of the observer are obtained as

$$\left.\begin{aligned} \xi &= \rho \cos \varphi' \sin h \\ \eta &= \rho \sin \varphi' \cos \delta - \rho \cos \varphi' \sin \delta \cos h \\ \zeta &= \rho \sin \varphi' \sin \delta + \rho \cos \varphi' \cos \delta \cos h \end{aligned}\right\} \qquad (12.37)$$

where h is the hour angle of the star and may be computed from one of the following expressions

$$h = \theta - \alpha = h^G + \lambda = \theta_G - \alpha + \lambda$$

where h^G is the Greenwich hour angle of the star.

The geocentric coordinates $\rho \sin\varphi'$ or $\rho \cos\varphi'$ may be calculated as before from

$$\rho \cos \varphi' = (N + H) \cos \varphi,$$
$$\rho \sin \varphi' = [N(1 - e^2) + H] \sin \varphi.$$

(2) The variations of the Besselian coordinates x', y', ξ', and η' may be computed similarly to the methods described above. The variations x', y' may be determined either from pretabulated values of x and y as described in section 12.31 or from analytical expressions. Equations (12.36) may be differentiated with respect to the time in order to obtain x' and y'. The work is rather elementary but the resulting formulas are long because α_M, δ_M, and π_M are all variables. However, relatively simple expressions may be derived for the time of conjunction because in this case $\alpha_M = \alpha$. For this instant

$$\left.\begin{aligned} x' &= \frac{\partial \alpha}{\partial h} \cos \delta_M \operatorname{cosec} \pi_M \,, \\ y' &= \frac{\partial \delta}{\partial h} \cos (\delta_M - \delta) \operatorname{cosec} \pi_M \end{aligned}\right. \qquad (12.38)$$

These quantities may be used with sufficient accuracy for the times of contact also. The values $\partial\alpha/\partial h$ and $\partial\delta/\partial h$ must be deduced from the hourly differences in the ephemerides. Further approximation is possible by assuming that $\cos(\delta_M - \delta) = 1$.

The variations ξ' and η' may be calculated from equations (12.25) by proper replacement of elements and by ignoring d':

$$\left.\begin{aligned} \xi' &= h'\rho \cos \varphi' \cos h = h'(-\eta \sin \delta + \zeta \cos \delta) \\ \eta' &= h'\xi \sin \delta \\ \zeta' &= -h'\xi \cos \delta \end{aligned}\right\} . \qquad (12.39)$$

(3) After the Besselian coordinates and their variations are at hand for the approximate time of the occultation T_0, the auxiliary quantities m, n, M and N may be calculated from equations (12.27) and (12.28).

(4) Having these values, the difference between the assumed and the correct times of contact τ is given by (12.30):

$$\begin{aligned} \sin \psi &= \frac{m}{k_M} \sin (M - N) \,, \\ \tau &= \frac{k_M}{n} \cos \psi - \frac{m}{n} \cos (M - N) \,, \end{aligned} \qquad (12.40)$$

where, as in the case of eclipses, $\cos\psi$ must be taken with a negative sign for immersions and with a positive sign for emersions. The time of contact will be

$$T = T_0 + \tau.$$

For greater accuracy, a second approximation should be performed where formula (12.30a) may be more convenient.

Instead of a second approximation, the following correction may be added to τ calculated in the first approximation:

$$d\tau = \frac{\tau^2}{n \cos \psi} [\bar{\eta} \cos (N + \psi) - \xi \sin (N + \psi)] \qquad (12.41)$$

where $\bar{\eta} = \rho \cos\varphi' \sin\delta \cos h$. In this case, the corrected time will be

$$T = T_0 + \tau + d\tau.$$

The determination of the approximate time of contact T_0 will be described in the next section.

(5) The computation of the position angle of the contact measured from the north point of the lunar limb is the same as in the case of solar eclipses, and it may be performed by equation (12.31). However, since there the angle was measured on the sun's limb, 180° should be added; also instead of a second approximation, the following correction may be used:

$$dP = \tau^2 (\bar{\eta} \sin N + \xi \cos N)/\cos\psi \qquad (12.42)$$

This way, formula (12.31) will be replaced by

$$P = N + \psi + dP + 180^\circ \qquad (12.43)$$

In all these computations the quadrant of the angle must be considered very carefully. It may be determined from the sign of $\sin\psi$ in the formula (12.40) and from the rule that $\cos\psi$ is negative for immersions and positive for emersions.

An example for an occultation prediction is given below.

EXAMPLE 12.3

Prediction of the Immersion of 24 Psc, January 20, 1961

Computations for the immersion of the star 24 Psc on January 20, 1961, at the 'Physics' station of The Ohio State University are included in this example. The geocentric coordinates of this station are

$$\rho \sin \varphi' = 0.6394072,$$
$$\rho \cos\varphi' = 0.7671179.$$

The star 24 Psc is not listed in 'The Apparent Places of Fundamental Stars,' so its apparent place was calculated from the 'General Catalogue' using the method shown in Examples 4.1 and 4.2. The results for $23^h 07^m$ UT are the following:

$$\alpha = 23^h 50^m 54\overset{s}{.}406,$$
$$\delta = 3^\circ 22' 23\overset{''}{.}53$$

The apparent place and parallax of the moon were taken from the 'American Ephemeris and Nautical Almanac' at the whole minute preceding and following the expected time of immersion by interpolating to the first, second, and third differences using Bessel's interpolation formula

$$f_p = f_0 + pd^1_{\frac{1}{2}} + B_2\,(d^2_0 + d^2_1) + B_3\,d^3_{\frac{1}{2}} + \dots$$

where p is the interpolating factor, B_2 and B_3 are Bessel's coefficients, and d^1, d^2, and d^3 are the first, second, and third differences. Since the moon's coordinates and parallax are tabulated in ephemeris time, a correction of 34 seconds was added to the universal time before interpolating. Computations are included in the table below which is followed by the computation for time of immersion of 24 Psc. $T_0 = 23^h07^m5^s.8$ UT was used as the time of the first approximation. The difference between the observed and the predicted times was $0^s.7$.

Computation of the Apparent Place and Parallax
of the Moon for Immersion of 24 Psc, January 20, 1961

Apparent Place of the Moon

ET	α_M	d^1	d^2	δ_M	d^1	d^2
22^h	$23^h48^m56^s.271$			$-3°05'57''.21$		
		134.112			673.42	
23^h	23 51 10.383		-0.232	-2 54 43 .79		0.04
		133.880			673.46	
24^h	23 53 24.262		-0.230	-2 43 30 .33		-0.03
		133.650			673.43	
01^h	23 55 37.913			-2 32 16 .96		

	α_M		δ_M	
	$23^h07^m34^s$	$23^h08^m34^s$	$23^h07^m34^s$	$23^h08^m34^s$
p	0.1261	0.1428	0.1261	0.1428
pd^1	$16^s.8837$	$19^s.1150$	$84''.9307$	$96''.1546$
$B_2(d^2_0+d^2_1)$	$0^s.0129$	$0^s.0143$	$0''.0003$	- 0 .0003
α_M and δ_M	$23^h51^m27^s.280$	$23^h51^m29^s.512$	$-2°53'18''.86$	$-2°53'07''.64$

Parallax of the Moon

ET	π_M	d^1	d^2
Jan 20.0	$60'01''.973$		
		-26.323	
Jan 20.5	59 35 .650		-1.527
		-27.850	
Jan 21.0	59 07 .800		-0.820
		-28.670	
Jan 21.5	58 39 .130		

	$23^h07^m34^s$	$23^h08^m34^s$
p	0. 9272	0. 9286
pd^1	$-25''.8218$	$-25''.8605$
$B_2(d^2_0+d^2_1)$	$0''.0399$	$0''.0399$
π_M	$59'9''.868$	$59'9''.829$

Immersion Prediction at "Physics" for
24 Psc, January 20, 1961

	23^h07^m UT	23^h08^m UT
α	357°43'36".09	357°43'36".09
δ	- 3°22'23".53	- 3°22'23".53
α_M	357°51'49".20	357°52'22".68
δ_M	- 2°53'18".86	- 2°53'07".64
π_M	59'09".868	59'09".829
sin δ	-0.05883949	-0.05883949
cos δ	0.99826746	0.99826746
sin δ_M	-0.05039375	-0.05033942
cos δ_M	0.99872943	0.99873217
$(\alpha_M - \alpha)$	8'13".11	8'46".59
sin $(\alpha_M - \alpha)$	0.00239066	0.00255298
cos $(\alpha_M - \alpha)$	0.99999714	0.99999675
sin π_M	0.01720940	0.01720911
x	0.13873941	0.14816241
y	0.49148262	0.49465039
h	26°05'35".43	26°20'37".89
sin h	0.43983219	0.44375730
cos h	0.89808000	0.89614701
ρ sin φ'	0.6394072	0.6394072
ρ cos φ'	0.7671179	0.7671179
ξ	0.33740297	0.34041399
η	0.67883572	0.67874851
$x - \xi$	-0.19866356	-0.19225158
$y - \eta$	-0.18735310	-0.18409812
	1st Approx.	2nd Approx.
UT	$23^h07^m05^s.80$	$23^h07^m05^s.00$
$(x' - \xi')$	0.384719	0.384719
$(y' - \eta')$	0.195299	0.195299
$x - \xi$	-0.19814383	-0.19812960
$y - \eta$	-0.18703846	-0.18711431
tan M	1.05884014	1.05886930
M	226°38'13".29	226°38'16".12
sin M	-0.72701852	-0.72702794
m	0.27240548	0.27251991
tan N	1.97015471	1.97015471
N	63°05'19".70	63°05'19".70
sin N	0.89170911	0.89170911
n	0.43144110	0.43144110

Immersion Prediction at "Physics" for 24 Psc, January 20, 1961 (continued)		
	1st Approx.	2nd Approx.
M - N	163°32'53".59	163°32'56".40
sin (M - N)	0.28320832	0.28318525
cos (M - N)	-0.95905842	-0.95906234
sin k_M	0.00468961	0.00468961
k''_M	967".305	967".305
ρ'' sin π_M	3549".6935	3549".6935
$k_M = k''_M/\rho''$ sin π_M	0.2725038	0.2725038
sin ψ	0.28310614	0.28320199
ψ	163°33'15".56	163°32'54".95
cos ψ	-0.95908858	-0.95906029
τ	$-0^h.000238$	$0^h.000037$
τ	$-0^s.86$	$0^s.13$
contact	$23^h07^m05^s.00$	$23^h07^m05^s.13$
P		46°38'14".65

12.42 Method of the 'American Ephemeris and Nautical Almanac' and of the 'Astronomical Ephemeris'

Although since 1960 the 'Ephemeris' does not contain any information about the circumstances of occultations, the occultation program of H. M. Nautical Almanac Office continues unchanged as a commitment. In this program, predictions are made for about 90 central stations together with latitude and longitude coefficients which enable observers in the vicinity of these stations to derive times for their own positions. Arrangements for the publication of these predictions in several periodicals have been made. The complete list of the central stations and the names of the periodicals in which the predictions are published may be found in [Nautical Almanac Offices, 1961, Section 10B]. Computer printed copies of predictions for any of the central stations, or of the elements of occultations, may be obtained upon request from H.M. Nautical Almanac Office (Royal Greenwich Observatory, Herstmonceux Castle, Hailsham, Sussex, England). The annual predictions for stations in North America are published in the November or December issues of 'Sky and Telescope' (Sky Publishing Company, Harvard College Observatory, Cambridge, Massachusetts).

The stars for which these predictions are made are those of magnitude 7.5 and brighter contained in the 'Catalog of 3539 Zodiacal Stars for the Equinox 1950.0' published in volume X/II of the 'Astronomical Papers prepared for the use of The American Ephemeris and Nautical Almanac' in 1940. At the bright limb, disappearances are given for stars of magnitude 4.5, and reappearances are given for stars of magnitude

3.5 or brighter. On the dark limb reappearances are given for stars of magnitude 6.5 and brighter. For further details of this prediction program, such as the preparation of the list of conjunctions and the calculation of the data published, the reader is referred to [Nautical Almanac Offices, 1961, Section 10B].

12.421 Tables of 'Sky and Telescope.' The tables published yearly in this magazine provide approximate occultation predictions for 15 central stations in North America shown in Fig. 12.11.

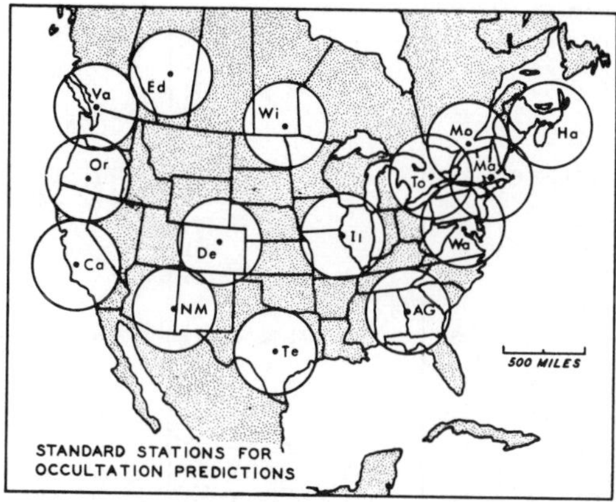

Fig. 12.11 Central stations for occultation predictions in North America

A sample of the tables is shown in Table 12.1. The columns give the date for each occultation, Zodiacal Catalogue number of the star, its magnitude, the phenomenon (1 = immersion, 2 = emersion), and the age of the moon in days. Under each station are given the universal time of the event to $0^m.1$, a and b latitude and longitude coefficients in minutes of time, and the position angle P to 1°. The a and b factors are changes in central-station-predicted times per degree of longitude and latitude respectively (see end of section 12.511). They enable fairly accurate computation of time for each local station (φ, λ) within 200–300 miles of a central station (φ', λ'). The local station time of approximate contact is

$$T_0 = T_0' - a(\lambda - \lambda') + b(\varphi - \varphi'),$$

with due regard being given to the arithmetic signs. The latitudes and longitudes are to be expressed in degrees. T_0' is the approximate contact time at the central station as given in the tables. In Fig. 12.11, the central stations are at the centers of circles of 250-mile radius, and in these areas the a and b factors should be quite useful. For po-

TABLE 12.1 'Sky and Telescope' Occultation Predictions for 1968 (Sample)

Date	Z.C. No.	Mag.	Ph.	Age of Moon	Wa WASHINGTON, D.C. W. 77°·065, N. 38°·920 U.T.	a	b	P	AG ALABAMA—GEORGIA W. 85°·000, N. 33°·000 U.T.	a	b	P
				d	h m	m	m	°	h m	m	m	°
Nov. 11	1211	6·2	2	20·5	10 03·2	·	·	233		N		
15	1644	4·1	1	24·6	9 55·9	−1·6	+1·3	90	9 42·4	−1·0	−0·1	119
15	1644	4·1	2	24·6	10 56·3	−0·7	−2·2	344	10 56·4	−1·4	−0·9	311
25	3037	7·3	1	4·7		A			1 21·1	−0·1	+1·0	27
29	26	7·0	1	8·9		N			6 31·1	−0·1	+3·1	2
30	132	6·9	1	9·9	5 45·0	−0·7	+0·3	48	5 37·3	−1·2	−0·1	65
Dec. 2	371	6·4	1	11·9	7 11·4	−1·0	+0·9	37	6 59·5	−1·4	+0·1	61
8	1169	5·4	2	18·0	8 26·9	−3·2	+1·9	238		N		
11	1504	5·7	2	21·1	10 08·2	·	·	9	10 22·8	−1·3	−2·4	328
12	1609	4·7	2	22·1		S				S		
13	1712	3·8	1	23·1	10 01·9	·	·	59	9 32·7	−1·7	+0·2	105
13	1712	3·8	2	23·1	10 25·8	·	·	23	10 44·2	−0·9	−2·0	336
15	1925	1·2	1	25·1	8 48·6	·	·	199		N		
15	1925	1·2	2	25·1	9 06·0	·	·	231		N		
22	3116	6·7	1	3·2	22 46·5	−1·2	−0·4	69		S		
25	3391	6·8	1	5·2	0 46·8	−0·2	+1·7	13	0 34·0	−0·5	+1·8	18
25	3394	7·4	1	5·3	1 39·1	−0·1	+1·6	15	1 28·5	−0·4	+1·4	24
28	209	7·2	1	8·3	2 19·2	−1·9	−0·5	78	2 04·2	−2·7	−0·5	87
31	545	4·2	1	11·3	1 02·9	−0·9	+3·0	30	0 38·8	−0·7	+2·9	32
31	550	6·8	1	11·3	1 24·8	−2·3	+0·6	87	1 01·8	−2·4	+0·8	90
31	551	7·1	1	11·3	1 32·1	−1·9	+1·5	65	1 07·9	−1·8	+1·6	68
31	552	3·0	1	11·3	2 00·9	−1·1	+3·5	24	1 32·9	−1·0	+3·1	30
31	552	3·0	2	11·3	3 08·8	−2·7	−1·8	289	2 52·0	−3·0	−0·3	275
31	559	6·6	1	11·4	2 44·3	−3·1	−2·4	117		G		
31	560	3·8	1	11·4	2 50·4	−2·3	+0·5	74	2 27·7	−2·7	+0·5	84
31	561	5·2	1	11·4	2 57·8	−2·0	+1·3	57	2 32·8	−2·3	+1·3	67
31	562	6·6	1	11·4		N			3 17·9	·	·	13
31	567	6·8	1	11·4		N				N		
31	570	6·8	1	11·4	4 20·9	·	·	138		N		
31	587	6·4	1	11·5	8 02·8	−0·3	−0·8	73	8 07·1	−0·3	−1·5	101

sitions outside these circles the factors should be interpolated. The letters inserted in the places of omitted predictions have the following meanings: G, grazing occultation; A, below or too near the horizon; N, no occultation; S, sunlight interferes.

12.422 Computer-Printed Tables of H. M. Nautical Almanac Office. These tables, obtainable upon request, contain the following information (in their notation): Zodiacal Catalogue number of the star occulted; magnitude; apparent right ascension and declination; the date and time of geocentric conjunction in right ascension T_0; the common Greenwich geocentric hour angle of the star and of the moon, H (positive to the west); the coordinate y of the shadow axis at the geocentric conjunction, Y; the variations of x and y in one hour mean time, x' and y'. It should be noted that at the time of geocentric conjunction in right ascension, the coordinate x of the shadow axis is zero.

The approximate time of contact at any given place by means of these tables may be calculated as follows:

(1) Compute the approximate instant of apparent conjunction of moon and star as seen from the place. This may be deduced from the time of the geocentric conjunction by the application of an approximate correction τ'. It will have the same sign as the local hour angle of geocentric conjunction, $h_0 = H + \lambda$. This correction may be computed from the formula,

$$\tau' = \frac{\xi_0}{x' - \xi'}$$

where $\xi_0 = \rho \cos\varphi' \sin h_0$, ξ' should be computed from (12.39), and x' is to be taken from the table.

The local time of conjunction will be $T_0 + \tau'$.

(2) Using the local time of conjunction as approximate time of the contact, compute the Besselian coordinates of the observer and of the shadow axis and their variations for this instant as follows:

$$\begin{aligned}
\xi &= \rho \cos\varphi' \sin(h_0 + \tau'), \\
\eta &= \rho \sin\varphi' \cos\delta - \rho \cos\varphi' \sin\delta \cos(h_0 + \tau'), \\
\xi' &= h' \rho \cos\varphi' \cos(h_0 + \tau'), \\
\eta' &= h' \xi \sin\delta, \\
x &= x'\tau', \\
y &= Y + y'\tau'.
\end{aligned}$$

In the first four equations, τ' should be used in sidereal units, and in the last two τ' should be used in mean time intervals. x' and y' are taken from the table.

(3) Using these quantities, equations (12.27) - (12.28) will give the auxiliaries, m, n, M, and N; and the angle ψ and the correction τ may be calculated from expression (12.40). The time of contact will be

$$T = T_0 + \tau' + \tau + d\tau$$

where $d\tau$ is given by equation (12.41).

(4) The position angles may be computed as described before by means of equations (12.43), (12.32), and (12.33).

12.43 Limits of an Occultation

Up to this point, it has been assumed that the occultation will take place at the given station. The limits where an occultation may be observed could be determined from occultation maps similar to those prepared for solar eclipses in the 'American Ephemeris and Nautical Almanac.' However, because of the large number of occultations, it is impractical to design and publish a map for each occultation, and the complete coverage provided by the predictions described in the previous section now makes them unnecessary.

Another method of determining the limits of an occultation is the use of an occultation machine. Such a machine is employed by H.M. Nautical Almanac Office for the purpose of determining for each selected conjunction those stations for which detailed investigation of possible occultations is to be made by means of electronic computers. In the machine the earth is represented by a globe and the moon-star shadow system by a cylindrical beam of light of correct radius. The initial setting of the machine is for the time of conjunction by setting the Besselian elements on appropriate scales. The globe and the shadow are moved by a driving mechanism in such a way that the machine will continue to present a picture of the actual circumstances. The path of the shadow thus is visualized, and the time intervals from conjunction to disappearances and reappearances and the corresponding position angles may be read off for all stations in the path. A description of the machine is given in the supplement to the 'Nautical Almanac' for 1938. The method of reduction was published in [Comrie, 1937].

A third method to determine the limits of an occultation together with approximate times of the contacts is the graphical method. The outline of the projection of the moon's disc on the fundamental plane is represented by a circle of radius k_M and the projections of the positions of the station are plotted for times before and after conjunction. If a point inside the circle is joined to one outside by a straight line, the point at which this line intersects the circle gives a good approximation of the projection of the station at the times of contacts. The preliminary time is obtained by assuming that the projection of the station moves with uniform speed along the straight line. For an illustration see [Lambert, 1949].

Of the various methods that can be devised for obtaining the limits of an occultation observation, the method of determining the limiting parallels is described here [Chauvenet, 1863]. This method could be used in combination with the computer-printed tables mentioned above. Compute the following quantities:

$$\left.\begin{aligned} \cos\gamma_1 &= Y\sin N \pm k_M && (\gamma < 180^\circ)\\ \cos\gamma_2 &= Y\sin N \mp k_M &&\\ \sin\beta &= \sin N\cos\delta && (\beta < 90^\circ)\\ \varphi_1 &= \beta \pm \gamma_1 &&\\ \sin\varphi_2 &= \sin(N \mp \gamma_2)\cos\delta && (N < 90^\circ) \end{aligned}\right\}$$

The upper signs should be used if δ is a northern declination; the lower should be used if it is a southern declination. When the declination is north φ_1 will be the northern limit and φ_2 the southern. The reverse is true for southern declinations. The quantities Y and δ may be computed or taken from the tables, and N may be computed from

$$\tan N = \frac{x'}{y'}$$

where x' and y' are given in the tables or may be computed in the usual way, always with positive signs. N is supposed to be less than 90°. When $\cos\gamma_1$ is imaginary, $\varphi_1 = \pm 90°$; when $\cos\gamma_2$ is imaginary, $\varphi_2 = \delta \mp 90°$. If $\varphi_1 = \beta \pm \gamma_1$ exceeds 90°, then φ_1 is either $180° - (\beta \pm \gamma_1)$ or $-180° - (\beta \pm \gamma_1)$.

The limiting parallels for occulted stars were given in the tables of the 'Ephemeris' before 1960.

If the observer is situated within the limiting parallels determined above, he will see the occultation if the quantity $H+\lambda$ taken without regard to sign is less than the semidiurnal arc of the star by at least one hour and the sun is not much more than an hour above the horizon at the local time $T_0 + \lambda$. For more details of the determination of the limits at a given occultation, the reader is referred to [Chauvenet, 1863, Chapter X] or to [Nautical Almanac Offices, 1961, Section 10B].

12.44 Prediction of the Isolimb of an Occultation

Occultation observations for geodetic purposes sometimes demand the prediction of a so-called equal limb line, or isolimb, on the surface of the earth. Observers on this line may observe a given occultation at the same selenographic latitude. For different observers to observe an occultation at the same point on the limb, i.e., at the same selenographic latitude and longitude, is impossible because of the libration of the moon.

As will be seen later, this method has great importance in eliminating the difficulties arising from the inaccurate knowledge of the profile of the moon.

The method was developed by H. Hirose of the Toyko Astronomical Observatory, and the calculations are as follows [Hirose, 1953]:

(1) Predict the occultation for a given base station determining the contact time T_1 and the angle Q by means of equation (12.23) substituting $k_M = L = \ell - i\zeta$,

$$\begin{aligned} k_M \sin Q &= x_1 - \xi_1, \\ k_M \cos Q &= y_1 - \eta_1. \end{aligned} \qquad (12.44)$$

The Besselian coordinates x_1, y_1, ξ_1, and η_1 should be computed for the time T_1.

(2) The Besselian coordinates of a second observer, ξ_2 and η_2, on the isolimb may be computed from the same equation keeping Q constant:

$$\begin{aligned} \xi_2 &= x_2 - k_M \sin Q, \\ \eta_2 &= y_2 - k_M \cos Q, \end{aligned} \qquad (12.45)$$

where x_2 and y_2 should be calculated for an assigned time T_2.

The evaluation of the coordinate ζ_2 requires the value of the distance ρ from the center of the earth. This distance is not known a priori because it is a function of the latitude of the point. The problem of determining ζ_2 could be solved by assuming that the base and the second point are situated on some geometrical surface, or by successive approximations, or by a direct procedure devised by Bessel given in [Nautical Almanac Offices, 1961, Section 9B].

The coordinate ζ_2 is calculated from the assumption that the base point and the second point are situated on the surface of an ellipsoid of the same eccentricity as that of the reference ellipsoid. The major and minor semiaxes of this ellipsoid are approximately given by

$$\bar{a} = a + \frac{a}{N} H,$$

$$\bar{b} = \bar{a}\sqrt{1 - e^2},$$

where a is the major semiaxis of the reference ellipsoid, H is the elevation of the second point above the ellipsoid, and N is its radius of curvature in the prime vertical plane.

The equation of this ellipsoid is

$$\frac{x^2 + q^2}{\bar{a}^2} + \frac{p^2}{\bar{b}^2} = 1 \tag{12.46}$$

where the coordinate axes x and q are in the equatorial plane (x coincides with the Besselian coordinate axis x), the axis p is the rotation axis of the earth. This system, which in fact is identical to the one used in section 12.222 (see Fig. 12.4), is obtained by rotating the Besselian coordinate system about its axis x by the declination of the shadow axis δ. The transformation equations between these two systems are

$$\left.\begin{aligned} x &= \xi \\ q &= -\eta \sin\delta + \zeta \cos\delta \\ p &= -\eta \cos\delta - \zeta \sin\delta \end{aligned}\right\}. \tag{12.47}$$

Substituting (12.47) in (12.46), the following quadratic equation for ζ is obtained

$$A\zeta^2 + 2B\zeta + C = 0 \tag{12.48}$$

where

$$\begin{aligned} A &= (1 + e^2 \sin^2\delta)\,\bar{a}^{-2}, \\ B &= \eta e^2 \sin\delta \cos\delta\,\bar{a}^{-2}, \\ C &= [\xi^2 + \eta^2 (1 + e^2\cos^2\delta)]\,\bar{a}^{-2} - 1. \end{aligned}$$

Using the quantities ζ_2 and η_2 from (12.45), equation (12.48) will give the coordinate ζ_2. The positive root should be used.

(3) The next step is to transform these Besselian coordinates into ordinary geodetic coordinates. This may be done by means of the following equations:

$$\begin{aligned} \tan\lambda &= \frac{v}{u}, \\ \tan\varphi &= \frac{w + Ne^2 \sin\varphi}{u\cos\lambda + v\sin\lambda}, \end{aligned} \tag{12.49}$$

where u, v, and w are the conventional geocentric rectangular coordinates of the observer (see equations (12.61)), which may also be calculated from

$$\left.\begin{aligned} u &= \xi \sin h^G - \cos h^G (\eta \sin \delta - \zeta \cos \delta) \\ v &= \xi \cos h^G + \sin h^G (\eta \sin \delta - \zeta \cos \delta) \\ w &= \eta \cos \delta + \zeta \sin \delta \end{aligned}\right\}, \qquad (12.50)$$

where h^G and δ are the Greenwich hour angle and apparent declination of the shadow axis, i. e., of the star, respectively.

Other points on the isolimb may be determined similarly. Hirose suggests that a second approximation may be useful in which the variations of the angle Q should be considered. These variations are due to the libration and to the topocentric variation of the quantity P - Q. For further discussion see the end of section 12.72.

12.5 Geodetic Applications of Occultations

The geodetic applications of occultations and solar eclipses are the following:

(1) Establishing geodetic ties between remote points, i.e., determining the geodetic coordinates of the points in the same geodetic system.

(2) Determining the equatorial radius of the earth.

(3) Determining the longitude of the station.

(4) Determining the flattening of the earth.

The latter two applications are out of date, therefore they will not be discussed here. The longitude of the station had been determined by means of eclipses before the radio or the telegraph was invented. The flattening of the earth may be determined more accurately by gravimetric methods or from the observations of artificial satellites than eclipse observations.

The general method of geodetic applications is the following: Suppose that the *prediction* of the contact times had been made on the basis of the assumed geodetic position of the observer (φ, λ, ρ), of the motion and parameters of the moon (α_M, δ_M, k_M, π_M), and of the sun (α_S, δ_S, k_S, π_S) as given in the tables of the ephemerides. The assumed geodetic positions will depend on the parameters of the reference ellipsoid used (a, f) and on the existing geodetic datum. In case of occultations the data of the sun will be replaced by those of the occulted star.

After observing the eclipse, generally it is found that even if the observation is errorless, the observed time of contact differs from the predicted time by $\delta\tau$. The difference is due to errors in the assumed data. In order to get the correct data, of which the geodetic positions of the observers situated at remote stations interests us most, corrections must be computed from equations like

$$\delta\tau = \bar{a}\delta\lambda + \bar{b}\delta\varphi + \bar{c}\delta\alpha_M + \bar{d}\delta\delta_M + \bar{e}\delta\pi_M + \bar{f}\delta k_M + \bar{g}\delta\alpha_S + \\ + \bar{h}\delta\delta_S + \bar{i}\delta k_S + \bar{j}\delta\pi_S + \bar{k}\delta\rho$$

where the left side is the difference between the observed and predicted times; on the right side $\delta\lambda$, $\delta\varphi$, etc., are the unknown corrections to the assumed data, and the coefficients $\bar{a}$ through $\bar{k}$ may be calculated from the observed and assumed data as will be shown later. Each observation provides an equation of this kind. From a sufficient number of equations, the unknowns may be calculated by the methods of adjustment computation. As will be seen later, some of the unknowns may be eliminated by various methods. In case of an occultation the corrections to the data of the sun $\delta\alpha_S$, $\delta\delta_S$, δk_S, and $\delta\pi_S$ will be replaced by corrections to the right ascension and declination of the star, which are negligible if the apparent position of the star has been properly computed from the star catalogue. In this case the equation above will take the form

$$\delta\tau = a\delta\lambda + b\delta\varphi + c\delta\alpha_M + d\delta\delta_M + e\delta\pi_M + f\delta k_M + g\delta\rho \ .$$

The corrections to the observer's position $\delta\lambda$ and $\delta\varphi$ represent either errors due to a possible difference in geodetic datums, or, since time is the only observation, they are the negative deflection of the vertical components when astronomic coordinates are used as assumed coordinates. The value $\delta\rho$ is the correction to the altitude of the observer above the ellipsoid; therefore, even if the correct height above sea level, corrected for refraction, has been used, it will be present and will represent the geoid-ellipsoid separation, the geoid undulation. The quantities $\delta\alpha_M$, $\delta\delta_M$, δk_M, and $\delta\pi_M$ are corrections to the tabular values of the moon's parameters. These equations may be set up in many different forms which could be necessary when not the time but, for instance, the position angle is observed, as is done sometimes in the case of solar eclipses or when the goal is not the determination of correct geodetic positions but, for instance, the equatorial semidiameter of the earth.

The calculations of the coefficients in the observation equations and the solution of the normal equations will be discussed next. They will be shown for occultations only and not for solar eclipses because of their lost geodetic significance. The method generally used in connection with eclipses is described in [Berroth and Hofmann, 1960; Mueller, 1964].

12.51 Position Determination

12.511 Effect of Errors in the Assumed Data on the Predicted Time of Contact. Substituting $k_M = L$ in equations (12.26), the equations corresponding to occultations are obtained,

$$\begin{aligned} k_M \sin Q &= x - \xi + (x' - \xi')\,\tau, \\ k_M \cos Q &= y - \eta + (y' - \eta')\,\tau. \end{aligned} \qquad (12.51)$$

Suppose k_M, Q, x, y, ξ, and η are all affected by errors δk_M, δQ, δx,

δy, $\delta\xi$, and $\delta\eta$, thus causing an error $\delta\tau$ in time. Assume furthermore that τ is so small (after second approximation) that it will not be necessary to include corrections for x', y', ξ', and η'. The assumed smallness is chiefly necessary in connection with the Besselian coordinates of the observer and their variations. The Besselian coordinates of the shadow axis x and y and their variations x' and y' are well determined for a certain period; in any case, x' and y' vary very little during an occultation. Putting all the increments into equations (12.51), one gets

$$\begin{aligned} (k_M + \delta k_M) \sin (Q + \delta Q) &= x + \delta x - (\xi + \delta\xi) + (x' - \xi')(\tau + \delta\tau), \\ (k_M + \delta k_M) \cos (Q + \delta Q) &= y + \delta y - (\eta + \delta\eta) + (y' - \eta')(\tau + \delta\tau). \end{aligned} \tag{12.52}$$

After developing and neglecting small quantities of the second order and subtracting (12.51) from (12.52), the relations yield

$$\begin{aligned} \delta k_M \sin Q + k_M \cos Q\, \delta Q &= \delta x - \delta\xi + (x' - \xi')\delta\tau, \\ \delta k_M \cos Q - k_M \sin Q\, \delta Q &= \delta y - \delta\eta + (y' - \eta')\delta\tau. \end{aligned}$$

The process of multiplying the first equation by $\sin Q$, the second by $\cos Q$, then adding them together, eliminates the correction δQ. Solving the result for $\delta\tau$, substituting $x' - \xi' = n \sin N$, $y' - \eta' = n \cos N$, and noting that $Q = N + \psi$, results in

$$\delta\tau = -\frac{\delta x - \delta\xi}{n \cos\psi} \sin (N + \psi) - \frac{\delta y - \delta\eta}{n \cos\psi} \cos (N + \psi) + \frac{\delta k_M}{n \cos\psi}. \tag{12.53}$$

As may be seen from equations (12.36), the quantities δx and δy will depend on errors in the right ascension, declination, and parallax of the moon, $\delta\alpha_M$, $\delta\delta_M$, and $\delta\pi_M$. The right ascension and declination of the star, α and δ, are considered to be accurately determined. The total derivatives of x and y therefore will be

$$\begin{aligned} \delta x &= \frac{\partial x}{\partial \alpha_M} \delta\alpha_M + \frac{\partial x}{\partial \delta_M} \delta\delta_M + \frac{\partial x}{\partial \pi_M} \delta\pi_M , \\ \delta y &= \frac{\partial y}{\partial \alpha_M} \delta\alpha_M + \frac{\partial y}{\partial \delta_M} \delta\delta_M + \frac{\partial y}{\partial \pi_M} \delta\pi_M , \end{aligned} \tag{12.54}$$

where the coefficients from differentiating equations (12.36) are

$$\begin{aligned} \frac{\partial x}{\partial \alpha_M} &= \cos \delta_M \cos (\alpha_M - \alpha) \operatorname{cosec} \pi_M, \\ \frac{\partial x}{\partial \delta_M} &= -\sin \delta_M \sin (\alpha_M - \alpha) \operatorname{cosec} \pi_M, \\ \frac{\partial x}{\partial \pi_M} &= -\cos \delta_M \sin (\alpha_M - \alpha) \frac{\cos \pi_M}{\sin^2 \pi_M} = -x \cot \pi_M, \\ \frac{\partial y}{\partial \alpha_M} &= \cos \delta_M \sin \delta \sin (\alpha_M - \alpha) \operatorname{cosec} \pi_M, \end{aligned}$$

$$\frac{\partial y}{\partial \delta_M} = [\cos \delta_M \cos \delta + \sin \delta_M \sin \delta \cos (\alpha_M - \alpha)] \operatorname{cosec} \pi_M,$$

$$\frac{\partial y}{\partial \pi_M} = -[\sin \delta_M \cos \delta - \cos \delta_M \sin \delta \cos (\alpha_M - \alpha)] \frac{\cos \pi_M}{\sin^2 \pi_M} = -y \cot \pi_M.$$

Because of the smallness of the quantities $(\alpha_M - \alpha)$ and $(\delta_M - \delta)$ at the time of occultation and the moderate requirements of accuracy, $\cos(\alpha_M - \alpha) = 1$, $\sin(\alpha_M - \alpha) = 0$, and $\delta_M = \delta$ may be substituted in the expressions above. With these simplifications, the increments in x and y from (12.54) will be

$$\begin{aligned} \delta x &= \delta\alpha_M \cos \delta_M \operatorname{cosec} \pi_M - \delta\pi_M x \cot \pi_M, \\ \delta y &= \delta\delta_M \operatorname{cosec} \pi_M - \delta\pi_M y \cot \pi_M. \end{aligned} \tag{12.55}$$

Similar treatment of equations (12.37) is necessary to obtain the increments in ξ and η. The variables there are φ', ρ, and $h = h^G + \lambda$, i.e., λ. It should be noted again that ρ includes the altitude above the ellipsoid and the fictitious height ΔH due to the refraction, as given in section 12.34. If the refraction correction and the height above sea level are included, then the increment in ρ will be the geoid undulation. The total derivative of ξ and η therefore will be

$$\begin{aligned} \delta\xi &= \frac{\partial \xi}{\partial \lambda} \delta\lambda + \frac{\partial \xi}{\partial \varphi'} \delta\varphi' + \frac{\partial \xi}{\partial \rho} \delta\rho, \\ \delta\eta &= \frac{\partial \eta}{\partial \lambda} \delta\lambda + \frac{\partial \eta}{\partial \varphi'} \delta\varphi' + \frac{\partial \eta}{\partial \rho} \delta\rho, \end{aligned} \tag{12.56}$$

where the coefficients from differentiating equations (12.37) are

$$\begin{aligned} \frac{\partial \xi}{\partial \lambda} &= \rho \cos \varphi' \cos h &&= p, \\ \frac{\partial \xi}{\partial \varphi'} &= -\rho \sin \varphi' \sin h &&= q, \\ \frac{\partial \xi}{\partial \rho} &= \cos \varphi' \sin h &&= \frac{\xi}{\rho}, \\ \frac{\partial \eta}{\partial \lambda} &= \rho \cos \varphi' \sin \delta \sin h &&= r, \\ \frac{\partial \eta}{\partial \varphi'} &= \rho \cos \varphi' \cos \delta + \rho \sin \varphi' \sin \delta \cos h &&= s, \\ \frac{\partial \eta}{\partial \rho} &= &&= \frac{\eta}{\rho}. \end{aligned}$$

Neglecting the difference between geodetic and geocentric latitudes in the correction $\delta\varphi'$ yields

$$\begin{aligned} \delta\xi &= p\delta\lambda + q\delta\varphi + \frac{\xi}{\rho} \delta\rho, \\ \delta\eta &= r\delta\lambda + s\delta\varphi + \frac{\eta}{\rho} \delta\rho. \end{aligned} \tag{12.57}$$

Substituting (12.55) without the simplifications and (12.57) into (12.53), and collecting the coefficients, the following equation is obtained:

$$\delta\tau = a\delta\lambda + b\delta\varphi + c\delta\alpha_M + d\delta\delta_M + e\delta\pi_M + f\delta k_M + g\delta\rho \qquad (12.58)$$

where

$$a = \frac{1.0472}{n\cos\psi}\left[p\sin(N+\psi) + r\cos(N+\psi)\right],$$

$$b = \frac{1.0472}{n\cos\psi}\left[q\sin(N+\psi) + s\cos(N+\psi)\right],$$

$$c = -\frac{0.017453}{n\cos\psi}\left[\frac{\sin(N+\psi)\cos\delta_M}{\sin\pi_M} + \frac{\cos\delta_M\sin\delta\sin(\alpha_M-\alpha)}{\sin\pi_M}\cos(N+\psi)\right],$$

$$d = -\frac{0.017453}{n\cos\psi}\left[\frac{\cos(N+\psi)}{\sin\pi_M} + \frac{\sin\delta_M\sin(\alpha_M-\alpha)}{\sin\pi_M}\sin(N+\psi)\right],$$

$$e = \frac{0.017453}{n\cos\psi}\,\frac{x\sin(N+\psi) + y\cos(N+\psi)}{\tan\pi_M},$$

$$f = \frac{0.36}{n\cos\psi},$$

$$g = \frac{0.5644}{n\cos\psi}\left[\xi\sin(N+\psi) + \eta\cos(N+\psi)\right].$$

The constants in these coefficients have been selected in such a way that if $\delta\lambda$ and $\delta\varphi$ are substituted in minutes of arc, $\delta\alpha_M$, $\delta\delta_M$, and $\delta\pi_M$ in seconds of arc, $\delta\rho$ in kilometers, and δk_M in units of 10^{-4}, then $\delta\tau$ will be in seconds of time. This device brings the first significant figure of the coefficients in the units place or in the first decimal place. In these equations n is based on the mean solar hour as unit, e.g., ξ' and η' must be computed accordingly. The unit of distance is the earth's equatorial radius.

In the coefficients c and d, the second terms in the brackets should be used only when extra accuracy is needed.

Using the first two and the last terms in (12.58) only, the equation may be used to extrapolate an occultation prediction in the vicinity of the base point. In this case, the coefficients a and b are identical to the longitude and latitude coefficients described in section 12.421.

12.512 On the Possibility of Determining the Errors in the Assumed Data. If enough equations of the type (12.58), one per station per contact, could be obtained, they might be solved to determine the various unknown errors in the assumed data. Suppose there are o occultations, each on a different night and each observed at s stations. Then each station will have three unknowns peculiar to it, $\delta\varphi$, $\delta\lambda$, and $\delta\rho$, or 3s unknowns for all the stations. Two unknowns, being corrections to the parameters of the moon, $\delta\pi_M$ and δk_M, will remain the same for all occultations anywhere. Each occultation will have the two unknowns $\delta\alpha_M$ and $\delta\delta_M$ peculiar to itself. The number of unknowns is thus

$$3s + 2o + 2.$$

Against this, there are 2so observations of contact on both limbs. With a large number of observations at many stations, it becomes possible to solve for the unknowns by the method of least squares.

This, however, is a simplified case for several reasons. Generally it will not be possible to observe both contacts every time. On the other hand, if the moon occults several stars during the same night, the corrections $\delta\alpha_M$ and $\delta\delta_M$ may be assumed the same for all occultations that night. Furthermore, $\delta\alpha_M$ and $\delta\delta_M$ may be determined separately from observing the occultations from stations at known positions or from meridian observations at observatories or by neglecting them by taking the position of the moon from refined ephemerides (see section 12.73). If these two unknowns are determined from separate sources, the rest of the unknowns for one station may be determined by continuing the observations only there.

To establish a geodetic tie, a minimum of two stations is needed. In this case, $2o + 8 = 4o$, i.e., a minimum of four occultations or eight contacts, must be observed at each station. If the corrections $\delta\alpha_M$ and $\delta\delta_M$ are determined from a different source, then the observation of two occultations or four contacts at each station would suffice.

The best program for tying several remote stations together would be the systematic observation of all available occultations at all the stations over a long period and the solution of the resulting equations by the method of least squares.

Solving the equations, the correct geodetic position of the observer may be calculated by adding the corrections to the assumed data:

$$\varphi + \delta\varphi,$$
$$\lambda + \delta\lambda,$$
$$H + \delta\rho.$$

The corrections $\delta\varphi$, $\delta\lambda$, and $\delta\rho$ include possible effects due to the difference in positions of the centers of the reference ellipsoids to which the stations are referenced, errors in the orientation and scale in the triangulation system, the components of the deflection of the vertical, etc.

An example of the computations for a single station is given on the next page.

EXAMPLE 12.4

Position of Williston Observatory, Massachusetts, from Occultations

Computer: J. H. St. Clair

Each year from 1925 through 1957, occultation observations made at Williston Observatory at Mount Holyoke College, Massachusetts, were published in 'The Astronomical Journal.' These observations were made under the direction of Professor Alice Farnsworth and were all immersions of stars at the dark limb of the moon.

Six observations made on the night of December 27, 1952, and three observations made March 3, 1955, which were published in 1954 and 1957, are used as observational data in a practical application of the preceding formulas. All observations were made with an 8-inch equatorial refractor with astronomical coordinates of

$$\Phi = +42^\circ 15' 18''.2,$$
$$\Lambda = -\ 4^h 50^m 18^s.99,$$

at an elevation of 76 m. Timing of contacts was accomplished with a chronograph and a short-wave radio tuned to WWV [Farnsworth, 1953 and 1956].

The nine observations listed in the table below produce nine observation equations. For the first six observations, all on the same night, the

Immersions Observed at Williston Observatory, Massachusetts

Star	Mag.	Date	Predicted Universal Time	Observed Universal Time
16 Tauri	5.4	27 Dec. 52	$23^h 40^m 11^s.83$	$23^h 40^m 18^s.8$
q Tauri	4.4	27 Dec. 52	23 45 47.75	23 45 48.4
21 Tauri	5.8	28 Dec. 52	0 10 52.04	0 10 54.0
22 Tauri	6.5	28 Dec. 52	0 15 49.13	0 15 51.2
20 Tauri	4.0	28 Dec. 52	0 19 25.62	0 19 29.6
BD+24°.571	6.8	28 Dec. 52	1 56 11.66	1 56 13.4
3 Gem.	5.8	3 Mar.55	1 22 47.71	1 22 46.6
4 Gem.	6.7	3 Mar.55	1 55 45.09	1 55 46.8
6 Gem.	6.3	3 Mar.55	2 53 59.23	2 53 56.8

errors in the tabulated coordinates of the moon can be considered as constants so the unknowns in the six equations will be $\delta\varphi$, $\delta\lambda$, $\delta\rho$, $\delta\alpha_M^1$, $\delta\delta_M^1$, $\delta\pi_M$, δk_M. The last three observations add two more unknowns, $\delta\alpha_M^2$

and $\delta\delta_M^2$. There are then nine equations and nine unknowns, and a unique solution is possible.

Predictions were made for the immersion of each star as described in section 12.41. The results of these predictions are listed below (T, P). Right ascension and declination at the beginning of the Besselian year for each star were given in the 'American Ephemeris and Nautical Almanac' since the observations were made before occultation data was deleted from that publication. Besselian day numbers were applied to find positions of the stars at contact (α, δ). The position (α_M, δ_M) and parallax (π_M) of the moon were taken from the 'Improved Lunar Ephemeris, 1952-59' by interpolating to the first, second, and third differences. Values were found for these quantities at the even minute preceding and following the observed time of immersion of each star.

Data on Immersion Prediction for the Stars and the Moon

	α	δ	α_M	δ_M	π_M	T	P
16 Tauri	55°30'31".785	24°08'47".58	54°49'54".615	24°39'08".79	56'47".892	$23^h40^m11^s.83$	127°.9
q Tauri	55 36 33 .030	24 19 29 .00	54 57 50 .505	24 39 42 .55	56 47 .790	23 45 47.75	89.4
21 Tauri	55 46 59 .970	24 24 45 .44	55 07 30 .255	24 42 30 .32	56 47 .282	0 10 52.04	84.1
22 Tauri	55 49 07 .335	24 23 10 .21	55 10 26 .235	24 43 03 .78	56 47 .180	0 15 49.13	92.6
20 Tauri	55 45 50 .175	24 13 32 .48	55 12 47 .040	24 43 30 .44	56 47 .099	0 19 25.62	133.8
BD+24°.571	56 19 54 .600	24 50 54 .98	56 09 44".955	24 54 05 .64	56 45 .139	1 56 11.66	27.7
3 Gem	91 45 35 .985	23 07 21 .67	91 43 26 .910	23 34 08 .64	58 51 .272	1 22 47.71	116.0
4 Gem	91 57 07 .665	23 00 28 .33	92 04 18 .030	23 32 11 .70	58 50 .904	1 55 45.09	133.4
6 Gem	92 24 26 .895	22 55 13 .68	92 40 54 .840	23 28 39 .35	58 50 .254	2 53 59.23	130.6

k_M = 0.272496

Computations involved in the application of occultations are quite long and laborious and as many checks for computational error as possible should be made to avoid time-consuming recomputation. The method of observation at one station affords many checks since several stars are predicted in a short time interval. Terms such as right ascension, declination, and parallax of the moon plus their trigonometric functions all increase or decrease in chronological order. The local hour angle of the stars and the observer's coordinates also vary chronologically affording an easy way to detect blunders. In successive approximations for the time of contact, each value for τ should be smaller than the preceding one. The value of m, the distance from the observer to the center of the shadow, should approach closer to the value of the assumed shadow radius (in this example, $k_M = 0.272496$) with each approximation.

A good check on the predictions can be made by plotting positions of contact around a circle representing the moon's limb. Stars observed from one station during a short period of time should trace parallel paths across the moon's surface (Fig. 12.12).

Formula (12.58) was used to form the observation equations. In applying these formulas, the values of h, x, y, ξ, η, and π_M must be taken at the observed time of contact rather than the computed. The number of places with which coefficients must be computed depends upon the precision of the timing. Since timing was to 0.ˢ1, there will be only two significant numbers in $\delta\tau$ and coefficients should be carried to four places at the most. With photoelectric observations (see section 12.62), the coefficients should be carried to five or six places.

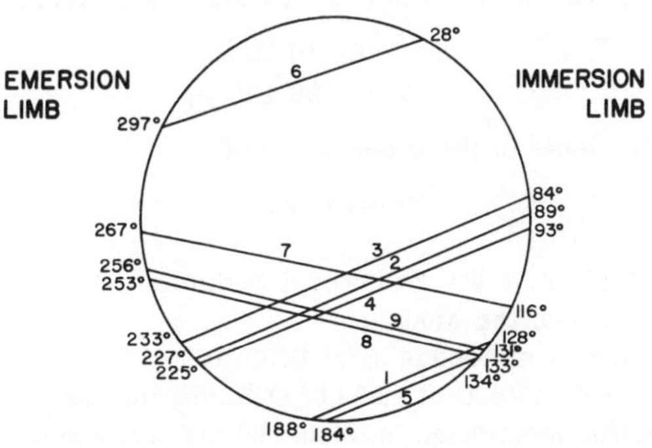

Fig. 12.12 Diagram of the path of occulted stars observed at Williston Observatory

The undulation term was neglected; the resulting observation equations for the nine contacts are tabulated below. Note that coefficients in equations (2), (3), and (4) are of the same order for each of the unknowns. Equations (7), (8), and (9) and equations (1) and (5) are also similar. Equation (6) differs from all the others. A look at Fig. 12.12 will explain these similarities in the coefficients since the paths of stars represented by similar equations are parallel and close together. The comparison of coefficients can be used as a rough check for errors in forming the observation equations.

Coefficients of Observation Equations

	$\delta\alpha_M^2$	$\delta\delta_M^2$	$\delta\alpha_M^1$	$\delta\delta_M^1$	$\delta\pi_M$	δk_M	$\delta\lambda$	$\delta\varphi$	$\delta\tau$
1.	0.0000	0.0000	-0.6082	-4.7390	2.0823	-1.6310	-0.4408	4.4726	0.697
2.	0.0000	0.0000	-1.7084	-1.7880	-0.6834	-0.8861	0.8166	2.3074	0.065
3.	0.0000	0.0000	-1.8769	-1.5562	-0.8305	-0.8829	1.1040	2.0767	0.196
4.	0.0000	0.0000	-1.7493	-1.9606	-0.5139	-0.9363	0.9768	2.3784	0.207
5.	0.0000	0.0000	-0.3640	-5.9624	2.9356	-2.0353	-1.5366	5.5464	0.398
6.	0.0000	0.0000	0.4343	-3.8097	0.2916	-1.3074	-0.4201	3.5724	0.174
7.	-2.3285	-0.1454	0.0000	0.0000	-0.0269	-0.9024	1.8758	-0.3615	-0.111
8.	-2.4406	-0.8880	0.0000	0.0000	0.7693	-0.9903	1.9532	0.0754	0.171
9.	-2.2619	-0.5548	0.0000	0.0000	0.9439	-0.8916	-1.5672	-0.5614	-0.243

The solution gives the following corrections to the assumed data:

$$\delta\varphi = +5''.13, \quad \delta\alpha_M^1 = +0''.06,$$
$$\delta\lambda = +5''.14, \quad \delta\delta_M^1 = +0''.12,$$
$$\delta k_M = -0.000012, \quad \delta\alpha_M^2 = +0''.10,$$
$$\delta\pi_M = +0''.06, \quad \delta\delta_M^2 = +0''.09.$$

The geodetic coordinates of the observation station therefore are

$$\varphi = +42°15'23''.33,$$
$$\lambda = -\ 4^h50^m19^s.33,$$

and the semidiameter of the moon is

$$k_M = 0.272484 \text{ earth radius.}$$

12.52 Determination of the Equatorial Semidiameter of the Earth and the Parallax of the Moon

The theory of this method is a modification of the type of observation equation described in section 12.511 [O'Keefe and Anderson, 1952].

The time of the occultation is observed and the distance between the observer and the center of the shadow Δ is computed by means of equation (12.15):

$$\Delta^2 = (x - \xi)^2 + (y - \eta)^2.$$

The distance between the observer and the edge of the shadow,

$$\delta\sigma = \Delta - k_M, \tag{12.59}$$

is then calculated and is considered the quantity observed. Instead of expression (12.58), a different type of observation equation is derived which is of the form

$$\delta\sigma = b_1\delta u + b_2\delta v + b_3\delta w + b_4\delta\alpha_M + b_5\delta\delta_M + b_6\delta r_M + b_7\delta k_M \tag{12.60}$$

where the coefficients b_1 through b_7 were obtained by collecting the differential coefficients in the derivation, similar to the procedure in section 12.511. The coefficients are

$$b_1 = -\sin h^G \sin Q + \sin\delta \cos h^G \cos Q,$$

$$b_2 = -\cos h^G \sin Q - \sin\delta \sin h^G \cos Q,$$

$$b_3 = -\cos\delta \cos Q,$$

$$b_4 = r_M \cos\delta_M \sin Q,$$

$$b_5 = r_M \cos Q,$$

$$b_6 = \frac{x \sin Q + y \cos Q}{r_M},$$

$$b_7 = +1.$$

The unknowns δu, δv, δw in (12.60) are corrections to the rectangular geocentric coordinates of the observer computed from

$$\left.\begin{aligned} u &= \rho \cos\varphi' \cos\lambda = (N + H) \cos\varphi \cos\lambda \\ v &= \rho \cos\varphi' \sin\lambda = (N + H) \cos\varphi \sin\lambda \\ w &= \rho \sin\varphi' \quad = [N(1 - e^2) + H]\sin\varphi \end{aligned}\right\} . \qquad (12.61)$$

The corrections δu, δv, δw, for the purposes of the calculations, are assumed to be constant in a certain triangulation system. The other unknowns in (12.60) were discussed before.

For a simple occultation of a star observed at several places, the quantities h^G, x, and y vary markedly; the values r_M, δ_M, δ vary only slightly. Therefore, if the observations are conducted along an isolimb, the quantity

$$U = b_3\delta w + b_4\delta\alpha_M + b_5\delta\delta_M + b_7\delta k_M \qquad (12.62)$$

is a constant provided that δw, $\delta\alpha_M$, $\delta\delta_M$, and δk_M are constants for the occultation. The δw is assumed to be constant, as mentioned before; the corrections to the coordinates of the moon $\delta\alpha_M$ and $\delta\delta_M$ are sufficiently constant for one occultation. The quantity δk_M would be constant if libration would not affect the lunar radius at the point on the limb where the occultation takes place. This effect is present; however, it is still assumed that δk_M is also a constant.

With these assumptions, equations (12.60) will take the form of

$$\delta\sigma = b_1\delta u + b_2\delta v + b_6\delta r_M + U. \qquad (12.63)$$

It may be shown that each of the three corrections above depends in turn on corrections to the earth's equatorial radius, to the geodetic latitude and longitude, and to the equatorial horizontal parallax of the moon, δa_e, $\delta\varphi$, $\delta\lambda$, and $\delta\pi_M$ respectively. The corresponding formulas are the following:

$$\left.\begin{aligned} \delta u &= \frac{u}{a_e}\,\delta a_e - v\delta\lambda - u\tan\varphi\,\delta u \\ \delta v &= \frac{v}{a_e}\,\delta a_e + u\delta\lambda - v\tan\varphi\,\delta v \\ \delta r_M &= \operatorname{cosec}\pi_M\,\delta a_e - a_e\cot\pi_M\operatorname{cosec}\pi_M\,\delta\pi_M \end{aligned}\right\} . \qquad (12.64)$$

It may be shown again that with the exception of the first terms in these equations, the rest is constant with a reasonable accuracy in a triangulation system. O'Keefe and Anderson computed these constants for the United States by substituting the data of Meades Ranch ($\varphi = 39^\circ 13' 26''.686$, $\lambda = -98^\circ 32' 30''.506$, $H = 0$, $\delta\varphi = -1''.2$, $\delta\lambda = -0''.5$) in (12.64). The second term in the third equation has been calculated from the expression of Lambert for the dynamic parallax, which is

$$\delta\pi_M = \frac{\tan\pi_M}{3a_e}\,\delta a_e + \frac{\tan\pi_M}{3\gamma_e}\,\delta\gamma_e - \frac{\tan\pi_M}{3\frac{1}{f}(\frac{1}{f} - 1)}\,\delta(\frac{1}{f}) + \frac{\tan\pi_M}{\frac{3}{\mu}(\frac{1}{\mu} + 1)}\,\delta(\frac{1}{\mu})$$

where γ_e is the mean equatorial gravity, f is the flattening of the ellipsoid, and μ is the mass of the moon in units of the earth's mass. Using the values of

$$a_e = 6378388 \text{ m},$$
$$f = 1/297.0,$$
$$\gamma_e = 978.052 \text{ cm s}^{-2},$$
$$\mu = 1/81.53,$$
$$\pi_M = 3422''.682,$$

Lambert found the following equation:

$$\delta\pi_M = 0.179\delta a_e - 1.17\delta\gamma_e - 0.0310\delta(\frac{1}{f}) + 0.170\delta(\frac{1}{\mu}),$$

which, if δa_e is substituted in kilometers and $\delta\gamma_e$ in cm s^{-2}, will give $\delta\pi_M$ in seconds of arc.

Estimating the most probable values of the last three terms in the equation above, the expression used in [O'Keefe and Anderson, 1952] is

$$\delta\pi_M = \frac{\tan \pi_M}{3a_e \sin 1''} \delta a_e - 0.025. \tag{12.65}$$

After substitutions, equations (12.64) for Meades Ranch take the form of

$$\delta u = -0.1151\delta a_e - 14.94 \text{ m},$$
$$\delta v = -0.767\delta a_e - 21.45 \text{ m},$$
$$\delta r_M = \frac{2}{3}\delta a_e \operatorname{cosec} \pi_M + 0.7731 \cot\pi_M \operatorname{cosec}\pi_M .$$

Using these expressions in (12.63) yields

$$\delta\sigma' = B\delta a_e + U \tag{12.66}$$

where

$$\delta\sigma' = \delta\sigma + 14.94b_1 + 21.45b_2 - 0.7731 \cot\pi_M \operatorname{cosec}\pi_M b_6,$$
$$B = -0.1151b_1 - 0.767b_2 + \frac{2}{3}\frac{x \sin Q + y \cos Q}{a_e}.$$

Equation (12.66) was employed in the solution. Solving it for δa, $\delta\pi_M$ may be calculated from (12.65). Obviously, in a triangulation system with different parameters than the above, the constants in (12.66) will have different values.

O'Keefe and Anderson, using nine stations and four occultations in 1949-1950 in the southwestern United States, obtained the following results:

$$a_e = 6378448 \pm 169 \text{ m},$$
$$\pi_M = 3422''.686 \pm 0''.03 .$$

Keeping the equatorial semiaxis of the international ellipsoid unchanged ($\delta a_e = 0$), and also assuming that $\delta\varphi = \delta\lambda = \delta\alpha_M = \delta\delta_M = \delta k_M = 0$, equation (12.60) yields

$$\delta\sigma = b_6\delta r_M .$$

Using this type of observation equations, the solution is

$$r_M = 384407.6 \pm 4.7\ \text{km},$$

which, through

$$\sin \pi_M = \frac{a_e}{r_M},$$

yields

$$\pi_M = 3422''.662 \pm 0''.042.$$

As follows from the derivation, the assumption included in these results is that the dynamic and geometric values of the moon's parallax agree.

In the solutions above, the elevations of the stations above the International ellipsoid were determined by using Hayford's geoid contours (U.S. Coast and Geodetic Survey, Special Publication No. 82, 1909) for the Clarke 1866 ellipsoid, and by adding the distances between the Clarke and the International ellipsoids, assuming that they are tangent at Meades Ranch. Using her geoidal map of North America (Bulletin Géodésique, No. 57, 1960), Irene Fischer recomputed the values above [Fischer, 1962]. Keeping the two ellipsoids tangent at Meades Ranch and the observation equations in the form $\delta\sigma = b_6 \delta r_M$, her results are

$$r_M = 384404.1\ \text{km},$$
$$\pi_M = 3422''.693.$$

Using her astrogeodetic world datum (Journal of Geophysical Research, Vol. 65, No. 7, 1960) ($1/f = 298.3$, $a_e = 6378166$ m), the results are

$$r_M = 384400.9\ \text{km},$$
$$\pi_M = 3422''.603.$$

The last quantities are in very good agreement with the dynamic mean values ($r_M = 384400 \pm 2$ km, $\pi_M = 3422''.610 \pm 0''.013$, referred to the same system) and also with the recent radar measurements of the moon's distance by Yaplee ($r_M = 384400 \pm 1.2$ km), as modified by -2 km by Fischer for the lunar radius used in the occultation calculations (1738 km). The uncertainty in the value quoted for the dynamic parallax is due to an estimate of the uncertainty in the mass of the moon. The most quoted values for the mass of the moon are those of Spencer Jones ($\mu = 1/81.27$), of Delano (1/81.219), of Rabe (1/81.375), the uncertanity being about 1/500. For comparison, the IAU-adopted (1964) values for these constants are [Nautical Almanac Offices, 1966]

$$r_M = 384400.2\ \text{km},$$
$$\pi_M = 3422''.608,$$
$$\mu = 1/81.30.$$

12.6 Observations of Occultations

As has been mentioned earlier, the solar eclipse is a relatively rare phenomenon which, moreover, may be observed only under good weather conditions. For this reason, and because its mission may be accomplished by other more flexible methods from the geodetic point of view, it has little practical significance today; thus the observation technique is not treated here. Instead, the reader is referred to [Berroth and Hofmann, 1960; Mueller, 1964].

12.61 Optical Methods

All that is needed in occultation observation is the recording of time of disappearance or reappearance of the star.

With visual observations it is not possible to fully exploit the potential accuracy of the method. At best the recorded time cannot be better than $0^s.2$ plus the effect of the personal response time. It is especially difficult to time the emersion of the star from behind the dark limb of the moon. The star seems to jump out from behind the moon and catches the observer by surprise, i.e., the observer has the feeling that the emersion occurred before he noticed it. This is especially the case when the occulted star is faint.

The use of photography in the observation could take care of the effect mentioned above, and it has been attempted at Harvard University and in France, but without real success. The main reason is that the photograph is not efficient at collecting light over short periods of time.

12.62 Photoelectric Method

The most effective tool for observing occultations appears to be the photoelectric cell. With this instrument, coupled with a telescope and an accurate time-recording device, it is possible to record the occultation of a ninth-magnitude star with a precision of $0^s.01$.

This method has been tested in several projects conducted by the U.S. Army Map Service in the United States, Morocco, and in the Azores in the years between 1949 and 1951. The purpose of the first project in 1949-1950 was the determination of the earth's equatorial radius and the distance to the moon. The theory and the results have been discussed in section 12.52. The occultations were observed at nine stations each equipped with a 12-inch Cassegrain reflective telescope on equatorial mounting. The telescope was connected to a photoelectric system consisting of a photocell, an amplifier, and a recorder. On the telescope a very small focal plane diaphragm was used to permit the light of the star to fall on the photocell while obscuring the light of the moon. The amplified output of the photocell together with the time signals were recorded on the recorder-oscillograph. The whole system was designed by D. D. Mears with the aid of A. E. Whitford of the University of Wisconsin [Mears, 1949; Whitford, 1939 and 1947]. Fig. 12.13 shows the

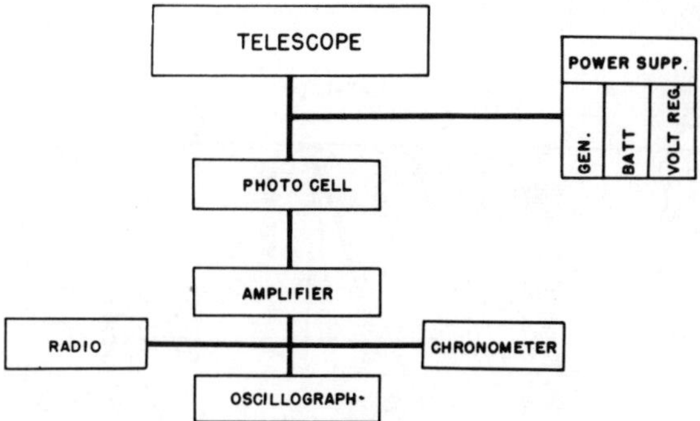

Fig. 12.13 Diagram of the photoelectric occultation equipment

diagram of the occultation equipment. A sample recording is shown in Fig. 12.14. The upper channel is the recording of the WWV time signals; the lower channel is the photocell output increasing downward.

Another major project was undertaken by the Army Map Service in the Pacific Star Occultation Program. Many occultation pairs were observed in areas shown in Fig. 12.15. As a result, direct occultation ties stretch all the way from Japan, the Philippines, and Okinawa to Hawaii and the United States. The stations were situated along isolimbs to avoid the effect of the lunar topography. Partial results giving the connection between the Philippines and Palau Island in the Carolines were published in [Henriksen et al., 1957] while the theory was published in [Henriksen, 1962].

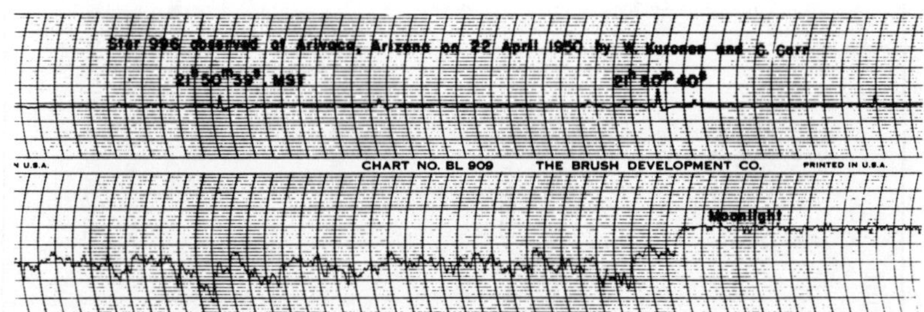

Fig. 12.14 Record from the photoelectric occultation equipment

In connection with the Pacific project, research and development began in 1948 at the Tokyo Astronomical Observatory. The first photoelectric apparatus using the Mears-Whitford plans was tested in early 1950. In the next two years the equipment was further tested and redesigned. Six sets of modified equipment were built by the Nippon Kogaku Company. This instrument had portable steel piers and was lighter in

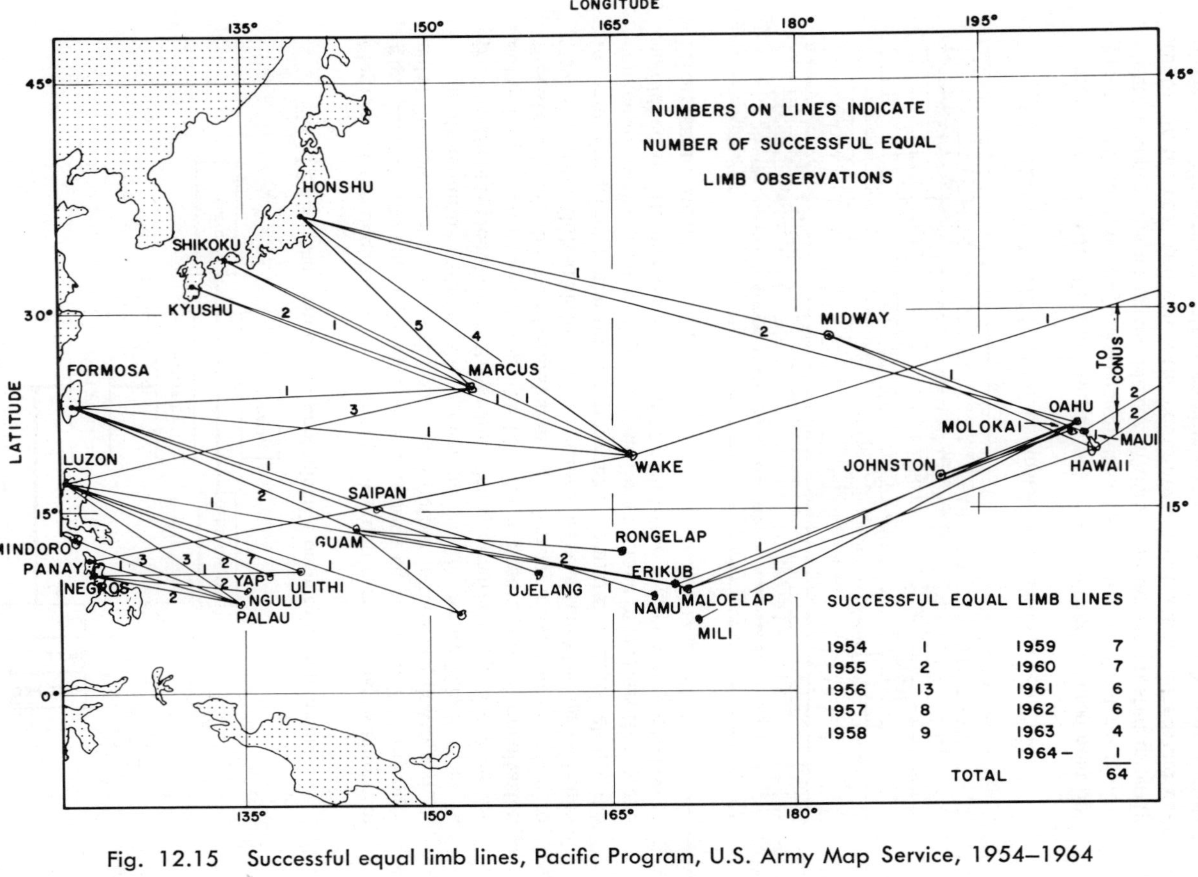

Fig. 12.15 Successful equal limb lines, Pacific Program, U.S. Army Map Service, 1954–1964

weight (Fig. 12.16). It was later supplemented by better equipment manufactured by the Sokkisha Company, also in Japan.

Fig. 12.16 Nippon-Kogaku occultation telescope

12.7 Major Factors Affecting the Accuracy of the Occultation Results

The accuracy of the geodetic information obtained from observations of occultations depends upon the accuracy with which the lunar motions and other parameters which enter the calculations are known and upon the accuracy of the observations.

12.71 Topography of the Lunar Limb

In order to obtain sufficiently accurate results, the radius of the moon should be known to within some hundred meters. The moon's limb being quite irregular, there is in a sense no simple lunar radius. Depending upon the location, values differing by as much as 2" or 6000 m may be found; therefore, the knowledge of the topography on and around the lunar limb is essential.

Of early attempts to chart the moon, perhaps the best known and most widely used is that of F. Hayn [Hayn, 1902-14]. A reexamination of these charts, however, showed that they were based on too few measurements spaced too far apart; therefore their accuracy for geodetic purposes is questionable. T. Weimer has also published profiles of the lunar limb which are perhaps more accurate but too sparse to be usable for interpolation [Weimer, 1952]. It is estimated that the maps of Hayn give errors of about 0".5, equivalent to 900 m in position. The charts of A. A. Nefedyev should also be mentioned here.

A new survey of the marginal zone of the moon has been carried out at the U.S. Naval Observatory by C. B. Watts. Some 500 lunar profiles were measured from several hundred photographs obtained at different observatories, and a topographic map was made. This improved compilation has been published in Volume 17 of the 'Astronomical Papers of the American Ephemeris' and is estimated to give errors of 0".07. The improvement is due not only to the many lunar profiles used but also to a more elaborate measurement control system. The new charts represent the irregularities of the moon's limb for various combinations of the librations. The limb corrections are indicated by lines of equal value at intervals of 0".2 as shown in Fig. 12.17. The position angle from the moon's axis is given in the upper right-hand corner of each frame. The horizontal and vertical arguments are the topocentric librations in longitude and latitude respectively. The corrections are for the mean distance of the moon.

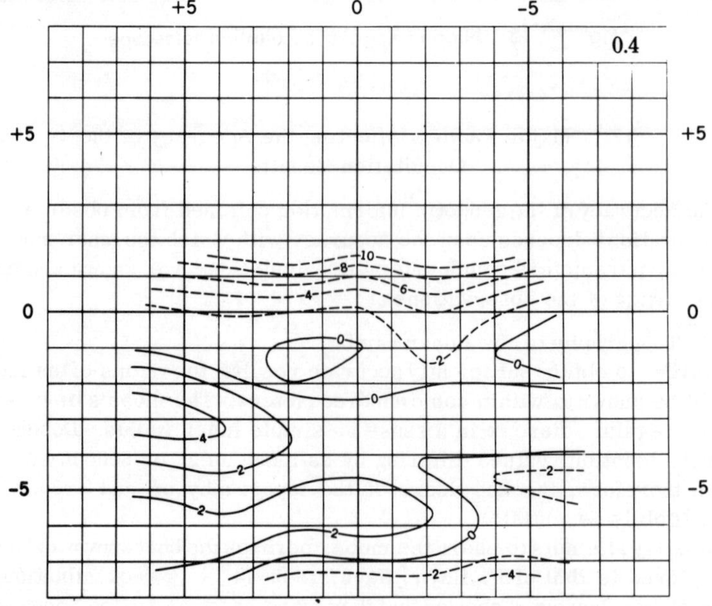

Fig. 12.17 Watts lunar limb-correction and libration chart. Sample

This whole problem is made somewhat simpler by the fact that the concern is with the difference in elevation only. Therefore the effect of the lunar topography may be minimized by observing the same lunar features at all stations in case of solar eclipses, or by observing from stations for which the occultations of a star take place behind the same point on the lunar limb. These stations, situated on an isolimb, may be selected as described in section 12.44. In this case the only effect that must be accounted for is the libration of the moon.

12.72 Libration

The moon does not always show the same limb to the observer because of libration. Due to this effect the features on the limb have a periodic motion. There are three kinds of librations, mainly for the following reasons:

—the eccentricity of the moon's orbit and the inclination of its axis of rotation with respect to the plane of orbit (optical libration),
—dynamic reasons (physical or dynamic libration),
—the rotation of the observer with the earth (diurnal or apparent libration).

The magnitudes of these librations are different. The first one is the largest, its maximum being about 8°; the second type is small, in the order of 0°.03; the diurnal libration may amount to 1°.

The calculation of the optical libration is based on the relationships of J. Cassini which are as follows:

(1) The descending node of the moon's equator on the ecliptic coincides with the ascending node of the orbit.

(2) The inclination of the equator of the moon with respect to the ecliptic is constant (1°32'.1).

(3) The period of axis rotation is equal to the mean sidereal period of one revolution of the moon in its orbital flight.

The 'Ephemeris,' in its sections for physical observations, contains the geocentric optical libration in longitude ℓ and latitude b (as the earth's selenographic longitude and latitude), the position angle C of the moon's axis (the angle that the lunar meridian through the apparent central point of the disc toward the north lunar pole forms with the declination circle through the central point), and the physical libration. The tabular librations and position angles of axis should be reduced to apparent or topocentric values. For this purpose, the following differential corrections derived by Atkinson may be used:

$$\Delta\ell = -\pi'_M \sin(Q - C) \sec b,$$
$$\Delta b = \pi'_M \cos(Q - C),$$
$$\Delta C = \sin(b + \Delta b)\,\Delta\ell - \pi'_M \sin Q \tan \delta_M,$$

where Q is the parallactic angle, and π'_M is the topocentric parallax. They may be calculated as follows:

$$\sin Q = \sin h_M \cos \varphi \operatorname{cosec} z_M,$$

$$\cos Q = \frac{\sin \varphi - \cos z_M \sin \delta_M}{\sin z_M \cos \delta_M},$$

$$\pi'_M = \pi_M (\sin z_M + 0.0084 \sin 2z_M),$$

where h_M and δ_M are the local hour angle and geocentric declination of the moon respectively, φ is the latitude of the observer. The geocentric zenith distance of the moon z_M may be computed from

$$\cos z_M = \sin\varphi \sin\delta_M + \cos\varphi \cos \delta_M \cos h_M.$$

The tabular values should be interpolated to the time of observation with second differences.

Section 10C in [Nautical Almanac Offices, 1961] gives somewhat different formulas to calculate the topocentric librations from the tabulated geocentric librations.

In case of observing an occultation from stations situated on an isolimb, further simplification is possible. S. W. Henriksen has proved that the velocity of the position angle is the same for all points on the moon's limb and is given by

$$\frac{dQ}{dt} = \frac{2\pi}{T} \sin i$$

where T is the moon's sidereal period, and i is the inclination of the moon's axis to the fundamental plane.

Considering this effect in the calculations of the isolimbs, it may be achieved that the occultation is always produced by the same feature of the limb.

12.73 Coordinates of the Moon

If corrections to the coordinates of the moon and the sun are not included in the observation equations, the position of these bodies should be known to about 0\".01. The presently accepted theory of the moon's orbit is that of E. Brown, who published his investigation in the 'Memoirs of the Royal Astronomical Society' between 1901 and 1908. He has formulated and published special tables for the computations of the position of the moon [Brown, 1919]. These tables were used in the determination of the coordinates of the moon until 1952 in the national ephemerides. The coordinates were given for every hour in these publications to 0\".1 in declination and $0^s.01$ in right ascension.

Later, with the improvement of electronic computers, it became possible to compute the coordinates of the moon directly from Brown's theory without using his tables. The first published result of this undertaking was the 'Improved Lunar Ephemeris, 1952-1959,' published by the U. S. Naval Observatory and the Royal Greenwich Observatory [Eckert et al., 1954]. In this publication the position of the moon is given to 0\".01 in declination, $0^s.001$ in right ascension, and 0\".001 in parallax. Since 1960, the improved lunar coordinates have been part of both the 'Astronomical Ephemeris' and the 'American Ephemeris and Nautical Almanac.'

The 'Ephemeris' contains the most precise material available at present for the position of the moon. However, since the adopted constant of sine parallax (3422".451) may be wrong, an unknown scale factor $(1+\varkappa)$ could be introduced in the formula expressing the moon's distance from the earth:

$$r_M = (1+\varkappa)\operatorname{cosec}\pi_M$$

where π_M is to be taken from the tables.

12.74 Parallax of the Moon

As has been mentioned in the preceding sections, the adopted value for the mean parallax of the moon is uncertain. The required accuracy is about the same as that of the coordinates of the moon, i.e., about 0".01. O'Keefe and Anderson and Fischer approached this precision when they determined the lunar parallax as described in section 12.52. It is a question, however, whether a once observed parallax can be used again or reduced to another occultation with the required accuracy; therefore, a correction for the parallax should always be included in the observation equations.

References

Berroth, A. and W. Hofmann. (1960). Kosmische Geodäsie. G. Brown, Karlsruhe.

Bessel, F. W. (1876). "Analyse der Finsternisse." Abhandlungen, 3, pp. 369-428. Rudolph Engelmann, Leipzig.

Brown, E. W. (1919). Tables of the Motion of the Moon. Yale University Press.

Buchanan, R. (1904). The Mathematical Theory of Eclipses According to Chauvenet's Transformation of Bessel's Method. Philadelphia.

Chauvenet, W. (1863). A Manual of Spherical and Practical Astronomy. Philadelphia. Reprinted by Dover Publications, Inc., New York, 1960.

Comrie, L. J. (1937). "The Reduction of Lunar Occultations." The Astronomical Journal, 46, pp. 61-67.

Eckert, W. J. (1958). "Improvement by Numerical Methods of Brown's Expressions for the Coordinates of the Moon." The Astronomical Journal, 63, pp. 415-418.

Eckert, W.J., R.Jones, and H.K.Clark. (1954). Improved Lunar Ephemeris, 1952-1959. U. S. Naval Observatory.

Farnsworth, A. (1953). "Occultations of Stars by the Moon Observed During 1952." The Astronomical Journal, 58, p. 175.

Farnsworth, A. (1956). "Occultations of Stars by the Moon Observed During 1955." The Astronomical Journal, 61, p. 360.

Fischer, I. (1962). "The Parallax of the Moon in Terms of a World Geodetic System." The Astronomical Journal, 67, pp. 373-378. Presented in a shorter form under the title "The Lunar Distance" at the 43rd Meeting of the American Geophysical Union, Washington, D. C., April, 1962.

Hayn, F. (1902, 1904, 1907, 1914). "Selenographische Koordinaten, I-IV." Abhandlungen, Sächsischen Akademie der Wissenschaften, 27 (9), 29 (1), 30 (1), 33 (1). Leipzig.

Henriksen, S. W. (1962). "The Application of Occultations to Geodesy." Army Map Service Technical Report, 46.

Henriksen, S.W., S.H. Genatt, C.D. Batchlor, and M.Q. Marchant. (1958). "Surveying by Occultations." Transactions of the American Geophysical Union, 38, 5, 1957; also, The Astronomical Journal, 63, pp. 291-295.

Hirose, H. (1953). "On the Prediction of the Equal Limb Line for an Occultation." The Annals of the Tokyo Astronomical Observatory, Second Series, III, 4.

Lambert, W.D. (1949). "Geodetic Applications of Eclipses and Occultations." Bulletin Géodésique, 13.

Mears, D. D. (1949). "Construction and Testing of Equipment for the Photoelectric Observations of Occultations." Army Map Service Technical Report, 3.

Mueller, I.I. (1964). Introduction to Satellite Geodesy. Frederick Ungar Publishing Co., Inc., New York.

Nautical Almanac Office, H. M. (1938). The Prediction and Reduction of Occultations, 1937. A Supplement to the Nautical Almanac for 1938. H. M. Stationery Office, London.

Nautical Almanac Offices of the United Kingdom and the United States of America. (1961). Explanatory Supplement to the Astronomical Ephemeris and the American Ephemeris and Nautical Almanac. H.M. Stationery Office, London. (The introduction of the IAU system of astronomical constants (1964) requires changes in this reference. They may be found in the next reference.)

Nautical Almanac Offices of the United Kingdom and the United States of America. (1966). The Introduction of the IAU System of Astronomical Constants into the Astronomical Ephemeris and into the American Ephemeris and Nautical Almanac, Supplement to the American Ephemeris, 1968.

O'Keefe, J. (1950). "A New Determination of the Lunar Parallax." The Astronomical Journal, 55, pp. 177-178.

O'Keefe, J. (1958). "The Occultation Method of Long Line Measurements." Bulletin Géodésique, 49.

O'Keefe, J.A. and J.P. Anderson. (1952). "The Earth's Equatorial Radius and the Distance of the Moon." The Astronomical Journal, 57, pp. 108-121; also, Bulletin Géodésique, 29, 1953.

O'Keefe, J. A. and D. D. Mears. (1954). "The 800-Inch Telescope." Journal of the Royal Astronomical Society of Canada, 48, 1.

Thomas, P. D. (1962). "Geodetic Positioning of the Hawaiian Islands." Surveying and Mapping, 22, 1.

Watts, C. B. and A. N. Adams. (1950). "Photographic and Photoelectric Technique for Mapping the Marginal Zone of the Moon." The Astronomical Journal, 55, pp. 81-82.

Weimer, T. (1952). Atlas de Profils Lunaries. Observatoire de Paris.

Whitford, A. E. (1939). "Photoelectric Observations of Diffraction at the Moon's Limb." Astrophysical Journal, 89, p. 351.

Whitford, A.E. (1947). "Angular Diameters of Stars from Occultations by the Moon." The Astronomical Journal, 52, p. 131.

INDEX

Aberration, 59, 88, 91, 116
  annual, see Annual aberration
  constant of, 92, 93, 102
  day numbers of, 93, 94
  diurnal, 96, 116, 484, 516
  law of, 91
  light time correction, 89
  planetary, 88
  secular, 89
  stellar, 89
Adjustment of optical instruments, 273, 290, 292
  effects of imperfections in, 305
Almucantar, 29
Altitude, 33
Annual aberration, 92
  e-terms of, 96
  perigee terms of, 96
Aphelion, 143
Apparent plane (eclipse), 557
Apparent position (place), 102, 116
Apparent time
  sidereal, 139
  solar, 145
Areal velocity, 143
Astrogeodetic deflections, 26, 425
Astrolabe, 456, 500, 518
  Danjon, 183, 301, 518, 536
  pendulum, 294
  prismatic, 291
Astronomic coordinates, 1, 19
  average, 21
  ordinary, 21
  reduced, 21
Astronomic triangle, 30
Astronomische Gesellschaft, 193, 195, 215, 218
Astronomisches Rechen - Institut, 192, 202, 213, 218, 222
Atomic clock (scale), 320, 334
  A.1, 175, 334
  A3, 175, 342, 343, 350, 354
  IATS, 354
  NBS-A, 175, 335, 355
  NBS-III, 175
  NBS(SA), 335, 357
  SAT, 342
  UTC(NBS), 335, 355
  UTC(USNO)'Master Clock,' 334, 367, 374
Atomic oscillator, 322
  cesium beam, 322, 326
  hydrogen maser, 322, 324, 326
  rubidium cell, 325, 326
Atomic time, see Time, atomic
Auwers, Artur, 193
Azimuth
  astronomic, 19
  of celestial body, 33
  corrections to, 428

determination of, 401
geodetic, 17
of normal section, 16
Azimuth setting error 306
determination of, 315
effect of, 310

Besselian day numbers, 75, 93, 102, 119
Besselian elements of an eclipse, 564
Besselian year, see Year, Besselian
Bessel's formula, 311
Black's method, 424
Boss, Benjamin, 202
Boss catalogue, see Star catalogues
Brosche's method, 205, 214
Bulletin Horaire, 345
Bureau International de l'Heure, 81, 163, 164, 336, 519

Calendar
astronomic, 153
civil, 152
ephemeris, 172
Calibration of optical instruments, 278, 290, 294
impersonal micrometer, 284
levels, 278, 282
Catalogues, see Star catalogues
Celestial coordinate systems, 32
ecliptic, 34
horizon, 33
hour angle, 33
right ascension, 34
transformations between, 37
Celestial sphere, 29
Cesium beam oscillator, see Atomic oscillator
Chandler period, 80
Chronograph, 391, 479, 504
Chronometer, 319, 326, 331, 393, 396; see also Clock, flying
Chronochord, 334
Chronofax (Newtek), 333
Hewlett-Packard, 325
Nardin, 327
Sulzer (Tracor), 330
Circumpolar star, 51
Clairaut's theorem, 10
Clock, 319, 326; see also Chronometer
atomic, see Atomic clock (scale)
'flying' (transportable), 323, 370, 372, 387
Shortt, 327
Codeclination, see Polar distance
Co-latitude, 37
Collimation
adjustment of, 275
determination of, 313
effect of, 305, 516
Colure
equinoctial, 32
solstitial, 32
Conventional International Origin (CIO), see Terrestrial pole, average
Coordinate systems
astronomic, 1, 18
celestial, see Celestial coordinate systems
geodetic, 1, 13
geographic, 1
spherical, 11
terrestrial, 21, 80
Culmination, 54
Curvature correction (star path)
for azimuth (horizontal circle) observations, 405, 410
for zenith distance observations, 419, 443, 483
Curvature of the plumb line, 21, 430, 471, 520

Danjon astrolabe, see Astrolabe, Danjon
Date
astronomic, 153
Besselian, 174
Greenwich sidereal, 141
Julian, 153
Julian ephemeris, 174
Date line, 150
Datum, geodetic, 23

Day
ephemeris, 168
sidereal, 140, 158
solar, 150, 158
see also Date
Day numbers
Besselian, see Besselian day numbers
independent, see Independent day numbers
second-order, see Second-order day numbers
Declination, 34
determination of, 182
Deflection of the vertical, 26
Delambre correction (curvature of star path), 443
Diurnal aberration, 96, 116, 425, 470, 484, 516
Doellen method, 516

Eccentric station correction
to azimuth, 432
to latitude, 472
to longitude, 520
Eccentricity
of ellipsoid, 14
first, 14
of orbit, 143
second, 14
Eclipse, 551
annular, 556
lunar, 551
partial, 556
prediction of, 552, 565
solar, 551
total, 556
Ecliptic, 31, 183
coordinate system, 35
latitude, 35, 184
longitude, 35
meridian, 31
obliquity of, 32, 183
parallel (of latitude), 31
pole, 31
Ellipsoid, 9, 10, 13
earth, 9
reference, 10, 11, 21, 23, 25
rotational, 13
triaxial, 13
Elongation, 56
Emersion, 568, 574
Ephemeris
date, 172
day, 168
hour, 168
hour angle, 172
meridian, 170
second, 168
year, 168
Ephemeris time, 138, 166, 545
determination of, 305, 545
equation of, 172
Epoch
atomic, 175
ephemeris, 168
sidereal, 139
solar, 145
Equal limb line, 583, 603
Equation of ephemeris time, 172
Equation of the equinox, 139
Equation of time, 147
Equations of optical instruments
horizontal angle measurement, 306
transit time measurement, 310
vertical angle measurement, 308
Equator
astronomic, 19
celestial, 29, 30, 75
geodetic, 14
terrestrial, 20, 21, 22
Equatorial semidiameter of earth, 10, 11
Equinoctial colure, 32
Equinox, 31
autumnal, 32
equation of the, 139
vernal, 31
Equipotential surface, 7
earth-spherop, 9
geoid, 7
geop, 7
geopotential surface, 7
spheroid, 9
spherop, 9

spheropotential surface, 9
Eulerian motion, 61, 80
Euler's theorem, 16

First-order astronomic position determination, 249, 401
azimuth, 401, 423
latitude, 401, 459
longitude, 401, 504
Flattening (ellipsoid), 10, 14
Floating zenith telescope, 297
Frequency standard, 320
NBS-III, 175, 323, 335
primary frequency standard, 321
Fundamental plane (eclipse), 557

Geodesic, 17
Geodetic datum, 23
Geoid, 7, 23, 27
reduction of observations to, 21, 430, 471, 520
Geoid undulation, 26, 27
Geop, 7
Geopotential
function, 6
(mass) coefficients, 7, 8, 9
surface, 7
Gravitational constant, 3, 11, 144
Gravity, 3
horizontal gradient of, 431
normal, 8
Gregorian calendar, 152

Hayn's lunar limb charts, 601
Height
geodetic, 21
orthometric, 21
Horizon
astronomic, 19
celestial, 29
coordinate system, 33
geodetic, 16
Horrebow-Talcott method, 459
Hour angle
coordinate system, 33
ephemeris, 172
Hour circle, 29
Hunter shutter eyepiece, 273

Hydrogen maser frequency standard, see Atomic oscillator

Immersion, 568, 574
Impersonal micrometer, 252, 481
calibration of, 284
of the Kern DKM3-A, 265
of the Wild T4, 258, 459, 504
Inclination of horizontal axis
adjustment, 275
determination of, 312, 316
effect of, 305
Inclination of vertical axis
adjustment of, 275
determination of, 316
effect of, 305
Independent day numbers, 76, 94, 119
Index correction
horizontal, 308, 314
vertical, see Zenith point of an instrument
International Association of Geodesy, 80
International Astronomical Union, 81, 164, 168, 202
International Latitude Service, 80
International Polar Motion Service, 80, 351
International Union of Geodesy and Geophysics, 82
Isolimb (occultation), 583, 603

Julian
calendar, 152
century, 152
date, 153
day number, 153
ephemeris date, 172
ephemeris day number, 172
year, 152

Kepler's laws, 141

Laplace
condition, 23
equation (azimuth), 23, 423
station (point), 424

Latitude
astronomic, 18, 20
determination of, 437
ecliptic, 35
geocentric, 15
geodetic, 15
reduced, 15
variation of, see Polar motion
Legendre polynomials, 7
Level
calibration of, 278
Horrebow, 259, 266, 459
striding, 259, 266, 275, 312, 316, 502, 505
Libration (moon), 548, 603
Light-time correction, see Aberration
Line of position, 525
Local sidereal time, 39
determination of, 478
see also, Longitude, determination of
Longitude
astronomic, 19
determination of, 476
ecliptic, 35
geodetic, 15
Loran-C, see Time signals
Lunar limb charts
of Hayn, 601
of Watts, 548, 602

Markowitz moon camera, 303, 546
Mayer's formula, 311, 505
Mayer's method, 504
Mean motion, 144
Mean Observatory, 337, 342, 343, 519
Mean position (place), 63, 116
Mean time
sidereal, 139
solar, 147
Meridian
astronomic, 19
average astronomic, 21, 80, 519
celestial, 29
ecliptic, 31
ephemeris, 170
geodetic, 14
Greenwich mean astronomic, 166, 170, 343
standard, 148
Meridian circle, 182, 250
Micrometer, see Impersonal micrometer
Moon
mass of, 597
parallax of, 573, 592, 596, 597
Moon camera, see Markowitz moon camera

Nadir, 29
National Bureau of Standards, time services, 355, 362
NBS-III Atomic Frequency Standard, 175, 323, 335
Newcomb, Simon, 63, 66, 80, 168
Normal gravity, 8
Normal section, 16
azimuth of, 16
Nušl-Fric circumzenithal, 294
Nutation, 59, 68, 116, 139
astronomic, 60
constant of, 70
in longitude, 69
long-period terms of, 70
in obliquity, 69
in right ascension, see Equation of the equinox
series of, 72
short-period terms of, 70, 79

Obliquity of the ecliptic, see Ecliptic
Obscuration, degree of (eclipse), 570
Observed position (place), 115
Occultation, 551
geodetic applications of, 585
observations of, 598
prediction of, 572
standard (central) stations for occultation predictions, 578
Omega navigational system, see Time signals
Optical instruments, 249, 598

adjustment of, 273
calibration of, 278
Oscillator, 319; see also Atomic oscillator; see also Quartz crystal oscillator

Parallactic angle, 37
Parallax, 59, 89, 98, 116
annual, 100
equatorial horizontal, 103
geocentric, 102
horizontal, 102
stellar, 100
Parallel
astronomic, 19
celestial, 29
ecliptic, 31
geodetic, 14
Passage instrument, 250
Penumbra, 556
Perihelion, 143
Personal equation, 271, 427, 484
Pevtsov's method, 469
Photographic determination of star positions, 187
Photographic zenith tube, 298, 538
Place
apparent, 102, 116
mean, 63, 116
observed, 115
true, 69, 116
Plate constants
in photographic determination of star positions, 188
Turner's, 188, 223, 230
Poisson's equation, 7
Polar distance, 34
ecliptic, 35
Polar motion, 61, 80, 164, 337, 338, 351, 519
effect on azimuth, 88, 429
effect on latitude, 87, 471
effect on longitude and time, 87, 140, 477, 519
Pole
average terrestrial, see Terrestrial pole, average
celestial, 29, 63, 69, 80, 85, 115
instantaneous terrestrial (true celestial), 21, 80, 85
Position
absolute, 21
apparent (celestial), 102, 116
astronomic, 1, 21
geodetic, 1, 15, 21
geographic, 1
mean (celestial), 63, 116
observed (celestial), 115
relative, 21
true (celestial), 69, 116
Position angle (eclipse), 568, 575
Position line (circle), 525
Potential function, 3, 6
spherical harmonic expansion of, 7
Precession, 59, 116, 118
annual, 65
constant of, 66
effect on time, 140
elements of, 62, 64, 65
general, 61, 62, 65
lunisolar, 60
planetary, 60, 65
in star catalogues, 181, 191, 193, 194, 208, 229
Prime vertical, 16, 30, 56
Proper motion, 59, 109, 118, 181

Quartz crystal oscillator, 321, 322, 328

Rapid Latitude Service, 81, 338
Reduction to central thread (transit), 480
Reference ellipsoid, see Ellipsoid
Refraction, 59, 89, 103, 115, 570
astronomic, 105
atmospheric, 105
mean, 106
Right ascension, 34
coordinate system, 34
determination of, 183
Rotation of the earth
irregularities in, 163
period of, 140
Rotation of the ecliptic, 60

axis of, 66
rate of, 66
Rubidium cell oscillator, see Atomic oscillator

Second of time
atomic, 176
ephemeris, 169
mean sidereal, 140, 158
mean solar, 158
Second-order astronomic position determination, 249, 401
azimuth, 402
latitude, 438
longitude, 481
Second-order day numbers, 120
Sidereal calendar, see Date, Greenwich sidereal
Sidereal time, see Time, sidereal
Sidereal year, 150
Simultaneous astronomic position determinations
latitude and azimuth, 472
latitude and longitude, 520
latitude, longitude, and azimuth, 541
Skew normal correction, 432
Solstice, 32, 184
Solstitial colure, 32
Spheroid, 9
Spherop, 9
earth-spherop, 9
Spheropotential
function, 9
surface, 9
Standard star coordinates, 188, 228
Star catalogues, 179
absolute, 180, 182
compilational, 181
fundamental, 181, 189
observational, 182
relationship between, 211, 509
relative, 180, 186
types of, 180
Star catalogues
Apparent Places of Fundamental Stars, 195, 203, 451, 452
Astrographic Catalogue, 237, 241
AGK1, 193, 218, 220
AGK2, 219, 222
AGK2R, 222
AGK3, 224
AGK3R, 224
Boss, see General Catalogue
Cape Photographic Catalogue, 229, 231
FC, 193, 195
FK3, 195, 214
FK4, 192, 195, 197, 200, 214
FK4 Sup, 195
General Catalogue, 192, 202, 206, 214
NFK, 194, 195
N30, 192, 209, 214
Preliminary General Catalogue, 204
SAO, 231, 234
Yale Catalogue, 225
Star constants, Besselian, 76, 93, 119
Star position, see Position, (celestial)
Star place, see Place
Stepped atomic time (SAT), 342
Sterneck method, 470
Sun
fictitious, 147
fictitious mean, 169
motion of, 141

Telescope
occultation, 598, 601
universal, 252, 269, 271
zenith, 252, 297
Terrestrial pole, average, 21, 80
of Cecchini, 353
Conventional International Origin, 351, 519
of epoch, 84, 338, 345, 519
IPMS 1900-05, 83, 84, 351
Theodolite
Askania TPR, 271
geodetic, 289, 292
Kern DKM3, 289
Kern DKM3-A, 260
one-second, 289